Birkhäuser

Cornerstones

J.J. Duistermaat
J.A.C. Kolk

Distributions

Theory and Applications

Translated from Dutch by J.P. van Braam Houckgeest

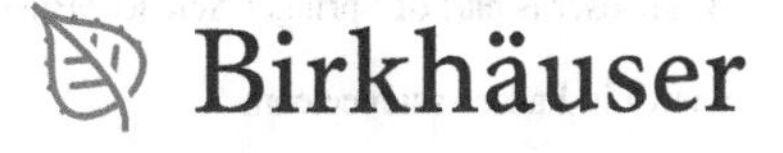

J.J. Duistermaat[†]
Mathematical Institute
Utrecht University

J.A.C. Kolk
Mathematical Institute
Utrecht University
P.O. Box 80.010
3508 TA Utrecht
The Netherlands
j.a.c.kolk@uu.nl

Translated from Dutch by J.P. van Braam Houckgeest

ISBN 978-0-8176-4672-1 e-ISBN 978-0-8176-4675-2
DOI 10.1007/978-0-8176-4675-2
Springer New York Dordrecht Heidelberg London

Library of Congress Control Number: 2010932757

Mathematics Subject Classification (2010): 46-01, 42-01, 35-01, 28-01, 34-01, 26-01

Printed on acid-free paper

Birkhäuser is part of Springer Science+Business Media

www.birkhauser-science.com

To V. S. Varadarajan

A True Friend and Source of Inspiration

Contents

Preface ... ix

Standard Notation ... xv

1 **Motivation** ... 1
Problems ... 13

2 **Test Functions** ... 17
Problems ... 31

3 **Distributions** ... 33
Problems ... 44

4 **Differentiation of Distributions** ... 45
Problems ... 48

5 **Convergence of Distributions** ... 51
Problems ... 54

6 **Taylor Expansion in Several Variables** ... 59
Problems ... 63

7 **Localization** ... 65
Problems ... 69

8 **Distributions with Compact Support** ... 71
Problems ... 81

9 **Multiplication by Functions** ... 83
Problems ... 88

10 **Transposition: Pullback and Pushforward** ... 91
Problems ... 106

11 Convolution of Distributions 115
Problems 132

12 Fundamental Solutions 137
Problems 146

13 Fractional Integration and Differentiation 153
13.1 The Case of Dimension One 153
13.2 Wave Family 157
13.3 Appendix: Euler's Gamma Function 164
Problems 168

14 Fourier Transform 177
Problems 199

15 Distribution Kernels 221
Problems 232

16 Fourier Series 237
Problems 260

17 Fundamental Solutions and Fourier Transform 271
17.1 Appendix: Fundamental Solution of $(I - \Delta)^k$ 279
Problems 282

18 Supports and Fourier Transform 287
Problems 306

19 Sobolev Spaces 311
Problems 317

20 Appendix: Integration 321

21 Solutions to Selected Problems 349

References 433

Index of Notation 435

Index 437

Preface

I am sure that something must be found. There must exist a notion of generalized functions which are to functions what the real numbers are to the rationals (G. Peano, 1912)

Not that much effort is needed, for it is such a smooth and simple theory (F. Trèves, 1975)

In undergraduate physics a lecturer will be tempted to say on certain occasions: "Let $\delta(x)$ be a function on the line that equals 0 away from 0 and is infinite at 0 in such a way that its total integral is 1. The most important property of $\delta(x)$ is exemplified by the identity

$$\int_{-\infty}^{\infty} \phi(x)\delta(x)\,dx = \phi(0),$$

where ϕ is any continuous function of x." Such a function $\delta(x)$ is an object that one frequently would like to use, but of course there is no such function, because a function that is 0 everywhere except at one point has integral 0. All the same, it is important to realize what our lecturer is trying to accomplish: **to describe an object in terms of the way it behaves when integrated against a function**. It is for such purposes that the theory of distributions, or "generalized functions," was created. It can be formulated in all dimensions, its mathematical scope is vast, and it has revolutionized modern analysis.

One way to elaborate on the distributional point of view[1] is to note that a pointwise definition of functions is not very relevant to many situations arising in engineering or physics. This is due to the fact that physical observations often do not represent sharp computations at a single point in space-time but rather averages of fluctuations in small but finite regions in space-time. This is an essential point in signal theory, where there are limitations to the determination of pulse lengths, and

[1] Here we follow the masterly exposition of Varadarajan [22, p. 185].

in quantum theory, where the electromagnetic fields of elementary particles cannot be measured unless one uses a macroscopic test body. From the mathematical point of view one can say that a measurement of a physical quantity by means of a test body yields an average of the values of that quantity in a very small region, the latter being represented by a smooth function that is zero outside a small domain. One replaces the test bodies by these functions, which are naturally called test functions. The value thus measured is a function on the space of test functions, and the interpretation of the measurement as an average makes it clear that this function must be linear. Thus, if T is the space of test functions (unspecified at this point), physical quantities assign real or complex values to functions in T. In keeping with our idea that measurements are averages, we recognize that sometimes things are not so bad and that actual point measurements are possible. Thus ordinary functions are also allowed to be viewed as functionals on T. If f is such an ordinary function, it represents the following functional on T:

$$\phi \mapsto \int f(x)\phi(x)\,dx \in \mathbf{C}.$$

However, since we admit measurements that are too singular to be represented by ordinary functions, we refer to the general functionals on T as generalized functions or distributions. We have been vague about what the space is in which we are operating and also what functions are chosen as test functions. This actually is a great strength of these ideas, because the methods evidently apply without any restriction on the nature or the dimensions of the space. In this book, however, we restrict ourselves to the most important case, that of open subsets of the Euclidean spaces $\mathbf{R}^n$.

Distributions are to functions what the real numbers $\mathbf{R}$ are to the rational numbers $\mathbf{Q}$. In $\mathbf{R}$, the cube root of any number also belongs to $\mathbf{R}$, as does the logarithm of the absolute value of a nonzero number; by contrast, $\sqrt[3]{2}$ and $\log 2$ do not belong to $\mathbf{Q}$. Moreover, $\mathbf{R}$ is the smallest extension of $\mathbf{Q}$ having such properties, while every real number can be approximated by rationals with arbitrary precision. Similarly, distributions are always infinitely differentiable, which is not true of all functions. Here, too, distributions are the smallest possible extension of the test functions satisfying this property, while every distribution can be approximated in the appropriate sense by test functions with arbitrary precision. Continuing the analogy, we mention that differential equations may have distributional solutions in situations where there are no classical solutions, that is, given by differentiable functions. In numerous problems it is of great advantage that solutions exist, even at the penalty of introducing new objects such as distributions, because the solutions can be subject to further study. The theory of distributions provides many tools for the investigation of these so-called weak solutions; for example, these tools enable one to determine when and where distributions are actually functions. One of the early triumphs of distribution theory was the result that every partial differential equation with constant coefficients has a fundamental solution in the sense of distributions: classically, nothing comparable is available.

Fourier theory is another branch of analysis in which a suitable subclass of all distributions helps to clarify many issues. This theory is a far-reaching generalization of writing a vector $x = (x_1, \dots, x_n)$ in $\mathbf{R}^n$ as

$$x = \sum_{k=1}^{n} x_k \, e_k,$$

that is, as a superposition of a finite sum of multiples of the basis vectors e_k. Analogously, in Fourier analysis one attempts to write functions or even distributions as superpositions of basic functions. In this case, finitely many functions do not suffice, but the collection of all bounded exponential functions turns out to be a good choice: bounded, because unbounded exponentials grow too fast at infinity, and exponential, because such functions are simultaneous eigenvectors of all partial derivatives.

The sense in which the infinite superposition represents the original object then becomes an important issue: is the convergence pointwise or uniform, or in a smeared sense? Fourier analysis in the distributional setting enables one to handle problems that classically were out of reach, as well as many new ones. So one obtains, working modulo 2π,

$$\delta(x) = \frac{1}{2\pi} \sum_{k=-\infty}^{\infty} e^{ikx}.$$

This formula goes back to Euler, except that he found the sum to be equal to 0 when x is away from 0.

Hörmander's monumental treatise [11] on linear partial differential equations and Harish-Chandra's pioneering work [10] on harmonic analysis on semisimple Lie groups over the fields of real, complex, or p-adic numbers are but two of the rich fruits borne by Schwartz's text [20], which gave birth to the theory of distributions.

This book aims to be a thorough, yet concise and application-oriented, introduction to the theory of distributions that can be covered in one semester. These constraints forced us to make choices: we try to be rigorous but do not construct a complete theory that prepares the reader for all aspects and applications of distributions. It supplies a certain degree of rigor for a kind of calculation that people long ago did completely heuristically, and it establishes what is legitimate and what is not. The amount of functional analysis that is needed in our treatment is reduced to a bare minimum: only the principle of uniform boundedness is used, while the Hahn–Banach theorems are applied to give alternative proofs, with one exception, of results obtained by different methods. On the other hand, in our exposition of the theory and, in particular, in the problems, we stress applications and interactions with other parts of mathematics.

As a result of this approach our text is complementary to the books [13] and [14] by A.W. Knapp, also published in the Cornerstones series. Building on firm foundations in functional analysis and measure theory, Knapp develops the theory rigorously and in greater depth and wider context than we do, by treating pseudodifferential operators on manifolds, for instance. In many ways our text is introductory;

on the other hand, it presents students of (theoretical) physics or electrical engineering with an idea of what distributions are all about from the mathematical point of view, while giving applied or pure mathematicians a taste of the power of distributions as a natural method in analysis. Our aim is to make the reader familiar with the essentials of the theory in an efficient and fairly rigorous way, while emphasizing the applications.

Solutions of important ordinary and partial differential equations, such as the equation for an electrical LRC network, those of Cauchy–Riemann, Laplace, and Helmholtz and the heat and wave equations, are studied in great detail. Tools for the investigation of the regularity of the solution, that is, its smoothness, are developed. Topics in signal reconstruction have also been treated, such as the mathematical theory underlying CT (= computed tomography) scanners as well as results on band-limited functions. The fundamentals of the theory of complex-analytic functions in one variable are efficiently derived in the context of distributions. In order to make the book self-contained, various results on special functions that are used in our treatment are deduced as consequences of the theory itself, wherever possible.

A large number of problems is included; they are found at the end of each chapter. Some of these illustrate the theory itself, while others explore its relevance to other parts of mathematics. They vary from straightforward applications of the theory to theorems or projects examining a topic in some depth. In particular, important aspects of multidimensional real analysis are studied from the point of view of distributions. Complete solutions to 146 of the 281 problems are provided; problems for which solutions are available are marked by the symbol *. A great number of the remaining exercises are supplied with copious hints, and many of the more difficult problems have been tested in take-home examinations.

In more technical terms, the first eleven chapters cover the basics of general distributions. Specifically, Chap. 10 presents a systematic calculus of pullback and pushforward for the transformation of distributions under a change of variables, whereas Chap. 13 considers complex-analytic one-parameter families of distributions with the aim of obtaining fundamental solutions of certain partial differential operators. Chap. 14 then goes on to treat the Fourier transform of the subclass of tempered distributions in the general, aperiodic case, which is of fundamental importance for the subsequent Chaps. 15–19. Chap. 15 discusses the notion of a distribution kernel of a continuous linear mapping. This notion enables an elegant verification of many properties of such mappings. More generally, it enables aspects of the theory of distributions to be surveyed from a fresh and unifying point of view, as is exemplified by many of the problems in the chapter. The Fourier inversion formula is used in a novel proof of the Kernel Theorem. The Fourier transform is applied in Chap. 16 to study the periodic case and in Chap. 17 to construct additional fundamental solutions. Chap. 18 deals with the Fourier transforms of compactly supported distributions, and Chap. 19 considers rudiments of the theory of Sobolev spaces.

Mathematically sophisticated readers, having perused the first ten chapters, might prefer to proceed immediately to Chaps. 14 and 15.

Important characteristics of the present treatment of the theory of distributions are the following. The theory as presented provides a highly coherent context with a strong potential for unification of seemingly distant parts of analysis. A systematic use of the operations of pullback and pushforward enables the development of a very clean and concise notation. A survey of distribution theory in the framework of distribution kernels allows a description that is algebraic rather than analytic in nature, and makes it possible to study distributions with a minimal use of test functions. In particular, within this framework some more advanced aspects of distribution theory can be developed in a highly efficient manner and transparent proofs can be given. The treatment emphasizes the role of symmetry in obtaining short arguments. In addition, distributions invariant under the actions of various groups of transformations are investigated.

Our preferred theory of integration is that of Riemann, because it will be more familiar to most readers than that of Lebesgue. In some instances, however, our arguments might be slightly shortened by the use of Lebesgue's theory. In the very limited number of cases in which Lebesgue integration is essential, we mention this explicitly. The reader who is not familiar with measure theory may safely skip these passages.

On the other hand, in the theory of distributions Radon measures arise naturally as linear forms defined on compactly supported continuous functions, and therefore the Daniell approach to the theory of integration, which emphasizes linear forms acting on functions instead of functions acting on sets, is very natural in this setting. In the Appendix, Chap. 20, we survey the theory of Lebesgue integration with respect to a measure from this point of view. Although the approach seems very appropriate in our context, we are aware of the fact that it is of limited value to the mathematical probabilist, who primarily requires a theory of integration on function spaces, which are not usually locally compact.

We strongly feel that a mathematical style of writing is appropriate for our purposes, so the book contains a certain amount of theorem–proof text. The reader of a text at this level of mathematical sophistication rightly expects to find all the information needed to follow the argument as well as clear expositions of difficult points, and the theorem–proof format is a time-honored vehicle for conveying these. Furthermore, in theorems one summarizes useful information for future application. Important results (for instance, the Fourier inversion formula) often get several proofs; in this manner different aspects or unexpected relations are brought to the fore.

The present text has evolved from a set of notes for courses taught at Utrecht University over the last twenty years, mainly to bachelor-degree students in their third year of theoretical physics and/or mathematics. In those courses, familiarity with measure theory, functional analysis, or even some of the more theoretical aspects of real analysis, such as compactness, could not be assumed. Since this book addresses the same type of audience, the present text was therefore designed to be essentially self-contained: the reader is assumed to have merely a working knowledge of linear algebra and of multidimensional real analysis (see [7], for instance), while only a few of the problems also require some acquaintance with the residue calculus from complex analysis in one variable. In some cases, the notion of a group will be encountered, mainly in the form of a (one-parameter) group of transformations acting on $\mathbf{R}^n$.

Each time the course was taught, the notes were corrected and refined, with the help of the students; we are grateful to them for their remarks. In particular, J.J. Kuit made a considerable number of original contributions and we benefitted from fruitful discussions with him. M.A. de Reus suggested many improvements. Also, we express our gratitude to our colleagues E.P. van den Ban, for making available the notes for his course in 1987 on distributions and Fourier transform and for very constructive criticism of a preliminary draft, and R.W. Bruggeman, for the improvements and additional problems that he contributed over the past few years. In addition, T.H. Koornwinder read substantial parts of the manuscript with great care when preparing a course on distributions, and contributed significantly, by many valuable queries and comments, to the accuracy of the final version.

Furthermore, we wish to acknowledge our indebtedness to A.W. Knapp, who played an essential role in the publication of this book, for his generous advice and encouragement. The enthusiasm and wisdom of Ann Kostant, our editor at Birkhäuser, made it all possible, and we are very grateful to her for this. Jessica Belanger saw the manuscript through its final stages of production. The original Dutch text has been translated with meticulous care by J.P. van Braam Houckgeest. In addition, his comments have led to considerable improvement in formulation. The second author is very thankful to M.J. Suttorp and H.W.M. Plokker, cardiologists, and their teams: their intervention was essential for the completion of this book.

The responsibility for any imprecisions remains entirely ours; we would be grateful to be told of them, at `j.a.c.kolk@uu.nl`.

Utrecht,
February 2010

Hans Duistermaat
Johan Kolk

Standard Notation

The symbol □ set against the right margin signifies the end of a proof. Furthermore, the symbol ⊘ marks the end of a definition, example or remark.

Item	Meaning
$\emptyset$	empty set
o, O	little and big O symbol of Landau
i	$\sqrt{-1}$
$x \in X$ or $X \ni x$	x an element of X
$x \notin X$	x not an element of X
$\{ x \in X \mid P \}$	the set of x in X such that P holds
∂X, $\overline{X}$	boundary and closure of the set X
$X \subset Y$ or $Y \supset X$	X a subset of Y
$X \cup Y$, $X \cap Y$, $X \setminus Y$	union, intersection, difference of sets
$X \times Y$	Cartesian product of sets
$f : X \to Y$, $x \mapsto f(x)$	mapping, effect of mapping
$f(\cdot, y)$	mapping $x \mapsto f(x, y)$
$f(X)$, $f^{-1}(X)$	direct and inverse image under f of the set X
$f\|_X$	restriction to X
$g \circ f$	composition of f and g, or of g following f
$\mathbf{Z}$, $\mathbf{Q}$, $\mathbf{R}$, $\mathbf{C}$	integers, rationals, reals, complex numbers
$\mathbf{Z}_{\geq a}$	integers greater than or equal to a
$\mathbf{N}$	$= \mathbf{Z}_{\geq 1}$, natural numbers
$\mathbf{R}_{>a}$	reals larger than a
$\|x\|$	absolute value of $x \in \mathbf{R}$
$[x]$	greatest integer $\leq x$
$\operatorname{sgn} x$	sign of x
$]\,a, b\,[$	open interval from a to b
$[\,a, b\,]$	closed interval from a to b
$[\,a, b\,[$, $]\,a, b\,]$	half-open intervals
(a, b)	column vector in $\mathbf{R}^2$
$(x_1, \ldots, x_n)$	column vector

$\|x\|$	norm of vector x
$\langle x, y \rangle$	inner product of vectors x and y
$\mathbf{R}^n$, $\mathbf{C}^n$	spaces of column vectors
$\operatorname{Re} z$, $\operatorname{Im} z$	real and imaginary parts of complex z
$\bar{z}$	complex conjugate of z
$\|z\|$	absolute value of $z \in \mathbf{C}$
$f(a\,\cdot)$	function $f : \mathbf{R}^n \to \mathbf{C}$ given by $x \mapsto f(ax)$
f', f''	derivative and second derivative of $f : \mathbf{R} \to \mathbf{R}$
$f^{(k)}$	kth-order derivative of f
Df	(total) derivative of mapping $f : \mathbf{R}^n \to \mathbf{R}^p$
$\partial_j f$	partial derivative of f with respect to jth variable
$\epsilon \downarrow 0$	ϵ approaches 0 through positive values
$\sum$, $\prod$	sum and product, possibly with a limit operation
I	identity matrix or operator
$\det A$	determinant of matrix or operator A
${}^tA$	transpose of matrix or operator A
$\simeq$	is isomorphic to, is equivalent to

Chapter 1
Motivation

Distributions form a class of objects that contains the continuous functions as a subset. Conversely, every distribution can be approximated by infinitely differentiable functions, and for that reason one also uses the term "generalized functions" instead of distributions. Even so, not every distribution is a function.

In several respects, the calculus of distributions can be developed more readily than the theory of continuous functions. For example, every distribution has a derivative, which itself is also a distribution (see Chap. 4). Hence, every continuous function considered as a distribution has derivatives of all orders. Conversely, we shall prove that every distribution can locally be written as a linear combination of derivatives of some continuous function (see Theorem 13.1 or Example18.2). If every continuous function is to be infinitely differentiable as a distribution, no proper subset of the space of distributions can therefore be adequate. In this sense, the extension of the concept of functions to that of distributions is as economical as it possibly can be.

This may be compared with the extension of the system $\mathbf{Z}$ of integers to the system $\mathbf{Q}$ of rational numbers, where to any x and $y \in \mathbf{Z}$ with $y \neq 0$ corresponds the quotient $\frac{x}{y} \in \mathbf{Q}$. In this case, too, $\mathbf{Q}$ is the smallest extension of $\mathbf{Z}$ having the desired properties.

We now discuss some more concrete types of problem and show how they are solved by the calculus of distributions. We also indicate some typical contexts in which these questions arise. It should be pointed out that the reader will not be assumed to be familiar with the nonmathematical concepts used in those contexts. Likewise, in Examples 1.1 through 1.5 we will occasionally use some mathematical tools that are not yet assumed to be known by the reader. The point of these examples is to provide insight into where the subject is going, rather than to give all details about each example.

Example 1.1. Here we consider the second-order derivative of a function that is nondifferentiable at one point.

The function f defined by $f(x) = |x|$ for $x \in \mathbf{R}$ is continuous on $\mathbf{R}$. It is differentiable on $\mathbf{R}_{<0}$ and on $\mathbf{R}_{>0}$, with derivative equaling -1 and $+1$ on these

J.J. Duistermaat and J.A.C. Kolk, *Distributions: Theory and Applications*,
Cornerstones, DOI 10.1007/978-0-8176-4675-2_1,

intervals, respectively. f is not differentiable at 0. So it seems natural to say that the derivative $f'(x)$ equals the sign $\operatorname{sgn}(x)$ of x, the value of $f'(0)$ not being defined and, intuitively speaking, being of little importance.

But beware, we obviously require that the second-order derivative $f''(x)$ equal 0 for $x < 0$ and for $x > 0$, while $f''(x)$ must have an essential contribution at $x = 0$. Indeed, if $f''(x) \equiv 0$, the conclusion is that $f'(x) \equiv c$ for a constant $c \in \mathbf{R}$; in other words, $f(x) = c\,x$ for all $x \in \mathbf{R}$, which is different from the function $x \mapsto |x|$, whatever choice is made for c. The correct description turns out to be that $f'' = 2\delta$, with δ as in the preface; see Problem 4.1 for more details. ⊘

Example 1.2. Now we are concerned with the electrical field of a point charge. In Maxwell's theory of electromagnetism there are physical difficulties with the concept of a point charge, and in its mathematical description a problem occurs as well.

Let $v : \mathbf{R}^3 \to \mathbf{R}^3$ be a continuously differentiable mapping, interpreted as a vector field on $\mathbf{R}^3$. Further, let

$$x \mapsto \operatorname{div} v(x) = \sum_{j=1}^{3} \frac{\partial v_j(x)}{\partial x_j}$$

be the *divergence* of v; this is a continuous real-valued function on $\mathbf{R}^3$. Suppose that X is a bounded and open set in $\mathbf{R}^3$ having a smooth *boundary* ∂X and lying at one side of ∂X; we write the outer normal to ∂X at the point $y \in \partial X$ as $\nu(y)$. The *Divergence Theorem* then asserts that

$$\int_X \operatorname{div} v(x)\,dx = \int_{\partial X} \langle v(y),\, \nu(y)\rangle\,dy; \tag{1.1}$$

see for instance Duistermaat–Kolk [7, Theorem 7.8.5]. Here the right-hand side is interpreted as an amount of volume that flows outward across the boundary, while $\operatorname{div} v$ is rather like a local expansion (= source strength) in a motion whose velocity field equals v.

Traditionally, one also wishes to allow *point (mass) sources* at a point p, for which $\int_X \operatorname{div} v(x)\,dx = c$ if $p \in X$ and $\int_X \operatorname{div} v(x)\,dx = 0$ if $p \notin \overline{X}$, where $\overline{X}$ is the *closure* of X in $\mathbf{R}^3$. (We make no statement for the case that $p \in \partial X$.) Here c is a positive constant, the strength of the point source in p. These conditions cannot be realized by a function $\operatorname{div} v$ continuous everywhere on $\mathbf{R}^3$.

More specifically, the divergence of the special vector field

$$v(x) = \frac{1}{\|x\|^3}\,x \tag{1.2}$$

vanishes at every point $x \neq 0$; verify this. This implies that the left-hand side of (1.1) equals 0 if $0 \notin \overline{X}$ and 4π if $0 \in X$; the latter result is obtained when we replace the set X by $X \setminus B$, where B is a closed ball around 0, having a radius sufficiently small that $B \subset X$, and then compute the right-hand side of (1.1) (see [7, Example 7.9.4]). Thus we would like to conclude that $\operatorname{div} v$ in this case equals the point source at the point 0 with strength 4π; in mathematical terms, $\operatorname{div} v = 4\pi\,\delta$, with δ the generalization to $\mathbf{R}^3$ of the δ from the preface (see Problem 4.6 and its solution for the details). ⊘

Example 1.3. The underlying mathematical background of this example is the theory of the Hilbert transform $\mathcal{H}$, which gives a way of describing a negative phase shift of signals by 90° (see Problem 14.52), that is,

$$(\mathcal{H}\cos)(x) = \sin x = \cos\left(x - \frac{\pi}{2}\right), \qquad (\mathcal{H}\sin)(x) = -\cos x = \sin\left(x - \frac{\pi}{2}\right).$$

In addition, in the example we come across interesting new distributions that play a role in quantum field theory.

The function $x \mapsto \frac{1}{x}$ is not absolutely integrable on any bounded interval around 0, so it is not immediately clear what

$$\int_{\mathbf{R}} \frac{\phi(x)}{x}\, dx$$

is to mean if ϕ is a continuous function that vanishes outside a bounded interval.

Even if $\phi(0) = 0$, the integrand is not necessarily absolutely integrable. For example, let $0 < \epsilon < c < 1$ and define (see Fig. 1.1)

$$\phi(x) = \begin{cases} 0 & \text{if} \quad x \le 0, \\ \dfrac{1}{|\log x|} & \text{if} \quad 0 < x \le c. \end{cases}$$

This function ϕ is continuous on $]-\infty, c\,]$ and can be extended to a continuous

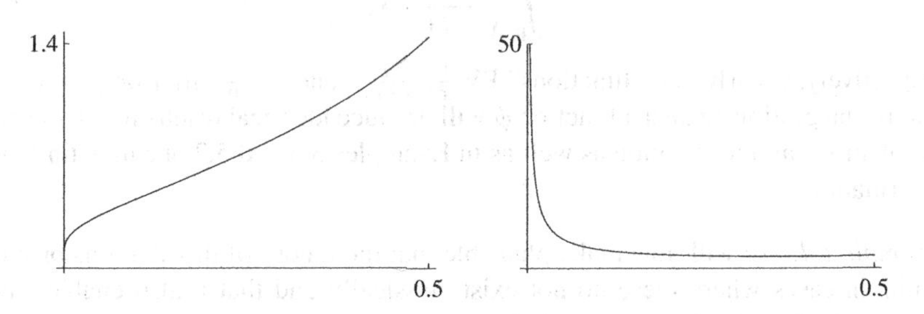

Fig. 1.1 Graphs of ϕ and $x \mapsto \frac{\phi(x)}{x}$

function on $\mathbf{R}$ that vanishes outside a bounded interval. Then

$$\int_{\epsilon}^{c} \frac{\phi(x)}{x}\, dx = \log|\log\epsilon| - \log|\log c|,$$

and the right-hand side converges to ∞ as $\epsilon \downarrow 0$.

But if ϕ is continuously differentiable and vanishes outside a bounded interval, integration by parts and the estimate $\phi(\epsilon) - \phi(-\epsilon) = O(\epsilon)$ as $\epsilon \downarrow 0$, which is a consequence of the Mean Value Theorem, give

$$\lim_{\epsilon\downarrow 0}\int_{\mathbf{R}\setminus[-\epsilon,\epsilon]}\frac{\phi(x)}{x}\,dx=-\int_{\mathbf{R}}\phi'(x)\,\log|x|\,dx. \tag{1.3}$$

Note the importance of the excluded intervals being symmetric about the origin. The left-hand side is called the *principal value* of the integral of $x\mapsto\frac{\phi(x)}{x}$ and is also written as

$$\mathrm{PV}\int_{\mathbf{R}}\frac{\phi(x)}{x}\,dx=:\Big(\mathrm{PV}\frac{1}{x}\Big)(\phi). \tag{1.4}$$

More generally, if $c\in\mathbf{R}$ and ϕ is a continuous function on $\mathbf{R}$, define

$$\Big(\mathrm{PV}\frac{1}{x-c}\Big)(\phi)=\lim_{\epsilon\downarrow 0}\int_{\mathbf{R}\setminus[c-\epsilon,c+\epsilon]}\frac{\phi(x)}{x-c}\,dx,$$

provided that this limit exists.

Other, equally natural, propositions can also be made. Indeed, always assuming ϕ to be continuously differentiable and to vanish outside a bounded interval,

$$\int_{\mathbf{R}}\frac{\phi(x)}{x+i\,\epsilon}\,dx$$

converges as $\epsilon\downarrow 0$, or $\epsilon\uparrow 0$, respectively; see Problem 1.3. The limit is denoted by

$$\int_{\mathbf{R}}\frac{\phi(x)}{x+i\,0}\,dx, \tag{1.5}$$

or

$$\int_{\mathbf{R}}\frac{\phi(x)}{x-i\,0}\,dx, \tag{1.6}$$

respectively. Clearly, the "functions" $\mathrm{PV}\,\frac{1}{x}$, $\frac{1}{x+i\,0}$, and $\frac{1}{x-i\,0}$ differ only at $x=0$, that is, integration against a function ϕ will produce identical results if $\phi(0)=0$. In Problem 1.3 and its solution as well as in Examples 3.3 and 5.7 one may find more information. ⊘

Example 1.4. We will now make plausible that the theory of distributions provides limits in cases where these do not exist classically and that it also enables more freedom in the interchange of analytic operations.

For $0\le r<1$ and $x\in\mathbf{R}$, summation of the geometric series leads to $\sum_{n\in\mathbf{Z}_{\geq 0}}(re^{i\,x})^n=\frac{1}{1-re^{i\,x}}$. By taking the real parts in this identity we obtain

$$\sum_{n\in\mathbf{Z}_{\geq 0}}r^n\cos nx=\frac{1-r\cos x}{1+r^2-2r\cos x}=:A_r(x).$$

Our interest is in the behavior of the preceding identity when $r=1$ or under taking limits for $r\uparrow 1$. First consider the series for $r=1$. Then it is clearly divergent, to ∞, if $x\in 2\pi\mathbf{Z}$. In fact, the series is divergent everywhere on $\mathbf{R}$. This follows from the fact that for no $x \in \mathbf{R}$ do we have $\cos nx \to 0$ as $n \to \infty$. Indeed, otherwise we would have $\sin^2 nx = 1 - \cos^2 nx \to 1$, but also $\sin^2 nx = \frac{1}{2}(1 - \cos 2nx) \to \frac{1}{2}$.

On the other hand,

$$\lim_{r \uparrow 1} A_r(x) = \frac{1}{2} \qquad (x \in \mathbf{R} \setminus 2\pi\mathbf{Z}) \qquad \text{and} \qquad A_r(x) = \frac{1}{1-r} \qquad (x \in 2\pi\mathbf{Z}).$$

Abel's Theorem (see [13, Theorem 1.48]), which would imply $\sum_{n \in \mathbf{Z}_{\geq 0}} \cos nx = \frac{1}{2}$ for $x \in \mathbf{R} \setminus 2\pi\mathbf{Z}$, does not apply, because of the divergence everywhere of the series.

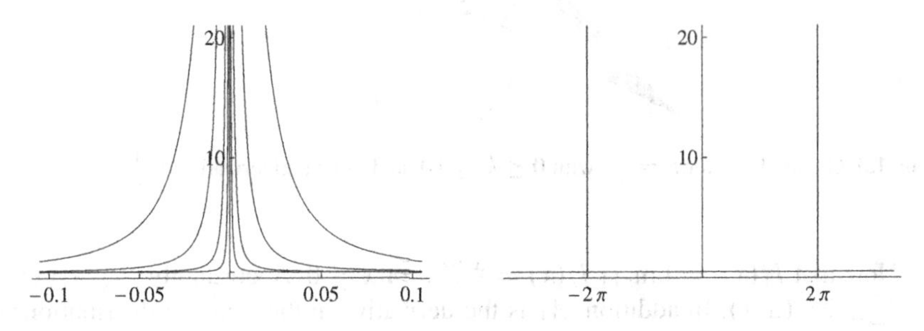

Fig. 1.2 Graph of A_r, for $r = 9\sum_{j=1}^{k} 10^{-j}$ with $2 \leq k \leq 5$, and of $A_{0.99999}$

Nevertheless the numerical evidence in Fig. 1.2 above strongly suggests that the sum of the series is given by the function having value $\frac{1}{2}$ on $\mathbf{R} \setminus 2\pi\mathbf{Z}$ and ∞ on $2\pi\mathbf{Z}$. However, in the theory of integration as discussed in Chap. 20 such a function would be identified with the constant function $\frac{1}{2}$ on $\mathbf{R}$, which ignores the serious divergence of the series on $2\pi\mathbf{Z}$. Therefore, it might be more reasonable to describe the limit of the series as $A_1 := \frac{1}{2} + c\sum_{k \in \mathbf{Z}} \delta_{2\pi k}$ on $\mathbf{R}$. Here $c \in \mathbf{C}$ is a suitable constant and $\delta_{2\pi k}$ denotes the Dirac function located at $2\pi k$. We determine c by demanding

$$\lim_{r \uparrow 1} \int_{-\pi}^{\pi} A_r(x)\, dx = \int_{-\pi}^{\pi} A_1(x)\, dx.$$

For $0 \leq r < 1$, an antiderivative I_r of A_r is given by (see Fig. 1.3 below)

$$I_r(x) = \frac{x}{2} + \arctan\left(\frac{1+r}{1-r} \tan\frac{x}{2}\right); \qquad \text{so} \qquad \int_{-\pi}^{\pi} A_r(x)\, dx = \sum_{\pm} \pm I_r(\pm\pi) = 2\pi,$$

which is also directly obvious by termwise integration of the series. On the other hand, $\int_{-\pi}^{\pi} A_1(x)\, dx = \pi + c$, and so $c = \pi$; phrased differently,

$$\sum_{n \in \mathbf{Z}_{\geq 0}} \cos n \cdot = \frac{1}{2} + \pi \sum_{k \in \mathbf{Z}} \delta_{2\pi k} \qquad \text{on} \quad \mathbf{R}.$$

The validity of this equality in the sense of distributions is rigorously verified in Problem 16.8.

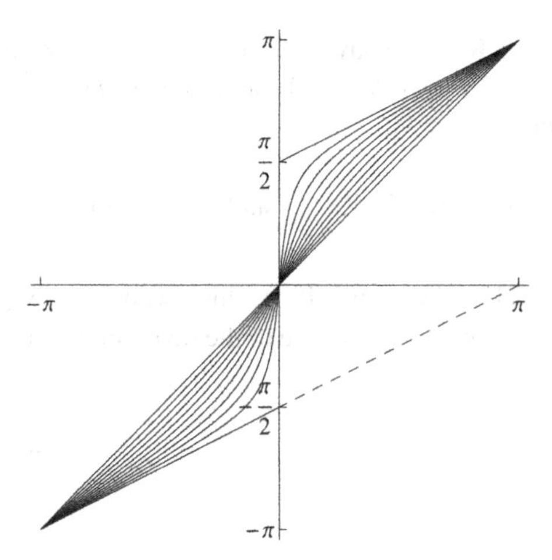

Fig. 1.3 Graph of I_r, for $r = \frac{k}{10}$ with $0 \le k \le 10$, and of an antiderivative of $\frac{1}{2}$

Note that $I_1(x) := \lim_{r\uparrow 1} I_r(x) = \frac{x \mp \pi}{2}$, for $x \lessgtr 0$. Accordingly $\int_{-\pi}^{\pi} A_1(x)\,dx = \sum_{\pm} \pm I_1(\pm\pi)$. In addition, A_1 is the derivative in the sense of distributions of I_1 (classically the latter is nondifferentiable at 0), as will be shown in Example 4.2.

According to the theory of integration we would have the limit of functions $\lim_{r\uparrow 1} A_r = \frac{1}{2}$ (the union of the lower solid line segment and the dashed line segment in Fig. 1.3 is the graph of an antiderivative of $\frac{1}{2}$) and then

$$\lim_{r\uparrow 1} \int_{-\pi}^{\pi} A_r(x)\,dx = 2\pi \neq \pi = \int_{-\pi}^{\pi} \lim_{r\uparrow 1} A_r(x)\,dx. \tag{1.7}$$

In view of the *Dominated Convergence Theorem*, Theorem 20.26.(iv), of *Lebesgue* or that of *Arzelà* (see [7, Theorem 6.12.3]), the fact that interchange of limit and integration in (1.7) is invalid means that the family $(A_r)_{0\le r<1}$ does not admit an integrable majorant on $[-\pi, \pi]$; see Fig. 1.4. In other words, there exists no function g on $\mathbf{R}$ that is integrable on $[-\pi, \pi]$, while $|A_r(x)| \le g(x)$, for all $0 \le r < 1$ and $x \in [-\pi, \pi]$. This is corroborated by the fact that near 0 the envelope, see [7, Exercise 5.38], of the graphs of the A_r, for $0 \le r < 1$, is given by the graph of $x \mapsto \frac{1}{2}(1 + \frac{1}{|\sin x|})$; this function is not integrable on bounded intervals containing 0.

Furthermore, with similar arguments as above, one derives the equality of distributions on $\mathbf{R}$

$$\sum_{n\in\mathbf{Z}} e^{inx} = 2\pi \sum_{k\in\mathbf{Z}} \delta_{2\pi k}(x).$$

This is the correct form of Euler's formula from the preface and gives in essence the basic result in the theory of Fourier series; see (16.7).

In addition, we have the following equality of distributions on a sufficiently small open neighborhood of 0, where we use the principal value from Example 1.3:

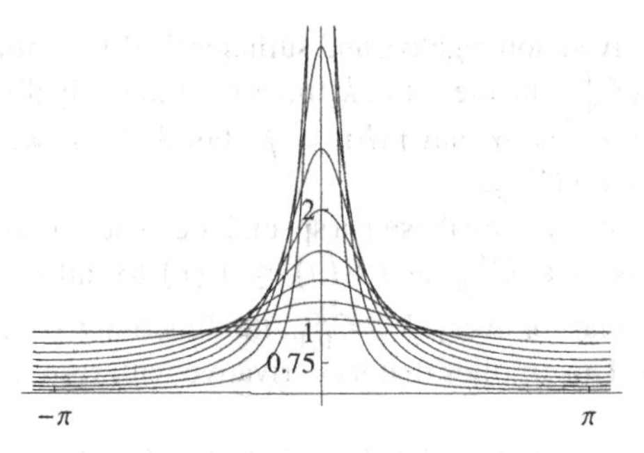

Fig. 1.4 Graph of A_r, for $r = \frac{k}{10}$ with $0 \le k \le 9$

$$\sum_{n \in \mathbf{N}} \sin nx = \frac{x \cos \frac{x}{2}}{2 \sin \frac{x}{2}} \Big(\mathrm{PV} \frac{1}{x} \Big) = \Big(\frac{1}{2} \log(1 - \cos x) \Big)'.$$

Note that the function in front of PV has limit 1 as $x \to 0$ and that $\log(1 - \cos)$ is integrable near 0, while $x \mapsto \frac{1}{x}$ is not. Further, $\lim_{n \to \infty} \sin nx = 0$ if and only if $x \in \pi\mathbf{Z}$. Indeed, under the assumption we obtain $2 \sin x \cos nx = \sin(n+1)x - \sin(n-1)x \to 0$, that is, $\sin x = 0$ by the preceding result about cos.

Finally, we observe that the family of functions (A_r) is closely related to the Poisson kernel (P_r) (see (16.12)), which plays an important role in boundary value problems for the Laplace equation. ⊘

Example 1.5. Another motivation, significant also for historical reasons, has its root in the *calculus of variations*, the theory of finding optimal solutions. An idea shared by most craftsmen, artists, engineers, and scientists is the principle of economy of means. Mathematically, this is the principle of least action and the theory of the associated differential equations of Euler–Lagrange. These variational equations form the basis for many mathematical models in the sciences and in economics. Solving them can be a daunting task; in addition, in the nineteenth century doubts arose about the existence of solutions in the general case. One might think that nature does not pose "artificial" problems and that the applied mathematician therefore need not worry about these matters. A physical theory, however, is not a description of nature but a model of nature that may well be troubled by mathematical difficulties. For instance, the description of the electric field near a very sharp charged needle poses problems both mathematically and physically: the actual experiment produces sparking.

The starting point for the rather lengthy discussion is an obvious calculation. This is then followed by an existence theorem concerning minima of functions, the full proof of which is not allowed by the present context, however. The discussion ends in the statement of a problem that will be solved by means of distribution theory at a later stage.

Consider

$$F(v) = \frac{1}{2} \int_a^b \big(p(x)\, v'(x)^2 + q(x)\, v(x)^2 \big)\, dx,$$

where p and q are given nonnegative and sufficiently differentiable functions on the interval $[a, b]$. Let $C^k_{\alpha,\beta}$ be the set of k times continuously differentiable functions v on $[a, b]$ with $v(a) = \alpha$ and $v(b) = \beta$. For $k \geq 1$, we consider F to be a real-valued function on $C^k_{\alpha,\beta}$.

We now ask whether among these v a special u can be found for which F reaches its minimum, that is, $u \in C^k_{\alpha,\beta}$ and $F(u) \leq F(v)$ for all $v \in C^k_{\alpha,\beta}$. If such u is obtained, one finds that for every $\phi \in C^k_{0,0}$, the function $t \mapsto F(u + t\phi)$ attains its minimum at $t = 0$. This implies that its derivative with respect to t at $t = 0$ equals 0, or

$$\int_a^b \big(p(x)\, u'(x)\, \phi'(x) + q(x)\, u(x)\, \phi(x)\big)\, dx = 0.$$

If $u \in C^2$, integration by parts gives

$$\int_a^b \Big(-\frac{d}{dx}(p(x)\, u'(x)) + q(x)\, u(x)\Big)\, \phi(x)\, dx = 0.$$

Since this must hold for all $\phi \in C^k_{0,0}$, we conclude that u must satisfy the second-order differential equation

$$(Lu)(x) := -\frac{d}{dx}(p(x)\, u') + q(x)\, u = 0. \tag{1.8}$$

This procedure may be applied to much more general functionals (functions on spaces of functions) F; the differential equations that one obtains for the stationary point u of F are called the *Euler–Lagrange equations*.

Until the middle of the nineteenth century the existence of a *minimizing* $u \in C^2$ was taken for granted. Weierstrass then brought up the seemingly innocuous example $a = -1$, $b = 1$, $\alpha = -1$, $\beta = 1$, $p(x) = x^2$, and $q(x) = 0$. One may then consider, for any $\epsilon > 0$, the function

$$v_\epsilon(x) = \frac{\arctan(x/\epsilon)}{\arctan(1/\epsilon)},$$

for $x \in [-1, 1]$. The denominator has been included in order to guarantee that $v_\epsilon(\pm 1) = \pm 1$. For $x < 0$, or $x > 0$, we have that $\arctan(x/\epsilon)$ converges to $-\pi/2$, or $+\pi/2$, respectively, as $\epsilon \downarrow 0$. Therefore, v_ϵ converges to the sign function sgn as $\epsilon \downarrow 0$.

To study the behavior of $F(v_\epsilon)$ we write

$$v'_\epsilon(x) = \frac{1}{\epsilon \arctan(1/\epsilon)}\, \frac{1}{1 + (x/\epsilon)^2}.$$

The change of variables $x = \epsilon\, y$ leads to

$$F(v_\epsilon) = \epsilon\, \frac{1}{\arctan^2(1/\epsilon)} \int_{-1/\epsilon}^{1/\epsilon} \frac{y^2}{2(1 + y^2)^2}\, dy.$$

Note that in this expression the factor $\frac{1}{\arctan^2(1/\epsilon)}$ converges to $4/\pi^2$ as $\epsilon \downarrow 0$. One has

$$\frac{2y^2}{(1+y^2)^2} = \phi'(y) \qquad \text{if} \qquad \phi(y) = -\frac{y}{1+y^2} + \arctan y.$$

That makes the integral equal to $(\phi(1/\epsilon) - \phi(-1/\epsilon))/4$, from which we can see that the integral converges to $\pi/4$ when $\epsilon \downarrow 0$. The conclusion is that $F(v_\epsilon) = \epsilon\, \psi(\epsilon)$, where $\psi(\epsilon)$ converges to $1/\pi$ as $\epsilon \downarrow 0$. In particular, $F(v_\epsilon)$ converges to zero as $\epsilon \downarrow 0$.

Thus we see that the infimum of F on $C^1_{-1,1}$ equals zero; indeed, even the infimum on the subspace $C^\infty_{-1,1}$ equals zero. However, if u is a C^1 function with $F(u) = 0$, we have $\frac{du}{dx}(x) \equiv 0$, which means that u is constant. But then u cannot satisfy the boundary conditions $u(-1) = -1$ and $u(1) = 1$. In other words, the restriction of F to the space $C^1_{-1,1}$ does not attain its minimum in this example.

In the beginning of the twentieth century the following discovery was made. Let $H_{(1)}$ be the space of the square-integrable functions v on $]\,a,\, b\,[$ whose derivatives v' are also square-integrable on $]\,a,\, b\,[$. Actually, this is not so easy to define. A correct definition is given in Chap. 19: $v \in H_{(1)}$ if and only if v is square-integrable and the distribution v' is also square-integrable. In order to understand this definition, we have to know how a square-integrable function can be interpreted as a distribution. Next, we use that the derivative of any distribution is another distribution, which may or may not equal a square-integrable function.

If v is a continuously differentiable function on $[a,\, b]$, application of the *Cauchy–Schwarz inequality* (see [7, Exercise 6.72]) gives

$$\begin{aligned} |v(x) - v(y)| &= \left| \int_y^x v'(z)\, dz \right| \leq \left(\int_y^x v'(z)^2\, dz \right)^{1/2} \left(\int_y^x dz \right)^{1/2} \\ &\leq \|v'\|_{L^2}\, |x-y|^{1/2}, \end{aligned} \tag{1.9}$$

where $\|v'\|_{L^2}$ is the L^2 norm of v'. This can be used to prove that every $v \in H_{(1)}$ can be interpreted as a continuous function on $[\,a,\, b\,]$ (also compare Example 19.3), with the same estimate

$$|v(x) - v(y)| \leq \|v'\|_{L^2}\, |x-y|^{1/2}.$$

The continuity of the functions $v \in H_{(1)}$ implies that one can meaningfully speak of the subspace $H^{\alpha,\beta}_{(1)}$ of the $v \in H_{(1)}$ for which $v(a) = \alpha$ and $v(b) = \beta$. Also, for every $v \in H_{(1)}$ the number $F(v)$ is well-defined.

Now assume that $p(x) > 0$ for every $x \in [a,\, b]$; this excludes the example of Weierstrass. The assumption implies the existence of a constant c with the property

$$\|v'\|^2_{L^2} \leq c\, F(v),$$

for all $v \in H_{(1)}$. In combination with the estimate for $|v(x) - v(y)|$ this tells us that every sequence $(v_j)_{j \in \mathbb{N}}$ in $H^{\alpha,\beta}_{(1)}$ with bounded values $F(v_j)$ is an equicontinuous and uniformly bounded sequence of continuous functions. By the *Arzelà–Ascoli Theorem* (see Knapp [13, Theorem 10.48]), a subsequence $(v_{j(k)})_{k\in\mathbf{N}}$ then converges uniformly to a continuous function u as $k \to \infty$. A second fact here offered without proof is that $u \in H^{\alpha,\beta}_{(1)}$ and that the values $F(v_{j(k)})$ converge to $F(u)$ as $k \to \infty$. This is now applied to a sequence of v_j for which $F(v_j)$ converges to the infimum i of F on $H^{\alpha,\beta}_{(1)}$. Thus one can show the existence of a $u \in H^{\alpha,\beta}_{(1)}$ with $F(u) = i$. In other words, F attains its minimum on $H^{\alpha,\beta}_{(1)}$.

This looked promising, but one then ran into the problem that initially all one could say about this minimizing u was that u' is square-integrable. This does not even imply that u is differentiable under the classical definition that the limit of the difference quotients exists. Because so far we do not even know that $u \in C^2$, the integration by parts is problematic and, as a consequence, so is the conclusion that u is a solution of the Euler–Lagrange equation.

What we can do is to integrate by parts with the roles of u and ϕ interchanged and thereby conclude that

$$\int_a^b u(x)\,(L\phi)(x)\,dx = 0, \tag{1.10}$$

for every $\phi \in C^\infty$ that vanishes identically in a neighborhood of the boundary points a and b. For this statement to be meaningful, u need only be a locally integrable function on the interval $I = \,]\,a, b\,[$. In that case the function u is said to satisfy the differential equation $Lu = 0$ *in a distributional sense*. Historically, a somewhat older term is *in a weak sense*, but this is not very specific.

Assume that p and q are sufficiently differentiable and that p has no zeros in the interval I. In this text we will show by means of distribution theory that if u is a locally integrable function and satisfies the equation $Lu = 0$ in the distributional sense, u is in fact infinitely differentiable in I and satisfies the equation $Lu = 0$ on I in the usual sense. See Theorem 9.4.

In this way, distribution theory makes a contribution to the calculus of variations: by application of the Arzelà–Ascoli Theorem it demonstrates the existence of a minimizing function $u \in H^{\alpha,\beta}_{(1)}$. Every minimizing function $u \in H^{\alpha,\beta}_{(1)}$ satisfies the differential equation $Lu = 0$ in the distributional sense; distribution theory yields the result that u is in fact infinitely differentiable and satisfies the differential equation $Lu = 0$ in the classical sense. This application may be extended to a very broad class of variational problems, also including functions of several variables, for which the Euler–Lagrange variation equation then becomes a partial differential equation. ⊘

Some of the interesting phenomena in the preceding examples form our starting point for the development of the theory of distributions. The estimation result from the next lemma, Lemma 1.6, will play an important role in what follows.

The functions in Examples 1.1, 1.2, and 1.3 are not continuous, or differentiable, respectively, at a special point. Singularities in functions can be mitigated by translating the function f back and forth and *averaging* the functions thus obtained with a *weight function* $\phi(y)$ that depends on the translation y applied to the original function. Let us assume that ϕ is sufficiently differentiable on $\mathbf{R}$, that $\phi(x) \geq 0$ for all $x \in \mathbf{R}$, that a constant $m > 0$ exists such that $\phi(x) = 0$ if $|x| \geq m$, and finally, that

$$\int_{\mathbf{R}} \phi(x)\,dx = 1. \tag{1.11}$$

For the existence of such ϕ, see Problem 1.4. The averaging procedure is described by the formula

$$(f * \phi)(x) = \int_{\mathbf{R}} f(x-y)\,\phi(y)\,dy = \int_{\mathbf{R}} f(z)\,\phi(x-z)\,dz. \tag{1.12}$$

The minus sign is used to obtain symmetric formulas; in particular, $f * \phi = \phi * f$. The function $f * \phi$ is called the *convolution* of f and ϕ, because one of the functions is reflected and translated, then multiplied by the other one, following which the result is integrated. Another interpretation is that of a measuring device recording a signal f around the position x, where $\phi(y)$ represents the sensitivity of the device at displacement y. In practice, this ϕ is never completely concentrated at $y = 0$; because of built-in inertia, $\phi(y)$ will have one or more bounded derivatives.

Yet another interpretation is obtained by defining $T_y f$, the *function translated by* y, via

$$(T_y f)(x) := f(x-y). \tag{1.13}$$

Here we use the rule that $(T_y f)(x+y)$, the value of the translated function at the translated point, equals $f(x)$, the value of the function at the original point. In other words, under T_y the graph translates to the right if $y > 0$. If we now read the first equality in (1.12) as an identity between functions of x, we have

$$f * \phi = \int_{\mathbf{R}} \phi(y)\,T_y f\,dy. \tag{1.14}$$

Here the right-hand side is defined as the limit of Riemann sums in the space of continuous functions (of x), where the limit is taken with respect to the supremum norm. Thus, the functions f translated by y are superposed, with application of a weight function $\phi(y)$, similar to a photograph that becomes softer (blurred) if the camera is moved during the exposure.

Indeed, differentiation with respect to x under the integral sign in the right-hand side in (1.12) yields, even in the case that f is merely continuous, that $f * \phi$ is differentiable, with derivative

$$(f * \phi)' = f * \phi'.$$

In obtaining this result, we have not used the normalization (1.11). We can therefore repeat this and conclude that $f * \phi$ is equally often continuously differentiable as ϕ. For more details, see the proof of Lemma 2.18 below.

How closely does the smoothed signal $f * \phi$ approximate the true signal f? If $a \leq f(z) \leq b$ for all $z \in [\,x-m,\, x+m\,]$, we conclude from (1.12) and (1.11) that $a \leq (f * \phi)(x) \leq b$ as well. This can be improved upon if we can bring the positive number m closer to 0. To achieve this, we replace the function ϕ by ϕ_ϵ (see Fig. 1.5), for an arbitrary constant $\epsilon > 0$, with

$$\phi_\epsilon(x) = \frac{1}{\epsilon}\,\phi\Big(\frac{x}{\epsilon}\Big). \tag{1.15}$$

Furthermore, ϕ_ϵ is equally often continuously differentiable as ϕ.

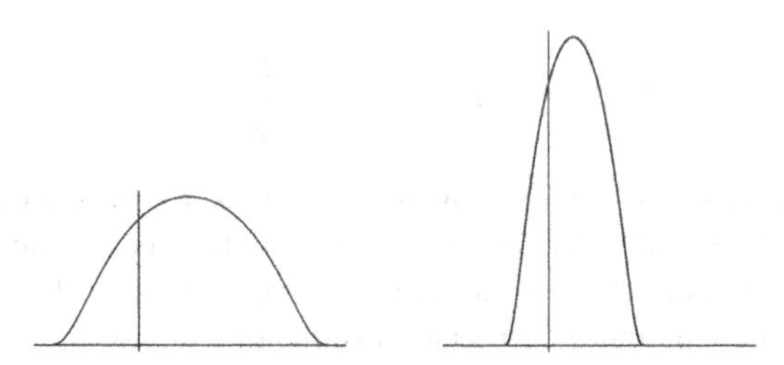

Fig. 1.5 Graph of ϕ_ϵ as in (1.15) with ϵ equal to 1 and 1/2, respectively

Lemma 1.6. *If f is continuous on* $\mathbf{R}$*, the function $f * \phi_\epsilon$ converges uniformly to f on every bounded interval $[\,a,\ b\,]$ as $\epsilon \downarrow 0$. And for every $\epsilon > 0$, the function $f * \phi_\epsilon$ on* $\mathbf{R}$ *is equally often continuously differentiable as ϕ.*

Proof. We have

$$\int_{\mathbf{R}} \phi_\epsilon(y)\,dy = \int_{\mathbf{R}} \phi\Big(\frac{y}{\epsilon}\Big)\,\frac{dy}{\epsilon} = \int_{\mathbf{R}} \phi(z)\,dz = 1,$$

from which

$$\begin{aligned}(f * \phi_\epsilon)(x) - f(x) &= \int_{\mathbf{R}} \big(f(x-y) - f(x)\big)\,\phi_\epsilon(y)\,dy\\ &= \int_{-\epsilon\,m}^{\epsilon\,m} \big(f(x-y) - f(x)\big)\,\phi_\epsilon(y)\,dy,\end{aligned}$$

where in the second identity we have used $\phi_\epsilon(y) = 0$ if $|y| > \epsilon\,m$. This leads to the estimate

$$\begin{aligned}|(f * \phi_\epsilon)(x) - f(x)| &\leq \int_{-\epsilon\,m}^{\epsilon\,m} |f(x-y) - f(x)|\,\phi_\epsilon(y)\,dy\\ &\leq \sup_{|y|\leq\epsilon\,m} |f(x-y) - f(x)|,\end{aligned}$$

where in the first inequality we have applied $\phi_\epsilon(y) \geq 0$ and in the second inequality we have once again used the fact that the integral of ϕ_ϵ equals 1.

The continuity of f gives that for every $\delta_0 > 0$ the function f is uniformly continuous on the bounded interval $[\,a - \delta_0,\ b + \delta_0\,]$ (if necessary, see [7, Theorem

1.8.15] taken in conjunction with Theorem 2.2 below). This implies that for every $\eta > 0$ there exists a $0 < \delta \le \delta_0$ with the property that $|f(x-y) - f(x)| < \eta$ if $x \in [a, b]$ and $|y| < \delta$. From this we may conclude $|(f * \phi_\epsilon)(x) - f(x)| \le \eta$, if $x \in [a, b]$ and $0 < \epsilon < \delta/m$. □

Bochner [3] has called the mapping $f \mapsto f * \phi_\epsilon$ an *approximate identity*, and Weyl [24], a *mollifier*.

Problems

1.1. For $a < c < b$, the integral $\int_a^b \frac{1}{x-c}\,dx$ is divergent. Prove

$$\mathrm{PV}\int_a^b \frac{1}{x-c}\,dx = \log\frac{b-c}{c-a}.$$

1.2. Calculate $\mathrm{PV}\int_{\mathbf{R}} \frac{\phi(x)}{x}\,dx$, for the following choices of ϕ:

$$\phi_1(x) = \frac{x}{1+x^2} \quad \text{and} \quad \phi_2(x) = \frac{1}{1+x^2}.$$

Which of these two integrals converges absolutely as an improper integral?

1.3.* Determine the difference between (1.4) and (1.5), and between (1.5) and (1.6). Each is a complex multiple of $\phi(0)$. See Example 14.30 and Problem 12.14 for different approaches.

1.4.* Determine a polynomial function p on $\mathbf{R}$ of degree six for which $p(a) =$

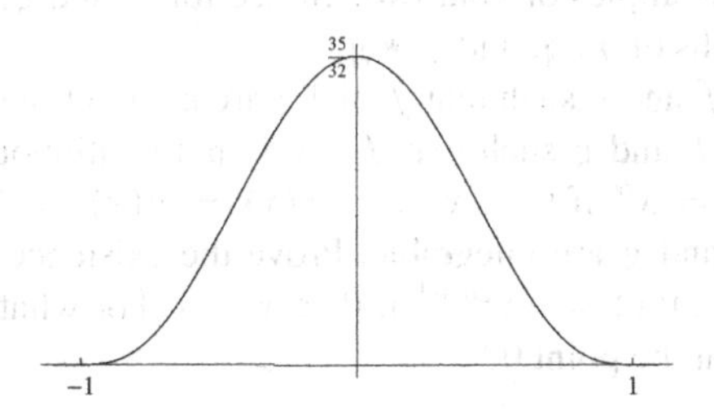

Fig. 1.6 Illustration for Problem 1.4

$p'(a) = p''(a) = 0$ for $a = \pm 1$, while in addition, $\int_{-1}^1 p(x)\,dx = 1$. Define $\phi(x) = p(x)$ for $|x| \le 1$ and $\phi(x) = 0$ for $|x| > 1$. Prove that ϕ is twice continuously differentiable. Sketch the graph of ϕ (see Fig. 1.6).

1.5.* Let ψ be twice continuously differentiable on $\mathbf{R}$ and let ψ equal 0 outside a bounded interval. Set $f(x) = |x|$. Calculate the second-order derivative of $g =$

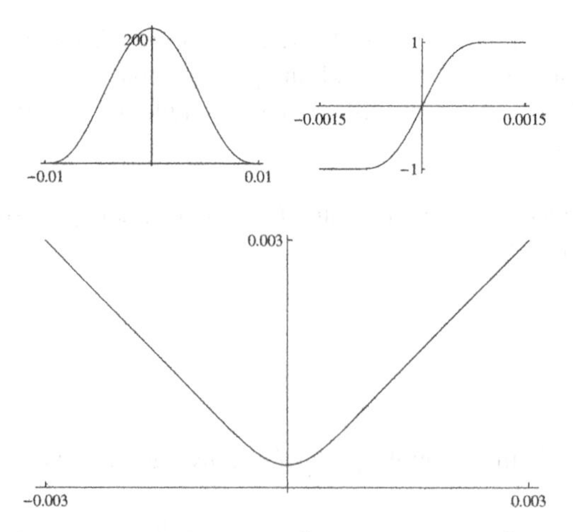

Fig. 1.7 Illustration for Problem 1.5. Graphs of $g'' = 2p_\epsilon$, for $\epsilon = 1/100$, and of g' and g, for $\epsilon = 1/1000$

$f * \psi$ by first differentiating under the integral sign, then splitting the integration at the singular point of f, and finally eliminating the differentiations in every sub-integral.

Now take ψ equal to ϕ_ϵ with ϕ as in Problem 1.4 and ϕ_ϵ as in (1.15). Draw a sketch of g'' for small ϵ, and, by finding antiderivatives, of g' and g. Show the sketches of g, g', g'' next to those of f, f', and f'' (?), respectively (see Fig. 1.7).

1.6.* We consider integrable functions f and g on $\mathbf{R}$ that vanish outside the interval $[-1, 1]$.

(i) Determine the interval outside which the convolution $f * g$ certainly vanishes.
(ii) Using simple examples of your own choice for f and g, calculate $f * g$, and sketch the graphs of f, g, and $f * g$.
(iii) Try to choose f and g such that f and g are not continuous while $f * g$ is.
(iv) Try to choose f and g such that $f * g$ is not continuous. Hint: let $\alpha > -1$, $f(x) = g(x) = x^\alpha$ if $0 < x < 1$, $f(x) = g(x) = 0$ if $x \leq 0$ or $x \geq 1$. Verify that f and g are integrable. Prove the existence of a constant $c > 0$ such that $(f * g)(x) = c\, x^{2\alpha+1}$ if $0 < x \leq 1$. For what values of α is $f * g$ discontinuous at the point 0?

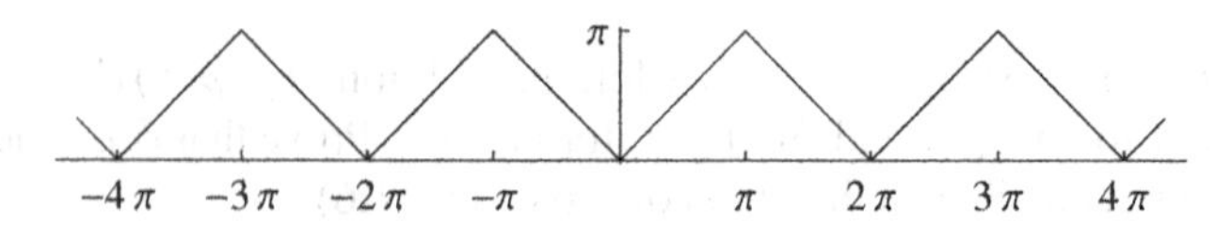

Fig. 1.8 Illustration for Problem 1.7. Graph of arccos ∘ cos

1.7. Set $I(x) = x$, for all $|x| \leq 1$, and let triangle : $\mathbf{R} \to \mathbf{R}$ denote the *unit triangle function* given by triangle$(x) = 1-|x|$, for $|x| \leq 1$ and triangle$(x) = 0$, for $|x| > 1$. Prove (see Fig. 1.8) in the notation of Definition 2.17 that

$$\cos \circ \arccos = I \quad \text{on} \quad [-1,1],$$
$$\arccos \circ \cos = \pi \sum_{k \in \mathbf{Z}} T_{(2k+1)\pi} \operatorname{triangle}\left(\frac{\cdot}{\pi}\right) = \sum_{k \in \mathbf{Z}} T_{2k\pi}(1_{[0,\pi]} * 1_{[0,\pi]}) \text{ on } \mathbf{R}.$$

Hint: show that arccos ∘ cos is continuous on $\mathbf{R}$, while on $\mathbf{R} \setminus \pi\, \mathbf{Z}$ one has

$$(\arccos \circ \cos)' = \frac{\sin}{|\sin|} = \sum_{k \in \mathbf{Z}} (-1)^k T_{k\pi} 1_{]0,\pi[}.$$

Chapter 2
Test Functions

We will now introduce test functions and do so by specializing the testing of f as in (1.12). If we set $x = 0$ and replace $\phi(y)$ by $\phi(-y)$, the result of testing f by means of the weight function ϕ becomes equal to the "integral inner product"

$$\langle f, \phi \rangle = \int_{\mathbf{R}} f(x)\,\phi(x)\,dx. \tag{2.1}$$

(For real-valued functions this is in fact an inner product; for complex-valued functions one uses the Hermitian inner product $\langle f, \overline{\phi} \rangle$.) In Chap. 1 we went on to vary ϕ, by translating and rescaling.

The idea behind the definition of distributions is that we consider (2.1) as a function of all possible test functions ϕ, in other words, we will be considering the mapping

$$\text{test}\, f : \phi \mapsto \int_{\mathbf{R}} f(x)\,\phi(x) \quad x.$$

Before we can do so, we first have to specify what functions will be allowed as test functions. The first requirement is that all these functions be **complex-valued**. Definition 2.5 below, of test functions, refers to compact sets. In this text we will be frequently encountering such sets; therefore we begin by collecting some information on them.

Definition 2.1. An *open cover* of a set K in $\mathbf{R}^n$ is a collection $\mathcal{U}$ of open sets in $\mathbf{R}^n$ such that their union contains K. That is, for every $x \in K$ there exists a $U \in \mathcal{U}$ with $x \in U$. A *subcover* is a subcollection $\mathcal{E}$ of $\mathcal{U}$ still covering K. In other words, $\mathcal{E} \subset \mathcal{U}$ and K is contained in the union of the sets U with $U \in \mathcal{E}$. The set K is said to be *compact* if every open cover of K has a finite subcover. This concept is applicable in very general topological spaces.

Next, recall the concept of a *subsequence* of an infinite sequence $(x(j))_{j \in \mathbf{N}}$. This is a sequence having terms of the form $y(j) = x(i(j))$ where $i(1) < i(2) < \cdots$; in particular, $\lim_{j\to\infty} i(j) = \infty$. Note that if the sequence $(x(j))_{j \in \mathbf{N}}$ converges to x, every subsequence of this sequence also converges to x. ⊘

J.J. Duistermaat and J.A.C. Kolk, *Distributions: Theory and Applications*,
Cornerstones, DOI 10.1007/978-0-8176-4675-2_2,

For the sake of completeness we prove the following theorem, which is known from analysis (see [7, Sect. 1.8]).

Theorem 2.2. *For a subset K of $\mathbf{R}^n$ the following properties (a) – (c) are equivalent.*

(a) K is bounded and closed.
(b) Every infinite sequence in K has a subsequence that converges to a point of K.
(c) K is compact.

Proof. **(a) ⇒ (c).** We begin by proving that a cube $B = \prod_{j=1}^{n} I_j$ is compact. Here I_j denotes a closed interval in $\mathbf{R}$ of length l, for every $1 \le j \le n$. Let $\mathcal{U}$ be an open cover of B; we assume that it does not contain a finite cover of B and will show that this assumption leads to a contradiction.

When we bisect a closed interval I of length l, we obtain $I = I^{(l)} \cup I^{(r)}$, where $I^{(l)}$ and $I^{(r)}$ are closed intervals of length $l/2$. Consider the cubes of the form $B' = \prod_{j=1}^{n} I'_j$, where for every $1 \le j \le n$ we have made a choice $I'_j = I_j^{(l)}$ or $I'_j = I_j^{(r)}$. Then B equals the union of the 2^n subcubes B'. If it were possible to cover each of these by a finite subcollection $\mathcal{E}$ of $\mathcal{U}$, the union of these $\mathcal{E}$ would be a finite subcollection of $\mathcal{U}$ covering B, in contradiction to the assumption. We conclude that there is a B' that is not covered by a finite subcollection of $\mathcal{U}$.

Applying mathematical induction, we thus obtain a sequence $(B^{(t)})_{t\in\mathbf{N}}$ of cubes with the following properties:

(i) $B^{(1)} = B$ and $B^{(t)} \subset B^{(t-1)}$ for every $t \in \mathbf{Z}_{\ge 2}$.
(ii) $B^{(t)} = \prod_{j=1}^{n} I_j^{(t)}$, where $I_j^{(t)}$ denotes a closed interval of length $2^{-t}\, l$.
(iii) $B^{(t)}$ is not covered by a finite subcollection of $\mathcal{U}$.

From (i) we now have, for every j, $I_j^{(t)} \subset I_j^{(t-1)}$, that is, the left endpoints $l_j^{(t)}$ of the $I_j^{(t)}$, considered as a function of t, form a monotonically nondecreasing sequence in $\mathbf{R}$. This sequence is bounded; indeed, $l_j^{(t)} \in I_j^{(s)}$ when $t \ge s$. As $t \to \infty$, the sequence therefore converges to an $l_j \in \mathbf{R}$; we have $l_j \in I_j^{(s)}$ because $I_j^{(s)}$ is closed. Conclusion: the limit point $l := (l_1, \dots, l_n)$ belongs to $B^{(s)}$, for every $s \in \mathbf{N}$.

Because $\mathcal{U}$ is a cover of B and $l \in B$, there exists a $U \in \mathcal{U}$ for which $l \in U$. Since U is open, there exists an $\epsilon > 0$ such that $x \in \mathbf{R}^n$ and $|x_j - l_j| < \epsilon$ for all j implies that $x \in U$. Choose $s \in \mathbf{N}$ with $2^{-s} < \epsilon$. Because $l \in B^{(s)}$, the fact that $x \in B^{(s)}$ implies that $|x_j - l_j| \le 2^{-s} < \epsilon$ for all j; therefore $x \in U$. As a consequence, $B^{(s)} \subset U$, in contradiction to the assumption that $B^{(s)}$ was not covered by a finite subcollection of $\mathcal{U}$.

Now let K be an arbitrary bounded and closed subset of $\mathbf{R}^n$ and $\mathcal{U}$ an open cover of K. Because K is bounded, there exists a closed cube B that contains K. Because K is closed, the complement $C := \mathbf{R}^n \setminus K$ of K is open. The collection $\widetilde{\mathcal{U}} := \mathcal{U} \cup \{C\}$ covers K and C, and therefore $\mathbf{R}^n$, and certainly B. In view of the foregoing, B is covered by a finite subcollection $\widetilde{\mathcal{E}}$ of $\widetilde{\mathcal{U}}$. Removing C from $\widetilde{\mathcal{E}}$, we obtain a finite subcollection $\mathcal{E}$ of $\mathcal{U}$; this covers K. Indeed, if $x \in K$, there exists

$U \in \widetilde{\mathcal{E}}$ with $x \in U$. Since U cannot equal C, we have $U \in \mathcal{E}$.
(c) ⇒ (b). Suppose that $(x(j))$ is an infinite sequence in K that has no subsequence converging in K. This means that for every $x \in K$ there exist an $\epsilon(x) > 0$ and an $N(x)$ for which $\|x - x(j)\| \geq \epsilon(x)$ whenever $j > N(x)$. Let

$$U(x) = \{ y \in K \mid \|y - x\| < \epsilon(x) \}.$$

The $U(x)$ with $x \in K$ form an open cover of K; condition (c) implies the existence of a finite subset F of K such that for every $x \in K$ there is an $f \in F$ with $x \in U(f)$. Let N be the maximum of the $N(f)$ with $f \in F$; then N is well-defined because F is finite. For every j we find that an $f \in F$ exists with $x(j) \in U(f)$, and therefore $j \leq N(f) \leq N$. This is in contradiction to the unboundedness of the indices j.
(b) ⇒ (a). Suppose that K satisfies (b). If K is not bounded, we can find a sequence $(x(j))_{j\in\mathbf{N}}$ with $\|x(j)\| \geq j$ for all j. There is a subsequence $(x(j(k)))_{k\in\mathbf{N}}$ that converges and that is therefore bounded, in contradiction to $\|x(j(k))\| \geq j(k) \geq k$ for all k. In order to prove that K is closed, suppose $\lim_{j\to\infty} x(j) = x$ for a sequence $(x(j))$ in K. This contains a subsequence that converges to a point $y \in K$. But the subsequence also converges to x, and in view of the uniqueness of limits we conclude that $x = y \in K$. □

The preceding theorem contains the *Bolzano–Weierstrass Theorem*, which states that every bounded sequence in $\mathbf{R}^n$ has a convergent subsequence; see [7, Theorem 1.6.3]. The implication (a) ⇒ (c) is also referred to as the *Heine–Borel Theorem*; see [7, Theorem 1.8.18]. However, linear spaces consisting of functions are usually of infinite dimension. In normed linear spaces of infinite dimension, "compact" is a much stronger condition than "bounded and closed," while in such spaces (b) and (c) are still equivalent.

As a first application of compactness we obtain conditions that guarantee that disjoint closed sets in $\mathbf{R}^n$ possess disjoint open neighborhoods; see Lemma 2.3 below and its corollary. To do so, we need some definitions, which are of independent interest.

Introduce the *set of sums* $A + B$ of two subsets A and B of $\mathbf{R}^n$ by means of

$$A + B := \{ a + b \mid a \in A,\ b \in B \}. \tag{2.2}$$

It is clear that $A + B$ is bounded if A and B are bounded. Also, $A + B$ is closed whenever A is closed and B compact. Indeed, suppose that the sequence $(c_j)_{j\in\mathbf{N}}$ in $A + B$ converges in $\mathbf{R}^n$ to c. One then has $c_j = a_j + b_j$ for some $a_j \in A$, $b_j \in B$. By the compactness of B, a subsequence $(b_{j(k)})_{k\in\mathbf{N}}$ converges to a $b \in B$. Consequently, the sequence with terms $a_{j(k)} = c_{j(k)} - b_{j(k)}$ converges to $a := c - b$ as $k \to \infty$. Because A is closed, a lies in A. The conclusion is that $c \in A + B$. In particular, $A + B$ is compact whenever A and B are both compact.

An example of two closed subsets A and B of $\mathbf{R}$ for which $A + B$ is not closed is the pair $A = \mathbf{Z}_{<0}$ and $B = \{ n + 1/n \mid n \in \mathbf{Z}_{\geq 2} \}$. Clearly, A and B are closed and $A + B$ does not contain any integer. On the other hand, for every $m \in \mathbf{Z}$ the numbers $m + 1/n = (m - n) + (n + 1/n)$ belong to $A + B$ if $n \in \mathbf{Z}_{\geq 2}$ and $n > m$, while $m + 1/n$ converges to m as $n \to \infty$.

Furthermore, the *distance* $d(x, U)$ from a point $x \in \mathbf{R}^n$ to a set $U \subset \mathbf{R}^n$ is defined by

$$d(x, U) = \inf\{ \|x - u\| \mid u \in U \}. \tag{2.3}$$

Note that $d(x, U) = 0$ if and only if $x \in \overline{U}$, the closure of U in $\mathbf{R}^n$. The δ-*neighborhood* U_δ of U is given by (see Fig. 2.1)

$$U_\delta = \{ x \in \mathbf{R}^n \mid d(x, U) < \delta \}. \tag{2.4}$$

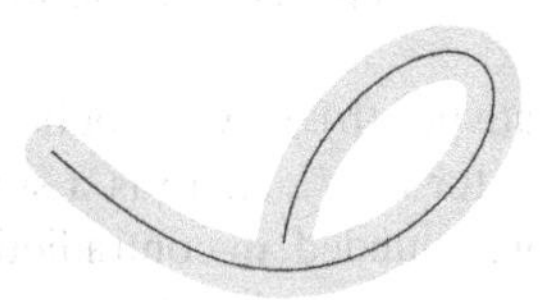

Fig. 2.1 Example of a δ-neighborhood

Observe that $x \in U_\delta$ if and only if a $u \in U$ exists with $\|x - u\| < \delta$. Using the notation $B(u; \delta)$ for the *open ball* of center u and radius δ, this gives

$$U_\delta = \bigcup_{u \in U} B(u; \delta),$$

which implies that U_δ is an open set. Also, $B(u; \delta) = \{u\} + B(0; \delta)$, and therefore

$$U_\delta = U + B(0; \delta).$$

Finally, we define $U_{-\delta}$ as the set of all $x \in U$ for which the δ-neighborhood of x is contained in U. Note that $U_{-\delta}$ equals the complement of $(\mathbf{R}^n \setminus U)_\delta$ and that consequently, $U_{-\delta}$ is a closed set.

Now we are prepared enough to obtain the following two results on separation of sets.

Lemma 2.3. *Let $K \subset \mathbf{R}^n$ be compact and $A \subset \mathbf{R}^n$ closed, while $K \cap A = \emptyset$. Then there exists $\delta > 0$ such that $K_\delta \cap A_\delta = \emptyset$.*

Proof. Assume the negation of the conclusion. Then there exists an element $x(j) \in K_{1/j} \cap A_{1/j}$, for every $j \in \mathbf{N}$. Therefore, one can select $y(j) \in K$ and $a(j) \in A$ satisfying

$$\|y(j) - x(j)\| < \frac{1}{j} \quad \text{and} \quad \|x(j) - a(j)\| < \frac{1}{j}; \qquad \text{so} \qquad \|y(j) - a(j)\| < \frac{2}{j}.$$

By passing to a subsequence, one may assume that the $y(j)$ converge to some $y \in K$ in view of criterion (b) in Theorem 2.2 for compactness. Hence $\|a(j) - y\| \to 0$, in other words, $a(j) \to y$ as $j \to \infty$. Since A is closed, this leads to $y \in A$; therefore $y \in K \cap A$, which is a contradiction. □

Corollary 2.4. *Consider $K \subset X \subset \mathbf{R}^n$ with K compact and X open. Then there exists a $\delta_0 > 0$ with the following property. For every $0 < \delta \leq \delta_0$ there is a compact set C such that*

$$K \subset K_\delta \subset C \subset C_\delta \subset X.$$

Proof. The set $A = \mathbf{R}^n \setminus X$ is closed and $K \cap A = \emptyset$. On account of Lemma 2.3 there is $\delta_0 > 0$ such that $K_{3\delta_0} \cap A = \emptyset$. Define $C = K + \overline{B(0;\delta)}$. Then C is compact as the set of sums of two compact sets; further, $C \subset K_{2\delta}$; hence $C_\delta \subset K_{3\delta} \subset K_{3\delta_0}$. This leads to $C_\delta \cap A = \emptyset$, and so $C_\delta \subset X$. □

After this longish intermezzo we next come to the definition of the space of test functions, one of the most important notions in the theory.

Definition 2.5. Let X be an open subset of $\mathbf{R}^n$. For $\phi : X \to \mathbf{C}$ the *support* of ϕ, written supp ϕ, is defined as the closure in X of the set of the $x \in X$ for which $\phi(x) \neq 0$. A *test function* on X is an infinitely differentiable complex-valued function on X whose support is a compact subset of X. (That is, supp ϕ is a compact subset of $\mathbf{R}^n$ and supp $\phi \subset X$.) The space of all test functions on X is designated as $C_0^\infty(X)$. (The subscript 0 is a reminder of the fact that the function vanishes on the complement of a compact subset, and thus in a sense on the largest part of the space.) It is a straightforward verification that $C_0^\infty(X)$ is a linear space under pointwise addition and multiplication by scalars of functions. ⊘

If we extend $\phi \in C_0^\infty(X)$ to a function on $\mathbf{R}^n$ by means of the definition $\phi(x) = 0$ for $x \in \mathbf{R}^n \setminus X$, we obtain a C^∞ function on $\mathbf{R}^n$. Indeed, $\mathbf{R}^n$ equals the union of the open sets $\mathbf{R}^n \setminus$ supp ϕ and X. On both these sets we have that ϕ is of class C^∞. The support of the extension equals the original support of ϕ. Stated differently, we may interpret $C_0^\infty(X)$ as the space of all $\phi \in C_0^\infty(\mathbf{R}^n)$ with supp $\phi \subset X$; with this interpretation we have $C_0^\infty(U) \subset C_0^\infty(V)$ if $U \subset V$ are open subsets of $\mathbf{R}^n$.

In the vast majority of cases the test functions need only be k times continuously differentiable, with k finite and sufficiently large. To avoid having to keep track of the degree of differentiability, one prefers to work with C_0^∞ rather than the space C_0^k of compactly supported C^k functions.

The question arises whether the combination of the requirements "compactly supported" and "infinitely differentiable" might not be so restrictive as to be satisfied only by the zero function. Indeed, if we were to replace the requirement that ϕ be infinitely differentiable by the requirement that ϕ be analytic, we would obtain only the zero function. Here we recall that a function ϕ is said to be *analytic* on X if for every $a \in X$, ϕ is given by a power series about a that is convergent on some neighborhood of a. This implies that ϕ is of class C^∞ and that the power series of ϕ about a equals the Taylor series of ϕ at a.

Furthermore, an open set X in $\mathbf{R}^n$ is said to be *connected* if X is **not** the union of two disjoint nonempty open subsets of X (for more details, refer to [7, Sect. 1.9]).

Lemma 2.6. *Let X be a connected open subset of $\mathbf{R}^n$ and ϕ an analytic function on X. Then either $\phi = 0$ on X or* $\operatorname{supp} \phi = X$. *In the latter case* $\operatorname{supp} \phi$ *is not compact, provided that X is not empty.*

Proof. Consider the set $U = \{ x \in X \mid \phi = 0 \text{ in a neighborhood of } x \}$; this definition implies that U is open in X. Now select $x \in X \setminus U$. Since ϕ equals its convergent power series in a neighborhood of x, there exists a (possibly higher-order) partial derivative of ϕ, say ψ, with $\psi(x) \neq 0$. Because ψ is continuous, there is a neighborhood V of x on which ψ differs from 0. Hence, $V \subset X \setminus U$, in other words, $X \setminus U$ is open in X. From the connectivity of X we conclude that either $U = X$, in which case $\phi = 0$ on X, or $U = \emptyset$, and in that case $\operatorname{supp} \phi = X$. □

Next we show that $C_0^\infty(X)$ is sufficiently rich. We fabricate the desired functions step by step.

Lemma 2.7. *Define the function $\alpha : \mathbf{R} \to \mathbf{R}$ by $\alpha(x) = e^{-\frac{1}{x}}$ for $x > 0$ and $\alpha(x) = 0$ for $x \leq 0$. Then $\alpha \in C^\infty(\mathbf{R})$ with $\alpha(x) > 0$ for $x > 0$, and* $\operatorname{supp} \alpha = \mathbf{R}_{\geq 0}$.

Proof. The only problem is the differentiability at 0; see Fig. 2.2. From the power series for the exponential function one obtains, for every $n \in \mathbf{N}$, the estimate $e^y \geq \frac{y^n}{n!}$ for all $y \geq 0$. Hence

$$\alpha(x) = \frac{1}{e^{1/x}} \leq \frac{n!}{1/x^n} = n!\, x^n \qquad (x > 0).$$

This tells us that α is differentiable at 0, with $\alpha'(0) = 0$.

As regards the higher-order derivatives, we note that for $x > 0$ the function α satisfies the differential equation

$$\alpha'(x) = \frac{\alpha(x)}{x^2}.$$

By applying this in the induction step we obtain, with mathematical induction on k,

$$\alpha^{(k)}(x) = p_k\Big(\frac{1}{x}\Big)\, \alpha(x),$$

where the p_k are polynomial functions inductively determined by

$$p_0(y) = 1 \qquad \text{and} \qquad p_{k+1}(y) = \big(p_k(y) - p_k{}'(y)\big)\, y^2.$$

In particular, p_k is of degree $2k$ and therefore satisfies an estimate of the form

$$|p_k(y)| \leq c(k)\, y^{2k} \qquad (y \geq 1).$$

From this we derive the estimate

$$|\alpha^{(k)}(x)| \le c(k)\, n!\, x^{n-2k} \qquad (0 < x \le 1).$$

If we then choose $n \ge 2k + 2$, we obtain, with mathematical induction on k, that $\alpha \in C^k(\mathbf{R})$ and $\alpha^{(k)}(0) = 0$. □

Lemma 2.8. *Let $\alpha \in C^\infty(\mathbf{R})$ be as in the preceding lemma. Let a and $b \in \mathbf{R}$ with $a < b$. Define the function $\beta = \beta_{a,b}$ by*

$$\beta(x) = \beta_{a,b}(x) = \alpha(x - a)\, \alpha(b - x).$$

One then has $\beta \in C^\infty(\mathbf{R})$ with $\beta > 0$ on $]\,a,\, b\,[$ and $\operatorname{supp} \beta = [\,a,\, b\,]$. *Furthermore,*

$$I(\beta) := \int_{\mathbf{R}} \beta(x)\, dx > 0.$$

The function $\gamma = \gamma_{a,b} := \frac{1}{I(\beta)}\, \beta$ has the same properties as β (see Fig. 2.2), while $\int_{\mathbf{R}} \gamma(x)\, dx = 1$.

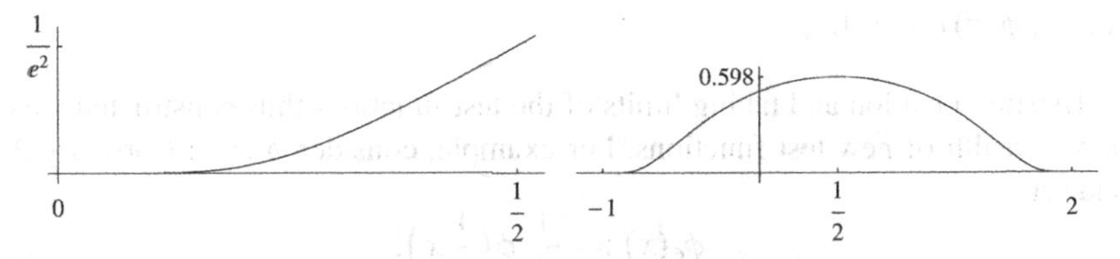

Fig. 2.2 Graphs of α as in Lemma 2.7 on $[\,0,\, 1/2\,]$ and of $\gamma_{-1,2}$ as in Lemma 2.8, with the scales adjusted

Lemma 2.9. *Let a_j and $b_j \in \mathbf{R}$ with $a_j < b_j$ and define $\gamma_{a_j,b_j} \in C_0^\infty(\mathbf{R})$ as in the preceding lemma, for $1 \le j \le n$. Write $x = (x_1, \dots, x_n) \in \mathbf{R}^n$. For a and $b \in \mathbf{R}^n$, define the function $\Gamma_{a,b} : \mathbf{R}^n \to \mathbf{R}$ by (see Fig. 2.3)*

$$\Gamma_{a,b}(x) = \prod_{j=1}^{n} \gamma_{a_j,b_j}(x_j).$$

Then we have

$$\Gamma_{a,b} \in C^\infty(\mathbf{R}^n), \qquad \Gamma_{a,b} > 0 \quad \text{on} \quad \prod_{j=1}^{n} \,]\,a_j,\, b_j\,[\,,$$

$$\operatorname{supp} \Gamma_{a,b} = \prod_{j=1}^{n} [\,a_j,\, b_j\,], \qquad \int_{\mathbf{R}^n} \Gamma_{a,b}(x)\, dx = 1.$$

For a complex number c, the notation $c \ge 0$ means that c is a nonnegative real number. For a complex-valued function f, $f \ge 0$ means that $f(x) \ge 0$ for every x

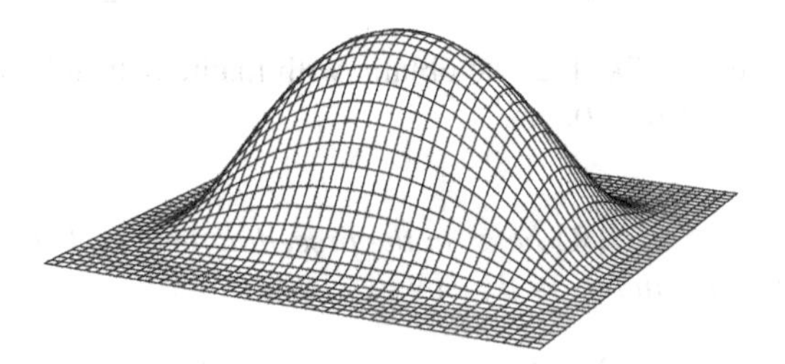

Fig. 2.3 Graph of $\Gamma_{(-1,2),(2,3)}$ as in Lemma 2.9

in the domain space of f. If g is another function, one writes $f \geq g$ or $g \leq f$ if $f - g \geq 0$.

Corollary 2.10. *For every point* $p \in \mathbf{R}^n$ *and every neighborhood* U *of* p *in* $\mathbf{R}^n$ *there exists a* $\phi \in C_0^\infty(\mathbf{R}^n)$ *with the following properties:*

(a) $\phi \geq 0$ *and* $\phi(p) > 0$.
(b) $\operatorname{supp} \phi \subset U$.
(c) $\int_{\mathbf{R}^n} \phi(x)\, dx = 1$.

By superposition and taking limits of the test functions thus constructed we obtain a wealth of new test functions. For example, consider ϕ as in Corollary 2.10 and set

$$\phi_\epsilon(x) := \frac{1}{\epsilon^n}\, \phi\Big(\frac{1}{\epsilon}\, x\Big). \tag{2.5}$$

Further, let f be an arbitrary function in $C_0(\mathbf{R}^n)$, the space of all *continuous functions* on $\mathbf{R}^n$ *with compact support*; these are easily constructed in abundance. By straightforward generalization of Lemma 1.6 to $\mathbf{R}^n$, the functions $f_\epsilon := f * \phi_\epsilon$ converge uniformly on $\mathbf{R}^n$ to f, as $\epsilon \downarrow 0$. The f_ϵ are test functions, in other words,

$$f_\epsilon \in C_0^\infty(\mathbf{R}^n), \tag{2.6}$$

as one can see from Lemma 2.18 below. Consequently, for every $f \in C_0(\mathbf{R}^n)$ there exists a family of functions in $C_0^\infty(\mathbf{R}^n)$ that converges to f uniformly on compact subsets. We say that $C_0^\infty(\mathbf{R}^n)$ is *dense* in $C_0(\mathbf{R}^n)$; see Definition 8.3 below for the general definition of dense sets.

Lemma 2.11. *For every* $a \in \mathbf{R}^n$ *and* $r > 0$ *there exists* $\phi \in C_0^\infty(\mathbf{R}^n)$ *satisfying*

$$\operatorname{supp} \phi \subset B(a; 2r), \qquad 0 \leq \phi \leq 1, \qquad \phi = 1 \quad on \quad B(a; r).$$

Proof. By translation and rescaling we see that it is sufficient to prove the assertion for $a = 0$ and $r = 1$. By Lemma 2.8 we can find $\beta \in C^\infty(\mathbf{R})$ such that $\beta > 0$ on $]\,1, 3\,[$ and $\operatorname{supp} \beta = [\,1, 3\,]$, while $I = \int_1^3 \beta(x)\, dx > 0$. Hence we may write

$$\eta(x) := \frac{1}{I}\int_x^3 \beta(t)\,dt.$$

Then $\eta \in C^\infty(\mathbf{R})$, $0 \leq \eta \leq 1$, while $\eta = 1$ on $]-\infty, 1]$ and $\eta = 0$ on $[3, \infty[$. Now set $\phi(x) = \eta(\|x\|^2) = \eta(x_1^2 + \cdots + x_n^2)$. □

We now review notation that will be needed for Definition 2.13 and Lemma 2.18 below, among other things. In this text we use the following notation for higher-order derivatives. A *multi-index* is a sequence

$$\alpha = (\alpha_1, \dots, \alpha_n) \in (\mathbf{Z}_{\geq 0})^n$$

of n nonnegative integers. The sum

$$|\alpha| := \sum_{j=1}^n \alpha_j$$

is called the *order* of the multi-index α. For every multi-index α we write

$$\partial_x^\alpha := \frac{\partial^\alpha}{\partial x^\alpha} := \partial_1^{\alpha_1} \circ \cdots \circ \partial_n^{\alpha_n}, \qquad \text{where} \qquad \partial_j := \frac{\partial}{\partial x_j}. \tag{2.7}$$

Furthermore, we use the shorthand notation

$$\partial^\alpha = \frac{\partial^\alpha}{\partial x^\alpha}$$

when we want to differentiate only with respect to the variables x_j. The crux is that the *Theorem on the interchangeability of the order of differentiation* (see for instance [7, Theorem 2.7.2]), which holds for functions sufficiently often differentiable, allows us to write every higher-order derivative in the form (2.7); also refer to the introduction to Chap. 6. Finally, in the case of $n = 1$, we define ∂ as $\partial^{(1)}$.

Remark 2.12. In (2.7) we defined the partial derivatives $\partial^\alpha f$ of arbitrary order of a function f depending on an arbitrary number of variables. For the kth-order derivatives of the product $f\,g$ of two functions f and g that are k times continuously differentiable, we have *Leibniz's formula*:

$$\partial^\alpha(f\,g) = \sum_{\beta \leq \alpha} \binom{\alpha}{\beta} \partial^{\alpha-\beta} f\, \partial^\beta g, \tag{2.8}$$

for $|\alpha| = k$. Here $\alpha = (\alpha_1, \dots, \alpha_n)$ and $\beta = (\beta_1, \dots, \beta_n)$ are multi-indices, while $\beta \leq \alpha$ means that for every $1 \leq j \leq n$ one has $\beta_j \leq \alpha_j$. The n-dimensional *binomial coefficients* in (2.8) are given by

$$\binom{\alpha}{\beta} := \prod_{j=1}^n \binom{\alpha_j}{\beta_j}, \qquad \text{where} \qquad \binom{p}{q} = \frac{p!}{(p-q)!\,q!},$$

for p and $q \in \mathbf{Z}$ with $0 \leq q \leq p$. Formula (2.8) is obtained with mathematical induction on the order $k = |\alpha|$ of differentiation, using *Leibniz's rule*

$$\partial_j(f\,g) = g\,\partial_j f + f\,\partial_j g \tag{2.9}$$

in the induction step. ⊘

Definition 2.5 is supplemented by the following, which introduces a notion of convergence in the infinite-dimensional linear space $C_0^\infty(X)$:

Definition 2.13. Let ϕ_j and $\phi \in C_0^\infty(X)$, for $j \in \mathbf{N}$ and X an open subset of $\mathbf{R}^n$. The sequence $(\phi_j)_{j\in\mathbf{N}}$ is said to converge to ϕ *in the space $C_0^\infty(X)$ of test functions* as $j \to \infty$, notation

$$\lim_{j\to\infty} \phi_j = \phi \quad \text{in} \quad C_0^\infty(X),$$

if the following two conditions are both met:

(a) there exists a compact subset K of X such that $\operatorname{supp} \phi_j \subset K$ for all j;
(b) for every multi-index α the sequence $(\partial^\alpha \phi_j)_{j\in\mathbf{N}}$ converges uniformly on X to $\partial^\alpha \phi$. ⊘

Observe that the data above imply that $\operatorname{supp} \phi \subset K$. The notion of convergence introduced in the definition above is very strong. The stronger the convergence, the fewer convergent sequences there are, and the more readily a function defined on $C_0^\infty(X)$ will be continuous.

Now we combine compactness and test functions in order to introduce the useful technical tool of a partition of unity over a compact set.

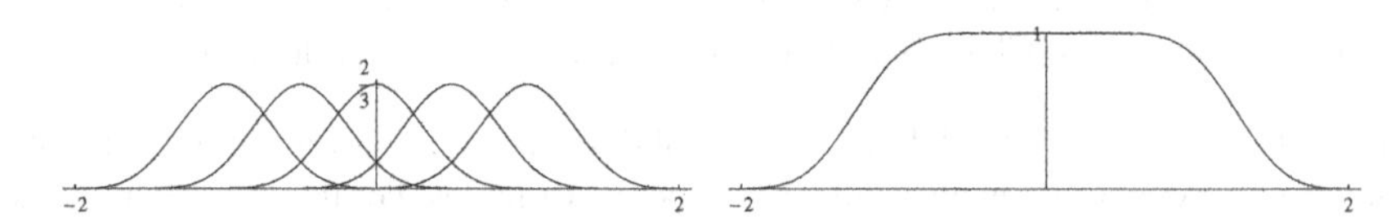

Fig. 2.4 Example of a partition of unity

Definition 2.14. Let K be a compact subset of an open subset X of $\mathbf{R}^n$ and $\mathcal{U}$ an open cover of K. A *$C_0^\infty(X)$ partition of unity over K subordinate to $\mathcal{U}$* is a finite sequence $\psi_1, \dots, \psi_l \in C_0^\infty(X)$ with the following properties (see Fig. 2.4):

(i) $\psi_j \geq 0$, for every $1 \leq j \leq l$, and $\sum_{j=1}^{l} \psi_j \leq 1$ on X;
(ii) there exists a neighborhood V of K in X with $\sum_{j=1}^{l} \psi_j(x) = 1$, for all $x \in V$;
(iii) for every j there is a $U = U(j) \in \mathcal{U}$ for which $\operatorname{supp} \psi_j \subset U$. ⊘

Given a function f on X, write $f_j = \psi_j\, f$ in the notation above. Then we obtain functions f_j with compact support contained in $U(j)$, while $f = \sum_{j=1}^{l} f_j$ on V. Furthermore, all $f_j \in C^k$ if $f \in C^k$. In the applications, the $U \in \mathcal{U}$ are small neighborhoods of points of K with the property that we can reach certain desired conclusions for functions with support in U. For example, partitions of unity were used in this way in [7, Theorem 7.6.1] to prove the integral theorems for open sets $X \subset \mathbf{R}^n$ with C^1 boundary.

Theorem 2.15. *For every compact set K contained in an open subset X of $\mathbf{R}^n$ and every open cover $\mathcal{U}$ of K there exists a $C_0^\infty(X)$ partition of unity over K subordinate to $\mathcal{U}$.*

Proof. For every $a \in K$ there exists an open set $U_a \in \mathcal{U}$ such that $a \in U_a$. Select $r_a > 0$ such that $B(a; 2r_a) \Subset U_a \cap X$. By criterion (c) in Theorem 2.2 for compactness, there exist finitely many $a(1), \dots, a(l)$ such that K is contained in the union V of the $B(a(j),\ r_{a(j)})$, for $1 \le j \le l$. Now select the corresponding $\phi_j \in C_0^\infty(X)$ as in Lemma 2.11 and set

$$\psi_1 = \phi_1; \qquad \psi_{j+1} = \phi_{j+1} \prod_{i=1}^{j} (1 - \phi_i) \qquad (1 \le j < l). \tag{2.10}$$

Then the conditions (i) and (iii) for a $C_0^\infty(X)$ partition of unity subordinate to $\mathcal{U}$ are satisfied by the $\psi_1, \dots, \psi_l$. The relation

$$\sum_{i=1}^{j} \psi_i = 1 - \prod_{i=1}^{j} (1 - \phi_i) \tag{2.11}$$

is trivial for $j = 1$. If (2.11) is true for $j < l$, then summing (2.10) and (2.11) yields (2.11) for $j + 1$. Consequently (2.11) is valid for $j = l$, and this implies that the $\psi_1, \dots, \psi_l$ satisfy condition (ii) for a partition of unity with V as defined above. □

Corollary 2.16. *Let K be a compact subset in $\mathbf{R}^n$. For every open neighborhood X of K in $\mathbf{R}^n$ there exists a $\chi \in C_0^\infty(\mathbf{R}^n)$ with $0 \le \chi \le 1$, $\operatorname{supp} \chi \subset X$ and $\chi = 1$ on an open neighborhood of K. In particular, for $\delta > 0$ sufficiently small, we can find such a function χ with $\chi = 1$ on K_δ.*

Proof. Consider the open cover $\{X\}$ of K and let $\psi_1, \dots, \psi_l$ be a subordinate partition of unity over K as in the preceding theorem. Then $\chi = \sum_j \psi_j$ satisfies all requirements. For the second assertion, apply Corollary 2.4 and the preceding result with K replaced by C as in the corollary. □

The function χ is said to be a *cut-off function* for the compact subset K of $\mathbf{R}^n$. Through multiplication by χ we can replace a function f defined on X by a function g with compact support contained in X. Here $g = f$ on a neighborhood of K and $g \in C^k$ if $f \in C^k$.

We still have to verify the claim in (2.6); it follows from Lemma 2.18 below. In the case of k equal to ∞, another proof will be given in Theorem 11.2. Later on, in demonstrating Theorem 11.22, we will need an analog of Corollary 2.16 in the case of not necessarily compact sets. To that end, we derive Lemma 2.19 below. In preparation, we introduce some concepts that are useful in their own right.

Definition 2.17. Let $X \subset \mathbf{R}^n$ be an open subset. A function $f : X \to \mathbf{C}$ is said to be *locally integrable* if for every $a \in X$, there exists an open rectangle $B \subset X$ with the properties that $a \in B$ and that f is integrable on B.

The *characteristic function* or *indicator function* 1_U of a subset U of $\mathbf{R}^n$ is defined by

$$1_U(x) = 1 \quad \text{if} \quad x \in U, \qquad 1_U(x) = 0 \quad \text{if} \quad x \in \mathbf{R}^n \setminus U.$$

U is said to be *measurable* if 1_U is locally integrable. ⊘

For the purposes of this book it will almost invariably be sufficient to interpret the concept of integrability, as we use it here, in the sense of *Riemann*. However, for distributions it is common to work with *Lebesgue integration*, which leads to a more comprehensive theory. Loosely speaking, Lebesgue's theory is more powerful than Riemann's, in the sense that it leads to a process of integration for more functions and to a simpler treatment of singular behavior of functions. On the other hand, a thorough treatment of Lebesgue integration is technically more demanding than that of Riemann integration. The distinction between the two concepts rarely arises in the case of the functions that will be encountered in this text. It is primarily in the description of spaces of **all** functions satisfying certain properties that the difference becomes important.

Readers who are not familiar with Lebesgue integration can find a way around this by restricting themselves to locally integrable functions with an absolute value whose improper Riemann integral exists, and otherwise taking our assertions about Lebesgue integration for granted. Some of these assertions do not apply to Riemann integration, but this need not be a reason for serious concern; we will discuss this issue when the need arises.

Nonetheless, for the benefit of readers who are interested in the relation between the theory of distributions and that of (Lebesgue) integration we concisely but fairly completely discuss integration in Chap. 20. In particular, local integrability is introduced in Definition 20.37.

Lemma 2.18. *Let f be locally integrable on $\mathbf{R}^n$ and $g \in C_0^k(\mathbf{R}^n)$. Then $f * g \in C^k(\mathbf{R}^n)$ and*

$$\operatorname{supp}(f * g) \subset \operatorname{supp} f + \operatorname{supp} g.$$

Here $\operatorname{supp} f + \operatorname{supp} g$ *is a closed subset of $\mathbf{R}^n$, compact if f, too, has compact support; in that case $f * g \in C_0^k(\mathbf{R}^n)$.*

Proof. We study $(f * g)(x)$ for $x \in U$, where $U \subset \mathbf{R}^n$ is bounded and open. Define $h(x, y) := f(y)\, g(x - y)$. Then the function $x \mapsto h(x, y)$ belongs to $C^k(U)$ for every $y \in \mathbf{R}^n$, because for every multi-index $\alpha \in (\mathbf{Z}_{\geq 0})^n$ with $|\alpha| \leq k$,

$$\frac{\partial^\alpha h}{\partial x^\alpha}(x, y) = f(y)\, \partial^\alpha g(x - y).$$

Let $B(r)$ be a ball about 0 of radius $r > 0$ such that supp $g \subset B(r)$. Then there exists an $r' > 0$ with $B(r) + U \subset B(r')$; furthermore, the characteristic function χ of $B(r')$ is integrable on $\mathbf{R}^n$. For every $x \in U$ the function $\frac{\partial^\alpha h}{\partial x^\alpha}(x, \cdot)$ vanishes outside $B(r')$; consequently, the latter function does not change upon multiplication by χ. In addition, we have

$$\left|\frac{\partial^\alpha h}{\partial x^\alpha}(x, y)\right| \le \sup_{x \in \mathbf{R}^n} |\partial^\alpha g(x)|\, |f(y)|\, \chi(y) \qquad ((x, y) \in U \times \mathbf{R}^n),$$

where $|f|\,\chi$ is an absolutely integrable function on $\mathbf{R}^n$. In view of a well-known theorem on changing the order of differentiation and integration (in the context of Riemann integration, see [7, Theorem 6.12.4]) we then know that $\int_{\mathbf{R}^n} h(x, y)\, dy$ is a C^k function of x whose derivatives equal the integral with respect to y of the corresponding derivatives according to x of the integrand $h(x, y)$.

Furthermore, $h(x, y) = 0$ if $x \in U$ and $y \notin K_U$, where

$$K_U := (\text{supp } f) \cap (\overline{U} + (-\text{supp } g)).$$

Now suppose $u \notin$ supp f + supp g. Then there exists a neighborhood U of u in $\mathbf{R}^n$ such that $x \notin$ supp f + supp g for all $x \in \overline{U}$, because the complement of supp f + supp g is open. But this means $K_U = \emptyset$, which implies that $(f * g)(x) = 0$ for all $x \in U$. □

Lemma 2.19. *Let $\phi \in C_0^\infty(\mathbf{R}^n)$, $\phi \ge 0$, $\int \phi(x)\, dx = 1$, and $\|x\| \le 1$ if $x \in$ supp ϕ. Suppose that the subset U of $\mathbf{R}^n$ is measurable; see Definition 2.17. Select $\delta > 0$ arbitrarily and define, for $0 < \epsilon < \delta$,*

$$\phi_\epsilon(x) = \frac{1}{\epsilon^n}\, \phi\Big(\frac{1}{\epsilon} x\Big) \qquad \textit{and} \qquad \chi_{U,\epsilon} := 1_U * \phi_\epsilon.$$

Then

$$\chi_{U,\epsilon} \in C^\infty(\mathbf{R}^n), \qquad 0 \le \chi_{U,\epsilon} \le 1, \qquad \text{supp } \chi_{U,\epsilon} \subset U_\delta.$$

Finally, $\chi_{U,\epsilon} = 1$ on a neighborhood of $U_{-\delta}$.

Proof. We have $\chi_{U,\epsilon} \in C^\infty(\mathbf{R}^n)$ by Lemma 2.18. Because $\phi \ge 0$, we obtain

$$0 = 0 * \phi_\epsilon \le 1_U * \phi_\epsilon \le 1 * \phi_\epsilon = 1(\phi_\epsilon) = 1.$$

Furthermore, if B_ϵ denotes the ϵ-neighborhood of 0, the support of $\chi_{U,\epsilon}$ is contained in supp 1_U + supp $\phi_\epsilon \subset \overline{U} + B_\epsilon$, and therefore also in the δ-neighborhood of U as $\epsilon < \delta$. The latter conclusion is reached when we replace U by $V = \mathbf{R}^n \setminus U$; note that $1 - \chi_{U,\epsilon} = 1 * \phi_\epsilon - 1_U * \phi_\epsilon = (1 - 1_U) * \phi_\epsilon = \chi_{V,\epsilon}$. □

Usually in applications of Lemma 2.19, the set U is either open or closed, but even then its characteristic function 1_U will not always be locally integrable in the sense of Riemann (see [7, Exercise 6.1]), whereas it is in the sense of Lebesgue; see Proposition 20.36. The only occasion in the text where this issue might play a role is in the proof of Theorem 11.22.

Finally, there are many situations in which one prefers to use, instead of $\phi \in C_0^\infty(\mathbf{R}^n)$, functions like (see Fig. 2.5)

$$\gamma(x) = \gamma_n(x) = \pi^{-\frac{n}{2}} \, e^{-\|x\|^2}. \tag{2.12}$$

The numerical factor is chosen such that the integral of γ over $\mathbf{R}^n$ equals 1; this γ

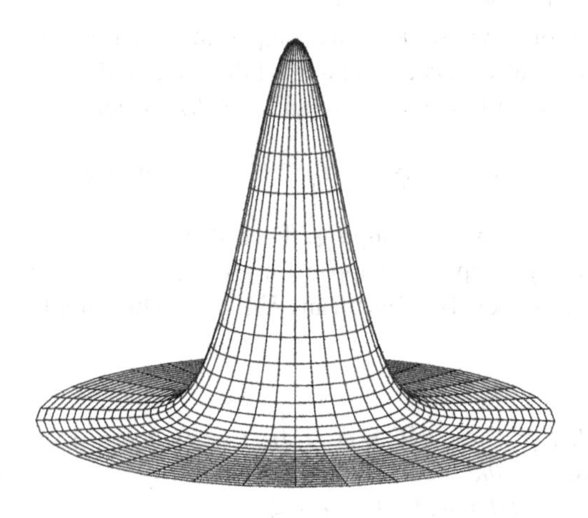

Fig. 2.5 Graph of γ_2, with different horizontal and vertical scales

is the *Gaussian density* or the *probability density of the normal distribution*, with expectation 0 and variance

$$\int_{\mathbf{R}^n} \|x\|^2 \, \gamma_n(x) \, dx = \frac{n}{2}. \tag{2.13}$$

For larger values of $\|x\|$ the values $\gamma(x)$ are so extremely small that in many situations we may just as well consider γ as having compact support. Naturally, this is only relative: if we were to use γ to test a function that grows at least like $e^{\|x\|^2}$ as $\|x\| \to \infty$, this would utterly fail.

For the sake of completeness we recall the calculation of $I_n := \int \gamma_n(x) \, dx$. Because $\gamma_n(x) = \prod_{j=1}^n \gamma_1(x_j)$, we have $I_n = (I_1)^n$. The change of variables $x = r(\cos\alpha, \, \sin\alpha)$ in a dense open subset of $\mathbf{R}^2$ now yields

$$I_2 = \frac{1}{\pi}\int_{\mathbf{R}^2} e^{-(x_1^2+x_2^2)}\,d(x_1,x_2) = \frac{1}{\pi}\int_{\mathbf{R}_{>0}}\int_{-\pi}^{\pi} e^{-r^2}\,r\,d\alpha\,dr = 1,$$

or

$$\int_{\mathbf{R}} e^{-x^2}\,dx = \sqrt{\pi}. \tag{2.14}$$

We refer to [7, Exercises 2.73 and 6.15, or 6.41] for other proofs of this identity. For the computation of (2.13) introduce spherical coordinates in $\mathbf{R}^n$ by $x = r\omega$ with $r > 0$ and ω belonging to the unit sphere $\{\, x \in \mathbf{R}^n \mid \|x\| = 1 \,\}$ in $\mathbf{R}^n$; see [7, Example 7.4.12] and (13.37). Next use the substitution $r^2 = s$ and formulas (13.30) and (13.31) below.

Problems

2.1.* Let $U \subset \mathbf{R}^n$ be a closed set. Prove that the corresponding distance function satisfies $|d(x, U) - d(y, U)| \leq \|x - y\|$, for all x and $y \in \mathbf{R}^n$.

2.2.* Let $\phi \in C_0^\infty(\mathbf{R})$, $\phi \neq 0$, and $0 \notin \operatorname{supp} \phi$. Decide whether the sequence $(\phi_j)_{j\in\mathbf{N}}$ converges to 0 in $C_0^\infty(\mathbf{R})$ if:

(i) $\phi_j(x) = j^{-1}\,\phi(x - j)$.
(ii) $\phi_j(x) = j^{-p}\,\phi(j\,x)$. Here p is a given positive integer.
(iii) $\phi_j(x) = e^{-j}\,\phi(j\,x)$.

In each of these cases verify that for every $x \in \mathbf{R}$ and every $k \in \mathbf{Z}_{\geq 0}$, the sequence $(\phi_j^{(k)}(x))_{j\in\mathbf{N}}$ converges to 0, and in addition, that in case (i) the convergence is even uniform on $\mathbf{R}$.

2.3.* Let ϕ and ϕ_ϵ be as in Lemma 2.19. Prove that for every $\psi \in C_0^\infty(X)$, the function $\psi * \phi_\epsilon$ converges to ψ in $C_0^\infty(X)$ as $\epsilon \downarrow 0$.

2.4. Consider $\beta \in C_0^\infty(\mathbf{R})$ with $\beta \geq 0$ and $\beta(x) = 0$ if and only if $|x| \geq 1$. Further assume that $\int \beta(x)\,dx = 1$. Let $0 < \epsilon < 1$, $\beta_\epsilon(x) = \frac{1}{\epsilon}\beta(\frac{x}{\epsilon})$, $I = [-1, 1] \subset \mathbf{R}$, and let $\psi = 1_I * \beta_\epsilon$. Determine where one has $\psi = 0$, where $0 < \psi < 1$, and where $\psi = 1$, and in addition, where $\psi' = 0$, $\psi' > 0$, and $\psi' < 0$, respectively.

Now let $\phi(x) = \beta(x_1)\,\beta(x_2)$ for $x \in \mathbf{R}^2$ and let $U = I \times I$, a square in the plane. Consider $\chi = \chi_{U,\epsilon}$ as in Lemma 2.19. Prove that $\chi(x) = \psi(x_1)\,\psi(x_2)$. Determine where one has $\chi = 0$, or $0 < \chi < 1$, or $\chi = 1$, and in addition, for $j = 1$ and 2, where $\partial_j\chi = 0$, $\partial_j\chi > 0$, $\partial_j\chi < 0$. Verify that if $0 < \chi < 1$, there is a j such that $\partial_j\chi \neq 0$. Prove by the *Submersion Theorem* (see [7, Theorem 4.5.2]) that for every $0 < c < 1$ the level set $N(c) := \{\, x \in \mathbf{R}^2 \mid \chi(x) = c \,\}$ is a C^∞ curve in the plane. Is this also true for the boundary of the support of χ and of $1 - \chi$? Give a description, as detailed as possible, of the level curves of χ, including a sketch.

2.5.* For $\epsilon > 0$, define $\gamma_\epsilon \in C^\infty(\mathbf{R})$ by

$$\gamma_\epsilon(x) = \frac{1}{\epsilon\sqrt{\pi}}\,e^{-x^2/\epsilon^2}.$$

Calculate $|\cdot| * \gamma_\epsilon$. Prove that this function is analytic on $\mathbf{R}$ and examine how closely it approximates the function $|\cdot|$. Also calculate its derivatives of first and second order. See Fig. 2.6.

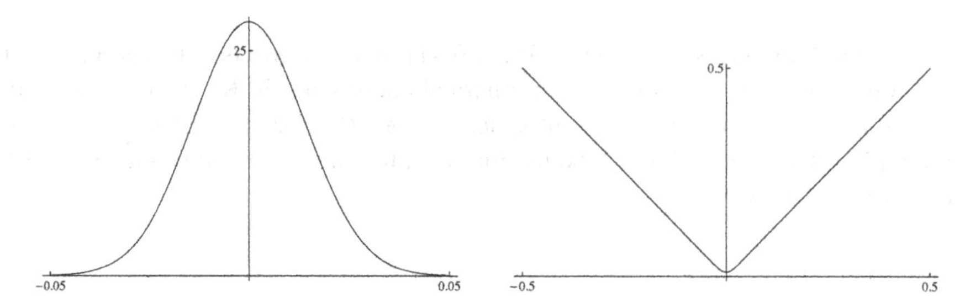

Fig. 2.6 Illustration for Problem 2.5. Graphs of γ_ϵ and $|\cdot| * \gamma_\epsilon$ with $\epsilon = 1/50$

2.6. Let U be a proper open subset of $\mathbf{R}^n$. Let $\gamma(x)$ be as in (2.12) and $\gamma_\epsilon(x) = \frac{1}{\epsilon^n}\gamma(\frac{1}{\epsilon}x)$, for $\epsilon > 0$. Denote the "probability of distance to 0 larger than r" by

$$\rho(r) = \int_{\|x\|>r} \gamma(x)\, dx.$$

Give an estimate of the r for which $\rho(r) < 10^{-6}$. Prove that $\chi_\epsilon := 1_U * \gamma_\epsilon$ is analytic and that $0 < \chi_\epsilon < 1$. Further prove that $\chi_\epsilon(x) \le \rho(\delta/\epsilon)$ if $d(x, U) = \delta > 0$; finally, show that $\chi_\epsilon(x) \ge 1 - \rho(\delta/\epsilon)$ if $d(x, \mathbf{R}^n \setminus U) = \delta > 0$.

2.7. Show, for $a > 0$, that

$$\int_{\mathbf{R}^n} e^{-a\|x\|^2/2}\, dx = \left(\frac{2\pi}{a}\right)^{\frac{n}{2}} \quad \text{and} \quad \int_{\mathbf{R}^n} \|x\|^2\, e^{-a\|x\|^2/2}\, dx = \frac{n}{a}\left(\frac{2\pi}{a}\right)^{\frac{n}{2}}.$$

Chapter 3
Distributions

For an arbitrary linear space E over $\mathbf{C}$, a $\mathbf{C}$-linear mapping : $E \to \mathbf{C}$ is also called a *linear form* or *linear functional* on E.

Definition 3.1. Let X be an open subset of $\mathbf{R}^n$. A *distribution on* X is a linear form u on $C_0^\infty(X)$ that is also *continuous* in the sense that

$$\lim_{j\to\infty} u(\phi_j) = u(\phi) \qquad \text{as} \qquad \lim_{j\to\infty} \phi_j = \phi \quad \text{in} \quad C_0^\infty(X).$$

Phrased differently, in this case continuity means preservation of convergence of sequences. The space of all distributions on X is denoted by $\mathcal{D}'(X)$. ⊘

The notation derives from the notation $\mathcal{D}(X)$ used by Schwartz for the space $C_0^\infty(X)$ of test functions equipped with the notion of convergence from Definition 2.13. (The letter $\mathcal{D}$ denotes differentiable.)

By the linearity of u, the assertion $u(\phi_j) \to u(\phi)$ is equivalent to $u(\phi_j - \phi) = u(\phi_j) - u(\phi) \to 0$, while $\phi_j \to \phi$ is equivalent to $\phi_j - \phi \to 0$. This implies that the continuity of a linear form u is equivalent to the assertion

$$\lim_{j\to\infty} u(\psi_j) = 0 \qquad \text{as} \qquad \lim_{j\to\infty} \psi_j = 0 \quad \text{in} \quad C_0^\infty(X).$$

Example 3.2. We have $u \in \mathcal{D}'(\mathbf{R})$ if $u(\phi) = \int_{\mathbf{R}} \phi(x)\,dx$, for all $\phi \in C_0^\infty(\mathbf{R})$. Indeed, u is well-defined because ϕ is continuous with compact support; the linearity of u is well-known and its continuity can be proved as follows. If $\lim_{j\to\infty} \phi_j = 0$ in $C_0^\infty(\mathbf{R})$, then there exists $m > 0$ such that $\operatorname{supp} \phi_j \subset [-m, m]$ for all j, while the convergence of the ϕ_j to the zero function is uniform on $[-m, m]$. Therefore we may interchange taking the limit and integration to obtain

$$\lim_{j\to\infty} u(\phi_j) = \lim_{j\to\infty} \int_{-m}^{m} \phi_j(x)\,dx = \int_{-m}^{m} \lim_{j\to\infty} \phi_j(x)\,dx = \int_{-m}^{m} 0\,dx = 0. \quad ⊘$$

J.J. Duistermaat and J.A.C. Kolk, *Distributions: Theory and Applications*,
Cornerstones, DOI 10.1007/978-0-8176-4675-2_3,

Example 3.3. We have $\mathrm{PV}\,\frac{1}{x} \in \mathcal{D}'(\mathbf{R})$ if we define this linear form, in the notation of Example 1.3, by

$$\mathrm{PV}\,\frac{1}{x} : C_0^\infty(\mathbf{R}) \to \mathbf{C} \qquad \text{with} \qquad \Big(\mathrm{PV}\,\frac{1}{x}\Big)(\phi) = \mathrm{PV}\int_{\mathbf{R}} \frac{\phi(x)}{x}\,dx.$$

Indeed, consider arbitrary $\phi \in C_0^\infty(\mathbf{R})$. Then there exists $m > 0$ with $\operatorname{supp}\phi \subset [-m, m]$. We may write $\int_{-m}^{m} |\log|x||\,dx =: c(m) > 0$ on account of the convergence of this integral. Hence, (1.3) leads to

$$\Big|\Big(\mathrm{PV}\,\frac{1}{x}\Big)(\phi)\Big| \leq c(m) \sup_{|x|\leq m} |\phi'(x)|,$$

and this implies that $\mathrm{PV}\,\frac{1}{x}$ is a continuous linear form on $C_0^\infty(\mathbf{R})$. See Example 5.7 for another proof. ⊘

For every complex linear space E with a notion of convergence, it is customary to denote the space of continuous linear forms on E by E'; this space is also referred to as the *topological dual* of E. If E is of finite dimension, every linear form on E is automatically continuous and E' is a complex linear space of the same dimension as E. For function spaces E of infinite dimension, this does not apply, and it is therefore sensible also to require continuity of the linear forms. This will open up a multitude of conclusions that one could not obtain otherwise, while the condition remains sufficiently weak to allow a large space of linear forms.

Remark 3.4. The study of general linear spaces E with a notion of convergence is referred to as *functional analysis*; this is outside the scope of this text. For the benefit of readers who (justifiably) find the preceding paragraph too vague, we add some additional clarification.

We require that addition and scalar multiplication in E be continuous with respect to the notion of convergence in E. That is, $x_j + y_j \to x + y$ and $c_j\,x_j \to c\,x$ if $x_j, x, y_j, y \in E$, $c_j \in \mathbf{C}$, and $x_j \to x$, $y_j \to y$ and $c_j \to c$ as $j \to \infty$. Furthermore, we impose the condition that limits of convergent sequences be uniquely determined. (The latter is a consequence of the usual requirement that E be a topological space having the Hausdorff property, which means that each two distinct points have nonintersecting neighborhoods.) In that case, E with its notion of convergence is also known as a *topological linear space*. A linear mapping $u : E \to \mathbf{C}$ is said to be continuous if $u\left(x_j\right) \to u(x)$ as $x_j \to x$ in E.

If E is of finite dimension, there exists a basis $(e_i)_{1\leq i\leq n}$ of E, where n is the dimension of E. The mapping that assigns to $x \in E$ the coordinates $(x_1, \dots, x_n) \in \mathbf{C}^n$ for which $x = \sum_{i=1}^{n} x_i\,e_i$ is a linear isomorphism from E to $\mathbf{C}^n$. The assertion is that via this linear isomorphism, convergence in E is equivalent to the usual coordinatewise convergence in $\mathbf{C}^n$.

Proof. The property that E is a topological linear space immediately leads to the conclusion that coordinatewise convergence implies convergence in E. We now prove the converse by mathematical induction on n.

Let $x_j \to x$ in E as $j \to \infty$. Select $1 \leq i \leq n$ and let c_j, or c, be the ith coordinate of x_j, or x, respectively. Suppose that the complex numbers c_j do not converge to c as $j \to \infty$; we will show that this assumption leads to a contradiction. By passing to a subsequence if necessary, we can arrange for the existence of a $\delta > 0$ with $|c_j - c| \geq \delta$ for all j. This implies that the sequence with terms $d_j = 1/\left(c_j - c\right)$ is bounded. Passing to a subsequence once again if necessary, we can arrange that there is a $d \in \mathbf{C}$ for which $d_j \to d$ in $\mathbf{C}$; here we apply Theorem 2.2.(b). With respect to the E-convergence, this leads to $y_j := d_j\,(x_j - x) \to d\,0 = 0$ as $j \to \infty$.

On the other hand, for every j, the ith coordinate of y_j equals 1. This means that for every j and k, the vector $y_j - y_k$ lies in the $(n-1)$-dimensional linear subspace E_0 of E consisting of the elements of E whose ith coordinates vanish. With respect to the E-convergence in E_0, the $y_j - y_k$ converge to zero if both j and k go to infinity; therefore, by the induction hypothesis, the $y_j - y_k$ converge coordinatewise to zero as $j, k \to \infty$. In other words, the y_j form a Cauchy sequence in $E \simeq \mathbf{C}^n$ with respect to coordinatewise convergence. This implies that $y \in E$ exists for which $y_j \to y$ by coordinates as $j \to \infty$, which in turn implies that $y_j \to y$ in E. Because we had already obtained that $y_j \to 0$ in E, it follows from the uniqueness of limits that $y = 0$. This leads to a contradiction to the fact that the y_j converge coordinatewise to y, because the ith coordinate of y_j equals 1 and the ith coordinate of $y = 0$ equals 0.

This negates the assumption that the complex numbers c_j do not converge to c. The conclusion therefore is that for every $1 \leq i \leq n$, the ith coordinate of x_j converges to the ith coordinate of x as $j \to \infty$. In other words, $x_j \to x$ in E implies the coordinatewise convergence of the x_j to x as $j \to \infty$. □

For an arbitrary linear function u on a linear space E with basis $(e_i)_{1 \leq i \leq n}$, one has

$$u(x) = u\left(\sum_{i=1}^{n} x_i\, e_i\right) = \sum_{i=1}^{n} x_i\, u\,(e_i)\,.$$

From this we can see that u is continuous with respect to coordinatewise convergence and is determined by its "matrix coefficients" $u_i = u\,(e_i)$, for $1 \leq i \leq n$. The conclusion is that for a topological linear space E of finite dimension, E' equals the space of **all** linear forms on E and this is a complex-linear space of the same dimension as E. ⊘

$\mathcal{D}'(X)$ is a complex-linear space with respect to addition and scalar multiplication defined by

$$(u+v)(\phi) = u(\phi) + v(\phi) \qquad \text{and} \qquad (\lambda\, u)(\phi) = \lambda\, u(\phi),$$

for u and $v \in \mathcal{D}'(X)$, $\phi \in C_0^\infty(X)$ and $\lambda \in \mathbf{C}$. It is a mere matter of writing out the definitions to show that $u + v$ and $\lambda\, u \in \mathcal{D}'(X)$ if $u, v \in \mathcal{D}'(X)$ and $\lambda \in \mathbf{C}$, and likewise, that with these definitions $\mathcal{D}'(X)$ does indeed become a linear space over the complex numbers $\mathbf{C}$.

Many functions can be interpreted as distributions. To explain this, we recall Definition 2.17 of local integrability of a function.

Theorem 3.5. *For every locally integrable function f on X,*

$$(\text{test } f)(\phi) = \int_X f(x)\,\phi(x)\,dx \qquad (\phi \in C_0^\infty(X)) \tag{3.1}$$

defines a distribution $u = \text{test } f$ on X.

Indeed, let $\phi_j \in C_0^\infty(X)$, $K \subset X$ compact, and $\operatorname{supp} \phi_j \subset K$ for all j. Then $\lim_{j\to\infty} u(\phi_j) = 0$ if $\lim_{j\to\infty} \phi_j = 0$ uniformly on K.

Proof. First we show that u is well-defined. By the compactness of $K := \operatorname{supp} \phi$, there are finitely many $a(i) \in X$ such that K is contained in the union of the rectangles $B_{a(i)}$. This means that the sum of the characteristic functions of the $B_{a(i)}$ is a majorant of the characteristic function of K. Using the identity $\int_K g(x)\,dx = \int 1_K(x)\,g(x)\,dx$, we deduce the absolute convergence of the integral in (3.1) from

$$\int_K |f(x)\,\phi(x)|\,dx \le \sum_i \int_{B_{a(i)}} |f(x)|\,|\phi(x)|\,dx \le c \sup_{x\in K} |\phi(x)|,$$

where

$$c = \sum_i \int_{B_{a(i)}} |f(x)|\,dx > 0.$$

The continuity of u now follows from the fact that for all j, $|u(\phi_j)|$ can be estimated by a constant times the supremum norm of ϕ_j. Indeed, the supports of the ϕ_j are contained in a fixed compact $K \subset X$ as $\phi_j \to 0$ in $C_0^\infty(X)$. □

Lemma 3.6. *The mapping $f \mapsto \text{test } f$ is linear and injective from $C(X)$, the linear space of all continuous functions on X, to $\mathcal{D}'(X)$, the space of distributions on X.*

Proof. By writing out the definitions one immediately verifies that the mapping is linear. A linear mapping is known to be injective if and only if its null space equals 0. In other words, assuming $(\text{test } f)(\phi) = 0$ for all test functions ϕ, we have to prove that $f(x) = 0$ for all $x \in X$.

With ϕ_ϵ as in (2.5), we obtain, for $x \in X$ and $\epsilon > 0$ sufficiently small,

$$(f * \phi_\epsilon)(x) = (\text{test } f)(T_x S\phi_\epsilon). \tag{3.2}$$

In the right-hand side the *reflection* S of a function ψ is defined by

$$(S\psi)(z) = \psi(-z).$$

We have already come across the translation T_x, in (1.13). This now yields

$$(T_x S\phi_\epsilon)(y) = (S\phi_\epsilon)(y - x) = \phi_\epsilon(x - y),$$

which enables us to see that (3.2) is another way of writing (1.12), with ϕ replaced by ϕ_ϵ.

The assumption $(\text{test } f)(\phi) = 0$, for all test functions ϕ, now implies that we obtain $(f * \phi_\epsilon)(x) = 0$, for every $\epsilon > 0$ and for all x. But in the text preceding (2.6) we have already seen that $f * \phi_\epsilon$ converges uniformly on compact sets to f as $\epsilon \downarrow 0$; and certainly, therefore,

$$f(x) = \lim_{\epsilon \downarrow 0} (f * \phi_\epsilon)(x) = 0. \qquad \square$$

Remark 3.7. This lemma justifies the usual identification of the continuous function f with the distribution test f. In other words, $C(X)$ is identified with the linear subspace $\{ \text{test } f \mid f \in C(X) \}$ of $\mathcal{D}'(X)$, and the premodifier "test" preceding the continuous function f is omitted when f is regarded as a distribution. A small amusement: $1 = \text{test } 1$ is integration of test functions. In this case the left-hand side is identified with the function that constantly equals 1; when we are piling identifications on top of each other, we should not be surprised that the notation becomes ambiguous.

For arbitrary locally integrable functions f the situation is a little subtler. A function f is said to be equal to g *almost everywhere* if the set $\{ x \in X \mid f(x) \neq g(x) \}$ has Lebesgue measure equal to 0; see Definition 20.14. In this case test $f =$ test g, while $f = g$ is not necessarily true in the strict sense that $f(x) = g(x)$ for all $x \in X$. Conversely, if test $f =$ test g, the functions f and g are equal almost everywhere. Below we outline the proof. $\oslash$

Proof. Write $h = f - g$. If $\phi \in C_0(X)$ there exists a sequence $\phi_j \in C_0^\infty(X)$ that converges uniformly to ϕ. This gives

$$\int h(x)\,\phi(x)\,dx = \lim_{j\to\infty} (\text{test } h)(\phi_j) = 0.$$

From the theory of Lebesgue integration it is known that the validity of this identity, for all $\phi \in C_0(X)$, implies that h vanishes almost everywhere; see Theorem 20.38 and Corollary 20.39.

While not all readers may be familiar with the latter result, it will be realized that h vanishes almost everywhere if the integral of h over any rectangle U equals 0. If h is tested with the functions $\chi_{U,\epsilon} \in C_0^\infty(X)$ from Lemma 2.19, we obtain that $\int_U h(x)\,dx = 0$ by taking the limit as $\epsilon \downarrow 0$. $\square$

It is customary in the theory of Lebesgue integration, see (20.18), to identify functions with each other when they are equal almost everywhere. We can therefore say that this custom amounts to **interpreting locally integrable functions f as distributions**, via the mapping $f \mapsto \text{test } f$. Again, the word "test" is omitted.

An example of a distribution **not** of the form test f for a locally integrable function f is the *Dirac function*, or *point measure*, δ_a, situated at the point $a \in X$. This is defined by

$$\delta_a(\phi) := \phi(a) \qquad (\phi \in C_0^\infty(X)),$$

in other words, by *evaluating* the test function ϕ at the point a. When $a = 0$, one simply uses the term Dirac function, without further specification, denoting it by δ. If $\delta_a = \text{test}\, f$, for a locally integrable function f, the restriction of f to $X \setminus \{a\}$ equals 0 almost everywhere. Because $\{a\}$ has measure 0, we conclude that f equals 0 almost everywhere; thus, test $f = 0$. This leads to a contradiction, because there exists a test function ϕ with $\phi(a) \neq 0$; consequently $\delta_a \neq 0$.

If one equates locally integrable functions with distributions, one should not be fussy about using function notation for distributions. In particular, for arbitrary $u \in \mathcal{D}'(X)$ one encounters the notation

$$u(\phi) = \int_X u(x)\,\phi(x)\,dx = \langle u, \phi \rangle = \langle \phi, u \rangle \qquad (\phi \in C_0^\infty(X)). \tag{3.3}$$

This will not lead to problems, provided one does not conclude that there is any meaning in the phrase "the value $u(x)$ of the distribution u at the point x." In the case of the Dirac function, for example, this certainly entails some problems at $x = 0$. So much for comments on the *Dirac notation*

$$\phi(a) = \int_{\mathbf{R}^n} \delta(x-a)\,\phi(x)\,dx = \int_{\mathbf{R}^n} \delta(x)\,\phi(x+a)\,dx.$$

(This formula describes the *sifting property* of the Dirac function.)

Normally, the continuity of a linear form u on $C_0^\infty(X)$ is most easily proved by giving an estimate for $u(\phi)$ in terms of a so-called C^k norm of $\phi \in C_0^\infty(X)$; see Example 3.3, for instance. In order to formulate this, we introduce, for every compact subset K of X, the linear space $C_0^\infty(K)$ consisting of all $\phi \in C_0^\infty(X)$ with supp $\phi \subset K$. On this space we have, for every $k \in \mathbf{Z}_{\geq 0}$, the C^k *norm*, defined by

$$\|\phi\|_{C^k} = \sup_{|\alpha| \leq k,\, x \in X} |\partial^\alpha \phi(x)|.$$

In Chap. 8 we establish, in a general context, a relation between the continuity of a linear mapping and estimates in terms of (semi)norms. For linear forms on $C_0^\infty(X)$ this looks as follows.

Theorem 3.8. *Let X be an open subset of $\mathbf{R}^n$. A linear form u on $C_0^\infty(X)$ belongs to $\mathcal{D}'(X)$ if and only if for every compact subset K of X there exist a constant $c > 0$ and an order of differentiation $k \in \mathbf{Z}_{\geq 0}$ with the property*

$$|u(\phi)| \leq c\, \|\phi\|_{C^k} \qquad (\phi \in C_0^\infty(K)). \tag{3.4}$$

Proof. $\Leftarrow$. $\phi_j \to \phi$ in $C_0^\infty(X)$ means that a compact subset K of X exists for which ϕ_j and $\phi \in C_0^\infty(K)$ for all j, while $\lim_{j\to\infty} \|\phi_j - \phi\|_{C^k} = 0$ for all k. If u satisfies (3.4), it then follows that $u(\phi_j) - u(\phi) = u(\phi_j - \phi)$ converges to 0 as $j \to \infty$.

$\Rightarrow$. Now suppose that u does not satisfy condition (3.4). This means that a compact subset K of X exists such that for every $c > 0$ and $k \in \mathbf{Z}_{\geq 0}$, there is a $\phi_{c,k} \in C_0^\infty(K)$ for which $|u(\phi_{c,k})| > c\, \|\phi_{c,k}\|_{C^k}$. This implies

$$\|\psi_{c,k}\|_{C^k} < \frac{1}{c} \qquad \text{and} \qquad |u(\psi_{c,k})| = 1,$$

if we take $\psi_{c,k} = \lambda\, \phi_{c,k}$ and $\lambda = 1/|u(\phi_{c,k})|$. The sequence $(\psi_{k,k})_{k\in\mathbf{N}}$ converges to 0 in $C_0^\infty(X)$, while $u(\psi_{k,k})$ does not converge to 0; therefore, u is not a distribution, which is a contradiction. $\square$

Theorem 3.8, which is concerned with the topological dual $\mathcal{D}'(X)$ of $C_0^\infty(X)$, is dual to Definition 2.13, which is about $C_0^\infty(X)$. Both assertions contain two quantifiers: the universal "for every" and the existential "there exists." The duality is reflected in the fact that the first quantifier in one assertion is replaced by the second quantifier in the other one. This fact is helpful for remembering the formulation of Theorem 3.8.

Example 3.9. From Example 3.3 one immediately derives estimates as in Theorem 3.8 for the linear form PV $\frac{1}{x}$ on $C_0^\infty(\mathbf{R})$. ⊘

Example 3.10. We have $u \in \mathcal{D}'(\mathbf{R})$ if

$$u(\phi) = \int_{\mathbf{R}^2} |x_1 x_2|\, \phi''(\|x\|)\, dx \qquad (\phi \in C_0^\infty(\mathbf{R})).$$

Indeed, u is a well-defined linear form on $C_0^\infty(\mathbf{R})$. Furthermore, if K is a compact subset of $\mathbf{R}$, there exists $m > 0$ such that $K \subset [-m, m]$. If $x \in \mathbf{R}^2$ satisfies $\|x\| \in K$, then $|x_i| \leq \|x\| \leq m$, for $1 \leq i \leq 2$. Since $\int_{-m}^{m} |x|\, dx = m^2$, we get

$$|u(\phi)| \leq m^4 \sup_{t\in K,\, |j|\leq 2} |\phi^{(j)}(t)| = m^4\, \|\phi\|_{C^2} \qquad (\phi \in C_0^\infty(K)).$$

The assertion now follows from Theorem 3.8.

Suppose that u were defined without absolute signs. Then we would have $u = 0$. This is directly verified by introducing polar coordinates in $\mathbf{R}^2$ (see Example 3.14 below). ⊘

Example 3.11. If V is a closed k-dimensional C^1 submanifold contained in an open subset X of $\mathbf{R}^n$ and f a continuous function on V, then

$$f\,\delta_V(\phi) = \int_V f(y)\,\phi(y)\, dy \qquad (\phi \in C_0^\infty(X))$$

defines a distribution $f\,\delta_V \in \mathcal{D}'(X)$. Here the integration over V is Euclidean k-dimensional integration, introduced in [7, Sect. 7.3], for example.

Indeed, if K is an arbitrary compact subset of X, there exists a closed ball B in $\mathbf{R}^n$ containing K. Then $m \in \mathbf{R}$ if $m = \sup_{y\in V\cap B} |f(y)|$ on account of the continuity of f, and also $\int_{V\cap B} dy < \infty$. Hence

$$|f\,\delta_V(\phi)| \leq m\,\|\phi\|_{C^0} \int_{V\cap B} dy \qquad (\phi \in C_0^\infty(K)). \tag{3.5}$$

If $f = 1$ we write $f\,\delta_V = \delta_V$. For $V = \{p\}$ this notation is consistent with the use of δ that we have seen before. If $k = n$, the set V is open in X and $\delta_V = 1_V$, the characteristic function of V. For $k = n - 1$, the distribution $f\,\delta_V$ is said to be a *layer* in X, with density function f. For $n = 3$ this may be visualized as a charge distribution over a surface V. ⊘

Definition 3.12. If $u \in \mathcal{D}'(X)$, the minimal $k \in \mathbf{Z}_{\geq 0}$ for which (3.4) holds, for certain $c > 0$, is known as the *order* of u on K. The supremum over all compact subsets K of X of the orders of u on K is called the *order* of the distribution u on X. In other words, u is of order k if and only if for every compact subset K of X there exists a constant $c = c(K) > 0$ such that

$$|u(\phi)| \leq c\,\|\phi\|_{C^k} \qquad (\phi \in C_0^\infty(K)). \tag{3.6}$$

⊘

Example 3.13. The estimate (3.5) implies that the order of $f\,\delta_V \in \mathcal{D}'(X)$ as in Example 3.11 equals 0. Both PV $\frac{1}{x} \in \mathcal{D}'(\mathbf{R})$ from Example 3.9 (use Problem 3.3) and $u \in \mathcal{D}'(\mathbf{R}^2)$ from Problem 3.5.(ii) are distributions of order 1. Furthermore, see Problem 3.2 for an example of a distribution on $\mathbf{R}$ of order k, for arbitrary $k \in \mathbf{Z}_{\geq 0}$. The order of a distribution u can be ∞, as for the distribution u on $\mathbf{R}$ defined by

$$u(\phi) = \sum_{j\in\mathbf{Z}_{\geq 0}} (-1)^j\,\phi^{(j)}(j) \qquad (\phi \in C_0^\infty(\mathbf{R})). \tag{3.7}$$

To see why this is so, use Problem 3.2 again. ⊘

Example 3.14. For $a > -1$, consider

$$u_a(\phi) = \int_{\mathbf{R}^2} |x_1 x_2|^a\,\phi''(\|x\|)\,dx \qquad (\phi \in C_0^\infty(\mathbf{R})).$$

Then u_a is a well-defined distribution, because $\int_{\mathbf{R}^2} |x_1x_2|^a\,dx = (\int_{\mathbf{R}} |x|^a\,dx)^2$ is convergent. In addition, u_a is of order ≤ 2, see also Example 3.10.

By introducing polar coordinates in $\mathbf{R}^2$, we obtain that

$$u_a(\phi) = \int_{-\pi}^{\pi} \frac{1}{2}|\sin 2\alpha|^a\,d\alpha \int_{\mathbf{R}_{>0}} r^{2a+1}\phi''(r)\,dr.$$

Let c_a denote a negative constant depending on a. We may then write $u_{-\frac{1}{2}}(\phi) = c_{-\frac{1}{2}}\,\phi'(0)$. If $a > -\frac{1}{2}$, integration by parts is permitted. This leads to

$$u_a(\phi) = c_a \int_{\mathbf{R}_{>0}} r^{2a}\phi'(r)\,dr \qquad \left(a > -\frac{1}{2}\right).$$

In particular, then, u_a is of order ≤ 1, for $a \geq -\frac{1}{2}$. Furthermore we obtain $u_0 = \pi\,\delta$, which is of order 0. Finally, suppose that $a > 0$. Upon integrating by parts once more, we see that also in this case u_a is of order 0. ⊘

In the next theorem we demonstrate that distributions of finite order k can be identified with the continuous linear forms on the space $C_0^k(X)$ of C^k functions with compact support in X. In this context, the convergence in $C_0^k(X)$ is that of Definition 2.13, where condition (b) is required only for all multi-indices α with $|\alpha| \leq k$. Note that $C_0^\infty(X) \subset C_0^k(X) \subset C_0^l(X)$ if $k > l$, in which there are no equalities. Furthermore, if $\infty \geq k > l$ and the sequence (ϕ_j) in $C_0^k(X)$ converges in $C_0^k(X)$ to ϕ, it also converges in $C_0^l(X)$. This implies that the restriction to $C_0^k(X)$ of a continuous linear form on $C_0^l(X)$ defines a continuous linear form on $C_0^k(X)$. In particular, the restriction to $C_0^\infty(X)$ of a continuous linear form on $C_0^l(X)$ is a distribution on X of order $\leq l$. A converse assertion can also be proved:

Theorem 3.15. *Let X be an open subset of $\mathbf{R}^n$, $k \in \mathbf{Z}_{\geq 0}$, and u a distribution on X of order $\leq k$. Then u has a unique extension to a continuous linear form v on $C_0^k(X)$.*

Proof. Let $f \in C_0^k(X)$. With the functions ϕ_ϵ as in (2.5) write $f_\epsilon = f * \phi_\epsilon$. We obtain, for every multi-index α with $|\alpha| \leq k$, that $\partial^\alpha f_\epsilon = (\partial^\alpha f) * \phi_\epsilon$ converges uniformly to $\partial^\alpha f$ as $\epsilon \downarrow 0$. From Lemma 2.18 it follows that for sufficiently small $\epsilon > 0$, the supports of the $f_\epsilon \in C_0^\infty(\mathbf{R}^n)$ are contained in a fixed compact subset of X. For these ϵ, therefore, $f_\epsilon \in C_0^\infty(X)$ and these functions converge to f in $C_0^k(X)$ as $\epsilon \downarrow 0$. Thus, if v exists at all, it is given by

$$v(f) = \lim_{\epsilon \downarrow 0} u(f_\epsilon).$$

This implies the uniqueness of a continuous linear extension of u to $C_0^k(X)$. From this uniqueness it follows in turn that the right-hand side (provided it exists) does not depend on the choice of the family of functions ϕ_ϵ as in (2.5).

Now the right-hand side in

$$|u(f_\epsilon) - u(f_\eta)| = |u(f_\epsilon - f_\eta)| \leq c\, \|f_\epsilon - f_\eta\|_{C^k}$$

converges to 0 as ϵ and $\eta \downarrow 0$. This implies that $(u(f_\epsilon))_{\epsilon>0}$ is a Cauchy sequence in $\mathbf{C}$ (with ϵ running through $\frac{1}{n}$ for $n \in \mathbf{N}$), and therefore converges to a complex number that we will denote by $v(f)$. Because the mapping $f \mapsto f * \phi_\epsilon$ is linear, we find that this defines a linear form v on $C_0^k(X)$. Furthermore, taking the limit as $\epsilon \downarrow 0$ in

$$\begin{aligned} |v(f)| &\leq |v(f) - u(f_\epsilon)| + |u(f_\epsilon)| \leq |v(f) - u(f_\epsilon)| + c\, \|f_\epsilon\|_{C^k} \\ &\leq |v(f) - u(f_\epsilon)| + c\, \|f_\epsilon - f\|_{C^k} + c\, \|f\|_{C^k} \end{aligned}$$

yields the result that $|v(f)| \leq c \, \|f\|_{C^k}$ for all $f \in C_0^k(X)$ with support in a fixed compact subset of X, which implies that v is continuous on $C_0^k(X)$. This also gives $v(f) = u(f)$ if $f \in C_0^\infty(X)$. □

It is common to write $v = u$, in other words, to identify u with its continuous extension to $C_0^k(X)$. A continuous linear form on $C_0(X) = C_0^0(X)$, the space of continuous functions with compact support in X, is also known as a *measure* on X, or preferably, a *Radon measure* on X, because one wants to distinguish it from the general set-theoretic concept of measure. Thus, distributions of order 0 are identified with Radon measures; for a more detailed discussion of this matter as well as a succinct but fairly complete discussion of integration theory, we refer to Chap. 20.

Example 3.16. If f is a locally integrable function on X, the second assertion in Theorem 3.5 implies that test f is a distribution on X of order 0. We conclude that test f has a unique extension to a Radon measure μ on X. This μ is called the measure with *density function* f; but in view of the identifications made, we can also write $f = \text{test}\, f = \mu$. An example of a Radon measure without locally integrable density function is the Dirac function, which is therefore also called the *Dirac measure*. The estimate (3.5) for $f\,\delta_V \in \mathcal{D}'(X)$ shows that $f\,\delta_V$ is a Radon measure on X. Because a finite linear combination of Radon measures is again a Radon measure, one may obtain examples of Radon measures for which some summands possess a density function and others not. ⊘

Definition 3.17. A distribution $u \in \mathcal{D}'(X)$ is said to be *positive* if $\phi \in C_0^\infty(X)$ and $\phi \geq 0$ imply that $u(\phi) \geq 0$. In this case we write $u \geq 0$. Here we use the convention that for a complex number c the notation $c \geq 0$ means that $c \in \mathbf{R}$ and $c \geq 0$.

A Radon measure u on X is said to be *positive* if $u(f) \geq 0$ for every nonnegative $f \in C_0(X)$. ⊘

Theorem 3.18. *Every positive distribution is a positive Radon measure.*

Proof. Suppose $u \in \mathcal{D}'(X)$ and $u \geq 0$. We begin by showing that u has order 0; on account of Theorem 3.15 it then has an extension to a Radon measure on X. Let K be a compact subset of X. Corollary 2.16 yields a $\chi \in C_0^\infty(X)$ with $0 \leq \chi \leq 1$ and $\chi = 1$ on K. This implies that $u(\chi) \geq 0$; in particular, $u(\chi)$ is real. Let $\phi \in C_0^\infty(K)$ be real-valued. With the notation $c = \|\phi\| = \sup_x |\phi(x)|$ we then get $c\,\chi - \phi \geq 0$, and therefore

$$c\,u(\chi) - u(\phi) = u(c\,\chi - \phi) \geq 0.$$

This implies that $c\,u(\chi) - u(\phi)$ is a nonnegative real number, so $u(\phi)$ is a real number and $u(\phi) \leq u(\chi)\,\|\phi\|$.

Now let $\phi \in C_0^\infty(K)$ be complex-valued. Then

$$u(\phi) = u(\mathrm{Re}\,\phi + i\,\mathrm{Im}\,\phi) = u(\mathrm{Re}\,\phi) + i\,u(\mathrm{Im}\,\phi)$$

with $u(\operatorname{Re}\phi)$ and $u(\operatorname{Im}\phi) \in \mathbf{R}$. In particular, $u(\phi) = u(\operatorname{Re}\phi)$ if $u(\phi) \in \mathbf{R}$. Using the notation

$$\alpha = \arg u(\phi) \qquad \text{and} \qquad \psi = \operatorname{Re}(e^{-i\alpha}\,\phi)$$

we now obtain

$$|u(\phi)| = e^{-i\alpha}\,u(\phi) = u(e^{-i\alpha}\,\phi) = u(\psi) \leq u(\chi)\,\|\psi\| \leq u(\chi)\,\|\phi\|. \tag{3.8}$$

This shows that u is of order 0.

If $f \in C_0(X)$ and $f \geq 0$, then also $f_\epsilon = f * \phi_\epsilon \geq 0$ because $\phi_\epsilon \geq 0$. Therefore $u(f_\epsilon) \geq 0$ and, consequently, also $u(f) = \lim_{\epsilon \downarrow 0} u(f_\epsilon) \geq 0$. Hence, the Radon measure u is positive. □

Remark 3.19. In constructing the proof we have derived that $u \geq 0$ implies that $u(\phi) \in \mathbf{R}$, for every continuous real-valued function ϕ with compact support. If both ϕ and ψ are real-valued and $\phi(x) \leq \psi(x)$ for all x, then $\psi - \phi \geq 0$, whence $u(\psi) - u(\phi) = u(\psi - \phi) \geq 0$, or $u(\phi) \leq u(\psi)$. The conclusion is that the linear form u is *monotone* in the sense that $u(\phi) \leq u(\psi)$, for ϕ and ψ real-valued and $\phi \leq \psi$, if and only if $u \geq 0$. ⊘

Example 3.20. A function f may be written as $f = f_+ - f_-$ with $f_\pm = \frac{1}{2}(|f| \pm f) \geq 0$, and these functions are continuous if f is. In particular, the distributions $f_\pm\,\delta_V$ as in Example 3.11 are positive. Theorem 3.18 then implies that they are positive Radon measures. This confirms that fact known from Example 3.16 that $f\,\delta_V$ is a Radon measure. Note that the argument does not require an estimate, like the one in (3.5). ⊘

Remark 3.21. One writes $\phi_j \uparrow 1$ as $j \to \infty$ for a sequence of functions ϕ_j if for every x we have $\phi_j(x) \uparrow 1$ as $j \to \infty$. A positive Radon measure μ is said to be a *probability measure* if a sequence of test functions ϕ_j exists with the property that $\phi_j \uparrow 1$ and $\mu(\phi_j) \uparrow 1$ as $j \to \infty$. This implies that for every sequence of test functions ψ_j with $\psi_j \uparrow 1$ as $j \to \infty$, one has $\mu(\psi_j) \uparrow 1$ as $j \to \infty$; one simply writes $\mu(1) = 1$.

For a (test) function ϕ, this $\mu(\phi)$ is called the *expectation* of ϕ with respect to the probability measure μ. If f is an integrable function with $f \geq 0$ and $\int f(x)\,dx = 1$, then $\mu = \text{test}\, f$ is a probability measure. In this case f is called a *probability density*. A finite linear combination

$$\mu = \sum_{j=1}^{l} p_j\,\delta_{a(j)}$$

of point measures, where the $a(j)$ are different points in $\mathbf{R}^n$, is a probability measure if and only if $p_j \geq 0$ for all j and $\sum_{j=1}^{l} p_j = 1$. The number p_j is then called the probability of the alternative $a(j)$.

Thus we arrive at a distributional interpretation of the calculus of probabilities. One should remain aware, however, that many results for probability measures do not hold for more general distributions. ⊘

Problems

3.1. Prove that

$$u(\phi) = \phi^{(k)}(0) \qquad (\phi \in C_0^\infty(\mathbf{R}))$$

defines a distribution $u \in \mathcal{D}'(\mathbf{R})$ of order $\leq k \in \mathbf{Z}_{\geq 0}$.

3.2.* Show that the distribution u from the preceding problem is not of order strictly lower than k on any neighborhood of 0. Hint: highly oscillatory test functions have large derivatives.

3.3. Show that PV $\frac{1}{x}$, $\frac{1}{x+i0}$ and $\frac{1}{x-i0}$ all are distributions of order 1. Hint: use Problem 1.3.

3.4. Verify that u, v, and w below are distributions on $\mathbf{R}^2$:

(i) $u(\phi) = \partial^\alpha \phi(x)$, where α is the multi-index $(1, 1)$ and x the point $(1, 1)$.
(ii) $v(\phi) = \int_{\mathbf{R}} \phi(t, 0)\, dt$.
(iii) $w(\phi) = \int_{\mathbf{R}^2} e^{\|x\|^2} \phi(x)\, dx$.

3.5.* We consider functions and distributions on $\mathbf{R}^2$.

(i) $r(x) = \|x\|$ defines a function on $\mathbf{R}^2$. Verify that $\log r$ and $1/r$ define distributions on $\mathbf{R}^2$. What is the order of these distributions?
(ii) (a) Define

$$u(\phi) = \int_0^\pi (\cos t\, \partial_1 + \sin t\, \partial_2)\phi(\cos t, \sin t)\, dt.$$

Show that this defines a distribution u on $\mathbf{R}^2$. What can you say about the order of u?

(b) The same questions for

$$v(\phi) = \int_0^\pi (-\sin t\, \partial_1 + \cos t\, \partial_2)\phi(\cos t, \sin t)\, dt.$$

Observe that $\cos t\, \partial_1 + \sin t\, \partial_2$ and $-\sin t\, \partial_1 + \cos t\, \partial_2$ is a directional derivative in a direction perpendicular and tangential, respectively, to the curve $t \mapsto (\cos, \sin t)$. In terms of vector analysis, u and v represent the flux of a gradient vector field across the curve and the work along it, respectively.

3.6. Demonstrate that for a continuous function f one has $f \geq 0$ as a distribution if and only if $f \geq 0$ as a function.

Chapter 4
Differentiation of Distributions

If f is continuously differentiable on an open set X in $\mathbf{R}^n$, one obtains by means of integration by parts that for every test function ϕ the following holds:

$$(\text{test}\, \partial_j f)(\phi) = -(\text{test}\, f)(\partial_j \phi).$$

Note that the boundary term on the right-hand side is absent because $\phi(x) = 0$ for x sufficiently large; see [7, Corollary 7.6.2].

For an arbitrary distribution u on X we now **define**

$$(\partial_j u)(\phi) = -u(\partial_j \phi) \qquad (1 \leq j \leq n,\ \phi \in C_0^\infty(X)).$$

Writing out the definition gives $\partial_j u \in \mathcal{D}'(X)$, which is called the *distributional derivative* of u with respect to the jth variable. The form of the definition is such that

$$\partial_j(\text{test}\, f) = \text{test}(\partial_j f)$$

for every continuously differentiable function f, so that we do not run into difficulties when we simply speak of "functions" when referring to distributions that are of the form test f. By mathematical induction on the number of derivatives we find that partial derivatives of u of arbitrary order are also distributions. As in the case of C^∞ functions, the order of differentiation may be changed arbitrarily:

Lemma 4.1. *For every distribution u on the open subset X of $\mathbf{R}^n$ and for every pair of indices $1 \leq j, k \leq n$, one has $\partial_j(\partial_k u) = \partial_k(\partial_j u)$.*

Proof. For every $\phi \in C_0^\infty(X)$ one obtains

$$\begin{aligned}(\partial_j \circ \partial_k u)(\phi) &= -\partial_k u(\partial_j \phi) = u(\partial_k \circ \partial_j \phi) = u(\partial_j \circ \partial_k \phi) = -\partial_j u(\partial_k \phi) \\ &= (\partial_k \circ \partial_j u)(\phi).\end{aligned}$$

□

With respect to the examples of continuous functions u for which $\partial_j\, \partial_k u \neq \partial_k\, \partial_j u$ we may therefore note that the two sides are, in fact, distributionally equal. It is interesting to see an example where this happens. For instance, consider

J.J. Duistermaat and J.A.C. Kolk, *Distributions: Theory and Applications*,
Cornerstones, DOI 10.1007/978-0-8176-4675-2_4,

$$u : \mathbf{R}^2 \to \mathbf{R} \qquad \text{given by} \qquad u(x) = x_1 x_2 \, \frac{x_1^2 - x_2^2}{x_1^2 + x_2^2}.$$

From [7, Theorem 2.7.2 and Example 2.7.1] it follows that $\partial_1 \partial_2 u(x) = \partial_2 \partial_1 u(x)$, except if $x = 0$. However, $\partial_1 \partial_2 u$ and $\partial_2 \partial_1 u$ coincide as locally integrable functions on $\mathbf{R}^2$, and accordingly as distributions.

Lemma 4.1 makes it possible to write all higher-order derivatives in the form

$$\partial^\alpha u := \frac{\partial^\alpha u}{\partial x^\alpha}.$$

Note that this distribution is of order $\leq k + m$ when u is of order $\leq k$ and $|\alpha| = m$. If U is open in $\mathbf{R}$ and $u \in \mathcal{D}'(U)$, we use the shorthand u' for the distributional derivative ∂u, and $u^{(k)}$ for $\partial^k u := \partial^{(k)} u$, where $k \in \mathbf{Z}_{>1}$.

In particular, every continuous function considered as a distribution has partial derivatives of all orders. Conversely, we shall prove in Examples 13.1 or 18.2 below that every distribution can locally be written as a linear combination of derivatives of some continuous function. If every continuous function is to be infinitely differentiable as a distribution, no proper subset of the space of distributions can therefore be adequate. In this sense, the distribution extension of the function concept is as economical as it possibly can be.

Example 4.2. $H := 1_{[0,\infty[}$, the characteristic function of $\mathbf{R}_{\geq 0}$, when interpreted as a distribution on $\mathbf{R}$, is known as the *Heaviside function.* (At the end of the nineteenth century, Heaviside introduced a kind of distribution calculus for computations on electrical networks. H may be interpreted in terms of the sudden switching on of a current.) One has $H' = \delta$. Indeed, for every $\phi \in C_0^\infty(\mathbf{R})$,

$$H'(\phi) = -H(\phi') = -\int_{\mathbf{R}_{>0}} \phi'(x)\, dx = \phi(0) = \delta(\phi).$$

More generally, for $a < b$, the characteristic function $1_{[a,b[}$ of the interval $[a, b[$ satisfies, in the notation (1.13),

$$1_{[a,b[} = T_a H - T_b H, \qquad \text{and so} \qquad 1'_{[a,b[} = \delta_a - \delta_b.$$

For $x \in \mathbf{R}$, we define $[x]$ as the integer such that $x - 1 < [x] \leq x$, the *integer part of* x. Then the *sawtooth function*

$$s = \frac{1}{2} + \frac{1}{\pi} \arctan\Big(\tan\Big(\pi\Big(\cdot - \frac{1}{2}\Big)\Big)\Big) : x \mapsto x - [x]$$

is locally integrable on $\mathbf{R}$ and defines $s \in \mathcal{D}'(\mathbf{R})$ (note the slight abuse of notation: the equality is not one of functions, because the right-hand side is not defined everywhere). With the notation $I(x) = x$, we have $s = I - \sum_{k \in \mathbf{Z}} k\, 1_{[k,k+1[}$, and therefore we have the following identity in $\mathcal{D}'(\mathbf{R})$:

$$s' = 1 - \sum_{k \in \mathbf{Z}} k(\delta_k - \delta_{k+1}) = 1 - \sum_{k \in \mathbf{Z}} \delta_k.$$

In Example 16.24 below this formula will lead to the Fourier series of every polynomial function made periodic with period 1. ⊘

The next theorem asserts that every distribution on $\mathbf{R}$ has a distributional antiderivative and that this is uniquely determined up to an additive constant.

Theorem 4.3. *Let I be an open interval in $\mathbf{R}$ and $f \in \mathcal{D}'(I)$. Then there exists $u \in \mathcal{D}'(I)$ for which $u' = f$. If, in addition, $v \in \mathcal{D}'(I)$ and $v' = f$, then $v = u + c$ for some constant $c \in \mathbf{C}$. In particular, $v \in \mathcal{D}'(I)$ and $v' = 0$ imply that v equals a constant function.*

Proof. Choose $\chi \in C_0^\infty(I)$ such that $1(\chi) = \int_I \chi(x)\,dx = 1$. For $\phi \in C_0^\infty(I)$, define the function $p(\phi)$ on $\mathbf{R}$ by

$$p(\phi)(x) := \int_{-\infty}^{x} \phi(t)\,dt - 1(\phi) \int_{-\infty}^{x} \chi(t)\,dt.$$

Then $p(\phi) \in C_0^\infty(I)$. Indeed, to see that $p(\phi)$ has compact support we note that there exist a and $b \in I$ for which $\phi(t) = \chi(t) = 0$ if $t < a$ or if $t > b$. This immediately implies that $p(\phi)(x) = 0$ if $x < a$, while in the case $x > b$ one has $p(\phi)(x) = 1(\phi) - 1(\phi)\,1 = 0$.

Denoting the derivative of ϕ by ϕ', we have $1(\phi') = 0$ and therefore $p(\phi') = \phi$. In other words, the mapping $p : C_0^\infty(I) \to C_0^\infty(I)$ is a left inverse of differentiation.

Furthermore, p is linear and sequentially continuous, that is,

$$\lim_{j\to\infty} p(\phi_j) = 0 \quad \text{in} \quad C_0^\infty(I) \quad \text{if} \quad \lim_{j\to\infty} \phi_j = 0 \quad \text{in} \quad C_0^\infty(I).$$

This implies that $u(\phi) := -f(p(\phi))$, for $\phi \in C_0^\infty(I)$, defines a distribution u on I. That distribution satisfies $u' = f$, because

$$u'(\phi) = -u(\phi') = f(p(\phi')) = f(\phi) \qquad (\phi \in C_0^\infty(I)).$$

To prove the second assertion in the theorem, observe that $w := v - u$ satisfies $w' = 0$, or $w(\phi') = 0$ for all $\phi \in C_0^\infty(I)$. In particular,

$$0 = w(p(\phi)') = w(\phi - 1(\phi)\,\chi) = w(\phi) - 1(\phi)\,w(\chi),$$

for all $\phi \in C_0^\infty(I)$, which implies that $w = w(\chi)$ test 1 $=$ test $w(\chi)$. Here we denote a constant function with value c by the same symbol c. □

Example 4.4. Let I be an open interval in $\mathbf{R}$ and $m \in \mathbf{Z}_{\geq 0}$. Then $u \in \mathcal{D}'(I)$ satisfies $u^{(m+1)} = 0$ if and only if there exists a polynomial function p on I of degree $\leq m$ such that $u = p$.

The proof is by means of Theorem 4.3 and mathematical induction on m. Thus, we obtain that $0 = u^{(m+1)} = (u^{(m)})'$ iff there exists $c \in \mathbf{C}$ such that $u^{(m)} = c$ iff $(u - c\frac{x^m}{m!})^{(m)} = 0$ iff there exists a polynomial function q of degree $\le m - 1$ such that $u - c\frac{x^m}{m!} = q$ iff $u = p$ with $p = c\frac{x^m}{m!} + q$. ⊘

Problems

4.1.* Prove that $|\cdot|' = \mathrm{sgn}$ and that $|\cdot|'' = 2\,\delta$ in $\mathcal{D}'(\mathbf{R})$.

4.2.* Show that $(\log|\cdot|)' = \mathrm{PV}\,\frac{1}{x}$ in $\mathcal{D}'(\mathbf{R})$. (For $(\log|\cdot|)''$, see Problem 13.6.)

4.3. Verify $(\tan\circ\arctan)' = 1$ and $(\arctan\circ\tan)' = 1 - \pi\sum_{k\in\mathbf{Z}}\delta_{(k+\frac{1}{2})\pi}$ in $\mathcal{D}'(\mathbf{R})$. Compute $(\arccos\circ\cos)'$; see Problem 1.7.

4.4. Let $\lambda \in \mathbf{C}$ and define $f(x) = e^{\lambda x}$ for $x > 0$ and $f(x) = 0$ for $x \le 0$. Prove that the derivatives of f satisfy

$$f^{(k)} = \lambda^k f + \sum_{j=0}^{k-1}\lambda^{k-1-j}\,\delta^{(j)} \qquad (k \in \mathbf{Z}_{\ge 0}).$$

Now let p be a polynomial of degree $m > 0$ and $p(\lambda) = 0$. Does $p(\partial)f$ vanish? Calculate the order of $p(\partial)f$.

4.5. Determine a continuous function f on $\mathbf{R}^n$ and a multi-index α for which $\partial^\alpha f = \delta$. Establish how much smaller α may be chosen if f is merely required to be locally integrable.

4.6.* Let $p \in \mathbf{R}^n$ and $v_j(x) = \|x - p\|^{-n}\,(x_j - p_j)$, for $1 \le j \le n$. For $n = 3$ this is the vector field v in (1.2). Verify that the v_j are locally integrable on $\mathbf{R}^n$ and thus define distributions on $\mathbf{R}^n$. Prove

$$\operatorname{div} v := \sum_{j=1}^{n}\partial_j v_j = c_n\,\delta_p,$$

where c_n denotes the $(n-1)$-dimensional volume of the sphere $S^{n-1} := \{\, x \in \mathbf{R}^n \mid \|x\| = 1\,\}$ in $\mathbf{R}^n$. (See (13.37) for an explicit formula for c_n.)

4.7.* (Sequel to Problem 4.6.) For $x \in \mathbf{R}^n \setminus \{0\}$, define

$$E(x) = \begin{cases} \dfrac{1}{(2-n)\,c_n\,\|x\|^{n-2}} & \text{if} \quad n \ne 2, \\[2ex] \dfrac{1}{2\pi}\log\|x\| & \text{if} \quad n = 2. \end{cases}$$

Prove that E is locally integrable on $\mathbf{R}^n$ and therefore defines a distribution on $\mathbf{R}^n$. Demonstrate the existence of a constant c for which $\partial_j E = c\,v_j$ as distributions. Then show that $\Delta E = \delta$, where

$$\Delta = \sum_{j=1}^{n} \partial_j^2 = \operatorname{div} \circ \operatorname{grad}$$

denotes the *Laplace operator* in $\mathbf{R}^n$.

4.8. Suppose $c_1, \ldots, c_q \in \mathbf{C}$ and $a_1 < \cdots < a_q \in \mathbf{R}$. Find the solutions $u \in \mathcal{D}'(I)$ of $u' = \sum_{j=1}^{q} c_j \, \delta_{a_j}$. In which case does one find $u = 1_I$ as a solution, for an interval I in $\mathbf{R}$?

Chapter 5
Convergence of Distributions

Definition 5.1. We now introduce a notion of convergence in the linear space $\mathcal{D}'(X)$ of distributions on an open set X in $\mathbf{R}^n$. Let $(u_j)_{j\in\mathbf{N}}$ be an infinite sequence in $\mathcal{D}'(X)$ and let $u \in \mathcal{D}'(X)$. One then writes

$$\lim_{j\to\infty} u_j = u \quad \text{in} \quad \mathcal{D}'(X) \qquad \text{if} \qquad \lim_{j\to\infty} u_j(\phi) = u(\phi) \qquad (\phi \in C_0^\infty(X)). \tag{5.1}$$

This is also referred to as *weak convergence* or *convergence of distributions*.

Instead of a sequence we can also take a family u_ϵ of distributions that depend on one or more real-valued parameters ϵ. We then say that u_ϵ converges in $\mathcal{D}'(X)$ to u as ϵ tends to a special value ϵ_0 if for every test function ϕ the complex numbers $u_\epsilon(\phi)$ converge to $u(\phi)$ as $\epsilon \to \epsilon_0$. ⊘

Example 5.2. Let $(f_j)_{j\in\mathbf{N}}$ be an infinite sequence of locally integrable functions on X and suppose that f is a locally integrable function on X with the property that for every compact subset K of X, one has

$$\lim_{j\to\infty} \int_K |f_j(x) - f(x)|\, dx = 0.$$

Then $f_j \to f$ in $\mathcal{D}'(X)$, in the sense that test $f_j \to$ test f in $\mathcal{D}'(X)$ as $j \to \infty$. Indeed, if $\phi \in C_0^\infty(X)$, application of standard inequalities for integrals yields

$$|\left(\text{test}\, f_j - \text{test}\, f\right)(\phi)| \leq \int_{\text{supp}\,\phi} |f_j(x) - f(x)|\, dx\, \sup_x |\phi(x)|.$$

In fact, this leads to convergence in the space of distributions of order 0, also referred to as *weak convergence of measures*. ⊘

Example 5.3. For $t > 0$ and $x \in \mathbf{R}$, set $u_t(x) = t\, H(x)\, e^{itx}$ with H as in Example 4.2. Then

$$u_t(\phi) = \int_{\mathbf{R}_{>0}} t\, e^{itx}\phi(x)\, dx = i\,\phi(0) + i \int_{\mathbf{R}_{>0}} e^{itx}\phi'(x)\, dx$$

J.J. Duistermaat and J.A.C. Kolk, *Distributions: Theory and Applications*,
Cornerstones, DOI 10.1007/978-0-8176-4675-2_5,

$$= i\,\phi(0) - \frac{\phi'(0)}{t} - \frac{1}{t}\int_{\mathbf{R}_{>0}} e^{itx}\phi''(x)\,dx \to i\,\phi(0) \quad \text{as} \quad t \to \infty,$$

for $\phi \in C_0^\infty(\mathbf{R})$. Hence $\lim_{t\to\infty} u_t = i\,\delta$ in $\mathcal{D}'(\mathbf{R})$. ⊘

The following *principle of uniform boundedness* will not be proved here. It is based on the *Banach–Steinhaus Theorem*, applied to the Fréchet space $C_0^\infty(K)$. See for example Rudin [19, Theorem 2.6] or Bourbaki [4, Livre V, Chap. III, §3, N°6 and §1, N°1, Corollaire]. Maybe the reader will see this principle explained as part of a course in functional analysis.

Lemma 5.4. *Let $(u_j)_{j\in\mathbf{N}}$ be a sequence in $\mathcal{D}'(X)$ with the property that for every $\phi \in C_0^\infty(X)$, the sequence $(u_j(\phi))_{j\in\mathbf{N}}$ in $\mathbf{C}$ is bounded. Then, for every compact subset K of X, there exist constants $c > 0$ and $k \in \mathbf{Z}_{\geq 0}$ such that*

$$|u_j(\phi)| \leq c\,\|\phi\|_{C^k} \qquad (j \in \mathbf{N},\ \phi \in C_0^\infty(K)). \tag{5.2}$$

Note the analogy in formulation with Theorem 3.8. The lemma above leads to the following property of *completeness* of the space of distributions. For the benefit of the reader who is not familiar with the principle of uniform boundedness, the proof of Theorem 5.5 is followed by a second proof of its part (i) that is based on the *method of the gliding hump*.

Theorem 5.5. *Let X be open in $\mathbf{R}^n$ and $(u_j)_{j\in\mathbf{N}}$ a sequence in $\mathcal{D}'(X)$ with the property that for every $\phi \in C_0^\infty(X)$ the sequence $(u_j(\phi))_{j\in\mathbf{N}}$ converges in $\mathbf{C}$ as $j \to \infty$; denote the limit by $u(\phi)$.*

(i) Then $u : \phi \mapsto u(\phi)$ defines a distribution on X. Furthermore, $\lim_{j\to\infty} u_j = u$ in $\mathcal{D}'(X)$.
(ii) If $\phi_j \to \phi$ in $C_0^\infty(X)$, then $\lim_{j\to\infty} u_j(\phi_j) = u(\phi)$ in $\mathbf{C}$.

Proof. Writing out the definitions, we find that u defines a linear form on $C_0^\infty(X)$. From the starting assumption it follows that the sequence $(u_j(\phi))$ is bounded for every $\phi \in C_0^\infty(X)$, and thus we obtain, for every compact $K \subset X$, an estimate of the form (5.2). Taking the limit in

$$|u(\phi)| \leq |u(\phi) - u_j(\phi)| + |u_j(\phi)| \leq |u(\phi) - u_j(\phi)| + c\,\|\phi\|_{C^k}$$

as $j \to \infty$, we get $|u(\phi)| \leq c\,\|\phi\|_{C^k}$ for all $\phi \in C_0^\infty(K)$. According to Theorem 3.8 this proves that $u \in \mathcal{D}'(X)$, and $u_j \to u$ in $\mathcal{D}'(X)$ now holds by definition.

Regarding the last assertion we observe that if $\phi_j \to \phi$ in $C_0^\infty(X)$, there exists a compact set $K \subset X$ such that ϕ_j and $\phi \in C_0^\infty(K)$ for all j. Applying Lemma 5.4 once again, we obtain from this

$$|u_j(\phi_j) - u(\phi)| \leq |u_j(\phi_j - \phi)| + |u_j(\phi) - u(\phi)| \leq c\,\|\phi_j - \phi\|_{C^k} + |u_j(\phi) - u(\phi)|,$$

which converges to 0 as $j \to \infty$. □

Proof. Suppose that u does not belong to $\mathcal{D}'(X)$. Then there exists a sequence (ϕ_j) in $C_0^\infty(X)$ that converges to 0 in $C_0^\infty(X)$, while $(u(\phi_j))$ does not converge to 0 as $j \to \infty$. Hence, by passing to a subsequence if necessary, we can arrange that there exists $c > 0$ such that $|u(\phi_j)| \geq c$. Recall from Definition 2.13 that $\phi_j \to 0$ in $C_0^\infty(X)$ is the case if there exists a compact set $K \subset X$ such that $\operatorname{supp} \phi_j \subset K$, while $\|\phi_j\|_{C^j} \leq \frac{1}{4^j}$ if we replace (ϕ_j) by a suitable subsequence if necessary. Accordingly, upon writing ϕ_j for $2^j \phi_j$, we obtain that $\phi_j \to 0$ in $C_0^\infty(X)$, while $|u(\phi_j)| \to \infty$ as $j \to \infty$.

Next, we define a subsequence of (ϕ_j), say (ψ_j) in $C_0^\infty(X)$, and a subsequence of (u_j), say (v_j) in $\mathcal{D}'(X)$, as follows. Select ψ_1 such that $|u(\psi_1)| > 2$. As $u_j(\psi_1) \to u(\psi_1)$, we may choose v_1 such that $|v_1(\psi_1)| > 2$. Now proceed by mathematical induction on j. Thus, assume that ψ_k and v_k have been chosen, for $1 \leq k < j$. Then select ψ_j from the sequence (ϕ_j) such that

$$\begin{aligned} &\text{(i)} \quad \|\psi_j\|_{C^j} < \frac{1}{2^j}, \qquad \text{(ii)} \quad |v_k(\psi_j)| < \frac{1}{2^{j-k}} \qquad (1 \leq k < j), \\ &\qquad \text{(iii)} \quad |u(\psi_j)| > \sum_{1 \leq k < j} |u(\psi_k)| + j + 1. \end{aligned} \tag{5.3}$$

Condition (i) can be satisfied, because of the properties of the ϕ_j; and (ii) because of $\phi_j \to 0$ in $C_0^\infty(X)$ and all v_k belong to $\mathcal{D}'(X)$, for $1 \leq j < k$; whereas (iii) holds because $|u(\phi_j)| \to \infty$. In addition, since $\lim_{j\to\infty} u_j(\psi) = u(\psi)$, for all $\psi \in C_0^\infty(X)$, condition (iii) implies that we may select v_j from the sequence (u_j) such that

$$|v_j(\psi_j)| > \sum_{1 \leq k < j} |v_j(\psi_k)| + j + 1. \tag{5.4}$$

Now, set $\psi = \sum_{k\in\mathbf{N}} \psi_k$. According to (5.3).(i) the series on the right-hand side converges in $C_0^\infty(X)$, which leads to $\psi \in C_0^\infty(X)$. Obviously, for any j,

$$v_j(\psi) = \sum_{1 \leq k < j} v_j(\psi_k) + v_j(\psi_j) + \sum_{j<k} v_j(\psi_k);$$

hence

$$|v_j(\psi)| \geq |v_j(\psi_j)| - \sum_{1 \leq k < j} |v_j(\psi_k)| - \sum_{j<k} |v_j(\psi_k)| > j + 1 - 1 = j,$$

on account of (5.4) and (5.3).(ii). On the other hand, (v_j) being a subsequence of (u_j) implies $\lim_{j\to\infty} v_j(\psi) = u(\psi)$. Summarizing, we have arrived at a contradiction. □

Remark 5.6. In applications one often meets families of distributions u_t that depend on one- or multi-dimensional real or complex parameters t, such that for each test function ϕ the complex numbers $u_t(\phi)$ converge as $t \to t_0$, where $t_0 = \pm\infty$ is allowed. Denote the limit by $u(\phi)$. Replacing u_t by u_{t_j}, where t_j is an infinite sequence such that $t_j \to t_0$ as $j \to \infty$, Theorem 5.5 implies that $u : \phi \mapsto u(\phi)$ is a distribution. The proof of the theorem may also be adapted to such families $(u_t)_t$. ⊘

Example 5.7. For $\epsilon > 0$, define $u_\epsilon : \mathbf{R} \to \mathbf{R}$ by $u_\epsilon(x) = \frac{1}{x} 1_{\mathbf{R}\setminus[-\epsilon,\epsilon]}(x)$. Then the u_ϵ are locally integrable functions; hence $u_\epsilon \in \mathcal{D}'(\mathbf{R})$, while $\lim_{\epsilon\downarrow 0} u_\epsilon(\phi) = \mathrm{PV}\,\frac{1}{x}(\phi)$ in the notation from Example 1.3, for every $\phi \in C_0^\infty(\mathbf{R})$. The preceding theorem then entails that $\mathrm{PV}\,\frac{1}{x} \in \mathcal{D}'(\mathbf{R})$; compare with Example 3.3. ⊘

Example 5.8. Let $(u_t)_{t\in\mathbf{R}}$ be a family of distributions on an open set X in $\mathbf{R}^n$. Assume that the function $t \mapsto u_t(\phi)$ is differentiable on $\mathbf{R}$, for every $\phi \in C_0^\infty(X)$. For every $t \in \mathbf{R}$ we then have the existence of

$$\lim_{h\to 0} \frac{1}{h}(u_{t+h} - u_t)(\phi) = \lim_{h\to 0} \frac{u_{t+h}(\phi) - u_t(\phi)}{h} = \frac{d}{dt}(u_t(\phi)) =: \Big(\frac{d}{dt}u_t\Big)(\phi).$$

By Theorem 5.5.(i), then, $\frac{d}{dt}u_t$ is a distribution on X. ⊘

The following lemma proves to be useful, for example in the theory of Fourier series from Chap. 16.

Lemma 5.9. *Let X be open in $\mathbf{R}^n$ and suppose that u_k and $u \in \mathcal{D}'(X)$ satisfy $\lim_{k\to\infty} u_k = u$ in $\mathcal{D}'(X)$. Then, for every multi-index α, one also has $\lim_{k\to\infty} \partial^\alpha u_k = \partial^\alpha u$ in $\mathcal{D}'(X)$.*

Proof. Let $\phi \in C_0^\infty(X)$ and $1 \le j \le n$ be arbitrary. Then

$$\partial_j u_k(\phi) = -u_k(\partial_j\phi) \to -u(\partial_j\phi) = \partial_j u(\phi) \qquad \text{as} \qquad k \to \infty. \qquad \square$$

Remark 5.10. An alternative definition of the (sequential) *completeness* of a space of distributions is that every Cauchy sequence in it converges to a limit contained in the space. Here a sequence u_j in $\mathcal{D}'(X)$ is a Cauchy sequence in $\mathcal{D}'(X)$ if and only if $u_j(\phi)$ is a Cauchy sequence in $\mathbf{C}$, for every $\phi \in C_0^\infty(X)$. In view of the completeness of $\mathbf{C}$ this definition is equivalent to the one from Theorem 5.5. ⊘

Problems

5.1.* Let $(f_j)_{j\in\mathbf{N}}$ be a sequence of nonnegative integrable functions on $\mathbf{R}^n$ with the following properties:

(a) For every j one has $\int_{\mathbf{R}^n} f_j(x)\,dx = 1$.
(b) For every $r > 0$ one has $\lim_{j\to\infty} \int_{\|x\|\ge r} f_j(x)\,dx = 0$.

Prove that $\lim_{j\to\infty} f_j = \delta$ in $\mathcal{D}'(\mathbf{R}^n)$.

5.2.* Let f be an integrable function on $\mathbf{R}^n$ and define $f_\epsilon(x) := \frac{1}{\epsilon^n} f(\frac{1}{\epsilon} x)$. Prove that

$$\lim_{\epsilon \downarrow 0} f_\epsilon = c\,\delta \quad \text{in} \quad \mathcal{D}'(\mathbf{R}^n) \qquad \text{with} \qquad c = \int_{\mathbf{R}^n} f(x)\,dx \in \mathbf{C}.$$

5.3. Using the results from Problem 1.3, prove that (see Problem 14.51 for another proof and more background and also Fig. 5.1)

$$\lim_{\epsilon \downarrow 0} \frac{x}{x^2+\epsilon^2} = \text{PV}\,\frac{1}{x} \qquad \text{and} \qquad \lim_{\epsilon \downarrow 0} \frac{\epsilon}{x^2+\epsilon^2} = \pi\,\delta \qquad \text{in} \qquad \mathcal{D}'(\mathbf{R}).$$

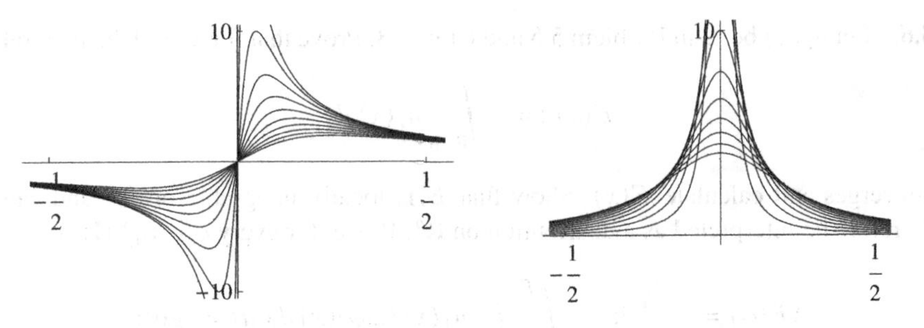

Fig. 5.1 Graphs of $x \mapsto \frac{x}{x^2+\epsilon^2}$ and $x \mapsto \frac{\epsilon}{x^2+\epsilon^2}$, for $\epsilon = \frac{k}{40}$ with $1 \leq k \leq 10$

5.4. If $e(j)$ denotes the jth standard basis vector in $\mathbf{R}^n$, show that

$$\lim_{t \to 0} \frac{1}{t}\,(\delta_{a-t\,e(j)} - \delta_a) = \partial_j \delta_a \quad \text{in} \quad \mathcal{D}'(\mathbf{R}^n).$$

5.5.* *(Heat or diffusion equation.)* Consider, for $t > 0$, the following function on $\mathbf{R}^n$ (compare with Example 14.15):

$$u_t(x) = \frac{1}{(4\pi t)^{\frac{n}{2}}}\, e^{-\frac{\|x\|^2}{4t}}.$$

Set $n = 1$, then make a sketch of this function and its first two derivatives, for a small positive value of t (see Fig. 5.2).
Verify that $u_t(x)$ satisfies the partial differential equation

$$\frac{d}{dt} u_t = \Delta u_t$$

on $\mathbf{R}^n$, the *heat equation* or *diffusion equation*. Calculate the limit in $\mathcal{D}'(\mathbf{R}^n)$ as $t \downarrow 0$ of u_t and of $\frac{d}{dt} u_t$. See Problems 14.32 and 17.1 for information from a different perspective.

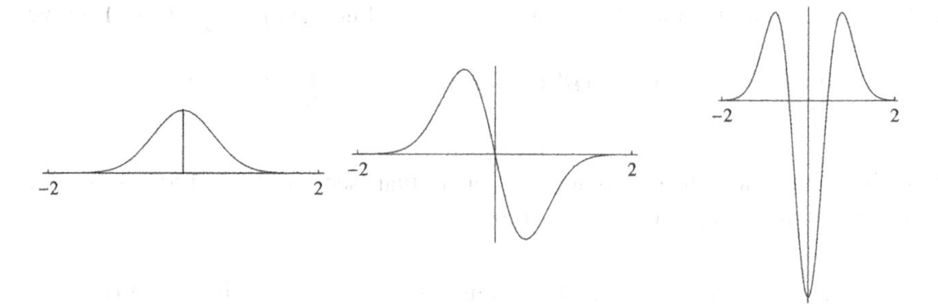

Fig. 5.2 Graphs of u_t and its first two derivatives, for $n = 1$ and $t = \frac{1}{10}$

5.6.* Let $u_t(x)$ be as in Problem 5.5 and let $n \geq 3$. Prove that for $x \neq 0$ the integral

$$E(x) := -\int_{\mathbf{R}_{>0}} u_t(x)\, dt$$

converges and calculate $E(x)$. Show that E is locally integrable on $\mathbf{R}^n$ and can therefore be interpreted as a distribution on $\mathbf{R}^n$. Prove, for every $\phi \in C_0^\infty(\mathbf{R}^n)$,

$$\Delta E(\phi) = -\lim_{s\downarrow 0,\, T\uparrow\infty} \int_s^T \int_{\mathbf{R}^n} u_t(x)\,(\Delta\phi)(x)\, dx\, dt = \phi(0),$$

and therefore $\Delta E = \delta$. Verify that E equals the E from Problem 4.7.

5.7. Define $f_t(x) = t^a\, e^{itx}$. For what $a \in \mathbf{R}$ does f_t converge to 0 as $t \to \infty$

(i) with respect to the C^k norm?
(ii) in $\mathcal{D}'(\mathbf{R})$?

5.8. Define $f_t(x) = t\, e^{itx} \log|x|$, for $x \in \mathbf{R}$. Calculate the limit of f_t in $\mathcal{D}'(\mathbf{R})$ as $t \to \infty$.

5.9. *(Riemann's "nondifferentiable" function.)* Let $f : \mathbf{R} \to \mathbf{R}$ be defined by

$$f(x) = \sum_{n\in\mathbf{N}} \frac{\sin(n^2 x)}{n^2}.$$

Weierstrass reported that Riemann suggested f as an example of a continuous function that is nowhere differentiable. (Actually, f is differentiable at points of the form $p\pi/q$ where p and q both are odd integers, with derivative equal to $-1/2$, but at no other points; see [5], for instance.) In Fig. 5.3, the first ten thousand terms have been summed for ten thousand different values of $0 \leq x \leq \pi$. Prove that

$$\lim_{N\to\infty} \sum_{n=1}^{N} \cos(n^2 x) = f'(x) \quad \text{in} \quad \mathcal{D}'(\mathbf{R}).$$

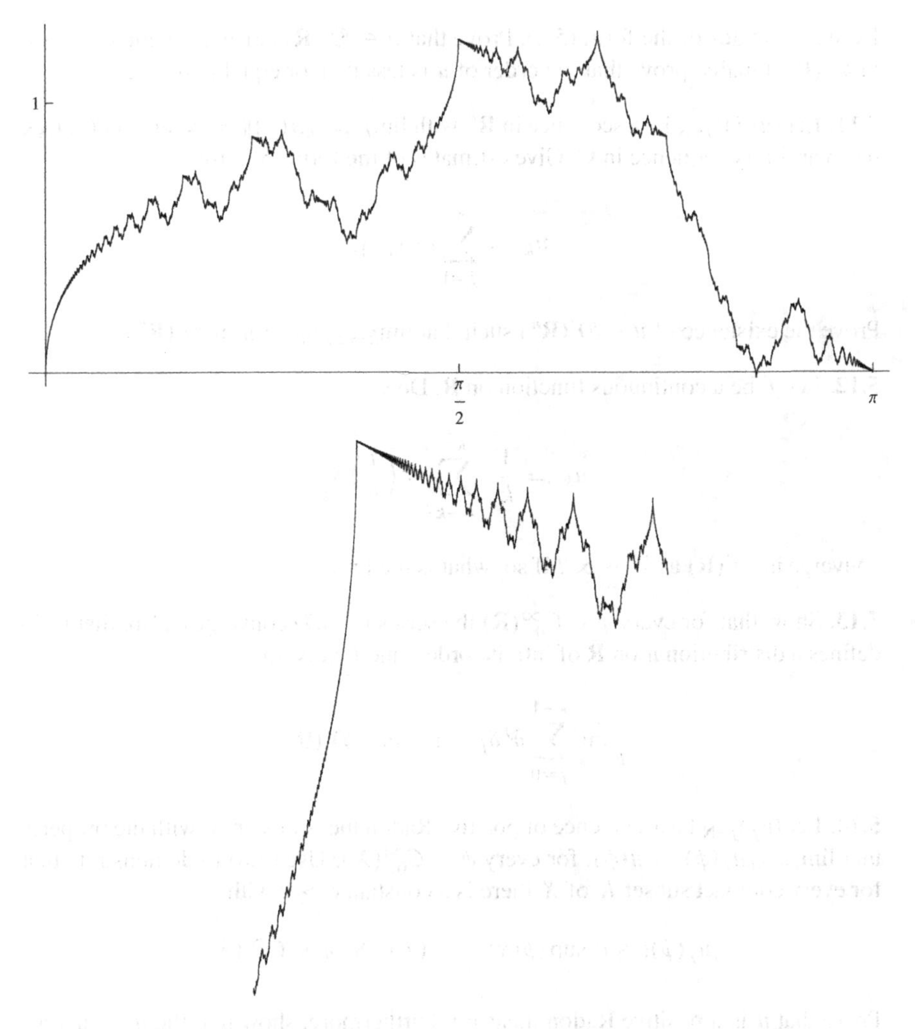

Fig. 5.3 The second graph is an enlargement of the first one, in a neighborhood of $\frac{\pi}{2}$

In the following problems the reader is asked to derive estimates of the form (5.2), so that we can use a fully proved variant of Theorem 5.5.

5.10. Let $c_n \in \mathbf{C}$ for all $n \in \mathbf{Z}$ and suppose that positive constants c and m exist such that

$$|c_n| \leq c\,|n|^m \qquad (n \in \mathbf{Z} \setminus \{0\}).$$

Let $\omega > 0$ and write

$$u_j(x) = \sum_{n=-j}^{j} c_n\, e^{in\omega x} \qquad (j \in \mathbf{N},\ x \in \mathbf{R}).$$

Derive estimates of the form (5.2). Prove that $u \in \mathcal{D}'(\mathbf{R})$ exists for which $u_j \to u$ in $\mathcal{D}'(\mathbf{R})$. Finally, prove that the order of u is less than or equal to $m+2$.

5.11. Let $(a(j))_{j\in\mathbf{N}}$ be a sequence in $\mathbf{R}^n$ with $\lim_{j\to\infty} \|a(j)\| = \infty$ and let $(c_j)_{j\in\mathbf{N}}$ be an arbitrary sequence in $\mathbf{C}$. Give estimates of the form (5.2) for

$$u_k := \sum_{j=1}^{k} c_j\, \delta_{a(j)}.$$

Prove the existence of $u \in \mathcal{D}'(\mathbf{R}^n)$ such that $\lim_{k\to\infty} u_k = u$ in $\mathcal{D}'(\mathbf{R}^n)$.

5.12. Let f be a continuous function on $\mathbf{R}$. Does

$$u_k := \frac{1}{k} \sum_{j=-k^2}^{k^2} f\Big(\frac{j}{k}\Big)\, \delta_{\frac{j}{k}}$$

converge in $\mathcal{D}'(\mathbf{R})$ as $k \to \infty$? If so, what is the limit?

5.13. Show that for every $\phi \in C_0^\infty(\mathbf{R})$ the series in (3.7) converges; also, that (3.7) defines a distribution u on $\mathbf{R}$ of infinite order; and finally, that

$$\lim_{k\to\infty} \sum_{j=0}^{k-1} \partial^j \delta_j = u \quad \text{in} \quad \mathcal{D}'(\mathbf{R}).$$

5.14. Let $(u_j)_{j\in\mathbf{N}}$ be a sequence of positive Radon measures on X with the property that $\lim_{j\to\infty} u_j(\phi) = u(\phi)$, for every $\phi \in C_0^\infty(X)$. Use (3.8) to demonstrate that for every compact subset K of X there is a constant $c > 0$ with

$$|u_j(\phi)| \le c \sup_x |\phi(x)| \qquad (j \in \mathbf{N},\ \phi \in C_0^\infty(K)).$$

Prove that u is a positive Radon measure. Furthermore, show that the u_j converge weakly as measures to u. That is, $\lim_{j\to\infty} u_j(f) = u(f)$, for every $f \in C_0(X)$.

Chapter 6
Taylor Expansion in Several Variables

Many classical asymptotic expansions imply interesting distributional limits. An example is the Taylor expansion of functions of several variables. Because we will be using this expansion elsewhere in this text, we begin by repeating its basic properties.

A mapping f from an open subset U of $\mathbf{R}^n$ to $\mathbf{R}^p$ is given by p real-valued functions of n variables, that is,

$$f(x) = f(x_1, \dots, x_n) = (f_1(x_1, \dots, x_n), \dots, f_p(x_1, \dots, x_n)).$$

Many properties of vector-valued functions can be derived from those of real-valued functions by proving them for each of the coordinate functions $f_i(x_1, \dots, x_n)$, with $1 \leq i \leq p$.

For instance, the mapping f is continuously differentiable if for every i and every j, the function f_i is partially differentiable with respect to the jth variable and if the function $\partial_j f_i$ is continuous on U. In that case one writes $f \in C^1(U, \mathbf{R}^p)$, or $f \in C^1(U, V)$ if $f(U) \subset V$.

The $\partial_j f_i(x)$, for $1 \leq j \leq n$ and $1 \leq i \leq p$, form the matrix of the linear mapping

$$Df(x) : v \mapsto \sum_{j=1}^{n} \partial_j f(x)\, v_j : \mathbf{R}^n \to \mathbf{R}^p.$$

This is the linear approximation at $h = 0$ of the increase $h \mapsto f(x+h) - f(x)$ of f and is also known as the *total derivative* of f at the point x. If $p = 1$, one normally writes $\mathrm{d}f(x)$ instead of $Df(x)$.

The functions $\partial_j f_i$ taken together can again be interpreted as a vector-valued function. This allows us to define, with mathematical induction on k, that $f \in C^k$ if $D^{k-1} f \in C^1$, where $D^k f$ is inductively defined by $D^k f(x) = D(D^{k-1} f)(x)$.

A theorem already mentioned is that $\partial_i(\partial_j f) = \partial_j(\partial_i f)$ if $f \in C^2$. For $f \in C^k$ this enables every higher-order derivative of f, where differentiation with respect to x_j takes place α_j times in total, to be rearranged in the form $\partial^\alpha f$, as in (2.7).

J.J. Duistermaat and J.A.C. Kolk, *Distributions: Theory and Applications*,
Cornerstones, DOI 10.1007/978-0-8176-4675-2_6,

An important result that we will often be using is the *chain rule*. For describing this, consider $f \in C^1(U, \mathbf{R}^p)$ and $g \in C^1(V, \mathbf{R}^q)$ where $U \subset \mathbf{R}^n$ and $V \subset \mathbf{R}^p$ are open subsets. The *composition* $g \circ f$ of f and g is defined by $(g \circ f)(x) = g(f(x))$ for all $x \in U$ with $f(x) \in V$; the continuity of f guarantees that the set of these x forms an open subset $U \cap f^{-1}(V)$ of $\mathbf{R}^n$. The chain rule asserts that we have $g \circ f \in C^1(U \cap f^{-1}(V), \mathbf{R}^q)$, while

$$D(g \circ f)(x) = Dg(f(x)) \circ Df(x); \qquad \text{or} \qquad D(g \circ f) = (Dg) \circ f\ Df.$$

With mathematical induction on k one finds that $g \circ f \in C^k$ if $f \in C^k$ and $g \in C^k$; for the higher-order derivatives inductive formulas are obtained that rapidly become very cumbersome to write out in explicit form. Considering the matrix coefficients occurring in the chain rule, one obtains the following identity of functions on $U \cap f^{-1}(V)$, for $1 \le j \le n$ and $1 \le i \le q$:

$$\partial_j (g_i \circ f) = \sum_{k=1}^{p} (\partial_k g_i) \circ f\ \partial_j f_k = \sum_{k=1}^{p} \partial_j f_k\ (\partial_k g_i) \circ f. \tag{6.1}$$

A practical way of studying $f(x)$ for x near a point $a \in U$ is to consider the function on $[0, 1] \subset \mathbf{R}$ given by

$$t \mapsto g(t) := f(a + t\,(x - a)).$$

We have $g(0) = f(a)$ and $g(1) = f(x)$; what we actually do is to consider f on the straight line from a to x. The difference vector $h = x - a$ is usually small. In the following lemma we use the abbreviations

$$x^\alpha = \prod_{j=1}^{n} x_j^{\alpha_j} \qquad \text{and} \qquad \alpha! = \prod_{j=1}^{n} \alpha_j!.$$

Lemma 6.1. *If $f \in C^k(U, \mathbf{R}^p)$, we have, for $0 \le j \le k$ and $a + th \in U$,*

$$\frac{1}{j!}\frac{d^j}{dt^j} f(a + th) = \sum_{|\alpha| = j} \frac{h^\alpha}{\alpha!}\, \partial^\alpha f(a + th). \tag{6.2}$$

Proof. Using the chain rule we obtain

$$\frac{d}{dt} f(a + th) = \sum_{j=1}^{n} h_j\, \partial_j f(a + th).$$

By mathematical induction on j this yields

$$\frac{d^j}{dt^j} f(a + th) = \sum_{|\alpha| = j} m_\alpha h^\alpha\, \partial^\alpha f(a + th), \tag{6.3}$$

where the $m_\alpha \in \mathbf{Z}$ are determined by recurrence relations; in particular, they are independent of the choices made for f, a, and t. To determine the m_α, we take $f(x) = x^\beta$ with $|\beta| = j$, $a = 0$ and differentiate at the point $t = 0$. Then $f(a+th) = (t\,h)^\beta = t^{|\beta|}\,h^\beta = t^j\,h^\beta$; the jth-order derivative with respect to t of this function equals $j!h^\beta$. But we also have, for every α with $|\alpha| = j$,

$$\partial^\alpha x^\beta = \begin{cases} \beta! & \text{if } \alpha = \beta, \\ 0 & \text{if } \alpha \neq \beta. \end{cases} \tag{6.4}$$

This proves that $m_\beta = j!/\beta!$.

Actually, the combinatorics associated with the recurrence relation are simple. Hence, one may alternatively finish the proof by noting that according to the induction hypothesis the right-hand side of (6.3) after division by $(j-1)!$ can be written as

$$\sum_{|\beta|=j-1} \sum_{i=1}^{n} \frac{h_i h^\beta}{\beta!} \partial_i \partial^\beta f(a+th) = \sum_{|\alpha|=j} \frac{h^\alpha \sum_{i=1}^{n} \alpha_i}{\alpha!} \partial^\alpha f(a+th). \qquad \square$$

For $f \in C^k(U, \mathbf{R}^p)$, the *Taylor polynomial* $T = T^k_{f,a}$ *of order* k at the point $a \in U$ is defined by

$$T(x) = \sum_{|\alpha| \le k} \frac{(x-a)^\alpha}{\alpha!} \partial^\alpha f(a). \tag{6.5}$$

The *remainder* $R = R^k_{f,a}$ *of order* k is defined as $R = f - T$. Then, by definition, $f = T + R$; we want to find out what information can be obtained about the remainder. The following formulation may be somewhat heavier than necessary for most applications, but it follows directly from the integral formula used for the remainder.

Theorem 6.2. *Let f be a C^l mapping defined on an open ball U in $\mathbf{R}^n$. For every $0 \le k \le l$, the mapping $(a, x) \mapsto R^k_{f,a}(x)$ is a C^{l-k} mapping on $U \times U$. Furthermore, for every compact $K \subset U$ and every $\epsilon > 0$, there exists a $\delta > 0$ such that*

$$\|R^k_{f,a}(x)\| \le \epsilon\,\|x-a\|^k \qquad \textit{if} \qquad a, x \in K \quad \textit{and} \quad \|x-a\| < \delta.$$

Finally, $R = R^k_{f,a}$ is of class C^l and $\partial^\alpha R(a) = 0$, for all multi-indices α with $|\alpha| \le k$.

Proof. The Taylor expansion of a C^k function g on an interval I containing 0 reads

$$g(t) = \sum_{j=0}^{k-1} \frac{t^j}{j!} g^{(j)}(0) + \int_0^t \frac{(t-s)^{k-1}}{(k-1)!} g^{(k)}(s)\,ds \qquad (t \in I);$$

the proof is obtained by mathematical induction on k. When we apply this formula, taking $g(t) = f(a + t\,(x-a))$ and substituting (6.2) for the derivatives, we obtain, for $t = 1$, the identity $f = T + R$, with $R = R^k_{f,a}$ and

$$R^k_{f,a}(x) = \sum_{|\alpha|=k} \frac{(x-a)^\alpha}{\alpha!} \int_0^1 k\,(1-s)^{k-1} \big(\partial^\alpha f(a + s\,(x-a)) - \partial^\alpha f(a)\big)\, ds.$$

On account of the locally uniform continuity of the kth-order derivatives of f, the integral converges locally uniformly to 0 as $\|x-a\| \to 0$. The fact that the mapping $(a, x) \mapsto R^k_{f,a}(x)$ is of class C^{l-k} follows by means of differentiation under the integral sign.

The function $R^k_{f,a}$ is of class C^l, being the difference of f and the Taylor polynomial $T := T^k_{f,a}$. From (6.4) we see that $\partial^\alpha f(a) = \partial^\alpha T(a)$ for all α with $|\alpha| \le k$. □

By way of application we give a "higher-order version" of Problem 5.2.

Proposition 6.3. *Let f be an integrable function on $\mathbf{R}^n$ with compact support. Let $a \in \mathbf{R}^n$ and set*

$$f_\epsilon(x) = \frac{1}{\epsilon^n}\, f\Big(\frac{1}{\epsilon}\,(x-a)\Big),$$

for $\epsilon > 0$. Furthermore, we define, for each multi-index α, the coefficient $c_\alpha \in \mathbf{C}$ by

$$c_\alpha = \int_{\mathbf{R}^n} x^\alpha\, f(x)\, dx,$$

and we introduce, for every $j \in \mathbf{Z}_{\ge 0}$, the distribution $u_j \in \mathcal{D}'(\mathbf{R}^n)$ by

$$u_j = \sum_{|\alpha|=j} \frac{c_\alpha}{\alpha!}\, \partial^\alpha \delta_a.$$

We then have, for all $k \in \mathbf{Z}_{\ge 0}$,

$$\lim_{\epsilon \downarrow 0} \epsilon^{-k}\, \big(f_\epsilon - \sum_{j=0}^{k} (-\epsilon)^j\, u_j\big) = 0 \quad \textit{in} \quad \mathcal{D}'(\mathbf{R}^n).$$

Proof. For every $\phi \in C_0^\infty(\mathbf{R}^n)$ we have

$$f_\epsilon(\phi) = \int_{\mathbf{R}^n} \epsilon^{-n}\, f\Big(\frac{1}{\epsilon}\,(x-a)\Big)\, \phi(x)\, dx = \int_{\mathbf{R}^n} f(h)\, \phi(a + \epsilon h)\, dh.$$

The assertion now follows by substituting the Taylor expansion of ϕ at the point a and of order k. □

The coefficient c_α is known as the αth *moment* of the function f. A consequence of the proposition is that $\epsilon^{-k} f_\epsilon$ converges distributionally to $(-1)^k u_k$ as $\epsilon \downarrow 0$ if $c_\alpha = 0$ for all multi-indices α with $|\alpha| < k$.

Problems

6.1. Formulate a variant of Proposition 6.3 in which f is not required to have compact support.

6.2.* Let u_t be as in Problem 5.5. Prove, for every $k \in \mathbf{Z}_{\geq 0}$,

$$\lim_{t \downarrow 0} \frac{1}{t^k} \left(u_t - \sum_{j=0}^{k-1} \frac{t^j}{j!} \Delta^j \delta \right) = \frac{1}{k!} \Delta^k \delta \quad \text{in} \quad \mathcal{D}'(\mathbf{R}^n).$$

Discuss the notation $u_t = e^{t\,\Delta}\,\delta$, for $t > 0$.

6.3.* (Sequel to Problem 6.2.) For $\phi \in C_0^\infty(\mathbf{R}^n)$ and $r \geq 0$, define the mean value $S_\phi(r)$ of ϕ over the sphere of radius r about 0 by

$$S_\phi(r) = \frac{1}{c_n} \int_S \phi(r\,y)\,dy.$$

Here S is the unit sphere in $\mathbf{R}^n$ and c_n its Euclidean $(n-1)$-dimensional volume, which is evaluated in (13.37).

(i) Show that $S_\phi \in C^\infty(\mathbf{R})$ and that

$$u_t(\phi) = c_n (4\pi t)^{-\frac{n}{2}} \int_{\mathbf{R}_{>0}} e^{-\frac{r^2}{4t}} r^{n-1}\, S_\phi(r)\,dr \qquad (t > 0).$$

(ii) Substitute the Taylor series of S_ϕ at 0 in this identity and express the coefficients of the series in terms of the $\Delta^j \phi(0)$. Under the assumption of convergence of the series, deduce *Pizzetti's formula*, for $r \geq 0$,

$$\begin{aligned} S_\phi(r) &= \Gamma\Big(\frac{n}{2}\Big) \sum_{k \in \mathbf{Z}_{\geq 0}} \frac{\Delta^k f(0)}{k!\,\Gamma(\frac{n}{2}+k)} \Big(\frac{r}{2}\Big)^{2k} \\ &= \Gamma\Big(\frac{n}{2}\Big) \Big(\frac{r\sqrt{-\Delta}}{2}\Big)^{1-\frac{n}{2}} J_{\frac{n}{2}-1}(r\sqrt{-\Delta}) f(0). \end{aligned}$$

Here Γ denotes the Gamma function as in Sect. 13.3 below, Δ denotes the Laplace operator, and $J_{\frac{n}{2}-1}$ the Bessel function of order $\frac{n}{2}-1$ as in [7, Exercise 6.66] for instance. Furthermore, the last equality is obtained by a formal substitution in the power series for the Bessel function. See [7, Exercises 7.22 and 7.54] for proofs along classical lines.]

(iii) Suppose that $\alpha_j \in 2\mathbf{Z}_{\geq 0}$, for $1 \leq j \leq n$ and consider the monomial function $y \mapsto y^\alpha = \prod_{1 \leq j \leq n} y_j^{\alpha_j}$. Define $(2a-1)!! = 1 \cdot 3 \cdots (2a-1)$, for $a \in \mathbf{N}$ and furthermore $(-1)!! = 1$. Prove (compare with [7, Exercise 7.21.(x)])

$$\frac{1}{c_n} \int_S y^\alpha\,dy = \frac{\prod_{1 \leq j \leq n} (\alpha_j - 1)!!}{\prod_{0 \leq l < \frac{|\alpha|}{2}} (n + 2l)} \in \mathbf{Q}.$$

Chapter 7
Localization

Generally speaking, distributions cannot be restricted to a point, that is, evaluated at that point. Restriction to an open set, however, is possible.

Following Definition 2.5 we have discussed how, for open subsets U and V of $\mathbf{R}^n$ with $U \subset V$, the space $C_0^\infty(U)$ is interpreted as a linear subspace of $C_0^\infty(V)$. More explicitly, we have the linear injection

$$\iota_{VU} : C_0^\infty(U) \to C_0^\infty(V). \tag{7.1}$$

Suppose that ϕ_j and ϕ belong to $C_0^\infty(U)$. Since a compact subset K of U is a compact subset of V too, it follows that

$$\lim_{j\to\infty} \phi_j = \phi \quad \text{in} \quad C_0^\infty(U) \quad \Longrightarrow \quad \lim_{j\to\infty} \iota_{VU}\phi_j = \iota_{VU}\phi \quad \text{in} \quad C_0^\infty(V).$$

Phrased differently, ι_{VU} preserves convergence of sequences. Now consider v in $\mathcal{D}'(V)$ and define

$$\rho_{UV}\, v : C_0^\infty(U) \to \mathbf{C} \qquad \text{by} \qquad (\rho_{UV}\, v)(\phi) = v(\iota_{VU}\phi) \qquad (\phi \in C_0^\infty(U)).$$

Then it follows that $\rho_{UV}\, v$ is a continuous linear form on $C_0^\infty(U)$. In other words, $\rho_{UV}\, v \in \mathcal{D}'(U)$; this distribution is said to be the *restriction* of v to U. This restriction of distributions defines a linear mapping

$$\rho_{UV} : \mathcal{D}'(V) \to \mathcal{D}'(U), \tag{7.2}$$

the *mapping induced by the inclusion* $C_0^\infty(U) \subset C_0^\infty(V)$. Note that upon changing to the dual spaces the sense of the arrows is reversed, as is visible in the following diagram:

$$\begin{array}{ccc} U & \longrightarrow & V \\ C_0^\infty(U) & \xrightarrow{\iota_{VU}} & C_0^\infty(V) \\ \mathcal{D}'(U) & \xleftarrow{\rho_{UV}} & \mathcal{D}'(V) \end{array}$$

J.J. Duistermaat and J.A.C. Kolk, *Distributions: Theory and Applications*,
Cornerstones, DOI 10.1007/978-0-8176-4675-2_7,

The distribution v is said to be of class C^k, or of order k, respectively, *on* U if $\rho_{UV}\, v$ has this property. If $w \in \mathcal{D}'(W)$ and $U \subset W$, we say that $v = w$ *on* U if $\rho_{UV}\, v = \rho_{UW}\, w$.

Theorem 7.1. *Let X be an open subset of $\mathbf{R}^n$ and $u \in \mathcal{D}'(X)$. Suppose that for every $x \in X$, there exists an open neighborhood U_x of x in X such that $u = 0$ on U_x. Then we have $u = 0$ (on X).*

Proof. Let $\phi \in C_0^\infty(X)$. The U_x with $x \in X$ form an open cover of the compact set $\operatorname{supp} \phi$. Theorem 2.15 yields $\psi_1, \dots, \psi_l \in C_0^\infty(X)$ with $\operatorname{supp} \psi_j \subset U_{x(j)}$ and $\sum_j \psi_j = 1$ on $\operatorname{supp} \phi$. Writing $\phi_j = \psi_j\, \phi$, we have $\phi_j \in C_0^\infty(U_{x(j)})$, so $u(\phi_j) = 0$ and $\phi = \sum_j \phi_j$; therefore $u(\phi) = \sum_j u(\phi_j) = 0$. □

Applying Theorem 7.1 to $u - v$, we further obtain that $u = v$ on X if for every $x \in X$, there exists an open neighborhood U of x in X such that $u = v$ on U.

Definition 7.2. The *support* of $u \in \mathcal{D}'(X)$, denoted by $\operatorname{supp} u$, is the set of all $x \in X$ for which there exists **no** open neighborhood U of x in X such that $u = 0$ on U. ⊘

Let $V := X \setminus \operatorname{supp} u$ be the complement of $\operatorname{supp} u$ in X. This is an open subset of X. Indeed, by definition, for every $x \in V$ there exists an open neighborhood U_x of x in X with $u = 0$ on U_x, which implies $U_x \subset V$. Theorem 7.1 now implies that $u = 0$ on V. Furthermore, if U is open in X and $u = 0$ on U, then $U \cap \operatorname{supp} u = \emptyset$, which is equivalent to $U \subset V$. In other words, V is the largest open subset of X on which $u = 0$. One has

$$u(\phi) = 0 \qquad \text{if} \qquad u \in \mathcal{D}'(X),\ \phi \in C_0^\infty(X) \quad \text{and} \quad \operatorname{supp} u \cap \operatorname{supp} \phi = \emptyset. \tag{7.3}$$

This characterization of $\operatorname{supp} u$ immediately implies that $\operatorname{supp} \partial^\alpha u \subset \operatorname{supp} u$, for every multi-index $\alpha \in (\mathbf{Z}_{\geq 0})^n$.

Example 7.3. Let V be a closed k-dimensional C^1 submanifold contained in an open subset X of $\mathbf{R}^n$ and f a continuous function on V. Consider the Radon measure $f\, \delta_V$ in $\mathcal{D}'(X)$ as in Example 3.11; then $\operatorname{supp} f\, \delta_V = V \cap \operatorname{supp} f = \operatorname{supp} f$, because f is a function on V. Furthermore, $\operatorname{supp} \partial^\alpha (f\, \delta_V) \subset V$, even though some of the differentiations in ∂^α might be in directions that are not tangential to V. ⊘

Example 7.4. Let $U \subset X$ be open subsets of $\mathbf{R}^n$ and $u \in \mathcal{D}'(X)$. Suppose $u(\phi) = 0$, for all $\phi \in C_0^\infty(X)$ satisfying $\operatorname{supp} \phi \cap U = \emptyset$. Then $\operatorname{supp} u \subset \overline{U}$ while in general it is not true that $\operatorname{supp} u \subset U$. For a counterexample, consider $U = \mathbf{R}_{>0} \subset \mathbf{R}$ and $u = 1_U \in \mathcal{D}'(\mathbf{R})$. Then $\operatorname{supp} u = \mathbf{R}_{\geq 0}$ and not $\operatorname{supp} u \subset \mathbf{R}_{>0}$. ⊘

If $U \subset X$ and $U \neq X$, the restriction mapping ρ_{UX} is not injective. Indeed, if $a \in X \setminus U$ and $u = \delta_a$, we have $u = 0$ on U.

Surjectivity of ρ_{UX} would mean that every distribution u on U is the restriction to U of a distribution v on X. If $\rho_{UX}\, v = u$, then v is said to be a *distributional extension* of u to X. An example: for every $k \in \mathbf{N}$, the function $x \mapsto x^{-k}$ on $\mathbf{R}_{>0}$, which is locally integrable on $\mathbf{R}_{>0}$ but not on all of $\mathbf{R}$, has the extension

$$\frac{(-1)^{k-1}}{(k-1)!}\,\partial^k(\log|\cdot|) \in \mathcal{D}'(\mathbf{R}).$$

However, there are many cases in which extension is impossible.

Example 7.5. The function $f : x \mapsto e^{1/x}$ on $\mathbf{R}_{>0}$ has no distributional extension to any neighborhood of 0 in $\mathbf{R}$. Suppose $\phi \in C_0^\infty(\mathbf{R}_{>0})$ satisfies $\phi \geq 0$ and $1(\phi) > 0$ and set $\phi_\epsilon(x) = \frac{1}{\epsilon}\,\phi(\frac{x}{\epsilon})$. We see that for every $N \in \mathbf{N}$ there exists a constant $c > 0$ such that $f(\phi_\epsilon) \geq c\,\epsilon^{-N}$, for all $0 < \epsilon \leq 1$. (Hint: use the fact that $e^{1/x} \geq x^{-N}/N!$ if $x > 0$ and apply the change of variables $x = \epsilon\, y$ in the integral.)

On the other hand, if $u \in \mathcal{D}'(\,]-\delta, \infty[\,)$ with $\delta > 0$ were a distributional extension of f, Theorem 3.8 asserts that there would exist c' and $c'' > 0$ and $k \in \mathbf{Z}_{\geq 0}$ with the property that $|u(\phi_\epsilon)| \leq c'\,\|\phi_\epsilon\|_{C^k} \leq c''\,\epsilon^{-k-1}$ for all $0 < \epsilon \leq 1$. Thus, the assumption that $u = f$ on $\mathbf{R}_{>0}$ leads to a contradiction. ⊘

There exists a counterpart of Theorem 7.1 in the form of an existence theorem:

Theorem 7.6. *Let $\mathcal{U}$ be a collection of open subsets of $\mathbf{R}^n$ with union X. Assume that for every $U \in \mathcal{U}$ a distribution $u_U \in \mathcal{D}'(U)$ is given and that $u_U = u_V$ on $U \cap V$, for all $U, V \in \mathcal{U}$. Then there exists a unique $u \in \mathcal{D}'(X)$ with the property that $u = u_U$ on U, for every $U \in \mathcal{U}$.*

Proof. Let K be a compact subset of X and ψ_j a partition of unity over K subordinate to the cover $\mathcal{U}$. In particular, there exists $U(j) \in \mathcal{U}$ such that supp $\psi_j \subset U(j) \in \mathcal{U}$. If $\phi \in C_0^\infty(X)$ and supp $\phi \subset K$, then ϕ equals the sum of the finitely many $\psi_j\,\phi \in C_0^\infty(X)$. We therefore have to take

$$u(\phi) = \sum_{j=1}^{l} u_{U(j)}(\psi_j\,\phi). \tag{7.4}$$

The uniqueness of u follows.

Before continuing we first prove that the right-hand side of (7.4) is independent of the choice of the partition ψ_j of unity over K subordinate to $\mathcal{U}$. Suppose χ_k is another partition of unity over K with supp $\chi_k \subset V(k) \in \mathcal{U}$. Then

$$\operatorname{supp}(\psi_j\,\chi_k\,\phi) \subset U(j) \cap V(k).$$

Consequently, $u_{U(j)}$ and $u_{V(k)}$ assume the same value on $\psi_j\,\chi_k\,\phi$. Therefore

$$\begin{aligned}\sum_{j=1}^{l} u_{U(j)}(\psi_j\,\phi) &= \sum_{j,k=1}^{l} u_{U(j)}(\psi_j\,\chi_k\,\phi) = \sum_{j,k=1}^{l} u_{V(k)}(\psi_j\,\chi_k\,\phi)\\ &= \sum_{k=1}^{l} u_{V(k)}(\chi_k\,\phi).\end{aligned}$$

We now interpret the right-hand side in (7.4) as the definition of u. If $\alpha, \beta \in C_0^\infty(X)$ and $a, b \in \mathbf{C}$, we take $K = \operatorname{supp} \alpha \cup \operatorname{supp} \beta$ above and thus obtain that $u(\alpha)$, $u(\beta)$, or $u(a\,\alpha + b\,\beta)$ equals the right-hand side in (7.4), replacing ϕ by α, β, or $a\,\alpha + b\,\beta$, respectively. From this it is immediate that $u(a\,\alpha + b\,\beta) = a\,u(\alpha) + b\,u(\beta)$. In other words, u is a linear form on $C_0^\infty(X)$.

If $\phi_k \to \phi$ in $C_0^\infty(X)$, there exists a compact subset K of X with the property that for every k one has that $\operatorname{supp} \phi_k \subset K$; this also implies that $\operatorname{supp} \phi \subset K$. Then, for every k, the $u(\phi_k)$ equal the right-hand side of (7.4) with ϕ replaced by ϕ_k. From the continuity of the $u_{U(j)}$ it now follows that $u(\phi_k) \to u(\phi)$ as $k \to \infty$. We have now proved that (7.4) defines a continuous linear form u on $C_0^\infty(X)$.

Finally, if $\operatorname{supp} \phi \subset U \in \mathcal{U}$, we have $\operatorname{supp}(\psi_j\,\phi) \subset U(j) \cap U$, and therefore $u_{U(j)}(\psi_j\,\phi) = u_U(\psi_j\,\phi)$. Summation over j gives $u(\phi) = u_U(\phi)$. □

Remark 7.7. Thanks to their properties described in Theorem 7.6, the linear spaces $\mathcal{D}'(U)$, as functions of the open subsets U of X, form a system known in the literature as a *(pre)sheaf*. In many cases, sheaves consist of spaces of functions on U, with the usual restriction of functions to open subsets. One example is the sheaf $U \mapsto C^\infty(U)$ of the infinitely differentiable functions. For these, the sheaf properties are a direct consequence of the local nature of the definition of the functions in the sheaf. Distributions form an example of a sheaf in which the linear spaces contain elements that are not functions. ⊘

Definition 7.8. Let $u \in \mathcal{D}'(X)$. The *singular support* of u, denoted by sing supp u, is the set of all $x \in X$ that do **not** possess any open neighborhood U in X such that $\rho_{UX}\, u \in C^\infty(U)$. Here ρ_{UX} is as in (7.2). ⊘

The complement C of sing supp u in X has an open cover $\mathcal{U}$ such that $\rho_{UX}\, u = u_U \in C^\infty(U)$, for every $U \in \mathcal{U}$. This follows from the definition of sing supp. If U and $V \in \mathcal{U}$, then $u_U = u_V$ on $U \cap V$, which implies that the u_U possess a common extension to a C^∞ function f on C. In view of Theorem 7.1 we have $u = f$ on C. In other words, u is of class C^∞ on C; every open subset of X on which u is of class C^∞ is contained in C. Obviously, sing supp $u \subset$ supp u.

In Problem 7.3 we use the following definition. In subsequent parts of this text we will repeatedly encounter linear partial differential operators with constant coefficients.

Definition 7.9. Let

$$P(\xi) = \sum_{|\alpha| \le m} c_\alpha \xi^\alpha$$

be a polynomial in n variables of degree m with coefficients $c_\alpha \in \mathbf{C}$. Replacing ξ by $\partial = \partial_x$ we obtain a differential operator

$$P(\partial) = P(\partial_x) = \sum_{|\alpha| \le m} c_\alpha\, \partial^\alpha = \sum_{|\alpha| \le m} c_\alpha \frac{\partial^\alpha}{\partial x^\alpha}, \tag{7.5}$$

a *linear partial differential operator* of *order m* and with *constant coefficients*. ⊘

Note that $P(\partial)$ maps $\mathcal{D}'(X)$ into itself, for any open subset X of $\mathbf{R}^n$; see also Problem 7.3.

Problems

7.1. Show that for a continuous function the support equals the support in the distributional sense.

7.2. Determine the supports of the Dirac function, the Heaviside function, and PV $\frac{1}{x}$.

7.3. Prove that supp $P(\partial)\, u \subset$ supp u, for every $u \in \mathcal{D}'(X)$ and every linear partial differential operator with constant coefficients $P(\partial)$.

7.4.* *(Integration of partial derivative.)* Let U be a measurable subset of X. Prove that the support of $\partial_j 1_U$ is contained in the boundary $\partial U = X \cap \overline{U} \setminus U^{\text{int}}$ of U in X. Here U^{int} is the set of the interior points of U. Now, let U be an open subset of X with C^1 boundary ∂U and lying at one side of ∂U. Prove that $\partial_j 1_U = -\nu_j\, \delta_{\partial U}$. Here ν_j denotes the jth component of the outer normal to ∂U. Also discuss the case $n = 1$.

7.5.* Let a and $f \in C^\infty(X)$ and let f be real-valued. A point $p \in X$ is said to be a *stationary point* of f if $\partial_j f(p) = 0$, for all $1 \le j \le n$; let S_f be the set of the stationary points of f in X. Prove that $X \setminus S_f$ is an open subset of X and that for every $m \in \mathbf{R}$,

$$\lim_{t\to\infty} t^m\, a\, e^{itf} = 0 \quad \text{in} \quad \mathcal{D}'(X \setminus S_f).$$

Hint: use the formula $e^{itf} = \partial_j(e^{itf})/(it\partial_j f)$, wherever $\partial_j f \neq 0$.

7.6.* Let $a \in \mathbf{C}$ and define $x_+^a = x^a = e^{a \log x}$ if $x > 0$ and $x_+^a = 0$ if $x < 0$. Prove the following assertions:

(i) If $\operatorname{Re} a > -1$, then x_+^a is locally integrable in $\mathbf{R}$ and can therefore be interpreted as a distribution.

(ii) If a is not a negative integer,

$$x_+^a = \frac{1}{(a+k)\cdots(a+1)}\,\partial_x^k x_+^{a+k} \qquad \text{with} \qquad k > -\operatorname{Re} a - 1$$

defines a distribution on $\mathbf{R}$ that is an extension of x_+^a on $\mathbf{R}\setminus\{0\}$. This distribution is independent of the choice of k. Determine the support and the order of this distribution. What happens if the real part of a is a negative integer (and the imaginary part does not vanish)?

(iii) Let $l_+(x) = \log x$ if $x > 0$ and $l_+(x) = 0$ if $x < 0$. Then l_+ is locally integrable on $\mathbf{R}$ and thus defines a distribution on $\mathbf{R}$. For every $k \in \mathbf{N}$,

$$\frac{(-1)^{k-1}}{(k-1)!}\,\partial^k l_+$$

is a distribution on $\mathbf{R}$ that forms an extension of x_+^{-k} on $\mathbf{R} \setminus \{0\}$. Again, determine its support and its order.

(iv) Formulate similar results starting from $(-x)_+^a$ or $l_+(-x)$.

7.7. Show that for every C^1 function f on $\mathbf{R}_{>0}$ a real-valued C^1 function g on $\mathbf{R}_{>0}$ can be found with the property that $f\, e^{ig}$ has an extension to a distribution of order ≤ 1 on $\mathbf{R}$. Hint: show that there exists a real-valued C^1 function g on $\mathbf{R}_{>0}$ such that $g'(x) \neq 0$, $f(x)/g'(x)$ converges to zero as $x \downarrow 0$, and f'/g' is absolutely integrable on a neighborhood of 0. Prove that

$$u(\phi) = i \int_{\mathbf{R}_{>0}} e^{i\, g(x)} \frac{d}{dx}\Big(\frac{f(x)\,\phi(x)}{g'(x)}\Big)\, dx \qquad (\phi \in C_0^\infty(\mathbf{R}))$$

defines a distribution u on $\mathbf{R}$ of order ≤ 1 and that $u = f\, e^{ig}$ on $\mathbf{R}_{>0}$.

7.8. A subset D of X is said to be *discrete* if for every $a \in D$, there is an open neighborhood U_a of a in X such that $D \cap U_a = \{a\}$. Let X be an open subset of $\mathbf{R}^n$ and D a discrete and closed subset of X. For every $a \in D$ let there be a partial differential operator $P_a(\partial)$ of order m_a. Write $u_a = P_a(\partial)\,\delta_a$. Demonstrate the following:

(i) There exists exactly one distribution u in X with $\operatorname{supp} u \subset D$ such that for every $a \in D$ there is an open neighborhood U_a of a in X on which $u = u_a$. One writes $u = \sum_{a \in D} u_a$.
(ii) $\operatorname{supp} u = D$, and the order of u equals the supremum of the m_a, for $a \in D$.
(iii) D is countable and for every enumeration $j \mapsto a(j)$ of D we have that $u = \lim_{k \to \infty} \sum_{j=1}^{k} u_{a(j)}$.

7.9. Let $(u_j)_{j \in \mathbf{N}}$ be a sequence in $\mathcal{D}'(X)$. Suppose that for every $x \in X$ there is an open neighborhood U of x in X with the property that the sequence $(u_j(\phi))_{j \in \mathbf{N}}$ in $\mathbf{C}$ converges for every $\phi \in C_0^\infty(U)$ as $j \to \infty$. Show that there exists $u \in \mathcal{D}'(X)$ such that $\lim_{j \to \infty} u_j = u$ in $\mathcal{D}'(X)$.

7.10. Determine the singular support of $|\cdot|$, the Dirac function, the Heaviside function, PV $\frac{1}{x}$, $1/(x + i0)$, and $1/(x - i0)$. And also of the components v_j of the vector field v in (1.2).

7.11. Prove that $\operatorname{sing\,supp}(\partial_j u) \subset \operatorname{sing\,supp} u$, for $1 \leq j \leq n$ and $u \in \mathcal{D}'(X)$.

Chapter 8
Distributions with Compact Support

If u is locally integrable on an open set X in $\mathbf{R}^n$ and has compact support, the integral

$$u(\phi) = \int_X u(x)\,\phi(x)\,dx$$

is absolutely convergent for every $\phi \in C^\infty(X)$, as follows from a slight adaptation of the proof of Theorem 3.5. More generally, every distribution with compact support can be extended to a continuous linear form on $C^\infty(X)$. Before we can give a precise formulation of this result, we have to define convergence in $C^\infty(X)$; clearly, Definition 2.13 has to be modified.

Definition 8.1. Let $(\phi_j)_{j\in\mathbf{N}}$ be a sequence in $C^\infty(X)$ and let $\phi \in C^\infty(X)$. We say that

$$\lim_{j\to\infty} \phi_j = \phi \quad \text{in} \quad C^\infty(X)$$

if for every multi-index α and for every compact subset K of X, the sequence $(\partial^\alpha \phi_j)_{j\in\mathbf{N}}$ converges uniformly on K to $\partial^\alpha \phi$ as $j \to \infty$.

A linear form u on $C^\infty(X)$ is said to be *continuous* if $u(\phi_j) \to u(\phi)$ whenever $\phi_j \to \phi$ in $C^\infty(X)$ as $j \to \infty$. The space of continuous linear forms on $C^\infty(X)$ is denoted by $\mathcal{E}'(X)$. One says that for $(u_j)_{j\in\mathbf{N}}$ and u in $\mathcal{E}'(X)$,

$$\lim_{j\to\infty} u_j = u \quad \text{in} \quad \mathcal{E}'(X)$$

if $\lim_{j\to\infty} u_j(\phi) = u(\phi)$ in $\mathbf{C}$, for every $\phi \in C^\infty(X)$ (compare with Definition 5.1). ⊘

The notation $\mathcal{E}'(X)$ echoes the notation $\mathcal{E}(X)$ used by Schwartz to indicate the space $C^\infty(X)$ endowed with the notion of convergence introduced in Definition 8.1.

Note that for a linear form u on $C^\infty(X)$ to be continuous it is sufficient that $u(\phi_j) \to 0$ whenever $\phi_j \to 0$ in $C^\infty(X)$. Furthermore, if $\phi_j \to 0$ in $C_0^\infty(X)$, then certainly $\phi_j \to 0$ in $C^\infty(X)$. It follows that for every $u \in \mathcal{E}'(X)$, the restriction ρu of u to the linear subspace $C_0^\infty(X)$ of $C^\infty(X)$ belongs to $\mathcal{D}'(X)$.

J.J. Duistermaat and J.A.C. Kolk, *Distributions: Theory and Applications*,
Cornerstones, DOI 10.1007/978-0-8176-4675-2_8,

A sequence $(K_j)_{j\in\mathbf{N}}$ of compact subsets of X is said to *absorb* the set X if for every compact subset K of X there exists an index j with $K \subset K_j$. The sequence is said to be *increasing* if $K_j \subset K_{j+1}$ for all j.

Lemma 8.2. *Let X be an open subset of $\mathbf{R}^n$. Then the following hold:*

(a) There exists an increasing sequence $(K_j)_{j\in\mathbf{N}}$ of compact subsets of X that absorbs X.
(b) There is a sequence $(\chi_j)_{j\in\mathbf{N}}$ in $C_0^\infty(X)$ with the property that for every $\phi \in C^\infty(X)$, the sequence with terms $\phi_j := \chi_j\,\phi \in C_0^\infty(X)$ converges in $C^\infty(X)$ to ϕ as $j \to \infty$.
(c) The restriction mapping $\rho : \mathcal{E}'(X) \to \mathcal{D}'(X)$ is injective.

Proof. **(a)**. Using the notation $C := \mathbf{R}^n \setminus X$, the complement of X in $\mathbf{R}^n$, we define

$$K_j = \left\{ x \in X \mid \|x\| \le j \ \text{ and } \ d(x, C) \ge \frac{1}{j} \right\} \qquad (j \in \mathbf{N}).$$

These are bounded and closed subsets of $\mathbf{R}^n$, and therefore compact on account of Theorem 2.2. Furthermore, every $K_j \subset K_{j+1} \subset X$. If K is a compact subset of X, there exists an R with $\|x\| \le R$ for all $x \in K$. In addition, the δ-neighborhood of K is contained in X for sufficiently small $\delta > 0$ in view of Corollary 2.4; this implies $d(x, C) \ge \delta$, for all $x \in K$. If we choose $j \in \mathbf{N}$ sufficiently large that $R \le j$ and $\delta \ge \frac{1}{j}$, then $K \subset K_j$.
(b). On the strength of Corollary 2.16, we can find $\chi_j \in C_0^\infty(X)$ with $\chi_j = 1$ on an open neighborhood of K_j. Let $\phi \in C^\infty(X)$. For every compact subset K of X there is a $j(K)$ such that $K \subset K_{j(K)}$; this implies $K \subset K_j$ if $j \ge j(K)$. But then one has $\phi_j = \chi_j\,\phi = \phi$ on an open neighborhood of K, and therefore also $\partial^\alpha \phi_j = \partial^\alpha \phi$ on K, for every multi-index α. It follows that certainly $\phi_j \to \phi$ in $C^\infty(X)$ as $j \to \infty$.
(c). This is a consequence of (b). Indeed, consider u and $v \in \mathcal{E}'(X)$ with $\rho\,u = \rho\,v$. For every $\phi \in C^\infty(X)$ we have

$$u(\phi) = \lim_{j\to\infty} u(\phi_j) = \lim_{j\to\infty} v(\phi_j) = v(\phi),$$

where $\phi_j \in C_0^\infty(X)$ and $\phi_j \to \phi$ in $C^\infty(X)$. □

Definition 8.3. A subset D of a topological space E is said to be *dense* in E if the closure of D is equal to E. ⊘

Lemma 8.2.(b) implies that $C_0^\infty(X)$ is dense in $C^\infty(X)$. Lemma 8.2.(c) is a special case of the general principle that a continuous extension to the closure of a set is uniquely determined.

Because of the injectivity of ρ, every $u \in \mathcal{E}'(X)$ can be identified with the distribution $\rho\,u \in \mathcal{D}'(X)$; accordingly, we will write $\rho\,u = u$ and consider $\mathcal{E}'(X)$ as a linear subspace of $\mathcal{D}'(X)$. Further note that $u_j \to u$ in $\mathcal{E}'(X)$ implies that $u_j \to u$ in $\mathcal{D}'(X)$.

In order to determine which elements in $\mathcal{D}'(X)$ belong to $\mathcal{E}'(X)$, we first take a closer look at convergence in $C^\infty(X)$.

Definition 8.4. For every $k \in \mathbf{Z}_{\geq 0}$ and compact subset K of X, we define

$$\|\phi\|_{C^k,K} = \sup_{|\alpha|\leq k,\, x\in K} |\partial^\alpha \phi(x)| \qquad (\phi \in C^k(X)). \tag{8.1}$$

For a sequence $(\phi_j)_{j\in\mathbf{N}}$ and ϕ in $C^k(X)$, we say that $\lim_{j\to\infty} \phi_j = \phi$ in $C^k(X)$ if we have $\lim_{j\to\infty} \|\phi_j - \phi\|_{C^k,K} = 0$, for every compact subset K of X. With this definition, $\phi_j \to \phi$ in $C^\infty(X)$ is equivalent to the assertion that $\phi_j \to \phi$ in $C^k(X)$ for every $k \in \mathbf{Z}_{\geq 0}$. ⊘

Because (8.1) is not subject to the condition supp $\phi \subset K$, (8.1) does not define a norm on either $C^k(X)$ or $C^\infty(X)$; for every k and K there exist $\phi \in C^\infty(X)$ with $\|\phi\|_{C^k,K} = 0$ and $\phi \neq 0$. However, the $n_{k,K} : \phi \mapsto \|\phi\|_{C^k,K}$ do form a separating collection of seminorms. We now discuss the definition and properties of seminorms for arbitrary linear spaces. In Theorem 8.8 we will resume the characterization of the distributions in $\mathcal{E}'(X)$.

Definition 8.5. For a linear space E over $\mathbf{C}$, a *seminorm* on E is defined as a function $n : E \to \mathbf{R}$ with the following properties. For every x and $y \in E$ and $c \in \mathbf{C}$, one has

(a) $n(x) \geq 0$,
(b) $n(x + y) \leq n(x) + n(y)$ (*subadditivity*),
(c) $n(c\,x) = |c|\, n(x)$ (*absolute homogeneity*).

In other words, n has all properties of a norm, except that there is no requirement that $n(x) = 0$ implies $x = 0$. (We note that condition (a) follows from (b) and (c).) Instead of a norm on E, one may consider a collection $\mathcal{N}$ of seminorms on E that is *separating* in the following sense:

(d) if $n(x) = 0$ for all $n \in \mathcal{N}$, then $x = 0$,
(e) if n and $m \in \mathcal{N}$ there exists $p \in \mathcal{N}$ such that

$$n(x) \leq p(x) \quad \text{and} \quad m(x) \leq p(x) \qquad (x \in E).$$ ⊘

An example is $E = C^k(X)$ with the collection of seminorms

$$\mathcal{N}^k := \{\, n_{k,K} \mid K \subset X,\ K \text{ compact } \}.$$

On $C^\infty(X)$ we use $\mathcal{N} = \bigcup_{k\in\mathbf{Z}_{\geq 0}} \mathcal{N}^k$.

A pair $(E, \mathcal{N})$ where E is a linear space and $\mathcal{N}$ a separating collection of seminorms on E is said to be a *locally convex topological linear space*. A subset U of E is said to be a *neighborhood* of a in $(E, \mathcal{N})$ if there exist $n \in \mathcal{N}$ and $\epsilon > 0$ such that

$$B(n, a, \epsilon) := \{\, x \in E \mid n(x - a) < \epsilon \,\} \subset U.$$

The word "topological" relates to "neighborhoods." Furthermore, a subset C of E is said to be *convex* if for every x and $y \in C$ the line segment from x to y, the set of $x + t\,(y - x)$ with $0 \le t \le 1$, is contained in C. For every $n \in \mathcal{N}$ and $\epsilon > 0$, one has that $B(n, a, \epsilon)$ is convex; this is the origin of the term "locally convex." The assertion about convexity follows from

$$n(x + t(y - x) - a) = n((1 - t)(x - a) + t(y - a)) \le (1 - t)\,n(x - a) + t\,n(y - a).$$

Readers familiar with topology will have observed that the $B(n, a, \epsilon)$ form a *basis*. A subset U of E is said to be *open* if U is a neighborhood of all of its elements. The open sets form a *topology* in E. For a deeper study of distribution theory we recommend that the reader peruse the general theory of locally convex topological linear spaces, as can be found in Bourbaki [4], for example.

Let x_j, for $j \in \mathbf{N}$, and $x \in E$. One then says that $\lim_{j\to\infty} x_j = x$ in $(E, \mathcal{N})$ if

$$\lim_{j\to\infty} n(x_j - x) = 0 \qquad (n \in \mathcal{N}).$$

If $(F, \mathcal{M})$ is another locally convex topological linear space, a linear mapping A from E to F is said to be *sequentially continuous* from $(E, \mathcal{N})$ to $(F, \mathcal{M})$ if $\lim_{j\to\infty} A(x_j) = A(x)$ in $(F, \mathcal{M})$ whenever $\lim_{j\to\infty} x_j = x$ in $(E, \mathcal{N})$.

A mapping $A : E \to F$ is said to be *continuous* from $(E, \mathcal{N})$ to $(F, \mathcal{M})$ if for every $a \in E$ and every neighborhood V of $A(a)$ in $(F, \mathcal{M})$ there exists a neighborhood U of a in $(E, \mathcal{N})$ such that $A(U) \subset V$. For a **linear** mapping this is equivalent to the assertion that for every $m \in \mathcal{M}$ there exist a constant $c > 0$ and $n \in \mathcal{N}$ with

$$m(A(x)) \le c\,n(x) \qquad (x \in E). \tag{8.2}$$

For the sake of completeness we give a proof of this general result from functional analysis.

Proof. Let $a \in E$ and let V be a neighborhood of $A(a)$ in $(F, \mathcal{M})$, that is, there exist $m \in \mathcal{M}$ and $\epsilon > 0$ such that $B(m, A(a), \epsilon) \subset V$. Suppose that there are a constant $c > 0$ and $n \in \mathcal{N}$ such that (8.2) holds. In that case, $x \in B(n, a, \delta)$ implies

$$m(A(x) - A(a)) = m(A(x - a)) \le c\,n(x - a) < c\,\delta.$$

Choosing $\delta = \epsilon/c$, we have $\delta > 0$, and we see that A maps the neighborhood $B(n, a, \delta)$ of a into $B(m, A(a), \epsilon)$, and therefore into V. From this we conclude that assertion (8.2) implies the continuity of A.

Conversely, suppose that A is continuous and that $m \in \mathcal{M}$. From the continuity of A we use only that there is some $a \in E$, that there are constants $\epsilon > 0$ and $\delta > 0$, and that there exists $n \in \mathcal{N}$ with the property that $n(x - a) < \delta$ implies $m(A(x) - A(a)) < \epsilon$. (This is a weak form of the continuity of the mapping A at the point a.) Because $A(x) - A(a) = A(x - a)$, we see that this is equivalent to the property that $n(y) < \delta$ implies $m(A(y)) < \epsilon$. (This is a weak form of the continuity of the mapping A at the point 0.) Now define $c = 2\epsilon/\delta$; we are going to prove that for every $x \in E$ one has $m(A(x)) \le c\,n(x)$. If $n(x) > 0$, one has, for $y = \lambda\,x$

where $\lambda = \delta/(2n(x))$,

$$n(y) = n(\lambda\, x) = \lambda\, n(x) = \delta/2 < \delta;$$

consequently

$$\epsilon > m(A(y)) = m(A(\lambda\, x)) = m(\lambda\, A(x)) = \lambda\, m(A(x)),$$

which implies $m(A(x)) < c\, n(x)$. If, on the other hand, $n(x) = 0$, then for every $\lambda > 0$ one has $n(y) = n(\lambda\, x) = \lambda\, n(x) = 0 < \delta$; it follows that $\epsilon > m(A(y)) = \lambda\, m(A(x))$, which implies $m(A(x)) \leq 0 \leq c\, n(x)$. □

Example 8.6. Consider open sets $X \subset \mathbf{R}^n$ and $Y \subset \mathbf{R}^m$, and a linear mapping $A : C_0^\infty(X) \to C^\infty(Y)$. According to (8.2), Definition 2.13, and (8.1), the mapping A is continuous if for any two compact sets $K \subset X$ and $K' \subset Y$ and any order of differentiation $k' \in \mathbf{Z}_{\geq 0}$, there exist a constant $c > 0$ and an order of differentiation $k \in \mathbf{Z}_{\geq 0}$ such that

$$\|A\phi\|_{C^{k'},\,K'} \leq c\, \|\phi\|_{C^k} \qquad (\phi \in C_0^\infty(K)). \qquad \oslash$$

For E as in the theory above and $F = \mathbf{C}$, with the absolute value as the only (semi)norm, we thus obtain the continuous linear forms on E, together forming the *topological dual* E' of E. The definition as given here is more exact than that in Chap. 3, because the definition there speaks rather vaguely about a "notion of convergence" in E. In E' we use finite sums of the seminorms $u \mapsto |u(x)|$ for all $x \in E$. For these, $\lim_{j\to\infty} u_j = u$ in E' is equivalent to $\lim_{j\to\infty} u_j(x) = u(x)$, for every $x \in E$; this is the *weak convergence* in E' that we introduced before, in Definitions 5.1 and 8.1.

If the separating collection $\mathcal{N}$ of seminorms is finite, we may, as far as the notion of convergence is concerned, as well replace the collection $\mathcal{N}$ by the unique norm given by the maximum of the $n \in \mathcal{N}$. We then find ourselves in the familiar context of linear spaces endowed with a norm, where linear mappings that satisfy estimates as in (8.2) are also said to be *bounded* linear mappings. For $C^k(X)$ and $C^\infty(X)$, the collection $\mathcal{N}$ cannot be replaced by a single norm. It is possible, however, without changing the notion of convergence, to replace the collection of seminorms by the increasing countable sequence with terms $n_i := n_{i,K_i}$, where (K_i) is an increasing sequence of compact subsets of X that absorbs X. Indeed, for every compact $K \subset X$ there exists a j with $K \subset K_j$; therefore $n_{k,K} \leq n_i$ if we choose i to be the maximum of k and j.

The following lemma generalizes a familiar result about linear mappings between linear spaces endowed with a norm.

Lemma 8.7. *Let $(E, \mathcal{N})$ and $(F, \mathcal{M})$ be locally convex topological linear spaces. Then every continuous linear mapping $A : E \to F$ is sequentially continuous. If $\mathcal{N}$*

is countable, the converse is also true, that is, every sequentially continuous linear mapping $A : E \to F$ is continuous.

Proof. Suppose that A is continuous and $x_j \to x$ in E. For every $m \in \mathcal{M}$ we then have, on account of (8.2),

$$m(A(x_j) - A(x)) = m(A(x_j - x)) \leq c\, n(x_j - x) \to 0 \qquad \text{as} \qquad j \to \infty.$$

Now suppose that $(n_j)_{j \in \mathbf{N}}$ is an enumeration of $\mathcal{N}$. By changing to

$$\nu_k := \max_{1 \leq j \leq k} n_j \qquad (k \in \mathbf{N}),$$

we obtain an increasing sequence of seminorms ν_k that defines the same notion of neighborhood in E.

Suppose A is not continuous. Then there exist an $m \in \mathcal{M}$ and, for every $c > 0$ and $k \in \mathbf{N}$, an $x = x_{c,k} \in E$ such that $m(A(x)) > c\, \nu_k(x)$. Next, take $c = k$ and introduce $y_k = \lambda\, x$ with $\lambda = 1/m(A(x))$. We then find that $1 = m(A(y_k)) > k\, \nu_k(y_k)$, from which we conclude that $y_k \to 0$ in $(E,\, \mathcal{N})$, without $A(y_k) \to 0$ in $(F,\, \mathcal{M})$. It follows that A is not sequentially continuous. □

The proof of Theorem 3.8 is the special case of the proof of the second assertion in Lemma 8.7 with $E = C_0^\infty(K)$ and $F = \mathbf{C}$. We have formulated Lemma 8.7 in this general form to avoid having to rewrite the same argument each time we work with another collection of seminorms.

Theorem 8.8. *Consider $u \in \mathcal{D}'(X)$. One has $u \in \mathcal{E}'(X)$ if and only if* supp u *is a compact subset of X. If that is the case, u is of finite order and there exist constants $c > 0$ and $k \in \mathbf{Z}_{\geq 0}$ satisfying*

$$|u(\phi)| \leq c\, \|\chi\, \phi\|_{C^k} \tag{8.3}$$

for every $\phi \in C^\infty(X)$ and every $\chi \in C_0^\infty(X)$ such that $\chi = 1$ on an open neighborhood of supp u.

Proof. Let $u \in \mathcal{E}'(X)$. By Lemma 8.7, with $E = C^\infty(X)$ and $F = \mathbf{C}$, it follows from the sequential continuity of u that u is continuous in the sense of (8.2). That is, there exist constants $c > 0$ and $k \in \mathbf{Z}_{\geq 0}$ and a compact subset K of X such that

$$|u(\phi)| \leq c\, \|\phi\|_{C^k,\, K} \qquad (\phi \in C^\infty(X)). \tag{8.4}$$

This implies, in particular, that $u(\phi) = 0$ if $\phi \in C_0^\infty(X \setminus K)$, in other words, supp $u \subset K$. The conclusion is that supp u is a compact subset of X and that u is of finite order; see Definition 3.12. The estimate (8.3) now follows from (8.4). Indeed, supp $(\phi - \chi\, \phi) \cap$ supp $u = \emptyset$ implies $u(\phi - \chi\, \phi) = 0$; in turn, this gives

$$|u(\phi)| = |u(\chi\, \phi)| \leq c\, \|\chi\, \phi\|_{C^k,\, K} \leq c\, \|\chi\, \phi\|_{C^k}.$$

Next, suppose that supp u is compact. Corollary 2.16 yields a $\chi \in C_0^\infty(X)$ with $\chi = 1$ on a neighborhood of supp u. Define $v(\phi) = u(\chi\,\phi)$ for every $\phi \in C^\infty(X)$; then v is a linear form on $C^\infty(X)$. Now, $\phi_j \to 0$ in $C^\infty(X)$ implies that $\chi\,\phi_j \to 0$ in $C_0^\infty(X)$ and therefore also $v(\phi_j) \to 0$. As a result, $v \in \mathcal{E}'(X)$. Finally, if $\phi \in C_0^\infty(X)$, then supp $(\phi - \chi\,\phi) \cap$ supp $u = \emptyset$, and so $0 = u(\phi - \chi\,\phi) = u(\phi) - v(\phi)$. This means that $u = \rho\, v$. □

In view of Theorem 8.8, $\mathcal{E}'(X)$ is said to be the *space of distributions with compact support* in X.

Lemma 8.9. *Let U and V be open subsets of $\mathbf{R}^n$ with $U \subset V$. For every $u \in \mathcal{E}'(U)$ there exists exactly one $v = i(u) \in \mathcal{E}'(V)$ such that $v = u$ on U and* supp $v =$ supp u. *This defines an injective continuous linear mapping i from $\mathcal{E}'(U)$ to $\mathcal{E}'(V)$. For $v \in \mathcal{E}'(V)$ one has that $v \in i(\mathcal{E}'(U))$ if and only if* supp $v \subset U$. *The mapping i is used to identify $\mathcal{E}'(U)$ with the space of the $v \in \mathcal{E}'(V)$ such that* supp $v \subset U$, *with the notation $i(u) = u$.*

Proof. The mapping ρ_{UV} from (7.2) defines a continuous linear mapping from $C^\infty(V)$ to $C^\infty(U)$. Thus, $v := u \circ \rho_{UV} \in \mathcal{E}'(V)$. It will be clear that $v = u$ on U, and that $v = 0$ on $C := V \setminus$ supp u. Because $V = U \cup C$, we conclude that $w = v$ whenever $w \in \mathcal{D}'(V)$, $w = u$ on U and $w = 0$ on C, which proves the uniqueness. It is also evident that $u = 0$ if $v = 0$, which yields the injectivity of the linear mapping i. □

The estimate (8.4) is of the same form as (3.4) in Theorem 3.8; however, in the case of (3.4) the test functions were restricted to $C_0^\infty(K)$. On $C_0^\infty(K)$, sequential continuity of linear forms is equivalent to continuity, and Theorem 3.8 asserts that the distributions on X are precisely those linear forms u on $C_0^\infty(X)$ for which the restriction of u to $C_0^\infty(K)$, for every compact subset K of X, is continuous from $C_0^\infty(K)$ to $\mathbf{C}$. This is the reason that we took the liberty, in discussing the definition of distributions, to speak of continuous linear forms, although in fact we had introduced them as sequentially continuous linear forms. We may add that it is also possible to endow $C_0^\infty(X)$ with an (uncountable) collection of seminorms $\mathcal{N}$, in such a way that the distributions are precisely the continuous linear forms on $C_0^\infty(X)$ with respect to $\mathcal{N}$; see for example Hörmander [12, Thm. 2.1.5].

Contrary to what might perhaps be expected, (8.4) does not generally hold for $K =$ supp u; a counterexample is given in Hörmander [12, Example 2.3.2]; see Problem 8.3. But on the other hand, we have the following result.

Theorem 8.10. *Let $a \in \mathbf{R}^n$ and let U be an open neighborhood of a in $\mathbf{R}^n$. Consider $u \in \mathcal{D}'(U)$ with* supp $u = \{a\}$. *Then u is of finite order, say k, and one has*

$$u = \sum_{|\alpha| \leq k} c_\alpha\, \partial^\alpha \delta_a \qquad \text{with} \qquad c_\alpha = \frac{u(x \mapsto (a-x)^\alpha)}{\alpha!}.$$

In particular, a Radon measure on U with support in a is of the form $c\,\delta_a$ with $c = u(1)$. Finally, there exists $c > 0$ such that $|u(\phi)| \leq c\,\|\phi\|_{C^k,\{a\}}$, for all $\phi \in C_0^\infty(U)$.

Proof. Considering that the conclusion relates only to the behavior of u near a, we may assume that U is a ball with center a. Furthermore, we can reduce the problem to the case that $a = 0$, by means of a translation. Because $u \in \mathcal{E}'(U)$, it follows from Theorem 8.8 that u has a uniquely determined continuous extension to $C^\infty(U)$, again denoted by u. Based on the estimate (8.3), we will first prove

$$u(\phi) = 0 \qquad \text{if} \qquad \phi \in C^\infty(U) \qquad \text{and} \qquad \partial^\alpha \phi(0) = 0 \qquad (|\alpha| \leq k).$$

Applying Taylor expansion, with an estimate for the remainder according to Theorem 6.2, we obtain for such a function ϕ,

$$\lim_{\epsilon \downarrow 0} \frac{1}{\epsilon^k} \sup_{\|x\| \leq \epsilon} |\phi(x)| = 0.$$

At 0, all derivatives to order $k - |\gamma|$ of the function $\partial^\gamma \phi$ vanish, and so the preceding formula implies

$$\lim_{\epsilon \downarrow 0} \frac{1}{\epsilon^{k-|\gamma|}} \sup_{\|x\| \leq \epsilon} |\partial^\gamma \phi(x)| = 0. \tag{8.5}$$

Furthermore, we specify the functions χ to be used in (8.3). In fact, choose $\chi \in C_0^\infty(\mathbf{R}^n)$ with $\chi = 1$ on a neighborhood of 0 and $\chi(x) = 0$ if $\|x\| > 1$. Write $\chi_\epsilon(x) = \chi(\frac{1}{\epsilon} x)$, for $\epsilon > 0$. We then have $\chi_\epsilon = 1$ on a neighborhood of 0 and $\chi_\epsilon(x) = 0$ if $\|x\| > \epsilon$; also, for every multi-index β there exists a constant $c_\beta > 0$ with the property

$$|\partial^\beta \chi_\epsilon(x)| \leq \frac{c_\beta}{\epsilon^{|\beta|}} \qquad (x \in \mathbf{R}^n).$$

Since $0 \notin \operatorname{supp}((1-\chi_\epsilon)\phi)$, we have $u(\phi) = u(\chi_\epsilon\, \phi) + u((1-\chi_\epsilon)\phi) = u(\chi_\epsilon\, \phi)$. Next we substitute $\psi = \chi_\epsilon\, \phi$ into the estimate (8.3). In doing so, we write $\partial^\alpha \psi(x)$ using Leibniz's formula as a sum of a constant times terms of the form $\partial^\beta \chi_\epsilon(x)\, \partial^\gamma \phi(x)$, with $\beta + \gamma = \alpha$. Each of these terms can be estimated uniformly in x, as

$$\frac{c_\beta}{\epsilon^{|\beta|}} \sup_{\|x\| \leq \epsilon} |\partial^\gamma \phi(x)|,$$

which converges to 0 as $\epsilon \downarrow 0$ on account of (8.5), because $|\beta| = |\alpha| - |\gamma| \leq k - |\gamma|$. The conclusion is that $|u(\phi)|$ converges to 0 as $\epsilon \downarrow 0$. This is possible only if $u(\phi) = 0$.

For general $\phi \in C^\infty(U)$, Taylor expansion to order k at 0 now gives

$$\phi(x) = \sum_{|\alpha| \leq k} \frac{\partial^\alpha \phi(0)}{\alpha!} x^\alpha + R(x),$$

where $R \in C^\infty(U)$ and $\partial^\alpha R(0) = 0$ whenever $|\alpha| \leq k$. The latter implies $u(R) = 0$, and therefore

$$u(\phi) = \sum_{|\alpha| \leq k} \frac{\partial^\alpha \phi(0)}{\alpha!} u(x \mapsto x^\alpha).$$

Applying the definition of derivative of δ leads to the formula in the theorem. $\square$

The preceding theorem is definitely a local result. Still, its proof had to wait until the present chapter dealing with C^∞ functions, on account of the use of the Taylor expansion. Generally speaking, a remainder in the Taylor expansion of a function in $C_0^\infty(X)$, with X open in $\mathbf{R}^n$, does not belong to $C_0^\infty(X)$, because Taylor polynomials do not belong to that space.

A characteristic application of the theorem occurs in the initial part of the proof of Theorem 13.3 below.

An important role in the theory of locally convex topological linear spaces is played by the *Hahn–Banach Theorems*. The plural is used here because the terminology is customarily applied to several closely related results. One of these will now be used to derive an interesting property of distributions; the result is the following *Dominated Extension Theorem*; see [19, Theorem 3.3]. Its verification requires transfinite induction, and the theorem is therefore nonconstructive.

Theorem 8.11. *Suppose E is a linear space over $\mathbf{C}$ and $n : E \to \mathbf{R}_{\geq 0}$ a seminorm on E, while L is a linear subspace of E and v a linear form on L such that*

$$|v(y)| \leq n(y) \qquad (y \in L).$$

Then v extends to a linear form u on E (i.e., the restriction of u to L equals v) that satisfies

$$|u(x)| \leq n(x) \qquad (x \in E).$$

For later use, in Remark 11.12, Example 14.31, and the proof of Theorem 15.4, we also mention the following, closely related, result; see [19, Theorem 3.5].

Theorem 8.12. *Suppose E is a locally convex topological linear space and L is a linear subspace of E. Then L is a dense subspace of E if and only if every continuous linear form on E that vanishes on L also vanishes on E.*

Relative to the filtration of distributions by order, those of order 0, the Radon measures, are the least singular ones. In addition, under the operation of differentiation these generate all distributions of higher order.

Theorem 8.13. *Suppose $X \subset \mathbf{R}^n$ is an open set, $u \in \mathcal{D}'(X)$, and $k \in \mathbf{Z}_{\geq 0}$. Then u is of order $\leq k$ if and only if for every open neighborhood U of* supp u *in X, there exist Radon measures $u_\alpha \in \mathcal{D}'(X)$, with $|\alpha| \leq k$, such that*

$$u = \sum_{|\alpha| \leq k} \partial^\alpha u_\alpha \qquad \textit{and} \qquad \operatorname{supp} u_\alpha \subset U.$$

Proof. Only $\Rightarrow$ requires a proof. Let $m \in \mathbf{N}$ be the number of multi-indices $\alpha \in (\mathbf{Z}_{\geq 0})^n$ such that $|\alpha| \leq k$. Then there exists a natural injection ι of $C_0^k(X)$ into the product space $(C_0(X))^m$; more specifically, ι assigns to each $\phi \in C_0^k(X)$ the m-tuple $(\partial^\alpha \phi)_{|\alpha| \leq k}$ of all derivatives of ϕ of order $\leq k$. Obviously ι is linear and injective. For $k > 0$ it is, however, not surjective, on account of Problem 12.12 or the equality of mixed partial derivatives of a C^k function (see [7, Theorem 2.7.2]). The image of $C_0^k(X)$ under ι is a linear subspace, say L, of $(C_0(X))^m$. Then the mapping $C_0^k(X) \to L$ is an isomorphism of locally convex topological linear spaces. Indeed, the ϕ_j converge to 0 in $C_0^k(X)$ as $j \to \infty$ if and only if all $\partial^\alpha \phi_j$ converge to 0 in $C_0(X)$, for $|\alpha| \le k$.

We may now transfer any continuous linear form on $C_0^k(X)$ as a continuous linear form on L and then extend the latter as a continuous linear form on $(C_0(X))^m$ on account of the Dominated Extension Theorem, Theorem 8.11. But given a finite family of locally convex topological linear spaces $(E_1, \dots, E_m)$, the product $E_1' \times \cdots \times E_m'$ of the topological dual spaces is canonically isomorphic to the topological dual of the product, via the correspondence

$$E_1' \times \cdots \times E_m' \ni (v_1, \dots, v_m) \quad \longleftrightarrow \\ (E_1 \times \cdots \times E_m \ni (y_1, \dots, y_m) \mapsto \textstyle\sum_{j=1}^m v_j(y_j) \in \mathbf{C}) \in (E_1 \times \cdots \times E_m)'.$$

Applying this to the product $(C_0(X))^m$, we see that a continuous linear form on this product is an m-tuple of Radon measures $(v_\alpha)_{|\alpha| \le k}$ on X, operating in the following manner:

$$(v_\alpha)((\phi_\alpha)) = \sum_{|\alpha| \le k} v_\alpha(\phi_\alpha).$$

It suffices to assume that this linear form extends the linear form u transferred to L and to take $\phi_\alpha = \partial^\alpha \phi$, to see that

$$u = \sum_{|\alpha| \le k} (-1)^{|\alpha|} \partial^\alpha v_\alpha.$$

We must now verify the condition on the supports of the Radon measures u_α. Take a function χ equal to 1 in a neighborhood of supp u and to 0 outside some closed subset contained in U. Such a function exists in view of Corollary 2.16. First of all, $u = \chi u$ since for all test functions ϕ, we have $\chi u(\phi) = u(\chi \phi) = u(\phi)$, and supp $(1 - \chi)\phi$ is contained in the complement of supp u. Therefore we obtain

$$u = \sum_{|\alpha| \le k} (-1)^{|\alpha|} \chi\, \partial^\alpha v_\alpha, \qquad \text{with} \qquad \chi\, \partial^\alpha v_\alpha = \sum_{\beta \le \alpha} c_\beta\, \partial^\beta ((\partial^{\alpha-\beta} \chi)\, v_\alpha)$$

for suitable $c_\beta \in \mathbf{R}$, as follows from Leibniz's formula (2.8). Finally, observe that the supports of the Radon measures $(\partial^{\alpha-\beta} \chi)\, v_\alpha$ are contained in supp χ, which, in turn, is contained in U. □

Problems

8.1. Suppose that the singular support of $u \in \mathcal{D}'(X)$ is a compact subset of X. Prove that u is of finite order.

8.2.* Let $(u_j)_{j\in\mathbf{N}}$ be a sequence in $\mathcal{E}'(X)$ that converges in $\mathcal{D}'(X)$ to $u \in \mathcal{D}'(X)$. If there exists a compact subset K of X with $\operatorname{supp} u_j \subset K$ for all j, then $\operatorname{supp} u \subset K$ and $u_j \to u$ in $\mathcal{E}'(X)$, in the sense that $u_j(\phi) \to u(\phi)$ for every $\phi \in C^\infty(X)$. Prove this.

Let $(a(j))_{j\in\mathbf{N}}$ be a sequence in X with the property that $d(a(j), \mathbf{R}^n \setminus X) \to 0$ or $\|a(j)\| \to \infty$ as $j \to \infty$. Prove that $\lim_{j\to\infty} \delta_{a(j)} = 0$ in $\mathcal{D}'(X)$, but **not** in $\mathcal{E}'(X)$.

8.3. Let $\left(x^{(j)}\right)_{j\in\mathbf{N}}$ be a sequence in $\mathbf{R}^n$ that converges to $x \in \mathbf{R}^n$. In addition, assume that all $x^{(j)}$ and x differ from each other. Finally, let $(\epsilon_j)_{j\in\mathbf{N}}$ be a sequence in $\mathbf{R}_{>0}$ with the property

$$\sum_{j\in\mathbf{N}} \epsilon_j = \infty, \qquad \text{while also} \qquad \sum_{j\in\mathbf{N}} \epsilon_j \, \|x^{(j)} - x\| < \infty.$$

(i) Prove the existence of $x^{(j)}$, x, and ϵ_j with these properties, for $j \in \mathbf{N}$. Hint: let $(n(k))_{k\in\mathbf{N}}$ be a strictly increasing sequence satisfying $\|x^{(n(k))} - x\| \le \frac{1}{k}$. Then take $\epsilon_j = \frac{1}{k}$ if $j = n(k)$, and $\epsilon_j = 0$ otherwise.

(ii) Verify that

$$u(\phi) = \sum_{j\in\mathbf{N}} \epsilon_j \left(\phi(x^{(j)}) - \phi(x)\right) \qquad (\phi \in C_0^\infty(\mathbf{R}^n))$$

defines a distribution u of order ≤ 1. Prove

$$\operatorname{supp} u = \{\, x^{(j)} \mid j \in \mathbf{N} \,\} \cup \{x\}$$

and show that this is a compact set.

(iii) Verify that for every $l \in \mathbf{N}$ there exists $\phi_l \in C_0^\infty(\mathbf{R}^n)$ with $\phi_l = 1$ on a neighborhood of $x^{(j)}$ if $1 \le j \le l$, while $\phi_l = 0$ on a neighborhood of $x^{(j)}$ if $j > l$ and $\phi_l = 0$ on a neighborhood of x. Prove that for every $k \in \mathbf{Z}_{\ge 0}$,

$$\|\phi_l\|_{C^k,\, \operatorname{supp} u} = 1, \qquad \text{while} \qquad u\left(\phi_l\right) = \sum_{j=1}^{l} \epsilon_j.$$

(iv) Demonstrate that (8.4) does not hold for any $k \in \mathbf{Z}_{\ge 0}$ with $K = \operatorname{supp} u$.

8.4. Prove that for every $r > 0$,

$$u_r(\phi) = \int_0^{2\pi} \phi(r\cos t, r\sin t)\, dt \qquad (\phi \in C_0^\infty(\mathbf{R}^2))$$

defines a Radon measure u_r on $\mathbf{R}^2$ with compact support and determine its support. Is u_r a locally integrable function? Calculate $u_r(x \mapsto x^\alpha)$ for all α with $|\alpha| \leq 2$.

For what functions θ does one have $\lim_{r \downarrow 0} \theta(r)\, u_r = \delta$? And what condition has to be met by the functions α and β so that

$$\lim_{r \downarrow 0} \big(\alpha(r)\, u_r + \beta(r)\, \delta\big) = \Delta\delta?$$

Prove that if $\alpha(r)\, u_r + \beta(r)\, \delta$ converges to a distribution v of order ≤ 2, it follows that v must be a linear combination of δ and $\Delta\delta$.

8.5.* Consider the function E on $\mathbf{R}^{n+1} \simeq \mathbf{R}^n \times \mathbf{R}$ defined by $E(x,t) = u_t(x)$ for $t > 0$, with u_t as in Problem 5.5, and $E(x,t) = 0$ for $t \leq 0$. Prove that E is locally integrable, so that E can be interpreted as a distribution on $\mathbf{R}^{n+1}$. Determine sing supp E.

Let $v = \partial_t E - \Delta_x E \in \mathcal{D}'(\mathbf{R}^{n+1})$, where Δ_x denotes the Laplace operator with respect to only the x variables. Determine supp v and estimate the order of v.

Finally, calculate v using integration by parts, as in Problem 5.6. Note that in this case there is no requirement that $n \geq 3$. See Problem 17.2 for another method.

Chapter 9
Multiplication by Functions

If X is an open subset of $\mathbf{R}^n$, the function ψ belongs to $C^\infty(X)$, and f is locally integrable on X, one has, for every test function ϕ,

$$(\text{test}(\psi\, f))(\phi) = (\text{test}\, f)(\psi\, \phi).$$

For an arbitrary $u \in \mathcal{D}'(X)$ we now **define**

$$(\psi\, u)(\phi) = u(\psi\, \phi) \qquad (\phi \in C_0^\infty(X)).$$

This leads to a linear form $\psi\, u$ on $C_0^\infty(X)$. Using Leibniz's formula (2.8) we see that the C^k norm of $\psi\, \phi$ can be estimated by a constant times the C^k norm of ϕ; using Theorem 3.8 we therefore conclude that $\psi\, u \in \mathcal{D}'(X)$.

The definition is formulated such that

$$\text{test}(\psi\, f) = \psi\, (\text{test}\, f)$$

for every locally integrable function f; thus we do not immediately run into notation difficulties if we omit the premodifier "test." From formula (9.2) below it follows that the mapping "multiplication by ψ": $u \mapsto \psi\, u$ is linear and continuous from $\mathcal{D}'(X)$ to $\mathcal{D}'(X)$. One has $u = \psi\, u$ on U if $\psi(x) = 1$ for all x in the open subset U of X. Because $\psi\, \phi = 0$ if $\phi \in C_0^\infty(X)$ and $\text{supp}\, \psi \cap \text{supp}\, \phi = \emptyset$, we also obtain

$$\text{supp}\, (\psi\, u) \subset \text{supp}\, \psi \cap \text{supp}\, u.$$

Furthermore, $(\chi\, \psi)u = \chi(\psi\, u)$, for $\chi \in C^\infty(X)$.

Example 9.1. Suppose X is an open set in $\mathbf{R}^n$ and let $a \in X$ and $\psi \in C^\infty(X)$. Then $\psi\, \delta_a = \psi(a)\, \delta_a$ in $\mathcal{D}'(X)$. Indeed, for every $\phi \in C_0^\infty(X)$, we have

$$(\psi\, \delta_a)(\phi) = \delta_a(\psi\, \phi) = (\psi\, \phi)(a) = \psi(a)\, \phi(a) = \psi(a)\, \delta_a(\phi).$$

In particular, $x\, \delta = 0$ in $\mathcal{D}'(X)$ if $0 \in X$. In turn, this implies $x\, \delta' = -\delta$, because

$$x\, \delta'(\phi) = \delta'(x\, \phi) = -\delta(\partial_x(x\, \phi)) = -\delta(\phi) - x\, \delta(\phi') = -\delta(\phi).$$

J.J. Duistermaat and J.A.C. Kolk, *Distributions: Theory and Applications*,
Cornerstones, DOI 10.1007/978-0-8176-4675-2_9,

This leads to $x^2\,\delta' = 0$ in $\mathcal{D}'(X)$, on account of

$$x^2\,\delta'(\phi) = x\,\delta'(x\,\phi) = -\delta(x\,\phi) = -x\,\delta(\phi) = 0. \qquad \oslash$$

Example 9.2. We have $x\ \mathrm{PV}\,\frac{1}{x} = 1$ in $\mathcal{D}'(\mathbf{R})$. In fact, for $\phi \in C_0^\infty(\mathbf{R})$,

$$x\ \mathrm{PV}\,\frac{1}{x}(\phi) = \mathrm{PV}\,\frac{1}{x}(x\,\phi) = \lim_{\epsilon\downarrow 0}\int_{\mathbf{R}\setminus[-\epsilon,\,\epsilon]}\phi(x)\,dx = \int_{\mathbf{R}}\phi(x)\,dx = 1(\phi).$$

Using Problem 1.3 and Example 9.1 we now obtain the following identities in $\mathcal{D}'(\mathbf{R})$:

$$1 = x\ \mathrm{PV}\,\frac{1}{x} = x\,\frac{1}{x \pm i\,0} \pm \pi\,i\,x\,\delta = x\,\frac{1}{x \pm i\,0}. \qquad \oslash$$

A first application of multiplication by functions is the following:

Lemma 9.3. *Let $(\chi_j)_{j\in\mathbf{N}}$ be a sequence in $C_0^\infty(X)$ as in Lemma 8.2.(b). For every $u \in \mathcal{D}'(X)$, the sequence of terms $u_j := \chi_j\,u \in \mathcal{E}'(X)$ then converges in $\mathcal{D}'(X)$ to u, as $j \to \infty$. In other words, $\mathcal{E}'(X)$ is dense in $\mathcal{D}'(X)$.*

Proof. If $\phi \in C_0^\infty(X)$, then $\chi_j\phi \to \phi$ in $C_0^\infty(X)$ on account of Lemma 8.2.(b); therefore $(\chi_j\,u)(\phi) = u(\chi_j\,\phi) \to u(\phi)$ as $j \to \infty$. □

By Theorem 3.15 we can similarly define the product $\psi\,u$ if $\psi \in C^k(X)$ and the distribution u is of order $\leq k$; in that case, $\psi\,u$ is of order $\leq k$. Thus $\psi\,\mu$ is a Radon measure for every continuous function ψ and Radon measure μ.

The *Leibniz rule* (2.9), (2.8) holds if $f = \psi \in C^\infty(X)$ and $g = u \in \mathcal{D}'(X)$. Indeed, for every $1 \leq j \leq n$ and $\phi \in C_0^\infty(X)$,

$$\begin{aligned}(\partial_j(\psi\,u))(\phi) &= (\psi\,u)(-\partial_j\phi) = u(-\psi\,\partial_j\phi) = u(\partial_j\psi\ \phi - \partial_j(\psi\,\phi))\\ &= (\partial_j\psi\,u + \psi\,\partial_j u)(\phi),\end{aligned}$$

which leads to

$$\partial_j(\psi\,u) = \partial_j\psi\,u + \psi\,\partial_j u \qquad (\psi \in C^\infty(X),\ u \in \mathcal{D}'(X)). \tag{9.1}$$

In addition, the product $\psi\,u$ is continuous as a function of ψ and u together. By this we mean that

$$\begin{aligned}&\psi_j\,u_j \to \psi\,u \quad \text{in} \quad \mathcal{D}'(X) \qquad \text{if}\\ &\psi_j \to \psi \quad \text{in} \quad C^\infty(X) \quad \text{and} \quad u_j \to u \quad \text{in} \quad \mathcal{D}'(X).\end{aligned} \tag{9.2}$$

For the proof we write $(\psi_j\,u_j)(\phi) = u_j(\psi_j\,\phi)$, for $\phi \in C_0^\infty(X)$. Next, we observe that $\psi_j\,\phi \to \psi\,\phi$ in $C_0^\infty(X)$ and we use Theorem 5.5.(ii).

The mapping $u \mapsto \psi\, u$, of multiplication by $\psi \in C^\infty(X)$, is also written by means of the notation ψ. In quantum mechanics this is a familiar operation; multiplying by x_j is said to be the jth *position operator*. It will be necessary to recall the definition from time to time, in order to avoid confusion in the notation. If $\psi \in C_0^\infty(X)$, then ψ is a sequentially continuous linear mapping from $\mathcal{D}'(X)$ to $\mathcal{E}'(X)$.

It is not possible to extend the product $\psi\, u$ in a continuous manner to arbitrary distributions ψ and u. For example, let $u_t \in C^\infty(\mathbf{R}^n)$, for $t > 0$, be as in Problem 5.5. Then $u_t\, \delta = u_t(0)\, \delta = (4\pi t)^{-n/2}\, \delta$ according to Example 9.1, which does not converge in $\mathcal{D}'(\mathbf{R}^n)$ as $t \downarrow 0$. On the other hand, u_t does converge to δ in $\mathcal{D}'(\mathbf{R}^n)$ as $t \downarrow 0$; this forms an obstacle to a well-behaved definition of the product $\delta\, \delta$ as an element of $\mathcal{D}'(\mathbf{R}^n)$.

The product that maps both $(\psi,\, u) \in C^\infty(X) \times \mathcal{D}'(X)$ and $(u,\, \psi) \in \mathcal{D}'(X) \times C^\infty(X)$ to $\psi\, u \in \mathcal{D}'(X)$ is not associative. Indeed, $\frac{1}{x+i0}\,(x\,\delta) = \frac{1}{x+i0}\,0 = 0$ by Example 9.1, while Example 9.2 implies $(\frac{1}{x+i0}\,x)\,\delta = 1\,\delta = \delta \neq 0$. On the other hand, the more elementary case of associativity $\psi(\chi\, u) = (\psi\chi)\, u$, for ψ and $\chi \in C^\infty(X)$ and $u \in \mathcal{D}'(X)$, does hold.

Now that we have multiplication by functions available, we can study *linear partial differential equations with variable coefficients*, of the form

$$P(x, \partial)u = \sum_{|\alpha| \leq m} c_\alpha(x)\, \partial^\alpha u = f, \tag{9.3}$$

for given distributions f and desired distributions u. Here the coefficients c_α are required to be C^∞ functions on the domain space X of the distributions. Variants involving C^k coefficients with finite k are also possible, but the order of u must then be limited and one has to keep track of the degrees of differentiability and of the orders. By way of example we show how the theory of linear ordinary differential equations, the case of $n = 1$, is developed in the distributional context.

Theorem 9.4. *Let I be an open interval in $\mathbf{R}$, $c_j \in C^\infty(I)$ for $0 \leq j \leq m$, and $c_m(x) \neq 0$ for all $x \in I$. Then there exists, for every $f \in \mathcal{D}'(I)$, a solution $u \in \mathcal{D}'(I)$ of the equation*

$$Pu := \sum_{j=0}^{m} c_j(x)\, u^{(j)} = f. \tag{9.4}$$

The subset in $\mathcal{D}'(I)$ of all solutions consists of the $u + h$, with h denoting any solution of the homogeneous equation, given by $f = 0$. Every such h is a classical C^∞ solution, and together they form an m-dimensional linear space over $\mathbf{C}$. If $f \in C^k(I)$, then $u \in C^{m+k}(I)$ and u is a solution in the classical sense.

Proof. Because $1/c_m \in C^\infty(I)$, we can multiply the equation by this factor; thus we obtain an equivalent equation for u, with c_m replaced by 1, the c_j by c_j/c_m, and f by f/c_m. Consequently, in what follows we may assume that $c_m = 1$.

We now perform the usual reduction to an m-dimensional first-order system. To achieve this, we introduce $v_j := u^{(j-1)}$, for $1 \le j \le m$. Also, $g_j := 0$ if $0 \le j < m$ and $g_m := f$. Together these make (9.4) equivalent to a system of the form $v' = L(x)\,v + g$, where the $m \times m$ matrix $L(x)$ can be obtained from the explicit form

$$v_j' = v_{j+1} \qquad (1 \le j < m), \qquad v_m' = -\sum_{j=1}^{m} c_{j-1}(x)\,v_j \;+\; g_m.$$

Let $x \mapsto \Phi(x)$ be a classical *fundamental matrix* for the homogeneous system $v' = L(x)\,v$, that is, an $m \times m$ matrix whose columns form a basis for the linear space of solutions of $v' = L(x)\,v$. Its existence is proved in the theory of ordinary differential equations; the coefficients are shown to be C^∞ functions of x, and similarly those of the inverse matrix $\Phi(x)^{-1}$. Furthermore, Φ satisfies the matrix differential equation $\Phi'(x) = L(x) \circ \Phi(x)$.

We now apply, in the distributional context, Lagrange's method of variation of constants for solving the inhomogeneous equation. This consists in substituting $v = \Phi(x)\,w$ and then deriving the differential equation that w must satisfy to make v a solution of the inhomogeneous equation $v' = L(x)\,v + g$. Because a distribution may be multiplied by a C^∞ function, the substitution $v = \Phi(x)\,w$ yields a bijective relation between the vector distributions v and w on I. Using Leibniz's rule we obtain for w,

$$L(x) \circ \Phi(x)\,w + g = \partial_x(\Phi(x)\,w) = L(x) \circ \Phi(x)\,w + \Phi(x)\,w',$$

in other words, $w' = \Phi(x)^{-1}g$. The notation used is shorthand for a calculation by components. But for every component this is an equation like the one we have already studied in Theorem 4.3. This yields the existence of a solution $w \in (\mathcal{D}'(I))^n$ and therefore of a solution $v = \Phi(x)\,w \in (\mathcal{D}'(I))^n$ of $v' = L(x)\,v + g$, and consequently, of a solution $u = v_1 \in \mathcal{D}'(I)$ of (9.4).

Owing to the linearity of the operator P, the solution space of the inhomogeneous equation consists of the $u + h$ with h the solutions of the homogeneous equation. Now we have $g = 0$ if $f = 0$; this gives the equation $w' = 0$, the only solutions of which are of the form $w = c \in \mathbf{R}^n$ according to the last assertion in Theorem 4.3. But this implies that $v = \Phi(x)\,c$, which leads to a classical solution of the homogeneous equation.

Finally, if $f \in C^k(I)$, there exists a classical solution $u_0 \in C^{k+m}(I)$. Every solution $u \in \mathcal{D}'(I)$ of the inhomogeneous equation is of the form $u = u_0 + h$ with h a classical C^∞ solution of the homogeneous equation. It follows that $u \in C^{k+m}(I)$ is a classical solution. □

The condition that the highest-order coefficient c_m should have no zeros is essential. This is sufficiently demonstrated by the example of the equation $x\,u = 0$ in $\mathcal{D}'(\mathbf{R})$, which has the nonclassical solution $u = \delta$, as follows from Example 9.1.

In the next theorem we formulate a converse assertion, considering $\mathbf{R}^n$ straight away.

Theorem 9.5. *Let X be open in $\mathbf{R}^n$ and $u \in \mathcal{D}'(X)$. If $\psi \in C^\infty(X)$ and $\psi\, u = 0$, then* supp u *is contained in the zero-set of ψ.*

Consider real-valued $\psi_j \in C^\infty(X)$, *for $1 \le j \le n$, and $a \in X$. Further suppose that $\psi_j(a) = 0$ and, additionally, that the total derivatives $D\psi_j(a)$ of the ψ_j at a are linearly independent, with $1 \le j \le n$. Then $\psi_j\, u = 0$, for all j, implies the existence of $c \in \mathbf{C}$ and an open neighborhood U of a in X satisfying $u = c\,\delta_a$ on the set U.*

Proof. Let $C = \{x \in X \mid \psi(x) \ne 0\}$. Then C is an open subset of X and $\frac{1}{\psi} \in C^\infty(C)$; therefore $u = \frac{1}{\psi}(\psi\, u) = \frac{1}{\psi}0 = 0$ on C. This proves that supp $u \subset X \setminus C$, the zero-set of ψ.

Denote by $\psi : X \to \mathbf{R}^n$ the C^∞ mapping having the ψ_j as component functions. Since $\psi(a) = 0$, we may write, for x sufficiently close to a,

$$\begin{aligned}\psi(x) = \psi(x) - \psi(a) &= \int_0^1 \frac{d}{dt}\psi(a + t(x-a))\,dt\\ &= \int_0^1 D\psi(a + t(x-a))\,dt\,(x-a) =: \Psi(x)\,(x-a),\end{aligned}$$

with Ψ an $n \times n$ matrix consisting of C^∞ functions. In particular, $\Psi(a) = D\psi(a)$; and by assumption this matrix is invertible. By continuity (for instance, consider the determinant of $\Psi(x)$), there exists an open ball U about a in $\mathbf{R}^n$ such that $\Psi(x)$ is invertible, for all $x \in U$. In view of Cramer's rule for inverse matrices, the matrix coefficients of Ψ^{-1} belong to $C^\infty(U)$. We now have

$$x - a = \Psi(x)^{-1}\,\psi(x).$$

By combining this equality with the identity of vectors $\psi(x)\,u = 0$ we obtain

$$(x-a)\,u = 0, \qquad \text{in other words,} \qquad (x_j - a_j)\,u = 0 \qquad (1 \le j \le n,\ x \in U).$$

In view of the assertion proved above we may conclude that $U \cap \operatorname{supp} u = \{a\}$; on U, therefore, we may consider u as an element of $\mathcal{E}'(U)$ on account of Theorem 8.8.

Taylor expansion of an arbitrary $\phi \in C^\infty(U)$ about a leads to

$$\phi(x) = \phi(a) + \sum_{j=1}^n (x_j - a_j)\,\phi_j(x) \qquad (x \in U),$$

for certain $\phi_j \in C^\infty(U)$. We obtain

$$\begin{aligned}u(\phi) &= u(\phi(a)) + \sum_{j=1}^n u((x_j - a_j)\,\phi_j) = \phi(a)\,u(1) + \sum_{j=1}^n (x_j - a_j)\,u(\phi_j)\\ &= u(1)\,\delta_a(\phi),\end{aligned}$$

and from this we deduce $u = u(1)\,\delta_a$ on U. □

Note that Example 9.1 shows that the nondegeneracy condition in the theorem, requiring that the derivatives be linearly independent, in general cannot be omitted.

Problems

9.1. Suppose $p \in C^1(]a, b[)$, $p(x) > 0$ for all $x \in]a, b[$ and q and r continuous on $]a, b[$. Prove that every Radon measure u that is a solution of the variational equation (1.10) for all $\phi \in C_0^\infty(]a, b[)$ is in fact C^2 and satisfies the Euler–Lagrange equation (1.8). Hint: use $v_2 = p(x)\, u'$.

9.2.* Let $\psi \in C^\infty(\mathbf{R}^n)$. Describe the distribution $\psi\, \partial^\alpha \delta$, for every multi-index α, as a linear combination of the $\partial^\beta \delta$ with $\beta \leq \alpha$.

9.3.* Determine all solutions $u \in \mathcal{D}'(\mathbf{R})$ of $x^k\, u = 0$, for $k \in \mathbf{N}$.

9.4.* Determine all solutions $u \in \mathcal{D}'(\mathbf{R})$ of $x\, u' = 0$. Show that these u form a linear space and determine its dimension.

9.5.* Observe that $u = \frac{1}{x+i0}$ is a solution of $x\, u = 1$, and prove that this implies $x\, u' = -u$.

Let $u \in \mathcal{D}'(\mathbf{R})$ be a solution of $x^k\, u = 1$. Applying the differential operator $x\, \partial_x$ to this identity, show that $v = \frac{-1}{k}\, u'$ is a solution of $x^{k+1}\, v = 1$. Finally, determine for every $k \in \mathbf{N}$, all solutions in $\mathcal{D}'(\mathbf{R})$ of the equation $x^k\, u = 1$.

9.6.* Prove

$$\lim_{t\to\infty} e^{itx}\, \mathrm{PV}\, \frac{1}{x} = \pi i\, \delta. \tag{9.5}$$

Deduce (refer to Problem 14.13 for another proof); see Fig. 9.1,

$$\lim_{t\to\infty} \frac{1}{2} \int_{-t}^{t} e^{ix\xi}\, d\xi = \lim_{t\to\infty} \frac{\sin tx}{x} = \pi\, \delta \qquad \text{and} \qquad \lim_{t\to\infty} \mathrm{PV}\, \frac{\cos tx}{x} = 0.$$

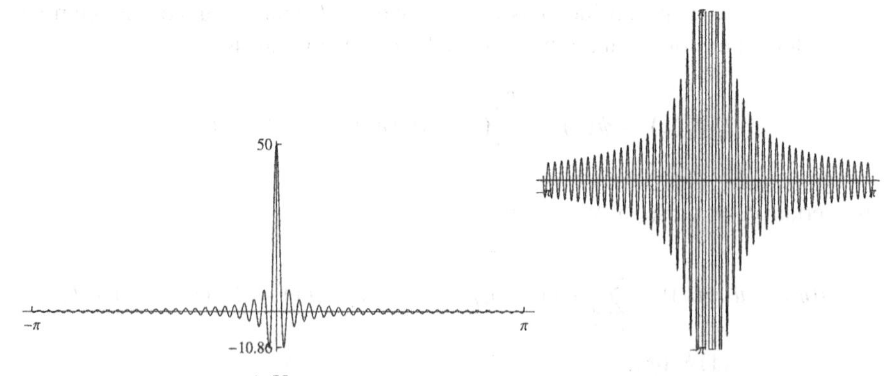

Fig. 9.1 Graphs of $x \mapsto \frac{\sin 50x}{x}$, one with scales adjusted and one with equal scales, but truncated

Next define $u_{t,\epsilon} \in C^\infty(\mathbf{R})$ by $u_{t,\epsilon}(x) = \frac{e^{itx}}{x+i\epsilon}$, where t and $\epsilon \in \mathbf{R}$ and $\epsilon \neq 0$. Using Problem 1.3 prove the following equalities in $\mathcal{D}'(\mathbf{R})$:

$$\text{(i)} \quad \lim_{t\to\infty}(\lim_{\epsilon\downarrow 0} u_{t,\epsilon}) = 0, \qquad \text{(ii)} \quad \lim_{t\to\infty}(\lim_{\epsilon\uparrow 0} u_{t,\epsilon}) = 2\pi i\,\delta,$$
$$\text{(iii)} \quad \lim_{\epsilon\to 0}(\lim_{t\to\infty} u_{t,\epsilon}) = 0.$$

In particular, limits in $\mathcal{D}'(\mathbf{R}^n)$ cannot be freely interchanged.

9.7.* Let $u \in \mathcal{D}'(\mathbf{R}^n)$ and $x_n\, u = 0$. Prove that there exists a uniquely determined $v \in \mathcal{D}'(\mathbf{R}^{n-1})$ such that $u(\phi) = v(\iota^*\phi)$ for every $\phi \in C_0^\infty(\mathbf{R}^n)$. Here

$$(\iota^*\phi)(x_1, \dots, x_{n-1}) = \phi(x_1, \dots, x_{n-1}, 0).$$

Hint: use $\chi \in C_0^\infty(\mathbf{R})$ with $\chi(x) = 1$ on a neighborhood of 0. For $\psi \in C_0^\infty(\mathbf{R}^{n-1})$, define the function $\psi \otimes \chi \in C_0^\infty(\mathbf{R}^n)$ by $(\psi \otimes \chi)(y, z) = \psi(y)\,\chi(z)$ and deduce that v must be defined by $v(\psi) := u(\psi \otimes \chi)$. See Problem 11.18 for a reformulation.

9.8. Prove that the differential equation

$$(1 - x^2)^2\, u' - 2x\, u = (1 - x^2)^2$$

has no solution $u \in \mathcal{D}'(I)$ if I is an open interval in $\mathbf{R}$ with $[-1,\ 1] \subset I$.

9.9. Let $a \in \mathbf{C}$. Show that the solutions $u \in \mathcal{D}'(\mathbf{R})$ of the differential equation $x\, u' = a\, u$ form a linear space $\mathcal{H}_a$. Demonstrate that multiplication by x, and the differentiation ∂_x, define linear mappings from $\mathcal{H}_a$ to $\mathcal{H}_{a+1}$ and $\mathcal{H}_{a-1}$, respectively. Determine $\mathcal{H}_a$. Hint: use Problem 7.6 and deal with the case $a \in \mathbf{Z}_{<0}$ separately. Decide whether the mappings above are injective or surjective, respectively.

9.10. Let $\xi \in \mathbf{C}$. Find the solutions $u \in \mathcal{D}'(\mathbf{R})$ of the equation $e^{ix\xi}\, u = u$. Distinguish between the cases $\xi = 0$, $\xi \in \mathbf{R} \setminus \{0\}$, and $\xi \in \mathbf{C} \setminus \mathbf{R}$.

9.11. Let $\omega(k)$, with $1 \leq k \leq n$, be a basis in $\mathbf{R}^n$. Let $b(j)$, with $1 \leq j \leq n$, be the *dual basis*, defined by $\langle b(j), \omega(k) \rangle = \delta_{jk}$, where $\delta_{jk} = 1$ if $j = k$ and $\delta_{jk} = 0$ if $j \neq k$. Define the lattice A in $\mathbf{R}^n$ by

$$A := \left\{ \sum_{j=1}^{n} 2\pi k_j\, b(j) \mid k \in \mathbf{Z}^n \right\}.$$

Show that $u \in \mathcal{D}'(\mathbf{R}^n)$ satisfies the equations $e^{i\langle x, \omega(j)\rangle}\, u = u$, for all $1 \leq j \leq n$, if and only if

$$u = \sum_{a \in A} c_a\, \delta_a \qquad (c_a \in \mathbf{C}).$$

9.12. (Sequel to Problem 7.4.) Let X be an open subset of $\mathbf{R}^n$ and let U be open in X with C^1 boundary ∂U. Let $v(x)$ be a continuous vector field on $\overline{U}$. Verify that $\sum_{j=1}^n v_j\, \partial_j 1_U = 0$ if and only if $v(y)$ is tangent to ∂U, for every $y \in \partial U$.

9.13. Determine all $u \in \mathcal{D}'(\mathbf{R}^2)$ such that $x_1 x_2 u = 0$ as well as $(x_1^2 - x_2^2)\, u = 0$. Hint: begin by determining the support of u. Then test u by means of $x_1 x_2 x^\alpha$ and $(x_1^2 - x_2^2)\, x^\beta$, for all multi-indices α and β.

9.14.* The ordinary differential equation with constant coefficients

$$L\, I'' + R\, I' + \frac{1}{C}\, I = \delta$$

in $\mathcal{D}'(\mathbf{R})$ describes the current I in an electrical network consisting of an inductor of inductance $L > 0$, a resistor of resistance $R \geq 0$, and a capacitor of capacitance $C > 0$ connected by ideal wires in which at time 0 a constant voltage of magnitude 1 is switched on. Prove that I is given by, for $t \in \mathbf{R}$ (see Fig. 9.2),

$$I(t) = \begin{cases} \dfrac{2}{\sqrt{R^2 - \frac{4L}{C}}}\, H(t)\, e^{-\frac{R}{2L}t} \sinh \dfrac{t}{2L}\sqrt{R^2 - \dfrac{4L}{C}}, & \text{if} \quad R^2 > \dfrac{4L}{C}; \\ \dfrac{1}{L}\, H(t)\, t\, e^{-\frac{R}{2L}t}, & \text{if} \quad R^2 = \dfrac{4L}{C}; \\ \dfrac{2}{\sqrt{\frac{4L}{C} - R^2}}\, H(t)\, e^{-\frac{R}{2L}t} \sin \dfrac{t}{2L}\sqrt{\dfrac{4L}{C} - R^2}, & \text{if} \quad R^2 < \dfrac{4L}{C}. \end{cases}$$

In the first case, the solution $I(t)$ is of exponential decay as $t \to \infty$, because

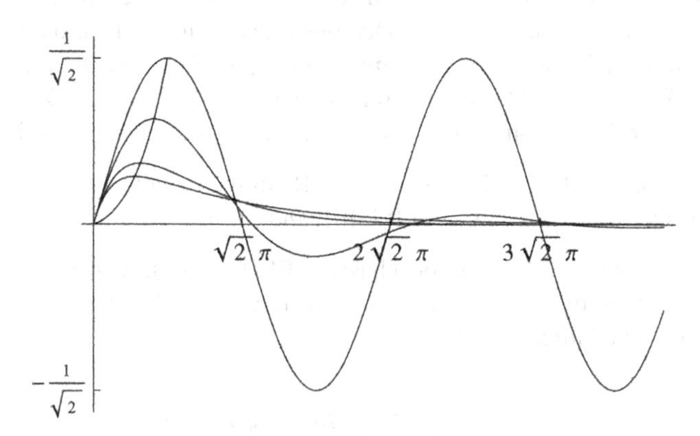

Fig. 9.2 Graphs of I for different values of R, while $L = 2$ and $C = 1$. Near 0 these correspond from top to bottom to $R = 0,\, 1,\, 2\sqrt{2}$, and 4. The vertical curve from $\frac{1}{2}\sqrt{2}(\pi, 1)$ to the origin passes through the maxima of the functions I for R running from 0 to ∞

$R > \sqrt{R^2 - \frac{4L}{C}}$; in the second, it is also of exponential decay; in the third, it is an exponentially damped sinusoidal oscillation. In this last case, under the additional condition of $R = 0$, the solution takes the form of an undamped sinusoidal oscillation $I(t) = \sqrt{\frac{C}{L}}\, H(t)\, \sin \frac{t}{\sqrt{LC}}$ (see Problem 14.28 for a different approach).

Chapter 10
Transposition: Pullback and Pushforward

We now consider the behavior of distributions under coordinate transformations and, more generally, under suitable C^∞ mappings.

Suppose that $(E, \mathcal{N})$ and $(F, \mathcal{M})$ are locally convex topological linear spaces and that A is a continuous linear mapping from E to F. (For the definitions of these notions, see Chap. 8.) For every v belonging to the topological dual F' of F, the mapping $v \circ A$, being a composition of continuous linear mappings $(E, \mathcal{N}) \xrightarrow{A} (F, \mathcal{M}) \xrightarrow{v} \mathbf{C}$, is a continuous linear form on E in view of a special case of Lemma 10.1 below. Therefore ${}^tA(v) := v \circ A$ is an element of the topological dual E' of E. Indeed,

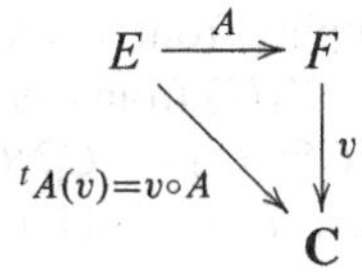

$$\begin{array}{ccc} E & \xrightarrow{A} & F \\ & {}_{{}^tA(v)=v\circ A}\searrow & \downarrow v \\ & & \mathbf{C} \end{array} \qquad \text{where} \qquad {}^tA(v) : E \ni x \mapsto v(A(x)) \in \mathbf{C}.$$

This defines a mapping ${}^tA : F' \to E'$, said to be the *transpose* of A, on account of the defining formula

$${}^tA(v)(x) := ({}^tA(v))(x) := v(A(x)) \qquad (v \in F',\ x \in E).$$

The situation may be summarized by means of the following diagram:

$$\begin{array}{ccc} E & \xrightarrow{A} & F \\ E' & \xleftarrow{{}^tA} & F' \end{array}$$

Note that ${}^tA$ is automatically continuous, and therefore sequentially continuous, from F' to E'. Indeed, if $x \in E$, one has, for every $v \in F'$, that $|{}^tA(v)(x)| = |v(A(x))|$, where the left-hand side is the x-seminorm of ${}^tA(v)$ and the right-hand side is the $A(x)$-seminorm of v.

J.J. Duistermaat and J.A.C. Kolk, *Distributions: Theory and Applications*,
Cornerstones, DOI 10.1007/978-0-8176-4675-2_10,

Lemma 10.1. *Let $(E, \mathcal{N})$, $(F, \mathcal{M})$, and $(G, \mathcal{L})$ be locally convex topological linear spaces. If A is a continuous linear mapping from $(E, \mathcal{N})$ to $(F, \mathcal{M})$ and B a continuous linear mapping from $(F, \mathcal{M})$ to $(G, \mathcal{L})$, then $B \circ A$ is a continuous linear mapping from $(E, \mathcal{N})$ to $(G, \mathcal{L})$. For its transpose one has ${}^t(B \circ A) = {}^tA \circ {}^tB$.*

Proof. Let $l \in \mathcal{L}$. By virtue of the continuity of B and of formula (8.2), there exist a constant $d > 0$ and $m \in \mathcal{M}$ with $l(B(y)) \leq d\, m(y)$ for all $y \in F$. On account of the continuity of A, there exist for this m a constant $c > 0$ and $n \in \mathcal{N}$ such that $m(A(x)) \leq c\, n(x)$ for all $x \in E$. Combining these two estimates, we obtain $l(B \circ A(x)) \leq d\, m(A(x)) \leq dc\, n(x)$ for all $x \in E$. Note that in the case of linear spaces endowed with a norm, this was the proof of the theorem that the operator norm of $B \circ A$ is less than or equal to the product of the operator norms of B and of A.

For the proof of the second assertion, consider $w \in G'$ and $x \in E$. Then

$$ {}^t(B \circ A)(w)(x) = w(B(A(x))) = {}^tB(w)(A(x)) = {}^tA({}^tB(w))(x). \qquad \square $$

We have, in fact, applied the principle of transposition several times before, when introducing operations involving distributions. For example, we defined $\partial_j : \mathcal{D}'(X) \to \mathcal{D}'(X)$ as the transpose of $-\partial_j : C_0^\infty(X) \to C_0^\infty(X)$; in other words, ${}^t\partial_j = -\partial_j$. Also, for $\psi \in C^\infty(X)$, we introduced the multiplication by $\psi : \mathcal{D}'(X) \to \mathcal{D}'(X)$ as the transpose of the multiplication by $\psi : C_0^\infty(X) \to C_0^\infty(X)$. If, in addition, ψ has compact support in X, multiplication by ψ is even continuous and linear from $C^\infty(X)$ to $C_0^\infty(X)$, which then immediately implies that multiplication by ψ, the transpose, defines a continuous linear mapping from $\mathcal{D}'(X)$ to $\mathcal{E}'(X)$. Furthermore, the restriction mapping $\rho_{UV} : \mathcal{D}'(V) \to \mathcal{D}'(U)$ from Chap. 7 is also a transpose, namely of the identical embedding $\iota_{VU} : C_0^\infty(U) \to C_0^\infty(V)$. We summarize the situation by the following diagram, where $v \in \mathcal{D}'(V)$ and $\phi \in C_0^\infty(U)$:

$$ \begin{array}{ccc} \iota_{VU} : C_0^\infty(U) & \longrightarrow & C_0^\infty(V) \\ \mathcal{D}'(U) & \longleftarrow & \mathcal{D}'(V) : \rho_{UV} \end{array} \qquad \text{with} \qquad (\rho_{UV}\, v)(\phi) = v(\iota_{VU}\phi). $$

Remark 10.2. For general continuous linear $A : C_0^\infty(U) \to C_0^\infty(V)$, it is customary to denote ${}^tA : \mathcal{D}'(V) \to \mathcal{D}'(U)$ by B if B is a previously introduced operator acting on functions and $B = {}^tA$ on the subspace $C_0^\infty(V)$ of $\mathcal{D}'(V)$. This is what we have done in the case of differentiation and multiplication by functions, and this will also be our guiding principle in the subsequent naming of transpose operators.

In Corollary 11.7 below we will see that $C_0^\infty(V)$ is dense in $\mathcal{D}'(V)$, which makes this procedure almost inevitable. Sometimes, however, the transpose operator turns out to be truly novel, for example when ${}^tA(\phi) \in \mathcal{D}'(U)$ is not a function for some $\phi \in C_0^\infty(V)$. In later parts of this chapter we will encounter further examples of this, for instance in Proposition 10.21. ⊘

Let X be an open subset of $\mathbf{R}^n$, Y an open subset of $\mathbf{R}^p$, and $\Phi : X \to Y$ a C^∞ mapping. Then $\psi \circ \Phi : x \mapsto \psi(\Phi(x))$ belongs to $C^\infty(X)$, for every $\psi \in C^\infty(Y)$. The mapping

$$\Phi^* : \psi \mapsto \psi \circ \Phi \quad : \quad C^\infty(Y) \to C^\infty(X)$$

is linear (albeit between infinite-dimensional spaces) and by means of the chain rule for differentiation it can be shown to be also continuous. Note that $(\Phi^*\psi)(x) = \psi(y)$ if $y = \Phi(x)$; thus, Φ^* is something like "the change of variables $y = \Phi(x)$ in functions of y." The operator Φ^* is said to be the *pullback* of functions on Y to functions on X, under the mapping $\Phi : X \to Y$. In fact, transposition is the pullback of continuous linear forms under continuous linear mappings, although the latter may have been defined on infinite-dimensional spaces. Furthermore, observe that for $X = \mathbf{R}$ and $Y = \{0\} \subset \mathbf{R}$, we get $(\Phi^*\psi)(x) = \psi(0)$, for $\psi \in C_0^\infty(Y)$ and $x \in \mathbf{R}$. In this case, therefore, we do not have $\Phi^* : C_0^\infty(Y) \to C_0^\infty(X)$.

Example 10.3. Let $\Phi : X \to Y$ and $\Psi : Y \to Z$ be C^∞ mappings, where $X \subset \mathbf{R}^n$, $Y \subset \mathbf{R}^p$, and $Z \subset \mathbf{R}^q$ are open sets. Then we have, on account of the chain rule (6.1),

$$(\partial_j \circ \Phi^*)\Psi_i = \partial_j(\Psi_i \circ \Phi) = \sum_{k=1}^{p} \partial_j \Phi_k \, (\Phi^* \circ \partial_k)\Psi_i,$$

for $1 \leq j \leq n$ and $1 \leq i \leq q$. In particular, if $q = 1$, this equality is valid for all $\Psi = \Psi_1 \in C^\infty(Y)$; therefore we obtain the following identity of continuous linear mappings, which describes *composition of partial differentiation and pullback*, for $1 \leq j \leq n$:

$$\partial_j \circ \Phi^* = \sum_{k=1}^{p} \partial_j \Phi_k \circ \Phi^* \circ \partial_k \quad : \quad C^\infty(Y) \to C^\infty(X). \tag{10.1}$$

On the right-hand side of this equality $\partial_j \Phi_k$ denotes the linear operator of multiplication by $\partial_j \Phi_k$ acting in $C^\infty(X)$.

In the special case of a linear mapping $\Phi = A : \mathbf{R}^n \to \mathbf{R}^n$ with matrix (a_{ij}), (10.1) takes the form

$$\partial_j \circ A^* = A^* \circ \sum_{k=1}^{n} a_{kj} \, \partial_k.$$

Hence

$$\partial \circ A^* = A^* \circ ({}^tA \, \partial), \qquad \text{where} \qquad \partial = (\partial_1, \ldots, \partial_n) \tag{10.2}$$

is regarded as a column vector and ${}^tA : \mathbf{R}^n \to \mathbf{R}^n$ denotes the transpose linear mapping. ⊘

The transpose of the continuous linear mapping $\Phi^* : C^\infty(Y) \to C^\infty(X)$ is a continuous linear mapping ${}^t(\Phi^*) : \mathcal{E}'(X) \to \mathcal{E}'(Y)$. This is written as ${}^t(\Phi^*) = \Phi_*$ and is said to be the *pushforward* of distributions on X with compact support to distributions on Y with compact support, under the mapping $\Phi : X \to Y$. The defining equation is

$$(\Phi_* u)(\psi) = u(\Phi^*\psi) = u(\psi \circ \Phi) \qquad (u \in \mathcal{E}'(X),\ \psi \in C^\infty(Y)). \tag{10.3}$$

We describe the preceding results by means of the following diagram:

$$X \xrightarrow{\Phi} Y$$

$$C^\infty(X) \xleftarrow{\Phi^*} C^\infty(Y)$$

$$\mathcal{E}'(X) \xrightarrow{\Phi_*} \mathcal{E}'(Y)$$

In mathematics, Φ_* is said to be *covariant* with respect to Φ, because Φ_* acts in the same direction as Φ itself, while Φ^* is said to be *contravariant* with respect to Φ, seeing that the sense of the arrow is reversed. Distribution theory parallels geometry regarding these conventions. In the latter, the underlying space and the points it contains are the initial (covariant) objects, the smooth functions on the space are dual (contravariant) objects, while compactly supported distributions are objects dual to smooth functions, hence covariant again. This is corroborated by the fact that the Dirac measures, which correspond to the points, form a subset of the distributions.

If Ψ is a C^∞ mapping from Y to an open subset Z of $\mathbf{R}^q$, then $\Psi \circ \Phi$ is a C^∞ mapping from X to Z. As in the proof of the second assertion in Lemma 10.1, we see that

$$(\Psi \circ \Phi)^* = \Phi^* \circ \Psi^* \quad : \quad C^\infty(Z) \to C^\infty(X).$$

Transposing this in turn, we obtain

$$(\Psi \circ \Phi)_* = \Psi_* \circ \Phi_* \quad : \quad \mathcal{E}'(X) \to \mathcal{E}'(Z). \tag{10.4}$$

Example 10.4. For every $x \in X$ one has $\Phi_*\delta_x = \delta_{\Phi(x)}$. Indeed, for every $\psi \in C^\infty(Y)$ we obtain $(\Phi_*\delta_x)(\psi) = \delta_x(\psi \circ \Phi) = \psi(\Phi(x)) = \delta_{\Phi(x)}(\psi)$. ⊘

Example 10.5. Set $V = \{ x \in \mathbf{R}^2 \mid \|x\| = 1 \}$ and consider the distribution $\Delta\,\delta_V \in \mathcal{E}'(\mathbf{R}^2)$ of order 2, where Δ denotes the Laplace operator as in Problem 4.7. Thus

$$(\Delta\,\delta_V)(\phi) = \int_V \Delta\phi(x)\,dx = \int_{-\pi}^{\pi} \Delta\phi(\cos\alpha, \sin\alpha)\,d\alpha \qquad (\phi \in C^\infty(\mathbf{R}^2)).$$

Define $\Phi : \mathbf{R}^2 \to \mathbf{R}$ by $\Phi(x) = \|x\|^2$. Then we have

$$\Phi_*\,\Delta\,\delta_V = 8\pi(\delta_1{}'' - \delta_1{}') \quad \text{in} \quad \mathcal{E}'(\mathbf{R}).$$

Indeed, using the chain rule repeatedly we obtain, for $\psi \in C^\infty(\mathbf{R})$,

$$\partial_1^2(\psi \circ \Phi)(x) = 4x_1^2\,\psi''(\|x\|^2) + 2\psi'(\|x\|^2),$$

and an analogous formula for differentiation with respect to x_2. It follows that

$$\begin{aligned}(\Phi_*\,\Delta\,\delta_V)(\psi) &= (\Delta\,\delta_V)(\Phi^*\psi) = \delta_V(\Delta(\psi \circ \Phi))\\ &= 4\int_{-\pi}^{\pi} ((\cos^2\alpha + \sin^2\alpha)\,\psi''(1) + \psi'(1))\,d\alpha = 8\pi(\delta_1{}'' - \delta_1{}')(\psi).\end{aligned}$$

See Problem 10.20.(vi) for another proof. ⊘

We are not allowed to conclude from the density of $\mathcal{E}'(X)$ in $\mathcal{D}'(X)$ and of $\mathcal{E}'(Y)$ in $\mathcal{D}'(Y)$ (see Lemma 9.3) that the continuous linear mapping $\Phi_* : \mathcal{E}'(X) \to \mathcal{E}'(Y)$ extends to a continuous linear mapping $\Phi_* : \mathcal{D}'(X) \to \mathcal{D}'(Y)$. This is because the topology of $\mathcal{E}'(X)$ is finer than the topology that $\mathcal{E}'(X)$ inherits as a linear subspace of $\mathcal{D}'(X)$. Nevertheless, in the following theorem Φ_* is extended, subject to additional conditions on the mapping Φ, to distributions that may not have compact support. We need some preparation.

For a closed subset A of X, we denote the space of the $u \in \mathcal{D}'(X)$ with $\text{supp}\, u \subset A$ by $\mathcal{D}'(X)_A$. We then say that $u_j \to u$ in $\mathcal{D}'(X)_A$ if $u_j \in \mathcal{D}'(X)_A$ for all j and $u_j \to u$ in $\mathcal{D}'(X)$ as $j \to \infty$. This implies that $u \in \mathcal{D}'(X)_A$.

The mapping $\Phi : X \to Y$ is said to be *proper* on A if for every compact subset L of Y, the set $\Phi^{-1}(L) \cap A$ is a compact subset of X. An important property of a proper and continuous mapping $\Phi : X \to Y$ is that it is *closed*, that is, the image under Φ of every closed set in X is closed in Y; see [7, Theorem 1.8.6].

Theorem 10.6. *Let Φ be a C^∞ mapping from the open subset X of $\mathbf{R}^n$ to the open subset Y of $\mathbf{R}^p$. Let A be a closed subset of X and suppose that Φ is proper on A.*

Then there exists a uniquely determined extension of $\Phi_* : \mathcal{E}'(X) \cap \mathcal{D}'(X)_A \to \mathcal{E}'(Y)$ to a linear mapping $\Phi_* : \mathcal{D}'(X)_A \to \mathcal{D}'(Y)$ with the property that $\Phi_* u_j \to \Phi_* u$ in $\mathcal{D}'(Y)$ if $u_j \to u$ in $\mathcal{D}'(X)_A$. Furthermore, $\Phi(\text{supp}\, u)$ is a closed subset of Y and

$$\text{supp}\,(\Phi_* u) \subset \Phi(\text{supp}\, u) \qquad (u \in \mathcal{D}'(X)_A). \tag{10.5}$$

In particular, if Φ is proper from X to Y, then $\Phi_* : \mathcal{E}'(X) \to \mathcal{E}'(Y)$ possesses an extension to a sequentially continuous linear mapping $\Phi_* : \mathcal{D}'(X) \to \mathcal{D}'(Y)$, and one has (10.5) for all $u \in \mathcal{D}'(X)$.

Proof. If $u \in \mathcal{D}'(X)_A$, the sequence of terms $u_j = \chi_j\, u \in \mathcal{E}'(X)$ from Lemma 9.3 converges to u in $\mathcal{D}'(X)_A$. This leads to the uniqueness: **if** the extension exists, we have to take

$$\Phi_* u = \lim_{j \to \infty} \Phi_* u_j .$$

Let L be an arbitrary compact subset of Y and $\chi \in C_0^\infty(X)$ with $\chi = 1$ on a neighborhood of the compact set $K = \Phi^{-1}(L) \cap A$; see Corollary 2.16. For all $\phi \in C_0^\infty(L)$, we define

$$v(\phi) = (\Phi_*(\chi\, u))(\phi) = u(\chi\, \Phi^* \phi).$$

Because the linear mapping Φ^* is continuous from $C_0^\infty(L)$ to $C^\infty(X)$ and multiplication by χ is continuous from $C^\infty(X)$ to $C_0^\infty(X)$, v is a continuous linear form on $C_0^\infty(L)$. Furthermore, if $u_j \to u$ in $\mathcal{D}'(X)_A$, then

$$(\Phi_* u_j)(\phi) = u_j(\Phi^* \phi) = (\chi\, u_j)(\Phi^* \phi) = u_j(\chi\, \Phi^* \phi)$$

converges to $v(\phi)$ as $j \to \infty$. From the resulting uniqueness it follows that this v has an extension to $C_0^\infty(Y)$; this is the $\Phi_* u$ with the desired continuity properties.

With respect to (10.5) we observe that $\phi \in C_0^\infty(Y)$ and supp $\phi \cap \Phi(\text{supp } u) = \emptyset$ imply that supp $(\Phi^*\phi) \subset \Phi^{-1}(\text{supp } \phi)$ is disjunct from supp u, and so

$$(\Phi_* u)(\phi) = \lim_{j\to\infty} (\Phi_* u_j)(\phi) = \lim_{j\to\infty} u_j(\Phi^*\phi) = 0.$$

The last equality follows from supp u_j = supp $\chi_j u \subset$ supp u, which is disjoint from supp $(\Phi^*\phi)$. Thus, we may conclude that supp $(\Phi_* u) \subset \overline{\Phi(\text{supp } u)} = \Phi(\text{supp } u)$, because the restriction of Φ to A is a closed mapping. □

Example 10.7. Denote by $1_X \in \mathcal{D}'(X)$ the characteristic function of an open subset X of $\mathbf{R}^n$. Write $\Delta : X \to X \times X$ for the *diagonal mapping*[1] $\Delta(x) = (x, x)$, which is proper. In this case, $\Delta(X) = \{ (x, x) \in \mathbf{R}^{2n} \mid x \in X \}$ is called the *diagonal* in $X \times X$. Next, consider the pushforward Radon measure

$$\Delta_* 1_X \in \mathcal{D}'(X \times X); \qquad \text{then} \qquad \Delta_* 1_X(\psi) = \int_X \psi(x, x)\, dx, \tag{10.6}$$

for $\psi \in C_0^\infty(X \times X)$. Note that supp $\Delta_* 1_X = \Delta(X)$, while $\Delta_* 1_X = 2^{-\frac{n}{2}} \delta_{\Delta(X)}$, where $\delta_{\Delta(X)}$ is the Radon measure on $X \times X$ as in Example 7.3. ⊘

We will now discuss the pushforward of distributions under C^∞ mappings having various additional properties. We begin by considering diffeomorphisms, and then, in Theorem 10.18 and Proposition 10.21, we study mappings between open sets of different dimensions.

A C^∞ *diffeomorphism* $\Phi : X \to Y$ is a C^∞ mapping from X to Y that is bijective and whose inverse mapping $\Psi := \Phi^{-1}$ is a C^∞ mapping from Y to X. First, we recall some basic facts about diffeomorphisms.

Applying the chain rule to $\Psi \circ \Phi(x) = x$ and to $\Phi \circ \Psi(y) = y$, we find, with the notation $y = \Phi(x)$, that $D\Psi(y) \circ D\Phi(x) = I$ and $D\Phi(x) \circ D\Psi(y) = I$. In other words, $D\Phi(x) : \mathbf{R}^n \to \mathbf{R}^p$ is a bijective linear mapping, with inverse $D\Psi(y)$; this also implies that $n = p$. Conversely, if Φ is an injective C^∞ mapping from an open subset X of $\mathbf{R}^n$ to $\mathbf{R}^n$ such that $D\Phi(x)$ is invertible for every $x \in X$, the *Inverse Function Theorem* asserts that $Y := \Phi(X)$ is an open subset of $\mathbf{R}^n$ and that Φ is a C^∞ diffeomorphism from X to Y. See, for example, [7, Theorem 3.2.8].

In the following theorem we use the notation

$$j_\Phi(x) = |\det D\Phi(x)| \qquad (x \in X) \tag{10.7}$$

for a differentiable mapping Φ from an open subset X of $\mathbf{R}^n$ to $\mathbf{R}^n$. The matrix of derivatives $D\Phi(x)$ is also referred to as the *Jacobi matrix* of Φ at the point x, and det $D\Phi(x)$ as the *Jacobi determinant* or *Jacobian*. Note that $j_\Phi \in C^\infty(X)$ if $\Phi : X \to \mathbf{R}^n$ is a C^∞ diffeomorphism.

---

[1] In this book, one can distinguish between the diagonal mapping and the Laplace operator on the basis of the context in which we make use of them.

Theorem 10.8. *Suppose that X and Y are open subsets of $\mathbf{R}^n$ and $\Phi : X \to Y$ a C^∞ diffeomorphism. Then Φ is proper and the pushforward $\Phi_* : \mathcal{D}'(X) \to \mathcal{D}'(Y)$ is a sequentially continuous linear mapping. In particular, for a locally integrable function f on X, $\Phi_* f$ is the locally integrable function on Y given by*

$$\Phi_* f = j_\Psi \, \Psi^* f = \Psi^*\Big(\frac{f}{j_\Phi}\Big), \qquad \textit{where} \qquad \Psi := \Phi^{-1}. \tag{10.8}$$

In particular, we have $j_\Phi \, \Phi^* \circ \Phi_* = I$ on $C_0^\infty(X)$.

Proof. The properness of Φ follows from the continuity of Ψ; accordingly, the existence of the pushforward $\Phi_* : \mathcal{D}'(X) \to \mathcal{D}'(Y)$ is a consequence of Theorem 10.6. For every $\psi \in C_0^\infty(Y)$ we obtain, by means of the change of variables $x = \Psi(y)$,

$$\begin{aligned}(\Phi_* f)(\psi) &= f(\Phi^* \psi) = f(\psi \circ \Phi) = \int_X f(x)\, \psi(\Phi(x))\, dx \\ &= \int_Y f(\Psi(y))\, \psi(y)\, j_\Psi(y)\, dy = (j_\Psi \, \Psi^* f)(\psi).\end{aligned} \tag{10.9}$$

See, for example, [7, Theorem 6.6.1] for the change of variables in an n-dimensional integral. □

Example 10.9. Fix $\epsilon > 0$. The (isotropic) positive *dilation* $\epsilon : \mathbf{R}^n \to \mathbf{R}^n$ given by $x \mapsto \epsilon\, x$ is a diffeomorphism. Hence we obtain for a locally integrable function f on $\mathbf{R}^n$ and $x \in \mathbf{R}^n$,

$$\epsilon_* f(x) = \epsilon^{-n} (\epsilon^{-1})^* f(x) = \frac{1}{\epsilon^n} f\Big(\frac{1}{\epsilon} x\Big) = f_\epsilon(x).$$

For the last function on the right-hand side we used the notation from Lemma 2.19. ⊘

Remark 10.10. In the case of a diffeomorphism, pushforward leaves the space of compactly supported functions of class C^∞ invariant, and therefore we can extend pullback of functions of class C^∞ to a mapping acting on distributions. More precisely, as a consequence of Theorem 10.8, the restriction of Φ_* to $C_0^\infty(X)$ is a continuous linear mapping A from $C_0^\infty(X)$ to $C_0^\infty(Y)$. Its transpose ${}^t A$ is a continuous linear mapping from $\mathcal{D}'(Y)$ to $\mathcal{D}'(X)$. Let $\psi \in C^\infty(Y)$. For every $\phi \in C_0^\infty(X)$ one has

$$({}^t A\, \psi)(\phi) = \psi(A\phi) = \psi(\Phi_* \phi) = (\Phi_* \phi)(\psi) = \phi(\Phi^* \psi) = (\Phi^* \psi)(\phi),$$

and therefore ${}^t A\, \psi = \Phi^* \psi$. In other words, ${}^t A : \mathcal{D}'(Y) \to \mathcal{D}'(X)$ is an extension of $\Phi^* : C^\infty(Y) \to C^\infty(X)$. This extension is also denoted by Φ^* in order to conform to the convention in Remark 10.2; it is said to be the *pullback under Φ of distributions on Y to distributions on X*. Because $C_0^\infty(Y)$ is dense in $\mathcal{D}'(Y)$, see Corollary 11.7 below, this continuous extension is uniquely determined and the naming is entirely natural. ⊘

The following result summarizes the results above.

Theorem 10.11. *Under the conditions of Theorem 10.8 the pullback $\Phi^* : \mathcal{D}'(Y) \to \mathcal{D}'(X)$ is a continuous linear mapping. For any $v \in \mathcal{D}'(Y)$ and $\phi \in C_0^\infty(X)$, it satisfies*

$$(\Phi^* v)(\phi) = j_\Psi \, v(\Psi^* \phi) = v\Big(\Psi^*\Big(\frac{\phi}{j_\Phi}\Big)\Big), \qquad \textit{where} \qquad \Psi := \Phi^{-1}.$$

Example 10.12. It is a direct application of Theorem 10.8 that $j_\Phi \, \Phi^* \circ \Phi_* = I$ on $\mathcal{D}'(X)$. In particular, Example 10.4 therefore implies

$$\Phi^* \delta_{\Phi(x)} = \frac{1}{j_\Phi(x)} \, \delta_x \qquad (x \in X). \qquad \oslash$$

From Theorem 10.8 it follows that for a diffeomorphism Φ, the pushforward under Φ and the pullback under the inverse Φ^{-1} of Φ are closely related to each other. Indeed, we have

$$\Phi_* \circ j_\Phi = (\Phi^{-1})^* \quad : \quad \mathcal{D}'(X) \to \mathcal{D}'(Y). \tag{10.10}$$

In particular, Φ_* and $(\Phi^{-1})^*$ are identical if and only if j_Φ equals 1, that is, if Φ is volume-preserving.

For $u \in \mathcal{D}'(X)$, the distribution $\Phi_* u \in \mathcal{D}'(Y)$ is also said to be the transform of u under Φ *as a distributional density*, while $(\Phi^{-1})^* u$ is said to be the transform of u *as a generalized function.*

Remark 10.13. The pullback of distributions under the diffeomorphisms that act as coordinate transformations enables one to define distributions on manifolds. See, for example, Hörmander [12, Sect. 6.3]. ⊘

Example 10.14. The reflection $S : x \mapsto -x$ in $\mathbf{R}^n$ about the origin satisfies $S^{-1} = S$ and $j_S = 1$, and therefore, in this case,

$$S_* = (S^{-1})^* = S^*.$$

This transformation of distributions on $\mathbf{R}^n$ is denoted by the same letter S; for every $u \in \mathcal{D}'(\mathbf{R}^n)$, Su is said to be the *reflected distribution.*

Somewhat more generally, if A is an invertible linear transformation of $\mathbf{R}^n$, then j_A equals the constant $|\det A|$, so that (10.10) leads to the following identity of continuous linear mappings on $\mathcal{D}'(\mathbf{R}^n)$:

$$|\det A| \, A_* = (A^{-1})^*. \qquad \oslash$$

Example 10.15. Let $T_h : x \mapsto x + h : \mathbf{R}^n \to \mathbf{R}^n$ be the *translation* in $\mathbf{R}^n$ by the vector $h \in \mathbf{R}^n$. This is evidently a diffeomorphism, whose Jacobi matrix equals the identity and whose Jacobi determinant therefore equals 1. Consequently, the mapping $T_h{}^* : C^\infty(\mathbf{R}^n) \to C^\infty(\mathbf{R}^n)$ has an extension to a continuous linear mapping

$$T_h{}^* = (T_h{}^{-1})_* = (T_{-h})_* : \mathcal{D}'(\mathbf{R}^n) \to \mathcal{D}'(\mathbf{R}^n). \tag{10.11}$$

If $e(j)$ denotes the jth basis vector in $\mathbf{R}^n$, one has, for every $\phi \in C_0^\infty(\mathbf{R}^n)$,

$$\begin{aligned}(T_{t\,e(j)}{}^*\phi - \phi)(x) &= \phi(x + te(j)) - \phi(x) = \int_0^1 \frac{d}{ds}\phi(x + s\,te(j))\,ds\\ &= t\int_0^1 \partial_j\phi(x + s\,te(j))\,ds.\end{aligned}$$

From this we obtain

$$\lim_{t\to 0}\frac{1}{t}\left(T_{t\,e(j)}{}^* - I\right)\phi = \partial_j\phi \quad \text{in} \quad C_0^\infty(\mathbf{R}^n). \tag{10.12}$$

In conjunction with (10.11), this implies that for $u \in \mathcal{D}'(\mathbf{R}^n)$ and $\phi \in C_0^\infty(\mathbf{R}^n)$,

$$\frac{1}{t}\left(T_{t\,e(j)}{}^*u - u\right)(\phi) = \frac{1}{t}\left((T_{-t\,e(j)})_*u - u\right)(\phi) = u\Big(\frac{1}{t}\left(T_{-t\,e(j)}{}^*\phi - \phi\right)\Big)$$

converges to $u(-\partial_j\phi) = \partial_j u(\phi)$ as $t \to 0$. In other words,

$$\lim_{t\to 0}\frac{1}{t}\left(T_{t\,e(j)}{}^* - I\right)u = \partial_j u \quad \text{in} \quad \mathcal{D}'(\mathbf{R}^n). \tag{10.13}$$

For example, if u is a layer on a surface V, as introduced in Example 7.3, one may regard $\partial_j u$ as having been obtained by translating $1/t$ times the charge distribution on V by $-t\,e(j)$, adding the opposite of this on V and then taking the limit as $t \to 0$. As long as one does not pass to the limit, this is similar to a *capacitor* from the theory of electricity. For this reason the derivative $\partial_j u$ of a layer u is also said to be a *double layer* and, more generally, the distribution $\partial^\alpha u$, with u a layer and $|\alpha| > 0$, is said to be a *multiple layer*. (The question whether a kth order derivative should be called a 2^k-fold layer or a $(k+1)$-fold layer is undecided.)

In (1.13) translation of functions was written more concisely as $T_h(f) := f \circ T_{-h} = (T_{-h})^*(f)$, in other words,

$$T_h = (T_{-h})^* = (T_h)_*. \tag{10.14}$$

For example, using this notation we obtain $T_h(\delta_a) = \delta_{a+h}$. The disadvantage of this notation is that now $\partial_j u$ is the derivative of $T_{-t\,e(j)}u$ with respect to t at $t = 0$. But that is the way it is: when the graph of an increasing function is translated to the right, it will lie lower.

Summarizing, we have the following identities of linear mappings acting in $\mathcal{D}'(\mathbf{R}^n)$, for $1 \le j \le n$:

$$\left.\frac{d}{dt} T_{t\,e(j)}\right|_{t=0} = -\partial_j \quad \text{and} \quad \left.\frac{d}{dt} T_{t\,e(j)}{}^*\right|_{t=0} = \partial_j. \tag{10.15}$$

In words, differentiation is the infinitesimal generator of translation; for more details, see [7, Sect. 5.9]. ⊘

Equation (10.12) can be generalized as follows. Let I be an open interval in $\mathbf{R}$ and let A_s be a continuous linear mapping from $(E, \mathcal{N})$ to $(F, \mathcal{M})$, for every $s \in I$. This one-parameter family of linear mappings is said to be *differentiable* at the point a if there exists a continuous linear mapping B from $(E, \mathcal{N})$ to $(F, \mathcal{M})$ such that for every $x \in E$,

$$\lim_{s \to a} \frac{1}{s-a} \left((A_s(x) - A_a(x)) \right) = B(x) \quad \text{in} \quad F.$$

In this case one writes $B = \frac{d}{ds} A_s|_{s=a}$. Writing out the definitions, one immediately sees that this implies that the one-parameter family $s \mapsto {}^t(A_s)$, of continuous linear mappings from F' to E', is also differentiable at $s = a$, while

$$\left.\frac{d}{ds} {}^t(A_s)\right|_{s=a} = {}^t\left(\left.\frac{d}{ds} A_s\right|_{s=a} \right).$$

The generalization to families of mappings that depend on more than one parameter is obvious.

Let X and Y be open subsets of $\mathbf{R}^n$ and $\mathbf{R}^p$, respectively. Let I be an open interval in $\mathbf{R}$ and let $\Phi : X \times I \to Y$ be a C^∞ mapping. Then $\Phi_t(x) = \Phi(x, t)$ defines, for every $t \in I$, a C^∞ mapping from X to Y. Denote by $\Phi_{t,k}$ the kth component function of Φ_t, for $1 \le k \le p$. Applying (10.1) with $j = n + 1$ to the Φ under consideration, one deduces that the one-parameter family $t \mapsto \Phi_t{}^*$, of continuous linear mappings from $C^\infty(Y)$ to $C^\infty(X)$, is differentiable and that one has the identity of continuous linear mappings

$$\frac{d}{dt} \Phi_t{}^* = \sum_{k=1}^{p} \frac{d}{dt} \Phi_{t,k} \circ \Phi_t{}^* \circ \partial_k \quad : \quad C^\infty(Y) \to C^\infty(X) \tag{10.16}$$

for the derivative with respect to t of the pullback. On the right-hand side of this equality $\frac{d}{dt} \Phi_{t,k}$ denotes the linear operator of multiplication by $\frac{d}{dt} \Phi_{t,k}$ acting in $C^\infty(X)$.

In particular, if the Φ_t are C^∞ diffeomorphisms and if we restrict the operators in this equality to $C_0^\infty(Y)$, we obtain an identity of continuous linear mappings sending $C_0^\infty(Y)$ to $C_0^\infty(X)$. Transposition and Lemma 10.1 as well as the fact that the transpose of ∂_k is $-\partial_k$ then immediately lead to the formula

$$\frac{d}{dt} (\Phi_t)_* = - \sum_{k=1}^{p} \partial_k \circ (\Phi_t)_* \circ \frac{d}{dt} \Phi_{t,k} \quad : \quad \mathcal{D}'(X) \to \mathcal{D}'(Y)$$

for the derivative with respect to t of the pushforward. In this case, $\frac{d}{dt}\Phi_{t,k}$ is the linear operator of multiplication by $\frac{d}{dt}\Phi_{t,k}$ acting in $\mathcal{D}'(X)$. If we restrict the operators in this formula to $C_0^\infty(X)$ and transpose once more, we find that (10.16) also holds when we have the continuous linear mappings act on $\mathcal{D}'(Y)$.

These formulas become simpler if $X = Y$, $0 \in I$, and $\Phi_0(x) = x$, for all $x \in X$; that is, $\Phi_0 = I$, the identity on X. In this case (compare with [7, Sect. 5.9])

$$x \mapsto v(x) = \frac{d}{dt}\Phi_t(x)\Big|_{t=0}$$

is said to be the *velocity vector field* of Φ at time $t = 0$. This defines a C^∞ vector field on X with components $v_j \in C^\infty(X)$; we now find for the pullback

$$\frac{d}{dt}{\Phi_t}^*\Big|_{t=0} = \sum_{j=1}^{n} v_j \circ \partial_j \quad : \quad \mathcal{D}'(X) \to \mathcal{D}'(X), \tag{10.17}$$

and for the pushforward

$$\frac{d}{dt}(\Phi_t)_*\Big|_{t=0} = -\sum_{j=1}^{n} \partial_j \circ v_j \quad : \quad \mathcal{D}'(X) \to \mathcal{D}'(X). \tag{10.18}$$

Conversely, for an arbitrary C^∞ vector field v on X, let $\Phi(x,t)$ be the solution $x(t)$ at time t of the system of differential equations

$$x'(t) = v(x(t)) \qquad \text{with initial condition} \qquad x(0) = x.$$

Suppose, for simplicity, that the solutions for all $(x,t) \in X \times \mathbf{R}$ exist. This is equivalent to the condition that none of the solutions will leave every compact subset of X in a finite period of time. According to the theory of ordinary differential equations, Φ in that case is a C^∞ mapping from $X \times \mathbf{R}$ to X and $\Phi_t : x \mapsto \Phi_t(x) = \Phi(x,t)$ is said to be the *flow over time t* with velocity vector field v. The latter satisfies the so-called *group law*

$$\Phi_{t+s} = \Phi_t \circ \Phi_s, \qquad \text{which also implies} \qquad {\Phi_{t+s}}^* = {\Phi_s}^* \circ {\Phi_t}^* \qquad (t,\ s \in \mathbf{R}).$$

In particular, all the Φ_t are C^∞ diffeomorphisms.

By differentiation with respect to t at $t = 0$ we find, in view of (10.17), that

$$\frac{d}{ds}{\Phi_s}^* = {\Phi_s}^* \circ \sum_{j=1}^{n} v_j \circ \partial_j. \tag{10.19}$$

Transposition gives

$$(\Phi_{t+s})_* = (\Phi_t)_* \circ (\Phi_s)_*,$$

from which we obtain, by differentiation with respect to s at $s = 0$, and on account of (10.18),

$$\frac{d}{dt}(\Phi_t)_* = -(\Phi_t)_* \circ \sum_{j=1}^{n} \partial_j \circ v_j. \tag{10.20}$$

The following theorem gives an application of this. The distribution $u \in \mathcal{D}'(X)$ is said to be *invariant* under the diffeomorphism $\Phi : X \to X$ *as a generalized function* if $(\Phi^{-1})^* u = u$. Applying Φ^*, we see that this is equivalent to $\Phi^* u = u$. This u is said to be *invariant* under Φ *as a distributional density* if $\Phi_* u = u$.

Theorem 10.16. *Let v be a C^∞ vector field on an open set X in $\mathbf{R}^n$ whose solutions are defined for all $(x, t) \in X \times \mathbf{R}$. Then $u \in \mathcal{D}'(X)$ is invariant as a generalized function, and as a distributional density, under the flow with velocity vector field v if and only if*

$$\sum_{j=1}^{n} v_j \circ \partial_j u = 0, \qquad \textit{and} \qquad \sum_{j=1}^{n} \partial_j \circ v_j\, u = 0, \qquad \textit{respectively}.$$

Proof. Because $\Phi_0 = I$, the identity in X, we have $\Phi_0{}^* u = I^* u = u$. Therefore, $\Phi_t{}^* u = u$ for all $t \in \mathbf{R}$ is equivalent to the assertion that $\Phi_t{}^* u$ is constant as a function of t. By testing with some $\phi \in C_0^\infty(X)$ and applying the theorem that a real-valued function of a real variable is constant if and only if its derivative vanishes, we find that the first assertion follows from (10.19). Likewise, the second assertion follows from (10.20), on account of $(\Phi_0)_* u = I_* u = u$. □

Homogeneity of functions and homogeneity of distributions can be discussed in analogous ways. For $c > 0$, we denote the mapping $x \mapsto c\,x : \mathbf{R}^n \to \mathbf{R}^n$ by c. Then $u \in \mathcal{D}'(\mathbf{R}^n)$ is said to be *homogeneous of degree* $a \in \mathbf{C}$ if for every $c > 0$,

$$c^* u = c^a\, u, \qquad \text{or equivalently} \qquad c^{n+a}\, c_* u = u.$$

Theorem 10.17. $u \in \mathcal{D}'(\mathbf{R}^n)$ *is homogeneous of degree* $a \in \mathbf{C}$ *if and only if it satisfies* Euler's differential equation:

$$\langle x, \operatorname{grad} u\rangle = \sum_{j=1}^{n} x_j\, \partial_j u = a\, u.$$

Proof. $x \mapsto e^t\, x$ is the flow of the vector field $v(x) = x$. Define $A_t(u) = e^{-at}\,(e^t)^* u$. Then $A_t \circ A_s = A_{t+s}$ and we obtain that u is homogeneous of degree $a \iff A_t(u) = u$ for all $t \in \mathbf{R} \iff \frac{d}{dt} A_t(u)|_{t=0} = 0$. By Leibniz's rule and (10.17), the left-hand side in the latter equality equals $-a\,u + \sum_j x_j\, \partial_j u$. □

Thus, Problem 9.9 amounted to determining all homogeneous distributions on $\mathbf{R}$.

As we have already established, pushforward under Φ of distributions with compact support by application of formula (10.3) is possible for arbitrary C^∞ mappings

Φ from an open subset X of $\mathbf{R}^n$ to an open subset Y of $\mathbf{R}^p$, even if $n > p$ or $n < p$. We conclude this chapter with a few remarks concerning these two cases under the assumption that Φ is generic, i.e., that all of its corresponding tangent mappings are of maximal rank. If $n > p$, and under the further assumption that Φ is a submersion, we can define pullback under Φ of distributions; this result requires the preceding theory for diffeomorphisms. In the case $n < p$, and under the further assumption of Φ being a proper immersion, we encounter a new phenomenon: the pushforward under Φ of a function is not necessarily a function anymore. This is related to $\Phi(X)$ not being an open subset of Y.

$n > p$ in case of a submersion. We first derive an intermediate result. Write $n = p + q$ with $q > 0$ and $x = (y, z)$ with $y \in \mathbf{R}^p$ and $z \in \mathbf{R}^q$. Let $\Pi : (y, z) \mapsto y : \mathbf{R}^n = \mathbf{R}^p \times \mathbf{R}^q \to \mathbf{R}^p$ be the projection onto the first p variables. This mapping is not proper, and therefore Π_* is defined only on $\mathcal{E}'(\mathbf{R}^n)$.

For a continuous function f on $\mathbf{R}^n$ with compact support, (10.3) now gives

$$(\Pi_* f)(\psi) = \int_{\mathbf{R}^p \times \mathbf{R}^q} f(y, z)\, \psi(y)\, d(y, z) = \int_{\mathbf{R}^p} \Big(\int_{\mathbf{R}^q} f(y, z)\, dz \Big) \psi(y)\, dy,$$

for all $\psi \in C^\infty(\mathbf{R}^p)$. This means that the distribution $\Pi_* f$ is given by the continuous function of y obtained from $f(x) = f(y, z)$ by integration over the *fiber* of the mapping Π over the value y, that is, by "integrating out" the z-variable:

$$(\Pi_* f)(y) = \int_{\mathbf{R}^q} f(y, z)\, dz. \tag{10.21}$$

On the strength of the Theorem about differentiation under the integral sign, we have $\Pi_* f \in C_0^\infty(\mathbf{R}^p)$ if $f \in C_0^\infty(\mathbf{R}^n)$. We conclude that the restriction of Π_* to $C_0^\infty(\mathbf{R}^n)$ is a continuous linear mapping A from $C_0^\infty(\mathbf{R}^n)$ to $C_0^\infty(\mathbf{R}^p)$.

Exactly as in Remark 10.10, we now obtain that $\Pi^* : C^\infty(\mathbf{R}^p) \to C^\infty(\mathbf{R}^n)$ has an extension to a continuous linear mapping $\Pi^* : \mathcal{D}'(\mathbf{R}^p) \to \mathcal{D}'(\mathbf{R}^n)$, defined as the transpose of the restriction to $C_0^\infty(\mathbf{R}^n)$ of Π_*.

Recall that $n > p$ and let Φ be a C^∞ mapping from an open $X \subset \mathbf{R}^n$ to an open $Y \subset \mathbf{R}^p$, with the property that its derivative $D\Phi(x) \in \mathrm{Lin}(\mathbf{R}^n, \mathbf{R}^p)$ at x is surjective, for every $x \in X$. One says that such a mapping is a *submersion*; see for example [7, Definition 4.2.6].

Theorem 10.18. *Let Φ be a C^∞ submersion from an open $X \subset \mathbf{R}^n$ to an open $Y \subset \mathbf{R}^p$. Then the restriction of Φ_* to $C_0^\infty(X)$ is a continuous linear mapping from $C_0^\infty(X)$ to $C_0^\infty(Y)$. Its transpose defines an extension of $\Phi^* : C^\infty(Y) \to C^\infty(X)$ to a continuous linear mapping from $\mathcal{D}'(Y)$ to $\mathcal{D}'(X)$; this is also denoted by Φ^* and is said to be the pullback of distributions under Φ.*

Proof. By means of the Submersion Theorem (see [7, Theorem 4.5.2], for example) one can find, for every $x \in X$, a neighborhood $U = U_x$ of x in X and a C^∞ diffeomorphism K from U onto an open subset V of $\mathbf{R}^n$ such that $\Phi = \Pi \circ K$ on

U. (Since K and Π map open sets to open sets, it now follows that this holds for Φ as well.) Applying this we obtain in $\mathcal{E}'(U)$,

$$\Phi_* = \Pi_* \circ K_* = \Pi_* \circ j_{\Lambda} \circ \Lambda^* \qquad \text{with} \qquad \Lambda := K^{-1}. \tag{10.22}$$

Here we have used (10.4) and Theorem 10.8. In particular, it now follows that the restriction of Φ_* to $C_0^\infty(U)$ is a continuous linear mapping from $C_0^\infty(U)$ to $C_0^\infty(Y)$. If for every compact subset K of X, we use a partition of unity over K subordinate to the cover by the U_x, for $x \in X$, we conclude that the restriction of Φ_* to $C_0^\infty(K)$ is a continuous linear mapping from $C_0^\infty(K)$ to $C_0^\infty(Y)$. □

Remark 10.19. As we have seen, $(\Pi_* g)(y)$ is obtained by integrating g over the linear manifold $\Pi^{-1}(\{y\})$, for a continuous function g with compact support. If f is a continuous function with compact support contained in U, one can see, by application of (10.22), that $(\Phi_* f)(y)$ is obtained by integrating $g := j_{\Lambda}\,(f \circ \Lambda)$ over $\Pi^{-1}(\{y\})$. But this amounts to an integration of f over the C^∞ submanifold

$$\Lambda(\Pi^{-1}(\{y\}) = K^{-1}(\Pi^{-1}(\{y\}) = \Phi^{-1}(\{y\}) \subset X,$$

the fiber of the mapping Φ over the value y.

If one chooses Euclidean $(n-p)$-dimensional integration over the inverse image $\Phi^{-1}(\{y\})$, then f must first be multiplied by a C^∞ factor that does not depend on f but only on the mapping Φ. We state here (see Problem 10.21 and its solution for a proof, or Problem 10.22.(iv) in the special case of $p = 1$) that this factor equals $\operatorname{gr}\Phi$ (with gr associated with gradient and Gramian), where

$$(\operatorname{gr}\Phi)(x) := \sqrt{\det(D\Phi(x) \circ {}^tD\Phi(x))} \qquad (x \in X),$$

which is the p-dimensional Euclidean volume of the parallelepiped in $\mathbf{R}^n$ spanned by the vectors $\operatorname{grad}\Phi_j(x)$, with $1 \le j \le p$. Here ${}^tD\Phi(x) \in \operatorname{Lin}(\mathbf{R}^p, \mathbf{R}^n)$ denotes the adjoint of $D\Phi(x)$, the matrix of which is given by the transpose of the matrix of $D\Phi(x)$. Observe that $\operatorname{gr}\Phi = \|\operatorname{grad}\Phi\|$ if $p = 1$. Summarizing, in the notation of Example 7.3 we have

$$(\Phi_* f)(y) = \int_{\Phi^{-1}(\{y\})} \frac{f(x)}{(\operatorname{gr}\Phi)(x)}\,dx = \frac{1}{\operatorname{gr}\Phi}\,\delta_{\Phi^{-1}(\{y\})}(f) \qquad (y \in Y). \tag{10.23}$$

⊘

Pushforward under a submersion Φ is not well-defined for all distributions, whereas pullback under Φ of distributions is. The latter will play an important role in Chap. 13.

Example 10.20. Define the *difference mapping*

$$d : \mathbf{R}^n \times \mathbf{R}^n \to \mathbf{R}^n \qquad \text{by} \qquad d(x, y) = x - y. \tag{10.24}$$

One directly verifies that d is a surjective linear mapping satisfying $(\operatorname{gr} d)(x,y) = 2^{\frac{n}{2}}$. (Note that $K : \mathbf{R}^n \times \mathbf{R}^n \to \mathbf{R}^n \times \mathbf{R}^n$ with $K(x,y) = (x-y,y)$ is the diffeomorphism such that $d = \Pi \circ K$ as in the proof of Theorem 10.18.) Now, in the notation of Example 10.7,

$$d^{-1}(\{0\}) = \{\,(x,z) \in \mathbf{R}^n \times \mathbf{R}^n \mid z = x\,\} = \Delta(\mathbf{R}^n),$$

the diagonal in $\mathbf{R}^n \times \mathbf{R}^n$. More generally, for $y \in \mathbf{R}^n$,

$$d^{-1}(\{y\}) = \{\,(x,z) \in \mathbf{R}^n \times \mathbf{R}^n \mid z = x - y\,\} = T_{(0,-y)}(\Delta(\mathbf{R}^n)).$$

Let $\delta_{\Delta(\mathbf{R}^n)}$ and $\Delta_* 1_{\mathbf{R}^n}$ be as in Example 10.7, while $\delta \in \mathcal{D}'(\mathbf{R}^n)$ denotes the Dirac measure. Then (10.23) implies, for $\chi \in C_0^\infty(\mathbf{R}^n \times \mathbf{R}^n)$,

$$\begin{aligned}(d_*\chi)(y) &= \frac{1}{2^{\frac{n}{2}}}\delta_{\Delta(\mathbf{R}^n)}(T_{(0,y)}\chi) = \Delta_* 1_{\mathbf{R}^n}(T_{(0,y)}\chi) = \int_{\mathbf{R}^n} \chi(x, x-y)\,dx,\\ &\text{in particular,} \qquad d^*\delta = \Delta_* 1_{\mathbf{R}^n} \in \mathcal{D}'(\mathbf{R}^n \times \mathbf{R}^n).\end{aligned} \tag{10.25}$$

See Problem 15.10 for another proof.

Using Dirac notation, one writes $d^*\delta(x,y) = \delta(x-y)$. Then the preceding identity leads to

$$\int_{\mathbf{R}^n \times \mathbf{R}^n} \delta(x-y)\chi(x,y)\,d(x,y) = \int_{\mathbf{R}^n} \chi(x,x)\,dx. \qquad \oslash$$

***$n < p$* in case of a proper embedding**. Suppose, for convenience, that Φ is proper. If, moreover, Φ is injective and $D\Phi(x) : \mathbf{R}^n \to \mathbf{R}^p$ is injective for every $x \in X$, then Φ is a proper C^∞ *embedding* of X onto the n-dimensional locally closed C^∞ submanifold $V = \Phi(X)$ of Y. The n-dimensional Euclidean integral of a continuous function f with compact support in V then equals the integral of $j_\Phi\, \Phi^* f$ over X, where

$$j_\Phi(x) = \sqrt{\det({}^t D\Phi(x) \circ D\Phi(x))} \qquad (x \in X)$$

now denotes the Euclidean n-dimensional volume in $\mathbf{R}^p$ of the parallelepiped spanned by the $D\Phi(x)(e(j)) \in \mathbf{R}^p$, $1 \le j \le n$. Here $e(j)$ denotes the jth standard basis vector in $\mathbf{R}^n$. See [7, Theorem 1.8.6, Exercise 4.20 and Sect. 7.3] among other references for more on these facts concerning proper embeddings and integration over submanifolds.

We further recall the distribution $g\,\delta_V$ of Euclidean integration over V with weight function g, as introduced in Example 7.3.

Proposition 10.21. *Let $n \le p$, suppose X is open in $\mathbf{R}^n$, and let Φ be a proper C^∞ embedding of X into the open subset Y of $\mathbf{R}^p$, having as its image the manifold $V = \Phi(X)$. Using the notation $\Psi = \Phi^{-1} : V \to X$ we obtain*

$$\Phi_* f = \Psi^*\Big(\frac{f}{j_\Phi}\Big)\,\delta_V \in \mathcal{D}'(Y), \tag{10.26}$$

for every continuous function f on X. Note that δ_V is a distribution of order 0, multiplied by the continuous function $\Psi^*(f/j_\Phi)$ on V.

Proof. This follows from a variant of (10.9). For $\psi \in C_0^\infty(Y)$ one has

$$\begin{aligned}(\Phi_* f)(\psi) &= \textstyle\int_X f(x)\,\psi(\Phi(x))\,dx = \int_X (f/j_\Phi)(x)\,\psi(\Phi(x))\,j_\Phi(x)\,dx\\ &= \int_V (f/j_\Phi)(\Psi(y))\,\psi(y)\,dy = \Psi^*(f/j_\Phi)\,\delta_V(\psi).\end{aligned}$$ □

Remark 10.22. Note the similarity in appearance between (10.26) and (10.8). Furthermore, observe that in this case of a proper embedding Φ, the pushforward Φ_* of distributions is well-defined on account of Φ being proper; however, Φ_* does not map functions to functions. Owing to the latter phenomenon, it is not possible to define the pullback Φ^* of distributions in a manner similar to the case of diffeomorphisms. ⊘

Problems

10.1.* *(Composition of pushforward and partial differentiation.)* Consider a C^∞ mapping Φ from an open subset X of $\mathbf{R}^n$ to an open subset Y of $\mathbf{R}^p$. Prove, for every $1 \le j \le n$, the following identity of continuous linear mappings:

$$\Phi_* \circ \partial_j = \sum_{k=1}^{p} \partial_k \circ \Phi_* \circ \partial_j \Phi_k \quad : \quad \mathcal{E}'(X) \to \mathcal{E}'(Y).$$

10.2. Let $X \subset \mathbf{R}^n$ be an open subset, μ a compactly supported Radon measure on X, and $f : X \to \mathbf{R}$ a C^∞ function. Prove that $f_*\mu$ is a Radon measure on $\mathbf{R}$ satisfying, in the notation of Theorem 20.34,

$$\int_{\mathbf{R}} \phi(t)\, f_*\mu(dt) = \int_X \phi \circ f(x)\,\mu(dx) \qquad (\phi \in C_0^\infty(\mathbf{R})).$$

In probability theory, $f_*\mu$ is said to be the *distribution of f under μ*. In that context, X, μ, f, and ϕ are usually more general than considered here.

10.3. Let u be a probability measure on X and $\Phi : X \to Y$ a continuous mapping. Discuss how to define $\Phi_* u$ so as to make it a probability measure on Y.

10.4. Let $\Phi : X \to Y$ be a C^∞ mapping that is not proper on the closed subset A of X. Demonstrate the existence of a sequence $(x(j))_{j\in\mathbf{N}}$ in A with the following properties:

(i) there exists $y \in Y$ such that $\lim_{j\to\infty} \Phi(x(j)) = y$;
(ii) for every compact subset K of X there exists a j_0 with $x(j) \notin K$, whenever $j \ge j_0$.

Prove that $\delta_{x(j)} \to 0$ in $\mathcal{D}'(X)_A$, while $\Phi_* \delta_{x(j)} \to \delta_y$ in $\mathcal{D}'(Y)$, as $j \to \infty$. Prove that the condition in Theorem 10.6, requiring Φ to be proper on A, is necessary for the conclusion in Theorem 10.6.

10.5. Let Φ be a C^∞ mapping from the open subset X of $\mathbf{R}^n$ to the open subset Y of $\mathbf{R}^p$. Prove supp $\Phi^*\phi \subset \Phi^{-1}(\text{supp}\ \phi)$. For a counterexample to the reverse inclusion, consider $X = Y = \mathbf{R}$, the function $\Phi(x) = x^4 - x^2$, and $\phi \in C_0^\infty(\mathbf{R})$ such that supp $\phi = [0, 2]$. Then supp $\Phi^*\phi = [-\sqrt{2}, -1] \cup [1, \sqrt{2}]$ and $\Phi^{-1}(\text{supp}\ \phi) = \{0\} \cup \text{supp}\ \Phi^*\phi$ (see Fig. 10.1).

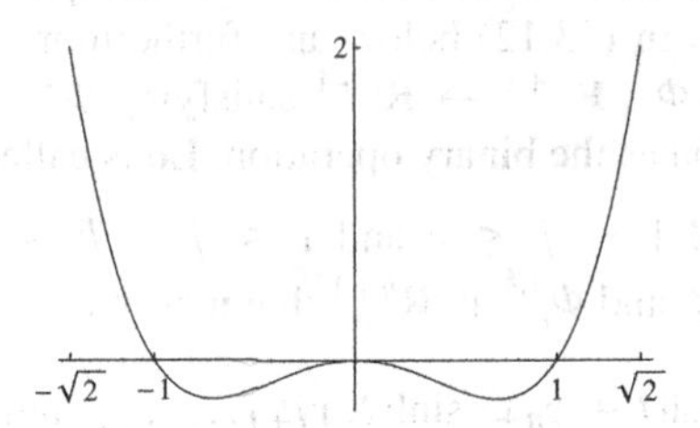

Fig. 10.1 Illustration for Problem 10.5. Graph of function Φ

10.6. *(Change of Variables Theorem.)* Let $\Psi : Y \to X$ be a C^∞ diffeomorphism of open subsets Y and X of $\mathbf{R}^n$. Show that $1_{\Psi(Y)} = \Psi_*(j_\Psi 1_Y)$ in $\mathcal{D}'(X)$.

10.7.* *(Composition of pullback and partial differentiation.)* Consider a C^∞ diffeomorphism $\Psi : Y \to X$ of open subsets Y and X of $\mathbf{R}^n$. Denote the inverse of the transpose of the Jacobi matrix of Ψ by $(\psi_{jk})_{1\le j,k\le n}$, that is, $({}^tD\Psi(y))^{-1} = (\psi_{jk}(y))_{1\le j,k\le n}$, for $y \in Y$. Derive the following identity of continuous linear mappings (compare with [7, Exercise 3.8]):

$$\Psi^* \circ \partial_j = \Big(\sum_{k=1}^{n} \psi_{jk} \circ \partial_k\Big) \circ \Psi^* \quad : \quad \mathcal{D}'(X) \to \mathcal{D}'(Y) \qquad (1 \le j \le n).$$

The change of variables $x = \Psi(y)$ relates the old variable x to the new variable y. The formula above expresses a partial derivative of a distribution with respect to the old variable in terms of partial derivatives of the transformed distribution with respect to the new variable.

10.8.* *(Interpretation of divergence.)* Suppose $\Phi_0 = I$. Verify, by combining Theorem 10.8 with (10.17) and (10.18), that

$$\frac{d}{dt} \det(D\Phi_t)\Big|_{t=0} = \sum_{j=1}^{n} \partial_j v_j = \operatorname{div} v.$$

10.9.* Consider a C^∞ vector field v on X with globally defined flow $(\Phi_t)_{t\in\mathbf{R}}$ and fix $x \in X$. Then prove the following: δ_x is invariant under Φ_t, for all $t \in \mathbf{R}$, as a distributional density $\iff \Phi_t(x) = x$ for all $t \in \mathbf{R} \iff v(x) = 0$. And also: δ_x is invariant under Φ_t, for all $t \in \mathbf{R}$, as a generalized function if and only if $v(x) = 0$ **and** $\operatorname{div} v(x) = 0$.

10.10.* The rotation in the plane $\mathbf{R}^2$ about the origin by the angle t is the linear mapping $(x_1, x_2) \mapsto (x_1 \cos t - x_2 \sin t,\ x_1 \sin t + x_2 \cos t)$. Demonstrate that $u \in \mathcal{D}'(\mathbf{R}^2)$ is invariant under all rotations if and only if $x_1\, \partial_2 u = x_2\, \partial_1 u$.

10.11. *(Lorentz-invariant distributions.)* Consider the quadratic form $q : \mathbf{R}^{n+1} \to \mathbf{R}$, of Lorentz type, as in (13.12) below and furthermore the collection $\mathbf{Lo}$ of all linear transformations $\Phi : \mathbf{R}^{n+1} \to \mathbf{R}^{n+1}$ satisfying $\Phi^* q = q$. Prove that $\mathbf{Lo}$ is a group with composition as the binary operation. $\mathbf{Lo}$ is called the *Lorentz group.*

(i) For all $t \in \mathbf{R}$ and $1 \le j \le n$ and $1 \le j < k \le n$, show that the linear transformations Φ_t^j and $\Phi_t^{j,k}$ in $\mathbf{R}^{n+1}$ that map y to

$$(y_1, \dots, y_{j-1}, y_j \cosh t + y_{n+1} \sinh t, y_{j+1}, \dots, y_j \sinh t + y_{n+1} \cosh t),$$
$$(y_1, \dots, y_{j-1}, y_j \cos t - y_k \sin t, y_{j+1}, \dots, y_{k-1}, y_j \sin t + y_k \cos t, \dots, y_{n+1}),$$

respectively, belong to $\mathbf{Lo}$. (In the case of $n = 3$, the Φ_t^j are *boosts* of rapidity t, which occur in *special relativity*.) Verify that the $(\Phi_t^j)_{t\in\mathbf{R}}$ and $(\Phi_t^{j,k})_{t\in\mathbf{R}}$ actually are one-parameter subgroups in $\mathbf{Lo}$ consisting of C^∞ diffeomorphisms of $\mathbf{R}^{n+1}$. Prove that the velocity vector fields corresponding to these one-parameter subgroups are

$$y \mapsto y_{n+1}\, e(j) + y_j\, e(n+1) \qquad \text{and} \qquad y \mapsto -y_k\, e(j) + y_j\, e(k),$$

respectively, where $(e(1), \dots, e(n+1))$ denotes the standard basis in $\mathbf{R}^{n+1}$. Furthermore, for any $u \in \mathcal{D}'(\mathbf{R}^{n+1})$, deduce

$$\frac{d}{dt}(\Phi_t^j)^* u\Big|_{t=0} = y_{n+1}\, \partial_j u + y_j\, \partial_{n+1} u = -\frac{d}{dt}(\Phi_t^j)_* u\Big|_{t=0},$$
$$\frac{d}{dt}(\Phi_t^{j,k})^* u\Big|_{t=0} = y_j\, \partial_k u - y_k\, \partial_j u = -\frac{d}{dt}(\Phi_t^{j,k})_* u\Big|_{t=0}.$$

(ii) Prove that $|\det \Phi| = 1$, for $\Phi \in \mathbf{Lo}$, and deduce that invariance of u under $\mathbf{Lo}$ as generalized function and as distributional density amount to the same thing. In particular, if u is invariant under $\mathbf{Lo}$, derive

$$(y_{n+1}\, \partial_j + y_j\, \partial_{n+1})u = 0 \qquad (1 \le j \le n),$$
$$(y_j\, \partial_k - y_k\, \partial_j)u = 0 \qquad (1 \le j < k \le n).$$

10.12. A distribution u in $\mathbf{R}$ is said to be *periodic with period* $a > 0$ if u is invariant under translation by a. Prove that every distributional Fourier series from Problem 5.10 is periodic with period $a = \frac{2\pi}{\omega}$.

10.13.* *(Poisson summation formula.)* For any $\lambda \in \mathbf{C}$, introduce $e_\lambda \in C^\infty(\mathbf{R})$ by $e_\lambda(x) = e^{\lambda x}$ and verify that $e_\lambda = \frac{1}{\lambda}\,\partial e_\lambda$. Fix $\omega > 0$. Now prove that the distributional Fourier series (see Definition 16.5 for more about this notion) given by

$$u = \sum_{n \in \mathbf{Z}} e_{in\omega}$$

defines a distribution on $\mathbf{R}$ having the following property. For every compact subset K of $\mathbf{R}$, there exists a constant $c > 0$ such that $|u(\phi)| \le c\,\|\phi\|_{C^2}$, for all $\phi \in C_0^\infty(K)$. Next show:

(i) $e_{i\omega} u = u$.
(ii) u is periodic with period $2\pi/\omega$.
(iii) There exists a constant $c \in \mathbf{C}$ such that

$$\sum_{n \in \mathbf{Z}} e_{in\omega} = c \sum_{k \in \mathbf{Z}} \delta_{k\,\frac{2\pi}{\omega}}.$$

(iv) The order of u is 0.

For the determination of $c = 2\pi/\omega$, see Example 16.8. This example and Problem 16.17 contain a different proof by means of Fourier transform. Furthermore, see Figs. 16.1 and 16.2.

10.14.* Let $0 < s < 1$. Prove the existence of $c = c(n, s) > 0$ such that

$$\int_{\mathbf{R}^n} \frac{|e^{i\langle x, \xi\rangle} - 1|^2}{\|x\|^{n+2s}}\,dx = \int_{\mathbf{R}^n} \frac{4\sin^2 \frac{\langle x, \xi\rangle}{2}}{\|x\|^{n+2s}}\,dx = c\,\|\xi\|^{2s} \qquad (\xi \in \mathbf{R}^n).$$

See Problem 14.40 for a computation of $c(1, s)$.

10.15.* *(Distributions homogeneous of fixed degree.)* Let $\mathcal{H}_a$ be the set of homogeneous distributions of degree $a \in \mathbf{C}$ in $\mathbf{R}^n$. Verify the following assertions.

(i) $\mathcal{H}_a$ is a linear subspace of $\mathcal{D}'(\mathbf{R}^n)$. Furthermore, $\mathcal{H}_a$ is a closed subspace of $\mathcal{D}'(\mathbf{R}^n)$, that is, if $u \in \mathcal{D}'(\mathbf{R}^n)$, $u_j \in \mathcal{H}_a$, and $u_j \to u$ in $\mathcal{D}'(\mathbf{R}^n)$ as $j \to \infty$, then $u \in \mathcal{H}_a$.
(ii) If $u \in \mathcal{H}_a$, one has $\partial_j u \in \mathcal{H}_{a-1}$ and $x_k\, u \in \mathcal{H}_{a+1}$.
(iii) If $\psi \in \mathcal{H}_a \cap C^\infty(\mathbf{R}^n)$, with $\psi \ne 0$, then $a \in \mathbf{Z}_{\ge 0}$ and ψ is a homogeneous polynomial function of degree a. Prove $\dim(\mathcal{H}_a \cap C^\infty(\mathbf{R}^n)) = \binom{a+n-1}{a}$. Hint: first prove that if the function $t \mapsto t^a : \mathbf{R}_{>0} \to \mathbf{C}$ possesses a C^∞ extension to $\mathbf{R}$, then $a \in \mathbf{Z}_{\ge 0}$, for example, by studying the behavior of the derivatives of this function as $t \downarrow 0$. Alternatively, a sufficiently high derivative would be homogeneous of negative degree and therefore not continuous at 0.
(iv) Every continuous function f on the sphere $\|x\| = 1$ possesses a uniquely determined extension to a homogeneous function g of degree a on $\mathbf{R}^n \setminus \{0\}$. This g is continuous. If $\operatorname{Re} a > -n$, then g is locally integrable on $\mathbf{R}^n$ and $g \in \mathcal{H}_a$.
(v) $\mathcal{H}_a$ is infinite-dimensional if $n > 1$ and $a \in \mathbf{C}$. Hint: use (iv) when $\operatorname{Re} a > -n$, and combine this with (ii) when $\operatorname{Re} a \le -n$.

10.16.* For all $\alpha \in (\mathbf{Z}_{\geq 0})^n$, prove that $\partial^\alpha \delta \in \mathcal{D}'(\mathbf{R}^n)$ is homogeneous of degree $-n - |\alpha|$.

10.17. Let $\Phi : X \to Y$ be a C^∞ submersion and let $K : X \to X$ be a C^∞ diffeomorphism such that $\Phi \circ K = \Phi$. Prove that for every $v \in \mathcal{D}'(Y)$ the distribution $u = \Phi^* v \in \mathcal{D}'(X)$ is invariant under K (as a generalized function).

10.18.* *(Composition of partial differentiation and pullback under submersion.)* Let X be open in $\mathbf{R}^n$, Y open in $\mathbf{R}^p$ and $\Phi : X \to Y$ a C^∞ submersion. Prove that (10.1) can be extended to an identity of continuous linear mappings: $\mathcal{D}'(Y) \to \mathcal{D}'(X)$.

10.19. Let $X = \mathbf{R}^2 \setminus (\mathbf{R}_{\leq 0} \times \{0\})$ and $Y = \mathbf{R}_{>0} \times \,]-\pi, \pi[$ and consider the C^∞ diffeomorphism $\Phi : X \to Y$ that assigns to $x = r(\cos\alpha, \sin\alpha) \in X$ its polar coordinates (r, α) in Y (see [7, Example 3.1.1]). For $u \in \mathcal{D}'(X)$, verify

$$\partial_r(\Phi_* u) = \frac{1}{r}\Phi_* u + \cos\alpha\, \Phi_*(\partial_1 u) + \sin\alpha\, \Phi_*(\partial_2 u) = \frac{1}{r}\Phi_* u + \sum_{k=1}^{2} \frac{x_k}{r}\, \Phi_*(\partial_k u).$$

10.20.* *(Composition of second-order partial differential operator with pullback under associated quadratic form.)* Consider the second-order linear partial differential operator with constant coefficients

$$P = \sum_{i,j=1}^{n} B_{ij}\, \partial_i \partial_j$$

in $\mathbf{R}^n$, where (B_{jk}) is an invertible symmetric real $n \times n$ matrix. Let (A_{kl}) be the inverse matrix and

$$\Phi : x \mapsto \sum_{k,l=1}^{n} A_{kl}\, x_k x_l \in \mathbf{R}$$

the corresponding quadratic form on $\mathbf{R}^n$. Verify the following assertions.

(i) The restriction of Φ to $\mathbf{R}^n \setminus \{0\}$ is a submersion from $\mathbf{R}^n \setminus \{0\}$ to $\mathbf{R}$.
(ii) One has the following identity of distributions in $\mathcal{D}'(\mathbf{R}^n \setminus \{0\})$:

$$(P \circ \Phi^*)v = \Phi^*(4y\, \partial_y^2 + 2n\, \partial_y)v \qquad (v \in \mathcal{D}'(\mathbf{R})).$$

Here $y = \Phi(x)$ denotes the variable in $\mathbf{R}$.
(iii) If $v \in \mathcal{D}'(\mathbf{R})$ is homogeneous of degree a, then $\Phi^* v$ is homogeneous of degree $2a$ on $\mathbf{R}^n \setminus \{0\}$.
(iv) If $\partial_y v$ is homogeneous of degree $-\frac{n}{2}$, then $u = \Phi^* v$ defines a solution on $\mathbf{R}^n \setminus \{0\}$ of the partial differential equation $Pu = 0$. When is u homogeneous? Of what degree?
(v) From part (ii) deduce the following two identities of continuous linear mappings:

$$\begin{aligned} P \circ \Phi^* &= \Phi^* \circ (4y\, \partial_y^2 + 2n\, \partial_y) &:&\quad C^\infty(\mathbf{R}) \to C^\infty(\mathbf{R}^n),\\ \Phi_* \circ P &= (4\partial_y^2 \circ y - 2n\, \partial_y) \circ \Phi_* &:&\quad \mathcal{E}'(\mathbf{R}^n) \to \mathcal{E}'(\mathbf{R}). \end{aligned}$$

(vi) Apply the latter equality in part (v) to derive the result in Example 10.5.

10.21.* Verify formula (10.23) using linear algebra to rewrite j_Λ, where j_Λ is as in (10.22). To this end, parametrize $U \cap \Phi^{-1}(\{y\})$ using Λ and show that the tangent space T of $\Phi^{-1}(\{y\})$ at $x = \Lambda(x')$ equals $\ker \phi$, where $\phi = D\Phi(x)$. Apply the chain rule to $\Phi \circ \Lambda = \Pi$ and deduce that T is spanned by certain column vectors λ_k that occur in the matrix of $\lambda := D\Lambda(x')$. Next, project the remaining column vectors in λ along T onto $T^\perp = \operatorname{im} {}^t\phi$ and express $(\det \lambda)^2 = \det({}^t\lambda\, \lambda)$ in terms of these projections and the λ_k spanning T. Finally, write the contribution coming from the projections in terms of the row vectors of ϕ.

10.22.* *(Pushforward of function under submersion to* **R** *as consequence of integration of a derivative.)* Let X be an open subset of $\mathbf{R}^n$ and $\Phi : X \to \mathbf{R}$ a C^∞ submersion. Denote by H the Heaviside function on $\mathbf{R}$.

(i) Show, for all $\phi \in C_0^\infty(X)$ and $y \in \mathbf{R}$, that

$$(\Phi_*\phi)(y) = (\Phi^*\delta_y)(\phi) = \Phi^*(\partial\, T_y H)(\phi)$$

and apply Problem 10.17 to derive

$$\partial_j\Phi\ \Phi^*\delta_y = \partial_j\Phi \circ \Phi^* \circ \partial(T_y H) = \partial_j(\Phi^*(T_y H)).$$

(ii) By approximating H by C^∞ functions, verify

$$(\Phi^*(T_y H))(\phi) = \int_{\Phi^{-1}(]\,y,\infty[)} \phi(x)\, dx.$$

(iii) Apply the Theorem on Integration of a Total Derivative (see [7, Theorem 7.6.1]) to deduce

$$\partial_j(\Phi^*(T_y H))(\phi) = \int_{\Phi^{-1}(\{y\})} \phi(x) \frac{\partial_j\Phi(x)}{\|\operatorname{grad}\Phi(x)\|}\, dx.$$

Here $\partial_j\Phi / \|\operatorname{grad}\Phi\|$ denotes the jth component of the normalized gradient vector field of Φ; the value of the vector field at x is the inner normal to $\Phi^{-1}(\{y\})$ at x. Furthermore, dx denotes Euclidean $(n-1)$-dimensional integration over $\Phi^{-1}(\{y\})$.

(iv) Combine the results from parts (i) and (iii), replace ϕ by $\phi\, \partial_j\Phi / \|\operatorname{grad}\Phi\|$, and sum over $1 \le j \le n$ to obtain formula (10.23):

$$(\Phi_*\phi)(y) = \int_{\Phi^{-1}(\{y\})} \frac{\phi(x)}{\|\operatorname{grad}\Phi(x)\|}\, dx \qquad (y \in Y).$$

10.23.* *(Pullback of Dirac measure under submersion.)* Let X be an open subset of $\mathbf{R}^n$ and consider a C^∞ submersion $\Phi : X \to \mathbf{R}$.

(i) Let $\phi \in C_0^\infty(X)$. Prove

$$(\Phi_*\phi)(y) = \frac{1}{\|\operatorname{grad} \Phi\|}\,\delta_{\Phi^{-1}(\{y\})}(\phi) \qquad (y \in \mathbf{R}).$$

Verify the following identity in $\mathcal{D}'(X)$ (compare with Example 10.12):

$$\Phi^*\delta_y = \frac{1}{\|\operatorname{grad} \Phi\|}\,\delta_{\Phi^{-1}(\{y\})} \qquad (y \in \mathbf{R}).$$

(ii) On the basis of part (i) conclude (compare with [7, Exercise 7.36]) that

$$1_{]-\infty,\,y[}(\Phi_*\phi) = 1_{\Phi^{-1}(]-\infty,\,y[)}(\phi) \qquad (y \in \mathbf{R});$$

and deduce from this, for all $y \in \mathbf{R}$,

$$(\Phi_*\phi)(y) = \partial_y \int_{\Phi^{-1}(]-\infty,\,y[)} \phi(x)\,dx, \qquad \Phi^*\delta_y = \partial_y \circ (1_{\Phi^{-1}(]-\infty,\,y[)}).$$

(iii) Give an independent proof of the second identity in part (ii) by means of successive application of the Fundamental Theorem of Integral Calculus on $\mathbf{R}$ (see [7, Theorem 2.10.1]), interchange of the order of integration, and integration by parts to $\psi(\Phi_*\phi)$, where $\psi \in C_0^\infty(\mathbf{R})$.

10.24.* *(Integration of total derivative.)* Let X be an open subset of $\mathbf{R}^n$ and $\Phi : X \to \mathbf{R}$ a C^∞ submersion. Define the open set $\Omega = \Phi^{-1}(\mathbf{R}_{>0})$, and suppose that Ω is nonempty and bounded. Note that the boundary $\partial\Omega$ of Ω equals $\Phi^{-1}(\{0\})$ and that this set is a C^∞ submanifold in $\mathbf{R}^n$ of dimension $n-1$. Derive from Problem 10.23.(i)

$$\Phi^*\delta = \frac{1}{\|\operatorname{grad} \Phi\|}\,\delta_{\partial\Omega} \quad \text{in} \quad \mathcal{D}'(X).$$

With H the Heaviside function on $\mathbf{R}$, prove on the basis of Problem 10.18 that

$$\partial_j(\Phi^* H) = \frac{\partial_j \Phi}{\|\operatorname{grad} \Phi\|}\,\delta_{\partial\Omega}.$$

Deduce that for all $\phi \in C_0^\infty(X)$,

$$\int_\Omega \partial_j\phi(x)\,dx = \int_{\partial\Omega} \phi(y)\,\nu_j(y)\,dy,$$

where $\nu(y)$ denotes the outer normal to $\partial\Omega$ at y and dy denotes Euclidean $(n-1)$-dimensional integration over $\partial\Omega$. Conclude that by this method we have proved the following *Theorem on Integration of a Total Derivative*; see [7, Theorem 7.6.1] and Problem 7.4:

$$\int_\Omega D\phi(x)\,dx = \int_{\partial\Omega} \phi(y)\,{}^t\nu(y)\,dy.$$

10.25. *(Wave equation.)* Denote the points in $\mathbf{R}^2$ by (x, t) and consider the mappings

$$\text{plus} : (x,t) \mapsto x + t \qquad \text{and} \qquad \text{minus} : (x,t) \mapsto x - t$$

from $\mathbf{R}^2$ to $\mathbf{R}$. Verify that these are submersions. Prove, for every pair of distributions $a \in \mathcal{D}'(\mathbf{R})$ and $b \in \mathcal{D}'(\mathbf{R})$, that the distribution $u := (\text{plus})^* a + (\text{minus})^* b$ on $\mathbf{R}^2$ satisfies the wave equation

$$\partial_t^2 u = \partial_x^2 u.$$

This gives a distributional answer to the classical question whether

$$(x,t) \mapsto a(x+t) + b(x-t)$$

is acceptable as a solution of the wave equation also when a and b are not C^2 functions, although continuous, for example. Describe $(\text{plus})^*\delta$ and $(\text{minus})^*\delta$.

10.26. Let Φ be a C^∞ mapping from an open subset X of $\mathbf{R}^n$ to an open subset Y of $\mathbf{R}^p$. Let C be the closure of $\Phi(X)$ in Y. Prove

$$\operatorname{supp} \Phi_* u \subset C \qquad (u \in \mathcal{E}'(X)).$$

Now assume that the p-dimensional measure of C equals 0, so that $\Phi_* u = 0$ if $\Phi_* u$ is locally integrable. Prove the existence of $\phi \in C_0^\infty(X)$ such that $\Phi_* \phi$ is not locally integrable. Hint: approximate δ_x by $\phi_j \in C_0^\infty(X)$.

10.27. Let Φ be a C^∞ mapping from an open subset X of $\mathbf{R}^n$ to an open subset Y of $\mathbf{R}^p$. Suppose that $\Psi : Y \to X$ is a C^∞ mapping such that $\Psi \circ \Phi(x) = x$, for all $x \in X$. Prove that Φ is a proper embedding.

Write $V := \Phi(X)$ and define

$$\mathcal{E}'(V) := \{\, v \in \mathcal{E}'(Y) \mid \psi\, v = 0 \text{ if } \psi \in C^\infty(Y) \text{ and } \psi = 0 \text{ on } V \,\}.$$

Prove that Φ_* is a bijective linear mapping from $\mathcal{E}'(X)$ to $\mathcal{E}'(V)$, with the restriction of Ψ_* to $\mathcal{E}'(V)$ as its inverse. Hint: use the fact that for every $\phi \in C^\infty(Y)$, the function $\psi = \phi - \Psi^* \circ \Phi^*(\phi)$ equals 0 on V.

Finally, let f be a real-valued C^∞ function on the open subset X of $\mathbf{R}^n$. Denote the points of $\mathbf{R}^{n+1}$ by (x, y), where $x \in \mathbf{R}^n$, $y \in \mathbf{R}$. Define $\Phi : X \to X \times \mathbf{R}$ by $\Phi(x) = (x,\ f(x))$, for $x \in X$. Let $v \in \mathcal{E}'(X \times \mathbf{R})$. Prove that $(y - f(x))\, v = 0$ if and only if there exists a $u \in \mathcal{E}'(X)$ with $v = \Phi_* u$.

Chapter 11
Convolution of Distributions

Convolution involves translation; that makes it difficult to define the former operation for functions or distributions supported by arbitrary open subsets in $\mathbf{R}^n$. Therefore we initially consider objects defined on all of $\mathbf{R}^n$.

In (3.2) we described the convolution $(f * \phi)(x)$ of a continuous function f on $\mathbf{R}^n$ and $\phi \in C_0^\infty(\mathbf{R}^n)$ as the testing of f with the function $T_x \circ S\phi : y \mapsto \phi(x-y)$. Here S and T_x are respectively the reflection, and the translation by the vector x, of functions, as described in Example 10.14 and Example 10.15. This suggests defining the *convolution product* $u * \phi$, for $u \in \mathcal{D}'(\mathbf{R}^n)$ and $\phi \in C_0^\infty(\mathbf{R}^n)$, as the function given by

$$(u * \phi)(x) = u(T_x \circ S\phi) \qquad (x \in \mathbf{R}^n). \tag{11.1}$$

Note that the definition also makes sense if $u \in \mathcal{E}'(\mathbf{R}^n)$ and $\phi \in C^\infty(\mathbf{R}^n)$. In Theorem 11.2 we will show that in both cases $u * \phi$ belongs to $C^\infty(\mathbf{R}^n)$. Furthermore, with a suitable choice of $\phi = \phi_\epsilon$ we can ensure that $u * \phi_\epsilon$ converges in $\mathcal{D}'(\mathbf{R}^n)$ to $u \in \mathcal{D}'(\mathbf{R}^n)$ for $\epsilon \downarrow 0$; see Lemma 11.6.

Example 11.1. For $a \in \mathbf{R}^n$ we have

$$(\delta_a * \phi)(x) = (T_x \circ S\phi)(a) = (S\phi)(a-x) = \phi(x-a).$$

Consequently, $\delta_a * \phi = T_a\phi$ for every $\phi \in C^\infty(\mathbf{R}^n)$, and in particular

$$\delta * \phi = \phi \qquad (\phi \in C^\infty(\mathbf{R}^n)). \tag{11.2}$$

In other words, δ acts as a unit element for the operation of convolution. ⊘

Theorem 11.2. *If $u \in \mathcal{D}'(\mathbf{R}^n)$ and $\phi \in C^\infty(\mathbf{R}^n)$ and at least one of the two sets* $\operatorname{supp} u$ *and* $\operatorname{supp} \phi$ *is compact, then $u * \phi \in C^\infty(\mathbf{R}^n)$. One has*

$$T_a(u * \phi) = (T_a u) * \phi = u * (T_a\phi), \tag{11.3}$$

for every $a \in \mathbf{R}^n$, and

$$\partial^\alpha(u * \phi) = (\partial^\alpha u) * \phi = u * (\partial^\alpha \phi), \tag{11.4}$$

J.J. Duistermaat and J.A.C. Kolk, *Distributions: Theory and Applications*,
Cornerstones, DOI 10.1007/978-0-8176-4675-2_11,

for every multi-index α. Furthermore,

$$\operatorname{supp}(u * \phi) \subset \operatorname{supp} u + \operatorname{supp} \phi. \tag{11.5}$$

Proof. We suppose that supp ϕ is compact; the other case is similar. Because $x \mapsto T_x \circ S\phi$ is continuous from $\mathbf{R}^n$ to $C_0^\infty(\mathbf{R}^n)$, the function $u * \phi$ is continuous on $\mathbf{R}^n$. Formula (11.3) follows when we compare

$$\begin{aligned}
(T_a(u * \phi))(x) &= (u * \phi)(x - a) &&= u(T_{x-a} \circ S\phi),\\
((T_a u) * \phi)(x) &= T_a u(T_x \circ S\phi) &&= u(T_{-a} \circ T_x \circ S\phi),\\
(u * (T_a\phi))(x) &= u(T_x \circ S \circ T_a\phi) &&= u(T_x \circ T_{-a} \circ S\phi).
\end{aligned}$$

Substituting $a = -t\, e(j)$ in (11.3), we obtain

$$\begin{aligned}
\frac{1}{t}\big((u * \phi)(x + t\, e(j)) - (u * \phi)(x)\big) &= \frac{1}{t}\big(T_{t\, e(j)}{}^* u - u\big)(T_x \circ S\phi)\\
&= u\big(T_x \circ S \circ \frac{1}{t}\big(T_{t\, e(j)}{}^*\phi - \phi\big)\big).
\end{aligned}$$

In view of (10.13) and (10.12) this converges as $t \to 0$. We see that $u * \phi$ is partially differentiable with respect to the jth variable, with derivative

$$\partial_j(u * \phi) = (\partial_j u) * \phi = u * (\partial_j \phi).$$

Since the partial derivatives are continuous, the conclusion is that $u * \phi$ is of class C^1. By mathematical induction on k one can then show that $u * \phi$ is of class C^k and that (11.4) holds for all multi-indices α with $|\alpha| = k$.

Here, $x \notin \operatorname{supp} u + \operatorname{supp} \phi$ means that supp u has an empty intersection with

$$x + (-\operatorname{supp} \phi) = x + \operatorname{supp} S\phi = \operatorname{supp}(T_x \circ S\phi),$$

which implies $(u * \phi)(x) = u(T_x \circ S\phi) = 0$. Because supp u is closed and supp ϕ compact, $C = \operatorname{supp} u + \operatorname{supp} \phi$ is a closed subset of $\mathbf{R}^n$; see the text following (2.2). Because $u * \phi = 0$ on the open subset $\mathbf{R}^n \setminus C$ of $\mathbf{R}^n$, the conclusion is that $\operatorname{supp}(u * \phi) \subset C$. □

A linear mapping $\mathcal{U}$ from functions on $\mathbf{R}^n$ to functions on $\mathbf{R}^n$ *commutes with all translations* if $\mathcal{U} \circ T_a = T_a \circ \mathcal{U}$, for all $a \in \mathbf{R}^n$. Convolution with u can be characterized in terms of such linear mappings:

Theorem 11.3. *Let $u \in \mathcal{D}'(\mathbf{R}^n)$. Then $u * : \phi \mapsto u * \phi$ is a continuous linear mapping from $C_0^\infty(\mathbf{R}^n)$ to $C^\infty(\mathbf{R}^n)$ that commutes with all translations. Conversely, if $\mathcal{U}$ is a continuous linear mapping from $C_0^\infty(\mathbf{R}^n)$ to $C^\infty(\mathbf{R}^n)$ that commutes with all translations, then there exists a uniquely determined $u \in \mathcal{D}'(\mathbf{R}^n)$ with $\mathcal{U}(\phi) =$*

$u * \phi$, for all $\phi \in C_0^\infty(\mathbf{R}^n)$. More specifically, $u = \delta \circ \mathcal{U} \circ S$, with δ the Dirac measure on $\mathbf{R}^n$.

Proof. It is evident that $u *$ is linear, and as (11.3) shows, $u *$ commutes with all translations. Since we have to verify that the conditions in Example 8.6 are satisfied, the proof of the continuity of $u * : C_0^\infty(\mathbf{R}^n) \to C^\infty(\mathbf{R}^n)$ is more involved.

Let A and B be compact subsets of $\mathbf{R}^n$. Then $K := A + (-B)$ is compact and according to Theorem 3.8 there exist a constant $c > 0$ and $k \in \mathbf{Z}_{\geq 0}$ such that

$$|u(\psi)| \leq c \, \|\psi\|_{C^k} \qquad (\psi \in C_0^\infty(K)).$$

Applying this to $\psi = T_x \circ S\phi$, with $x \in A$ and $\phi \in C_0^\infty(B)$, we get

$$\|u * \phi\|_{C^0, A} \leq c \, \|\phi\|_{C^k} \qquad (\phi \in C_0^\infty(B)).$$

Because $\partial^\alpha(u * \phi) = u * (\partial^\alpha \phi)$, it now follows that for every $m \in \mathbf{Z}_{\geq 0}$ there exists a constant $c' > 0$ with

$$\|u * \phi\|_{C^m, A} \leq c' \, \|\phi\|_{C^{k+m}} \qquad (\phi \in C_0^\infty(B)).$$

Because this holds for all A, B, and m, the desired conclusion is obtained.

Now consider $\mathcal{U}$ as described above. If $\mathcal{U}$ is of the form $u *$, then $(\mathcal{U}\psi)(0) = u(S\psi)$; thus, u is determined by $u(\psi) = (\mathcal{U} \circ S\psi)(0)$, for $\psi \in C_0^\infty(\mathbf{R}^n)$. If we now interpret this as the definition of u, we see that $u \in \mathcal{D}'(\mathbf{R}^n)$ and

$$\begin{aligned}(u * \phi)(x) &= u(T_x \circ S\phi) = (\mathcal{U} \circ S \circ T_x \circ S\phi)(0) = (\mathcal{U} \circ T_{-x}\phi)(0) \\ &= (T_{-x} \circ \mathcal{U}\phi)(0) = \mathcal{U}(\phi)(x),\end{aligned}$$

for every $x \in \mathbf{R}^n$; that is, $u * \phi = \mathcal{U}(\phi)$. □

The preceding theorem implies that any translation acting in $C_0^\infty(\mathbf{R}^n)$ has to be a convolution operator, as has been explicitly demonstrated in Example 11.1. Furthermore, see Problem 15.11 for another proof of this theorem.

In proving Theorem 11.5 we will use the following *principle of integration under the distribution sign.*

Lemma 11.4. *Let X be an open subset of $\mathbf{R}^n$ and suppose that the mapping $A : \mathbf{R}^m \to C_0^\infty(X)$ has the following properties:*

(a) There exists a compact subset T of $\mathbf{R}^m$ such that $A(t) = 0$ whenever $t \notin T$.
(b) There is a compact subset K of X such that supp $A(t) \subset K$ *for all $t \in \mathbf{R}^m$.*
(c) For every $k \in \mathbf{Z}_{\geq 0}$ and every $\epsilon > 0$ there exists $\delta > 0$ such that $\|s - t\| < \delta$ implies $\|A(s) - A(t)\|_{C^k} < \epsilon$.

Then

$$\Big(\int_{\mathbf{R}^m} A(t)\, dt\Big)(x) = \int_{\mathbf{R}^m} A(t)(x)\, dt$$

defines a function $\int A(t)\,dt$ *belonging to* $C_0^\infty(X)$. *For every* $u \in \mathcal{D}'(X)$, *one has that* $u \circ A : \mathbf{R}^m \to \mathbf{C}$ *is a continuous function with compact support, while*

$$u\Big(\int_{\mathbf{R}^m} A(t)\,dt\Big) = \int_{\mathbf{R}^m} (u \circ A)(t)\,dt.$$

Proof. It follows from the assumptions made that the (finite) Riemann sums

$$S_h := h^m \sum_{t \in \mathbf{Z}^m} A(h\,t) \qquad (h \in \mathbf{R}_{>0})$$

converge in $C_0^\infty(X)$ to $\int A(t)\,dt$ as $h \downarrow 0$. One begins by proving this in $C_0^0(X)$ and then uses the Theorem about differentiation under the integral sign to obtain the result in $C_0^\infty(X)$. From the continuity of A and of u it follows that $u \circ A$ is a continuous function. The complex number

$$u(S_h) = h^m \sum_{t \in \mathbf{Z}^m} u(A(h\,t)) = h^m \sum_{t \in \mathbf{Z}^m} (u \circ A)(h\,t)$$

is an approximating Riemann sum for the integral of $u \circ A$. Using the continuity of u once again, we now obtain

$$u\Big(\int_{\mathbf{R}^m} A(t)\,dt\Big) = u\big(\lim_{h\downarrow 0} S_h\big) = \lim_{h\downarrow 0} u(S_h) = \int_{\mathbf{R}^m} (u \circ A)(t)\,dt. \qquad \square$$

Theorem 11.5. *For* $u \in \mathcal{D}'(\mathbf{R}^n)$ *and* ϕ *and* ψ *in* $C_0^\infty(\mathbf{R}^n)$ *one has*

$$(u * \phi)(\psi) = u(S\phi * \psi). \tag{11.6}$$

Proof. The mapping

$$A : x \mapsto \psi(x)\,(T_x \circ S\phi) \quad : \quad \mathbf{R}^n \to C_0^\infty(\mathbf{R}^n)$$

satisfies the conditions of Lemma 11.4. Applying (1.14), we obtain

$$\begin{aligned}(u * \phi)(\psi) &= \int (u * \phi)(x)\,\psi(x)\,dx = \int u(T_x \circ S\phi)\,\psi(x)\,dx \\ &= \int u(\psi(x)\,(T_x \circ S\phi))\,dx = u\Big(\int \psi(x)\,(T_x \circ S\phi)\,dx\Big) \\ &= u(S\phi * \psi).\end{aligned} \qquad \square$$

Theorem 11.5 is useful in proving the following two results on approximation of distributions by functions.

Lemma 11.6. *Let* $\phi \in C_0^\infty(\mathbf{R}^n)$ *and* $1(\phi) = 1$. *Write* $\phi_\epsilon(x) = \epsilon^{-n}\,\phi(\frac{1}{\epsilon}\,x)$, *for* $\epsilon > 0$. *For every* $u \in \mathcal{D}'(\mathbf{R}^n)$ *we then have that* $u * \phi_\epsilon \in C^\infty(\mathbf{R}^n)$ *converges in* $\mathcal{D}'(\mathbf{R}^n)$ *to* u *as* $\epsilon \downarrow 0$.

Proof. Let $\psi \in C_0^\infty(\mathbf{R}^n)$. Because $1(S\phi) = (S1)(\phi) = 1(\phi) = 1$ and $S(\phi_\epsilon) = (S\phi)_\epsilon$, we find, in view of Lemma 1.6, which can be extended to $\mathbf{R}^n$ in a straightforward manner, that $S\phi_\epsilon * \psi \to \psi$ in $C_0^\infty(\mathbf{R}^n)$. In combination with Theorem 11.5 this implies

$$(u * \phi_\epsilon)(\psi) = u(S\phi_\epsilon * \psi) \to u(\psi)$$

as $\epsilon \downarrow 0$. Since this holds for every $\psi \in C_0^\infty(\mathbf{R}^n)$, we conclude that $u * \phi_\epsilon \to u$ in $\mathcal{D}'(\mathbf{R}^n)$. □

Next we extend a sharpening of this result to open subsets of $\mathbf{R}^n$ by "cutting and smoothing."

Corollary 11.7. *Let X be an open subset of $\mathbf{R}^n$. For every $u \in \mathcal{D}'(X)$ there exists a sequence $(u_j)_{j\in\mathbf{N}}$ in $C_0^\infty(X)$ with $\lim_{j\to\infty} u_j = u$ in $\mathcal{D}'(X)$.*

Proof. Choose an increasing sequence of compact sets K_j in X that absorbs X; see Lemma 8.2.(a). Further, choose cut-off functions $\chi_j \in C_0^\infty(X)$ such that $\chi_j = 1$ on an open neighborhood of K_j. Finally, choose $\epsilon(j) > 0$ such that $\lim_{j\to\infty} \epsilon(j) = 0$ and $\operatorname{supp} \chi_j + \operatorname{supp} \phi_j \subset X$ for every j. Here $\phi_j := \phi_{\epsilon(j)}$ with ϕ_ϵ as in Lemma 11.6. The third assumption is satisfied if $\epsilon(j)c < \delta_j$, where c is the supremum of the $\|x\|$ with $x \in \operatorname{supp} \phi$ and where $\delta_j > 0$ has been chosen such that the δ_j-neighborhood of $\operatorname{supp} \chi_j$ is contained in X; see Fig. 11.1. In addition, suppose that the δ_j converge to 0 as $j \to \infty$.

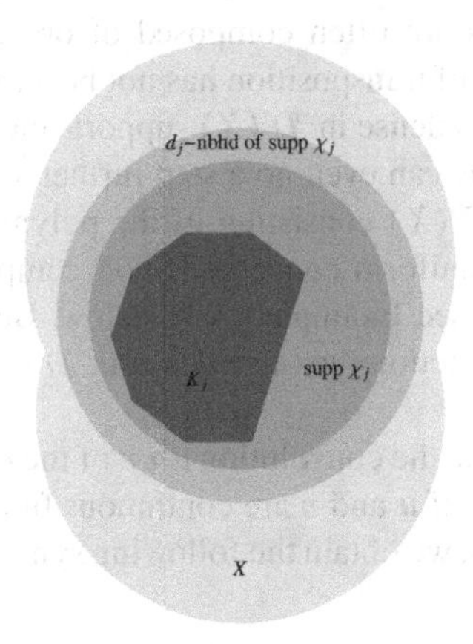

Fig. 11.1 Illustration of the proof of Corollary 11.7

Because $\chi_j u \in \mathcal{E}'(X)$, we can interpret it as an element of $\mathcal{E}'(\mathbf{R}^n) \subset \mathcal{D}'(\mathbf{R}^n)$; see Lemma 8.9. We now take

$$u_j := (\chi_j u) * \phi_j \in C^\infty(\mathbf{R}^n).$$

In view of (11.5), we find that $\operatorname{supp} u_j$ is a compact subset of X, and so $u_j \in C_0^\infty(X)$.

Let $\psi \in C_0^\infty(X) \subset C_0^\infty(\mathbf{R}^n)$. There exists a j_0 such that for all $j \geq j_0$, the δ_j-neighborhood of supp ψ is contained in K_j; consequently, $\chi_j\,(S\phi_j * \psi) = S\phi_j * \psi$. Thus, on the strength of Theorem 11.5 we get, for $j \geq j_0$,

$$u_j(\psi) = (\chi_j\, u)(S\phi_j * \psi) = u(\chi_j\,(S\phi_j * \psi)) = u(S\phi_j * \psi).$$

Because $S\phi_j * \psi \to \psi$ in $C_0^\infty(X)$ we find that $u_j(\psi) \to u(\psi)$ as $j \to \infty$ and conclude that $u_j \to u$ in $\mathcal{D}'(X)$. □

By Corollary 11.7, many identities that hold for test functions can be extended to distributions. The principle is that if A and B are sequentially continuous operators on $\mathcal{D}'(X)$ with $A(\phi) = B(\phi)$ for all $\phi \in C_0^\infty(X)$, then $A(u) = B(u)$ for all $u \in \mathcal{D}'(X)$. Indeed, with $(u_j)_{j\in\mathbf{N}}$ in $C_0^\infty(X)$ as in Corollary 11.7 we obtain

$$A(u) = \lim_{j\to\infty} A(u_j) = \lim_{j\to\infty} B(u_j) = B(u).$$

(The principle as introduced here is the same as the "principle of uniqueness of continuous extensions to the closure.") This was the basis for conjecturing (9.1), (11.3), (11.4), and (11.6), for instance, and even (11.5). The proof by transposition was then a matter of writing out the definitions.

It is a very fruitful principle, if only for generating conjectures, but it should not be overlooked that its application presupposes that A and B have already been shown to be sequentially continuous extensions to $\mathcal{D}'(X)$ of $A = B$ on $C_0^\infty(X)$. Because these extensions are often composed of operators defined by means of transposition, the method of transposition has not become redundant.

The fact that $C_0^\infty(X)$ is dense in $\mathcal{D}'(X)$ supports the concept of distributions as generalized functions. One can even go a step further, observing that, for example, the linear subspace in $C^\infty(X)$ consisting of the polynomial functions is dense in $C^\infty(X)$ with respect to uniform convergence on compacta. This is *Weierstrass's Approximation Theorem*; see Example 14.31 below. Because $C^\infty(X)$ is dense in $\mathcal{D}'(X)$, the polynomial functions are also dense in $\mathcal{D}'(X)$.

Our next aim is to define the convolution $u * v$ of the most general distributions u and v in $\mathcal{D}'(\mathbf{R}^n)$ possible. If u and v are continuous functions and if at least one of them has compact support, we obtain the following symmetric expression for testing $u * v$ with $\phi \in C_0^\infty(\mathbf{R}^n)$:

$$(u * v)(\phi) = \iint u(x)\, v(z-x)\, dx\, \phi(z)\, dz = \iint u(x)\, v(y)\, \phi(x+y)\, dx\, dy. \tag{11.7}$$

Actually, the left-hand side should read test$(u * v)$, instead of $u * v$. In the integration over z we have applied the change of variables $z = x + y$.

In the right-hand side of (11.7), the function

$$u \otimes v : (x, y) \mapsto u(x)\, v(y) \tag{11.8}$$

on $\mathbf{R}^n \times \mathbf{R}^n$ is tested against the function

$$\Sigma^*\phi : (x, y) \mapsto \phi(x + y). \tag{11.9}$$

The function $u \otimes v$ is said to be the *tensor product* of u and v. In (11.9) we have used the *sum mapping*

$$\Sigma : (x, y) \mapsto x + y \quad : \quad \mathbf{R}^n \times \mathbf{R}^n \to \mathbf{R}^n, \tag{11.10}$$

by which $\phi \in C^\infty(\mathbf{R}^n)$ is pulled back to a function in $C^\infty(\mathbf{R}^n \times \mathbf{R}^n)$. Using the transpose Σ_* of Σ^* in a formal way, see (10.3), we can write

$$(u * v)(\phi) = (u \otimes v)(\Sigma^*\phi) = (\Sigma_*(u \otimes v))(\phi) \qquad (\phi \in C_0^\infty(\mathbf{R}^n)).$$

This suggests the following definition for the convolution product $u * v$ of the distributions u and v in $\mathbf{R}^n$:

$$u * v = \Sigma_*(u \otimes v), \tag{11.11}$$

the tensor product of u and v pushed forward under the sum mapping. In explicit form this becomes

$$(u * v)(\phi) = u(x \mapsto v(y \mapsto \phi(x + y))) \qquad (\phi \in C_0^\infty(\mathbf{R}^n)). \tag{11.12}$$

To justify this, we first consider the extension of the definition of the tensor product $u \otimes v$ to arbitrary distributions u and v. See Example 15.11 for another approach.

Theorem 11.8. *Let X and Y be open subsets of $\mathbf{R}^n$ and $\mathbf{R}^p$, respectively, and let $u \in \mathcal{D}'(X)$ and $v \in \mathcal{D}'(Y)$. Then there exists exactly one distribution on $X \times Y$, denoted by $u \otimes v$ and said to be the* tensor product *of u and v, with the property*

$$(u \otimes v)(\phi \otimes \psi) = u(\phi)\, v(\psi), \tag{11.13}$$

for all $\phi \in C_0^\infty(X)$ and $\psi \in C_0^\infty(Y)$. One has, for $\theta \in C_0^\infty(X \times Y)$,

$$(u \otimes v)(\theta) = u(x \mapsto v(y \mapsto \theta(x, y))) = v(y \mapsto u(x \mapsto \theta(x, y))). \tag{11.14}$$

Equation (11.14) also applies if $u \in \mathcal{E}'(X)$, $v \in \mathcal{E}'(Y)$, and $\theta \in C^\infty(X \times Y)$. The mapping

$$(u, v) \mapsto u \otimes v \quad : \quad \mathcal{D}'(X) \times \mathcal{D}'(Y) \to \mathcal{D}'(X \times Y) \tag{11.15}$$

is sequentially continuous with respect to each variable separately.

Proof. We begin by verifying that the right-hand side in (11.14) is well-defined and is continuously dependent on the test function θ. Given $x \in X$, introduce the partial function $\theta(x, \cdot) : y \mapsto \theta(x, y)$ in $C_0^\infty(Y)$; then $x \mapsto \theta(x, \cdot)$ is a continuous mapping: $X \to C_0^\infty(Y)$. Hence composition with the continuous $v : C_0^\infty(Y) \to \mathbf{C}$ implies that

$$(V\theta)(x) := v \circ \theta(x, \cdot)$$

defines a continuous function $V\theta$ on X with compact support. Using the linearity, and once again using the continuity of v, one can show that

$$\lim_{t \to 0} \frac{1}{t}\big((V\theta)(x + t\, e(j)) - (V\theta)(x)\big) = V(\partial_j \theta)(x).$$

By mathematical induction on k we find that $V\theta$ is of class C^k and that

$$\partial^\alpha (V\theta) = V(\partial^\alpha \theta),$$

for all multi-indices α with $|\alpha| \leq k$. Thus we see that $V\theta \in C_0^\infty(X)$; moreover, making use of the estimates for v of the form (3.4), we conclude that

$$V : \theta \mapsto V\theta \quad : \quad C_0^\infty(X \times Y) \to C_0^\infty(X)$$

defines a continuous linear mapping. Consequently, $u \otimes v \in \mathcal{D}'(X \times Y)$ if we set

$$u \otimes v = u \circ V \quad : \quad C_0^\infty(X \times Y) \to \mathbf{C}. \tag{11.16}$$

Because

$$(V(\phi \otimes \psi))(x) = v(\phi(x)\, \psi) = \phi(x)\, v(\psi) \qquad (x \in X),$$

it follows that $V(\phi \otimes \psi) = v(\psi)\, \phi$. If we then apply u to this, we come to the conclusion that (11.13) holds. This proves the existence part of the theorem.

Now suppose that $w \in \mathcal{D}'(X \times Y)$ and $w(\phi \otimes \psi) = u(\phi)\, v(\psi)$ whenever $\phi \in C_0^\infty(X)$ and $\psi \in C_0^\infty(Y)$. Then $\omega := w - (u \otimes v)$ has the property that $\omega(\phi \otimes \psi) = 0$ for all $\phi \in C_0^\infty(X)$ and $\psi \in C_0^\infty(Y)$. In Lemma 11.10 below, which is of independent interest, we will demonstrate that this implies $\omega = 0$. In that case, $w = u \otimes v$, and thus we will also have proved the uniqueness.

Furthermore, the right-hand side of the second identity in (11.14) defines a distribution that also satisfies (11.13); in view of the uniqueness just proved, this distribution also equals $u \otimes v$.

To prove the last assertion we observe that on the strength of (11.16), it follows from $\lim_{j \to \infty} u_j = 0$ in $\mathcal{D}'(X)$ that for every $\theta \in C_0^\infty(X \times Y)$,

$$\lim_{j \to \infty} u_j \otimes v(\theta) = \lim_{j \to \infty} u_j(V\theta) = 0,$$

because $V\theta \in C_0^\infty(X)$. The analogous result holds for $u \otimes v_j$. □

Example 11.9. Let the notation be as in Example 10.20. Now (10.25) takes the form, for ϕ and $\psi \in C_0^\infty(\mathbf{R}^n)$ and $y \in \mathbf{R}^n$,

$$d_*(\phi \otimes \psi)(y) = \int_{\mathbf{R}^n} \phi(x)\, S\psi(y - x)\, dx = \phi * S\psi(y), \quad \text{so} \quad d_*(\phi \otimes \psi) = \phi * S\psi.$$

Theorem 11.5 then implies the following result, which will be of use in Example 15.11 below:

$$d^*u(\phi \otimes \psi) = u(d_*(\phi \otimes \psi)) = u(S\psi * \phi) = u * \psi(\phi) \qquad (u \in \mathcal{D}'(\mathbf{R}^n)). \ \oslash$$

The previously defined tensor product of continuous functions u and v satisfies (11.13), and in view of the uniqueness it therefore equals the distributional tensor product of u and v. In other words, the tensor product of distributions is an extension of the tensor product of continuous functions.

The mapping in (11.15) is actually sequentially continuous with respect to all variables; this can be proved using Theorem 5.5.

Next we come to the lemma that was invoked in the proof of Theorem 11.8.

Lemma 11.10. *Let X and Y be open subsets of $\mathbf{R}^n$ and $\mathbf{R}^p$, respectively, and $\omega \in \mathcal{D}'(X \times Y)$. If $\omega(\phi \otimes \psi) = 0$, for all $\phi \in C_0^\infty(X)$ and $\psi \in C_0^\infty(Y)$, then $\omega = 0$.*

Proof. Let $(x_0, y_0) \in X \times Y$. We can find a corresponding open neighborhood U of (x_0, y_0) whose closure is a compact subset K of $X \times Y$. If $\chi \in C_0^\infty(X \times Y)$ equals 1 on an open neighborhood of K, one has $\chi\omega \in \mathcal{E}'(X \times Y) \subset \mathcal{D}'(\mathbf{R}^n \times \mathbf{R}^p)$. Then we get in the notation of Lemma 11.6 that, if $\epsilon > 0$ is sufficiently small,

$$((\chi\omega) * (\phi_\epsilon \otimes \psi_\epsilon))(x, y) = (\chi\omega)((T_x S\phi_\epsilon) \otimes (T_y S\psi_\epsilon)) = \omega((T_x S\phi_\epsilon) \otimes (T_y S\psi_\epsilon))$$

equals 0, for all $(x, y) \in K$, in view of $\operatorname{supp} T_z S\zeta_\epsilon = z + (-\epsilon \operatorname{supp} \zeta)$. In particular, we have $(\chi\omega) * (\phi_\epsilon \otimes \psi_\epsilon) = 0$ on U. On account of Lemma 11.6, the limit of the left-hand side, as $\epsilon \downarrow 0$, equals $\chi\omega$; thus we obtain $\chi\omega = 0$ on U, and so $\omega = 0$ on U. By applying Theorem 7.1 we see that this implies $\omega = 0$. □

This lemma has a useful application, as will be shown in Remark 11.12. In preparation we treat the following:

Example 11.11. Write $C_0^\infty(X) \otimes C_0^\infty(Y)$ for the linear subspace of $C_0^\infty(X \times Y)$ consisting of the finite linear combinations of the functions $\phi \otimes \psi$, where $\phi \in C_0^\infty(X)$ and $\psi \in C_0^\infty(Y)$. This linear subspace is a proper subset of $C_0^\infty(X \times Y)$. In order to verify this, note that the following assertions are equivalent, for any $\chi \in C_0^\infty(X \times Y)$:

(i) $\chi \in C_0^\infty(X) \otimes C_0^\infty(Y)$.

(ii) The linear subspace of $C_0^\infty(X)$ spanned by the set of functions $\{\, \chi(\cdot, y) \in C_0^\infty(X) \mid y \in Y \,\}$ is of finite dimension; here $\chi(\cdot, y)(x) = \chi(x, y)$.

(iii) The linear subspace of $C_0^\infty(Y)$ spanned by the set of functions $\{\, \chi(x, \cdot) \in C_0^\infty(Y) \mid x \in X \,\}$ is of finite dimension.

For an example of $\chi \in C_0^\infty(\mathbf{R} \times \mathbf{R})$ that does not satisfy condition (ii), consider (compare with Lemma 2.7)

$$\chi(x, y) = \begin{cases} e^{\frac{1}{\|(x,y)\|^2-1}} & \text{if} \quad \|(x, y)\| < 1, \\ 0 & \text{if} \quad \|(x, y)\| \geq 1. \end{cases}$$

Indeed, select $k \in \mathbf{Z}_{\geq 0}$ and $1 > y_1 > \cdots > y_k > 0$ arbitrarily. Then

$$\chi(x, y_j) \gtreqqless 0 \qquad \text{if} \qquad |x| \lesseqgtr \sqrt{1 - y_j^2}. \tag{11.17}$$

Suppose there exist $c_j \in \mathbf{C}$ such that $\sum_{1 \leq j \leq k} c_j \, \chi(x, y_j) = 0$, for all $x \in \mathbf{R}$. Then consider $x \in \mathbf{R}$ satisfying $\sqrt{1 - y_{k-1}^2} < |x| < \sqrt{1 - y_k^2}$ and apply (11.17) to find that $c_k = 0$. Continuing in this manner, we see that $c_j = 0$ for all j. This demonstrates the linear independence of the functions $\chi(\cdot, y_j)$, for $1 \leq j \leq k$. ⊘

Remark 11.12. Notwithstanding that $C_0^\infty(X) \otimes C_0^\infty(Y)$ is a proper linear subspace of $C_0^\infty(X \times Y)$, it is a dense subset of the latter space. Indeed, using the Hahn–Banach Theorem, Theorem 8.12, one directly deduces the claim from Lemma 11.10. Phrased differently, test functions of both the variables x and y can be approximated by linear combinations of products of test functions of x and y, respectively. We will give a direct proof in Theorem 16.17 below and another one in Theorem 15.4. ⊘

Example 11.13. If we denote the variable in X or Y by x or y, respectively, then $u \in \mathcal{D}'(X)$ and $v \in \mathcal{D}'(Y)$ can also be denoted by the function symbols $u(x)$ and $v(y)$, respectively. In that case it is usual to use the function notation $(u \otimes v)(x, y) = u(x)\, v(y)$ also for distributions $u \in \mathcal{D}'(X)$ and $v \in \mathcal{D}'(Y)$.

If $a \in X$ and $b \in Y$ are given points, then

$$(\delta_a \otimes \delta_b)\,(\phi \otimes \psi) = \delta_a(\phi)\,\delta_b(\psi) = \phi(a)\,\psi(b) = \delta_{(a,b)}(\phi \otimes \psi),$$

from which we see that $\delta_a \otimes \delta_b = \delta_{(a,b)}$. In function notation: $\delta_a(x)\,\delta_b(y) = \delta_{(a,b)}(x, y)$. For example, if $X = \mathbf{R}^n$ and $Y = \mathbf{R}^p$, then $\delta(x)\,\delta(y) = \delta(x, y)$, a formula popular in the physics literature.

Now let $X = Y = \mathbf{R}^n$. Then $\delta(x + y)\,\delta(x - y) = \frac{1}{2}\,\delta(x, y)$ in $\mathcal{D}'(\mathbf{R}^n \times \mathbf{R}^n)$. Indeed, $\delta(x+y)\,\delta(x-y)$ corresponds to $\Phi^*(\delta \otimes \delta) \in \mathcal{D}'(\mathbf{R}^n \times \mathbf{R}^n)$, with $\Phi(x, y) = (x + y, x - y)$. Now we have, for $\phi \in C_0^\infty(\mathbf{R}^n \times \mathbf{R}^n)$,

$$\Phi^*(\delta \otimes \delta)(\phi) = (\delta \otimes \delta)(\Phi_*\phi) = j_{\Phi^{-1}}(\delta \otimes \delta)((\Phi^{-1})^*\phi) = \frac{1}{2}\,\phi(0, 0) = \frac{1}{2}\delta(\phi),$$

where in order to establish the second identity, we have used Theorem 10.8, and where we have observed that $\Phi^{-1} = \frac{1}{2}\,\Phi$ and $j_{\Phi^{-1}} = \frac{1}{2}$ for the third. ⊘

We now want to define the convolution $u * v$ of the distributions u and v belonging to $\mathcal{D}'(\mathbf{R}^n)$ as the tensor product, pushed forward under the sum mapping, as in (11.11). In the pushing forward we intend to use Theorem 10.6. We are, however, faced with the problem that the sum mapping is not a proper mapping from $\mathbf{R}^n \times \mathbf{R}^n$ to $\mathbf{R}^n$, unless $n = 0$. Indeed, the inverse image of $z \in \mathbf{R}^n$ consists of the set of all $(x, z - x)$ with $x \in \mathbf{R}^n$. In the following lemma we use sets of sums as defined in (2.2).

Lemma 11.14. *Let A and B be closed subsets of $\mathbf{R}^n$. Then the following assertions are equivalent:*

(a) The sum mapping (11.10) is proper on $A \times B$.
(b) For every compact subset L of $\mathbf{R}^n$, $A \cap (L + (-B))$ is bounded in $\mathbf{R}^n$.
(c) For every compact subset L of $\mathbf{R}^n$, $B \cap (L + (-A))$ is bounded in $\mathbf{R}^n$.

Assertions (a)–(c) obtain if either A or B is compact.

Proof. (a) means that for every compact subset L of $\mathbf{R}^n$ the set V of the $(x, y) \in A \times B$, with $x + y \in L$, is compact. Because this set is closed, we may replace the word "compact" by "bounded." Furthermore, the projection of V onto the first component equals $A \cap (L + (-B))$ and the projection onto the second component is $B \cap (L + (-A))$. This proves (a) $\Rightarrow$ (b) and (a) $\Rightarrow$ (c).

On the other hand,

$$\Sigma^{-1}(L) \subset \big(A \cap (L + (-B))\big) \times \big(L + \big(-\big(A \cap (L + (-B))\big)\big)\big).$$

This implies (b) $\Rightarrow$ (a) and by an analogous argument one shows that (c) $\Rightarrow$ (a). □

Definition 11.15. For every u and v in $\mathcal{D}'(\mathbf{R}^n)$ such that the sum mapping (11.10) is proper on $\operatorname{supp} u \times \operatorname{supp} v$, (11.11) defines a distribution $u * v \in \mathcal{D}'(\mathbf{R}^n)$, said to be the *convolution* of u and v. In particular, $u * v$ is defined if at least one of the two distributions u and v has compact support. ⊘

Example 11.16. By substituting a Dirac measure into (11.12) we obtain

$$u * \delta_a = \delta_a * u = T_a u \qquad (u \in \mathcal{D}'(\mathbf{R}^n),\ a \in \mathbf{R}^n). \tag{11.18}$$

In particular (compare with (11.2)),

$$u * \delta = \delta * u = u \qquad (u \in \mathcal{D}'(\mathbf{R}^n)). \tag{11.19}$$

⊘

Let A be a closed subset of $\mathbf{R}^n$. In the following theorem, as in Theorem 10.6, we use the notation $\mathcal{D}'(\mathbf{R}^n)_A$ for the set of $u \in \mathcal{D}'(\mathbf{R}^n)$ such that $\operatorname{supp} u \subset A$. We say that $u_j \to u$ in $\mathcal{D}'(\mathbf{R}^n)_A$ if $u_j \in \mathcal{D}'(\mathbf{R}^n)_A$ for all j and $u_j \to u$ in $\mathcal{D}'(\mathbf{R}^n)$. In that case, it automatically follows that $u \in \mathcal{D}'(\mathbf{R}^n)_A$. For the properness of the sum mapping, see Lemma 11.14.

Theorem 11.17. *Let A and B be closed subsets of $\mathbf{R}^n$ such that the sum mapping (11.10) is proper on $A \times B$. The convolution product $(u, v) \mapsto u * v$ is the uniquely determined extension of the convolution product*

$$(u, \phi) \mapsto u * \phi \quad : \quad \mathcal{D}'(\mathbf{R}^n)_A \times C_0^\infty(\mathbf{R}^n)_B \to C^\infty(\mathbf{R}^n),$$

as introduced in (11.1), to a mapping from $\mathcal{D}'(\mathbf{R}^n)_A \times \mathcal{D}'(\mathbf{R}^n)_B$ *to* $\mathcal{D}'(\mathbf{R}^n)$ *that is sequentially continuous with respect to each variable separately.*

The convolution of distributions satisfies the following computational rules:

$$\operatorname{supp}(u * v) \subset \operatorname{supp} u + \operatorname{supp} v, \tag{11.20}$$

$$u * v = v * u \qquad \textit{(commutative rule)}, \tag{11.21}$$

$$(u * v) * w = u * (v * w) \qquad \textit{(associative rule)}, \tag{11.22}$$

$$(u * v)(\phi) = u(Sv * \phi) \qquad (\phi \in C_0^\infty(\mathbf{R}^n)), \tag{11.23}$$

$$T_a(u * v) = (T_a u) * v = u * (T_a v) \qquad (a \in \mathbf{R}^n), \tag{11.24}$$

$$\partial^\alpha(u * v) = (\partial^\alpha u) * v = u * (\partial^\alpha v) \qquad (\alpha \in (\mathbf{Z}_{\geq 0})^n). \tag{11.25}$$

Rule (11.22) holds if the sum mapping $\Sigma : (x, y, z) \mapsto x + y + z$ *is proper on* $\operatorname{supp} u \times \operatorname{supp} v \times \operatorname{supp} w$. *This certainly is the case if at least two out of the three distributions* u, v, *and* w *have compact support. The right-hand side in (11.23) has to be understood as* $u(\chi(Sv*\phi))$, *where* $\chi \in \mathcal{D}'(\mathbf{R}^n)$ *and* $\chi = 1$ *on a neighborhood of* $A \cap (L + (-B))$ *with* L *a compact subset of* $\mathbf{R}^n$ *containing* $\operatorname{supp} \phi$.

Proof. Regarding the extension of the convolution product to $\mathcal{D}'(\mathbf{R}^n)_A \times \mathcal{D}'(\mathbf{R}^n)_B$, we first note the following. For P and Q closed subsets of $\mathbf{R}^n$ with $P \cap Q$ compact, there exists $\chi \in C_0^\infty(\mathbf{R}^n)$ with $\chi = 1$ on a neighborhood of $P \cap Q$. Given any $u \in \mathcal{D}'(\mathbf{R}^n)_P$ and $\psi \in C^\infty(\mathbf{R}^n)_Q$, then $u(\chi\,\psi)$ is well-defined and independent of the choice of χ. In this manner, we extend u to a continuous linear functional on $C^\infty(\mathbf{R}^n)_Q$ and the extension is uniquely determined, because $C_0^\infty(\mathbf{R}^n)_Q$ is dense in the latter space.

Now consider $v \in \mathcal{D}'(\mathbf{R}^n)_B$. Then $Sv \in \mathcal{D}'(\mathbf{R}^n)_{-B}$, so we obtain, for any compact $L \subset \mathbf{R}^n$ and $\phi \in C_0^\infty(\mathbf{R}^n)_L$,

$$Sv * \phi \in C^\infty(\mathbf{R}^n)_{L+(-B)}.$$

Next apply the first result in the proof choosing $P = A$ and $Q = L + (-B)$. Then $P \cap Q$ is compact by Lemma 11.14.(b), and accordingly we may define, for $u \in \mathcal{D}'(\mathbf{R}^n)_A$,

$$(u \,\tilde{*}\, v)(\phi) = u(Sv * \phi).$$

If $v \in C_0^\infty(\mathbf{R}^n)$, then (11.6) implies that the right-hand side is equal to $(u * v)(\phi)$. This being the case for every $\phi \in C_0^\infty(\mathbf{R}^n)$, we conclude that $u \,\tilde{*}\, v = u * v$ if $v \in C_0^\infty(\mathbf{R}^n)$: the convolution product of distributions is an extension of the convolution product of a distribution and a test function as introduced in (11.1). Furthermore, (11.23) now follows on identifying $\tilde{*}$ with $*$. In addition, we have for $u \in \mathcal{D}'(\mathbf{R}^n)$ and v and $\phi \in C_0^\infty(\mathbf{R}^n)$ that $u(Sv * \phi) = \Sigma_*(u \otimes v)(\phi)$. This may be proved by observing that both sides depend continuously on $u \in \mathcal{D}'(\mathbf{R}^n)$ and that $\mathcal{E}'(\mathbf{R}^n)$ is dense in $\mathcal{D}'(\mathbf{R}^n)$. For $u \in \mathcal{E}'(\mathbf{R}^n)$ we obtain

$$\Sigma_*(u \otimes v)(\phi) = (u \otimes v)(\phi \circ \Sigma) = u(x \mapsto v(\phi(x + \cdot))) = u(Sv * \phi).$$

Using (10.5), we see that

$$\begin{aligned}\operatorname{supp}(u * v) &= \operatorname{supp} \Sigma_*(u \otimes v) \subset \Sigma(\operatorname{supp}(u \otimes v)) \\ &= \Sigma(\operatorname{supp} u \times \operatorname{supp} v) = \operatorname{supp} u + \operatorname{supp} v.\end{aligned}$$

Let $\tau : (x, y) \mapsto (y, x)$ be the interchanging of variables. The second identity in (11.14) can be written as $(u \otimes v)(\theta) = (v \otimes u)(\tau^*(\theta))$ for every test function θ, that is,

$$u \otimes v = \tau_*(v \otimes u).$$

The commutativity of addition means that $\Sigma \circ \tau = \Sigma$. Application of (10.4) in the situation at hand is admissible, because it can be shown that (10.4) remains valid for Φ a diffeomorphism and Ψ equal to Φ as in Theorem 10.5. Since τ is a diffeomorphism, this leads to

$$u * v = \Sigma_* \circ \tau_*(v \otimes u) = (\Sigma \circ \tau)_*(v \otimes u) = \Sigma_*(v \otimes u) = v * u.$$

Thus, the commutativity of convolution is seen to be a direct consequence of the commutativity of addition.

For arbitrary u, v, and $w \in \mathcal{E}'(\mathbf{R}^n)$ we find by writing out the definitions that $(u * (v * w))(\theta)$ equals

$$u(x \mapsto v(y \mapsto w(z \mapsto \theta(x + (y + z))))),$$

while $((u * v) * w)(\theta)$ is equal to

$$u(x \mapsto v(y \mapsto w(z \mapsto \theta((x + y) + z)))).$$

In this case it follows that (11.22) is a direct consequence of the associativity of addition. For $u \in \mathcal{D}'(\mathbf{R}^n)_A$, $v \in \mathcal{D}'(\mathbf{R}^n)_B$, and $w \in \mathcal{D}'(\mathbf{R}^n)_C$ with Σ proper on $A \times B \times C$, we get associativity because it already holds on a dense subset.

By virtue of Theorem 11.8 we conclude that $u \mapsto u * v = \Sigma_*(u \otimes v)$, being a composition of two such mappings, is sequentially continuous; using the commutativity of convolution, we find that $v \mapsto u * v$ is also sequentially continuous.

Finally, (11.24) and (11.25) can be proved by writing out the definitions and applying transposition. These rules can also be derived from the corresponding formulas for $v \in C_0^\infty(\mathbf{R}^n)$, approximating $v \in \mathcal{D}'(\mathbf{R}^n)$ by a sequence of test functions and using the sequential continuity of the convolution product. □

Example 11.18. Let $\mathcal{U} = u *$ as in Theorem 11.3. Then $\mathcal{U}$ has an extension to a sequentially continuous linear mapping $\mathcal{U} : \mathcal{E}'(\mathbf{R}^n) \to \mathcal{D}'(\mathbf{R}^n)$ satisfying $\mathcal{U}(v) = u * v$, for all $v \in \mathcal{E}'(\mathbf{R}^n)$. In particular, $u = \mathcal{U}(\delta)$. ⊘

Remark 11.19. The associativity of convolution implies that there is no need to use parentheses in the case of multiple convolution. The convolution $u_1 * u_2 * \cdots * u_k$ of k distributions $u_1, u_2, \ldots, u_k$ is well-defined if the sum mapping

$$\Sigma : (x^{(1)}, x^{(2)}, \ldots, x^{(k)}) \mapsto \sum_{j=1}^{k} x^{(j)}$$

is proper on $\operatorname{supp} u_1 \times \operatorname{supp} u_2 \times \cdots \times \operatorname{supp} u_k$. One then has

$$u_1 * u_2 * \cdots * u_k = \Sigma_*(u_1 \otimes u_2 \otimes \cdots \otimes u_k). \tag{11.26}$$

The notation for the repeated tensor product in the right-hand side of this formula suggests that the tensor product is associative as well. In fact, for test functions this is evident. Using Corollary 11.7 and the continuity of the tensor product, we then also obtain the associativity of the tensor product for arbitrary distributions. The proof of (11.26) can be given by mathematical induction on k. ⊘

Remark 11.20. Assume that u and $v \in \mathcal{D}'(\mathbf{R}^n)$ and that the sum mapping is proper on $\operatorname{supp} u \times \operatorname{supp} v$. If in this situation $v \in C^\infty(\mathbf{R}^n)$, then also $u * v \in C^\infty(\mathbf{R}^n)$, while

$$(u * v)(x) = u(T_x \circ S v) \qquad (x \in \mathbf{R}^n). \tag{11.27}$$

Although this is the same formula as (11.1), we first of all need to justify the right-hand side in (11.27), because we have not assumed u or v to have compact support. Consider a compact subset $L \subset \mathbf{R}^n$. On account of Lemma 11.14.(b) there exists a compact $K \subset \mathbf{R}^n$ with the property that $\operatorname{supp} u \cap (L + (-\operatorname{supp} v)) \subset K$. Next, write $A = L + (-\operatorname{supp} v)$. Then A is a closed subset of $\mathbf{R}^n$, while $\operatorname{supp} u \cap A$ is compact. Therefore there exists a uniquely determined sequentially continuous extension of u to the space of $\phi \in C^\infty(\mathbf{R}^n)$ with $\operatorname{supp} \phi \subset A$. This extension is also denoted by u. To verify its existence, we write $u(\phi) := u(\chi\,\phi)$, where $\chi \in C_0^\infty(\mathbf{R}^n)$ and $\chi = 1$ on a neighborhood of K (see Corollary 2.16). Now consider any $x \in L$ and write $\phi = T_x \circ S v$. Then $\operatorname{supp} \phi = \{x\} + (-\operatorname{supp} v) \subset A$, and it follows that $u(T_x \circ S v) = u(\phi)$ is well-defined.

We will now show that $u * v \in C^\infty(\mathbf{R}^n)$ and that (11.27) holds. Let U be an arbitrary bounded open subset of $\mathbf{R}^n$ with compact closure L. According to the assumption made,

$$K := \Sigma^{-1}(L) \cap (\operatorname{supp} u \times \operatorname{supp} v)$$

is a compact subset of $\mathbf{R}^n \times \mathbf{R}^n$. Let $\chi \in C_0^\infty(\mathbf{R}^n \times \mathbf{R}^n)$ with $\chi = 1$ on a neighborhood of K. The definition of distributions pushed forward, given in the proof of Theorem 10.6, leads to the formula

$$(u * v)(\theta) = (u \otimes v)(\chi \Sigma^*(\theta)),$$

for every $\theta \in C_0^\infty(U)$. In addition, we can take $\chi = \alpha \otimes \beta$, with α and $\beta \in C_0^\infty(\mathbf{R}^n)$ equal to 1 on a large bounded subset of $\mathbf{R}^n$. But this means that **on** U,

$$u * v = (\alpha u) * (\beta v). \tag{11.28}$$

Here $\beta v \in C_0^\infty(\mathbf{R}^n)$. This identity is derived from $(u \otimes v)((\alpha \otimes \beta)\Sigma^*(\theta)) = (\alpha u \otimes \beta v)(\Sigma^*(\theta))$, which follows from $(\psi u)(\phi) = \phi(\psi u)(1) = \phi u(\psi) = u(\psi\phi)$, for $\psi \in C_0^\infty$, $u \in \mathcal{D}'$, and $\phi \in C^\infty$. In Theorem 11.2 we have concluded that the right-hand side of (11.28) is a C^∞ function given by

$$((\alpha u) * (\beta v))(x) = (\alpha u)(T_x \circ S(\beta v)).$$

For $x \in U$ this is equal to the right-hand side of (11.27), with the interpretation given above.

More precisely, we claim that

$$u(\alpha\,\beta(x-\cdot)v(x-\cdot)) = u(\alpha\,v(x-\cdot)) \qquad (x \in U)$$

and that $\alpha = 1$ on a neighborhood of supp $u \cap (L + (-\operatorname{supp} v))$. Equivalently we may write the former identity as

$$u(\alpha(\beta(x-\cdot)-1)v(x-\cdot)) = 0 \qquad (x \in U). \tag{11.29}$$

For the proof of (11.29), fix $x \in U$, which implies $x \in L$. By the definition of K, we find for $y \in \mathbf{R}^n$ the following three possibilities: $y \notin \operatorname{supp} u$ or $x - y \notin \operatorname{supp} v$ or $(y, x-y) \in K$. But $K \subset U_1 \times U_2$, where U_1 and U_2 are open subsets of $\mathbf{R}^n$ on which α and β, respectively, are equal to 1. Hence, if $(y, x-y) \in K$, then $y \in U_1 \cap (x + (-U_2))$. Thus supp u is covered by the following open sets: the complement in $\mathbf{R}^n$ of $x + (-\operatorname{supp} v)$ and the set $U_1 \cap (x + (-U_2))$. For y belonging to either one of these sets we have $\alpha(y)(\beta(x-y)-1)v(x-y) = 0$; this leads to (11.29). Finally, we show that $\alpha = 1$ on a neighborhood of supp $u \cap (L + (-\operatorname{supp} v))$. Since $\chi = 1$ on an open neighborhood of K, we see that $\alpha = 1$ on an open neighborhood of the image of K under the projection $\mathbf{R}^n \times \mathbf{R}^n \to \mathbf{R}^n$ with $(y, z) \mapsto y$. By the definition of K this image equals supp $u \cap (L + (-\operatorname{supp} v))$. ⊘

Remark 11.21. For application in (12.10) below we derive a result that is of interest in its own right. Suppose f and g are locally integrable functions on $\mathbf{R}^n$ and Σ is proper on supp $f \times$ supp g. Then $f * g \in \mathcal{D}'(\mathbf{R}^n)$ as in the definition in (11.12) is actually a locally integrable function on $\mathbf{R}^n$ too. More precisely, let $L \subset \mathbf{R}^n$ be compact and set

$$K = \{\,(y,z) \in \operatorname{supp} f \times \operatorname{supp} g \mid y + z \in L\,\} \subset \mathbf{R}^n \times \mathbf{R}^n,$$

which is also compact on account of Σ being proper. Then we obtain, for almost all $x \in L$,

$$(f * g)(x) = \int_{\pi_1(K)\cap(L+(-\pi_2(K)))} f(y)\,g(x-y)\,dy.$$

Here $\pi_i : \mathbf{R}^n \times \mathbf{R}^n \to \mathbf{R}^n$ denotes the projection onto the ith factor, for $1 \le i \le 2$.

Indeed, consider $\phi \in C_0^\infty(\mathbf{R}^n)$ with supp $\phi \subset L$. Select $\chi \in C_0^\infty(\mathbf{R}^n \times \mathbf{R}^n)$ such that χ equals 1 on a neighborhood of K. Then we obtain

$$\begin{aligned}
(\operatorname{test} f * \operatorname{test} g)(\phi) &= (\operatorname{test} f \otimes \operatorname{test} g)(\chi(\phi \circ \Sigma)) \\
&= (\operatorname{test} f)(y \mapsto (\operatorname{test} g)(z \mapsto \chi(y,z)\,\phi(y+z))) \\
&= \int_{\mathbf{R}^n} f(y) \int_{\mathbf{R}^n} g(z)\,\chi(y,z)\,\phi(y+z)\,dz\,dy \\
&= \int_{\mathbf{R}^n} f(y) \int_{\mathbf{R}^n} g(x-y)\,\chi(y,x-y)\,\phi(x)\,dx\,dy
\end{aligned}$$

$$= \int_{\mathbf{R}^n} \int_{\mathbf{R}^n} f(y)\, g(x-y)\, \chi(y, x-y)\, dy\, \phi(x)\, dx$$
$$= \int_L \int_{\pi_1(K) \cap (L + (-\pi_2(K)))} f(y)\, g(x-y)\, \chi(y, x-y)\, dy\, \phi(x)\, dx$$
$$= \int_L \int_{\pi_1(K) \cap (L + (-\pi_2(K)))} f(y)\, g(x-y)\, dy\, \phi(x)\, dx.$$

In obtaining the last equality, we need the extra assumption that χ equals 1 on a neighborhood of the set

$$\pi_1(K) \times \big(L + \big(-\big(L + (-\pi_2(K))\big)\big)\big),$$

which contains K.

In the case of compact supp f, we may also write

$$(f * g)(x) = \int_{\mathrm{supp}\, f} f(y)\, g(x-y)\, dy. \tag{11.30}$$

If μ is a Radon measure on $\mathbf{R}^n$ of compact support and g is locally integrable on $\mathbf{R}^n$, then the proof above can be modified to give, in the notation of Chap. 20,

$$(\mu * g)(x) = \int_{\mathrm{supp}\, \mu} g(x-y)\, \mu(dy). \tag{11.31}$$

⊘

In general, we have the following estimate for the singular support of the convolution of two distributions.

Theorem 11.22. *Consider u and $v \in \mathcal{D}'(\mathbf{R}^n)$ and suppose that the sum mapping $(x, y) \mapsto x + y$ is proper on* supp $u \times$ supp v. *Then*

$$\mathrm{sing\,supp}\,(u * v) \subset \mathrm{sing\,supp}\, u + \mathrm{sing\,supp}\, v. \tag{11.32}$$

Proof. Let $A = \mathrm{sing\,supp}\, u$ and $B = \mathrm{sing\,supp}\, v$. For every $\delta > 0$ we can find a cut-off function $\alpha \in C^\infty(\mathbf{R}^n)$ with $\alpha = 1$ on an open neighborhood of A and such that supp α is contained in the δ-neighborhood A_δ of A. For this purpose we can choose the convolution of the characteristic function of the $\frac{1}{2}\,\delta$-neighborhood of A and ϕ_ϵ, for sufficiently small $\epsilon > 0$; see Lemma 2.19. Let β be an analogous cut-off function, having the same properties with respect to B instead of A.

Write $u_1 := \alpha\, u$ and $u_2 := (1-\alpha)\, u$, and analogously $v_1 := \beta\, v$ and $v_2 := (1-\beta)\, v$. Then $u = u_1 + u_2$ with supp $u_1 \subset A_\delta$, $u_2 \in C^\infty(\mathbf{R}^n)$. Furthermore, $v = v_1 + v_2$ with supp $v_1 \subset B_\delta$ and $v_2 \in C^\infty(\mathbf{R}^n)$. Out of the four terms in

$$u * v = u_1 * v_1 + u_1 * v_2 + u_2 * v_1 + u_2 * v_2,$$

the last three are of class C^∞ on account of Remark 11.20. This implies

$$\operatorname{sing\,supp} u*v = \operatorname{sing\,supp} u_1*v_1 \subset \operatorname{supp} u_1*v_1 \subset \operatorname{supp} u_1+\operatorname{supp} v_1 \subset A_\delta+B_\delta.$$

If we can demonstrate that for every $x \notin A + B$ there exists $\delta > 0$ with $x \notin A_\delta + B_\delta$, that is, $x \notin \operatorname{sing\,supp}(u * v)$, then we will have proved that

$$\operatorname{sing\,supp}(u * v) \subset A + B.$$

The mapping Σ being proper on $\operatorname{supp} u \times \operatorname{supp} v$, it follows from [7, Theorem 1.8.6] that $A + B$ is a closed subset of $\mathbf{R}^n$. Such a set equals the intersection of all of its δ-neighborhoods; accordingly, there exists $\delta > 0$ such that $x \notin (A + B)_{2\delta}$. The conclusion now follows because $A_\delta + B_\delta \subset (A + B)_{2\delta}$. □

Remark 11.23. It may also happen that the convolution $u * v$ can be formed without making any assumptions about the supports of u and of v. The space of all integrable functions on $\mathbf{R}^n$ is denoted by $L^1(\mathbf{R}^n)$, or simply L^1, and provided with the *integral norm*

$$\|v\|_{L^1} := \int_{\mathbf{R}^n} |v(x)|\, dx.$$

One of the main results of Lebesgue integration, see Theorem 20.40, asserts that L^1 is a *Banach space*, that is, a normed linear space that is complete.

Let E be a linear subspace of $\mathcal{D}'(\mathbf{R}^n)$ endowed with a norm $u \mapsto \|u\|$, with the following four properties:

(i) For every $a \in \mathbf{R}^n$ and $u \in E$, one has $T_a u \in E$ and $\|T_a u\| = \|u\|$.
(ii) E is complete.
(iii) $C_0^\infty(\mathbf{R}^n) \subset E$ and convergence in $C_0^\infty(\mathbf{R}^n)$ implies convergence in E.
(iv) $C_0^\infty(\mathbf{R}^n)$ is dense in E.

An example of such a space is $E = L^1$.

In that case, $y \mapsto \psi(y)\, T_y\phi$ is continuous from $\mathbf{R}^n$ to E, for every ϕ and $\psi \in C_0^\infty(\mathbf{R}^n)$. In view of (1.14), the integral of this function is equal to $\phi * \psi$. Applying the triangle inequality to the approximating Riemann sums, we obtain the inequality

$$\|\phi * \psi\| \leq \|\psi\|_{L^1}\, \|\phi\|.$$

Now let $u \in E$ and $v \in L^1$. Then there exist sequences $(u_j)_{j\in\mathbf{N}}$ and $(v_j)_{j\in\mathbf{N}}$ in $C_0^\infty(\mathbf{R}^n)$ with $\|u - u_j\| \to 0$ and $\|v - v_j\|_{L^1} \to 0$ as $j \to \infty$. The sequence $(u_j * v_j)_{j\in\mathbf{N}}$ is a Cauchy sequence in E. The limit, denoted by $u * v$, is independent of the choice of the sequences (u_j) and (v_j) and satisfies

$$\|u * v\| \leq \|v\|_{L^1}\, \|u\|.$$

The mapping $(u, v) \mapsto u * v$ thus defined is continuous from $E \times L^1$ to E and is the only continuous extension to $E \times L^1$ of the convolution product on $C_0^\infty(\mathbf{R}^n) \times$

$C_0^\infty(\mathbf{R}^n)$. It is left to the reader as a problem to prove these assertions concerning $u * v$.

Applying this for $E = L^1$, one can prove by means of the theory of Lebesgue integration on product spaces that for every u and $v \in L^1$ the integral

$$\int_{\mathbf{R}^n} u(x-y)\, v(y)\, dy$$

converges to $(u * v)(x)$ for almost all x; see also Problem 11.22. The example $u = v = |x|^{-a}\, 1_{[-1,1]}$ in $L^1(\mathbf{R})$, for $\frac{1}{2} \leq a < 1$, shows that the integral need not converge for all x. ⊘

Problems

11.1.* For each of the following cases, find a combination of a distribution u and a test function ϕ on $\mathbf{R}$ that solves the equation:

(i) $u * \phi = 0$.
(ii) $u * \phi = 1$.
(iii) $u * \phi = x$.
(iv) $u * \phi = \sin$.

11.2. If $\lim_{j\to\infty} u_j = u$ in $\mathcal{D}'(\mathbf{R}^n)$ and $\lim_{j\to\infty} \phi_j = \phi$ in $C_0^\infty(\mathbf{R}^n)$, then

$$\lim_{j\to\infty} u_j * \phi_j = u * \phi \quad \text{in} \quad C^\infty(\mathbf{R}^n).$$

Prove this using the principle of uniform boundedness from Chap. 5.

11.3.* Given a mapping $\mathcal{A} : C_0^\infty(\mathbf{R}^n) \to C^\infty(\mathbf{R}^n)$, verify that the following assertions are equivalent.

(i) $\mathcal{A}$ is a continuous linear mapping and commutes with the partial differentiations ∂_j, for all $1 \leq j \leq n$.
(ii) There exists a $u \in \mathcal{D}'(\mathbf{R}^n)$ such that $\mathcal{A}(\phi) = u * \phi$, for every $\phi \in C_0^\infty(\mathbf{R}^n)$.

Hint: differentiate $a \mapsto T_a \circ \mathcal{A} \circ T_{-a}$.

11.4. If $t \mapsto u_t$ is a continuous mapping from $\mathbf{R}^m$ to $\mathcal{D}'(\mathbf{R}^n)$ with $u_t = 0$ for all t outside a bounded subset of $\mathbf{R}^m$, we define the integral $\int u_t\, dt \in \mathcal{D}'(\mathbf{R}^n)$ by

$$\Big(\int_{\mathbf{R}^m} u_t\, dt\Big)(\phi) := \int_{\mathbf{R}^m} u_t(\phi)\, dt \qquad (\phi \in C_0^\infty(\mathbf{R}^n)).$$

Now prove that for $u \in \mathcal{D}'(\mathbf{R}^n)$ and $\phi \in C_0^\infty(\mathbf{R}^n)$,

$$u * \phi = \int_{\mathbf{R}^n} \phi(x)\, T_x u\, dx.$$

11.5.* Let $u \in \mathcal{D}'(\mathbf{R}^n)$ with $\partial_j u = 0$, for all $1 \le j \le n$. Prove the existence of a constant $c \in \mathbf{C}$ such that $u = c$. See Problem 12.9 for a generalization.

11.6.* Fix $u \in \mathcal{D}'(\mathbf{R}^n)$ and set

$$L = \{ P(\partial)\, u \in \mathcal{D}'(\mathbf{R}^n) \mid P \text{ polynomial function on } \mathbf{R}^n \}.$$

Prove that the linear subspace L of $\mathcal{D}'(\mathbf{R}^n)$ is of finite dimension if and only if u is an *exponential polynomial* on $\mathbf{R}^n$, that is, a function of the form $u = \sum_{k=1}^{l} P_k\, e_{a_k}$, where P_k is a polynomial function on $\mathbf{R}^n$ and $a_k \in \mathbf{C}^n$, while e_{a_k} is defined in (14.1), for $1 \le k \le l$.

11.7.* Let f be a polynomial function on $\mathbf{R}$ of degree $\le m$ and $T \in \mathcal{E}'(\mathbf{R})$. Show that $f * T$ is a polynomial function of degree $\le m$.

11.8.* Calculate $\delta_a * \delta_b$, for a and $b \in \mathbf{R}^n$.

11.9. Suppose that f and $g \in C(\mathbf{R}^n)$ and that f has compact support. Verify that $(\text{test}\, f) * (\text{test}\, g) = \text{test}(f * g)$.

11.10.* Let P be a linear partial differential operator in $\mathbf{R}^n$ with constant coefficients. Prove that $Pu = (P\delta) * u$ for every $u \in \mathcal{D}'(\mathbf{R}^n)$.

Also prove that $P(u * v) = (Pu) * v = u * (Pv)$ if u and $v \in \mathcal{D}'(\mathbf{R}^n)$ and if the sum mapping is proper on $\operatorname{supp} u \times \operatorname{supp} v$.

11.11.* Let $v \in \mathcal{E}'(\mathbf{R}^n)$. Prove that the convolution mapping $Sv*$ sends $C_0^\infty(\mathbf{R}^n)$ to $C_0^\infty(\mathbf{R}^n)$ and is continuous linear. Consider the transpose

$${}^t(Sv*) : \mathcal{D}'(\mathbf{R}^n) \to \mathcal{D}'(\mathbf{R}^n).$$

Prove that ${}^t(Sv*)u = u * v$, for all $u \in \mathcal{D}'(\mathbf{R}^n)$. See Example 15.11 for another proof.

11.12. Let E be the linear subspace of $\mathcal{D}'(\mathbf{R}^n)$ consisting of the finite linear combinations of the δ_a with $a \in \mathbf{R}^n$. Prove that for every continuous function f in $\mathbf{R}^n$ there exists a sequence $(u_j)_{j\in\mathbf{N}}$ in E such that $\lim_{j\to\infty} u_j = f$ in $\mathcal{D}'(\mathbf{R}^n)$. Hint: see Problem 5.12.

Now go on to show that for every $u \in \mathcal{D}'(\mathbf{R}^n)$ there exists a sequence (u_j) in E with $\lim_{j\to\infty} u_j = u$ in $\mathcal{D}'(\mathbf{R}^n)$. Given that the convolution of distributions is continuous in each of the variables, use this to prove, once again, the properties (11.24), (11.25), (11.23), (11.22), and (11.21).

11.13. Define $\mathcal{D}'(\mathbf{R})_+$ as the union of the $\mathcal{D}'(\mathbf{R})_{[l,\infty[}$ over all $l \in \mathbf{R}$. We say that $u_j \to u$ in $\mathcal{D}'(\mathbf{R})_+$ if there exists an $l \in \mathbf{R}$ with $u_j \to u$ in $\mathcal{D}'(\mathbf{R})_{[l,\infty[}$. Prove that for every u and $v \in \mathcal{D}'(\mathbf{R})_+$ the sum mapping is proper on $\operatorname{supp} u \times \operatorname{supp} v$ and additionally, that the convolution product

$$(u, v) \mapsto u * v \quad : \quad \mathcal{D}'(\mathbf{R})_+ \times \mathcal{D}'(\mathbf{R})_+ \to \mathcal{D}'(\mathbf{R})_+$$

satisfies all computational rules for the convolution product. In assertions concerning sequential continuity, one only needs to replace convergence in $\mathcal{D}'(\mathbf{R})$ by convergence in $\mathcal{D}'(\mathbf{R})_+$.

11.14.* Define, by mathematical induction on k, χ_+^k by $\chi_+^1 = H$, the Heaviside function, and $\chi_+^k = H * \chi_+^{k-1}$ if $k > 1$. Calculate all χ_+^k and all derivatives $(\chi_+^k)^{(l)}$. Prove that for every $k \in \mathbf{N}$ and $f \in \mathcal{D}'(\mathbf{R})_+$, there exists exactly one $u \in \mathcal{D}'(\mathbf{R})_+$ with $u^{(k)} = f$.

For $\phi \in C^k(\mathbf{R})$, verify the formula

$$(\phi\, H)^{(k)} = \sum_{j=0}^{k-1} \phi^{(k-j-1)}(0)\, \delta^{(j)} + \phi^{(k)}\, H.$$

Write out the formula resulting from the convolution of the left-hand side and the right-hand side, respectively, and χ_+^k. Do you recognize the result?

11.15.* Calculate $(1 * \delta') * H$ and $1 * (\delta' * H)$.

11.16. Prove that the tensor product of two probability measures is also a probability measure. Verify that for two probability measures μ and ν on $\mathbf{R}^n$ the convolution $\mu * \nu$ is well-defined and defines a probability measure on $\mathbf{R}^n$. The convolution $\mu * \nu$ is said to be the *independent sum of the probability measures* μ and ν. Do you recognize this from probability theory? Finally, calculate the probability measure $\mu * \mu * \cdots * \mu$, the independent sum of N copies of μ, if $\mu = p\,\delta_1 + (1-p)\,\delta_0$ and $0 \le p \le 1$.

11.17. Calculate the distribution $\delta'(x+y)\,\delta(x-y)$ on $\mathbf{R}^2$. Express the result in terms of the Dirac measure at $0 \in \mathbf{R}^2$ and its derivatives.

11.18. Let $u \in \mathcal{D}'(\mathbf{R}^n)$ and $x_n\, u = 0$. Show that the result from Problem 9.7 can be rephrased as the existence of a uniquely determined $v \in \mathcal{D}'(\mathbf{R}^{n-1})$ such that $u = v \otimes \delta^{\mathbf{R}}$ in $\mathcal{D}'(\mathbf{R}^n)$, where $\delta^{\mathbf{R}}$ denotes the Dirac measure on $\mathbf{R}$. (See Problem 15.7.(ii) for a different proof.)

11.19.* For $t \in \mathbf{R}$, consider the translation T_t in $\mathbf{R}^n$ given by $T_t : x = (x', x_n) \mapsto (x', x_n + t) \in \mathbf{R}^n \times \mathbf{R}$. Write $\mathbf{T} = (T_t)_{t \in \mathbf{R}}$ for the corresponding one-parameter group of translations. Suppose that $u \in \mathcal{D}'(\mathbf{R}^n)$ is invariant under the action induced by $\mathbf{T}$; in other words, it satisfies $(T_t)_* u = u$, for all $t \in \mathbf{R}$.

(i) Show that this is the case if and only if $\partial_n\, u = 0$.
(ii) Let $\pi : \mathbf{R}^n \to \mathbf{R}^{n-1}$ be the orthogonal projection $\pi : (x', x_n) \mapsto x'$, which corresponds to the decomposition $\mathbf{R}^n = \mathbf{R}^{n-1} \times \mathbf{R}$. Prove the existence of a $v \in \mathcal{D}'(\mathbf{R}^{n-1})$ such that $u = \pi^* v = v \otimes 1_{\mathbf{R}}$. (See Problem 15.7.(i) for a different proof.)

Background. The geometry underlying this result is as follows. Every orbit in $\mathbf{R}^n$ under the action of $\mathbf{T}$ is of the form $\{\, T_t(x', 0) = (x', t) \in \mathbf{R}^n \mid t \in \mathbf{R} \}$ for some

$x' \in \mathbf{R}^{n-1}$, and this set equals the level set $\pi^{-1}(\{x'\})$; geometrically, it is a line perpendicular to the hyperplane $\mathbf{R}^{n-1} \times \{0\}$. In other words, the set of orbits under $\mathbf{T}$ coincides exactly with the set of level sets in $\mathbf{R}^n$ of the surjective submersion π. Therefore, distributions on $\mathbf{R}^n$ that are invariant under $\mathbf{T}$ factorize through π, that is, they are pullbacks under π of distributions on $\mathbf{R}^{n-1}$.

11.20.* *(Distribution supported by linear subspace.)* Corresponding to the decomposition $\mathbf{R}^n = \mathbf{R}^p \times \mathbf{R}^q$ for p and $q > 0$, write $x = (y, z)$. Suppose $u \in \mathcal{E}'(\mathbf{R}^n)$ satisfies $\operatorname{supp} u \subset \mathbf{R}^p \times \{0\} \subset \mathbf{R}^n$. Set $A = \{0\} \times (\mathbf{Z}_{\geq 0})^q \subset (\mathbf{Z}_{\geq 0})^n$, let $\iota : \mathbf{R}^p \to \mathbf{R}^n$ be the natural embedding with $\iota(y) = (y, 0)$, and $\delta^{\mathbf{R}^q}$ the Dirac measure supported at $0 \in \mathbf{R}^q$. Prove the existence of $u_\alpha \in \mathcal{E}'(\mathbf{R}^p)$, with $\alpha \in A$, such that

$$u = \sum_{\alpha \in A} (\partial^\alpha \circ \iota_*) u_\alpha = \sum_{\alpha \in (\mathbf{Z}_{\geq 0})^q} (I \otimes \partial^\alpha) \iota_* u_\alpha = \sum_{\alpha \in (\mathbf{Z}_{\geq 0})^q} u_\alpha \otimes \partial^\alpha \delta^{\mathbf{R}^q},$$

where the sum is actually finite. (See Problem 15.8 for another proof.)

In other words, u is a finite sum of *transversal derivatives* of compactly supported distributions on $\mathbf{R}^p$. Any linear subspace in $\mathbf{R}^n$ of dimension p can be transformed into $\mathbf{R}^p \times \{0\}$ by means of a rotation about the origin.

11.21. In this problem we use the notation from Example 14.22. Let X be an open subset of $\mathbf{R}^n$ and consider p and $p' \geq 1$ satisfying $1/p + 1/p' = 1$, on the understanding that $p' = \infty$ if $p = 1$ and $p' = 1$ if $p = \infty$. For all $f \in L^p(X)$ and $g \in L^{p'}(X)$, prove *Hölder's inequality*

$$\int_X |f(x) g(x)| \, dx \leq \|f\|_{L^p(X)} \, \|g\|_{L^{p'}(X)}.$$

Conclude that $fg \in L^1(X)$. Hint: see [7, Exercise 6.73.(i)].

11.22. *(Young's inequality.)* In this problem we use the notation from Example 14.22. Let p, q, and $r \geq 1$ and $1/p + 1/q + 1/r = 2$. Then we have *Young's inequality*, which asserts, for all f, g, and $h \in C_0(\mathbf{R}^n)$,

$$\left| \int_{\mathbf{R}^n} f(x) \, (g * h)(x) \, dx \right| \leq \|f\|_{L^p} \, \|g\|_{L^q} \, \|h\|_{L^r}.$$

Indeed, we are free to assume that f, g, and h are real and nonnegative. Introduce $p' \geq 1$ by $1/p + 1/p' = 1$, and similarly q' and r'. Write the integral on the left-hand side as

$$I = \int_{\mathbf{R}^n} \int_{\mathbf{R}^n} f(x) \, g(x-y) \, h(y) \, dy \, dx = \int_{\mathbf{R}^n} \int_{\mathbf{R}^n} a(x,y) \, b(x,y) \, c(x,y) \, dx \, dy$$

with

$$a(x,y) = f(x)^{p/r'} \, g(x-y)^{q/r'},$$
$$b(x,y) = g(x-y)^{q/p'} \, h(y)^{r/p'},$$
$$c(x,y) = f(x)^{p/q'} \, h(y)^{r/q'}.$$

Noting that $1/p' + 1/q' + 1/r' = 1$, we can use Hölder's inequality from Problem 11.21 for three functions to obtain $|I| \leq \|a\|_{L^{r'}} \|b\|_{L^{p'}} \|c\|_{L^{q'}}$. But

$$\|a\|_{L^{r'}} = \Big(\int_{\mathbf{R}^n}\int_{\mathbf{R}^n} f(x)^p\, g(x-y)^q\, dx\, dy\Big)^{1/r'} = \|f\|_{L^p}^{p/r'}\, \|g\|_{L^q}^{q/r'},$$

and similarly for b and c. The second equality above is a consequence of changing variables from y to $x - y$ and integrating first with respect to y. This leads to the desired inequality.

Now prove the following. For f and $g \in C_0(\mathbf{R}^n)$ and p, q and $r \geq 1$ satisfying $1 + 1/p = 1/q + 1/r$, one has the inequality

$$\|f * g\|_{L^p} \leq \|f\|_{L^q}\, \|g\|_{L^r}, \qquad \text{in particular} \qquad \|f * g\|_{L^p} \leq \|f\|_{L^p}\, \|g\|_{L^1}.$$

In order to obtain this estimate, apply Young's inequality with $f = s\,|g * h|^{p'/p}$, where s is the function defined by $s\,(g * h) = |g * h|$. This implies

$$\|g * h\|_{L^{p'}} \leq \|g\|_{L^q}\, \|h\|_{L^r}, \qquad \text{where} \qquad 1 + \frac{1}{p'} = \frac{1}{q} + \frac{1}{r}.$$

Then replace p' by p.

Finally, prove the validity of the estimates if the functions belong to the appropriate spaces of type $L^p(\mathbf{R}^n)$.

Chapter 12
Fundamental Solutions

Definition 12.1. Let $P = P(\partial) = \sum_{|\alpha| \leq m} c_\alpha \, \partial^\alpha$ be a linear partial differential operator in $\mathbf{R}^n$ with constant coefficients, as introduced in (7.5). A *fundamental solution* of P is a distribution $E \in \mathcal{D}'(\mathbf{R}^n)$ such that $PE = \delta$, the Dirac measure at the origin. ⊘

Every linear partial differential operator with constant coefficients (not all of them equal to 0) has a fundamental solution. For more on this, see below: Theorem 17.11, Remarks 17.12 and 18.9 for special cases and Theorem 18.4 for the general result. For linear partial differential operators with variable coefficients the existence of fundamental solutions need not be the case.

The importance of fundamental solutions lies in the following:

Theorem 12.2. *Suppose E is a fundamental solution of P. Then we have*

$$P(E * f) = f \qquad (f \in \mathcal{E}'(\mathbf{R}^n)), \tag{12.1}$$

$$u = E * Pu \qquad (u \in \mathcal{E}'(\mathbf{R}^n)). \tag{12.2}$$

Proof. Using (11.25) and (11.19), we obtain $P(E * f) = (PE) * f = \delta * f = f$, and therefore (12.1). Combination of (11.25) and (12.1) yields that $E * (Pu) = P(E * u) = u$, which implies (12.2). □

For every distribution f on $\mathbf{R}^n$ with compact support, formula (12.1) implies the existence of a distributional solution $u = E * f \in \mathcal{D}'(\mathbf{R}^n)$ of the inhomogeneous linear partial differential equation $Pu = f$. Additionally, under the assumption that the solution u of $Pu = f$ has compact support, (12.2) means that the solution is uniquely determined and given by $u = E * f$.

A word of warning: it may seem as if $E\,*$ is a two-sided inverse of P, which would imply that P is bijective. But that is not actually the case, because the domain spaces do not correspond: generally speaking, we do not know more about $E\,*$ than that it is a mapping from $\mathcal{E}'(\mathbf{R}^n)$ to $\mathcal{D}'(\mathbf{R}^n)$. The differential operator P maps $\mathcal{D}'(\mathbf{R}^n)$ to $\mathcal{D}'(\mathbf{R}^n)$ and $\mathcal{E}'(\mathbf{R}^n)$ to $\mathcal{E}'(\mathbf{R}^n)$, but not $\mathcal{D}'(\mathbf{R}^n)$ to $\mathcal{E}'(\mathbf{R}^n)$. In general,

J.J. Duistermaat and J.A.C. Kolk, *Distributions: Theory and Applications*,
Cornerstones, DOI 10.1007/978-0-8176-4675-2_12,

partial differential operators are far from injective, as we will see in the next example and in Problem 14.3 below.

Another consequence of this is that fundamental solutions are not uniquely determined. If E is a fundamental solution of P, then $\widetilde{E}$ is a fundamental solution of P if and only if $\widetilde{E} = E + u$, with $u \in \mathcal{D}'(\mathbf{R}^n)$ a solution of $Pu = 0$.

Example 12.3. Let $P = \Delta = \sum_{j=1}^n \partial_j^2$ be the *Laplace operator* in $\mathbf{R}^n$. A C^2 function u is said to be *harmonic* on the open subset U of $\mathbf{R}^n$ if $\Delta u = 0$ on U; this terminology is carried over to distributions $u \in \mathcal{D}'(U)$. In other words, the kernel of Δ consists of the harmonic functions, or distributions, respectively. If $n > 1$, then the harmonic polynomials on $\mathbf{R}^n$ form a linear space of infinite dimension. Indeed, the functions $x \mapsto (x_1 + i\, x_2)^k$ on $\mathbf{R}^n$ are harmonic, for all $k \in \mathbf{Z}_{\geq 0}$. (If $n = 1$ and U is an interval, then $u \in \mathcal{D}'(U)$ and $\Delta u = 0$ if and only if u on U is equal to a polynomial function of degree ≤ 1.)

In Problem 4.7 a fundamental solution of the Laplace operator was found to be the locally integrable function E on $\mathbf{R}^n$ that on $\mathbf{R}^n \setminus \{0\}$ is given by

$$E(x) = \begin{cases} \dfrac{1}{(2-n)\, c_n\, \|x\|^{n-2}} & \text{if} \quad n \neq 2, \\[2ex] \dfrac{1}{2\pi} \log \|x\| & \text{if} \quad n = 2. \end{cases} \tag{12.3}$$

For c_n, see also (13.37) below. For $f \in \mathcal{E}'(\mathbf{R}^n)$, the distribution $u = E * f$ is said to be the *potential* of the distribution f, a terminology that has its origin in the situation in which $n = 3$ and f denotes a mass or charge density. We conclude that the potential u of f satisfies *Poisson's equation*

$$\Delta u = f \quad \text{in} \quad \mathcal{D}'(\mathbf{R}^n);$$

and in particular, the potential u is harmonic on the complement of supp f, the largest open set on which $f = 0$. Furthermore, if u is a solution of Poisson's equation that has compact support, then u necessarily equals the potential of f.

The general fundamental solution of the Laplace operator is equal to the sum of the δ-potential and a harmonic function on $\mathbf{R}^n$. If $n \geq 3$, then E can be characterized as the fundamental solution of Δ that converges to 0 as $\|x\| \to \infty$. This can be proved by means of Fourier transform; see Problem 17.4 below. ⊘

As far as the study of the singular supports (see Definition 7.8) of solutions is concerned, addition of C^∞ functions is irrelevant. This can also be expressed by saying that calculations are performed *modulo* C^∞ when summands of class C^∞ are neglected. A distribution E on $\mathbf{R}^n$ is said to be a *parametrix* of P if there exists a $\psi \in C^\infty(\mathbf{R}^n)$ such that

$$PE = \delta + \psi, \tag{12.4}$$

in other words, if E modulo C^∞ satisfies the equation for a fundamental solution. For certain P a parametrix can be obtained by iterative methods.

Theorem 12.4. *Suppose that* P *possesses a parametrix* E *with* $\operatorname{sing\,supp} E = \{0\}$. *Then, for every open subset* X *of* $\mathbf{R}^n$,

$$\operatorname{sing\,supp} u = \operatorname{sing\,supp} Pu \qquad (u \in \mathcal{D}'(X)). \tag{12.5}$$

Proof. We see immediately that $\operatorname{sing\,supp} Pu \subset \operatorname{sing\,supp} u$, which is the case for every linear partial differential operator P in X with C^∞ coefficients. We will now prove the converse inclusion.

If u has compact support, we can interpret u as an element of $\mathcal{E}'(\mathbf{R}^n)$ according to Lemma 8.9. We have

$$E * Pu = (PE) * u = \delta * u + \psi * u.$$

Since $\delta * u = u$ and $\psi * u \in C^\infty(\mathbf{R}^n)$ by Theorem 11.2, we obtain, using (11.32),

$$\begin{aligned}\operatorname{sing\,supp} u &= \operatorname{sing\,supp} E * Pu \subset \operatorname{sing\,supp} E + \operatorname{sing\,supp} Pu \\ &= \{0\} + \operatorname{sing\,supp} Pu = \operatorname{sing\,supp} Pu.\end{aligned}$$

Now let $u \in \mathcal{D}'(X)$ be arbitrary and $x \in X \setminus \operatorname{sing\,supp} Pu$. Choose $\chi \in C_0^\infty(X)$ with $\chi = 1$ on an open neighborhood U of x. Then $P(\chi u) = Pu$ on U, and therefore $x \notin \operatorname{sing\,supp} P(\chi u)$. The above yields $\operatorname{sing\,supp} P(\chi u) = \operatorname{sing\,supp}(\chi u)$, because χu has compact support. Hence $x \notin \operatorname{sing\,supp}(\chi u)$, which in turn implies that $x \notin \operatorname{sing\,supp} u$, because $\chi = 1$ on a neighborhood of x. □

The theorem means that if $u \in \mathcal{D}'(X)$ is a solution of the partial differential equation $Pu = f$, then $u \in C^\infty$ on every open subset U of X where $f \in C^\infty$. A linear partial differential operator P with C^∞ coefficients is said to be *hypoelliptic* if P has this property, that is, if (12.5) holds. If E is a parametrix of a hypoelliptic operator P with constant coefficients, then necessarily $\operatorname{sing\,supp} E = \operatorname{sing\,supp} PE = \operatorname{sing\,supp} \delta = \{0\}$.

The term "hypoelliptic" conveys the fact that this condition is weaker than the condition that P is elliptic; see Theorem 17.6. For the definition of the term "elliptic," see Definition 17.2.

Remark 12.5. If all derivatives can be estimated such that convergence of each of the Taylor series is guaranteed, the proof of Theorem 12.4 can be modified so as to yield results on the real-analyticity of solutions. Thus one has that u is real-analytic wherever $Pu = 0$, if there exist $E \in \mathcal{D}'(\mathbf{R}^n)$ and an open neighborhood U of 0 in $\mathbf{R}^n$ with the following properties:

(a) E is real-analytic on $U \setminus \{0\}$,
(b) $PE - \delta$ is real-analytic on U.

Estimation of all derivatives can be circumvented by means of complex analysis; see Remark 12.13. This is then followed by the proof of this assertion. ⊘

Remark 12.6. For linear partial differential operators $P = P(x, \partial)$ with variable coefficients there is a well-developed theory in which the convolution operator $E\,*$, defined by

$$(E * f)(x) = \int E(x-y)\, f(y)\, dy,$$

is replaced by a *singular integral operator* K of the form

$$(Kf)(x) = \int k(x,y)\, f(y)\, dy.$$

Here the *integral kernel* $k(x, y)$ is a distribution on $X \times X$ if the open set X in $\mathbf{R}^n$ denotes the domain space of the functions f and Kf; see Chap. 15 below. The *operator* K is then said to be a *parametrix* of P if $P \circ K = I + R$ and $K \circ P = I + S$, where R and S are integral operators with a C^∞ kernel on $X \times X$. It is not difficult to prove that P is hypoelliptic when P has a parametrix K with

$$\operatorname{sing\,supp} k \subset \{\, (x,y) \in X \times X \mid x = y \,\}.$$

However, the **construction** of parametrices, for sufficiently general operators with variable coefficients, involves too much work to be dealt with in this text. ⊘

Example 12.7. Every ordinary differential operator (the case $n = 1$) is hypoelliptic in the open set where the coefficient of the highest-order term does not vanish. See Theorem 9.4. ⊘

Example 12.8. The fundamental solution E of the Laplace operator given in (12.3) satisfies $\operatorname{sing\,supp} E = \{0\}$; consequently, the Laplace operator is hypoelliptic. In particular, every harmonic distribution is a harmonic C^∞ function.

The solution E is not only C^∞, but even real-analytic on the complement of the origin. Remark 12.5 therefore implies that u is real-analytic wherever $\Delta u = 0$. In particular, every harmonic function and distribution is real-analytic. ⊘

A note on history: Weyl [24, Lemma 2] proved that every square-integrable function u such that $\int u(x)\, \Delta\phi(x)\, dx = 0$ for every $\phi \in C_0^2$ is in fact C^∞. This assertion is known as *Weyl's Lemma* and has become the prototype of *Regularity Theorems* like Theorem 12.4. Weyl's proof follows the same lines as the proof of Theorem 12.4 given here.

Example 12.9. In Problem 8.5 a fundamental solution E was found for the operator $P = \partial_t - \Delta_x$ in $(n+1)$-dimensional (x, t)-space. This, too, satisfies $\operatorname{sing\,supp} E = \{0\}$; consequently, the heat operator, too, is hypoelliptic. If the distribution u is a solution of the heat equation $Pu = 0$ on an open subset U of $\mathbf{R}^{n+1}$, then u is a C^∞ function on U. However, in the present case E is not analytic on the entire plane of the (x, t) with $t = 0$, and here it is not true that u is analytic wherever $Pu = 0$. P is called the *n-dimensional heat operator*, although this operator is defined in $\mathbf{R}^{n+1}$. This derives from the interpretation of $x = (x_1, \ldots, x_n)$ as the position coordinates and t as the time coordinate, which are often treated as playing different roles. ⊘

Example 12.10. Problems 10.25 and 12.7 describe $u \in \mathcal{D}'(\mathbf{R}^2)$ that satisfy the equation $(\partial_1^2 - \partial_2^2)u = 0$ in $\mathbf{R}^2$ but are far from C^∞ functions. This shows that the *wave operator* $\partial_1^2 - \partial_2^2$ is not hypoelliptic. Indeed, for every $n \geq 1$ the *n-dimensional wave operator* $\square := \partial_t^2 - \Delta_x$, with $x \in \mathbf{R}^n$, is not hypoelliptic; see Theorem 13.3. ⊘

Example 12.11. We now consider complex analysis from a distributional point of view.

A function f defined on an open subset V of $\mathbf{C}$ is called *complex-differentiable* at the point $z \in V$ when $\frac{f(z+h)-f(z)}{h}$ converges as $h \to 0$ in $\mathbf{C}$. The limit is said to be the *complex derivative* $f'(z) \in \mathbf{C}$ of f at z. If we apply the identification $\mathbf{C} \simeq \mathbf{R}^2$, by writing $z \in \mathbf{C}$ as $z = x + iy$ in the usual way, with x and $y \in \mathbf{R}$, then this condition is stronger than the condition that f is differentiable as a complex-valued function of two real variables (x, y). Indeed, when f depends on the combination $z = x + iy$, its partial derivatives satisfy

$$\partial_x f(z) = f'(z)\,\partial_x z = f'(z) \qquad \text{and} \qquad \partial_y f(z) = f'(z)\,\partial_y z = i\,f'(z).$$

This means that f is complex-differentiable if and only if f is real-differentiable and if, moreover,

$$\partial_y f = i\,\partial_x f \qquad \textit{(Cauchy–Riemann equation).} \tag{12.6}$$

It is then natural to call a distribution $u \in \mathcal{D}'(V)$ *complex-differentiable* if u satisfies (12.6); in other words, if $Pu = 0$, with P equal to the first-order linear partial differential operator with constant coefficients

$$P = i\,\partial_x - \partial_y = i(\partial_x + i\,\partial_y).$$

The formula for integration by parts in $\mathbf{R}^n$ reads, with $1 \leq j \leq n$,

$$\int_U f(x)\,\partial_j g(x)\,dx = -\int_U g(x)\,\partial_j f(x)\,dx + \int_{\partial U} f(y)\,g(y)\,\nu_j(y)\,d_{\mathrm{Eucl}}y. \tag{12.7}$$

Here U is an open subset of $\mathbf{R}^n$ with C^1 boundary ∂U and with the property that U lies at one side of ∂U. Furthermore, f and $g \in C^1(\overline{U})$ and $(\operatorname{supp} f) \cap \overline{U}$, for example, is compact. Finally, $\nu_j(y)$ is the jth component of the outer normal to ∂U at y. See, for example, [7, Corollary 7.6.2].

In $\mathbf{R}^2 \simeq \mathbf{C}$ we can write $i\,\nu_1 - \nu_2 = i(\nu_1 + i\,\nu_2) = i\,\nu$, where the vector $\nu(y)$ is now interpreted as a complex number. The vector $i\,\nu(y)$ is equal to the tangent vector to ∂U at y of length 1, oriented to have U to the left. If ∂U is locally parametrized by a C^1 curve $\gamma : [a, b] \to \mathbf{C}$, where $\gamma'(t)$ has the same orientation as $i\,\nu(\gamma(t))$, one has, for a continuous function f with support in $\gamma([a, b])$,

$$\int_{\partial U} f(z)\,i\,\nu(z)\,d_{\mathrm{Eucl}}z = \int_a^b f(\gamma(t))\,\gamma'(t)\,dt.$$

Here the vector $\gamma'(t)$ is also interpreted as a complex number. This is said to be the *complex line integral* of f over ∂U with respect to the orientation of ∂U described above, and is denoted by

$$\int_{\partial U} f(z)\,dz \qquad \text{with} \qquad dz = i\,\nu(z)\,d_{\mathrm{Eucl}}z.$$

Thus we obtain for the operator P the following transposition formula:

$$\int_U f(z)\,Pg(z)\,dz = -\int_U g(z)\,Pf(z)\,dz + \int_{\partial U} f(z)\,g(z)\,dz. \tag{12.8}$$

In this formula, U, f, and g are as above, with $n = 2$. The integrals over U are Euclidean 2-dimensional integrals, while that over ∂U is a complex line integral.

For a first application of (12.8) we assume that U is bounded, that f is also a complex-differentiable function on U, and that g equals the constant function 1. We then immediately get the following version of *Cauchy's Integral Theorem*:

$$\int_{\partial U} f(z)\,dz = 0. \tag{12.9}$$

In order to find a fundamental solution of P, we consider the function $z \mapsto \frac{1}{z}$ on $\mathbf{C} \setminus \{0\}$. This function is locally integrable on $\mathbf{C} = \mathbf{R}^2$ (use polar coordinates) and can therefore be interpreted as a distribution. From (12.6) it is evident that $P\frac{1}{z} = 0$ on $\mathbf{R}^2 \setminus \{0\}$. For $\phi \in C_0^\infty(\mathbf{R}^2)$ we obtain, applying (12.8),

$$\begin{aligned}\Big(P\frac{1}{z}\Big)(\phi) &= -\frac{1}{z}(P\phi) = -\lim_{\epsilon\downarrow 0}\int_{|z|>\epsilon}\frac{P\phi(z)}{z}\,dz = -\lim_{\epsilon\downarrow 0}\int_{|z|=\epsilon}\frac{\phi(z)}{z}\,dz\\ &= -\lim_{\epsilon\downarrow 0}\int_0^{2\pi}\frac{\phi(\epsilon\, e^{-it})}{\epsilon\, e^{-it}}\,\epsilon\, e^{-it}\,(-i)\,dt = 2\pi i\,\phi(0).\end{aligned}$$

Here we use that the outer normal to $\{\, z \in \mathbf{C} \mid |z| = \epsilon \,\}$ points toward the origin; therefore this circle has to be traversed clockwise, which is the case under the mapping $t \mapsto \epsilon\, e^{-it}$. We conclude that $P\frac{1}{z} = 2\pi i\,\delta$, in other words, $E(z) = \frac{1}{2\pi i\, z}$ defines a fundamental solution of P.

Because this fundamental solution is C^∞, and even complex-analytic, outside the origin, we conclude on the strength of Theorem 12.4 that P is hypoelliptic. It follows that in particular, **every complex-differentiable distribution on an open set V equals a C^∞ function on V** and as such is complex-differentiable in the classical sense.

Now let U be an open subset of V with C^1 boundary ∂U. Assume that $\overline{U}$ is a compact subset of V, $f \in C^1(V)$, and $g \in C_0^\infty(V)$. Interpreting the function g in (12.8) as a test function and writing the left-hand side as $(f\,1_U)(Pg) = -P(f\,1_U)(g)$, we can rewrite (12.8) as the identity

$$P(f\,1_U) = (Pf)\,1_U - f\,\delta_{\partial U}^{\mathrm{compl}}$$

in the space of distributions on V of order ≤ 1 with compact support, where $\delta_{\partial U}^{\text{compl}}$ denotes the complex line integration over the boundary. If we now apply the convolution operator $E\,*$ to this identity, and recall (12.2), the left-hand side becomes equal to $f\,1_U$; and thus we obtain *Pompeiu's integral formula*

$$f(\zeta) = \frac{1}{2\pi i}\int_U \frac{Pf(z)}{\zeta - z}\,dz + \frac{1}{2\pi i}\int_{\partial U} \frac{f(z)}{z-\zeta}\,dz \qquad (\zeta \in U). \tag{12.10}$$

Here we applied (11.30) and (11.31) for obtaining the first and second terms on the right-hand side, respectively. In particular, if f is complex-differentiable on V, that is, $Pf = 0$ and $f \in C^\infty(V)$, one obtains the well-known *Cauchy integral formula*

$$f(\zeta) = \frac{1}{2\pi i}\int_{\partial U} \frac{f(z)}{z-\zeta}\,dz \qquad (\zeta \in U). \tag{12.11}$$

Thus, f is expressed on U in terms of the restriction $f|_{\partial U}$ of f to the boundary ∂U. Not every analytic function g on ∂U is of the form $g = f|_{\partial U}$ for a complex-analytic function f on U; see Problem 16.13.

Cauchy's integral formula gives an arbitrary complex-differentiable function f as a "continuous linear combination" of the very simple complex-differentiable functions $z \mapsto \frac{1}{z-\zeta}$, where the variable z runs over the boundary ∂U. The singularity at $z = \zeta$ need not bother us here, provided we ensure that $\zeta \in U$ steers clear of ∂U. In particular, for a given point $a \in U$ and with ζ in a sufficiently small neighborhood $U(a)$ of a, we can substitute the power series

$$\frac{1}{z-\zeta} = \frac{1}{(z-a)-(\zeta-a)} = \frac{1}{z-a}\,\frac{1}{1-\frac{\zeta-a}{z-a}} = \sum_{k\in\mathbf{Z}_{\geq 0}} \frac{(\zeta-a)^k}{(z-a)^{k+1}}$$

into (12.11). This implies that the complex-differentiable f can be expressed by a convergent complex power series in a neighborhood of a. More precisely,

$$f(\zeta) = \sum_{k\in\mathbf{Z}_{\geq 0}} \Big(\frac{1}{2\pi i}\int_{\partial U} \frac{f(z)}{(z-a)^{k+1}}\,dz\Big)(\zeta-a)^k \qquad (\zeta \in U(a)).$$

Conversely, it is known that every convergent complex power series is complex-differentiable and even infinitely differentiable, and the power series is equal to the Taylor series. See any textbook about analysis in one variable. It follows that the conditions "complex-differentiable distribution" and "is locally equal to a convergent complex power series" are equivalent. In this case, f is said to be *complex-analytic*.

Instead of working with the operator $P = i\,\partial_x - \partial_y$, it is more usual in the literature to use

$$\partial_z := \frac{1}{2}\,(\partial_x - i\,\partial_y) \qquad \text{and} \qquad \partial_{\bar z} := \frac{1}{2}\,(\partial_x + i\,\partial_y) = \frac{1}{2i}\,P. \tag{12.12}$$

That makes (12.6) equivalent to $\partial_{\bar{z}} f = 0$. If this is the case, then $\frac{df}{dz} = \partial_z f$. The operator $\partial_{\bar{z}}$ is said to be the *Cauchy–Riemann operator;* it has the fundamental solution $\frac{1}{\pi z}$, and in the literature (12.10) is mostly written in the form

$$f(\zeta) = -\frac{1}{\pi} \int_U \frac{\partial_{\bar{z}} f(z)}{z - \zeta}\, dz + \frac{1}{2\pi i} \int_{\partial U} \frac{f(z)}{z - \zeta}\, dz \qquad (\zeta \in U). \tag{12.13}$$

In several variables one has an analogous theory: a function $z \mapsto f(z)$ on $\mathbf{C}^n$ is complex-differentiable if f is complex-differentiable as a function of each of the variables z_j, that is, if $\partial_{\bar{z}_j} f = 0$, for $1 \le j \le n$. These are referred to as the Cauchy–Riemann equations for f. It can be proved that a distributional solution of these equations is of class C^∞ and can locally be developed into power series, where the proof makes use of a higher-dimensional version of Cauchy's integral formula. These functions are said to be *complex-analytic in several variables.* ⊘

Remark 12.12. The notation becomes clear when we write $f(x, y) = F(z, \bar{z})$, with $z = x + iy$ and $\bar{z} = x - iy$. This yields

$$\partial_x f = \partial_z F + \partial_{\bar{z}} F \qquad \text{and} \qquad \partial_y f = i\, \partial_z F - i\, \partial_{\bar{z}} F.$$

If we now solve this system for $\partial_z F$ and $\partial_{\bar{z}} F$ in terms of $\partial_x f$ and $\partial_y f$, we obtain formulas that we recognize as the definitions of ∂_z and $\partial_{\bar{z}}$. ⊘

Remark 12.13. In Lemma 2.6, a function f on an open subset U of $\mathbf{R}^n$ was said to be analytic on U if f has local power series representations at all points of U. To distinguish it from the above, such f is also said to be *real-analytic* on U. By substituting complex values for the variables into the power series, we see that f has a complex-analytic extension to an open neighborhood V of U in $\mathbf{C}^n \simeq \mathbf{R}^{2n}$. Conversely, the restriction to $U := V \cap \mathbf{R}^n$ of a complex-analytic function on V is a real-analytic function on U.

It follows that a function f is real-analytic if f can be extended to a function on a complex neighborhood that satisfies the Cauchy–Riemann equations on that neighborhood. In many cases, this involves less effort than working out the estimates for all derivatives required to establish the convergence of the Taylor series. ⊘

Finally, we come back to the proof of the assertion in Remark 12.5. We observe that the *analytic singular support* $\operatorname{sing\,supp}_{\text{anal}}(u)$ may be defined in the obvious manner, for any $u \in \mathcal{D}'(\mathbf{R}^n)$.

Lemma 12.14. *Let $u \in \mathcal{D}'(\mathbf{R}^n)$ and $v \in \mathcal{E}'(\mathbf{R}^n)$. Then we have*

$$\operatorname{sing\,supp}_{\text{anal}}(u * v) \subset \operatorname{sing\,supp}_{\text{anal}} u + \operatorname{supp} v.$$

Proof. Set $U_0 = \mathbf{R}^n \setminus \operatorname{sing\,supp}_{\text{anal}} u$ and denote by u_0 the restriction of u to U_0. The function u_0 has a complex-analytic extension to an open subset $U \subset \mathbf{C}^n$

satisfying $U \cap \mathbf{R}^n = U_0$. Selecting $a \in \mathbf{R}^n$ with $a \notin \operatorname{sing\,supp}_{\text{anal}} u + \operatorname{supp} v$, we have $a + (-\operatorname{supp} v) \subset U_0$. Let V be an open neighborhood of a in $\mathbf{C}^n$ such that $V + (-\operatorname{supp} v) \subset U$. For $z \in V$, write

$$u_0 * v(z) = v(u_0(z - \cdot)) = \int_{\mathbf{R}^n} u_0(z - x)\, v(x)\, dx.$$

This equality should be interpreted in the sense of distributions, and it defines a complex-analytic function on V. We note that on $V \cap \mathbf{R}^n$, the function $u_0 * v$ coincides with the usual convolution product $u * v$ as in (11.27). Thus, the latter is analytic on the open neighborhood $V \cap \mathbf{R}^n$ of a in $\mathbf{R}^n$; this implies $a \notin \operatorname{sing\,supp}_{\text{anal}}(u * v)$. □

Theorem 12.15. *Let P be a linear partial differential operator in $\mathbf{R}^n$ with constant coefficients. Suppose there exist $E \in \mathcal{D}'(\mathbf{R}^n)$ and an open neighborhood U of 0 in $\mathbf{R}^n$ with the following properties:*

(a) E is analytic on $U \setminus \{0\}$,
(b) $F := PE - \delta$ is analytic on U.

We then have, for every open subset $X \subset \mathbf{R}^n$ and $u \in \mathcal{D}'(X)$, that u is analytic on X if $Pu = 0$ on X.

Proof. We fix a point $x \in X$ and will prove that u is analytic on a neighborhood of x in X. To this end, we select $\epsilon > 0$ such that the open ball $B(0;\epsilon)$ is contained in U, while $B(x;\epsilon) \subset X$. Furthermore, we consider $\chi \in C_0^\infty(B(x;\epsilon))$ satisfying $\chi = 1$ on an open neighborhood of x in X. Then we have $P(\chi u) = f$ with $f \in \mathcal{E}'(\mathbf{R}^n)$ and

$$\operatorname{supp} f \subset \operatorname{supp} D\chi \subset \mathring{B}(x;\epsilon) := B(x;\epsilon) \setminus \{x\}. \tag{12.14}$$

We obtain

$$E * f = E * P(\chi u) = PE * (\chi u) = F * (\chi u) + \chi u,$$

which shows that it is sufficient to verify

(i) $x \notin \operatorname{sing\,supp}_{\text{anal}}(E * f)$,
(ii) $x \notin \operatorname{sing\,supp}_{\text{anal}}(F * (\chi u))$.

(i). On account of Lemma 12.14 and (12.14) we have

$$\operatorname{sing\,supp}_{\text{anal}}(E * f) \subset \operatorname{sing\,supp}_{\text{anal}} E + \operatorname{supp} f \subset \operatorname{sing\,supp}_{\text{anal}} E + \mathring{B}(x;\epsilon).$$

Now $\operatorname{sing\,supp}_{\text{anal}} E \cap \mathring{B}(0;\epsilon) = \emptyset$ on the strength of condition (a); therefore $0 \notin \operatorname{sing\,supp}_{\text{anal}} E + \mathring{B}(0;\epsilon)$, which implies $x \notin \operatorname{sing\,supp}_{\text{anal}} E + \mathring{B}(x;\epsilon)$. In other words, (i) is valid.

(ii). We note that again on account of Lemma 12.14,

$$\operatorname{sing\,supp}_{\text{anal}}(F * (\chi u)) \subset \operatorname{sing\,supp}_{\text{anal}} F + \operatorname{supp} \chi \subset \operatorname{sing\,supp}_{\text{anal}} F + B(x;\epsilon).$$

Furthermore, condition (b) gives $\operatorname{sing\,supp}_{\text{anal}} F \cap B(0;\epsilon) = \emptyset$, which leads to $0 \notin \operatorname{sing\,supp}_{\text{anal}} F + B(0;\epsilon)$; hence $x \notin \operatorname{sing\,supp}_{\text{anal}} F + B(x;\epsilon)$ and we conclude that (ii) holds too. □

Problems

12.1. Determine all fundamental solutions $E_k \in \mathcal{D}'(\mathbf{R})$ of $P = \partial^k$, for $k \in \mathbf{N}$. Which of these are homogeneous? Of what degree?

12.2.* *(Dipoles.)* The distribution $\sum_j v_j\, \partial_j \delta_a$ is said to be a *dipole* at a point a with *dipole vector* v. Calculate the potential in $\mathbf{R}^3$ of the dipole at the origin, with dipole vector equal to the first basis vector. Sketch the corresponding level curves in the (x_1, x_2)-plane. What do the level curves of the potential of δ look like? Also work out this problem for the potential in $\mathbf{R}^2$. See Fig. 12.1.

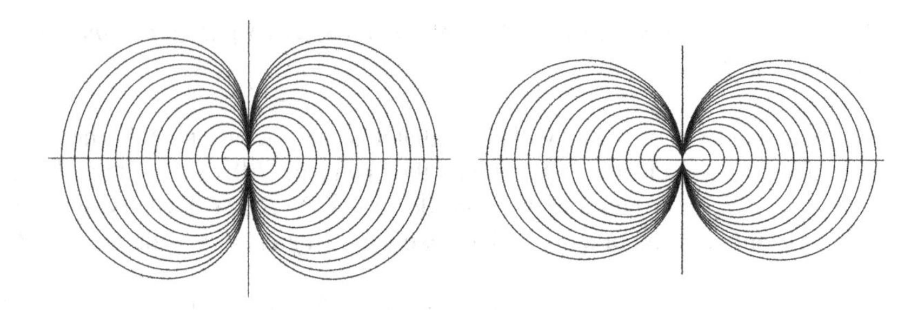

Fig. 12.1 Illustration for Problem 12.2. Equipotential curves in the (x_1, x_2)-plane in $\mathbf{R}^3$ and in $\mathbf{R}^2$, respectively

12.3.* *(Potential of a bar.)* Calculate the potential u in $\mathbf{R}^3$ of the distribution f defined by

$$f(\phi) = \int_{-a}^{a} \phi(x_1, 0, 0)\, dx_1 \qquad (\phi \in C_0^\infty(\mathbf{R}^3)),$$

a bar of length $2a$. Indicate where the distribution u is harmonic. Is u a (locally integrable) function? Determine $\operatorname{sing\,supp} u$. How does u behave as $a \to \infty$?

12.4.* *(Green's formula and Dirichlet's problem.)* Suppose $E \in \mathcal{D}'(\mathbf{R}^n)$ is a fundamental solution of the Laplace operator Δ and consider a harmonic function u on the open subset X of $\mathbf{R}^n$. Let U be an open and $\overline{U}$ a compact subset of X. Then prove that $v := \Delta(u\, 1_U)$ is a distribution on X with $\operatorname{supp} v \subset \partial U$, the boundary of U. Also prove that $v \in \mathcal{E}'(X)$ and $u\, 1_U = E * v$.

Now assume that the boundary ∂U is of class C^1 and denote the outer normal to the boundary at the point $y \in \partial U$ by $\nu(y)$. For a C^1 function f on a neighborhood of ∂U, the *normal derivative* $\partial_\nu f$ of f is defined by

$$\partial_\nu f(y) := \sum_{j=1}^{n} \nu_j(y)\, \partial_j f(y) \qquad (y \in \partial U).$$

This is a continuous function on ∂U. Now prove *Green's formula*

$$u(x) = \int_{\partial U} \big(u(y)\, \partial_\nu(y \mapsto E(x-y)) - E(x-y)\, \partial_\nu u(y)\big)\, dy \qquad (x \in U);$$

and furthermore, that the right-hand side vanishes for $x \in X \setminus \overline{U}$.

Observe that this formula expresses u in U in terms of u and $\partial_\nu u$ on the boundary ∂U. However, that is not the whole story, for it is known that u is completely determined on U by $u|_{\partial U}$. This is the so-called *Dirichlet problem*, which is not discussed in the present book; see, however, Example 16.15 and Problems 14.38, 14.54, 16.14, and 18.8.

12.5. *(Mean Value Theorem and maximum principle.)* Let the notation be as in Problem 12.4. Apply Green's second identity with u and with $v = 1$ to obtain $\int_{\partial U} \frac{\partial u}{\partial \nu}(y)\, dy = 0$. Next consider $x \in X$ and $r > 0$ such that $B(x;r) \subset X$ and apply Green's formula from Problem 12.4 with $U = B(x;r)$ and E as in (12.3) to obtain the following *Mean Value Theorem* for harmonic functions:

$$u(x) = \frac{1}{\mathrm{vol}_{n-1}(\partial B(x;r))} \int_{\partial B(x;r)} u(y)\, dy.$$

Deduce

$$u(x) = \frac{1}{\mathrm{vol}_n(B(x;r))} \int_{B(x;r)} u(y)\, dy.$$

Derive the *maximum principle*: if u is a harmonic function on X, then u cannot have a local maximum or minimum in X. More precisely, if $u(x) \gtreqless u(y)$ for all $y \in B(x;r)$, where $r > 0$, then u is constant on the component of X containing x.

12.6.* Let P be a harmonic polynomial function on $\mathbf{R}^n$ and $x \in \mathbf{R}^n$. Verify

$$\int_{\mathbf{R}^n} P(x+y)\, e^{-\frac{1}{2}\|y\|^2}\, dy = (2\pi)^{\frac{n}{2}}\, P(x).$$

12.7.* *(One-dimensional wave operator.)* Define the open set $V = \{\, (x,t) \in \mathbf{R}^2 \mid |x| < t \,\}$, see Fig. 12.2. Compute a constant $a \in \mathbf{C}$ such that $E = a\, 1_V$ is a fundamental solution of the wave operator $\Box = {\partial_t}^2 - {\partial_x}^2$. Hint: for the computation one has to evaluate an integral. The evaluation of this integral can be simplified in several ways. One possibility is to apply the change of variables $(x,t) = \Psi(y)$, where Ψ is the rotation in $\mathbf{R}^2$ about the origin by the angle $\pi/4$. Alternatively, one can apply Green's Integral Theorem (see [7, Theorem 8.3.5]). In this case, given a test function ϕ, determine a vector field v such that $\Box\phi = \operatorname{curl} v = \partial_x v_2 - \partial_t v_1$.

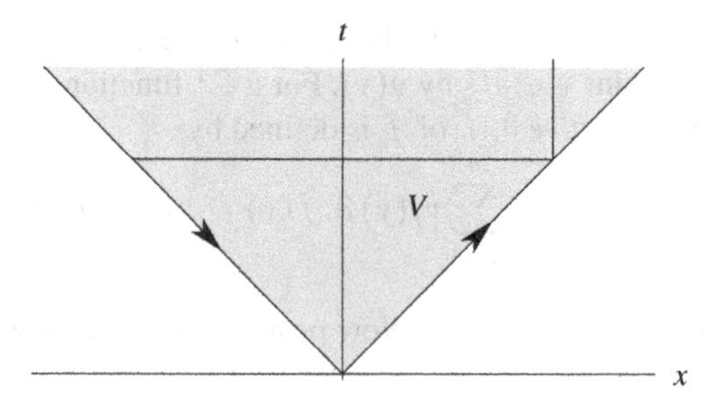

Fig. 12.2 Illustration for Problem 12.7. The set V

Determine $\operatorname{supp} E$ and $\operatorname{sing\,supp} E$. Show that $u = E * f$ is a well-defined distribution on $\mathbf{R}^2$ if the support of $f \in \mathcal{D}'(\mathbf{R}^2)$ is contained in a half-space of the form $H = \{ (x,t) \mid t \geq t_0 \}$, for some $t_0 \in \mathbf{R}$. Prove that u is a solution of the inhomogeneous wave equation $\square\, u = f$.

12.8. *(Three-dimensional wave operator.)* Denote the coordinates in $\mathbf{R}^4$ by (x,t), where $x \in \mathbf{R}^3$ and $t \in \mathbf{R}$, and the wave operator by $\square = \partial_t{}^2 - \Delta_x$. Let u be the characteristic function of $V = \{ (x,t) \mid \|x\| < t \}$, the interior of the forward light cone. Further, let $v = \square\, u$ and $w = \square\, v$. Write $q(x,t) = t^2 - \|x\|^2$ and $P = \{ (x,t) \mid t > 0 \}$, the positive half-space.

Verify that $u = q^* H$ on P. Use Problem 10.20 to determine v and w in P; describe v. Prove that v and w are homogeneous distributions on $\mathbf{R}^n$ and determine their degrees. Prove that $w = c\,\delta$ for some constant $c \in \mathbf{C}$. Determine c by testing with the function $t \mapsto e^{-t}$, and justify this procedure. Calculate a fundamental solution E of $\square$.

If $\sigma : (x,t) \mapsto (x,-t)$, then prove that $\sigma^*(E)$ is also a fundamental solution of $\square$ and that $U = E - \sigma^*(E)$ satisfies the wave equation $\square\, U = 0$. Determine $\operatorname{supp} U$ and $\operatorname{sing\,supp} U$.

12.9.* Prove the following generalization of Problem 11.5. Let X be a connected open subset of $\mathbf{R}^n$ and suppose $u \in \mathcal{D}'(X)$ satisfies $\partial_j u = 0$, for all $1 \leq j \leq n$. Prove the existence of a constant $c \in \mathbf{C}$ such that $u = c\, 1_X$.

12.10.* *(Distribution invariant under positive dilations and rotations.)* Let $u \in \mathcal{D}'(\mathbf{R}^n)$ be invariant under positive dilations and rotations. Verify that u equals a constant function on $\mathbf{R}^n$.

12.11.* Assume that P and Q are hypoelliptic operators and show in that case that the composition $P \circ Q$ is also a hypoelliptic operator. Calculate the composition of ∂_z and $\partial_{\bar{z}}$. Use this to determine, from the fundamental solution (12.3) of Δ, a fundamental solution of ∂_z, and also of $\partial_{\bar{z}}$.

12.12.* *(Determining an antiderivative if integrability conditions are satisfied.)* Let E be a fundamental solution of Δ in $\mathbf{R}^n$ and let g_j, for $1 \leq j \leq n$, be distributions with compact support in $\mathbf{R}^n$, with the property that $\partial_j g_k = \partial_k g_j$ for all $1 \le j,\, k \le n$. Prove that the distribution

$$f = \sum_{j=1}^{n} \partial_j E * g_j$$

satisfies the system of differential equations

$$\partial_j f = g_j \qquad (1 \le j \le n).$$

(The problem with the usual method of obtaining f, namely by integration of the g_j along lines, is that distributions on $\mathbf{R}^n$ cannot in all cases be restricted to curves.)

Now prove that, for arbitrary distributions g_j in $\mathbf{R}^n$, there exists a distribution f in $\mathbf{R}^n$ with $\partial_j f = g_j$ if and only if $\partial_j g_k = \partial_k g_j$, for all $1 \le j,\, k \le n$. Hint: in the formula above for f, replace the distributions g_j by $\phi\, g_j$, where ϕ denotes a cut-off function.

12.13.* *(Helmholtz's equation).* Let $k \in \mathbf{R}_{\ge 0}$ and define

$$E_\pm \in C^\infty(\mathbf{R}^3 \setminus \{0\}) \qquad \text{by} \qquad E_\pm(x) = \frac{e^{\pm ik\|x\|}}{\|x\|}.$$

(i) Prove that $E_\pm \in \mathcal{D}'(\mathbf{R}^3)$. Begin by writing, for $x \in \mathbf{R}^3 \setminus \{0\}$,

$$E_\pm(x) = \cos k\|x\| \, \frac{1}{\|x\|} \pm i\, \frac{\sin k\|x\|}{\|x\|} =: g(x)\, \frac{1}{\|x\|} \pm i\, h(x),$$

and show that both functions g and h belong to $C^\infty(\mathbf{R}^3)$.

Next one is asked to show, by two different methods, that $-\frac{1}{4\pi} E_\pm$ is a fundamental solution of *Helmholtz's differential equation* with parameter k, or, differently phrased, that in $\mathcal{D}'(\mathbf{R}^3)$ one has

$$(\Delta + k^2) E_\pm = -4\pi\delta. \tag{12.15}$$

The first approach is a verification that the distribution $E_\pm$ does in fact satisfy (12.15); the second one is an actual construction of the solution.

(ii) Verify the following equalities in $\mathcal{D}'(\mathbf{R}^3)$:

$$\operatorname{grad} g(x) = -k\, h(x)\, x, \qquad \operatorname{grad}\Big(\frac{1}{\|\cdot\|}\Big)(x) = -\frac{1}{\|x\|^3}\, x,$$
$$\Delta g = -k(k\, g + 2h).$$

Prove that in addition, one has in $\mathcal{D}'(\mathbf{R}^3)$,

$$\Delta\Big(g\, \frac{1}{\|\cdot\|}\Big) = g\, \Delta\Big(\frac{1}{\|\cdot\|}\Big) + 2\Big\langle \operatorname{grad} g,\, \operatorname{grad}\Big(\frac{1}{\|\cdot\|}\Big)\Big\rangle + \frac{1}{\|\cdot\|}\, \Delta g$$

$$= -k^2 \, g \frac{1}{\|\cdot\|} - 4\pi \, \delta,$$

and use these results to prove (12.15).

Now the second method. Suppose $f \in C^\infty(\mathbf{R}^3 \setminus \{0\})$ to be a radial function, that is, there exists $f_0 \in C^\infty(\mathbf{R}_{>0})$ having the property $f(x) = f_0(\|x\|)$.

(iii) Evaluate Δf in terms of derivatives of f_0.
(iv) Further assume that $(\Delta + k^2) f = 0$. Determine the differential equation satisfied by $\widetilde{f_0} \in C^\infty(\mathbf{R}_{>0})$, where $\widetilde{f_0}(r) = r f_0(r)$, and use it to prove, for a and $b \in \mathbf{C}$,

$$f(x) = a \, \frac{\cos k \|x\|}{\|x\|} + b \, \frac{\sin k \|x\|}{\|x\|} \qquad (x \in \mathbf{R}^3 \setminus \{0\}).$$

(v) Use Green's second identity (see for instance [7, Example 7.9.6]) to show that in $\mathcal{D}'(\mathbf{R}^3)$,

$$(\Delta + k^2) f = -4\pi a \, \delta.$$

(vi) Conclude that every rotation-invariant fundamental solution of Helmholtz's equation is given by

$$\sum_\pm c_\pm E_\pm, \qquad \text{where} \qquad c_\pm \in \mathbf{C}, \ \sum_\pm c_\pm = 1.$$

12.14.* *(Boundary values of complex-analytic functions.)* In this problem we construct generalizations of the distributions $\frac{1}{x \pm i0} \in \mathcal{D}'(\mathbf{R})$ from (1.5) and (1.6). We need some notation.

If U is an open subset of $\mathbf{C}$, denote by $\mathcal{O}(U)$ the linear space of complex-differentiable functions on U. Write $H_+ = \{ z = x + iy \in \mathbf{C} \mid y > 0 \}$. For $N \in \mathbf{Z}_{\geq 0}$, $a < b$, and $0 < h$, introduce a seminorm on $\mathcal{O}(H_+)$ and linear subspaces, respectively, by

$$\begin{aligned} n_{N,a,b,h}(f) &= \sup\{ y^N \, |f(x+iy)| \mid a \le x \le b, \, 0 < y \le h \}, \\ \mathcal{O}_N(H_+) &= \{ f \in \mathcal{O}(H_+) \mid n_{N,a,b,h}(f) < \infty, \ \text{for all } a < b, \, 0 < h \}, \\ \mathcal{O}_*(H_+) &= \bigcup_{N \in \mathbf{Z}_{\geq 0}} \mathcal{O}_N(H_+). \end{aligned}$$

Furthermore, let $G = \{ z \in \mathbf{C} \mid z \neq 0 \text{ and } -\pi < \arg z < \pi \}$ and define the (principal) branch $\log : G \to \mathbf{C}$ of the logarithm by putting $\log z = \log |z| + i \arg z$. Denote the restriction of log to H_+ by $\log_+ \in \mathcal{O}(H_+)$.

(i) Prove that both $\log_+$ and $z \mapsto \frac{1}{z}$ belong to $\mathcal{O}_1(H_+)$.

For every $\phi \in C^\infty(\mathbf{R})$ and $N \in \mathbf{Z}_{\geq 0}$, introduce $\widetilde{\phi}_N \in C^\infty(\mathbf{C})$ and $R_{\phi,N} \in C^\infty(\mathbf{R})$, respectively, by

$$\widetilde{\phi}_N(x+iy) = \sum_{k=0}^{N} \frac{\phi^{(k)}(x)}{k!}(iy)^k \qquad \text{and} \qquad R_{\phi,N}(x) = \frac{\phi^{(N+1)}(x)}{2N!}.$$

Here $\widetilde{\phi}_N$ is the Nth-order Taylor polynomial in y, which an analytic extension of ϕ to H_+ would have if it did exist.

(ii) Demonstrate that $\partial_{\bar{z}}\widetilde{\phi}_N(z) = R_{\phi,N}(x)\,(iy)^N$, for all $z \in \mathbf{C}$.
(iii) Show that for every $\psi \in C^\infty(H_+)$ and $f \in \mathcal{O}(H_+)$ we have $\partial_{\bar{z}}(\psi f) = (\partial_{\bar{z}}\psi) f$.
(iv) Select $a < b$ and $c < d$ such that $R = [a,b] \times [c,d] \subset H_+$. Verify, for all $g \in C^\infty(H_+)$,

$$\int_{\partial R} g(z)\,dz = 2i \int_R \partial_{\bar{z}} g(x+iy)\,dx\,dy.$$

From now on, suppose that $f \in \mathcal{O}_N(H_+)$ and write $f^\epsilon(z) := f(z+i\epsilon)$, for $\epsilon > 0$.

(v) Let $a < b$ and $0 < h$. Prove that for all $\phi \in C_0^\infty(\mathbf{R})$ with $\operatorname{supp}\phi \subset\,]\,a,b\,[$,

$$\int_a^b \phi(x)\,f^\epsilon(x)\,dx = \int_a^b \widetilde{\phi}_N(x+ih)\,f^\epsilon(x+ih)\,dx$$
$$+2i \int_0^h \int_a^b R_{\phi,N}(x)\,(iy)^N\,f^\epsilon(x+iy)\,dx\,dy.$$

(vi) Demonstrate that for every $\phi \in C_0^\infty(\mathbf{R})$ the limit

$$\beta_+(f)(\phi) := \lim_{\epsilon \downarrow 0} \int_{\mathbf{R}} \phi(x)\,f(x+i\epsilon)\,dx$$

exists. In fact, prove that for any $h > 0$,

$$\beta_+(f)(\phi) = \sum_{k=0}^{N} \frac{(ih)^k}{k!} \int_{\mathbf{R}} \phi^{(k)}(x)\,f(x+ih)\,dx$$
$$+ \frac{(ih)^{N+1}}{N!} \int_{\mathbf{R}} \phi^{(N+1)}(x) \int_0^1 t^N\,f(x+ith)\,dt\,dx.$$

Deduce that $\beta_+(f)$ is an element of $\mathcal{D}'(\mathbf{R})$ of order at most $N+1$.

The preceding results lead to a linear operator $\beta_+ : \mathcal{O}_*(H_+) \to \mathcal{D}'(\mathbf{R})$, which is called the *boundary value map*. Note that $\beta_+(\frac{1}{z}) = \frac{1}{x+i0}$. In analogy with this result we also write $f(x+i0)$ instead of $\beta_+(f)$.

(vii) Verify $\frac{1}{x+i\,0} = \mathrm{PV}\,\frac{1}{x} - \pi i\,\delta$.

By means of Cauchy's integral formula it is not difficult to prove that the complex differentiation ∂_z maps $\mathcal{O}_N(H_+)$ into $\mathcal{O}_{N+1}(H_+)$. As a consequence, ∂_z defines a linear operator from $\mathcal{O}_*(H_+)$ into itself.

(viii) Establish the identity of operators $\beta_+ \circ \partial_z = \partial_x \circ \beta_+$ on $\mathcal{O}_*(H_+)$.

Similarly as above, define $H_- = \{ z = x + iy \in \mathbf{C} \mid y < 0 \}$, the corresponding boundary value map $\beta_- : \mathcal{O}_*(H_-) \to \mathcal{D}'(\mathbf{R})$, as well as $\log_-$.

(ix) Demonstrate $\beta_+(\log_+) - \beta_-(\log_-) = 2\pi i \, SH$, where H denotes the Heaviside function and deduce

$$\frac{1}{(x+i0)^k} - \frac{1}{(x-i0)^k} = 2\pi i \, \frac{(-1)^k}{(k-1)!} \, \delta^{(k-1)} \qquad (k \in \mathbf{N}).$$

Here we have introduced, for $x + iy \in G$ and $a \in \mathbf{C}$,

$$(x+iy)^a = e^{a(\log|x+iy| + i \arg(x+iy))}.$$

For every $a \in \mathbf{C}$ we thus obtain a complex-analytic function on G. Note that we have obtained the *Plemelj–Sokhotsky jump relations* (see Example 14.30 or Problem 1.3 for other proofs in the case of $k = 1$). Furthermore, observe that the jump relations do not depend on the choice of the argument of the complex logarithm.

(x) Prove, for $x \in \mathbf{R}$ and $a \in \mathbf{C}$,

$$(x \pm i0)^a = \lim_{\pm y \downarrow 0} (x+iy)^a = x^a H(x) + e^{\pm i\pi a}(-x)^a H(-x).$$

(xi) Define

$$p : \mathbf{R} \times H_- \to \mathbf{C} \qquad \text{by} \qquad p(r,z) = r^2 - z^2.$$

Demonstrate that p takes its values in G. Introduce, for every $r \in \mathbf{R}$ and $a \in \mathbf{C}$,

$$f_{r,a} : H_- \to \mathbf{C} \qquad \text{satisfying} \qquad f_{r,a}(z) = p(r,z)^{-\frac{a}{2}}.$$

For $\operatorname{Re} a \leq 0$ and $x \in \mathbf{R}$, show that $f_{r,a}(x - i0) := \lim_{\epsilon \downarrow 0} f_{r,a}(x - i\epsilon)$ exists and defines a distribution on $\mathbf{R}$. In the case of $\operatorname{Re} a > 0$, prove that

$$f_{r,a}(x - i0) := \beta_-(f_{r,a}) \in \mathcal{D}'(\mathbf{R})$$

is well-defined. Give an estimate for the order of this boundary value and verify that its restriction to $\mathbf{R} \setminus \{\pm r\}$ equals the real-analytic function

$$f_{r,a}(x - i0) = \begin{cases} (r^2 - x^2)^{-\frac{a}{2}}, & |x| < |r|; \\ e^{\mp i \frac{\pi}{2} a}(x^2 - r^2)^{-\frac{a}{2}}, & \pm x > |r|. \end{cases}$$

Chapter 13
Fractional Integration and Differentiation

In this chapter we deal with "complex powers of the operator $\frac{d}{dx}$," a concept already found in the posthumous works of Riemann and elaborated by Marcel Riesz in the 1930s and 1940s. The relevant article [18] is lengthy, but with the help of a little distribution theory and complex analysis, all results can readily be proved. We will also deal with Riesz's treatment of the wave operator $\square = \partial_t^2 - \Delta_x$ in arbitrary dimension; thus we will obtain, among other things, a fundamental solution of $\square$.

13.1 The Case of Dimension One

Let $\mathcal{D}'(\mathbf{R})_+$ be the space of distributions $u \in \mathcal{D}'(\mathbf{R})$ such that there exists $l \in \mathbf{R}$ with $\operatorname{supp} u \subset [\,l, \infty\,[$. For u and $v \in \mathcal{D}'(\mathbf{R})_+$ the convolution product $u * v \in \mathcal{D}'(\mathbf{R})_+$ is well-defined; see Problem 11.13 or Theorem 11.17. Every $u \in \mathcal{D}'(\mathbf{R})_+$ possesses a uniquely determined kth-order antiderivative in $\mathcal{D}'(\mathbf{R})_+$, that is, $v \in \mathcal{D}'(\mathbf{R})_+$ with $\partial^k v = u$, for $k \in \mathbf{N}$. This is obtained via the definition (compare with [7, Exercise 2.75]), with H denoting the Heaviside function,

$$v = \chi_+^k * u, \qquad \text{where} \qquad \chi_+^k(x) = \frac{x^{k-1}}{(k-1)!}\,H(x) \qquad (x \in \mathbf{R}).$$

Indeed, by mathematical induction on $k \in \mathbf{N}$ one gets $\partial^k \chi_+^k = \delta$, since on account of the Leibniz rule and Examples 4.2 and 9.1,

$$\partial \chi_+^{k+1} = \partial\Big(\frac{x^k}{k!}\,H\Big) = \frac{x^{k-1}}{(k-1)!}\,H + \frac{x^k}{k!}\,\delta = \frac{x^{k-1}}{(k-1)!}\,H = \chi_+^k.$$

In other words, finding kth-order antiderivatives is equal to convolution with the function χ_+^k. Conversely, kth-order differentiation can also be regarded as a convolution operator, namely, as convolution with the distribution $\delta^{(k)}$.

It was Riesz who discovered that these operators can be embedded in an analytic way into a family of convolution operators $I_+^a = \chi_+^a\,*$ depending on $a \in \mathbf{C}$. Here

J.J. Duistermaat and J.A.C. Kolk, *Distributions: Theory and Applications*, Cornerstones, DOI 10.1007/978-0-8176-4675-2_13,

the χ_+^a have the following properties, for all a and $b \in \mathbf{C}$:

$$\chi_+^a \in \mathcal{D}'(\mathbf{R}) \quad \text{and} \quad \operatorname{supp} \chi_+^a \subset \mathbf{R}_{\geq 0}, \tag{13.1}$$

$$\chi_+^k(x) = \frac{x^{k-1}}{(k-1)!} H(x) \qquad (k \in \mathbf{N}), \tag{13.2}$$

$$\chi_+^{-k} = \delta^{(k)} \qquad (k \in \mathbf{Z}_{\geq 0}), \tag{13.3}$$

$$\chi_+^a * \chi_+^b = \chi_+^{a+b}. \tag{13.4}$$

By (13.1), the operator I_+^a of convolution with χ_+^a is a continuous linear operator from $\mathcal{D}'(\mathbf{R})_+$ to $\mathcal{D}'(\mathbf{R})_+$. On the strength of (13.2), I_+^k is equal to finding a kth-order antiderivative in $\mathcal{D}'(\mathbf{R})_+$, while according to (13.3), I_+^{-k} is equal to kth-order differentiation. Finally, (13.4) implies the validity of the group law $I_+^a \circ I_+^b = I_+^{a+b}$. Because $\chi_+^0 = \delta$, one has $I_+^0 = I$, the identity in $\mathcal{D}'(\mathbf{R})_+$. On account of the group law with $b = -a$ we also deduce from this that I_+^a is bijective from $\mathcal{D}'(\mathbf{R})_+$ to $\mathcal{D}'(\mathbf{R})_+$, with inverse I_+^{-a}. If one describes I_+^a as "finding antiderivatives of order $a \in \mathbf{C}$," then I_+^{-a} could be called "differentiation of order a."

The family $(\chi_+^a)_{a\in\mathbf{C}}$ forms a *complex-analytic family of distributions*, in the sense that $a \mapsto \chi_+^a(\phi)$ is a complex-analytic function on $\mathbf{C}$, for every $\phi \in C_0^\infty(\mathbf{R})$. In this case one also uses the term *distribution-valued complex-analytic function*. This property will be used to deduce the validity for all $a \in \mathbf{C}$ of numerous identities involving χ_+^a from the validity on an arbitrary nonempty open subset U of $\mathbf{C}$. Thus, we are free to choose U such that the desired identity can be verified on U by direct calculation.

This procedure involves the following principle. If f and g are analytic functions on a connected open subset V of $\mathbf{C}$, one has $f = g$ on V whenever $f = g$ on a nonempty open subset U of V. This follows by applying Lemma 2.6 to $\phi = f - g$. The functions f and g may be taken as the left-hand side and the right-hand side, or vice versa, of the identity involving χ_+^a, after testing with an arbitrary test function. This is called the *principle of analytic continuation of identities*.

We now give the definition of the family χ_+^a, followed by a summary of the properties. The starting point is the definition

$$\chi_+^a(x) = \frac{x^{a-1}}{\Gamma(a)} H(x) \qquad (x \in \mathbf{R}). \tag{13.5}$$

Here $x^c = e^{c \log x}$ for $x > 0$ and $c \in \mathbf{C}$. The function $\chi_+^a(x)$ is locally integrable on $\mathbf{R}$ if and only if $\operatorname{Re} a > 0$; for these values of a we interpret χ_+^a as an element of $\mathcal{D}'(\mathbf{R})$.

The factor $\Gamma(a)$ in the denominator is in terms of Euler's Gamma function, see the appendix in Sect. 13.3 below and in particular, Corollary 13.6. This is a complex-analytic function on $\mathbf{C} \setminus \mathbf{Z}_{\leq 0}$ **without zeros**. For every $k \in \mathbf{Z}_{\geq 0}$, Γ has a simple pole at $-k$, with residue $(-1)^k/k!$. Consequently, $1/\Gamma$ possesses an extension to a complex-analytic function on $\mathbf{C}$, with zeros only at the points $-k$, with $k \in \mathbf{Z}_{\geq 0}$, and derivative $(-1)^k\, k!$ at those points. Further, $\Gamma(k) = (k-1)!$ for $k \in \mathbf{N}$, which means that (13.2) is satisfied.

Integration by parts and the formula $\Gamma(a+1) = a\,\Gamma(a)$ imply the following identity in $\mathcal{D}'(\mathbf{R})$:

$$\partial \chi_+^{a+1} = \chi_+^a \qquad (\operatorname{Re} a > 0). \tag{13.6}$$

This enables us to **define** $\chi_+^a \in \mathcal{D}'(\mathbf{R})$ for every $a \in \mathbf{C}$, by means of

$$\chi_+^a = \partial^k \chi_+^{a+k} \qquad (k \in \mathbf{Z}_{\geq 0},\ \operatorname{Re}(a+k) > 0). \tag{13.7}$$

The right-hand side does not depend on the choice of k, while for $\operatorname{Re} a > 0$ the definition is identical to the χ_+^a that we have defined above. This also implies that (13.6) holds for all $a \in \mathbf{C}$. Furthermore, for every $\phi \in C_0^\infty(\mathbf{R})$, the following function is complex-analytic on $\mathbf{C}$:

$$a \mapsto \chi_+^a(\phi) = (-1)^k\, \chi_+^{a+k}(\phi^{(k)}),$$

as follows by interchanging the Cauchy–Riemann operator from (12.12) and integration.

For every $\phi \in C_0^\infty(\mathbf{R})$ with $\operatorname{supp} \phi \subset \mathbf{R}_{<0}$ one has $\chi_+^a(\phi) = 0$ if $\operatorname{Re} a > 0$; it follows by analytic continuation that this identity holds for all $a \in \mathbf{C}$; this proves (13.1). In the same way, analytic continuation yields the validity of (13.5) for all $a \in \mathbf{C}$.

From the analytic continuation of (13.6) and Example 4.2 we obtain

$$\chi_+^0 = \partial \chi_+^1 = \partial H = \delta. \tag{13.8}$$

By repeated differentiation of this we now get (13.3).

From (13.5) we see that

$$\operatorname{sing\,supp} \chi_+^a = \{0\} \tag{13.9}$$

if $a \notin \mathbf{Z}_{\leq 0}$. For $a \in \mathbf{Z}_{\leq 0}$, (13.5) implies that $\operatorname{supp} \chi_+^a \subset \{0\}$; in this case we conclude again that (13.9) is valid, using (13.3). Because χ_+^a has order 0 if and only if $\operatorname{Re} a > 0$, we see that the order of χ_+^a equals k if $-k < \operatorname{Re} a \leq -k+1$, with $k \in \mathbf{N}$. This can also be formulated by saying that the singularity of χ_+^a at 0 becomes worse and worse as $\operatorname{Re} a \to -\infty$.

If $\operatorname{Re} a > 0$ and $\operatorname{Re} b > 0$, we have, for $x > 0$,

$$\Gamma(a)\,\Gamma(b)\,(\chi_+^a * \chi_+^b)(x) = \int_0^x y^{a-1}\,(x-y)^{b-1}\,dy$$

$$= x^{a+b-1} \int_0^1 t^{a-1}\,(1-t)^{b-1}\,dt = \Gamma(a+b)\,B(a,b)\,\chi_+^{a+b}(x).$$

Here B is Euler's Beta function; see (13.34) below. In view of (13.33), this implies that (13.4) holds if $\operatorname{Re} a > 0$ and $\operatorname{Re} b > 0$. The identity (13.4) now follows by analytic continuation (after testing) for all $a \in \mathbf{C}$, for given $b \in \mathbf{C}$ with $\operatorname{Re} b > 0$, and then also, by analytic continuation with respect to the variable b, for all $a \in \mathbf{C}$ and $b \in \mathbf{C}$.

By analytic continuation one also proves the following identity in $\mathcal{D}'(\mathbf{R})$:

$$x\,\chi_+^a = a\,\chi_+^{a+1} \qquad (a \in \mathbf{C}).$$

Combining this with (13.6), we now obtain

$$x\,\partial\chi_+^a = (a-1)\,\chi_+^a. \tag{13.10}$$

In other words, for every $a \in \mathbf{C}$ we conclude, on the strength of Theorem 10.17, that χ_+^a is a distribution homogeneous of degree $a-1$.

Marcel Riesz called $I_+^a(u) = \chi_+^a * u$ a *Riemann–Liouville integral* of u; he had encountered it in his work on so-called Cesàro means.

For suitable functions ϕ and a in an appropriate subset of $\mathbf{C}$, one writes

$$\mathcal{M}\phi(a) := \Gamma(a)\,\chi_+^a(\phi) = \int_{\mathbf{R}_{>0}} x^{a-1}\,\phi(x)\,dx,$$

where $\mathcal{M} : \phi \mapsto \mathcal{M}\phi$ is said to be the *Mellin transform*; see Problem 15.6 below for more details. For the Mellin transform of $x \mapsto e^{-x}$, sin and cos, and $x \mapsto \frac{1}{x+1}$, see (13.30), Problem 13.13, and (16.32), respectively. In view of this, the χ_+^a are sometimes called *Mellin distributions* or *Mellin kernels*.

Fundamental solutions are a useful tool in proving results concerning the structure of distributions.

Theorem 13.1. *Let X be an open subset of $\mathbf{R}^n$. For every $u \in \mathcal{D}'(X)$ and bounded open subset U of X, there exists $f \in C(X)$ such that the restriction of u to U equals a derivative of finite order of f.*

Proof. Indeed, for $k \in \mathbf{Z}_{\geq 0}$, define $E_k = \chi_+^k \otimes \cdots \otimes \chi_+^k \in \mathcal{D}'(\mathbf{R}^n)$ and the partial differential operator $\widetilde{\partial}^k := \partial^{(k,\dots,k)}$ in $\mathbf{R}^n$. Then (13.6) and (13.8) lead to

$$\widetilde{\partial}^k E_k = \delta \in \mathcal{D}'(\mathbf{R}^n), \qquad \text{while} \qquad E_k \in C^{k-2}(\mathbf{R}^n) \qquad (k \geq 2).$$

In other words, E_k is a fundamental solution of $\widetilde{\partial}^k$. Next, select $\chi \in C_0^\infty(X)$ satisfying $\chi = 1$ on a neighborhood of U. Then $u = \chi u$ on U. Because $\chi u \in \mathcal{E}'(X) \subset \mathcal{E}'(\mathbf{R}^n)$, it is of finite order, say $k \in \mathbf{Z}_{\geq 0}$, according to Theorem 8.8. From Theorem 11.17 it follows that

$$\chi u = \chi u * \widetilde{\partial}^{k+2} E_{k+2} = \widetilde{\partial}^{k+2}(\chi u * E_{k+2}) \in \mathcal{D}'(\mathbf{R}^n).$$

It remains to prove that $\chi u * E_{k+2}$ is a continuous function. Define ϕ_ϵ as in Lemma 2.19 and apply Theorem 11.2 to get

$$f_\epsilon := (\chi u * E_{k+2}) * \phi_\epsilon \in C^\infty(\mathbf{R}^n).$$

On account of (11.22), the associative law applies, and so application of (11.1) with $\chi u \in \mathcal{E}'(\mathbf{R}^n)$ and $E_{k+2} * \phi_\epsilon \in C^\infty(\mathbf{R}^n)$ gives

$$f_\epsilon(x) = \chi u * (E_{k+2} * \phi_\epsilon)(x) = \chi u(T_x \circ S(E_{k+2} * \phi_\epsilon)). \tag{13.11}$$

As in (2.6) one shows that the $E_{k+2} * \phi_\epsilon$ converge to E_{k+2} in $C^k(\mathbf{R}^n)$ as $\epsilon \downarrow 0$. Since χu is of order k, application of (8.3) to the right-hand side of (13.11) implies that the continuous functions f_ϵ converge, uniformly on compact sets, to the function f defined by $x \mapsto \chi u(T_x \circ SE_{k+2})$ as $\epsilon \downarrow 0$. Hence f is also continuous on X. On the other hand, by Lemma 11.6 the f_ϵ converge to $\chi u * E_{k+2}$ in $\mathcal{D}'(X)$ as $\epsilon \downarrow 0$. It is clear that the limits are the same in this case, so that one has $\chi u * E_{k+2} = f$ in $C(X)$. See Example 18.2 for another proof. □

13.2 Wave Family

We now describe the version for the *wave operator*

$$\Box = \partial_t^2 - \Delta_x = \partial_{n+1}^2 - \sum_{j=1}^n \partial_j^2$$

in $\mathbf{R}^{n+1}$, whose points we denote by $y = (x, t)$. Here $(x_1, \dots, x_n) \in \mathbf{R}^n$ are the position coordinates and $t \in \mathbf{R}$ is the time coordinate. (For generalization to multi-dimensional time, see Kolk–Varadarajan [15].) All definitions will be given in terms of the corresponding quadratic form q on $\mathbf{R}^{n+1}$, of Lorentz type, defined by

$$q(y) = q(x,t) = t^2 - \|x\|^2 = t^2 - \sum_{j=1}^n x_j^2 = {}^t(x,t)J(x,t), \tag{13.12}$$

where ^{t} denotes the transpose and J is the $(n+1) \times (n+1)$ diagonal matrix with coefficients $(-1, \dots, -1, 1)$. Further, we denote the interiors C_+ and C_- of the (solid) *forward* and the *backward cones* by

$$C_\pm = \{ y = (x,t) \in \mathbf{R}^{n+1} \mid q(y) > 0,\ t \gtrless 0 \}, \tag{13.13}$$

respectively (see Fig. 13.1). Then the boundary ∂C_+ of C_+ is the nappe of the cone

$$\partial C_+ = \{ (x,t) \in \mathbf{R}^{n+1} \mid \|x\| = t \}. \tag{13.14}$$

The point of departure this time is the function R_+^a on an open dense subset of $\mathbf{R}^{n+1}$, defined by

$$R_+^a(y) = \begin{cases} c(a)\, q(y)^{\frac{a-n-1}{2}} & \text{if } y \in C_+, \\ 0 & \text{if } y \in \mathbf{R}^{n+1} \setminus \overline{C_+}. \end{cases} \tag{13.15}$$

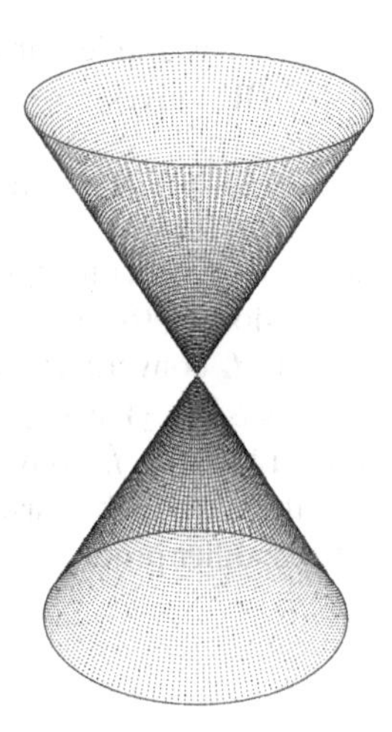

Fig. 13.1 Illustration for (13.13). C_+ and C_- are bounded by the upper and the lower nappes of the cone, respectively

Here $c(a) = c_n(a)$ is a constant to be determined; see Lemma 13.2 below. Note that for $\operatorname{Re} a < n + 1$ the function R_+^a is unbounded along ∂C_+. It is locally integrable on $\mathbf{R}^{n+1}$ if and only if $\operatorname{Re} a > n - 1$; if that condition is satisfied, we obtain an element of $\mathcal{D}'(\mathbf{R}^{n+1})$.

Lemma 13.2. *If we choose (for the second equality, see (13.39) below)*

$$c(a) = \frac{\Gamma(\frac{a+1}{2})}{\pi^{\frac{n}{2}}\,\Gamma(a)\,\Gamma(\frac{a-n+1}{2})} = \frac{1}{2^{a-1}\,\pi^{\frac{n-1}{2}}\,\Gamma(\frac{a}{2})\,\Gamma(\frac{a-n+1}{2})}, \tag{13.16}$$

we obtain, for $\operatorname{Re} a > n - 1$ *and* $\operatorname{Re} b > n - 1$,

$$R_+^a * R_+^b = R_+^{a+b}. \tag{13.17}$$

Proof. For the calculation of the function $R := R_+^a * R_+^b$ at $y = (x, t)$ we have to integrate

$$R_+^a(x - \xi, t - \tau)\, R_+^b(\xi, \tau)$$

over $\eta = (\xi, \tau) \in C_+$ with $y - \eta \in C_+$. The latter two relations imply that $\|\xi\| < \tau < t$, which means that this set of integration is uniformly bounded whenever t varies over a bounded set; thus, there are no problems with the convergence of the integral. Next, on account of $y - \eta \in \overline{C_+}$ and $\eta \in C_+$, we see that the sum y belongs to C_+, and therefore $\operatorname{supp} R \subset \overline{C_+}$. And finally, we observe that a change of variables $\eta = \rho\,\zeta$, with $\rho > 0$, results in $R(\rho\, y) = \rho^{2m}\, R(y)$, where

$$2m = (a - n - 1) + (b - n - 1) + (n + 1) = a + b - n - 1.$$

That is, R is homogeneous of degree $a + b - n - 1$.

The *Lorentz group* **Lo** is the collection of all linear transformations A in $\mathbf{R}^{n+1}$ satisfying $A^* q = q$; it is a group with composition as the binary operation. For the further determination of R we use the *connected component* $\mathbf{Lo}^\circ$ of **Lo**. This is the subgroup of all $A \in \mathbf{Lo}$ with $\det A = 1$ and $A(C_+) = C_+$ (see [7, Exercise 5.70]) and is called the *proper orthochronous Lorentz group*. It is clear that R_+^a is invariant under all $A \in \mathbf{Lo}^\circ$. Conversely: **if f is a function on C_+ that is invariant under $\mathbf{Lo}^\circ$ and homogeneous of degree $2m$, then there exists a constant $c \in \mathbf{C}$ with $f = c\, q^m$ on C_+.** Indeed, f is constant on all *orbits* $\{ Ay \mid A \in \mathbf{Lo}^\circ \}$ of points $y \in C_+$ under the action of $\mathbf{Lo}^\circ$. It is known that the orbits in C_+ under the action of $\mathbf{Lo}^\circ$ are equal to the level surfaces of q in C_+ (see Fig. 13.2),

$$\{ y \in C_+ \mid q(y) = \text{constant} \}. \tag{13.18}$$

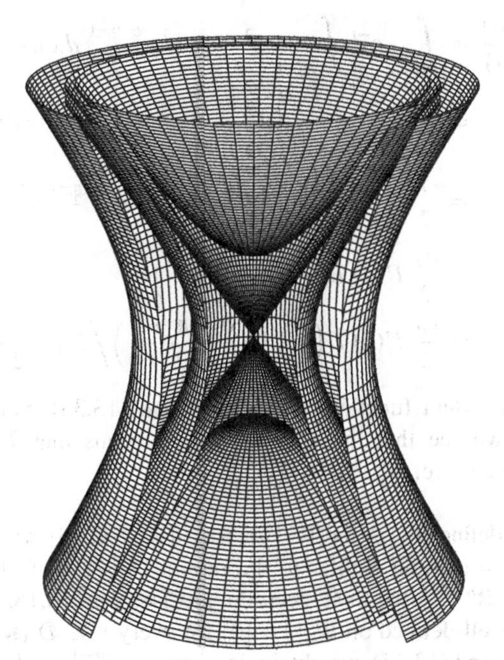

Fig. 13.2 Illustration of (13.18). Surfaces of the levels 4, 1, 0, and -1 of q on $\mathbf{R}^3$

This implies that there exists a function g on $\mathbf{R}_{>0}$ such that $f(y) = g(q(y))$ for all $y \in C_+$. From the homogeneity of f we deduce that g must be homogeneous of degree m on $\mathbf{R}_{>0}$.

For every $A \in \mathbf{Lo}^\circ$, the change of variables $\eta = A\,\zeta$ in the integral for $R = R_+^a * R_+^b$ yields that $R(A\,y) = R(y)$; in other words, R is invariant under all $A \in \mathbf{Lo}^\circ$. Applying the preceding characterization, we find a constant $c(a, b) \in \mathbf{C}$ such that

$$R_+^a * R_+^b = c(a,b)\, R_+^{a+b}. \tag{13.19}$$

For the calculation of $c(a,b)$ we test both sides of this identity in $\mathcal{D}'(\mathbf{R}^{n+1})$ with the function $(x,t) \mapsto e^{-t}$ in $C^\infty(\mathbf{R}^{n+1})$. On $\overline{C_+}$ this function decreases fast enough to ensure convergence of the integrals. If we now write

$$T(a) = \int_{\mathbf{R}} \int_{\mathbf{R}^n} e^{-t}\, R_+^a(x,t)\, dx\, dt,$$

and note that $e^t = e^{t-\tau}\, e^\tau$, we find that testing of the left-hand side of (13.19) equals $T(a)\, T(b)$; therefore, $c(a,b)$ is given by

$$T(a)\, T(b) = c(a,b)\, T(a+b).$$

On the other hand, with the changes of variables $\|x\| = r = t\,\sqrt{s}$ and the notation c_n for the Euclidean $(n-1)$-dimensional volume of the unit sphere in $\mathbf{R}^n$ we obtain

$$\begin{aligned}
\frac{T(a)}{c(a)} &= \int_{\mathbf{R}_{>0}} e^{-t} \int_{\|x\|<t} (t^2 - \|x\|^2)^{\frac{a-n-1}{2}}\, dx\, dt \\
&= c_n \int_{\mathbf{R}_{>0}} e^{-t} \int_0^t (t^2 - r^2)^{\frac{a-n-1}{2}}\, r^{n-1}\, dr\, dt \\
&= \frac{c_n}{2} \int_{\mathbf{R}_{>0}} e^{-t}\, t^{a-1}\, dt \int_0^1 (1-s)^{\frac{a-n-1}{2}}\, s^{\frac{n-2}{2}}\, ds \\
&= \frac{c_n}{2}\, \Gamma(a)\, B\Big(\frac{n}{2}, \frac{a-n+1}{2}\Big) \\
&= \frac{c_n}{2}\, \Gamma(a)\, \Gamma\Big(\frac{n}{2}\Big)\, \Gamma\Big(\frac{a-n+1}{2}\Big) \Big/ \Gamma\Big(\frac{a+1}{2}\Big).
\end{aligned}$$

Here B is Euler's Beta function, see (13.34) and (13.33). Substituting formula (13.37) for c_n, we see that the choice (13.16) means that $T(a) = T(b) = T(a+b) = 1$; therefore, $c(a,b) = 1$. □

Next we will define $\mathbf{R}_+^a \in \mathcal{D}'(\mathbf{R}^{n+1})$, for all $a \in \mathbf{C}$. To this end, we compute $\square\, R_+^a$ and apply a method similar to the one in (13.7). In fact, note that q is a submersion from $\mathbf{R}^{n+1} \setminus \{0\}$ to $\mathbf{R}$; according to Theorem 10.18, therefore, the distribution q^*v is well-defined on $\mathbf{R}^{n+1} \setminus \{0\}$ for every $v \in \mathcal{D}'(\mathbf{R})$. More precisely, comparing (13.5) and (13.15), we obtain, in view of $\frac{a-n-1}{2} = \frac{a-n+1}{2} - 1$,

$$R_+^a = d(a)\, q^*\big(\chi_+^{\frac{a-n+1}{2}}\big) \quad \text{on} \quad \mathbf{R}^{n+1} \setminus \overline{C_-}, \tag{13.20}$$

$$\text{where} \quad d(a) = \frac{\Gamma(\frac{a+1}{2})}{\pi^{\frac{n}{2}}\, \Gamma(a)} \quad (\operatorname{Re} a > n-1). \tag{13.21}$$

The specification "on $\mathbf{R}^{n+1} \setminus \overline{C_-}$" is necessary because $q^*\chi_+^c$ behaves identically on C_- and C_+. Indeed, the reflection $y \mapsto -y$ leaves $q^*\chi_+^c$ invariant and carries C_+ into C_-. However, $R_+^a = 0$ everywhere on $\mathbf{R}^{n+1} \setminus \overline{C_+}$, which contains $\overline{C_-} \setminus \{0\}$. On

$q^{-1}(\,]-\infty,\,0\,[\,)$, that is, the complement of $\overline{C_+} \cup \overline{C_-}$, equation (13.20) is correct, since both sides vanish there. Furthermore, if $\frac{a+1}{2} = -k$ with $k \in \mathbf{Z}_{\geq 0}$, then $a = -1-2k \in \mathbf{Z}_{<0}$; so the pole in the numerator in (13.21) is compensated by that of $\Gamma(a)$ in the denominator. As a result, $a \mapsto d(a)$ possesses an extension to a complex-analytic function on $\mathbf{C}$.

If the target variable of q in $\mathbf{R}$ is also denoted by q, we obtain from Problem 10.20.(ii) the following identity in $\mathcal{D}'(\mathbf{R}^{n+1} \setminus \{0\})$:

$$\Box(q^* v) = q^*(2(n+1)\,v' + 4q\,v'').$$

Apply this identity to $v = \chi_+^{\frac{a-n+1}{2}}$. Note that $v' = \chi_+^{\frac{a-n-1}{2}}$; see (13.6). This means that v' is homogeneous of degree $\frac{a-n-3}{2}$, see (13.10), so that Theorem 10.17 implies

$$q\,v'' = \frac{a-n-3}{2}\,v'; \qquad \text{therefore} \qquad 2(n+1)\,v' + 4q\,v'' = (2a-4)\,v'.$$

On $\mathbf{R}^{n+1} \setminus \overline{C_-}$ this yields

$$\Box\, R_+^a = d(a)\,(2a-4)\,q^*v' = \frac{d(a)\,(2a-4)}{d(a-2)}\,R_+^{a-2}.$$

Because $\Gamma(\frac{a+1}{2}) = \frac{a-1}{2}\Gamma(\frac{a-1}{2})$ and $\Gamma(a) = (a-1)(a-2)\,\Gamma(a-2)$, the end result is

$$\Box\, R_+^a = R_+^{a-2} \qquad (\operatorname{Re} a > n+1). \tag{13.22}$$

Initially this equality is valid in $\mathcal{D}'(\mathbf{R}^{n+1} \setminus \{0\})$, but because the partial derivatives of R_+^a vanish along ∂C_+ for $\operatorname{Re} a > n+1$, as does R_+^{a-2}, it is also valid in $\mathcal{D}'(\mathbf{R}^{n+1})$.

This now enables us to **define**, for all $a \in \mathbf{C}$, as we did for the χ_+^a, the distributions R_+^a by

$$R_+^a = \Box^k\, R_+^{a+2k} \qquad (k \in \mathbf{Z}_{\geq 0},\ \operatorname{Re} a + 2k > n-1).$$

The right-hand side is independent of k, and $a \mapsto R_+^a(\phi)$ is a complex-analytic function on $\mathbf{C}$, for every $\phi \in C_0^\infty(\mathbf{R}^{n+1})$. By analytic continuation, the identities (13.15), (13.17), (13.20), and (13.22) are now seen to be **valid for all** $a \in \mathbf{C}$**, in the distributional sense**. Sometimes the R_+^a are called *Riesz distributions* or *Riesz kernels*. With respect to the distribution R_+^a, (13.15) here means

$$\operatorname{supp} R_+^a \subset \overline{C_+}, \tag{13.23}$$

while on the open set C_+ the distribution R_+^a equals the real-analytic function $c(a)\,q^{\frac{a-n-1}{2}}$. This also implies in the notation of (13.14),

$$\operatorname{sing\,supp} R_+^a \subset \partial C_+. \tag{13.24}$$

We further mention the equation

$$\sum_{j=1}^{n+1} y_j\,\partial_j\,R_+^a = (a-n-1)\,R_+^a, \tag{13.25}$$

which says that for every $a \in \mathbf{C}$, the distribution R_+^a is homogeneous of degree $a-n-1$. Furthermore, by analytic continuation of identities this is seen to imply that for every $a \in \mathbf{C}$, the distribution R_+^a is invariant under the proper orthochronous Lorentz group $\mathbf{Lo}^\circ$ and satisfies

$$q\,R_+^a = a\,(a-n+1)\,R_+^{a+2}. \tag{13.26}$$

Theorem 13.3. *$R_+^0 = \delta$ and $E_+ := R_+^2$ is a fundamental solution of $\square$. For $n = 1$ or n even, one has* $\operatorname{supp} E_+ = \overline{C_+}$, *while* $\operatorname{supp} E_+ = \partial C_+$ *if $n > 1$ is odd. In all cases,* $\operatorname{sing\,supp} E_+ = \partial C_+$.

Proof. In (13.21) we have $\frac{1}{\Gamma(a)} = 0$ and $\Gamma(\frac{a+1}{2}) \neq 0$ at $a = 0$. Therefore it follows by analytic continuation of (13.20) to $a = 0$ that $R_+^0 = 0$ on $\mathbf{R}^{n+1} \setminus \overline{C_-}$. By combining this with (13.23), we can conclude that $\operatorname{supp} R_+^0 \subset \{0\}$. But on the strength of Theorem 8.10 this leads to

$$R_+^0 = \sum_{|\alpha| \le m} c_\alpha\,\partial^\alpha\delta,$$

for some $m \in \mathbf{Z}_{\ge 0}$ and $c_\alpha \in \mathbf{C}$. Now (13.25) implies that R_+^0 is homogeneous of degree $-n-1$, while $\partial^\alpha\delta$ is homogeneous of degree $-n-1-|\alpha|$ according to Problem 10.16. As a consequence we obtain

$$0 = \Big(n+1+\sum_{j=1}^{n+1} y_j\,\partial_j\Big)\,R_+^0 = -\sum_{|\alpha|\le m} c_\alpha\,|\alpha|\,\partial^\alpha\delta. \tag{13.27}$$

Further, testing with $y \mapsto y^\beta$ shows that the distributions $\partial^\alpha\delta$ are linearly independent in $\mathcal{D}'(\mathbf{R}^{n+1})$. This implies that (13.27) can be true only if $c_\alpha = 0$, for all $\alpha \neq 0$. This now proves that $R_+^0 = c\,\delta$, for some $c \in \mathbf{C}$.

But then (13.17) yields

$$R_+^a = R_+^{0+a} = R_+^0 * R_+^a = c\,\delta * R_+^a = c\,R_+^a,$$

which implies that $c = 1$, because there certainly exists an $a \in \mathbf{C}$ such that $R_+^a \neq 0$. This proves the first assertion; the second one follows from (13.22) for $a = 2$.

As regards the description of $\operatorname{supp} E_+$ and $\operatorname{sing\,supp} E_+$, we observe that (13.20) leads to

$$E_+ = \frac{1}{2\,\pi^{\frac{n-1}{2}}}\,q^*\big(\chi_+^{\frac{3-n}{2}}\big) \qquad \text{on} \qquad \mathbf{R}^{n+1} \setminus \overline{C_-}.$$

If $n = 1$, we have that $\chi_+^1 = H$ and therefore $E_+ = \frac{1}{2} 1_{C_+}$, a fundamental solution that we have encountered before, in Problem 12.7. For $n = 3$ we obtain $E_+ = \frac{1}{2\pi} q^*\delta$ on $\mathbf{R}^4 \setminus \overline{C_-}$; compare this with Problems 12.8, 13.10, and 13.11. More generally we get, for $n = 2k + 3$ with $k \in \mathbf{Z}_{\geq 0}$,

$$E_+ = \frac{1}{2\,\pi^{k+1}}\, q^*\delta^{(k)} \qquad \text{on} \qquad \mathbf{R}^{n+1} \setminus \overline{C_-}.$$

From this we can see that $\operatorname{supp} E_+ = \partial C_+$. If, however, n is even, then we have $\operatorname{supp} \chi_+^{(3-n)/2} = \mathbf{R}_{\geq 0}$, and so $\operatorname{supp} E_+ = \overline{C_+}$.

In (13.24) we already established that $\operatorname{sing\,supp} E_+ \subset \partial C_+$. For n odd, the equality of these sets follows from the foregoing, while if n is even, the equality follows from the observation that the $\frac{3-n}{2}$th power of q is not differentiable coming from within C_+ at points y satisfying $q(y) = 0$, that is, at points belonging to ∂C_+. □

Hadamard referred to the assertion $\operatorname{supp} E_+ = \partial C_+$ for $n > 1$ odd as the *Huygens principle*. In view of what Huygens himself wrote about wave propagation, a case can also be made for calling the assertion $\operatorname{sing\,supp} E_+ = \partial C_+$, which holds in every dimension, the Huygens principle.

Remark 13.4. The fact that $\operatorname{supp} E_+ \subset \overline{C_+}$ enables us to form $E_+ * u$ for every $u \in \mathcal{D}'(\mathbf{R}^{n+1})$ such that $\operatorname{supp} u$ is contained in the half-space $\{\, (x,t) \in \mathbf{R}^{n+1} \mid t \geq 0 \,\}$; see the first part of the proof of Lemma 13.2. If, moreover, the distribution u satisfies the wave equation $\square\, u = 0$, then $u = 0$. Indeed,

$$u = \delta * u = (\square\, E_+) * u = \square\, (E_+ * u) = E_+ * \square\, u = 0.$$

This implies that E_+ is the only fundamental solution E with support in the half-space $t \geq 0$. Indeed, in that case $u := E_+ - E$ also has support in $t \geq 0$, while $\square\, u = \square\, E_+ - \square\, E = \delta - \delta = 0$, and therefore $E_+ - E = u = 0$.

By elaborating this further we can also obtain the *uniqueness* of solutions $u = u(x,t)$ for $t > 0$ of the *Cauchy problem*

$$\square\, u = f(x,t), \qquad \lim_{t \downarrow 0} u(x,t) = a(x), \qquad \lim_{t \downarrow 0} \partial_t u(x,t) = b(x), \tag{13.28}$$

for a given inhomogeneous term f and given initial functions a and b.

To prove the uniqueness, we have to demonstrate that $u(x,t) = 0$ for $t > 0$, in the case that $\square\, u = 0$ and $u(x,t)$ and $\partial_t u(x,t)$ both converge to 0 as $t \downarrow 0$. On the strength of the foregoing we can indeed draw that conclusion if we can show that extension of u by $u(x,t) = 0$ for $t < 0$ yields a distribution u on $\mathbf{R}^{n+1}$, while additionally, this distribution satisfies $\square\, u = 0$ on $\mathbf{R}^{n+1}$.

We now consider this Cauchy problem for the partial differential operator $\square$ as an initial-value problem for the ordinary differential operator $\frac{d^2}{dt^2} - \Delta_x$ in the variable t, with integral curves of the form $t \mapsto u_t \in \mathcal{D}'(\mathbf{R}^n)$. If we write $u_t(x) = u(x,t)$, then u_t is a function on $\mathbf{R}^n$ for every $t > 0$ and we obtain, for every test function ϕ with support in $t > 0$,

$$u(\phi) = \int_{\mathbf{R}_{>0}} u_t(\phi_t)\,dt. \tag{13.29}$$

This leads to the idea of defining a *distributional solution of the Cauchy problem* (13.28) as a family $(u_t)_{t\in\mathbf{R}_{>0}}$ of distributions on $\mathbf{R}^n$ with the following properties:

(a) $t \mapsto u_t(\psi)$ is of class C^2, for every $\psi \in C_0^\infty(\mathbf{R}^n)$,
(b) $\frac{d^2}{dt^2}u_t - \Delta_x u_t = f_t$, for every $t > 0$,
(c) $u_t \to a$ and $\frac{d}{dt}u_t \to b$ in $\mathcal{D}'(\mathbf{R}^n)$, as $t \downarrow 0$.

Here we assume that $(f_t)_{t\in\mathbf{R}_{>0}}$ is a continuous family of distributions on $\mathbf{R}^n$, and that a and b belong to $\mathcal{D}'(\mathbf{R}^n)$. We obtain uniqueness for this Cauchy problem if $f_t \equiv 0$ and $a = b = 0$ imply that $u_t \equiv 0$. In the first instance one may think of C^2 functions u, instead of distributions.

The identity (13.29), now for all $\phi \in C_0^\infty(\mathbf{R}^{n+1})$ and without the restriction $t > 0$ for the support of ϕ, defines a distribution u on $\mathbf{R}^{n+1}$, with support in the half-space $t \geq 0$. This can be demonstrated by means of the principle of uniform boundedness, while also using the continuity of $t \mapsto u_t : \mathbf{R}_{\geq 0} \to \mathcal{D}'(\mathbf{R}^n)$. All we have to do is show that $\Box\, u = 0$. But this follows from

$$\begin{aligned}(\Box\, u)(\phi) = u(\Box\, \phi) &= \lim_{s\downarrow 0} \int_s^\infty u_t\Big(\frac{d^2}{dt^2}\phi_t - \Delta\phi_t\Big)\,dt \\ &= \lim_{s\downarrow 0}\Big(-u_s\Big(\frac{d}{ds}\phi_s\Big) + \frac{d}{ds}u_s(\phi_s) + \int_s^\infty \Big(\frac{d^2}{dt^2}u_t - \Delta u_t\Big)(\phi_t)\,dt\Big) = 0.\end{aligned}$$

For C^2 functions this is clear; for the justification in the distributional context, the principle of uniform boundedness is invoked once more.

An existence theorem for the Cauchy problem is given in Example 18.7 below, by means of Fourier transform. ⊘

13.3 Appendix: Euler's Gamma Function

Euler's Gamma function is defined by

$$\Gamma(a) := \int_{\mathbf{R}_{>0}} e^{-t}\,t^{a-1}\,dt. \tag{13.30}$$

This is an absolutely convergent integral if $a \in \mathbf{C}$ and $\operatorname{Re} a > 0$. Differentiation with respect to a under the integral sign leads to the conclusion that Γ satisfies the Cauchy–Riemann equation in the complex right half-plane, and therefore defines a complex-analytic function on that open set.

By writing $e^{-t} = -\frac{d}{dt}e^{-t}$ and integrating by parts, we obtain

$$\Gamma(a) = (a-1)\,\Gamma(a-1) \qquad (\operatorname{Re} a > 1). \tag{13.31}$$

Because $\Gamma(1) = 1$, it follows by mathematical induction on k that

$$\Gamma(k) = (k-1)! \qquad (k \in \mathbf{N}).$$

Euler's brilliant idea was now to define, for every $a \in \mathbf{C}$ with $a \notin \mathbf{Z}_{\leq 0}$,

$$\Gamma(a) := \frac{\Gamma(a+k)}{a(a+1)\cdots(a+k-1)} \qquad (k \in \mathbf{N},\ \mathrm{Re}\, a + k > 0). \tag{13.32}$$

Using (13.31), one sees that the right-hand side does not depend on the choice of k, while in the complex right half-plane the definition is identical to the $\Gamma(a)$ introduced before. In this way we obtain a complex-analytic extension of the Gamma function to $\mathbf{C} \setminus \mathbf{Z}_{\leq 0}$; see Fig. 13.3. Equation (13.31) holds for all $a \in \mathbf{C}$ where the left-hand side and the right-hand side are both defined.

For $\mathrm{Re}\, a > 0$ and $\mathrm{Re}\, b > 0$ we can write

$$\begin{aligned}
\Gamma(a)\,\Gamma(b) &= \int_{\mathbf{R}_{>0}} \int_{\mathbf{R}_{>0}} e^{-(t+u)}\, t^{a-1}\, u^{b-1}\, dt\, du \\
&= \int_{\mathbf{R}_{>0}} \int_0^s e^{-s}\, t^{a-1}\, (s-t)^{b-1}\, dt\, ds \\
&= \int_{\mathbf{R}_{>0}} e^{-s}\, s^{a+b-1}\, ds \int_0^1 r^{a-1}\, (1-r)^{b-1}\, dr.
\end{aligned}$$

This has been obtained by first substituting $u = s - t$ and then $t = r\,s$. In other words,

$$\Gamma(a)\,\Gamma(b) = \Gamma(a+b)\, B(a, b) \qquad (\mathrm{Re}\, a > 0,\ \mathrm{Re}\, b > 0), \tag{13.33}$$

where B is *Euler's Beta function*, defined by

$$B(a, b) = \int_0^1 t^{a-1}\, (1-t)^{b-1}\, dt. \tag{13.34}$$

The change of variables $t = \frac{u}{u+1}$ gives

$$B(a, b) = \int_{\mathbf{R}_{>0}} \frac{u^{a-1}}{(1+u)^{a+b}}\, du = 2 \int_{\mathbf{R}_{>0}} \frac{v^{2a-1}}{(1+v^2)^{a+b}}\, dv. \tag{13.35}$$

With $t = \cos^2 \alpha$, for $0 < \alpha < \frac{\pi}{2}$, we obtain

$$B(a, b) = 2 \int_0^{\frac{\pi}{2}} \cos^{2a-1} \alpha\, \sin^{2b-1} \alpha\, d\alpha. \tag{13.36}$$

In particular, taking $a = b = \frac{1}{2}$ in either one of the preceding formulas immediately leads to

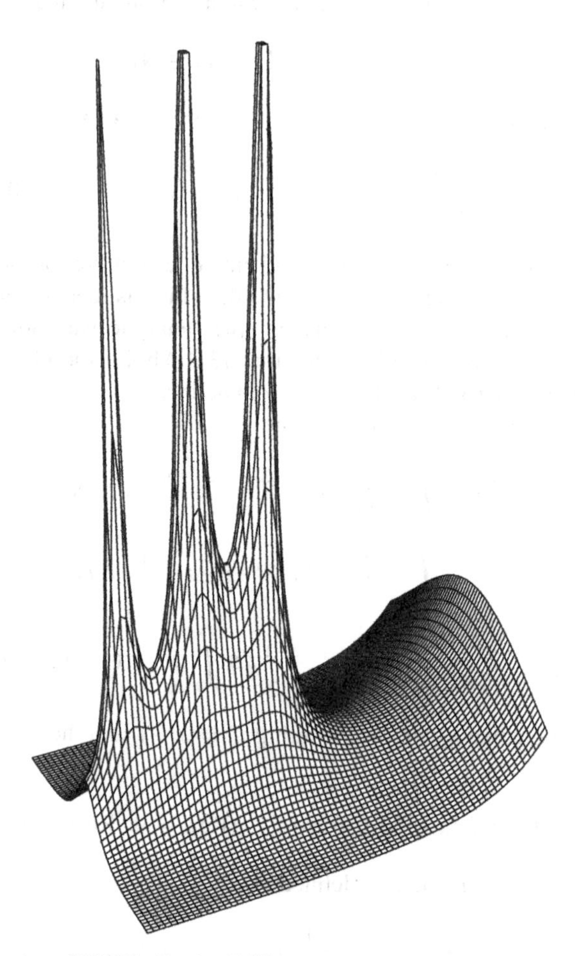

Fig. 13.3 Illustration of (13.32). Graph of $|\Gamma|$

$$\Gamma\Big(\frac{1}{2}\Big)^2 = 2\int_0^{\frac{\pi}{2}} d\alpha = \pi, \qquad \text{that is,} \qquad \Gamma\Big(\frac{1}{2}\Big) = \sqrt{\pi}.$$

This reconfirms (2.14), because of

$$\Gamma\Big(\frac{1}{2}\Big) = \int_{\mathbf{R}_{>0}} e^{-t}\, t^{-\frac{1}{2}}\, dt = 2\int_{\mathbf{R}_{>0}} e^{-x^2}\, dx = \int_{\mathbf{R}} e^{-x^2}\, dx.$$

A formula for the Euclidean $(n-1)$-dimensional volume c_n of the unit sphere $\{ x \in \mathbf{R}^n \mid \|x\| = 1 \}$ in $\mathbf{R}^n$ now follows from

$$\pi^{\frac{n}{2}} = \int_{\mathbf{R}^n} e^{-\|x\|^2}\, dx = c_n \int_{\mathbf{R}_{>0}} e^{-r^2}\, r^{n-1}\, dr$$

$$= c_n \int_{\mathbf{R}_{>0}} e^{-t}\, t^{\frac{n-1}{2}}\, \frac{1}{2} t^{-\frac{1}{2}}\, dt = c_n\, \frac{1}{2}\, \Gamma\Big(\frac{n}{2}\Big).$$

This yields (see [7, Example 7.9.1 or Exercise 7.21.(viii)] for alternative proofs)

$$c_n = \frac{2\,\pi^{\frac{n}{2}}}{\Gamma(\frac{n}{2})}. \tag{13.37}$$

With $n = 2m$ even, one has $\Gamma(\frac{n}{2}) = \Gamma(m) = (m-1)!$. With $n = 2m+1$ odd, one obtains

$$\Gamma\Big(\frac{n}{2}\Big) = \Gamma\Big(\frac{1}{2} + m\Big) = \pi^{\frac{1}{2}} \prod_{j=1}^{m} \Big(j - \frac{1}{2}\Big).$$

Note that $\sin 2\alpha = \sin 2(\frac{\pi}{2} - \alpha)$ and apply the change of variables $\sin^2 2\alpha = t$. This leads to $4 \sin\alpha \cos\alpha\, d\alpha = \frac{1}{2\sqrt{1-t}}\, dt$, and thus we obtain from (13.36)

$$\begin{aligned} B(a,\,a) &= 4 \int_0^{\frac{\pi}{4}} (\cos\alpha\ \sin\alpha)^{2a-1}\, d\alpha = 2^{1-2a} \int_0^1 t^{a-1}(1-t)^{-\frac{1}{2}}\, dt \\ &= 2^{1-2a} B\Big(a,\,\frac{1}{2}\Big). \end{aligned} \tag{13.38}$$

The following *duplication formula of Legendre* is an immediate consequence (see [7, Exercise 6.53] for a different proof):

$$\Gamma(2a)\Gamma\Big(\frac{1}{2}\Big) = 2^{2a-1}\, \Gamma(a)\, \Gamma\Big(a + \frac{1}{2}\Big). \tag{13.39}$$

By analytic continuation, the formula is valid for $a \in \mathbf{C} \setminus \frac{1}{2}\mathbf{Z}_{\leq 0}$.

Lemma 13.5. *For* $0 < \operatorname{Re} a < 1$, *one has the* reflection formula *for the Gamma function*

$$\Gamma(a)\,\Gamma(1-a) = B(a, 1-a) = \frac{\pi}{\sin \pi a}.$$

Proof. The first identity follows from $\Gamma(a+b) = \Gamma(1) = 1$ if $a + b = 1$.

For $0 < \operatorname{Re} a < 1$, we obtain from (13.36)

$$\Gamma(a)\,\Gamma(1-a) = B(1-a,\,a) = 2 \int_0^{\frac{\pi}{2}} \tan^{2a-1} \alpha\, d\alpha.$$

Set $\tan\alpha = x$. Then $(1 + \tan^2\alpha)\, d\alpha = dx$, and so we obtain from (16.32) below,

$$\Gamma(a)\,\Gamma(1-a) = 2 \int_0^\infty \frac{x^{2a-1}}{1+x^2}\, dx = \frac{\pi}{\sin \pi a}. \tag{13.40}$$

See (14.45) in Problem 14.39 or [7, Exercises 6.58 or 6.59] for other proofs. □

Corollary 13.6. *We have $\Gamma(a)\,\Gamma(1-a) = \frac{\pi}{\sin \pi a}$, for all $a \in \mathbf{C} \setminus \mathbf{Z}$. For every $a \in \mathbf{C}$ with $a \notin \mathbf{Z}_{\leq 0}$ we have $\Gamma(a) \neq 0$. If $k \in \mathbf{Z}_{\geq 0}$, then $\lim_{a \to -k}(a+k)\,\Gamma(a) = (-1)^k/k!$. The function $\frac{1}{\Gamma}$ possesses an extension to an entire analytic function on $\mathbf{C}$, with zeros only at $-k$, for $k \in \mathbf{Z}_{\geq 0}$, and with derivative at those points equal to $(-1)^k\,k!$; see Fig. 13.4.*

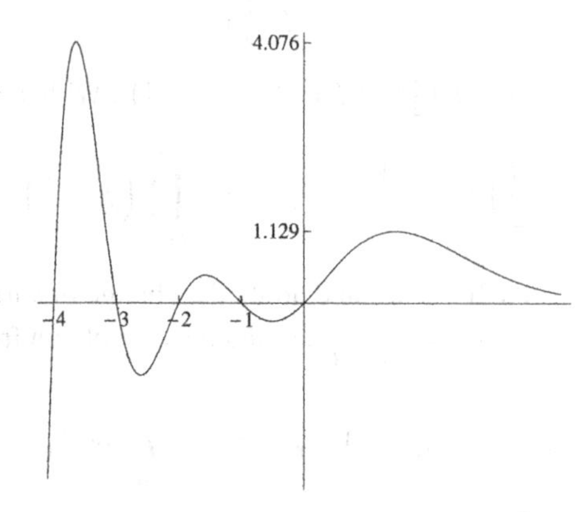

Fig. 13.4 Illustration of Corollary 13.6. Graph of the restriction of $\frac{1}{\Gamma}$ to a segment in $\mathbf{R}$

Proof. The first assertion follows by analytic continuation. The assertion that Γ has simple poles at the points $-k$ for $k \geq 0$, with residue $(-1)^k/k!$ at those points, follows from this formula, or directly from (13.32). □

Alternatively, for obtaining the analytic continuation and the properties of the poles one may use

$$\Gamma(a) = \int_{\mathbf{R}_{>0}} e^{-t}\,t^{a-1}\,dt = \int_0^1 e^{-t}\,t^{a-1}\,dt + \int_1^\infty e^{-t}\,t^{a-1}\,dt$$
$$= \sum_{k \in \mathbf{Z}_{\geq 0}} \frac{(-1)^k}{k!\,(a+k)} + \int_1^\infty e^{-t}\,t^{a-1}\,dt,$$

where the second integral on the right-hand side defines a complex-analytic function.

Problems

13.1.* *(Abel's integral equation.)* Given $0 < a < 1$ and $\phi \in C_0^\infty(\mathbf{R}_{>0})$, show that the solution $u \in \mathcal{D}'(\mathbf{R})_+$ to the integral equation

$$\phi(x) = \int_0^x \frac{u(y)}{(x-y)^a}\,dy$$

is given by

$$u(x) = \frac{\sin \pi a}{\pi} \int_0^x \frac{\phi'(y)}{(x-y)^{1-a}}\, dy$$

and belongs to $C^\infty(\mathbf{R}_{>0})$. The special case of $a = \frac{1}{2}$ is called *Abel's integral equation.*

13.2.* For $a \in \mathbf{C}$ with $\operatorname{Re} a > 0$, define $\chi_-^a \in \mathcal{D}'(\mathbf{R})$ by

$$\chi_-^a(\phi) = \int_{\mathbf{R}_{<0}} \frac{(-x)^{a-1}}{\Gamma(a)} \phi(x)\, dx \qquad (\phi \in C_0^\infty(\mathbf{R})).$$

Furthermore, set $|\chi|^a = \sum_\pm \chi_\pm^a$. Copy the theory about χ_+^a for the case of χ_-^a, by proving $\chi_-^a(\phi) = \chi_+^a(S\phi)$, and also for the case of $|\chi|^a$. In particular, show that

$$|\chi|^{-k} = \begin{cases} 2\delta^{(k)} & \text{if} \quad k \in \mathbf{Z}_{\geq 0} \text{ even,} \\ 0 & \text{if} \quad k \in \mathbf{N} \text{ odd.} \end{cases}$$

13.3. In the notation of Problems 12.14.(x) and 13.2, prove, for $a \in \mathbf{C}$ with $\operatorname{Re} a > 0$,

$$\frac{(x \pm i0)^a}{\Gamma(a+1)} = \chi_+^{a+1} + e^{\pm i\pi a} \chi_-^{a+1}.$$

Derive this equality for every $a \in \mathbf{C}$ and conclude that each of $a \mapsto (x \pm i0)^a$ is a complex-analytic family of distributions on $\mathbf{R}$.

13.4.* For $-1 < \operatorname{Re} a < 0$ and $\phi \in C_0^\infty(\mathbf{R})$, prove that the function $x \mapsto x^{a-1}(\phi(x) - \phi(0))$ is integrable on $\mathbf{R}_{\geq 0}$ and that

$$\chi_+^a(\phi) = \int_{\mathbf{R}_{>0}} \frac{x^{a-1}}{\Gamma(a)} (\phi(x) - \phi(0))\, dx.$$

In particular, for these values of a it is not true that $\chi_+^a = \text{test}\, \frac{x^{a-1}}{\Gamma(a)} H$.

More generally, show that for $k \in \mathbf{Z}_{\geq 0}$ and $-k-1 < \operatorname{Re} a < -k$,

$$\chi_+^a(\phi) = \int_{\mathbf{R}_{>0}} \frac{x^{a-1}}{\Gamma(a)} \Big(\phi(x) - \sum_{j=0}^k \frac{x^j}{j!} \phi^{(j)}(0)\Big)\, dx$$

$$= \frac{1}{\Gamma(a)} \lim_{\epsilon \downarrow 0} \Big(\int_\epsilon^\infty x^{a-1} \phi(x)\, dx + \sum_{j=0}^k \frac{\phi^{(j)}(0)}{j!\,(a+j)} \epsilon^{a+j} \Big).$$

Conclude that upon restriction to the linear subspace $L \subset C_0^\infty(\mathbf{R})$ consisting of the $\phi \in C_0^\infty(\mathbf{R})$ with $0 \notin \operatorname{supp} \phi$, one actually has

$$\chi_+^a(\phi) = \int_{\mathbf{R}_{>0}} \frac{x^{a-1}}{\Gamma(a)} \phi(x)\, dx, \qquad \text{that is,} \qquad \chi_+^a\big|_L = \Big(\text{test}\, \frac{x^{a-1}}{\Gamma(a)} H\Big)\Big|_L.$$

Verify that the last identity holds for all $a \in \mathbf{C}$.

13.5. Similarly to Problem 13.4, introduce $x_\pm^a \in \mathcal{D}'(\mathbf{R})$, for $-k-1 < \operatorname{Re} a < -k$ with $k \in \mathbf{Z}_{\geq 0}$, by setting

$$x_\pm^a(\phi) = \int_{\mathbf{R}_{>0}} x^a\Big(\phi(\pm x) - \sum_{j=0}^{k-1} \frac{\phi^{(j)}(0)}{j!}(\pm x)^j\Big)\,dx \qquad (\phi \in C_0^\infty(\mathbf{R})).$$

Then $x_\pm^a$ is well-defined. Next, let $|x|^a = \sum_\pm x_\pm^a \in \mathcal{D}'(\mathbf{R})$. Replace k by $2k$ in the formulas above, then demonstrate that $|x|^a$ is well-defined for $-2k-1 < \operatorname{Re} a < -2k+1$ and satisfies

$$|x|^a(\phi) = \int_{\mathbf{R}_{>0}} x^a\Big(\sum_\pm \phi(\pm x) - 2\sum_{j=0}^{k-1} \frac{\phi^{(2j)}(0)}{(2j)!}x^{2j}\Big)\,dx \qquad (\phi \in C_0^\infty(\mathbf{R})).$$

In particular, for $k \in \mathbf{Z}_{\geq 0}$ even, deduce that

$$|x|^{-k}(\phi) = \int_{\mathbf{R}_{>0}} x^{-k}\Big(\sum_\pm \phi(\pm x) - 2\sum_{j=0}^{\frac{k}{2}-1} \frac{\phi^{(2j)}(0)}{(2j)!}x^{2j}\Big)\,dx \qquad (\phi \in C_0^\infty(\mathbf{R})).$$

13.6.* Prove $\partial_x \,\mathrm{PV}\,\frac{1}{x} = -|x|^{-2}$ in $\mathcal{D}'(\mathbf{R})$ in the notation of Problem 13.5.

13.7. Let $a \in \mathbf{C}$, $k \in \mathbf{Z}$, and $0 < \operatorname{Re} a < k$. Prove that for every $f \in C^k(\mathbf{R})$ with $\operatorname{supp} f \subset [\,l, \infty\,[$ there exists exactly one continuous function u in $\mathbf{R}$ such that $\operatorname{supp} u \subset [\,l, \infty\,[$ and

$$\int_l^x (x-y)^{a-1}\,u(y)\,dy = f(x) \qquad (x \geq 0).$$

Derive a formula for $u(x)$ in terms of f, for every $x \geq 0$. How does $u = u_j$ behave as $j \to \infty$, if $f = f_j$ converges to the Heaviside function in $\mathcal{D}'(\mathbf{R})_+$?

13.8. Prove

$$R_+^{-2k} = \square^k \delta \qquad (k \in \mathbf{Z}_{\geq 0}).$$

For what $a \in \mathbf{C}$ does one have that $\operatorname{supp} R_+^a \subset \partial C_+$?

13.9.* Consider the distribution

$$\rho := \frac{d}{da} R_+^a\Big|_{a=0}.$$

Prove

$$E_+ = \frac{1}{(1-n)}\,q\,\rho \quad \text{if} \quad n > 1, \qquad \text{and} \qquad \sum_{j=1}^{n+1} y_j\,\partial_j \rho = -(n+1)\,\rho + \delta.$$

Is ρ homogeneous?

13.10.* *(Wave operator.)* In this problem, we directly compute a fundamental solution in $\mathcal{D}'(\mathbf{R}^4)$ of the wave operator $\square = \partial_t^2 - \Delta_x$, where $(x,t) \in \mathbf{R}^3 \times \mathbf{R} \simeq \mathbf{R}^4$. As usual, we suppose that $q : \mathbf{R}^4 \to \mathbf{R}$ is given by $q(x,t) = t^2 - \|x\|^2$; write $X = \mathbf{R}^4 \setminus \{0\}$.

(i) Let $\phi \in C_0^\infty(X)$ and prove by means of Problem 10.23.(ii) that

$$
\begin{aligned}
(q^*\delta)(\phi) &= \frac{d}{dy}\Big|_{y=0} \int_{\mathbf{R}^3} \int_{-\sqrt{\|x\|^2+y}}^{\sqrt{\|x\|^2+y}} \phi(x,t)\,dt\,dx \\
&= \sum_{\pm} \frac{1}{2} \int_{\mathbf{R}^3} \frac{\phi(x, \pm\|x\|)}{\|x\|}\,dx =: \sum_{\pm} \delta_\pm(\phi).
\end{aligned}
$$

(ii) Verify that the reflection in the hyperplane $\{ (x,0) \in \mathbf{R}^4 \mid x \in \mathbf{R}^3 \}$ transforms each of the distributions $\delta_\pm \in \mathcal{D}'(X)$ into the other. Use positivity to show that δ_+ is a Radon measure on X having the forward cone ∂C_+ as its support, and prove that δ_+ can be extended to a Radon measure on $\mathbf{R}^4$.

The next goal is a verification that $\square\,\delta_\pm = 2\pi\,\delta$ in $\mathcal{D}'(\mathbf{R}^4)$.

(iii) Write q for the variable in $\mathbf{R}$ and in particular, $v' = \partial_q v$ for $v \in \mathcal{D}'(\mathbf{R})$. Show that we have the following identity in $\mathcal{D}'(X)$ (compare with Problem 10.20.(ii)):

$$\square(q^*v) = 4q^*((q\,\partial_q + 2)v').$$

Conclude by means of the homogeneity of $\delta \in \mathcal{D}'(\mathbf{R})$ that $\square\,(q^*\delta) = 0$ on X. The supports of δ_+ and δ_- being disjoint, we obtain $\square\,\delta_\pm = 0$ on X.

(iv) Next prove the existence of $k \in \mathbf{Z}_{\geq 0}$ and $c_\alpha \in \mathbf{C}$ such that in $\mathcal{D}'(\mathbf{R}^4)$,

$$\square\,\delta_+ = \sum_{|\alpha| \leq k} c_\alpha \partial^\alpha \delta.$$

Show that δ_+ is homogeneous of degree -2 and conclude that $\square\,\delta_+$ is homogeneous of degree -4. Verify that $\partial^\alpha\delta$ is homogeneous of degree $-4 - |\alpha|$. Deduce the existence of $c \in \mathbf{C}$ satisfying $\square\,\delta_+ = c\,\delta$.

(v) Verify that the definition of $\delta_+(\phi)$ also applies with $\phi \in C^\infty(\mathbf{R}^4)$ for which the support intersects ∂C_+ in a compact set. Hence, choose ϕ of the form $\phi(x,t) = \psi(t)$ with $\psi \in C_0^\infty(\mathbf{R})$. Now use spherical coordinates in $\mathbf{R}^3$ in order to prove $c = 2\pi$. Conclude that

$$\square\,E_+ = \delta \qquad \text{if} \qquad E_+(\phi) = \frac{1}{4\pi} \int_{\mathbf{R}^3} \frac{\phi(x, \|x\|)}{\|x\|}\,dx.$$

(vi) Given $f \in C_0^\infty(\mathbf{R}^4)$, verify that a solution $u \in C^\infty(\mathbf{R}^4)$ of the inhomogeneous wave equation $\square\,u = f$ is provided by the *retarded potential*

$$u(x,t) = \frac{1}{4\pi} \int_{\mathbf{R}^3} \frac{f(y, t - \|x-y\|)}{\|x-y\|}\,dy.$$

13.11. In the case of $n = 3$ and $n = 2$, according to (13.24) a fundamental solution E_+ for the wave operator is given on $\mathbf{R}^{n+1} \setminus \overline{C_-}$ by, respectively,

$$E_+ = \frac{1}{2\pi} q^* \delta \qquad \text{and} \qquad E_+ = \frac{1}{2\sqrt{\pi}} q^* \chi_+^{\frac{1}{2}}.$$

We will derive more explicit forms of these fundamental solutions. Actually, with $n = 3$ and the sphere $S(t) = \{ x \in \mathbf{R}^3 \mid \|x\| = t \}$ we have, for all $\phi \in C_0^\infty(\mathbf{R}^4)$,

$$E_+(\phi) = \int_{\mathbf{R}_{>0}} \frac{1}{4\pi t} \int_{S(t)} \phi(x,t)\, dx\, dt. \tag{13.41}$$

With $n = 2$ and the solid disk $B(t) = \{ x \in \mathbf{R}^2 \mid \|x\| < t \}$ we have, for all $\phi \in C_0^\infty(\mathbf{R}^3)$,

$$E_+(\phi) = \frac{1}{2\pi} \int_{\mathbf{R}_{>0}} \int_{B(t)} \frac{\phi(x,t)}{\sqrt{t^2 - \|x\|^2}}\, dx\, dt. \tag{13.42}$$

First we consider the case of $n = 3$.

(i) Prove
$$q_*\phi(0) = \frac{1}{2} \sum_{\pm} \int_{\mathbf{R}^3} \frac{\phi(x, \pm\|x\|)}{\|x\|}\, dx =: \sum_{\pm} \delta_\pm(\phi).$$

Hint: use that $q : \mathbf{R}^4 \setminus \{0\}$ is a submersion and note that $q^{-1}(\{0\}) = \bigcup_\pm \partial C_\pm$, while $\partial C_\pm = \operatorname{graph}(h_\pm)$ with

$$h_\pm : \mathbf{R}^3 \to \mathbf{R} \qquad \text{defined by} \qquad h_\pm(x) = \pm\|x\|.$$

Next apply [7, Special case in Sect. 7.4.III]. Finally, note that $\|\operatorname{grad} q(x,t)\| = 2\sqrt{2}\,\|x\|$, for $(x,t) \in q^{-1}(\{0\})$.

(ii) Show that $\delta_\pm \in \mathcal{D}'(\mathbf{R}^4)$ is homogeneous of degree -2 by verification of the definition, by application of a result in the theory and in a problem, respectively. Prove that $\delta_\pm$ is a Radon measure invariant under $\mathbf{Lo}^\circ$ and supported by $\partial C_\pm$.

(iii) Verify (13.41) by introducing spherical coordinates in $\mathbf{R}^3$.

Next we derive (13.42) from (13.41) by means of *Hadamard's method of descent*. In this method solutions of a partial differential equation are obtained by considering them as special solutions of another equation that involves more variables and can be solved. Accordingly, consider $\phi \in C_0^\infty(\mathbf{R}^3)$ and define $\widetilde{\phi} \in C_0^\infty(\mathbf{R}^4)$ by

$$\widetilde{\phi}(x, x_3, t) = \phi(x,t) \qquad (x \in \mathbf{R}^2,\ x_3 \in \mathbf{R},\ t \in \mathbf{R}).$$

(iv) The sphere $S(t)$ minus its equator is the union of the graphs of the functions

$$h_\pm : B(t) \to \mathbf{R} \qquad \text{given by} \qquad h_\pm(x) = \pm\sqrt{t^2 - \|x\|^2}.$$

Imitate the method from [7, Example 7.4.10] to prove

$$\int_{S(t)} \widetilde{\phi}(x,t)\,dx = 2t \int_{B(t)} \frac{\phi(x,t)}{\sqrt{t^2 - \|x\|^2}}\,dx \qquad (t \in \mathbf{R}_{>0}).$$

Now prove (13.42).

Formula (13.42) can also be proved directly as follows.

(v) For $p > 0$, show that

$$F(p) := q^{-1}(\{p^2\}) \cap C_+ = \{\, (x,t) \in \mathbf{R}^3 \mid t^2 - \|x\|^2 = p^2,\ t > 0 \,\},$$

which is one sheet of a two-sheeted hyperboloid. Deduce $F(p) = \mathrm{graph}(h(p,\cdot))$, where

$$h : \mathbf{R} \times \mathbf{R}^2 \to \mathbf{R} \qquad \text{is given by} \qquad h(p,x) = \sqrt{p^2 + \|x\|^2}.$$

Derive, for $\phi \in C_0^\infty(\mathbf{R}^3 \setminus \overline{C_-})$,

$$q_*\phi(p^2) = \frac{1}{2} \int_{\mathbf{R}^2} \frac{\phi(x,\, h(p,x))}{h(p,x)}\,dx.$$

(vi) Using a change of variables in $\mathbf{R}_{>0}$, verify that for $\psi \in C_0^\infty(\mathbf{R})$,

$$\frac{1}{2\sqrt{\pi}}\, \chi_+^{\frac{1}{2}}(\psi) = \frac{1}{\pi} \int_{\mathbf{R}_{>0}} \psi(p^2)\,dp.$$

(vii) Combining parts (v) and (vi), deduce, for $\phi \in C_0^\infty(\mathbf{R}^3 \setminus \overline{C_-})$,

$$\frac{1}{2\sqrt{\pi}}\, (q^* \chi_+^{\frac{1}{2}})(\phi) = \frac{1}{2\pi} \int_{\mathbf{R}^2} \int_{\mathbf{R}_{>0}} \frac{\phi(x,\, h(p,x))}{h(p,x)}\,dp\,dx.$$

(viii) Next introduce the change of variables $h(p,x) = t$ in $\mathbf{R}_{>0}$, that is, $p = \sqrt{t^2 - \|x\|^2}$. Then $p > 0$ implies $t > \|x\|$. Deduce (see Fig. 13.5)

$$E_+(\phi) = \frac{1}{2\pi} \int_{\mathbf{R}^2} \int_{\|x\|}^{\infty} \frac{\phi(x,t)}{\sqrt{t^2 - \|x\|^2}}\,dt\,dx.$$

Furthermore, interchange the order of integration to obtain (13.42).

Finally we give an application of the results obtained.

(ix) Given $f \in C_0^\infty(\mathbf{R}^4)$, verify that a solution $u \in C^\infty(\mathbf{R}^4)$ of the inhomogeneous wave equation $\square u = f$ is provided by the *retarded potential*

$$u(x,t) = \frac{1}{4\pi} \int_{\mathbf{R}^3} \frac{f(y,\, t - \|x - y\|)}{\|x - y\|}\,dy.$$

13.12. *(Lorentz-invariant distribution on* $\mathbf{R}^{n+1} \setminus \{0\}$ *is a pullback under the quadratic form* q.*)* Let $q : \mathbf{R}^{n+1} \to \mathbf{R}$ be the quadratic form from (13.12) and define

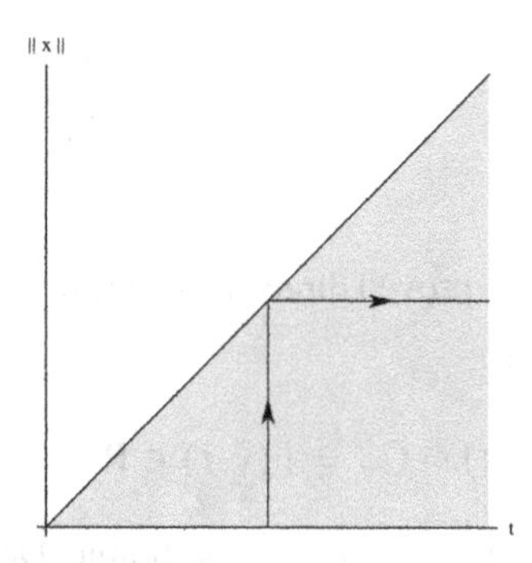

Fig. 13.5 Domain of integration in Problem 13.11.(viii)

$$\Phi_\pm : \mathbf{R}^n \times \mathbf{R}_{\gtrless 0} \to \mathbf{R}^{n+1}_+ := \{\, z \in \mathbf{R}^{n+1} \mid \textstyle\sum_{j=1}^n z_j^2 + z_{n+1} > 0 \,\}$$
$$\text{by} \qquad \Phi_\pm(y) = (y_1, \dots, y_n, q(y)) = z.$$

Prove that $z \mapsto (z_1, \dots, z_n, \pm\sqrt{\sum_{j=1}^n z_j^2 + z_{n+1}})$ are the inverses of $\Phi_\pm$ and that the $\Phi_\pm$ are C^∞ diffeomorphisms. Show that

$$^t(D\Phi_\pm(y)^{-1}) = \begin{pmatrix} 1\, 0 \dots 0 & \dfrac{y_1}{y_{n+1}} \\ \vdots & \vdots \\ 0\, 0 \dots 1 & \dfrac{y_n}{y_{n+1}} \\ 0\, 0 \dots 0 & \dfrac{1}{2y_{n+1}} \end{pmatrix}.$$

Given $u \in \mathcal{D}'(\mathbf{R}^{n+1} \setminus \{0\})$, consider its restrictions $u_\pm$ to the open subsets $\mathbf{R}^n \times \mathbf{R}_{\gtrless 0}$. Write $v_\pm = (\Phi_\pm^{-1})^* u_\pm \in \mathcal{D}'(\mathbf{R}^{n+1}_+)$. Derive from Problem 10.7 the following identity, for $1 \le j \le n$:

$$\Phi_\pm^*(\partial_j v_\pm) = \frac{1}{y_{n+1}}(y_{n+1}\, \partial_j + y_j\, \partial_{n+1}) u_\pm \quad \text{in} \quad \mathcal{D}'(\mathbf{R}^n \times \mathbf{R}_{\gtrless 0}).$$

In the particular case of u being invariant under the Lorentz group **Lo**, deduce from Problem 10.11.(ii) that $\partial_j v_\pm = 0$, for $1 \le j \le n$. Next apply Problem 11.19 repeatedly to obtain $w_\pm \in \mathcal{D}'(\mathbf{R})$ such that $v_\pm = 1_{\mathbf{R}^n} \otimes w_\pm$. Denote by $\Pi : \mathbf{R}^{n+1} \to \mathbf{R}$ the projection onto the last factor. Then $\Pi \circ \Phi_\pm = q$. Deduce

$$v_\pm = 1_{\mathbf{R}^n} \otimes w_\pm = \Pi^* w_\pm = (\Phi_\pm^{-1})^* q^* w_\pm, \qquad \text{that is,} \qquad u_\pm = q^* w_\pm \tag{13.43}$$

on $\mathbf{R}^n \times \mathbf{R}_{\gtrless 0}$.

In the arguments above, the coordinate y_{n+1} had a special role. Mutatis mutandis, by means of the equations from Problem 10.11.(ii) the conclusion in (13.43) can be drawn for every coordinate y_j, with $1 \le j \le n$, instead of y_{n+1}. Furthermore, use the Lorentz-invariance of u and the fact that the Lorentz group acts transitively on the level sets of q to conclude that w is independent of the choices made. Deduce that there exists a unique $w \in \mathcal{D}'(\mathbf{R})$ such $u = q^* w$ on

$$\bigcup_{j=1}^{n+1} \{ y \in \mathbf{R}^{n+1} \mid y_j \neq 0 \} = \mathbf{R}^{n+1} \setminus \{0\}.$$

The mapping q is not submersive on all of $\mathbf{R}^{n+1}$. As a consequence, a treatment of Lorentz-invariant distributions on $\mathbf{R}^{n+1}$ requires a more elaborate study.

13.13. We shall prove (see Problem 14.39 or [7, Exercise 6.60.(iii)] for another demonstration), for $a \in \mathbf{C}$,

$$\int_{\mathbf{R}_{>0}} x^{a-1} \begin{Bmatrix} \sin \\ \cos \end{Bmatrix} x \, dx = \Gamma(a) \begin{Bmatrix} \sin \\ \cos \end{Bmatrix} \Big(a \frac{\pi}{2} \Big) \qquad \begin{matrix} (-1 < \mathrm{Re}\, a < 1); \\ (0 < \mathrm{Re}\, a < 1). \end{matrix} \tag{13.44}$$

Let $t > 0$ and apply Cauchy's Integral Theorem (12.9) to the function $f(z) = e^{-z} z^{a-1}$ and the open set $U \subset \mathbf{C}$ that equals the quarter-circle of radius t with vertices 0, t, and it, of which the vertex 0, however, is cut off by a small quarter-circle of radius ϵ with $0 < \epsilon < t$. Show that, for $0 < \mathrm{Re}\, a < 1$,

$$\begin{aligned} 0 = \int_\epsilon^t e^{-x} x^{a-1} \, dx + i \int_0^{\frac{\pi}{2}} e^{-te^{i\phi}} t^a e^{ia\phi} \, d\phi + i \int_t^\epsilon e^{-iy} (iy)^{a-1} \, dy \\ + i \int_{\frac{\pi}{2}}^0 e^{-\epsilon e^{i\phi}} \epsilon^a e^{ia\phi} \, d\phi =: \sum_{j=1}^4 I_j. \end{aligned}$$

For $0 \leq \phi \leq \frac{\pi}{2}$, we have $\cos\phi \geq 1 - \frac{2}{\pi}\phi$ and therefore

$$\int_0^{\frac{\pi}{2}} e^{-t\cos\phi} \, d\phi \leq \frac{\pi}{2} \frac{1}{t}; \qquad \text{hence} \qquad |I_2| \leq \frac{\pi}{2} t^{\mathrm{Re}\, a - 1} e^{|\mathrm{Im}\, a| \frac{\pi}{2}}.$$

Furthermore,

$$|I_4| \leq \frac{\pi}{2} \epsilon^{\mathrm{Re}\, a} e^{|\mathrm{Im}\, a| \frac{\pi}{2}}.$$

Conclude by taking limits for $\epsilon \downarrow 0$ and $t \to \infty$, for $0 < \mathrm{Re}\, a < 1$, that

$$\Gamma(a) = \int_{\mathbf{R}_{>0}} e^{-x} x^{a-1} \, dx = e^{ia\frac{\pi}{2}} \int_{\mathbf{R}_{>0}} e^{-iy} y^{a-1} \, dy.$$

If we carry out the same reasoning as above but using the quarter-circle of radius t of vertices 0, t, and $-it$, the quarter-circle again being indented at 0 by a quarter-circle of radius ϵ, we obtain

$$\Gamma(a) = e^{-ia\frac{\pi}{2}} \int_{\mathbf{R}_{>0}} e^{iy} y^{a-1} \, dy.$$

Deduce the formulas in (13.44) for $0 < \operatorname{Re} a < 1$ and verify that the first equation in (13.44) is valid for $-1 < \operatorname{Re} a < 1$.

Chapter 14
Fourier Transform

Let $\lambda \in \mathbf{C}^n$. The function

$$e_\lambda : x \mapsto e^{\langle x, \lambda \rangle} = e^{x_1 \lambda_1 + \cdots + x_n \lambda_n} \tag{14.1}$$

in $C^\infty(\mathbf{R}^n)$ has the remarkable property that

$$\partial^\alpha e_\lambda = \lambda^\alpha e_\lambda \qquad (\alpha \in (\mathbf{Z}_{\geq 0})^n).$$

As a consequence, for every linear partial differential operator with constant coefficients

$$P = P(\partial) = \sum_{|\alpha| \leq m} c_\alpha \, \partial^\alpha \quad : \quad C^\infty(\mathbf{R}^n) \to C^\infty(\mathbf{R}^n) \tag{14.2}$$

one has

$$P(\partial) \, e_\lambda = P(\lambda) \, e_\lambda \qquad \text{with} \qquad P(\lambda) = \sum_{|\alpha| \leq m} c_\alpha \, \lambda^\alpha.$$

In other words, e_λ is an eigenvector of all of the operators $P(\partial)$ simultaneously. We note in passing that e_λ is also an eigenvector of all translations T_a, for $a \in \mathbf{R}^n$, with eigenvalue $e_\lambda(-a)$.

The idea behind the Fourier transform is that the operators P can better be understood if the vectors on which they act can be written as linear combinations of eigenvectors; indeed, P then acts as a scalar multiplication on each summand, the scalar being the eigenvalue corresponding to the eigenvector in question. A particularity is that here the eigenvalues do not form a discrete set but an entire continuum, namely $\mathbf{C}^n$. For this reason it is natural to replace the linear combinations by integrals over λ, where a weight function plays the role of the coefficients in the integration. The approximating Riemann sums (in the λ-space) then become finite linear combinations.

As will appear later on, for a large subspace $\mathcal{S}(\mathbf{R}^n)$ of $C^\infty(\mathbf{R}^n)$, every function in $\mathcal{S}(\mathbf{R}^n)$ can be written as an integral of this type over the set $i\,\mathbf{R}^n = \{\, i\xi \in \mathbf{C}^n \mid \xi \in \mathbf{R}^n \,\}$. The function $e_{i\xi}$ is constant on the planes $\{\, x \in \mathbf{R}^n \mid \langle x, \xi \rangle = c \,\}$, for $c \in \mathbf{R}$, and behaves like a harmonic oscillation transversely to them; accordingly,

J.J. Duistermaat and J.A.C. Kolk, *Distributions: Theory and Applications*,
Cornerstones, DOI 10.1007/978-0-8176-4675-2_14,

this is called a *plane harmonic wave*. These are precisely the e_λ, with $\lambda \in \mathbf{C}^n$, that are bounded on $\mathbf{R}^n$, a great help in obtaining estimates.

For a Fourier series on $\mathbf{R}$ of the form

$$\phi(x) = \sum_{n \in \mathbf{Z}} c_n \, e^{in\omega x},$$

the Fourier coefficients c_n are given in terms of the function ϕ by

$$c_n = \frac{\omega}{2\pi} \int_0^{2\pi/\omega} e^{-in\omega x} \, \phi(x) \, dx,$$

the average of $e_{-in\omega} \, \phi$ over one period of ϕ; compare with Chap. 16. If we cast off the shackles of periodic functions, the numerical factor and the interval of integration are no longer determined.

For an integrable function u on $\mathbf{R}^n$ one now defines the *Fourier transform* $\mathcal{F} u$ (often written as $\hat{u}$ in the literature) as

$$\mathcal{F} u(\xi) = \int_{\mathbf{R}^n} e^{-i\langle x, \xi \rangle} \, u(x) \, dx \qquad (\xi \in \mathbf{R}^n). \tag{14.3}$$

Several other conventions are current in the literature. This definition is fairly standard in the theory of linear partial differential equations and leads to simple formulas, but is has the disadvantage of not being unitary on square-integrable functions (see Theorem 14.32 below). Some authors use the $+$ instead of the $-$ sign, others add a factor 2π to the exponential or put $(2\pi)^{-n/2}$ in front of the integral. In our convention, the factor $(2\pi)^{-n}$ occurs only in the inversion formula for the Fourier transform; see Theorem 14.13 below.

Example 14.1. If $1_{[a,b]}$ denotes the characteristic function of the interval $[a, b]$, then

$$\mathcal{F} 1_{[a,b]}(\xi) = i \, \frac{e^{-ib\xi} - e^{-ia\xi}}{\xi}. \tag{14.4}$$

In particular, if $-a = b = 1$, the right-hand side takes the form

$$\frac{e^{i\xi} - e^{-i\xi}}{i\xi} = \frac{2 \sin \xi}{\xi} =: 2 \operatorname{sinc} \xi,$$

where the notation sinc originates from *sinus cardinalis*; see Fig. 14.1. ⊘

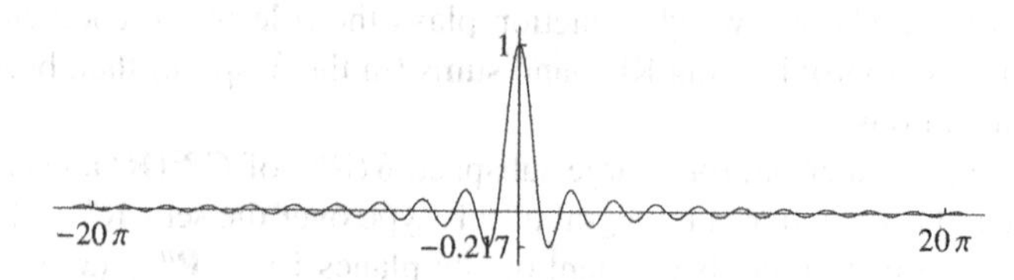

Fig. 14.1 Example 14.1. Graph of sinc

The following result is known as the *Riemann–Lebesgue Theorem*. In our development of the theory of distributions, it is the first instance in which certain properties of Lebesgue integrable functions have to be studied in some depth.

Theorem 14.2. *If u is a Lebesgue integrable function on $\mathbf{R}^n$, then $\mathcal{F}u$ is uniformly continuous on $\mathbf{R}^n$ and $\mathcal{F}u(\xi)$ converges to 0 if $\xi \in \mathbf{R}^n$ and $\|\xi\| \to \infty$. In particular, $\mathcal{F}u$ is bounded on $\mathbf{R}^n$ and satisfies*

$$|\mathcal{F}u(\xi)| \le \int_{\mathbf{R}^n} |u(x)|\,dx \qquad (\xi \in \mathbf{R}^n). \tag{14.5}$$

Proof. Since $|e^{-i\langle x,\xi\rangle}| = 1$, the estimate (14.5) follows from

$$\left|\int_{\mathbf{R}^n} e^{-i\langle x,\xi\rangle}\,u(x)\,dx\right| \le \int_{\mathbf{R}^n}\left|e^{-i\langle x,\xi\rangle}\,u(x)\right|\,dx = \int_{\mathbf{R}^n}|u(x)|\,dx.$$

For $0 \ne \xi \in \mathbf{R}^n$, we have

$$\begin{aligned}\mathcal{F}u(\xi) &= \int_{\mathbf{R}^n} e^{-i\langle x,\xi\rangle}\,u(x)\,dx = -\int_{\mathbf{R}^n} e^{-i\left\langle x+\frac{\pi}{\|\xi\|^2}\xi,\,\xi\right\rangle}\,u(x)\,dx\\ &= -\int_{\mathbf{R}^n} e^{-i\langle x,\xi\rangle}\,u\Big(x-\frac{\pi}{\|\xi\|^2}\xi\Big)\,dx.\end{aligned}$$

Hence

$$\mathcal{F}u = \frac{1}{2}\mathcal{F}\Big(I - T_{\frac{\pi}{\|\xi\|^2}\xi}\Big)u. \tag{14.6}$$

Lemma 20.44 with $p = 1$ now implies that $\mathcal{F}u(\xi) \to 0$ as $\|\xi\| \to \infty$.

It remains to show that $\mathcal{F}u$ is uniformly continuous. For arbitrary ξ and $\eta \in \mathbf{R}^n$, we obtain

$$|\mathcal{F}u(\xi+\eta) - \mathcal{F}u(\xi)| = \left|\int_{\mathbf{R}^n} e^{-i\langle x,\xi\rangle}(e^{-i\langle x,\eta\rangle}-1)\,u(x)\,dx\right| \le 2\int_{\mathbf{R}^n}|u(x)|\,dx.$$

For $x \in \mathbf{R}^n$, we have $\lim_{\eta\to 0}(e^{-i\langle x,\eta\rangle} - 1) = 0$. Therefore we see on account of *Lebesgue's Dominated Convergence Theorem*, see Theorem 20.26.(iv), that $\lim_{\eta\to 0}\mathcal{F}u(\xi+\eta) = \mathcal{F}u(\xi)$ uniformly in ξ. □

Remark 14.3. An alternative proof of Theorem 14.2 can be based on Example 14.1 and runs as follows.

Let $C(0)$ be the space of continuous functions f on $\mathbf{R}^n$ with the property that $f(\xi) \to 0$ as $\|\xi\| \to \infty$. If v is the characteristic function of an n-dimensional rectangle

$$\left\{x \in \mathbf{R}^n \mid a_j \le x_j \le b_j,\ 1 \le j \le n\right\},$$

we see from (14.4) that $\mathcal{F}v \in C(0)$. If v is a finite linear combination of characteristic functions of rectangles, then again $\mathcal{F}v \in C(0)$.

Furthermore, it is known from the theory of Lebesgue integration that for every Lebesgue integrable function u on $\mathbf{R}^n$ and every $\epsilon > 0$ there exists a finite linear combination v of characteristic functions of rectangles with $\int_{\mathbf{R}^n} |u(x) - v(x)|\, dx < \frac{\epsilon}{2}$. On account of (14.5), with u replaced by $u - v$, this implies that

$$\sup_{\xi \in \mathbf{R}^n} |\mathcal{F}u(\xi) - \mathcal{F}v(\xi)| < \frac{\epsilon}{2}.$$

Because $\mathcal{F}v \in C(0)$, there exists $R > 0$ with the property $|\mathcal{F}v(\xi)| < \frac{\epsilon}{2}$ whenever $\|\xi\| > R$. From this it follows that

$$|\mathcal{F}u(\xi)| \leq |\mathcal{F}u(\xi) - \mathcal{F}v(\xi)| + |\mathcal{F}v(\xi)| < \frac{\epsilon}{2} + \frac{\epsilon}{2} = \epsilon \qquad (\|\xi\| > R).$$

Thus we have shown that $\mathcal{F}u(\xi) \to 0$ as $\|\xi\| \to \infty$.

The limit of a uniformly convergent sequence of continuous functions is continuous. Because $\mathcal{F}u$ is the uniform limit of the above continuous functions $\mathcal{F}v$, we conclude that $\mathcal{F}u$ is continuous. This completes the proof that $\mathcal{F}u \in C(0)$. ⊘

If u has compact support, the definition of the Fourier transform can be extended to distributions and the Fourier transform is a complex-analytic function:

Lemma 14.4. *If u is an integrable function with compact support, then $\mathcal{F}u(\xi) = u(e_{-i\xi})$. For every $u \in \mathcal{E}'(\mathbf{R}^n)$ the function $\zeta \mapsto \mathcal{F}u(\zeta) := u(e_{-i\zeta})$ is complex-analytic on $\mathbf{C}^n$.*

Proof. The first assertion is evident. Let $u \in \mathcal{E}'(\mathbf{R}^n)$. Then, by the continuity of $\zeta \mapsto e_{-i\zeta} : \mathbf{C}^n \to C^\infty(\mathbf{R}^n)$, it follows that $\mathcal{F}u$ is continuous on $\mathbf{C}^n$. The difference quotient

$$\frac{1}{t}(e_{-i(\zeta + t\, e(j))} - e_{-i\zeta}) \qquad (t \in \mathbf{C})$$

converges to the function $x \mapsto -i x_j\, e_{-i\zeta}(x)$ in $C^\infty(\mathbf{R}^n)$ as $t \to 0$. From this we see that $\mathcal{F}u$ is complex-differentiable with respect to every variable, with partial derivative equal to the continuous function

$$\partial_j \mathcal{F}u(\zeta) = u(-i x_j\, e_{-i\zeta}) \qquad (\zeta \in \mathbf{C}^n). \tag{14.7}$$

On the strength of Cauchy's integral formula (12.11), applied to each variable, it now follows that $v = \mathcal{F}u$ is a complex-analytic function on $\mathbf{C}^n$. That is, for every $z \in \mathbf{C}^n$ the Taylor series of v,

$$\sum_\alpha \frac{\partial^\alpha v(z)}{\alpha!} (\zeta - z)^\alpha,$$

converges to $v(\zeta)$, for ζ in a suitable neighborhood in $\mathbf{C}^n$ of the point z. □

Example 14.5. The function sinc is complex-analytic on $\mathbf{C}$. This can also be seen from the power series expansion

$$\operatorname{sinc} z = \sum_{k \in \mathbf{Z}_{\geq 0}} \frac{(-1)^k}{(2k+1)!} z^{2k} \qquad (z \in \mathbf{C}). \qquad \oslash$$

For $\eta \in \mathbf{R}^n$ the function $u(e_{-\eta}) = \mathcal{F}u(-i\eta)$ is said to be the *Laplace transform* of u. The classical case is that in which $n = 1$, where u is a probability measure with support in $\mathbf{R}_{\geq 0}$, and $\eta > 0$. For this reason the complex-analytic function $\mathcal{F}u$ on $\mathbf{C}^n$, for $u \in \mathcal{E}'(\mathbf{R}^n)$, is said to be the *Fourier–Laplace transform* of u. In Theorem 18.1 below we will prove that the restriction of $\mathcal{F}u$ to $\mathbf{R}^n$ is a function that is at most of polynomial growth.

The analog of the Completeness Theorem for Fourier series is that "arbitrary" functions u can be written as continuous superpositions of the $e_{i\xi}$ with weight factors $c\,\mathcal{F}u(\xi)$, where c is a constant to be determined. Here we run into a problem, because $\mathcal{F}u$ is not necessarily integrable on $\mathbf{R}^n$ for every integrable function u. For example, if $1_{[a,b]}$ denotes the characteristic function of the interval $[a, b]$, then, by (14.4), $\mathcal{F}1_{[a,b]}$ is a real-analytic function on $\mathbf{R}$ that is not absolutely integrable on $\mathbf{R}$. The following space $\mathcal{S}$ of test functions, which lies between $C_0^\infty(\mathbf{R}^n)$ and $C^\infty(\mathbf{R}^n)$, proves to be very useful for constructing the theory.

Definition 14.6. A function ϕ on $\mathbf{R}^n$ is said to be *rapidly decreasing* if for every multi-index β, the function $x \mapsto x^\beta \phi(x)$ is bounded on $\mathbf{R}^n$. One defines $\mathcal{S} = \mathcal{S}(\mathbf{R}^n)$ as the space of all $\phi \in C^\infty(\mathbf{R}^n)$ such that $\partial^\alpha \phi$ is rapidly decreasing for every multi-index α.

If $(\phi_j)_{j\in\mathbf{N}}$ is a sequence in $\mathcal{S}$ and $\phi \in \mathcal{S}$, then ϕ_j is said to *converge to ϕ in $\mathcal{S}$*, notation $\lim_{j\to\infty} \phi_j = \phi$ in $\mathcal{S}$, if for all multi-indices α and β, the sequence of functions $(x^\beta\, \partial^\alpha \phi_j)_{j\in\mathbf{N}}$ converges uniformly on all of $\mathbf{R}^n$ to $x^\beta\, \partial^\alpha \phi$. $\oslash$

Two alternative formulations can be given: a function ϕ on $\mathbf{R}^n$ is rapidly decreasing if and only if for every $N > 0$ one has $\phi(x) = O\left(\|x\|^{-N}\right)$ as $\|x\| \to \infty$. We have $\phi \in \mathcal{S}$ if and only if $\phi \in C^\infty$ and $P(x, \partial)\phi$ is bounded on $\mathbf{R}^n$, for every linear partial differential operator $P(x, \partial)$ with polynomial coefficients.

$\mathcal{S} = \mathcal{S}(\mathbf{R}^n)$ is a linear subspace of $C^\infty = C^\infty(\mathbf{R}^n)$. When provided with the norms

$$\phi \mapsto \|\phi\|_{\mathcal{S}(k,N)} := \sup_{|\alpha|\leq k,\, |\beta|\leq N,\, x\in\mathbf{R}^n} |x^\beta\, \partial^\alpha \phi(x)| \tag{14.8}$$

it becomes a locally convex topological linear space in which convergence of sequences corresponds to that in Definition 14.6. Because it would be sufficient to consider the countable nondecreasing sequence of norms that is obtained by taking $N = k$, sequential continuity of linear operators on $\mathcal{S}$ is equivalent to continuity with respect to these norms; see Lemma 8.7.

We have $C_0^\infty \subset \mathcal{S} \subset C^\infty$. Furthermore, Definitions 2.13, 14.6, and 8.4 imply $\phi_j \to \phi$ in $C_0^\infty \ \Rightarrow\ \phi_j \to \phi$ in $\mathcal{S} \ \Rightarrow\ \phi_j \to \phi$ in C^∞. This is tantamount to the assertion that the identity, interpreted as a mapping from C_0^∞ to $\mathcal{S}$, is continuous, and the same applies to the identity from $\mathcal{S}$ to C^∞. Phrased differently, we have *continuous inclusions* in the following:

$$C_0^\infty(\mathbf{R}^n) \subset \mathcal{S}(\mathbf{R}^n) \subset C^\infty(\mathbf{R}^n). \tag{14.9}$$

By Lemma 8.2.(b), $C_0^\infty(\mathbf{R}^n)$ is dense in $C^\infty(\mathbf{R}^n)$ and therefore, a fortiori, $\mathcal{S}(\mathbf{R}^n)$ is dense in $C^\infty(\mathbf{R}^n)$. This implies that a continuous linear operator on $C^\infty(\mathbf{R}^n)$ is uniquely determined by its restriction to $\mathcal{S}(\mathbf{R}^n)$. The following lemma shows that likewise, every continuous linear operator on $\mathcal{S}(\mathbf{R}^n)$ is uniquely determined by its restriction to $C_0^\infty(\mathbf{R}^n)$.

Lemma 14.7. *Let $\chi \in C_0^\infty(\mathbf{R}^n)$ with $\chi(x) = 1$ for $\|x\| < 1$. Write $(\epsilon^*\chi)(x) = \chi(\epsilon\, x)$. For every $\phi \in \mathcal{S}(\mathbf{R}^n)$ and $\epsilon > 0$, one has that $\phi_\epsilon := (\epsilon^*\chi)\,\phi$ belongs to $C_0^\infty(\mathbf{R}^n)$ and converges in $\mathcal{S}(\mathbf{R}^n)$ to ϕ as $\epsilon \downarrow 0$.*

Proof. By application of Leibniz's formula we find that the value $x^\beta\, \partial^\alpha(\phi_\epsilon - \phi)(x)$ equals $(\chi(\epsilon\, x) - 1)\, x^\beta\, \partial^\alpha \phi(x)$ plus terms that can be estimated by a constant times $\epsilon^{|\gamma|}\, |x^\beta\, \partial^{\alpha-\gamma}\phi(x)|$, for $|\gamma| \neq 0$ and $\gamma \leq \alpha$. But the first term can be estimated by the supremum over the x with $\|x\| \geq 1/\epsilon$ of $|x^\beta\, \partial^\alpha \phi(x)|$. From this we see that $\lim_{\epsilon\downarrow 0} \phi_\epsilon \to \phi$ in $\mathcal{S}(\mathbf{R}^n)$. □

Theorem 14.8. *If P is a linear partial differential operator with polynomial coefficients, then P is a continuous linear mapping from $\mathcal{S}$ to $\mathcal{S}$.*

Proof. For the norms defined in (14.8) and $\phi \in \mathcal{S}$ we have the estimates

$$\|\partial_j\, \phi\|_{\mathcal{S}(k,N)} \leq \|\phi\|_{\mathcal{S}(k+1,N)}, \qquad \|x_j\, \phi\|_{\mathcal{S}(k,N)} \leq \|\phi\|_{\mathcal{S}(k,N+1)} + k\, \|\phi\|_{\mathcal{S}(k-1,N)},$$

where we have used that

$$\partial^\alpha\left(x_j\, \phi\right) = x_j\, \partial^\alpha \phi + \alpha_j\, \partial^{\alpha-\delta_j}\phi \qquad (1 \leq j \leq n).$$

Here δ_j is the multi-index having a 1 in the jth position and zeros everywhere else. From this follow the assertions for $P = x_j$ and for $P = \partial_j$. The general assertion then is obtained by mathematical induction on the order of the operator and the degree of the polynomials in the coefficients. □

For the next lemma the following notation proves useful:

$$D_j := \frac{1}{i}\partial_j \quad (1 \leq j \leq n).$$

The operator $\hbar\, D_j = \frac{\hbar}{i}\partial_j$ is known in quantum mechanics as the jth *momentum operator*. In this expression $\hbar = \frac{h}{2\pi}$, where h is Planck's constant, a quantity describing the atomic scale.

Lemma 14.9. *The Fourier transform $\mathcal{F} : u \mapsto \mathcal{F}u$ defines a continuous linear mapping from $\mathcal{S}$ to $\mathcal{S}$. For every $1 \leq j \leq n$, $\phi \in \mathcal{S}$, and ξ and $a \in \mathbf{R}^n$, one has*

$$\mathcal{F}(D_j\phi)(\xi) = \quad \xi_j\,(\mathcal{F}\phi)(\xi), \tag{14.10}$$
$$\mathcal{F}(x_j\,\phi)(\xi) = -D_j(\mathcal{F}\phi)(\xi), \tag{14.11}$$
$$\mathcal{F}(T_a{}^*\phi)(\xi) = \quad (e_{ia}\,\mathcal{F}\phi)(\xi), \tag{14.12}$$
$$\mathcal{F}(e_{ia}\phi)(\xi) = (T_{-a}{}^*\mathcal{F}\phi)(\xi). \tag{14.13}$$

Here $T_a{}^*$ is as in Example 10.15. In compact notation, we have the following identities of continuous linear mappings from $\mathcal{S}$ to $\mathcal{S}$:

$$\mathcal{F}\circ D_j = \xi_j\circ\mathcal{F} \quad \textit{and} \quad \mathcal{F}\circ x_j = -D_j\circ\mathcal{F},$$
$$\mathcal{F}\circ T_a{}^* = e_{ia}\circ\mathcal{F} \quad \textit{and} \quad \mathcal{F}\circ e_{ia} = T_{-a}{}^*\circ\mathcal{F}.$$

Proof. Let $\phi\in\mathcal{S}$. Then the estimate $|\phi(x)|\le c\,(1+\|x\|)^{-N}$ for all $x\in\mathbf{R}^n$ and some $c>0$ and $N>n$ implies that ϕ is integrable on $\mathbf{R}^n$. On account of the Riemann–Lebesgue Theorem, Theorem 14.2, it follows that $\mathcal{F}\phi$ is a bounded continuous function on $\mathbf{R}^n$. But $x_j\,\phi$ is also integrable and $D_j=\frac{1}{i}\partial_{\xi_j}$ acting on the integrand in (14.3) yields $-e^{-i\langle x,\xi\rangle}\,x_j\,\phi(x)$. Therefore, the theorem on differentiation under the integral sign implies that $\mathcal{F}\phi$ is differentiable with respect to ξ_j, with continuous derivative given by (14.11). By mathematical induction on k we find that $\mathcal{F}\phi\in C^k(\mathbf{R}^n)$, for all $k\in\mathbf{Z}_{\ge 0}$, and that all derivatives of $\mathcal{F}\phi$ are bounded.

We obtain (14.10) when we apply integration by parts to the approximating integral over a large rectangle. For every pair of multi-indices α and β the function

$$\xi^\beta\,D^\alpha(\mathcal{F}\phi)=\xi^\beta\,\mathcal{F}((-x)^\alpha\,\phi)=\mathcal{F}(D^\beta((-x)^\alpha\phi)) =: \mathcal{F}\psi \tag{14.14}$$

is bounded, because $\psi\in\mathcal{S}$ according to Theorem 14.8. Here we have used the following self-explanatory notation:

$$D^\beta=\prod_{j=1}^n D_j{}^{\beta_j}=(-i)^{|\beta|}\,\partial^\beta.$$

We conclude that $\mathcal{F}(\mathcal{S})\subset\mathcal{S}$.

For (14.12), use the substitution of variables $x=y-a$ to obtain

$$(\mathcal{F}\circ T_a{}^*)\phi(\xi)=\int_{\mathbf{R}^n}e^{-i\langle x,\xi\rangle}\phi(x+a)\,dx=e^{i\langle a,\xi\rangle}\int_{\mathbf{R}^n}e^{-i\langle y,\xi\rangle}\phi(y)\,dy$$
$$=(e_{ia}\circ\mathcal{F})\phi(\xi).$$

The remaining identity (14.13) has a similar proof.

Finally we prove the continuity of $\mathcal{F}$ on the basis of (8.2). In particular, it is sufficient to show, for all k and $N\in\mathbf{Z}_{\ge 0}$, the existence of $c=c(k,N)>0$ such that we have

$$\|\mathcal{F}\phi\|_{\mathcal{S}(k,N)}\le c\,\|\phi\|_{\mathcal{S}(N,k+n+1)} \qquad (\phi\in\mathcal{S}).$$

To this end we first apply (14.5) to ψ in (14.14), then use the estimate

$$\int_{\mathbf{R}^n} |\psi(x)|\,dx = \int_{\mathbf{R}^n} (1+\|x\|)^{-(n+1)}\,(1+\|x\|)^{n+1}\,|\psi(x)|\,dx$$
$$\leq c' \sup_{x\in\mathbf{R}^n} (1+\|x\|)^{n+1}\,|\psi(x)|,$$

where $c' = \int_{\mathbf{R}^n} (1+\|x\|)^{-(n+1)}\,dx < \infty$. Application of Leibniz's rule then leads to the desired estimate. □

Remark 14.10. Observe that (14.12) and (14.13) taken in conjunction with (10.15) imply (14.10) and (14.11), respectively. The reverse implications may be proved as in Problem 11.3. ⊘

Example 14.11. For every $a \in \mathbf{C}$ with $\operatorname{Re} a > 0$,

$$u_a(x) = e^{-a\,x^2/2} \qquad (x \in \mathbf{R})$$

defines a function $u_a \in \mathcal{S}(\mathbf{R})$; the proof is analogous to that of Lemma 2.7. Therefore, $\mathcal{F}u_a \in \mathcal{S}$. Now $Du_a = iax\,u_a$, and we obtain by Fourier transformation of both sides, also taking into account (14.10) and (14.11),

$$D\mathcal{F}u_a = i\frac{1}{a}\xi\,\mathcal{F}u_a.$$

This yields

$$\mathcal{F}u_a = c(a)\,u_{\frac{1}{a}} \qquad \text{with} \qquad c(a) = (\mathcal{F}u_a)(0) = \int_{\mathbf{R}} e^{-ax^2/2}\,dx.$$

For $a > 0$, the change of variables $x = (2/a)^{1/2}\,y$ leads to

$$c(a) = \sqrt{\frac{2\pi}{a}}, \tag{14.15}$$

where we have also used $\int_{\mathbf{R}} e^{-x^2}\,dx = \sqrt{\pi}$; see (2.14).

Now $a \mapsto c(a)$ is complex-analytic on $H = \{\, a \in \mathbf{C} \mid \operatorname{Re} a > 0 \,\}$; differentiation under the integral sign shows, for example, that c satisfies the Cauchy–Riemann equation. With the definition

$$a^{-\frac{1}{2}} = |a|^{-\frac{1}{2}}\,e^{-\frac{1}{2}i\,\arg a}, \qquad \text{where} \qquad -\frac{\pi}{2} < \arg a < \frac{\pi}{2},$$

the right-hand side of (14.15) is a complex-analytic function on H as well. The difference v between the two sides equals 0 on the positive real axis. Therefore, the Taylor series of v vanishes at every point on that axis, which implies that $v = 0$ on an open neighborhood in $\mathbf{C}$ of $\mathbf{R}_{>0}$. On account of Lemma 2.6 we conclude that $v = 0$ on H. In other words, (14.15) holds for all $a \in \mathbf{C}$ with $\operatorname{Re} a > 0$.

The preceding results can be generalized to $\mathbf{R}^n$ by

$$u_a(x) = e^{-\sum_{j=1}^n a_j\,x_j{}^2/2} \qquad (a \in \mathbf{C}^n,\ \operatorname{Re} a_j > 0,\ x \in \mathbf{R}^n).$$

Since

$$e^{-i\langle x,\xi\rangle} = \prod_{j=1}^{n} e^{-ix_j\,\xi_j} \qquad \text{and} \qquad u_a(x) = \prod_{j=1}^{n} u_{a_j}(x_j),$$

we conclude that

$$\mathcal{F} u_a = (2\pi)^{\frac{n}{2}} \prod_{j=1}^{n} \frac{1}{\sqrt{a_j}}\, u_{\frac{1}{a_j}}. \qquad \oslash$$

The following lemma prepares the theorem that every $\phi \in \mathcal{S}$ can be written as a continuous linear combination of the $e_{i\xi}$, with $\xi \in \mathbf{R}^n$ and $c\,\mathcal{F}\phi(\xi)$ as weight function, giving the "Fourier coefficients."

Lemma 14.12. *Define $\mathcal{G} = S \circ \mathcal{F} \circ \mathcal{F} : \mathcal{S}(\mathbf{R}^n) \to \mathcal{S}(\mathbf{R}^n)$. Then $\mathcal{G}$ is a continuous linear mapping that commutes with the position and momentum operators and also with all translation operators. More specifically, for $x \in \mathbf{R}^n$ and $1 \le j \le n$,*

$$\text{(i)}\quad \mathcal{G}\circ x_j = x_j\circ\mathcal{G}, \qquad \text{(ii)}\quad \mathcal{G}\circ D_j = D_j\circ\mathcal{G}, \qquad \text{(iii)}\quad \mathcal{G}\circ T_x{}^* = T_x{}^*\circ\mathcal{G}. \qquad (14.16)$$

Proof. From (14.11) and (14.10) we deduce

$$\begin{aligned}\mathcal{F}\circ\mathcal{F}\circ x_j &= \mathcal{F}\circ(-D_j)\circ\mathcal{F} = -x_j\circ\mathcal{F}\circ\mathcal{F},\\ \mathcal{F}\circ\mathcal{F}\circ D_j &= \mathcal{F}\circ x_j\circ\mathcal{F} = -D_j\circ\mathcal{F}\circ\mathcal{F}.\end{aligned}$$

For the reflection $S\phi(x) = \phi(-x)$ we have

$$S\circ(-x_j) = x_j\circ S \qquad \text{and} \qquad S\circ(-D_j) = D_j\circ S;$$

thus we conclude that (14.16).(i) and (ii) hold. Furthermore, (14.12) and (14.13) imply (14.16).(iii), for all $x \in \mathbf{R}^n$. □

Theorem 14.13. *The Fourier transform $\mathcal{F} : \mathcal{S}(\mathbf{R}^n) \to \mathcal{S}(\mathbf{R}^n)$ is bijective, with inverse $\mathcal{F}^{-1} = (2\pi)^{-n}\, S\circ\mathcal{F} = (2\pi)^{-n}\,\mathcal{F}\circ S$. This is expressed by the formula*

$$\phi(x) = \frac{1}{(2\pi)^n}\int_{\mathbf{R}^n} e^{i\langle x,\xi\rangle}\,\mathcal{F}\phi(\xi)\,d\xi \qquad (\phi \in \mathcal{S}(\mathbf{R}^n),\ x \in \mathbf{R}^n). \qquad (14.17)$$

Proof. With δ the Dirac measure on $\mathbf{R}^n$ and $\mathcal{G}$ as in Lemma 14.12, consider $u := \delta\circ\mathcal{G} \in \mathcal{D}'(\mathbf{R}^n)$. In view of (14.16).(i) we conclude

$$x_j\,u = u\circ x_j = \delta\circ\mathcal{G}\circ x_j = \delta\circ x_j\circ\mathcal{G} = 0 \qquad (1 \le j \le n).$$

Theorem 9.5 then implies the existence of $c \in \mathbf{C}$ such that $u = c\,\delta$. Furthermore, for $\phi \in C_0^\infty(\mathbf{R}^n)$ and $x \in \mathbf{R}^n$, we now obtain, on account of (14.16).(iii),

$$\mathcal{G}\phi(x) = T_x{}^*(\mathcal{G}\phi)(0) = \mathcal{G}(T_x{}^*\phi)(0) = u(T_x{}^*\phi) = c\,\delta(T_x{}^*\phi) = c\,\phi(x).$$

Hence $\mathcal{G} = c\,I$ on $C_0^\infty(\mathbf{R}^n)$ and Lemma 14.7 then implies that $\mathcal{G} = c\,I$ on $\mathcal{S}(\mathbf{R}^n)$.

To determine c, we can apply $\mathcal{G}$ to $\phi = e^{-\|\cdot\|^2/2}$, for example. Then we see from Example 14.11, with all $a_j = 1$, that $\mathcal{F}\phi = (2\pi)^{n/2}\,\phi$. Therefore $\mathcal{G}\phi = (2\pi)^n\phi$, that is, $c = (2\pi)^n$.

We can now conclude that $((2\pi)^{-n} S \circ \mathcal{F}) \circ \mathcal{F} = I$. This is formula (14.17); in other words, $\mathcal{F}$ has left-inverse $(2\pi)^{-n} S \circ \mathcal{F}$. It then follows that $\mathcal{F}$ is injective. A more symmetric form is obtained if we left-multiply by $(2\pi)^n S$:

$$\mathcal{F} \circ \mathcal{F} = (2\pi)^n\, S.$$

If we now right-multiply this in turn by $(2\pi)^{-n} S$, we see that $(2\pi)^{-n}\mathcal{F} \circ S$ is a right-inverse of $\mathcal{F}$; its existence implies that $\mathcal{F}$ is surjective. The end result is that $\mathcal{F}$ is bijective, with inverse as in the theorem. □

Identity (14.17) is known as the *Fourier inversion formula*. Remark 14.18, Example 14.26, and Problems 14.7 and 15.9 below contain different proofs. In the case $n = 1$ the constant in the formula can also be determined by means of Fourier series; see Example 16.8 below.

Theorem 14.24 below implies that $\mathcal{G}\delta$ is a well-defined distribution. On account of this fact as well as Example 11.18, the proof of (14.17) can be simplified as follows. The example immediately leads to

$$\mathcal{G} = (\mathcal{G}\delta)\,*, \qquad \text{while} \qquad x_j(\mathcal{G}\delta) = \mathcal{G}(x_j\delta) = \mathcal{G}\,0 = 0; \qquad \text{so} \qquad \mathcal{G} = c\,I.$$

Remark 14.14. The plane waves are characterized by the fact that they are exactly the bounded eigenfunctions of the differential operator $D = (D_1, \dots, D_n)$, that is, if $\psi \in C^\infty(\mathbf{R}^n)$ satisfies the eigenvalue equation $D\psi = \lambda\,\psi$ with $\lambda \in \mathbf{C}^n$, and if ψ is a bounded function, then $\lambda = \xi$ with $\xi \in \mathbf{R}^n$, and $\psi(x) = \psi(0)\,e^{i\langle x, \xi\rangle}$; see Problem 14.1. The arbitrary function in $\mathcal{S}(\mathbf{R}^n)$ can therefore be written as a superposition of bounded eigenfunctions of the differential operator D acting on $C^\infty(\mathbf{R}^n)$. The formula

$$(\mathcal{F} \circ D \circ \mathcal{F}^{-1})\phi = \xi\,\phi \qquad (\phi \in \mathcal{S}(\mathbf{R}^n))$$

shows that the differential operator D acting on $\mathcal{S}(\mathbf{R}^n)$ can be diagonalized by conjugation with the Fourier transform $\mathcal{F}$, and that the action in $\mathcal{S}(\mathbf{R}^n)$ of the operator obtained by conjugation is that of multiplication by the coordinate ξ.

Apparently, $\mathcal{F}$ is the unitary (modulo a factor) linear coordinate transform in $\mathcal{S}(\mathbf{R}^n)$ that diagonalizes the Hermitian linear operator D (see Problem 14.4) acting in the infinite-dimensional linear space $\mathcal{S}(\mathbf{R}^n)$. Note that this is a strong analog of the *Spectral Theorem* from linear algebra for Hermitian linear operators in a finite-dimensional linear space over $\mathbf{C}$ provided with an inner product. See Theorem 14.32 below for the sense in which $\mathcal{F}$ modulo a factor is unitary. In fact, this case provides a model for the generalization of the theorem to the infinite-dimensional setting. ⊘

Example 14.15. Combining Theorem 14.13 and Example 14.11 gives immediately (compare with Problem 5.5), for $t > 0$,

$$\mathcal{F}^{-1} e^{-t\|\cdot\|^2} = \frac{1}{(4\pi t)^{\frac{n}{2}}}\, e^{-\frac{\|\cdot\|^2}{4t}}. \qquad \oslash$$

In the following theorem we use the notation

$$\langle \phi, \psi \rangle = \int_{\mathbf{R}^n} \phi(x)\, \psi(x)\, dx,$$
$$(\phi, \psi) = \langle \phi, \overline{\psi} \rangle = \int_{\mathbf{R}^n} \phi(x)\, \overline{\psi(x)}\, dx. \tag{14.18}$$

Here the overline denotes complex conjugation; thus, (14.18) defines a Hermitian inner product in the space of functions. These expressions are meaningful if $\phi\, \psi$ is integrable.

Theorem 14.16. *For all* $\phi, \psi \in \mathcal{S}(\mathbf{R}^n)$ *we have*

$$\langle \mathcal{F}\phi, \psi \rangle = \langle \phi, \mathcal{F}\psi \rangle, \tag{14.19}$$
$$(\phi, \psi) = (2\pi)^{-n}\, (\mathcal{F}\phi, \mathcal{F}\psi) \qquad \textit{(Parseval's formula)}, \tag{14.20}$$
$$\phi * \psi \in \mathcal{S} \quad \textit{and} \quad \mathcal{F}(\phi * \psi) = \mathcal{F}\phi\, \mathcal{F}\psi, \tag{14.21}$$
$$\mathcal{F}(\phi\, \psi) = (2\pi)^{-n}\, \mathcal{F}\phi * \mathcal{F}\psi. \tag{14.22}$$

Proof. Both sides in (14.19) are equal to the double integral

$$\int_{\mathbf{R}^n} \int_{\mathbf{R}^n} \phi(x)\, e^{-i\langle x, \xi \rangle}\, \psi(\xi)\, dx\, d\xi,$$

where we also use $\langle x, \xi \rangle = \langle \xi, x \rangle$.

The proof of (14.20) begins with the observation that the complex conjugate of $e^{-i\langle x, \xi \rangle}$ equals $e^{i\langle x, \xi \rangle}$, which implies that $\overline{\mathcal{F}\psi} = S \circ \mathcal{F}\overline{\psi}$. Combining (14.19) and Theorem 14.13, we then conclude

$$\langle \mathcal{F}\phi, \overline{\mathcal{F}\psi} \rangle = \langle \mathcal{F}\phi, S \circ \mathcal{F}\overline{\psi} \rangle = \langle \phi, \mathcal{F} \circ S \circ \mathcal{F}\overline{\psi} \rangle = (2\pi)^n\, \langle \phi, \overline{\psi} \rangle.$$

The convolution $\phi * \psi$ is bounded; this merely requires one of the factors to be integrable and the other one to be bounded. Using

$$x^\beta = (x - y + y)^\beta = \sum_{\gamma \le \beta} \binom{\beta}{\gamma} (x - y)^{\beta - \gamma}\, y^\gamma$$

we see that

$$x^\beta\, (\phi * \psi) = \sum_{\gamma \le \beta} \binom{\beta}{\gamma} (x^{\beta - \gamma}\, \phi) * (x^\gamma\, \psi).$$

Combining this with (11.4), we find that $x^\beta\, \partial^\alpha(\phi * \psi)$ is bounded for all multi-indices α and β, which implies $\phi * \psi \in \mathcal{S}$. The formula $\mathcal{F}(\phi * \psi) = \mathcal{F}\phi\, \mathcal{F}\psi$ is obtained by substituting $e^{-i\langle x,\xi\rangle} = e^{-i\langle x-y,\xi\rangle}\, e^{-i\langle y,\xi\rangle}$ in the double integral that represents the left-hand side.

For (14.22) we replace ϕ and ψ in (14.21) by $\mathcal{F}\phi$ and $\mathcal{F}\psi$, respectively. Further, we apply $\mathcal{F}^{-1}$ to both sides and use $\mathcal{F}^2 = (2\pi)^n\, S$. Thus we get

$$\mathcal{F}\phi * \mathcal{F}\psi = \mathcal{F}^{-1}(\mathcal{F}^2\phi\, \mathcal{F}^2\psi) = (2\pi)^n \mathcal{F} \circ S(S\phi\, S\psi) = (2\pi)^n \mathcal{F}(\phi\, \psi). \qquad \square$$

Remark 14.17. According to (14.19) the restriction to $\mathcal{S}(\mathbf{R}^n)$ of the transpose of $\mathcal{F}$ is $\mathcal{F}$ again. Using transposition and ${}^tD_j = -D_j$, we see that the identities (14.10) and (14.11) actually are equivalent. Similarly, (10.11) leads to the equivalence of (14.12) and (14.13).

From (1.14) and (14.12) we obtain, for $\phi \in \mathcal{S}(\mathbf{R}^n)$, the following identity of linear mappings from $\mathcal{S}$ to $\mathcal{S}$:

$$\begin{aligned}\mathcal{F} \circ (\phi\, *) &= \mathcal{F} \circ \int_{\mathbf{R}^n} \phi(a)\, T_{-a}{}^*\, da = \int_{\mathbf{R}^n} \phi(a)\, \mathcal{F} \circ T_{-a}{}^*\, da \\ &= \int_{\mathbf{R}^n} \phi(a)\, e_{-ia} \circ \mathcal{F}\, da = \Big(\int_{\mathbf{R}^n} e_{-ia}\, \phi(a)\, da\Big) \circ \mathcal{F} = (\mathcal{F}\phi) \circ \mathcal{F}.\end{aligned}$$

Note that this provides an alternative proof of the equality in (14.21).

In view of Remark 14.10 and the proof of Theorem 14.13 it is obvious by now that all functorial properties of the Fourier transform are a consequence of the equalities (14.10) and ${}^t\mathcal{F} = \mathcal{F}$. In fact, in Problem 15.13 below it will be shown that these two equalities determine the Fourier transform up to a scalar factor. In Problem 15.5 we show how to obtain these functorial properties by one uniform method. $\oslash$

Remark 14.18. The identity (14.19) leads to an elegant proof of (14.17). To this end, apply (14.19) with ψ replaced by $\epsilon^*\psi$, for $\epsilon > 0$. Use a change of variables twice or Problem 14.30.(ii) as well as the *Dominated Convergence Theorem of Arzelà*, see [7, Theorem 6.12.3], or that of *Lebesgue*, see Theorem 20.26.(iv), when passing to the limit as $\epsilon \downarrow 0$. This yields

$$\psi(0) \int_{\mathbf{R}^n} \mathcal{F}\phi(\xi)\, d\xi = \phi(0) \int_{\mathbf{R}^n} \mathcal{F}\psi(\xi)\, d\xi. \qquad (14.23)$$

Next, take $\psi = e^{-\|\cdot\|^2/2}$ and apply Example 14.11 and Problem 2.7 to obtain (14.17) for $x = 0$. Finally, treat the case of arbitrary $x \in \mathbf{R}^n$ by invoking (14.12). $\oslash$

Definition 14.19. A *tempered distribution* on $\mathbf{R}^n$ is a (sequentially) continuous linear form on $\mathcal{S} = \mathcal{S}(\mathbf{R}^n)$. The space of tempered distributions is denoted by $\mathcal{S}'$ or $\mathcal{S}'(\mathbf{R}^n)$. The term originates from the French “distribution tempérée.”

One says that for a sequence $(u_j)_{j\in\mathbf{N}}$ and u in $\mathcal{S}'$,

$$\lim_{j\to\infty} u_j = u \quad \text{in} \quad \mathcal{S}'$$

if $\lim_{j\to\infty} u_j(\phi) = u(\phi)$ in $\mathbf{C}$, for every $\phi \in \mathcal{S}$ (compare with Definitions 5.1 and 8.1). $\oslash$

Restriction of $u \in \mathcal{E}' = \mathcal{E}'(\mathbf{R}^n)$ to $\mathcal{S}$ yields a $\rho u \in \mathcal{S}'$; restriction of $u \in \mathcal{S}'$ to C_0^∞ gives a $\rho u \in \mathcal{D}' = \mathcal{D}'(\mathbf{R}^n)$. The restriction mappings ρ from $\mathcal{E}'$ to $\mathcal{S}'$ and from $\mathcal{S}'$ to $\mathcal{D}'$ are continuous and linear. These mappings are also injective, on account of Lemmas 8.2.(c) and 14.7, respectively. In both cases ρu is identified with u; this leads to the continuous inclusions (compare with (14.9))

$$\mathcal{E}'(\mathbf{R}^n) \subset \mathcal{S}'(\mathbf{R}^n) \subset \mathcal{D}'(\mathbf{R}^n). \tag{14.24}$$

We further note that Lemma 14.7 implies that for every $u \in \mathcal{S}'$ the $(\epsilon^* \chi)\, u \in \mathcal{E}'$ converge in $\mathcal{S}'$ to u as $\epsilon \downarrow 0$. Hence, $\mathcal{E}'$ is dense in $\mathcal{S}'$. Furthermore, C_0^∞ is dense in $\mathcal{E}'$, see Corollary 11.7, and C_0^∞ is dense even in $\mathcal{S}'$, in the sense that for every $u \in \mathcal{S}'$ there exists a sequence $(u_j)_{j\in\mathbf{N}}$ in C_0^∞ such that $u_j \to u$ in $\mathcal{S}'$ as $j \to \infty$. For example, take

$$u_j = \phi_\epsilon * \left((\epsilon^* \chi)\, u\right),$$

with ϕ_ϵ as in Lemma 11.6, χ as in Lemma 14.7, and $\epsilon = 1/j$, and apply (11.20). Indeed, on account of Theorem 11.5 this boils down to $u_j(\psi) = u((\epsilon^*\chi)(S\phi_\epsilon * \psi))$ tending to $u(\psi)$, for every $\psi \in \mathcal{S}$. Thus we have to prove that $(\epsilon^*\chi)(S\phi_\epsilon * \psi)$ tends to ψ in $\mathcal{S}$. The details of the argument are the same as in the initial part of the proof of Theorem 14.33 below. Consequently, every continuous operator on $\mathcal{S}'$ is uniquely determined by its restriction to C_0^∞.

For $u \in \mathcal{D}'$, one has that $u \in \mathcal{S}'$ if and only if there exist a constant $c > 0$ and an $\mathcal{S}$ norm n such that

$$|u(\phi)| \leq c\, n(\phi) \qquad (\phi \in C_0^\infty). \tag{14.25}$$

The continuous extension of u to $\mathcal{S}$ is obtained by taking $u(\phi) = \lim_{\epsilon\downarrow 0} u(\phi_\epsilon)$, for $\phi \in \mathcal{S}$ and ϕ_ϵ as in Lemma 14.7.

Example 14.20. Define

$$u(\phi) = \int_{\mathbf{R}} \phi(x)\, e^x \cos e^x\, dx \qquad (\phi \in \mathcal{S}(\mathbf{R})).$$

In the given form, the integral is not absolutely convergent and has to be interpreted as $\lim_{b\to\infty} \int_{-\infty}^b \phi(x)\, e^x \cos e^x\, dx$. However, using integration by parts and $\lim_{x\to\pm\infty} |\phi(x)| = 0$, it can be rewritten as the absolutely convergent integral $-\int_{\mathbf{R}} \phi'(x) \sin e^x\, dx$. Hence

$$|u(\phi)| \leq \int_{\mathbf{R}} |\phi'(x)|\, dx \leq \int_{\mathbf{R}} \frac{1}{1+x^2}\, dx \sup_{0\leq j\leq 1,\, x\in\mathbf{R}} (1+x^2)|\phi^{(j)}(x)|.$$

This estimate establishes that $u \in \mathcal{S}'(\mathbf{R})$. Observe that there exists no polynomial p on $\mathbf{R}$ such that $|e^x \cos e^x| \le |p(x)|$, for all $x \in \mathbf{R}$. $\oslash$

The following is a direct consequence of Theorem 14.8.

Theorem 14.21. *If P is a linear partial differential operator with polynomial coefficients, then P is a continuous linear mapping from $\mathcal{S}'(\mathbf{R}^n)$ to $\mathcal{S}'(\mathbf{R}^n)$.*

Example 14.22. For u integrable on $\mathbf{R}^n$, we are going to prove that $u \in \mathcal{S}'$. Denoting the space of Lebesgue integrable functions on $\mathbf{R}^n$ by $L^1 = L^1(\mathbf{R}^n)$, see Theorem 20.40, we have the continuous inclusions

$$C_0^\infty(\mathbf{R}^n) \subset \mathcal{S}(\mathbf{R}^n) \subset L^1(\mathbf{R}^n) \subset \mathcal{S}'(\mathbf{R}^n).$$

More generally: if $1 \le p < \infty$, then L^p is defined as the space of the locally integrable functions u on $\mathbf{R}^n$, modulo functions that are equal to zero almost everywhere, such that $|u|^p$ is integrable; see Theorem 20.41 for more details. In L^p,

$$\|u\|_{L^p} := \Big(\int_{\mathbf{R}^n} |u(x)|^p \, dx\Big)^{1/p}$$

defines a norm; L^p is *complete* with respect to this norm. L^∞ denotes the space of the essentially bounded locally integrable functions on $\mathbf{R}^n$ (modulo functions that are equal to zero almost everywhere); this is a Banach space with respect to the norm

$$\|u\|_{L^\infty} := \operatorname{ess\,sup} |u|.$$

Here ess sup f, the *essential supremum* of a function f, is defined as the infimum of all $\sup g$ as g runs through the set of all functions that are equal to f almost everywhere.

Let q be the real number such that $\frac{1}{p} + \frac{1}{q} = 1$, on the understanding that $q = \infty$ if $p = 1$ and $q = 1$ if $p = \infty$. Hölder's inequality from Problem 11.21 then implies, for $u \in L^p$ and $\phi \in \mathcal{S}$,

$$|u(\phi)| \le \|u\|_{L^p} \, \|\phi\|_{L^q}.$$

Next, choose $N \in \mathbf{Z}_{\ge 0}$ such that $Nq > n$, with $N = 0$ if $q = \infty$, that is, $p = 1$. If we then write

$$\phi(x) = (1 + \|x\|)^{-N} \, (1 + \|x\|)^N \, \phi(x)$$

and use the integrability of $(1 + \|x\|)^{-qN}$, we conclude that there exists a constant $c_N > 0$ with the property that for every $u \in L^p$ and $\phi \in \mathcal{S}$,

$$|u(\phi)| \le \|u\|_{L^p} \, c_N \, \|\phi\|_{\mathcal{S}(0,N)}. \tag{14.26}$$

We also obtain that $\|\phi\|_{L^q} \le c_N \, \|\phi\|_{\mathcal{S}(0,N)}$. Furthermore, (14.26) proves that we have the following continuous inclusions:

$$C_0^\infty(\mathbf{R}^n) \subset \mathcal{S}(\mathbf{R}^n) \subset L^p(\mathbf{R}^n) \subset \mathcal{S}'(\mathbf{R}^n).$$

In other words, convergence with respect to the L^p norm implies convergence in $\mathcal{S}'$; see Remark 20.43 for more details.

Combining this with Theorem 14.21, we find that the space of tempered distributions is really quite large: for every $1 \leq p \leq \infty$ we can start with $u \in L^p$, then apply to it an arbitrary linear partial differential operator P with polynomial coefficients, and conclude that the result $P\,u$ is a tempered distribution. ⊘

Definition 14.23. If $u \in \mathcal{S}'(\mathbf{R}^n)$, its Fourier transform $\mathcal{F}u \in \mathcal{S}'(\mathbf{R}^n)$ is defined by

$$\mathcal{F}u(\phi) = u(\mathcal{F}\phi) \qquad (\phi \in \mathcal{S}(\mathbf{R}^n)).$$ ⊘

In view of (14.19) the new and the old definition of $\mathcal{F}\phi$ coincide for $\phi \in \mathcal{S}(\mathbf{R}^n) \subset \mathcal{S}'(\mathbf{R}^n)$.

The slightly frivolous notation (3.3) is continued to

$$\mathcal{F}u(\xi) = \int_{\mathbf{R}^n} e^{-i\langle x,\xi\rangle} u(x)\,dx, \tag{14.27}$$

where we are not allowed to give ξ a value, since the expression is meaningful only after integration over ξ, with a test function as weight function.

Theorem 14.24. *For every $u \in \mathcal{S}'(\mathbf{R}^n)$ we have $\mathcal{F}u \in \mathcal{S}'(\mathbf{R}^n)$. The mapping $\mathcal{F} : u \mapsto \mathcal{F}u$ is a continuous linear mapping from $\mathcal{S}'(\mathbf{R}^n)$ to $\mathcal{S}'(\mathbf{R}^n)$ that is a common extension of the Fourier transform on $\mathcal{S}(\mathbf{R}^n)$, on the space $L^1(\mathbf{R}^n)$ of Lebesgue integrable functions, or on the space $\mathcal{E}'(\mathbf{R}^n)$ of distributions with compact support, respectively.*

If $u \in \mathcal{E}'(\mathbf{R}^n)$, then $\mathcal{F}u \in \mathcal{S}'(\mathbf{R}^n)$ corresponds to the analytic function $\mathcal{F}u$ from Lemma 14.4, and therefore certainly $\mathcal{F}u \in C^\infty(\mathbf{R}^n)$.

For every $u \in \mathcal{S}'(\mathbf{R}^n)$ and $1 \leq j \leq n$ we have

$$\mathcal{F}(D_j u) = \xi_j\,\mathcal{F}u \qquad \textit{and} \qquad \mathcal{F}(x_j\,u) = -D_j\mathcal{F}u. \tag{14.28}$$

Finally, $\mathcal{F}$ is bijective from $\mathcal{S}'(\mathbf{R}^n)$ to $\mathcal{S}'(\mathbf{R}^n)$, with inverse equal to $(2\pi)^{-n}\,S \circ \mathcal{F} = (2\pi)^{-n}\,\mathcal{F} \circ S$.

Proof. These assertions follow from Lemma 14.9 and Theorem 14.13, where we recognize $\mathcal{F} : \mathcal{S}' \to \mathcal{S}'$ as the transpose of $\mathcal{F} : \mathcal{S} \to \mathcal{S}$. The assertion that $\mathcal{F}$ is an extension of the Fourier transform on $\mathcal{S}$ follows from (14.19). On L^1, $\mathcal{F}$ corresponds to the Fourier transform, because the Fourier transform on L^1 is also a continuous extension of that on $\mathcal{S}$ and (14.19) holds for $\phi \in L^1$ and $\psi \in \mathcal{S}$.

Now let $u \in \mathcal{E}'(\mathbf{R}^n)$. As we have seen above, $\mathcal{E}'(\mathbf{R}^n) \subset \mathcal{S}'(\mathbf{R}^n)$; therefore $u \in \mathcal{S}'(\mathbf{R}^n)$, and thus $\mathcal{F}u \in \mathcal{S}'(\mathbf{R}^n)$ is defined. In order to prove that this distributional Fourier transform $\mathcal{F}u$ is equal to the analytic function $\widehat{u} = \mathcal{F}u$ from Lemma 14.4, we write, for every $\phi \in C_0^\infty(\mathbf{R}^n)$,

$$\mathcal{F}u(\phi) = u(\mathcal{F}\phi) = u\Big(\int_{\mathbf{R}^n} e_{-i\xi}\,\phi(\xi)\,d\xi\Big) = \int_{\mathbf{R}^n} u(e_{-i\xi})\,\phi(\xi)\,d\xi$$

$$= \int_{\mathbf{R}^n} \widehat{u}(\xi)\,\phi(\xi)\,d\xi = (\text{test}\,\widehat{u})(\phi),$$

where for the third equality we have used Lemma 11.4 with $A(\xi) = \phi(\xi)\,e_{-i\xi}$ as well as the linearity of u. So as to satisfy condition (b) from that lemma, i.e., that the support of $A(\xi)$ is contained in a compact set for all $\xi \in \mathbf{R}^n$, we have, in fact, to perform the calculation above with u replaced by $\chi\, u$, where $\chi \in C_0^\infty(\mathbf{R}^n)$ is a test function that equals 1 on a neighborhood of supp u. Thus, we have shown that $\mathcal{F} u = \text{test}\,\widehat{u}$.

The formulas in (14.28) follow from those in $\mathcal{S}$ by continuous extension or by transposition. The same holds for the formula $\mathcal{F} \circ \mathcal{F} = (2\pi)^n\, S$ in $\mathcal{S}'$. □

Example 14.25. We have, for all $a \in \mathbf{R}^n$,

$$T_a \circ \mathcal{F} = \mathcal{F} \circ e_{ia} \qquad \text{and} \qquad \mathcal{F} \circ T_a = e_{-ia} \circ \mathcal{F} \quad \text{in} \quad \mathcal{S}'(\mathbf{R}^n). \tag{14.29}$$

Indeed, the two identities follow by transposition from (14.12) and (14.13). ⊘

Example 14.26. Because $\mathcal{F}\delta(\phi) = \delta(\mathcal{F}\phi) = \mathcal{F}\phi(0) = 1(\phi)$, for every $\phi \in \mathcal{S}(\mathbf{R}^n)$, we see that

$$\mathcal{F}\delta = 1 \quad \text{in} \quad \mathcal{S}'(\mathbf{R}^n). \tag{14.30}$$

Since $S1 = 1$, we find on account of the Fourier inversion formula from Theorem 14.24 that

$$\mathcal{F}1 = (2\pi)^n\,\delta. \tag{14.31}$$

This identity is equivalent to the inversion formula in the form (14.17). In order to see this, apply both sides of (14.30) to $T_x{}^*\phi$, with $x \in \mathbf{R}^n$, and use (14.29) to conclude

$$(2\pi)^n\,\phi(x) = \mathcal{F}1(T_x{}^*\phi) = 1(\mathcal{F} \circ T_x{}^*)(\phi) = 1(e_{ix} \circ \mathcal{F})(\phi)$$
$$= \int_{\mathbf{R}^n} e^{i\langle x,\xi\rangle}\,\mathcal{F}\phi(\xi)\,d\xi.$$

Significantly enough, (14.31) also may be obtained independently of the Fourier inversion formula (14.17) as follows. Observe that

$$0 = \mathcal{F}0 = \mathcal{F}(D_j 1) = \xi_j\,\mathcal{F}1 \qquad (1 \le j \le n).$$

Therefore $\mathcal{F}1 = c\,\delta$ on account of Theorem 9.5, and so $c = 1(\mathcal{F}e^{-\|\cdot\|^2/2}) = (2\pi)^n$. Probably, this is the most transparent proof of (14.17). Furthermore, this proof too is based on (14.10) (actually, its global form (14.12)) and ${}^t\mathcal{F} = \mathcal{F}$. See the natural extension of Problem 9.6 to $\mathbf{R}^n$ or Problems 14.31 and 14.32 for related proofs.

In the Dirac notation (14.27) the equality (14.31) looks rather spectacular:

$$\delta(x) = \frac{1}{(2\pi)^n} \int_{\mathbf{R}^n} e^{i\langle x,\xi\rangle}\,d\xi. \tag{14.32}$$

See (16.7) below for the analog in the periodic case

With d equal to the difference mapping from (10.24), formula (14.31) implies

$$(2\pi)^n\, d^*\delta = d^*\mathcal{F}1,$$

which in Dirac notation is known as the *Fourier–Gel'fand formula*

$$\delta(x-y) = \frac{1}{(2\pi)^n}\int_{\mathbf{R}^n} e^{i\langle x-y,\,\xi\rangle}\,d\xi.$$

In fact, it leads to yet another proof of (14.17), in view of

$$\phi(x) = \int_{\mathbf{R}^n}\phi(y)\,\delta(x-y)\,dy = \frac{1}{(2\pi)^n}\int_{\mathbf{R}^n}\int_{\mathbf{R}^n} e^{i\langle x-y,\,\xi\rangle}\,\phi(y)\,dy\,d\xi.$$

If P is a polynomial function in n variables, then $P = P\,1$, and so (14.28) implies

$$\mathcal{F}(P(D)\,\delta) = P \qquad \text{and} \qquad \mathcal{F}P = (2\pi)^n\, P(-D)\,\delta. \tag{14.33}$$

In particular, if Δ denotes the Laplace operator, then

$$\mathcal{F}(P(\Delta)\,\delta) = (-\|\cdot\|^2)^*\,P. \tag{14.34}$$

Furthermore, application of (14.29) implies, for every $a \in \mathbf{R}^n$,

$$\mathcal{F}\,\delta_a = \mathcal{F}\circ T_a\,\delta = e_{-ia}\circ\mathcal{F}\,\delta = e_{-ia}; \tag{14.35}$$

in turn, this entails

$$\mathcal{F}(P\,e_{ia}) = T_a(\mathcal{F}P) = (2\pi)^n\,T_a(P(i\,\partial)\,\delta) = (2\pi)^n\,P(i\,\partial)\,\delta_a. \tag{14.36}$$

On account of Theorem 8.10 we obtain that a distribution on $\mathbf{R}^n$ has finite support if and only if it equals the Fourier transform of an *exponential polynomial*, which is defined in Problem 11.6, on $\mathbf{R}^n$ that is a tempered distribution, or more precisely, the Fourier transform of a function of the form $u = \sum_{k=1}^{l} P_k\,e_{ia_k}$, where $l \in \mathbf{Z}_{\geq 0}$, P_k is a polynomial function on $\mathbf{R}^n$, and $a_k \in \mathbf{R}^n$, for $1 \leq k \leq l$ (compare with Problem 14.24 below). Note that the linear subspace $L = \{\, P(\partial)\,u \in \mathcal{S}'(\mathbf{R}^n) \mid P$ a polynomial function on $\mathbf{R}^n \,\}$ is of finite dimension and see Problem 11.6 for the converse assertion. Furthermore, similar to the problem of multiple eigenvalues in the theory of the indicial equation for nth-order ordinary differential equations, the occurrence of polynomials is related to confluence, but now of Dirac measures. ⊘

Remark 14.27. By now, we have collected abundant evidence for the fact that the smoother a tempered distribution u, the faster the decay at infinity of its Fourier transform $\mathcal{F}u$. Phrased differently, in the decomposition of a distribution into plane harmonic waves, many oscillations of high frequency are needed to account for the small-scale variations of a "rough" distribution, while fewer are needed for a "smooth" distribution. Indeed:

- If u is a derivative of a Dirac measure, then $\mathcal{F}u$ is an exponential polynomial that is tempered; see (14.36).
- If u is of class L^1, then $\mathcal{F}u$ is continuous and vanishes at infinity; see the Riemann–Lebesgue Theorem, Theorem 14.2.
- If u is an integrable function that is Hölder continuous of order $0 < \alpha < 1$, then $\mathcal{F}u(\xi) = O((1 + \|\xi\|)^{-\alpha})$ as $\|\xi\| \to \infty$; see Problem 14.11 below.
- If u is a function of class C^k with derivatives of sufficient decay, then $\mathcal{F}u(\xi) = O((1 + \|\xi\|)^{-k})$ as $\|\xi\| \to \infty$; use integration by parts.
- If u is a nontrivial polynomial, then $\operatorname{supp} \mathcal{F}u = \{0\}$; see (14.33). ⊘

Remark 14.28. Define $f : \mathbf{R} \to \mathbf{R}$ by $f(x) = e^x$. One might be tempted to argue that

$$\mathcal{F} f = \mathcal{F}\Big(\sum_{k \in \mathbf{Z}_{\geq 0}} \frac{x^k}{k!} \Big) = 2\pi \sum_{k \in \mathbf{Z}_{\geq 0}} \frac{i^k}{k!} \delta^{(k)}.$$

The last expression, however, does not define an element in $\mathcal{D}'(\mathbf{R})$: its support is $\{0\}$ but it is not of finite order. This means that f does not belong to $\mathcal{S}'(\mathbf{R})$, whereas the partial sums of the series for f do. This shows that convergence of a sequence of functions uniformly on compact sets in $\mathbf{R}$ need not imply convergence of that sequence in $\mathcal{S}'(\mathbf{R})$. ⊘

Example 14.29. According to (1.3) we get, for $\mathrm{PV}\,\frac{1}{x} \in \mathcal{D}'(\mathbf{R})$ and any $\phi \in C_0^\infty(\mathbf{R})$,

$$\mathrm{PV}\,\frac{1}{x}(\phi) = -\int_{\mathbf{R}} (1 + x^2)\phi'(x)\, \frac{\log|x|}{1 + x^2}\, dx,$$

and so

$$\Big|\mathrm{PV}\,\frac{1}{x}(\phi)\Big| \leq \int_{\mathbf{R}} \frac{\log|x|}{1 + x^2}\, dx \, \sup_{x \in \mathbf{R}} (1 + x^2)|\phi'(x)|.$$

This estimate proves that $\mathrm{PV}\,\frac{1}{x} \in \mathcal{S}'(\mathbf{R})$. On account of Example 9.2 we have $x\,\mathrm{PV}\,\frac{1}{x} = 1$ in $\mathcal{D}'(\mathbf{R})$ and so in $\mathcal{S}'(\mathbf{R})$; hence (14.28) implies

$$i\,\partial_x\Big(\mathcal{F}\,\mathrm{PV}\,\frac{1}{x}\Big) = \mathcal{F}\Big(x\,\mathrm{PV}\,\frac{1}{x}\Big) = \mathcal{F}1 = 2\pi\,\delta$$

in view of the preceding example. Example 4.2 and Theorem 4.3 then lead to $\mathcal{F}\,\mathrm{PV}\,\frac{1}{x} = -2\pi i\, H + c$, for some $c \in \mathbf{C}$, where H denotes the Heaviside function, which is a tempered distribution. Furthermore, $\mathrm{PV}\,\frac{1}{x}$ is an odd distribution and so is $\mathcal{F}\,\mathrm{PV}\,\frac{1}{x}$ in view of $S\mathcal{F} = \mathcal{F}S$; this implies $-2\pi i + c = -c$, which leads to $c = \pi i$. Therefore, with sgn denoting the sign function on $\mathbf{R}$,

$$\mathcal{F}\,\mathrm{PV}\,\frac{1}{x} = -\pi i\ \mathrm{sgn}; \qquad \text{and so} \qquad \mathcal{F}\,\mathrm{sgn} = -2i\ \mathrm{PV}\,\frac{1}{\xi}, \tag{14.37}$$

because $2\pi\ \mathrm{PV}\,\frac{1}{x} = \mathcal{F}S\mathcal{F}\,\mathrm{PV}\,\frac{1}{x} = -\pi i\,\mathcal{F}S\,\mathrm{sgn} = \pi i\,\mathcal{F}\,\mathrm{sgn}$.

For any $\phi \in \mathcal{S}(\mathbf{R})$, the latter identity in (14.37) entails

$$\int_{\mathbf{R}_{>0}} \int_{\mathbf{R}} \phi(x) \sin \xi x \, dx \, d\xi = \lim_{\epsilon \downarrow 0} \int_{\mathbf{R} \setminus [-\epsilon, \epsilon]} \frac{\phi(x)}{x} \, dx.$$

Indeed,

$$\begin{aligned}
\mathcal{F} \operatorname{sgn}(\phi) = \operatorname{sgn}(\mathcal{F}\phi) &= \int_{\mathbf{R}} \operatorname{sgn}(\xi) \int_{\mathbf{R}} e^{-ix\xi} \phi(x) \, dx \, d\xi \\
&= -\int_{\mathbf{R}_{>0}} \int_{\mathbf{R}} e^{i\xi x} \phi(x) \, dx \, d\xi + \int_{\mathbf{R}_{>0}} \int_{\mathbf{R}} e^{-i\xi x} \phi(x) \, dx \, d\xi \\
&= -2i \int_{\mathbf{R}_{>0}} \int_{\mathbf{R}} \phi(x) \sin \xi x \, dx \, d\xi.
\end{aligned}$$

Finally, note that $H = \frac{1}{2} + \frac{1}{2} \operatorname{sgn}$. Applying (14.37), we now see that

$$\mathcal{F} H = \pi \, \delta - i \, \mathrm{PV} \frac{1}{\xi}. \tag{14.38}$$

⊘

Example 14.30. According to Example 4.2 one has $\delta = \partial H$, and on account of (14.30) this implies

$$1 = \mathcal{F}\delta = \mathcal{F}(iDH) = i\xi \, \mathcal{F} H.$$

But this does not uniquely determine $\mathcal{F} H$, because for every constant $c \in \mathbf{C}$, the equation above is also satisfied by $\mathcal{F} H + c \, \delta$. (All solutions are described in Problem 9.5.) For every $\epsilon > 0$, however, $H_\epsilon(x) = e^{-\epsilon x} H(x)$ is a Lebesgue integrable function on $\mathbf{R}$, and therefore $\mathcal{F} H_\epsilon$ is a continuous function. Furthermore, $\partial H_\epsilon = -\epsilon \, H_\epsilon + \delta$, and so

$$(i\xi + \epsilon) \, \mathcal{F} H_\epsilon = 1; \qquad \text{that is,} \qquad \mathcal{F} H_\epsilon(\xi) = \frac{1}{i\xi + \epsilon}, \tag{14.39}$$

because $\mathcal{F} H_\epsilon$ is continuous. Since $\lim_{\epsilon \downarrow 0} H_\epsilon = H$ in $\mathcal{S}'$ we now conclude, on the strength of Theorem 14.24, that

$$\mathcal{F} H = \lim_{\epsilon \downarrow 0} \frac{1}{i\xi + \epsilon} = \frac{1}{i} \, \frac{1}{\xi - i0} \tag{14.40}$$

with convergence in $\mathcal{S}'(\mathbf{R})$. A combination of (14.38) and (14.40) now leads to the following *Plemelj–Sokhotsky jump relations* (compare with Problems 1.3 and 12.14):

$$\frac{1}{x \pm i\,0} \pm \pi i \, \delta = \mathrm{PV} \frac{1}{x}. \tag{14.41}$$

⊘

Example 14.31. Consider $u \in \mathcal{E}'(\mathbf{R}^n)$ satisfying $u(x \mapsto x^\alpha) = 0$, for every multi-index α. From (14.7) we derive $\partial^\alpha \mathcal{F} u(\zeta) = (-i)^{|\alpha|} u(x \mapsto x^\alpha e_{-i\zeta})$. Since $\mathcal{F} u$ is complex-analytic on $\mathbf{C}^n$ on account of Lemma 14.4, we obtain, by power series expansion about 0,

$$\mathcal{F}u(\zeta) = \sum_{\alpha} \frac{(-i)^{|\alpha|}}{\alpha!} u(x \mapsto x^{\alpha})\, \zeta^{\alpha} = 0 \qquad (\zeta \in \mathbf{C}^n).$$

In other words, $\mathcal{F}u = 0$, but then Theorem 14.24 implies $u = 0$.

Using the Hahn–Banach Theorem, Theorem 8.12, one now deduces that the linear subspace in $C^{\infty}(\mathbf{R}^n)$ consisting of the polynomial functions is dense in $C^{\infty}(\mathbf{R}^n)$. In other words, given a function $\phi \in C^{\infty}(\mathbf{R}^n)$, there exists a sequence $(p_j)_{j \in \mathbf{N}}$ of polynomial functions on $\mathbf{R}^n$ such that for every multi-index α and compact set K in $\mathbf{R}^n$, the sequence $(\partial^{\alpha} p_j)_{j \in \mathbf{N}}$ converges to $\partial^{\alpha}\phi$ uniformly on K as $j \to \infty$. This is *Weierstrass's Approximation Theorem* (see [7, Exercise 6.103] for a proof along classical lines and the corollary to Theorem 16.20 for still another); conversely, the vanishing of u is a direct consequence of this theorem. ⊘

Next, we generalize Parseval's formula (14.20). A function f on $\mathbf{R}^n$ is said to be *square-integrable* if f is a locally integrable function such that $\int_{\mathbf{R}^n} |f(x)|^2\, dx < \infty$. The space of square-integrable functions is denoted by $L^2 = L^2(\mathbf{R}^n)$. According to the Cauchy–Schwarz inequality, the so-called L^2 *inner product* (ϕ, ψ) from (14.18) is well-defined for all ϕ and $\psi \in L^2$. The corresponding norm

$$\|f\| = \|f\|_{L^2} := (f, f)^{1/2} = \Big(\int_{\mathbf{R}^n} |f(x)|^2\, dx\Big)^{1/2}$$

is said to be the L^2 *norm*. The space $L^2(\mathbf{R}^n)$ is with respect to this norm. In other words: $L^2(\mathbf{R}^n)$ is a *Hilbert space*, an inner product space that is complete with respect to the corresponding norm. Additionally, the space of continuous functions with compact support is dense in $L^2(\mathbf{R}^n)$ by Theorem 20.41 and the definition preceding it. From Example 14.22 we obtain the continuous inclusion $L^2 \subset \mathcal{S}'$.

For a general Hilbert space $\mathcal{H}$, a linear mapping $U : \mathcal{H} \to \mathcal{H}$ is said to be a *unitary isomorphism* if $U(\mathcal{H}) = \mathcal{H}$ and

$$(Uf, Ug) = (f, g) \qquad (f, g \in \mathcal{H}).$$

This implies that U is bijective and that U^{-1} from $\mathcal{H}$ to $\mathcal{H}$ is unitary as well. (Observe that in an infinite-dimensional Hilbert space an injective linear mapping is not necessarily surjective.)

Theorem 14.32. *If u belongs to $L^2(\mathbf{R}^n)$, this is also true of its Fourier transform $\mathcal{F}u$. Parseval's formula (14.20) applies to all ϕ and $\psi \in L^2(\mathbf{R}^n)$. It follows that the restriction of $\widetilde{\mathcal{F}} := (2\pi)^{-\frac{n}{2}}\, \mathcal{F}$ to $L^2(\mathbf{R}^n)$ defines a unitary isomorphism from $L^2(\mathbf{R}^n)$ onto $L^2(\mathbf{R}^n)$.*

Proof. Applying Remark 20.43 below, we obtain a sequence $(\phi_j)_{j \in \mathbf{N}}$ in $C_0^{\infty}(\mathbf{R}^n)$ with $\lim_{j \to \infty} \|u - \phi_j\| = 0$. On account of (14.20) this leads to

$$\|\widetilde{\mathcal{F}}\phi_j - \widetilde{\mathcal{F}}\phi_k\| = \|\phi_j - \phi_k\| \le \|\phi_j - u\| + \|u - \phi_k\| \to 0,$$

for $j, k \to \infty$. Therefore $(\widetilde{\mathcal{F}}\phi_j)_{j \in \mathbf{N}}$ is a Cauchy sequence in L^2, and in view of the completeness of L^2, see Theorem 20.41, there exists $v \in L^2$ satisfying

$\lim_{j\to\infty} \|v - \widetilde{\mathcal{F}}\phi_j\| = 0$. This implies that $\widetilde{\mathcal{F}}\phi_j \to v$ in $\mathcal{S}'$. Because, in addition, $\phi_j \to u$ in $\mathcal{S}'$, we obtain $\widetilde{\mathcal{F}}\phi_j \to \widetilde{\mathcal{F}}u$ in $\mathcal{S}'$, and on account of the uniqueness of limits we derive that $\widetilde{\mathcal{F}}u = v$. Because $v \in L^2$, the conclusion is that $\widetilde{\mathcal{F}}u \in L^2$.

Combining the estimates

$$\|\widetilde{\mathcal{F}}u\| \le \|\widetilde{\mathcal{F}}u - \widetilde{\mathcal{F}}\phi_j\| + \|\widetilde{\mathcal{F}}\phi_j\| \qquad \text{and} \qquad \|\widetilde{\mathcal{F}}\phi_j\| = \|\phi_j\| \le \|u\| + \|u - \phi_j\|$$

and taking the limit as $j \to \infty$, we also deduce

$$\|\widetilde{\mathcal{F}}u\| \le \|u\|.$$

This implies that $\widetilde{\mathcal{F}}$ is continuous with respect to the L^2 norm. It follows that the right-hand side in (14.20) is continuous with respect to both variables ϕ and ψ in L^2, as is the left-hand side. If we now approximate ϕ and ψ by $\phi_j \in C_0^\infty$ and $\psi_j \in C_0^\infty$, respectively, relative to the L^2 norm and use $(\phi_j, \psi_j) = (\widetilde{\mathcal{F}}\phi_j, \widetilde{\mathcal{F}}\psi_j)$, we obtain (14.20) for ϕ and ψ in L^2.

Finally we prove the surjectivity of $\widetilde{\mathcal{F}}$. According to Theorem 14.13, we have $\widetilde{\mathcal{F}}(\mathcal{S}) = \mathcal{S}$, while $\mathcal{S}$ is dense in L^2. It follows that the image of L^2 under $\widetilde{\mathcal{F}}$ is dense in L^2, and because $\widetilde{\mathcal{F}}$ is unitary the image is closed; therefore, the image is equal to $L^2(\mathbf{R}^n)$. Indeed, if $\widetilde{\mathcal{F}}u_j \to v$ in L^2, then $\|\widetilde{\mathcal{F}}u_j - \widetilde{\mathcal{F}}u_k\| = \|u_j - u_k\|$ shows that $(u_j)_{j\in\mathbf{N}}$ is a Cauchy sequence in L^2. Hence there exists $u \in L^2$ with $u_j \to u$ in L^2, and therefore $\widetilde{\mathcal{F}}u_j \to \widetilde{\mathcal{F}}u$ in L^2. But this leads to $v = \widetilde{\mathcal{F}}u$. □

We give an extension of (14.21) to distributions. Note that the product $\mathcal{F}u\,\mathcal{F}v$ is well-defined as an element of $\mathcal{D}'(\mathbf{R}^n)$ if $u \in \mathcal{S}'(\mathbf{R}^n)$ and $v \in \mathcal{E}'(\mathbf{R}^n)$, because then $\mathcal{F}v \in C^\infty(\mathbf{R}^n)$ according to Theorem 14.24.

Theorem 14.33. *If $u \in \mathcal{S}'(\mathbf{R}^n)$ and $v \in \mathcal{E}'(\mathbf{R}^n)$, then $u * v \in \mathcal{S}'(\mathbf{R}^n)$, while $\mathcal{F}(u * v) = \mathcal{F}u\,\mathcal{F}v$. In particular, $\mathcal{F}u\,\mathcal{F}v \in \mathcal{S}'(\mathbf{R}^n)$.*

Proof. On the strength of (11.23), we recognize that to prove $u * v \in \mathcal{S}'$ we only have to demonstrate that $\phi \mapsto Sv * \phi : C_0^\infty \to C^\infty$ possesses an extension to a continuous mapping from $\mathcal{S}$ to $\mathcal{S}$. This means that for every pair of multi-indices α and β, we have to estimate the number

$$x^\beta\, \partial^\alpha (Sv * \phi)(x) = x^\beta\, (Sv * \partial^\alpha \phi)(x)$$

uniformly in x by an $\mathcal{S}$ norm of ϕ; this requires only a uniform estimate by an $\mathcal{S}$ norm of $\psi = \partial^\alpha \phi$. In view of (11.1),

$$(Sv * \psi)(x) = Sv(T_x \circ S\psi) = v(S \circ T_x \circ S\psi) = v(T_{-x}\psi).$$

Using the continuity of v on $C^\infty(\mathbf{R}^n)$, we deduce from (8.4) that it is sufficient to show that for every multi-index γ and compact $K \subset \mathbf{R}^n$, the number

$$x^\beta\, \partial_y{}^\gamma (\psi(y + x)) = \bigl(x^\beta\, \partial^\gamma \psi\bigr)(y + x)$$

can be estimated uniformly in $x \in \mathbf{R}^n$ and in $y \in K$ by an $\mathcal{S}$ norm of ψ. The change of variables $x = z - y$ and the uniform estimate $|(z-y)^\beta| \le c\,(1 + \|z\|)^{|\beta|}$, for all $z \in \mathbf{R}^n$ and $y \in K$, then yield the desired result.

We now prove $\mathcal{F}(u * v) = \mathcal{F}u\,\mathcal{F}v$ by continuous extension of this identity for u and $v \in C_0^\infty$. We write, for $u \in \mathcal{S}'(\mathbf{R}^n)$, $v \in \mathcal{E}'(\mathbf{R}^n)$, and $\phi \in C_0^\infty$,

$$\mathcal{F}(u * v)(\phi) = (u * v)(\mathcal{F}\phi) = u(Sv * \mathcal{F}\phi).$$

Because $\mathcal{F}\phi \in \mathcal{S}$, we obtain from the above that $Sv * \mathcal{F}\phi \in \mathcal{S}$. Following Definition 14.19 we observed that for every $u \in \mathcal{S}'$ there exists a sequence (u_j) in C_0^∞ with the property that $u_j \to u$ in $\mathcal{S}'$ as $j \to \infty$. From this we now conclude that $\mathcal{F}(u_j * v) \to \mathcal{F}(u * v)$ in $\mathcal{D}'$. But we also have $\mathcal{F}v\,\phi \in C_0^\infty$, and therefore we deduce from

$$(\mathcal{F}u\,\mathcal{F}v)(\phi) = \mathcal{F}u(\mathcal{F}v\,\phi) = u(\mathcal{F}(\mathcal{F}v\,\phi))$$

that one also has $\mathcal{F}u_j\,\mathcal{F}v \to \mathcal{F}u\,\mathcal{F}v$ in $\mathcal{D}'$.

Thus we conclude that $\mathcal{F}(u * v) = \mathcal{F}u\,\mathcal{F}v$ for all $u \in \mathcal{S}'$, if it holds for all $u \in C_0^\infty$, for given $v \in \mathcal{E}'$. The commutativity of the convolution in turn implies that $\mathcal{F}(u * v) = \mathcal{F}u\,\mathcal{F}v$ for all $v \in \mathcal{E}'$, because it holds for all $v \in C_0^\infty$, for given $u \in C_0^\infty$. □

A result related to the theorem above is that $u * \phi \in C^\infty(\mathbf{R}^n)$ and $\mathcal{F}(u * \phi) = \mathcal{F}u\,\mathcal{F}\phi$ if $u \in \mathcal{S}'(\mathbf{R}^n)$ and $\phi \in \mathcal{S}(\mathbf{R}^n)$.

In specific examples of homogeneous distributions it is often straightforward that the distributions are tempered. More generally one has *Hörmander's Theorem* [12, Thm. 7.1.18]:

Theorem 14.34. *If $u \in \mathcal{D}'(\mathbf{R}^n)$ and the restriction of u to $\mathbf{R}^n \setminus \{0\}$ is homogeneous (of arbitrary degree $a \in \mathbf{C}$), then $u \in \mathcal{S}'(\mathbf{R}^n)$.*

Proof. We begin with a $\rho \in C_0^\infty(\mathbf{R}_{>0})$ such that $\rho \ge 0$ and ρ is not identically zero. Through multiplication by a suitable positive number we can ensure that $\int \rho(1/s)\,s^{-1}\,ds = 1$. Let $r > 0$. By the change of variables $s = t/r$ we deduce

$$\int_{\mathbf{R}_{>0}} \rho\Big(\frac{r}{t}\Big)\,\frac{dt}{t} = 1 \qquad (r > 0).$$

Now define

$$\psi(x) := 1 - \int_1^\infty \rho\Big(\frac{\|x\|}{t}\Big)\,\frac{dt}{t}.$$

If x is bounded, the interval of integration can be replaced by a bounded interval, and the theorem on differentiation under the integral sign gives $\psi \in C^\infty(\mathbf{R}^n)$. On the other hand, if $\|x\|$ is large, the interval of integration can be replaced by $\mathbf{R}_{>0}$, which implies that $\psi(x) = 0$. In conclusion, $\psi \in C_0^\infty(\mathbf{R}^n)$.

Using the notation $\chi(x) := \rho(\|x\|)$ and $t : x \mapsto t\,x$, we now conclude, for every $\phi \in C_0^\infty(\mathbf{R}^n)$,

$$u(\phi) = u(\psi\,\phi) + u\Big(\int_1^\infty (1/t)^*\chi\,t^{-1}\,dt\,\phi\Big) = u(\psi\,\phi)$$
$$+\int_1^\infty u\left((1/t)^*\chi\,\phi\right)\,t^{-1}\,dt = u(\psi\,\phi) + \int_1^\infty u\left(\chi\,t^*\phi\right)\,t^{a+n-1}\,dt,$$

where we have used the linearity of u, the identity $(1/t)^*\chi\,\phi = (1/t)^*\,(\chi\,t^*\phi)$, and $(1/t)_* = t^n\,t^*$ (see Theorem 10.8), and $t^*u = t^a\,u$ on $\mathbf{R}^n \setminus \{0\}$.

Because both $\psi\,u$ and $\chi\,u$ are distributions with compact support, we deduce the existence of a constant $c_1 > 0$, a $k \in \mathbf{Z}_{\geq 0}$, and compact subsets K of $\mathbf{R}^n$ and L of $\mathbf{R}^n \setminus \{0\}$ such that

$$|u(\phi)| \leq c_1\,\|\phi\|_{C^k,K} + c_1 \int_1^\infty \|t^*\phi\|_{C^k,L}\,t^{\operatorname{Re}a+n-1}\,dt;$$

see (8.4). Now

$$\|t^*\phi\|_{C^k,L} = t^k\,\|\phi\|_{C^k,t\,L}$$

and for every N there exists a constant $c_2 > 0$ such that

$$\|\phi\|_{C^k,t\,L} \leq c_2\,t^{-N}\,\|\phi\|_{\mathcal{S}(k,N)}, \tag{14.42}$$

with the notation of (14.8). Indeed, it is sufficient to prove the estimate for $k = 0$. Because $0 \notin L$, there is a constant $m > 0$ such that $m \leq \max_{1\leq j\leq n}|x_j|$ for all $x \in L$. Consequently, for $x \in t\,L$ there exists a j such that $x_j \geq t\,m$, and therefore

$$|\phi(x)| = |x_j^N|\,|\phi(x)|\,|x_j|^{-N} \leq |x_j^N|\,|\phi(x)|(tm)^{-N} \leq m^{-N}\,t^{-N}\,\|\phi\|_{\mathcal{S}(0,N)}.$$

Choosing $N > k+\operatorname{Re}a+n$ in (14.42), we obtain a constant $c > 0$ with the property

$$|u(\phi)| \leq c\,\|\phi\|_{\mathcal{S}(k,N)} \qquad (\phi \in C_0^\infty(\mathbf{R}^n)).$$

This implies that u can be extended to a continuous linear form on $\mathcal{S}(\mathbf{R}^n)$; on account of Lemma 14.7 this extension is uniquely determined. □

Problems

14.1. *(Characterization of exponentials.)* Suppose $u \in \mathcal{D}'(\mathbf{R}^n)$ and $\partial_j u = \lambda_j\,u$, for some $\lambda \in \mathbf{C}^n$ and all $1 \leq j \leq n$. Demonstrate the existence of a constant $c \in \mathbf{C}$ such that $u = c\,e_\lambda$.

14.2. Prove $e_{-\lambda} \circ P(\partial) \circ e_\lambda = P(\partial + \lambda)$, for $\lambda \in \mathbf{C}^n$ and $P(\partial)$ as in (14.2).

14.3.* Suppose that $n > 1$ and let P be a polynomial function on $\mathbf{C}^n$ of degree at least one. Show that the zero-set of P is infinite. Deduce that the space of complex-analytic solutions u on $\mathbf{C}^n$ of the linear partial differential equation with constant coefficients $P(\partial)u = 0$ is of infinite dimension.

14.4. Prove that D_j is a Hermitian linear operator acting in $\mathcal{S}(\mathbf{R}^n)$ when this linear space is provided with the Hermitian inner product from (14.18), that is, $(D_j\phi, \psi) = (\phi, D_j\psi)$, for all $1 \leq j \leq n$ and ϕ and $\psi \in \mathcal{S}(\mathbf{R}^n)$.

14.5.* Assume that $\phi_1, \ldots, \phi_n$ are integrable functions on $\mathbf{R}$. Using the notation (11.8) prove that $\phi = \otimes_{j=1}^n \phi_j$ defines an integrable function on $\mathbf{R}^n$ with the property $\mathcal{F}\phi = \otimes_{j=1}^n \mathcal{F}\phi_j$.

14.6. Suppose that u and $v := \mathcal{F}u$ are Lebesgue integrable on $\mathbf{R}^n$. Prove that u equals the continuous function $(2\pi)^{-n}\, S \circ \mathcal{F}v$ almost everywhere on $\mathbf{R}^n$.

14.7.* Let $L : \mathcal{S} \to \mathcal{S}$ be a linear mapping that commutes with the position and momentum operators. Show that there exists a constant $c \in \mathbf{C}$ such that $L = c\, I$, with I the identity on $\mathcal{S}$. Deduce the Fourier inversion formula (14.17).

14.8.* Let $\phi \in \mathcal{S}(\mathbf{R}^n)$. Prove one of the following assertions and then derive the other:

(i) If $\phi(0) = 0$, then we may write $\phi = \sum_{j=1}^n x_j\, \phi_j$ with $\phi_j \in \mathcal{S}(\mathbf{R}^n)$.
(ii) If $\int_{\mathbf{R}^n} \phi(x)\, dx = 0$, then we may write $\phi = \sum_{j=1}^n \partial_j \phi_j$ with $\phi_j \in \mathcal{S}(\mathbf{R}^n)$.

14.9.* *(Eigenvalues of Fourier transform.)* What are the possible eigenvalues of the Fourier transform acting on $\mathcal{S}(\mathbf{R})$? And when acting on $L^2(\mathbf{R})$? For three eigenvalues, try to find corresponding eigenfunctions. See Problem 14.42 for a different approach.

14.10.* Let P be a harmonic polynomial function on $\mathbf{C}^n$ and define $(-i)^* P(z) = P(-i\, z)$, for all $z \in \mathbf{C}^n$. Verify *Hecke's formula*

$$\mathcal{F}(e^{-\frac{1}{2}\|\cdot\|^2} P) = (2\pi)^{\frac{n}{2}} e^{-\frac{1}{2}\|\cdot\|^2} (-i)^* P.$$

14.11.* Suppose $f \in C_0(\mathbf{R}^n)$ is Hölder continuous of order $0 < \alpha \leq 1$ (for this notion, see Definition 19.4 below). Prove the existence of $c > 0$ such that

$$|\mathcal{F} f(\xi)| \leq \frac{c}{(1 + \|\xi\|)^\alpha} \qquad (\xi \in \mathbf{R}^n).$$

14.12.* *(L^1 Sobolev inequality in special case.)* For $f \in \mathcal{S}(\mathbf{R}^n)$, write $\|f\|_{L^\infty}$ and $\|f\|_{L^1}$ as in Example 14.22. Show that there exists a constant $c = c_n > 0$ such that for all $f \in \mathcal{S}(\mathbf{R}^n)$,

$$\|f\|_{L^\infty} \leq c_n \sum_{|\alpha| \leq n+1} \|D^\alpha f\|_{L^1}.$$

14.13.* Prove, for every $t > 0$ and $\phi \in \mathcal{S}(\mathbf{R})$, that

$$\int_{-t}^{t} \mathcal{F}\phi(\xi)\, d\xi = 2 \int_{\mathbf{R}} \phi(x) \frac{\sin tx}{x}\, dx$$

and deduce (compare with Problem 9.6)

$$\lim_{t\to\infty} \frac{\sin tx}{x} = \pi\,\delta \quad \text{in} \quad \mathcal{S}'(\mathbf{R}).$$

14.14.* Consider the following functions on $\mathbf{R}$:

$$a(x) = e^{-x^2+2x}, \qquad b(x) = e^{-x} H(x), \qquad c(x) = e^{-|x|}, \qquad d(x) = \frac{1}{1+x^2}.$$

(i) For each of these, sketch its graph.
(ii) Verify that these functions belong to $L^1 = L^1(\mathbf{R})$. Which of them belong to $\mathcal{S} = \mathcal{S}(\mathbf{R})$ and which to $L^2 = L^2(\mathbf{R})$?
(iii) Calculate the Fourier transforms of these functions. Deduce *Laplace's integral* (compare with Problem 18.8.(iii) or [7, Exercise 2.85])

$$\int_{\mathbf{R}_{>0}} \frac{\cos x\,\xi}{1+x^2}\, dx = \frac{\pi}{2} e^{-|\xi|} \qquad (\xi \in \mathbf{R}).$$

(iv) Sketch the graphs of the Fourier transforms of the functions b, c, and d; make separate sketches of the real and imaginary parts.
(v) For each of the Fourier transforms found, determine whether it is a function in $\mathcal{S}$, in L^1, or in L^2.

14.15.* Define $f = e^{-|\cdot|}$ on $\mathbf{R}$. Differentiate f twice and use the result to prove that $\frac{1}{2} f$ is a fundamental solution of the differential operator $D^2 + I$ in $\mathbf{R}$; in addition, calculate $\mathcal{F} f$. Derive $\mathcal{F} \arctan = \frac{\pi}{i}\,\mathrm{PV}\, \frac{e^{-|\xi|}}{\xi}$.

14.16. Suppose $a > 0$ and prove

$$\frac{1}{4}\mathcal{F}(1_{[-a,a]})\, \mathcal{F}(e^{-|\cdot|})(\xi) = \frac{\sin a\xi}{\xi(1+\xi^2)} =: g(\xi).$$

Show that $g \in \mathcal{S}'(\mathbf{R})$ and that $g \in L^1(\mathbf{R})$ actually. Verify that the solution $f \in \mathcal{S}'(\mathbf{R})$ of the equation $\mathcal{F} f = g$ is given by

$$2f(x) = \begin{cases} 1 - e^{-a}\cosh x & (|x| \le a), \\ e^{-|x|}\sinh a & (|x| \ge a). \end{cases}$$

Deduce that $f \in C(\mathbf{R}) \cap L^1(\mathbf{R})$ and verify that $\int_{\mathbf{R}_{>0}} g(\xi)\, \cos x\xi\, d\xi = \pi f(x)$.

14.17.* Use the function d from Problem 14.14 to prove (compare with Problem 19.8.(viii))

$$\mathcal{F} f_0(\xi) = -\pi i \ \mathrm{sgn}(\xi) e^{-|\xi|} \qquad \text{if} \qquad f_0(x) = \frac{x}{1+x^2}.$$

Next, suppose $f \in C(\mathbf{R})$ satisfies $f(x) = \frac{1}{x} + O(\frac{1}{x^2})$ as $|x| \to \infty$. Show that $\mathcal{F} f$ is a function that is continuous except at the origin, where left and right limits exist. Verify that the jump $\mathcal{F} f(0_+) - \mathcal{F} f(0_-)$ equals $-2\pi i$.

14.18.* *(Constant in Fourier inversion formula.)* Prove that in the context of Problem 14.5, one has $\phi \in \mathcal{S}(\mathbf{R}^n)$ if $\phi_j \in \mathcal{S}(\mathbf{R})$, for every $j = 1, \dots, n$. Use this to prove that $c_n = (c_1)^n$ if c_n is the constant in (14.17). Compute c_1 by calculating $\mathcal{G}(\phi)(0) = \int \mathcal{F}\phi(\xi)\, d\xi$ for $\phi = e^{-|\cdot|}$. A problem with this method is that ϕ does not belong to $\mathcal{S}(\mathbf{R})$. Argue that this can be overcome through approximating ϕ by a suitable sequence of functions in $\mathcal{S}(\mathbf{R})$.

14.19. Use Parseval's formula to verify

$$\int_{\mathbf{R}} \frac{1}{(1+x^2)^2}\, dx = \int_{\mathbf{R}} \frac{x^2}{(1+x^2)^2}\, dx = \frac{\pi}{2}.$$

14.20. Suppose a and $b > 0$. Prove

$$\int_{\mathbf{R}_{>0}} \frac{\sin ax \ \sin bx}{x^2}\, dx = \frac{\pi}{2} \min(a, b) = \frac{\pi}{4}(a + b - |a - b|).$$

14.21. Prove that (14.23) comes down to the equality $(\mathcal{F}1) \otimes \delta = \delta \otimes \mathcal{F}1$ in $\mathcal{S}'(\mathbf{R}^n \times \mathbf{R}^n)$, which in turn is equivalent to $\delta \otimes \delta = \delta \otimes \delta$.

14.22. Let u be a locally integrable function on $\mathbf{R}^n$ such that there exists $N \in \mathbf{R}$ with $u(x) = O(\|x\|^N)$ as $\|x\| \to \infty$. Prove that $u \in \mathcal{S}'$.

14.23. Prove that every tempered distribution is of finite order.

14.24.* *(Characterization of tempered exponentials.)* Suppose $u \in \mathcal{S}'(\mathbf{R}^n)$ and $\phi \in \mathcal{S}(\mathbf{R}^n)$. Prove that there exist constants $C > 0$ and $N \in \mathbf{Z}_{\geq 0}$ such that

$$|u(T_a\phi)| \leq C\,(1 + \|a\|)^N \qquad (a \in \mathbf{R}^n).$$

Now let $\xi \in \mathbf{C}^n$. Prove that the function $e_{i\xi}$ defines a tempered distribution on $\mathbf{R}^n$ if and only if $\xi \in \mathbf{R}^n$.

14.25. For what combinations of p and $q \in \mathbf{N}$ does $x \mapsto e^{x^p}\, e^{i\, e^{x^q}}$ define a tempered distribution on $\mathbf{R}$?

14.26. Let $a > 0$ be a constant. Calculate the Fourier transforms of the following distributions on $\mathbf{R}$:

(i) $x \mapsto e^{-ax} H(x)$.
(ii) $e^{-a|\cdot|}$.
(iii) $x \mapsto \frac{1}{x^2+a^2}$.

Discuss possible complex-analytic extensions of the Fourier transforms to open subsets of $\mathbf{C}$. Furthermore, try to establish whether the functions above converge in $\mathcal{S}'(\mathbf{R})$ as $a \downarrow 0$. If so, determine the Fourier transforms of their limits. Calculate $\mathcal{F}^{-1} H$.

14.27.* Determine the Fourier transforms of the functions: $\mathbf{R} \to \mathbf{R}$ given by the following formulas:

(i) cos, sin, $x \mapsto x \sin x$, $\cos^2$, $\cos^k$ for $k \in \mathbf{N}$.
(ii) sinc.
(iii) $x \mapsto x\, H(x)$ (use Problem 13.6), $|\cdot|$, $\sin|\cdot|$.

Note that only in case (ii) is the Fourier transform a function. In the notation of Problem 13.5, deduce from part (iii) that $\mathcal{F}|\cdot|^{-2} = -\pi\,|\cdot|$.

14.28.* Derive the results from Problem 9.14 by means of the Fourier transform.

The solutions of (systems of) linear ordinary differential equations with constant coefficients are at most of exponential growth at infinity, and for that reason might be nontempered. Therefore, *Laplace transform* is often used as a tool in situations like the present one; but it may be replaced by a combination of Fourier transform and analytic continuation.

14.29. Let $P(D) = \prod_{j=1}^{m}(D - \lambda_j)$ be a differential operator in $\mathbf{R}$, where $\lambda_j \in \mathbf{C}$ for $1 \leq j \leq m$. Prove that all fundamental solutions of $P(D)$ are tempered distributions if and only if $\lambda_j \in \mathbf{R}$ for all $1 \leq j \leq m$.

Now let $P(D)$ be a linear partial differential operator with constant coefficients on $\mathbf{R}^n$, with $n > 1$. Prove that there are always solutions $u \in \mathcal{D}'(\mathbf{R}^n)$ of $P(D)u = 0$ such that $u \notin \mathcal{S}'(\mathbf{R}^n)$. Conclude that not all fundamental solutions of $P(D)$ are tempered distributions (compare with Problem 14.3).

14.30.* *(Fourier transform, pullback, and pushforward.)* Let A be an invertible linear transformation of $\mathbf{R}^n$. Write the transpose of the inverse of A as B; then $B = ({}^tA)^{-1}$.

(i) Demonstrate the following identity of continuous linear operators in $\mathcal{S}'(\mathbf{R}^n)$:

$$\mathcal{F} \circ A^* = ({}^tA)_* \circ \mathcal{F} = |\det B|\, B^* \circ \mathcal{F}.$$

See Problem 15.5 for another proof.

(ii) For $c > 0$, denote the mapping $x \mapsto c\,x : \mathbf{R}^n \to \mathbf{R}^n$ by c. Deduce

$$\mathcal{F} \circ (c^{-1})^* = c^n\, c^* \circ \mathcal{F} \qquad \text{or equivalently,} \qquad \mathcal{F} \circ c_* = c^* \circ \mathcal{F}.$$

Prove that $\mathcal{F}u$ is homogeneous of degree $-n-a$ if $u \in \mathcal{S}'(\mathbf{R}^n)$ is homogeneous of degree $a \in \mathbf{C}$.

(iii) Verify that $u \in \mathcal{S}'(\mathbf{R}^n)$ is invariant under all orthogonal transformations if and only if $\mathcal{F}u$ has that property.

(iv) Let A be an invertible linear transformation of $\mathbf{R}^n$. Prove that A^* commutes with $\mathcal{F}$ if and only if A is orthogonal.

14.31. Here we describe another derivation of (14.31) that is independent of (14.17) (see the natural extension of Problem 9.6 to $\mathbf{R}^n$ or Problem 14.32 for related proofs).

Set $\phi = e^{-\|\cdot\|^2/2}$. From a slight modification of Lemma 14.7 it follows that $\lim_{\epsilon\downarrow 0} \epsilon^*\phi = 1_{\mathbf{R}^n}$ in $\mathcal{S}'(\mathbf{R}^n)$. Hence, successively using the continuity of $\mathcal{F}$, Problems 15.5 and 5.2, as well as the notation from Example 10.9, we obtain the following identities in $\mathcal{S}'(\mathbf{R}^n)$:

$$\mathcal{F}1_{\mathbf{R}^n} = \lim_{\epsilon\downarrow 0}(\mathcal{F}\circ\epsilon^*)\phi = \lim_{\epsilon\downarrow 0}(\epsilon_*\circ\mathcal{F})\phi = \lim_{\epsilon\downarrow 0}(\mathcal{F}\phi)_\epsilon = c\,\delta,$$
$$\text{where} \qquad c = 1_{\mathbf{R}^n}(\mathcal{F}\phi) = (2\pi)^n.$$

14.32.* In the notation of Example 14.11 consider $\phi = (2\pi)^{-n/2}u_{(1,\dots,1)} \in \mathcal{S}(\mathbf{R}^n)$ and $\epsilon_*\phi$, for $\epsilon > 0$, as in Example 10.9. Prove that $\epsilon^*\mathcal{F}\phi = e^{-\epsilon^2\|\cdot\|^2/2}$. Verify the equivalence of the following two assertions:

(i) $\lim_{\epsilon\downarrow 0}\epsilon^*\mathcal{F}\phi = 1$ in $\mathcal{S}'(\mathbf{R}^n)$,

(ii) $\lim_{\epsilon\downarrow 0}\epsilon_*\phi = \delta$ in $\mathcal{S}'(\mathbf{R}^n)$ (compare with Problem 5.5).

Prove one of the two statements.

14.33.* Let P be a polynomial function on $\mathbf{R}^n$ and suppose that $u = P(D)\,\delta \in \mathcal{D}'(\mathbf{R}^n)$ is invariant under all orthogonal transformations in $\mathbf{R}^n$. Prove the existence of a polynomial function P_0 on $\mathbf{R}$ such that $u = P_0(\Delta)\,\delta$, where Δ is the Laplacian.

14.34. Let $P(D)$ be a linear partial differential operator in $\mathbf{R}^n$ with variable coefficients. Prove that $P(D)\circ A^* = A^*\circ P(D)$ for all translations and orthogonal transformations A in $\mathbf{R}^n$ if and only if there exists a polynomial function P_0 on $\mathbf{R}$ such that $P(D) = P_0(\Delta)$, where Δ is the Laplacian.

14.35.* Define $\mathbf{GL}(n,\mathbf{R})$ as the group of invertible real $n\times n$ matrices and $\mathbf{SL}(n,\mathbf{R})$ as the subgroup of $\mathbf{GL}(n,\mathbf{R})$ consisting of the matrices with determinant 1. Suppose that $n > 1$ and that $u \in \mathcal{D}'(\mathbf{R}^n)$ is invariant under $\mathbf{SL}(n,\mathbf{R})$ as a generalized function. Prove the existence of c_1 and $c_2 \in \mathbf{C}$ such that $u = c_1 1_{\mathbf{R}^n} + c_2\,\delta$. If, in addition, u is invariant under $\mathbf{GL}(n,\mathbf{R})$, show that $u = c_1 1_{\mathbf{R}^n}$.

14.36.* *(Bochner's Theorem.)* Consider $p \in C(\mathbf{R}^n)$. Prove that the following conditions are equivalent:

(i) The convolution operator $\phi \mapsto \phi * p$ is positive on $C_0^\infty(\mathbf{R}^n)$, that is, we have $(p*\phi, \phi) \geq 0$ for all $\phi \in C_0^\infty(\mathbf{R}^n)$, where $(\cdot\,,\cdot)$ denotes the Hermitian inner product from (14.18).

(ii) p is *positive definite*, that is, $\sum_{l,m=1}^{k} c_l \overline{c_m}\, p(x_l - x_m) \geq 0$, for all $c_1, \dots, c_k \in \mathbf{C}$ and $x_1, \dots, x_k \in \mathbf{R}^n$, where $k \in \mathbf{N}$. In other words, $(p(x_l - x_m))_{1 \leq l,m \leq k}$ is a positive semidefinite matrix.

(iii) p equals $\mathcal{F}\mu$, where μ is a positive Radon measure on $\mathbf{R}^n$ of finite mass $\mu(1) = p(0)$; see Remark 3.21.

14.37. Prove that the following functions on $\mathbf{R}^n$ are positive definite: e_{ia}, for $a \in \mathbf{R}^n$; $e^{-a\|\cdot\|^2}$ and $e^{-a\|\cdot\|}$, both for $a > 0$ (for the latter; see Problem 18.8.(i)). Why is cos positive definite on $\mathbf{R}$ while sin is not?

14.38.* *(Dirichlet problem on* $[0,1]$ *and Green's function.)* Set $J = \,]0,1[$ and consider the pushforward Radon measure $\Delta_* 1_J$ on $\mathbf{R}^2$ as in Example 10.7. Let I be the identity mapping acting in the space of locally integrable functions. The function $g : \mathbf{R}^2 \to \mathbf{C}$ is said to be *Green's function* associated with the differential operator D^2 on J if g is continuous and vanishes on $\mathbf{R}^2 \setminus \overline{J}^2$, while

$$(D^2 \otimes I)g = \Delta_* 1_J \qquad \text{and} \qquad g(0,\xi) = g(1,\xi) = 0 \qquad (\xi \in J). \qquad (14.43)$$

Here the differential equation is an equality in $\mathcal{D}'(\mathbf{R}^2)$. Further, suppose that f is a locally integrable function on $\mathbf{R}$. Show that $u \in \mathcal{D}'(\mathbf{R})$ given by

$$u(x) = \int_0^1 g(x,\xi)\, f(\xi)\, d\xi = \int_{\mathbf{R}} g(x,\xi)\, f(\xi)\, d\xi \qquad (x \in J)$$

is a solution to the Dirichlet problem $D^2 u = f$ in $\mathcal{D}'(J)$ and $u(0) = u(1) = 0$.

Consider the particular case of $f \in C_0^\infty(\mathbf{R})$ and furthermore the continuous linear mapping $\otimes f : C_0^\infty(\mathbf{R}) \to C_0^\infty(\mathbf{R}^2)$ satisfying $\otimes f(\phi) = \phi \otimes f$. Then derive that $u = {}^t(\otimes f)g \in \mathcal{E}'(\mathbf{R})$.

Prove that $E = -\frac{1}{2}|\cdot|$ is a fundamental solution of the differential operator D^2 on $\mathbf{R}$ and deduce that $T_\xi E$ satisfies the inhomogeneous differential equation in (14.43). Note that $x \mapsto ax + b$ is a solution of the homogeneous differential equation and determine a and $b \in \mathbf{C}$ such that $g(x,\xi) = T_\xi E(x) + ax + b$. Verify that g is uniquely determined and that

$$g(x,\xi) = \begin{cases} \xi(1-x), & 0 \leq \xi \leq x \leq 1, \\ x(1-\xi), & 0 \leq x \leq \xi \leq 1. \end{cases}$$

Now let $p > 0$ and apply the same method to the differential operator $D^2 + p^2 I$ on J. Show that in this case,

$$E = \frac{e^{-p|\cdot|} - 1}{2p} \quad \text{and} \quad g(x,\xi) = \begin{cases} \dfrac{\sinh p\,\xi \,\sinh p\,(1-x)}{p \sinh p}, & 0 \leq \xi \leq x \leq 1, \\[2ex] \dfrac{\sinh p\,x \,\sinh p\,(1-\xi)}{p \sinh p}, & 0 \leq x \leq \xi \leq 1. \end{cases}$$

Observe that Taylor expansion of E and g with respect to the variable p shows that the corresponding distributions obtained before arise in the limit as $p \downarrow 0$.

Finally, note that in [7, Exercises 7.69 and 7.70] this technique is applied in higher-dimensional situations in a classical setting.

14.39. *(Fourier transform of $\chi^a_\pm$.)* Let χ^a_- be as in Problem 13.2. Prove that $\chi^a_\pm \in \mathcal{S}'(\mathbf{R})$ for every $a \in \mathbf{C}$, and that for every $\phi \in \mathcal{S}(\mathbf{R})$, $a \mapsto \chi^a_\pm(\phi)$ is a complex-analytic function on $\mathbf{C}$. Deduce that $a \mapsto \chi^a_\pm$ is a complex-analytic family of tempered distributions and that $a \mapsto \mathcal{F}\chi^a_\pm$ also has this property.

Suppose $0 < a < 1$. Show that $\chi^a_+ + \chi^a_-$ and $\chi^a_+ - \chi^a_-$ are distributions on $\mathbf{R}$ that are homogeneous of degree $a-1$, and furthermore even and odd, respectively. Derive the existence of $c_\pm(a) \in \mathbf{C}$ such that

$$\mathcal{F}(\chi^a_+ + \chi^a_-) = c_+(a)(\chi^{1-a}_+ + \chi^{1-a}_-), \qquad \mathcal{F}(\chi^a_+ - \chi^a_-) = c_-(a)(\chi^{1-a}_+ - \chi^{1-a}_-).$$

By means of testing these identities against the Schwartz functions $x \mapsto \phi(x) = e^{-x^2/2}$ and $x \mapsto x\phi(x)$, respectively, and using (13.39) and Corollary 13.6, compute

$$c_+(a) = 2\Gamma(1-a)\cos\frac{\pi}{2}a \qquad \text{and} \qquad c_-(a) = -2i\,\Gamma(1-a)\sin\frac{\pi}{2}a.$$

In addition, deduce (see Problem 13.13 or [7, Exercise 6.60.(iii)] for different proofs)

$$\int_{\mathbf{R}_{>0}} x^{a-1}\begin{Bmatrix}\cos\\ \sin\end{Bmatrix} x\,dx = \Gamma(a)\begin{Bmatrix}\cos\\ \sin\end{Bmatrix}\Big(\frac{\pi}{2}a\Big). \tag{14.44}$$

Conversely, one may use (14.44) and Theorem 14.24 to prove

$$c_+(a)\,c_+(1-a) = 2\pi, \qquad \text{that is,} \qquad \Gamma(a)\,\Gamma(1-a) = \frac{\pi}{\sin \pi a}, \tag{14.45}$$

for $a \in \mathbf{C} \setminus \mathbf{Z}$, the reflection formula from Corollary 13.6.

Furthermore, in the notation of Problem 13.3, show that

$$\mathcal{F}\chi^a_\pm = e^{-i\frac{\pi}{2}a}\Gamma(1-a)(\chi^{1-a}_\pm + e^{i\pi a}\chi^{1-a}_\mp) = \frac{e^{\mp i\frac{\pi}{2}a}}{(\xi \mp i0)^a}.$$

Prove the validity of these formulas for all $a \in \mathbf{C}$ using analytic continuation.

Deduce, in the notation of Problem 13.5, for $a \in \mathbf{C}$ not a positive even integer and not a negative odd integer, that

$$\begin{aligned}\mathcal{F}|\chi|^a &= \ \ 2\Gamma(1-a)\cos\frac{\pi}{2}a\,|\chi|^{1-a},\\ \mathcal{F}|\cdot|^a &= -2\Gamma(1+a)\sin\frac{\pi}{2}a\,|\cdot|^{-1-a},\end{aligned}$$

for $a \in \mathbf{C}$ not a nonnegative even integer and not a negative odd integer. Note that the case of $a = 1$ occurs in Problem 14.27.(iii), in the form $\mathcal{F}|\cdot| = -2|\cdot|^{-2}$. Use the reflection formula from Lemma 13.5 to conclude, for $a \in \mathbf{C}$ not a nonpositive even integer and not a positive odd integer, that

$$\cos\frac{\pi}{2}a\,\mathcal{F}|\cdot|^{-a} = \frac{\pi}{\Gamma(a)}|\cdot|^{a-1}. \tag{14.46}$$

Note that the case of $a = 2$ occurs in Problem 14.27, in the form $\mathcal{F}|\cdot|^{-2} = -\pi|\cdot|$.

14.40. For $0 < s < 1$, consider the constant $c(1,s)$ from Problem 10.14. Use integration by parts to show that $c(1,s) = -4\Gamma(-2s)\cos\pi s > 0$. In particular, show that $c(1,\frac{1}{2}) = 2\pi$ and $\lim_{s\downarrow 0} c(1,s) = \lim_{s\uparrow 1} c(1,s) = +\infty$.

14.41. For symmetric $A \in \mathbf{GL}(n,\mathbf{R})$, denote by $\operatorname{sgn} A$ the *signature* van A, i.e., the number of positive minus the number of negative eigenvalues of A, all counted with multiplicities. Prove by means of analytic continuation or (14.44) that

$$\mathcal{F}e^{\frac{i}{2}\langle A\cdot,\cdot\rangle} = \frac{(2\pi)^{\frac{n}{2}}}{\sqrt{|\det A|}}\,e^{i\left(\frac{\pi}{4}\operatorname{sgn} A - \frac{1}{2}\langle A^{-1}\cdot,\cdot\rangle\right)}.$$

14.42. *(Metaplectic representation.)* Let $\mathbf{SL}(2,\mathbf{R})$ be as in Problem 14.35 and write $\mathfrak{sl}(2,\mathbf{R})$ for the corresponding *Lie algebra*; see [7, Sect. 5.10]. Then the following matrices are a basis for $\mathfrak{sl}(2,\mathbf{R})$:

$$h = \begin{pmatrix} 1 & 0 \\ 0 & -1 \end{pmatrix}, \qquad e^+ = \begin{pmatrix} 0 & 1 \\ 0 & 0 \end{pmatrix}, \qquad e^- = \begin{pmatrix} 0 & 0 \\ 1 & 0 \end{pmatrix}.$$

For a and $b \in \mathfrak{sl}(2,\mathbf{R})$, define the *commutator* $[a,b]$ as $ab - ba$. Show that $[a,b]$ also belongs to $\mathfrak{sl}(2,\mathbf{R})$, and in addition, that

$$[h, e^{\pm}] = \pm 2\, e^{\pm}, \qquad [e^+, e^-] = h.$$

Next define the linear mapping ω from $\mathfrak{sl}(2,\mathbf{R})$ to the linear space of differential operators acting on $\mathcal{S}(\mathbf{R})$ by

$$\omega(h) = x\,\partial_x + \frac{1}{2}, \qquad \omega(e^+) = \frac{i}{2}x^2, \qquad \omega(e^-) = \frac{i}{2}\partial_x^2.$$

Verify that ω is a homomorphism of Lie algebras; more precisely, prove that ω maps into the linear space of skew-Hermitian operators with respect to the Hermitian inner product (14.18), and that ω preserves commutators. Furthermore, demonstrate that $\mathcal{F}\circ\omega(e^-) = -\omega(e^+)$. Define the following mappings from $\mathbf{R}$ to the group $\mathbf{GL}(\mathcal{S}(\mathbf{R}))$ of invertible linear operators from $\mathcal{S}(\mathbf{R})$ into itself:

$$t \mapsto \widetilde{\omega}(\exp(th)) := e^{\frac{t}{2}}(e^t)^*, \qquad t \mapsto \widetilde{\omega}(\exp(te^+)) := e^{\frac{i}{2}tx^2},$$

$$t \mapsto \widetilde{\omega}(\exp(te^-)) := \frac{1}{\sqrt{2\pi t}}\,e^{i\left(-\frac{\pi}{4}+\frac{x^2}{2t}\right)} * .$$

Here the last expression defines a convolution operator. Prove that we obtain one-parameter subgroups of $\mathbf{GL}(\mathcal{S}(\mathbf{R}))$. Show that these subgroups have $\omega(h)$, $\omega(e^+)$, and $\omega(e^-)$ as infinitesimal generators, respectively. Hint: in the third case, apply Problem 14.41 and compare with Problem 17.7. Verify that all the operators in these subgroups are unitary with respect to the Hermitian inner product on $L^2(\mathbf{R})$.

The *Theorem of Shale–Weil* asserts that there exists a homomorphism of groups $\widetilde{\omega}$ from $\widetilde{\mathbf{SL}}(2,\mathbf{R})$, the double covering group of $\mathbf{SL}(2,\mathbf{R})$, to the group of unitary linear operators acting in $L^2(\mathbf{R})$ having ω as its tangent mapping. In particular, therefore, $\widetilde{\omega}\circ\exp = \exp\circ\,\omega$ on $\mathfrak{sl}(2,\mathbf{R})$. The three one-parameter groups constructed above generate a group $\mathbf{G}$ of unitary operators in $L^2(\mathbf{R})$ that is isomorphic to $\widetilde{\mathbf{SL}}(2,\mathbf{R})$.

Now consider the following differential operators on $\mathcal{S}(\mathbf{R})$:

$$a^{\mp} = x \pm \partial_x.$$

Note that these operators are each others adjoints. Then we have the commutator relation

$$[a^-, a^+] = 2; \qquad \text{hence} \qquad [a^-, (a^+)^j] = 2j\,(a^+)^{j-1} \qquad (j \in \mathbf{Z}_{>0}).$$

Write

$$\phi_0(x) = e^{-\frac{1}{2}x^2} \qquad \text{and} \qquad \phi_j = (a^+)^j \phi_0 \qquad (j \in \mathbf{Z}_{>0}). \tag{14.47}$$

Then $a^-\phi_0 = 0$, and therefore, for $j \in \mathbf{Z}_{>0}$,

$$a^-\phi_j = a^-(a^+)^j\phi_0 = ([a^-, (a^+)^j] + (a^+)^j a^-)\phi_0 = 2j\,(a^+)^{j-1}\phi_0 = 2j\,\phi_{j-1}.$$

Consequently, a^+ is called a *creation operator* and similarly a^- an *annihilation operator*. In the notation (14.18), we now compute the inner products

$$\begin{aligned}(\phi_j, \phi_l) &= (a^+\phi_{j-1}, \phi_l) = (\phi_{j-1}, a^-\phi_l) = 2l(\phi_{j-1}, \phi_{l-1})\\ &= 2^l\, l!\,\delta_{jl}\,(\phi_0, \phi_0) = 2^l\, l!\,\sqrt{\pi}\,\delta_{jl}.\end{aligned}$$

In other words, $((2^j\, j!\sqrt{\pi})^{-\frac{1}{2}}\,\phi_j)_{j\in\mathbf{Z}_{\geq 0}}$ forms an orthonormal system in $L^2(\mathbf{R})$.

By means of mathematical induction it is easy to prove the existence of polynomial functions H_j of degree j such that $\phi_j = H_j\,\phi_0$. The H_j are the *Hermite polynomials*. For any $\phi \in \mathcal{S}(\mathbf{R})$, verify

$$a^+\phi(x) = -e^{\frac{1}{2}x^2}\partial_x(e^{-\frac{1}{2}x^2}\phi); \qquad \text{deduce} \qquad H_j(x) = (-1)^j e^{x^2}\partial_x^j e^{-x^2},$$

which is the *Rodrigues formula* for H_j. The *Hermite functions* ψ_j with

$$\psi_j(x) := \frac{1}{\sqrt{2^j\, j!\sqrt{\pi}}}\, H_j(x) e^{-\frac{1}{2}x^2} \qquad (j \in \mathbf{Z}_{\geq 0})$$

form an orthonormal basis for $L^2(\mathbf{R})$. For a proof of the completeness, suppose that $f \in L^2(\mathbf{R})$ is orthogonal to all of the ψ_j. Then define $F : \mathbf{C} \to \mathbf{C}$ by $F(\zeta) = \int_{\mathbf{R}} e^{-ix\zeta - x^2/2}\,\overline{f(x)}\,dx$. Show that F is a complex-analytic function on $\mathbf{C}$ satisfying

$$\partial^j F(0) = (-i)^j \int_{\mathbf{R}} x^j e^{-\frac{1}{2}x^2} \overline{f(x)}\, dx = 0 \qquad (j \in \mathbf{Z}_{\geq 0}).$$

Conclude that F must vanish identically on $\mathbf{C}$. For arbitrary $\xi \in \mathbf{R}$, this yields

$$0 = F(\xi) = \int_{\mathbf{R}} e^{-ix\xi} e^{-\frac{1}{2}x^2} \overline{f(x)}\, dx;$$

and hence $x \mapsto e^{-x^2/2} f(x) = 0$ in $L^2(\mathbf{R})$ on account of Theorem 14.32.

Set $k = i(e^- - e^+)$ and extend ω by linearity to $\mathfrak{sl}(2, \mathbf{C})$. The differential operator

$$2\omega(k) = x^2 - \partial_x^2$$

is known as the *Hermite operator*. The Hermite functions are eigenfunctions of the Hermite operator. In fact, the identities $x = \frac{1}{2}(a^- + a^+)$ and $\partial_x = \frac{1}{2}(a^- - a^+)$ imply that

$$\omega(e^{\pm}) = \frac{i}{8}(a^- \pm a^+)^2, \qquad \omega(k) = \frac{1}{4}(a^- a^+ + a^+ a^-).$$

This entails

$$\omega(k)\psi_j = \frac{1}{4}(a^- \psi_{j+1} + 2j\, a^+ \psi_{j-1}) = \frac{1}{4}(2j + 2 + 2j)\psi_j = \Big(j + \frac{1}{2}\Big)\psi_j.$$

Note that the Fourier transform commutes with the Hermite operator and that the eigenspaces of the Hermite operator are of dimension 1. Deduce that the Hermite functions are eigenfunctions of the Fourier transform (compare with Problem 14.9). More specifically, the identities $\widetilde{\mathcal{F}} \circ a^+ = -i a^+ \circ \widetilde{\mathcal{F}}$ and $\widetilde{\mathcal{F}} \psi_0 = \psi_0$ lead to

$$\widetilde{\mathcal{F}} \psi_j = \widetilde{\mathcal{F}} \circ (a^+)^j \psi_0 = (-i)^j (a^+)^j \widetilde{\mathcal{F}} \psi_0 = (-i)^j \psi_j \qquad (j \in \mathbf{Z}_{\geq 0}). \quad (14.48)$$

It is remarkable that the Fourier transform also belongs to the group $\mathbf{G}$. Indeed, we have

$$\exp\Big(-\frac{\pi i}{2}\omega(k)\Big)\psi_j = e^{-\frac{\pi i}{2}(j+\frac{1}{2})}\psi_j = (-i)^{j+\frac{1}{2}}\psi_j = (-i)^{\frac{1}{2}} \widetilde{\mathcal{F}} \psi_j;$$

hence

$$\widetilde{\mathcal{F}} = \widetilde{\omega}\Big(i^{\frac{1}{2}} \exp\Big(-\frac{\pi i}{2} k\Big)\Big).$$

In turn, this explains why the Fourier transform and the Hermite operator commute.

Once it is known that the collection of Hermite functions $(\psi)_{j \in \mathbf{Z}_{\geq 0}}$ is an orthonormal basis for $L^2(\mathbf{R})$, then one may define the normalized Fourier transform $\widetilde{\mathcal{F}}$ as the linear mapping : $L^2(\mathbf{R}) \to L^2(\mathbf{R})$ satisfying (14.48); see Problem 15.14.

The construction above is essentially the "algebraic" treatment of the harmonic oscillator in quantum mechanics. For this reason the homomorphism $\widetilde{\omega}$ (or ω) is called the *metaplectic representation*, *Segal–Shale–Weil representation* or *oscillator representation* of $\widetilde{\mathbf{SL}}(2, \mathbf{R})$; see [9, Chap. 4].

It is straightforward to generalize the treatment above to $\mathcal{S}(\mathbf{R}^n)$ by considering tensor products of Hermite functions, starting with the differential operators

$$\omega^n(h) = \sum_{j=1}^{n} x_j \partial_j + \frac{n}{2}, \qquad \omega^n(e^+) = \frac{i}{2}\|\cdot\|^2, \qquad \omega^n(e^-) = \frac{i}{2}\Delta.$$

14.43.* *(Fourier transform of R_+^a.)* Let $n \geq 2$. Consider the complex-analytic family of distributions $(R_+^a)_{a\in\mathbf{C}}$ from Chap. 13 and let $C_\pm$ be as in (13.13). Show that $R_+^a \in \mathcal{S}'(\mathbf{R}^{n+1})$, for all $a \in \mathbf{C}$. Prove that $(\mathcal{F} R_+^a)_{a\in\mathbf{C}}$ is a complex-analytic family of tempered distributions.

For $a \in \mathbf{C}$, apply Problem 12.14.(xi) to demonstrate the following identity in $\mathcal{S}'(\mathbf{R}^{n+1})$:

$$\mathcal{F} R_+^a = (\|\xi\|^2 - (\tau - i0)^2)^{-\frac{a}{2}} := \lim_{\epsilon\downarrow 0}((\xi,\tau) \mapsto \|\xi\|^2 - (\tau - i\epsilon)^2)^{-\frac{a}{2}}.$$

Verify that the restriction of $\mathcal{F} R_+^a$ to $\mathbf{R}^{n+1} \setminus \bigcup_\pm \partial C_\pm$ is the real-analytic function satisfying

$$(\xi,\tau) \mapsto \begin{cases} e^{\mp i\frac{\pi}{2}a}(\tau^2 - \|\xi\|^2)^{-\frac{a}{2}}, & (\xi,\tau) \in C_\pm; \\ (\|\xi\|^2 - \tau^2)^{-\frac{a}{2}}, & (\xi,\tau) \in \mathbf{R}^{n+1} \setminus \bigcup_\pm \overline{C_\pm}. \end{cases}$$

For $\mathrm{Re}\, a < 0$, show that $\mathcal{F} R_+^a$ is given by a locally integrable function on $\mathbf{R}^{n+1}$. In the case of $\mathrm{Re}\, a \geq 0$, prove that $\mathcal{F} R_+^a$ is a distribution of order at most $N + 1$ if $N \in \mathbf{Z}_{\geq 0}$ satisfies $N \geq \mathrm{Re}\, a$.

14.44.* Let $a > 0$. Compute $\mathcal{F}$ sinc and show that

$$\int_{\mathbf{R}} \mathrm{sinc}\, a\, x\, dx = \int_{\mathbf{R}} \mathrm{sinc}^2\, a\, x\, dx = \frac{\pi}{a}.$$

Determine $\mathcal{F}\,\mathrm{sinc}^2$ and obtain the value of the second integral once more. See Problem 16.19 or 16.21 and [7, Example 2.10.14 or Exercises 0.14, 6.60 or 8.19] for different methods.

14.45. Consider $u_a(x) = e^{-ax^2/2}$ as in Example 14.11, but now for $\mathrm{Re}\, a \geq 0$. Let $t \in \mathbf{R}$. Prove that $u_{it} \in \mathcal{S}'$ and that $\lim_{a\to it,\, \mathrm{Re}\, a>0} u_a = u_{it}$ in $\mathcal{S}'$. Calculate $\mathcal{F} u_{it}$.

14.46. For what $\lambda \in \mathbf{C}$ does the difference equation $T_1 u = \lambda u$ have a solution $u \in \mathcal{S}'(\mathbf{R})$ with $u \neq 0$? Describe the solution space for every such λ. The same questions for the difference equation $\frac{1}{2}(T_1 u + T_{-1} u) = \lambda u$.

14.47. Let $u \in \mathcal{S}'(\mathbf{R}^n)$ with $u \neq 0$ be a solution of the differential equation $\Delta u = \lambda u$, with $\lambda \in \mathbf{C}$. Prove that $\lambda \leq 0$ and that u has an extension to an entire analytic function on $\mathbf{C}^n$. Show that u is a polynomial if $\lambda = 0$. Discuss the rotationally symmetric solutions u, for every $\lambda \leq 0$.

14.48. Let μ be a probability measure on $\mathbf{R}^n$. Prove that $\mu \in \mathcal{S}'$ and that $\mathcal{F}\mu$ is a continuous function on $\mathbf{R}^n$, with $|\mathcal{F}\mu(\xi)| \le 1$ for all $\xi \in \mathbf{R}^n$ and $\mathcal{F}\mu(0) = 1$. Hint: take $\chi_r \in C_0^\infty, 0 \le \chi_r \le 1$ and $\chi_r(x) = 1$ whenever $\|x\| \le r$. Verify that $\mathcal{F}(\chi_r \mu)$ converges uniformly on $\mathbf{R}^n$ as $r \to \infty$. Also prove that $\mathcal{F}(\mu * \nu) = \mathcal{F}\mu\, \mathcal{F}\nu$ if μ and ν are probability measures.

14.49.* *(Hilbert transform.)* The *Hilbert transform* $\mathcal{H}\phi \in C^\infty(\mathbf{R})$ of $\phi \in C_0^\infty(\mathbf{R})$ is defined as

$$\mathcal{H}\phi(x) = \frac{1}{\pi}\Big(\phi * \mathrm{PV}\frac{1}{y}\Big)(x) = \lim_{\epsilon \downarrow 0} \frac{1}{\pi} \int_{|y|>\epsilon} \frac{\phi(x-y)}{y}\, dy \qquad (x \in \mathbf{R}).$$

(i) Show that $\mathcal{H}\phi \in \mathcal{S}'(\mathbf{R})$ and that we have the following identity of linear operators on $C_0^\infty(\mathbf{R})$:

$$\mathcal{F} \circ \mathcal{H} = -i\ \mathrm{sgn} \circ \mathcal{F}.$$

The question of natural domain and range spaces for $\mathcal{H}$ is subtle. For instance, extension of the definition of $\mathcal{H}$ to all of $\mathcal{S}'(\mathbf{R})$ by means of the formula above is impossible on account of the discontinuity of the sign function. Furthermore, if $\phi \in C_0^\infty(\mathbf{R})$ and $\mathcal{F}\phi(0) \neq 0$, then $\mathcal{H}\phi \notin L^1(\mathbf{R})$ because the Fourier transform of the latter function is discontinuous at 0. Nevertheless, there are positive results.

(ii) Prove that $\|\mathcal{H}\phi\| = \|\phi\|$, where $\|\cdot\|$ denotes the L^2 norm. Verify that $\mathcal{H}$ possesses an extension to a unitary isomorphism $\mathcal{H}$ from $L^2(\mathbf{R})$ onto $L^2(\mathbf{R})$ and show that $\mathcal{H}^2 = -I$.

(iii) Set $f(x) = \frac{1}{1+x^2}$ and $g(x) = \frac{x}{1+x^2}$. Prove $\mathcal{H}f = g$ and verify that the identity $\|f\| = \|g\|$ is corroborated by Problem 14.19.

(iv) Demonstrate that $\mathcal{H}\cos = \sin$ and $\mathcal{H}\sin = -\cos$ in two different ways. Deduce that $\mathcal{H}e_{\pm i} = \mp i\, e_{\pm i}$.

(v) For $a > 0$, consider $u = 1_{[-a,a]} \in L^2(\mathbf{R})$. Verify, for $x \in \mathbf{R}$,

$$\mathcal{H}u(x) = \frac{1}{\pi}\log\left|\frac{x+a}{x-a}\right|, \qquad (\mathcal{F}\circ\mathcal{H})u = -2i\,\frac{\sin a\,\cdot}{|\cdot|},$$

$$\int_{\mathbf{R}_{>0}} \log^2\left|\frac{x+a}{x-a}\right| dx = a\,\pi^2, \qquad \int_{\mathbf{R}} \mathrm{sinc}^2\, a\, dx = \frac{\pi}{a}.$$

(vi) Deduce, for $0 < b \neq a$, that

$$\mathcal{F}\Big(\frac{\sin a\,\cdot}{|\cdot|}\Big)(\xi) = i\,\log\left|\frac{\xi-a}{\xi+a}\right| \quad \text{and} \quad \int_{\mathbf{R}_{>0}} \frac{\sin ax \sin bx}{x}\, dx = \frac{1}{2}\log\left|\frac{a+b}{a-b}\right|.$$

(vii) Verify $\mathcal{H}(\frac{\sin a\,\cdot}{\cdot}) = \frac{1-\cos a\,\cdot}{\cdot}$ and $\mathcal{H}(\frac{1-\cos a\,\cdot}{\cdot}) = -\frac{\sin a\,\cdot}{\cdot}$.

14.50.* Consider a and $b \in \mathbf{C}$ with $a^2 + b^2 \neq 0$ and $\phi \in C_0^\infty(\mathbf{R})$. Show that $\psi \in C^\infty(\mathbf{R})$ satisfies *Cauchy's integral equation*

$$a\,\psi(x) + \frac{b}{\pi}\,\mathrm{PV}\int_{\mathbf{R}} \frac{\psi(y)}{x-y}\, dy = \phi(x) \qquad \text{if} \qquad \psi = \frac{1}{a^2+b^2}(a\,I - b\,\mathcal{H})\phi$$

in the notation of Problem 14.49.

14.51. *(Poisson kernel, associated Poisson kernel, and Hilbert transform.)* (See Problem 18.8 for a generalization in higher dimensions.) Let the notation be as in Problem 14.49. Define the *Poisson kernel* $(P_t)_{t>0}$ and the *associated Poisson kernel* $(Q_t)_{t>0}$, respectively, by setting, with $z = x + it \in \mathbf{C}$,

$$P_t(x) := P(x,t) = -\operatorname{Im}\frac{1}{\pi z} = \frac{1}{\pi}\frac{t}{x^2+t^2} = \frac{1}{\pi t}(t^{-1})^*\frac{1}{1+x^2},$$
$$Q_t(x) := Q(x,t) = \operatorname{Re}\frac{1}{\pi z} = \frac{1}{\pi}\frac{x}{x^2+t^2} = \frac{1}{\pi t}(t^{-1})^*\frac{x}{1+x^2}.$$

(i) Prove that P and Q are harmonic functions on $\{\,(x,t)\in\mathbf{R}^2 \mid t>0\,\}$ and that $P_t \in L^1(\mathbf{R})$ and $Q_t \in L^2(\mathbf{R})$, for all $t>0$.
(ii) Show that $\mathcal{F} P_t = e^{-t|\cdot|}$ and $\mathcal{F} Q_t = -i\,\operatorname{sgn} e^{-t|\cdot|}$, for $t>0$.
(iii) Conclude (compare with Problem 5.3)

$$\lim_{t\downarrow 0} P_t = \delta \qquad\text{and}\qquad \lim_{t\downarrow 0} Q_t = \frac{1}{\pi}\,\mathrm{PV}\,\frac{1}{x} \quad\text{in}\quad \mathcal{S}'(\mathbf{R}).$$

In particular, verify that we have the following identity of functions, for $\phi \in \mathcal{S}(\mathbf{R})$:

$$\lim_{t\downarrow 0} P_t * \phi = \phi \qquad\text{and}\qquad \lim_{t\downarrow 0} Q_t * \phi = \mathcal{H}\phi.$$

(iv) Demonstrate $\mathcal{H} P_t = Q_t$, $\|P_t\| = \|Q_t\|$ and $Q_t * \phi = P_t * \mathcal{H}\phi$, for $\phi \in \mathcal{S}(\mathbf{R})$.
(v) Furthermore, deduce, for t and $t' > 0$, that

$$1(P_t) = 1, \qquad P_t * P_{t'} = P_{t+t'}, \qquad Q_t * P_{t'} = Q_{t+t'}.$$

As a consequence, (P_t) is said to have the *semigroup property*. Young's inequality from Problem 11.22 may be used to show that the convolution involving Q_t is well-defined.

14.52.* *(Characterization of Hilbert transform.)* According to Problem 14.49.(i) and (iv) the Hilbert transform $\mathcal{H} : C_0^\infty(\mathbf{R}) \to \mathcal{S}'(\mathbf{R})$ satisfies

$$\mathcal{F}\circ\mathcal{H} = -i\,\operatorname{sgn}\circ\mathcal{F}, \qquad\text{and as a consequence,}\qquad \mathcal{H}\cos = \sin.$$

Without explicit computation of Fourier transforms we will show that these properties imply that $\mathcal{H}$ is given by $\mathcal{H}\phi = \phi * \frac{1}{\pi}\,\mathrm{PV}\,\frac{1}{x}$. (For a different proof, see Problem 15.15.)

To this end, show that $\mathcal{H}$ commutes with translations and deduce that there exists a uniquely determined $u \in \mathcal{D}'(\mathbf{R})$ such that $\mathcal{H} = u\,*$. Next, prove that $\mathcal{H}$ commutes with positive dilations and deduce that u is homogeneous of degree -1. Conclude the existence of a_1 and $a_2 \in \mathbf{C}$ such that $u = a_1\,\mathrm{PV}\,\frac{1}{x} + a_2\,\delta$ and finally prove that $\mathcal{H}$ anticommutes with reflection and deduce that $u = \frac{1}{\pi}\,\mathrm{PV}\,\frac{1}{x}$.

14.53. *(Complex-differentiable functions and Hilbert transform.)* Let $f : \mathbf{C} \to \mathbf{C}$ be complex-differentiable and $U(r) \subset \mathbf{C}$ the upper half of the disk of radius $r > 0$ with center 0. Suppose that f satisfies decay properties for $|z| \to \infty$ with $\pi < \arg z < 0$ that entail, for every w with $\operatorname{Im}(w) > 0$,

$$\lim_{r\to\infty} \int_{\partial U(r)} \frac{f(z)}{z-w}\,dz = \int_{\mathbf{R}} \frac{f(t)}{t-w}\,dt.$$

Application of Cauchy's integral formula (12.11) leads to, for $x \in \mathbf{R}$, $y > 0$, and r sufficiently large,

$$f(x+iy) = \frac{1}{2\pi i}\int_{\partial U(r)} \frac{f(z)}{z-x-iy}\,dz = \frac{1}{2\pi i}\int_{\mathbf{R}} \frac{f(t)}{t-x-iy}\,dt.$$

Consider x temporarily as a constant and $t \in \mathbf{R}$ as a variable. Then (14.41) implies

$$\frac{1}{t-x-i0} = \mathrm{PV}\,\frac{1}{t-x} + \pi i\,\delta_x.$$

As a consequence verify, denoting the restriction of f to $\mathbf{R}$ by $f_{\mathbf{R}}$, for $x \in \mathbf{R}$,

$$\begin{aligned} f_{\mathbf{R}}(x) &= \lim_{y\downarrow 0} f(x+iy) = \frac{1}{2\pi i}\frac{1}{t-x-i0}(f_{\mathbf{R}}) = \frac{1}{2\pi i}\,\mathrm{PV}\,\frac{1}{t-x}(f_{\mathbf{R}}) + \frac{1}{2}f_{\mathbf{R}}(x) \\ &= \frac{1}{2\pi i}\,\mathrm{PV}\int_{\mathbf{R}} \frac{f_{\mathbf{R}}(t)}{t-x}\,dt + \frac{1}{2}f_{\mathbf{R}}(x) = \frac{i}{2}\,\mathcal{H} f_{\mathbf{R}}(x) + \frac{1}{2}f_{\mathbf{R}}(x), \end{aligned}$$

with $\mathcal{H}$ the Hilbert transform from Problem 14.49. By taking real and imaginary parts of the resulting identity deduce the following identities of functions on $\mathbf{R}$:

$$\mathcal{H}\,\mathrm{Re}\,f_{\mathbf{R}} = \mathrm{Im}\,f_{\mathbf{R}} \qquad \text{and} \qquad \mathcal{H}\,\mathrm{Im}\,f_{\mathbf{R}} = -\mathrm{Re}\,f_{\mathbf{R}}.$$

Derive that $\mathcal{H}^2 = -I$ on functions of the form $f_{\mathbf{R}}$. Verify the computations of the Hilbert transforms in Problem 14.49.(iv) and (vii) by applying the preceding identities in the case of the complex-analytic functions $z \mapsto e^{iz}$ and $z \mapsto \frac{1-e^{iaz}}{z}$, respectively.

As a result, for a suitable class of complex-differentiable functions f, the real part of $f_{\mathbf{R}}$ determines the imaginary part of $f_{\mathbf{R}}$ via the Hilbert transform, and vice versa.

14.54.* *(Functions on $\mathbf{R}$ as "jump" of functions on $\mathbf{C}\setminus\mathbf{R}$.)* We use the notation from Problems 14.49 and 14.51. Set $H_\pm = \{\, z = x + it \in \mathbf{C} \mid \pm t > 0 \,\}$ and define the function H_{-iz}, for $z \in H_+$, as in Example 14.30.

(i) Prove, for $\phi \in \mathcal{S}(\mathbf{R})$ and $z \in H_+$, that

$$\frac{1}{i}\int_{\mathbf{R}} \frac{\phi(y)}{y-z}\,dy = \mathcal{F} H_{-iz}(\phi) = \int_{\mathbf{R}_{>0}} e^{iz\xi}\,\mathcal{F}\phi(\xi)\,d\xi.$$

(ii) More generally, show that

$$\Phi(z) := \frac{1}{i}\int_{\mathbf{R}} \frac{\phi(y)}{y-z}\,dy = \pm \int_{\mathbf{R}_{\gtrless 0}} e^{iz\xi}\,\mathcal{F}\phi(\xi)\,d\xi =: \Phi_\pm(z) \qquad (z \in H_\pm).$$

Deduce that Φ is complex-differentiable on $\mathbf{C} \setminus \mathbf{R}$ and that we may write $\Phi = \sum_\pm \Phi_\pm$ with supp $\Phi_\pm \subset \overline{H_\pm}$ and $\Phi_\pm$ complex-differentiable on $H_\pm$.

Define u_ϕ and $v_\phi : H_+ \to \mathbf{C}$ respectively, for $z \in H_+$, by

$$u_\phi(z) = \frac{1}{2\pi}(\Phi_+(z) - \Phi_-(\overline{z})) \qquad \text{and} \qquad v_\phi(x,t) = -\frac{i}{2\pi}(\Phi_+(z) + \Phi_-(\overline{z})).$$

(iii) Verify, for $x + it \in H_+$,

$$u_\phi(x+it) = P_t * \phi(x) = \frac{1}{2\pi}\int_{\mathbf{R}} e^{-t|\xi|}\,e^{ix\xi}\,\mathcal{F}\phi(\xi)\,d\xi,$$

$$v_\phi(x+it) = Q_t * \phi(x) = \frac{1}{2\pi}\int_{\mathbf{R}} e^{-t|\xi|}\,e^{ix\xi}\,\mathcal{F}(\mathcal{H}\phi)(\xi)\,d\xi,$$

and prove in two different manners that

$$\lim_{t\downarrow 0} u_\phi(\cdot + it) = \phi \quad \text{and} \quad \lim_{t\downarrow 0} v_\phi(\cdot + it) = \mathcal{H}\phi.$$

(iv) Suppose that ϕ is real-valued. Demonstrate the identities of functions $u_\phi = \frac{1}{\pi}\operatorname{Re}\Phi_+$ and $v_\phi = \frac{1}{\pi}\operatorname{Im}\Phi_+$ on H_+. Deduce that u_ϕ and v_ϕ are harmonic functions on H_+.

(v) Generalize the results above to the case of $f \in L^2(\mathbf{R})$ instead of $\phi \in \mathcal{S}(\mathbf{R})$. Define arg : $\mathbf{C} \setminus \mathbf{R} \to \mathbf{R}$ by $-\pi < \arg z < \pi$. Verify, for $a > 0$, $f = 1_{[-a,a]}$, and $z = a + it \in H_+$,

$$u_f(z) = \frac{1}{\pi}(\arg(z-a) - \arg(z+a)) = \frac{1}{\pi}\Big(\arctan\frac{a+x}{t} + \arctan\frac{a-x}{t}\Big),$$

$$v_f(z) = \frac{1}{\pi}\log\left|\frac{z+a}{z-a}\right|; \qquad \text{and} \qquad \lim_{t\downarrow 0} u_f(x+it) = 1_{[-a,a]}(x),$$

for $x \in \mathbf{R}$ and $t > 0$. Note the agreement with the results in Problem 14.49.(v). Furthermore, show that

$$\int_{\mathbf{R}_{>0}} e^{-t\xi}\cos x\xi\,\frac{\sin a\xi}{\xi}\,d\xi = \frac{1}{2}\Big(\arctan\frac{a+x}{t} + \arctan\frac{a-x}{t}\Big).$$

The complex-differentiable functions $\Phi_\pm$ on $H_\pm$ have extensions to continuous functions on $\overline{H_\pm}$. In part (iii) it is asserted that $\lim_{t\downarrow 0}(\Phi_+(x+it) - \Phi_-(x-it)) = 2\pi\,\phi(x)$, for $x \in \mathbf{R}$. In general therefore, these extensions do not coincide on $\mathbf{R}$. This means that it is impossible to "glue" the $\Phi_\pm$ to define a single complex-differentiable function on all of $\mathbf{C}$. This point of view is of importance in the theory of *hyperfunctions*.

On the positive side, u_ϕ is a solution to the Dirichlet problem $\Delta u = 0$ on H_+ and $u|_{\mathbf{R}} = \phi$, for $\phi \in \mathcal{S}(\mathbf{R})$; see Problem 12.4.

14.55. If $u \in L^1$, $v \in \mathcal{E}'$, and $u * v = 0$, then prove that $u = 0$ or $v = 0$. Calculate $1 * \partial_j \delta$, for $1 \leq j \leq n$.

14.56. Prove the following assertions, in which the convolution product is as defined in Remark 11.23:

(i) If $u, v \in L^1$ then $u * v \in L^1$ and $\mathcal{F}(u * v) = \mathcal{F}u\, \mathcal{F}v$.
(ii) If $u \in L^1$ and $u * u = u$, then $u = 0$.
(iii) If $u \in L^2$ and $v \in L^1$, then $u * v \in L^2$ and $\mathcal{F}(u * v) = \mathcal{F}u\, \mathcal{F}v$.
(iv) If $v \in L^1$, there exists $f \in L^2$ such that the equation $u * v = f$ has no solution $u \in L^2$.

14.57. Let $u_0 \in \mathcal{S}'(\mathbf{R}^n)$. Prove that there exists exactly one differentiable family

$$t \mapsto u_t : \mathbf{R}_{>0} \to \mathcal{S}'(\mathbf{R}^n)$$

such that $\frac{d}{dt} u_t = \Delta_x u_t$ and $\lim_{t \downarrow 0} u_t = u_0$ in $\mathcal{S}'(\mathbf{R}^n)$. Calculate $\mathcal{F}u_t$ and u_t.

14.58.* *(Homogeneous fundamental solution of Laplace operator.)* Let $n \geq 3$. Prove the following assertions:

(i) $v(\xi) = -\|\xi\|^{-2}$, for $\xi \in \mathbf{R}^n \setminus \{0\}$, defines a locally integrable function on $\mathbf{R}^n$, and therefore a distribution on $\mathbf{R}^n$, which we denote by v as well.
(ii) v is homogeneous of degree -2 and $v \in \mathcal{S}'(\mathbf{R}^n)$.
(iii) Define $E := \mathcal{F}^{-1}v$. Then $\Delta E = \delta$, $E \in \mathcal{S}'(\mathbf{R}^n)$, and E is homogeneous of degree $2 - n$.
(iv) Every homogeneous fundamental solution of the Laplace operator is equal to E.
(v) E is C^∞ on $\mathbf{R}^n \setminus \{0\}$ and invariant under all rotations.
(vi) There exists a constant c such that E is equal to the locally integrable function $x \mapsto c\, \|x\|^{2-n}$.
(vii) $c = \frac{1}{(2-n)c_n}$ with c_n as in (13.37).

14.59.* *(Cauchy–Riemann operator.)* In this problem the notation is as in Example 12.11. Our goal is to obtain a fundamental solution E of the *Cauchy–Riemann operator* $\frac{\partial}{\partial \bar{z}}$ by means of the Fourier transform. Differently phrased, we use that transform to solve the following partial differential equation for E on $\mathbf{R}^2$:

$$\frac{\partial E}{\partial x}(x, y) - \frac{1}{i}\frac{\partial E}{\partial y}(x, y) = 2\delta(x, y) = 2\delta(x)\delta(y).$$

To this end, suppose that $x \mapsto E(x, \cdot)$ is a C^1 family in $\mathcal{S}'(\mathbf{R})$ and deduce on the basis of the Fourier transform with respect to the other variable that under this assumption,

$$\frac{d\mathcal{F}E}{dx}(x, \eta) - \eta \mathcal{F}E(x, \eta) = 2\delta(x).$$

Here $\mathcal{F}E$ denotes the partial Fourier transform. Write H for the Heaviside function on $\mathbf{R}$ and prove that, for a function $\eta \mapsto c(\eta)$ still to be determined,

$$\mathcal{F} E(x,\eta) = 2(c(\eta) + H(x))e^{x\eta} \qquad ((x,\eta) \in \mathbf{R}^2).$$

Observe that $\eta \mapsto e^{x\eta}$ does not define a tempered distribution on $\mathbf{R}$ if $x\eta > 0$. Obviate this problem by means of the choice $c(\eta) = -H(\eta)$ and show that in this case

$$\mathcal{F} E(x,\eta) = -2\,\mathrm{sgn}(\eta)\, H(-x\,\mathrm{sgn}(\eta))\, e^{x\eta} \qquad ((x,\eta) \in \mathbf{R}^2).$$

Now prove (compare with the assertion preceding (12.13))

$$E(x,y) = \frac{1}{\pi}\frac{1}{x+iy}, \qquad \text{in other words,} \qquad E(z) = \frac{1}{\pi z},$$

and show that E is indeed a tempered distribution on $\mathbf{C}$. Finally, derive (14.39) from the preceding formulas.

14.60.* *(Laplace operator.)* In this problem we construct a fundamental solution of the *Laplace operator* Δ on $\mathbf{R}^n$ by a method analogous to the one in Chap. 13. To this end we define, for $a \in \mathbf{C}$ with $\mathrm{Re}\, a > 0$, the function $R^a : \mathbf{R}^n \to \mathbf{C}$ by

$$R^a(x) = c(a)\, \|x\|^{a-n}, \qquad \text{where} \qquad c(a) = \frac{2}{c_n \Gamma(\frac{a}{2})} = \frac{\Gamma(\frac{n}{2})}{\pi^{\frac{n}{2}}\Gamma(\frac{a}{2})}.$$

(i) Prove that R^a is a locally integrable function on $\mathbf{R}^n$ and in consequence defines a distribution in $\mathcal{D}'(\mathbf{R}^n)$. Prove that in fact $R^a \in \mathcal{S}'(\mathbf{R}^n)$ and verify

$$R^a(e^{-\|\cdot\|^2}) = 1.$$

(ii) Let $q\colon \mathbf{R}^n \to \mathbf{R}$ be the quadratic form $x \mapsto \|x\|^2$; then verify that $q\colon \mathbf{R}^n \setminus \{0\} \to \mathbf{R}$ is a C^∞ submersion. Now prove that on $\mathbf{R}^n \setminus \{0\}$ and for $\mathrm{Re}\, a > 0$ we have

$$R^a = d(a)\, q^*\Big(\chi_+^{\frac{a-n+2}{2}}\Big) \qquad \text{with} \qquad d(a) = \frac{2\Gamma(\frac{a-n+2}{2})}{c_n\Gamma(\frac{a}{2})} = \frac{\Gamma(\frac{n}{2})\Gamma(\frac{a-n+2}{2})}{\pi^{\frac{n}{2}}\Gamma(\frac{a}{2})}.$$

Show that $a \mapsto c(a)$ admits an extension to a complex-analytic function on $\mathbf{C}$ and recall that $a \mapsto \chi_+^a$ is a family of distributions on $\mathbf{R}$ that is complex-analytic on $\mathbf{C}$. Use this to verify that the formula above is valid on $\mathbf{R}^n \setminus \{0\}$ for all $a \in \mathbf{C}$.

(iii) Prove in two different ways that on $\mathbf{R}^n \setminus \{0\}$ and for $\mathrm{Re}\, a > 0$,

$$\Delta R^a = 2(a-n)\, R^{a-2}. \tag{14.49}$$

Now suppose $n \geq 3$. Show that the definition of $R^a \in \mathcal{S}'(\mathbf{R}^n)$ may be extended to all $a \in \mathbf{C}$ by setting, for $k \in \mathbf{Z}_{\geq 0}$ and $0 < \mathrm{Re}\, a + 2k \leq 2$,

$$R^a = \frac{1}{2^k \prod_{1\leq j\leq k}(a+2j-n)}\, \Delta^k R^{a+2k}. \tag{14.50}$$

(iv) Next prove that $R^0 = 0$ on $\mathbf{R}^n \setminus \{0\}$ and use positivity to deduce $R^0 = \delta$. Conclude that

$$\Delta\Big(\frac{\|\cdot\|^{2-n}}{(2-n)c_n}\Big) = \delta.$$

This is the way in which the fundamental solution of Δ from Problem 4.7 was obtained. More generally, demonstrate that

$$R^{-2k} = \frac{(-1)^k}{2^k \prod_{0\le j<k}(n+2j)}\,\Delta^k\delta \qquad (k\in\mathbf{N}).$$

(v) For all $a \in \mathbf{C}$, verify that R^a is homogeneous of degree $a-n$ and invariant under the orthogonal group of $\mathbf{R}^n$.

(vi) In the notation of Problem 6.3.(ii), prove that

$$S_\phi^{(2k)}(0) = \frac{(-1)^k\,(2k)!}{k!}\,R^{-2k}(\phi).$$

Background. The occurrence of numerical factors in (14.49) and (14.50) can be avoided by multiplying R^a by

$$\frac{\Gamma(\frac{n-a}{2})}{2^a\,\Gamma(\frac{n}{2})}, \qquad \text{that is, by introducing} \qquad \widetilde{R}^a = \frac{\Gamma(\frac{n-a}{2})}{2^a\pi^{\frac{n}{2}}\,\Gamma(\frac{a}{2})}\|\cdot\|^{a-n}. \tag{14.51}$$

This normalization results if we require that $\widetilde{R}^a * f = f$, where $f(x) = e^{ix_1}$. Observe that we do not get a complex-analytic family on all of $\mathbf{C}$, because $\widetilde{R}^a$ is not even well-defined if $a \in n + 2\mathbf{Z}_{\ge 0}$. On the other hand, we obtain, for all $k \in \mathbf{Z}_{\ge 0}$ and for $a \in \mathbf{C}$ satisfying $\operatorname{Re} a + k < n$,

$$\widetilde{R}^a = (-1)^k\Delta^k\,\widetilde{R}^{a+2k} \qquad \text{and} \qquad \widetilde{R}^{-2k} = (-1)^k\Delta^k\delta.$$

Furthermore, the group law $\widetilde{R}^a * \widetilde{R}^b = \widetilde{R}^{a+b}$ is valid, although with restrictions on a and b. This can be proved from the convolution identity above, or by means of Fourier transform.

(vii) Prove, for all $a \in \mathbf{C}$, the following identity in $\mathcal{S}'(\mathbf{R}^n)$:

$$\mathcal{F}R^a = \pi^{\frac{n}{2}}2^a\,R^{n-a}.$$

More explicitly, for $0 < \operatorname{Re} a < n$,

$$\Gamma\Big(\frac{a}{2}\Big)\,\mathcal{F}\frac{1}{\|\cdot\|^a} = \pi^{\frac{n}{2}}2^{n-a}\,\Gamma\Big(\frac{n-a}{2}\Big)\frac{1}{\|\cdot\|^{n-a}}. \tag{14.52}$$

In particular, for $n \ge 3$, we have $\mathcal{F}\|\cdot\|^{2-n} = (n-2)c_n\,\|\cdot\|^{-2}$, with c_n as in (13.37). According to Problem 14.27 the last formula is true even in the case of $n = 1$, while it is not for $n = 2$. In the case of $n = 1$, prove that (14.52) is equivalent to (14.46).

(viii) Deduce

$$\mathcal{F} R^{-2k} = \frac{1}{2^k \prod_{0 \le j < k}(n+2j)} \|\cdot\|^{2k} \qquad (k \in \mathbf{Z}_{\ge 0})$$

and verify that this identity agrees with the formulas for R^{-2k} in part (iv).

14.61.* *(Radon transform.)* For $\omega \in S^{n-1}$ and $t \in \mathbf{R}$, define

$$\Phi_\omega : \mathbf{R}^n \to \mathbf{R} \qquad \text{by} \qquad \Phi_\omega(x) = \langle \omega, x \rangle \qquad \text{and} \qquad N(\omega, t) = \Phi_\omega^{-1}(\{t\}).$$

Note that $N(\omega, t)$ equals the hyperplane $\{ x \in \mathbf{R}^n \mid \langle \omega, x \rangle = t \}$.

(i) Prove that Φ_ω satisfies $\| \operatorname{grad} \Phi_\omega(x) \| = 1$, for all $x \in \mathbf{R}^n$. Next, show (see Theorem 10.18) that for any $\phi \in C_0^\infty(\mathbf{R}^n)$ and $t \in \mathbf{R}$,

$$(\Phi_\omega)_* \phi(t) = \int_{N(\omega,t)} \phi(x)\, d_{n-1}x =: \mathcal{R}\phi(\omega, t),$$

where $d_{n-1}x$ denotes integration with respect to the Euclidean hyperarea on $N(\omega, t)$ and $\mathcal{R}\phi : S^{n-1} \times \mathbf{R} \to \mathbf{R}$ is called the *Radon transform* of ϕ.

(ii) Prove that

$$\mathcal{R}(T_h{}^* \phi)(\omega, t) = \mathcal{R}\phi(\omega, t + \langle \omega, h \rangle) \qquad (h \in \mathbf{R}^n)$$

and deduce, for $1 \le j \le n$ and with Δ as usual denoting the Laplacian acting on $\mathbf{R}^n$, that

$$\mathcal{R}(\partial_j \phi)(\omega, t) = \omega_j\, \partial_t \mathcal{R}\phi(\omega, t) \qquad \text{and} \qquad \mathcal{R}(\Delta\phi)(\omega, t) = \partial_t^2 \mathcal{R}\phi(\omega, t).$$

(iii) Suppose $n \in \mathbf{N}$ is odd. In the notation of Problem 13.2, prove

$$\Phi_\omega{}^*(|\chi|^{-n+1})(\phi) = 2\partial_t^{n-1} \mathcal{R}\phi\big|_{t=0} = 2\mathcal{R}(\Delta^{\frac{n-1}{2}} \phi)\big|_{t=0} \qquad (\phi \in C_0^\infty(\mathbf{R}^n)).$$

Let $a \in \mathbf{C}$ and let $d\omega$ denote integration with respect to the Euclidean hyperarea on S^{n-1}. Define $A^a \in \mathcal{D}'(\mathbf{R}^n)$ by

$$A^a(\phi) = \int_{S^{n-1}} \Phi_\omega{}^*(|\chi|^{a+1})(\phi)\, d\omega \qquad (\phi \in C_0^\infty(\mathbf{R}^n)).$$

(iv) Prove that $(A^a)_{a \in \mathbf{C}}$ is a complex-analytic family of distributions and that for $\operatorname{Re} a > 0$,

$$A^a(\phi) = \int_{\mathbf{R}^n} \int_{S^{n-1}} \frac{|\langle \omega, x \rangle|^a}{\Gamma(a+1)}\, d\omega\, \phi(x)\, dx.$$

In particular, A^a is a continuous function on $\mathbf{R}^n$ for $\operatorname{Re} a > 0$. Also show that

$$A^{-n}(\phi) = 2 \int_{S^{n-1}} \partial_t^{n-1} \mathcal{R}\phi(\omega, t)\big|_{t=0}\, d\omega.$$

(v) Prove that the function A^a is homogeneous of degree a and is invariant under rotations acting on $\mathbf{R}^n$, for all $\operatorname{Re} a > 0$. Deduce the existence of $d(a) \in \mathbf{C}$

such that $A^a = d(a)\, R^{a+n} \in \mathcal{D}'(\mathbf{R}^n)$ for all $a \in \mathbf{C}$, where $(R^a)_{a\in\mathbf{C}}$ is the complex-analytic family from Problem 14.60. Prove, for $a \in \mathbf{C} \setminus (-2\mathbf{N})$, that

$$d(a) = \frac{2^{1-a}\pi^n}{\Gamma(\frac{n}{2})\Gamma(\frac{a}{2}+1)} \qquad \text{and conclude that} \qquad R^{a+n} = \frac{1}{d(a)} A^a.$$

Note that it is sufficient to evaluate $A^a(e_n)$ for $\operatorname{Re} a > 0$ owing to the invariance under rotations, where e_n denotes the nth standard basis vector in $\mathbf{R}^n$. Furthermore, use Legendre's duplication formula (13.39).

(vi) Now assume that $n \in \mathbf{N}$ is odd. Apply Problem 14.60.(iv), part (ii), and the reflection formula from Lemma 13.5 to obtain, for $x \in \mathbf{R}^n$ and $\phi \in C_0^\infty(\mathbf{R}^n)$),

$$\phi(x) = \frac{(-1)^{\frac{n-1}{2}}}{2(2\pi)^{n-1}} \int_{S^{n-1}} \partial_t^{n-1} \mathcal{R}\phi(\omega, t)\big|_{t=\langle \omega, x\rangle}\, d\omega.$$

For n odd, the formula above is called *Radon's inversion formula*, that is, it recovers ϕ from its Radon transform $\mathcal{R}\phi$. Differently phrased: ϕ is known when its integrals over all hyperplanes in $\mathbf{R}^n$ are known. Note that $\{\, N(\omega, \langle \omega, x\rangle) \mid \omega \in S^{n-1}\,\}$ is the collection of all hyperplanes in $\mathbf{R}^n$ passing through x. Hence, this inversion formula is local, in the sense that only the integrals over hyperplanes close to x are needed.

If n is even, the constant $1/d(-n)$ is undefined, while the definition of A^{-n} involves the constant $1/\Gamma(1-n) = 0$. In this case, proceed as follows.

(vii) Assume n to be even. Apply Legendre's duplication formula twice, then apply the reflection formula in order to show that

$$\frac{1}{d(-n)\Gamma(1-n)} = \frac{(-1)^{\frac{n}{2}}(n-1)!}{(2\pi)^n}.$$

In the notation of Problem 13.5, deduce

$$\delta = \frac{(-1)^{\frac{n}{2}}(n-1)!}{(2\pi)^n} \int_{S^{n-1}} \Phi_\omega{}^*(|\cdot|^{-n})\, d\omega;$$

that is, for $\phi \in C_0^\infty(\mathbf{R}^n)$ and $x \in \mathbf{R}^n$,

$$\phi(x) = \frac{(-1)^{\frac{n}{2}}(n-1)!}{(2\pi)^n} \int_{S^{n-1}} \int_{\mathbf{R}_{>0}} t^{-n} \Big(\sum_{\pm} \mathcal{R}\phi(\omega, \langle \omega, x\rangle \pm t) - 2\sum_{k=0}^{\frac{n}{2}-1} \partial_s^{2k} \mathcal{R}\phi(\omega, s)\big|_{s=\langle \omega, x\rangle} \frac{t^{2k}}{(2k)!}\Big)\, dt\, d\omega.$$

Note that in the case that n is even, the inversion formula is not local. Consider the special case of $n = 2$ and define, for $x \in \mathbf{R}^2$ and $t \geq 0$,

$$\overline{\mathcal{R}}\phi(x,t) = \frac{1}{4\pi}\int_{S^1}\sum_{\pm}\mathcal{R}\phi(\omega,\langle\,\omega,\,x\,\rangle \pm t)\,d\omega.$$

Then $\overline{\mathcal{R}}\phi(x,t)$ is the average of $\mathcal{R}\phi$ over the set of all lines in $\mathbf{R}^2$ at a distance t from the point x (note that $N(\omega,\langle\,\omega,\,x\,\rangle + t) = N(-\omega,\langle\,-\omega,\,x\,\rangle - t))$. The inversion formula now takes the form

$$\begin{aligned}\phi(x) &= -\frac{1}{\pi}\int_{\mathbf{R}_{>0}}\frac{\overline{\mathcal{R}}\phi(x,t) - \overline{\mathcal{R}}\phi(x,0)}{t^2}\,dt = -\frac{1}{\pi}\int_{\mathbf{R}_{>0}}\frac{\partial_t\overline{\mathcal{R}}\phi(x,t)}{t}\,dt\\ &= -\frac{1}{\pi}\int_{\mathbf{R}_{>0}}\frac{d\,\overline{\mathcal{R}}\phi_x(t)}{t}.\end{aligned}$$

Here the second identity is obtained by means of a formal integration by parts. The last expression is a Riemann–Stieltjes integral and equals, apart from differences in notation, the formula given by Radon in his 1917 paper.

Chapter 15
Distribution Kernels

A very important result in distribution theory is the Schwartz Kernel Theorem, which can be efficiently proved by means of the Fourier transform.

Suppose $X \subset \mathbf{R}^n$ and $Y \subset \mathbf{R}^m$ are open subsets and k is a continuous function on $X \times Y$. Then

$$\mathcal{K} : \psi \mapsto \mathcal{K}\psi, \qquad \text{where} \qquad \mathcal{K}\psi(x) = \int_Y k(x, y)\psi(y)\, dy$$

defines a linear mapping $\mathcal{K}$ from the space of compactly supported continuous functions on Y to the space of continuous functions on X. The mapping $\mathcal{K}$ is said to be the linear *integral operator* or *integral transform* defined by the *integral kernel*[1] k and is often denoted by k_{op}. Integral transforms act in spaces of functions as changes of variables, which can be useful making problems more tractable. In the historical development of functional analysis, integral operators played an important role.

Further assume $\psi \in C_0^\infty(Y)$ and $\phi \in C_0^\infty(X)$. Integrating the function $\mathcal{K}\psi$ against the function ϕ gives

$$\int_{X \times Y} k(x, y)\, \phi(x)\, \psi(y)\, d(x, y),$$

which implies

$$\mathcal{K}\psi(\phi) = k(\phi \otimes \psi). \tag{15.1}$$

Here the tensor product $\phi \otimes \psi$ is defined similarly to (11.8). Since $\phi \otimes \psi$ belongs to $C_0^\infty(X \times Y)$, the function k in (15.1) may be replaced by an arbitrary element of $\mathcal{D}'(X \times Y)$. We also observe that $\phi \otimes \psi$ depends continuously on ϕ when ψ is fixed and depends continuously on ψ when ϕ is fixed. As a consequence, (15.1) defines a

$$\text{continuous linear operator} \quad \mathcal{K} : C_0^\infty(Y) \to \mathcal{D}'(X), \tag{15.2}$$

which is denoted by $\mathcal{K} = k_{\mathrm{op}}$. These observations lead to a generalization of the concept of linear integral operator with far-reaching implications.

---

[1] This notion of kernel is not to be confused with the one from linear algebra.

J.J. Duistermaat and J.A.C. Kolk, *Distributions: Theory and Applications*,
Cornerstones, DOI 10.1007/978-0-8176-4675-2_15,

Furthermore, the Kernel Theorem, Theorem 15.2, below asserts that for every $\mathcal{K}$ as in (15.2) there exists a unique $k \in \mathcal{D}'(X \times Y)$ such that $\mathcal{K} = k_{\mathrm{op}}$, that is, (15.1) applies. On account of its uniqueness, we may call k, or $k_{\mathcal{K}}$, the *distribution kernel*, or *Schwartz kernel*, of $\mathcal{K}$, while k_{op} is said to be the *operator generated* by k. Some authors use the same symbol for the kernel and for the mapping generated by it.

The Kernel Theorem enables the investigation of mappings from test functions to distributions by means of methods from distribution theory itself, through a study of the corresponding distribution kernels. Thus it is possible to obtain numerous properties of integral operators. In particular, this is the case for the Fourier transform: its intertwining properties with many operations on functions are easily obtained in this manner; see Problem 15.5 below. Kernels are important in the theory of linear partial differential equations; in particular, the Kernel Theorem unites differential and integral operators into a common framework.

The verification of the theorem begins by deriving a formula for k in terms of $\mathcal{K}$. This formula then implies the uniqueness of k but also provides a starting point for proving its existence. Principal ingredients in the proof are the basic property of Fourier decomposition, i.e., that a function can be represented as a continuous superposition of exponential functions, plus the fact that a multidimensional exponential function already equals a tensor product.

After this introduction, we now discuss the notion of kernel in greater detail. Consider open sets $X \subset \mathbf{R}^n$ and $Y \subset \mathbf{R}^m$, and let $k \in \mathcal{D}'(X \times Y)$. Then

$$(\phi, \psi) \mapsto k(\phi \otimes \psi) \qquad (\phi \in C_0^\infty(X), \psi \in C_0^\infty(Y)) \tag{15.3}$$

is a bilinear form on $C_0^\infty(X) \times C_0^\infty(Y)$. For fixed ψ it becomes a linear form on $C_0^\infty(X)$ and for fixed ϕ it becomes a linear form on $C_0^\infty(Y)$. Both of these are distributions. Indeed, given compact sets $K \subset X$ and $K' \subset Y$, Theorem 3.8 implies the existence of a constant $c > 0$ and an order of differentiation $N \in \mathbf{Z}_{\geq 0}$ such that $|k(\phi \otimes \psi)| \leq c \, \|\phi \otimes \psi\|_{C^N}$, for $\phi \in C_0^\infty(K)$ and $\psi \in C_0^\infty(K')$. It is helpful to replace this by the equivalent estimate

$$|k(\phi \otimes \psi)| \leq c \, \|\phi\|_{C^N} \, \|\psi\|_{C^N} \qquad (\phi \in C_0^\infty(K), \psi \in C_0^\infty(K')). \tag{15.4}$$

Write $\mathcal{K}\psi$ for the linear form $\phi \mapsto k(\phi \otimes \psi)$ on $C_0^\infty(X)$, that is,

$$\mathcal{K}\psi(\phi) = k(\phi \otimes \psi) \qquad (\phi \in C_0^\infty(X)). \tag{15.5}$$

The inequality (15.4) then gives an estimate as in Theorem 3.8 for $\mathcal{K}\psi$ when ψ is fixed, so that $\mathcal{K}\psi \in \mathcal{D}'(X)$. Moreover, (15.4) shows that the linear mapping

$$\mathcal{K} : C_0^\infty(Y) \to \mathcal{D}'(X) \qquad \text{given by} \qquad \psi \mapsto \mathcal{K}\psi$$

is sequentially continuous: if $\psi_j \to 0$ in $C_0^\infty(Y)$ as $j \to \infty$, then $\mathcal{K}\psi_j \to 0$ in $\mathcal{D}'(X)$. It is also obvious from (15.4) that the linear mapping

$$C_0^\infty(X) \to \mathcal{D}'(Y) \qquad \text{given by} \qquad \phi \mapsto (\psi \mapsto k(\phi \otimes \psi))$$

is sequentially continuous.

To write this mapping in a similar way, one can first define the *transposed kernel* $\tau_* k \in \mathcal{D}'(Y \times X)$ of k, as follows. The isomorphism $\tau : X \times Y \to Y \times X$ given by $\tau(x, y) = (y, x)$ induces isomorphisms of pullback τ^* and pushforward τ_* satisfying $\tau_* k(\chi) = k(\tau^* \chi)$, for every $\chi \in C_0^\infty(Y \times X)$. In particular, $\tau^*(\psi \otimes \phi) = \phi \otimes \psi$. It follows from (15.5) that the mapping $\mathcal{K}_{\tau_* k} : C_0^\infty(X) \to \mathcal{D}'(Y)$ generated by $\tau_* k$ has the property

$$\mathcal{K}_{\tau_* k}\phi(\psi) = \tau_* k(\psi \otimes \phi) = k(\phi \otimes \psi) \qquad (\psi \in C_0^\infty(Y)).$$

This notation thus makes it possible to discuss the two mappings associated with the bilinear form (15.3) on the same footing.

It is useful to observe that the restriction to $C_0^\infty(X)$ of the transpose mapping ${}^t\mathcal{K}$, and the mapping $\mathcal{K}_{\tau_* k}$ coincide, that is,

$$ {}^t\mathcal{K} = \mathcal{K}_{\tau_* k}. \tag{15.6}$$

In fact, one obtains ${}^t\mathcal{K}\phi(\psi) = \mathcal{K}\psi(\phi) = k(\phi \otimes \psi) = \mathcal{K}_{\tau_* k}\phi(\psi)$, for $\phi \in C_0^\infty(X)$ and $\psi \in C_0^\infty(Y)$. Furthermore, each of the mappings $\mathcal{K}$ and ${}^t\mathcal{K}$ determines the other one.

Let us consider two simple examples. If $k \in C^\infty(X \times Y)$, then (15.5) yields the classical integral operator

$$\mathcal{K}\psi(x) = \int_Y k(x, y)\psi(y)\, dy \qquad (\psi \in C_0^\infty(Y)),$$

and the transpose operator is

$$ {}^t\mathcal{K}\phi(y) = \int_X k(x, y)\phi(x)\, dx \qquad (\phi \in C_0^\infty(X)).$$

If $k = u \otimes v$, where $u \in \mathcal{D}'(X)$ and $v \in \mathcal{D}'(Y)$, then the ranges of both mappings are one-dimensional. Indeed,

$$\mathcal{K}\psi = v(\psi)u \qquad \text{and} \qquad {}^t\mathcal{K}\phi = u(\phi)v.$$

More generally, in the notation of Example 11.11, we note that the following assertions are equivalent, for any $k \in C_0^\infty(X \times Y)$:

(i) $k \in C_0^\infty(X) \otimes C_0^\infty(Y)$.
(ii) The integral operator $k_{\text{op}} : C_0^\infty(Y) \to C^\infty(X)$ is of finite rank, that is, its range space is of finite dimension.
(iii) The transpose of k_{op} is of finite rank.

Finally, consider the special case of a sequentially continuous linear mapping $\mathcal{K} : C_0^\infty(Y) \to C^\infty(X)$. In order to derive a formula for at least one distribution kernel, say k, of $\mathcal{K}$, suppose that $\chi = \phi \otimes \psi \in C_0^\infty(X) \otimes C_0^\infty(Y)$ and note that

$$\phi(x)\,\psi = \phi \otimes \psi(x, \cdot) = \chi(x, \cdot) \in C_0^\infty(Y) \qquad \text{and}$$

$$\mathcal{K}\psi(x)\,\phi(x) = \mathcal{K}(\phi(x)\,\psi)(x) = \mathcal{K}(\chi(x,\cdot))(x),$$

with $\chi(x,\cdot)(y) = \chi(x,y)$. Hence, (15.1) takes the form

$$k(\chi) = \int_X \mathcal{K}(\chi(x,\cdot))(x)\,dx. \tag{15.7}$$

In fact, $C_0^\infty(X) \otimes C_0^\infty(Y)$ is dense in $C_0^\infty(X \times Y)$ according to Remark 11.12 or Theorem 15.4 below; therefore (15.7) is valid for all $\chi \in C_0^\infty(X \times Y)$ and determines the unique kernel k of $\mathcal{K}$. See Theorem 15.5 below for a more precise formulation.

Example 15.1. Let $I : C_0^\infty(X) \to C_0^\infty(X)$ be the identity mapping and denote the kernel as given in (15.7) by $k_I \in \mathcal{D}'(X \times X)$. We then obtain, in the notation of Example 10.7 and in particular with the natural embedding $\Delta : X \to X \times X$,

$$k_I(\chi) = \int_X \chi(x,x)\,dx = \Delta_* 1_X(\chi) \qquad (\chi \in C_0^\infty(X \times X)).$$

Furthermore, consider the special case of $X = \mathbf{R}^n$ and denote by d the difference mapping from (10.24) and by δ the Dirac measure on $\mathbf{R}^n$. Using (10.25) one obtains

$$k_I = \Delta_* 1_{\mathbf{R}^n} = d^*\delta \in \mathcal{D}'(\mathbf{R}^n \times \mathbf{R}^n). \tag{15.8}$$

This example shows that operators with very good regularity properties, meaning that they preserve $C_0^\infty(X)$, may have distribution kernels that are quite singular distributions. It also demonstrates that not every continuous linear operator from $L^2(Y)$ to $L^2(X)$ is necessarily generated by a kernel belonging to $L^2(X \times Y)$. ⊘

Suppose now that one has an arbitrary sequentially continuous linear mapping $C_0^\infty(Y) \to \mathcal{D}'(X)$. Then it is natural to ask whether there exists a kernel in $\mathcal{D}'(X \times Y)$ that generates this mapping. The main assertion of the following *Kernel Theorem*, which is due to Schwartz, is that this is indeed the case.

Theorem 15.2. *Consider open sets $X \subset \mathbf{R}^n$ and $Y \subset \mathbf{R}^m$. A linear mapping $\mathcal{K} : C_0^\infty(Y) \to \mathcal{D}'(X)$ is sequentially continuous if and only if it is generated by a distribution kernel $k \in \mathcal{D}'(X \times Y)$; phrased differently,*

$$\mathcal{K}\psi(\phi) = k(\phi \otimes \psi) \qquad (\phi \in C_0^\infty(X),\, \psi \in C_0^\infty(Y)). \tag{15.9}$$

Moreover, the kernel k is uniquely determined by the mapping $\mathcal{K}$.

Proof. We begin by establishing uniqueness of k. This is the easy part of the proof; moreover, it suggests a way of verifying the existence of k.

To this end, suppose that $k \in \mathcal{D}'(X \times Y)$ is a distribution kernel for $\mathcal{K}$. Furthermore, let K be an arbitrary compact subset of $X \times Y$ and denote by K_1 and K_2 its orthogonal projections onto $\mathbf{R}^n$ and $\mathbf{R}^m$, respectively. Then K_1 and K_2 are compact subsets of X and Y and consequently there exist $\rho \in C_0^\infty(X)$ and $\sigma \in C_0^\infty(Y)$

such that $\rho = 1$ on K_1 and $\sigma = 1$ on K_2, respectively. Thus $\rho \otimes \sigma \in C_0^\infty(X \times Y)$, while $\rho \otimes \sigma = 1$ on $K \subset K_1 \times K_2$. Next, set $l = (\rho \otimes \sigma)k \in \mathcal{E}'(X \times Y)$. Then l and k coincide on $C_0^\infty(K)$, while $l \in \mathcal{E}'(\mathbf{R}^n \times \mathbf{R}^m) \subset \mathcal{S}'(\mathbf{R}^n \times \mathbf{R}^m)$ on account of Lemma 8.9 and (14.24). In view of (15.9) we now have, for $(\xi, \eta) \in \mathbf{R}^n \times \mathbf{R}^m$,

$$l(e_{i\xi} \otimes e_{i\eta}) = k((\rho \otimes \sigma)(e_{i\xi} \otimes e_{i\eta})) = k(\rho e_{i\xi} \otimes \sigma e_{i\eta}) = \mathcal{K}(\sigma e_{i\eta})(\rho e_{i\xi}).$$

We recall that l is continuous and linear. Therefore, considering the integral as a limit of Riemann sums and using (14.17), we obtain, for every $\chi \in C_0^\infty(K) \subset C_0^\infty(\mathbf{R}^n \times \mathbf{R}^m)$,

$$\begin{aligned} k(\chi) = l(\chi) &= l\left(\frac{1}{(2\pi)^{n+m}} \int_{\mathbf{R}^n \times \mathbf{R}^m} \mathcal{F}\chi(\xi, \eta)\, e_{i\xi} \otimes e_{i\eta}\, d(\xi, \eta)\right) \\ &= \frac{1}{(2\pi)^{n+m}} \int_{\mathbf{R}^n \times \mathbf{R}^m} \mathcal{F}\chi(\xi, \eta)\, l(e_{i\xi} \otimes e_{i\eta})\, d(\xi, \eta) \\ &= \frac{1}{(2\pi)^{n+m}} \int_{\mathbf{R}^n \times \mathbf{R}^m} \mathcal{F}\chi(\xi, \eta)\, \mathcal{K}(\sigma e_{i\eta})(\rho e_{i\xi})\, d(\xi, \eta). \end{aligned} \tag{15.10}$$

The identity (15.10) constitutes an explicit formula for the restriction of k to $C_0^\infty(K)$ in terms of $\mathcal{K}$, which proves that this restriction is uniquely determined by $\mathcal{K}$. Since K is an arbitrary compact subset of $X \times Y$, this implies that k is uniquely determined by $\mathcal{K}$.

Conversely, to prove the existence of k, let us assume that we are given a sequentially continuous linear mapping $\mathcal{K} : C_0^\infty(Y) \to \mathcal{D}'(X)$. Define the bilinear form

$$B : C_0^\infty(X) \times C_0^\infty(Y) \to \mathbf{C} \qquad \text{by setting} \qquad B(\phi, \psi) = \mathcal{K}\psi(\phi). \tag{15.11}$$

Our next step is to derive an estimate similar to (15.4). Let $K \subset X$ and $K' \subset Y$ be compact sets. Since $\mathcal{K}\psi \in \mathcal{D}'(X)$, there exists, on account of Theorem 3.8, an estimate

$$|B(\phi, \psi)| = |\mathcal{K}\psi(\phi)| \le c_\psi \, \|\phi\|_{C^{N_\psi}} \qquad (\phi \in C_0^\infty(K)).$$

Here both $c_\psi > 0$ and $N_\psi \in \mathbf{Z}_{\ge 0}$ depend on ψ. Furthermore, since $\mathcal{K}$ is sequentially continuous and linear, $\psi \mapsto B(\phi, \psi)$ is an element of $\mathcal{D}'(Y)$, for each $\phi \in C_0^\infty(X)$, and again on the basis of Theorem 3.8 we obtain

$$|B(\phi, \psi)| \le c'_\phi \, \|\psi\|_{C^{M_\phi}} \qquad (\psi \in C_0^\infty(K')).$$

These two inequalities show that the restriction of the bilinear form B to the linear subspace $C_0^\infty(K) \times C_0^\infty(K')$ is separately continuous, which in turn yields that it is also jointly continuous. The verification of this fact is deferred to Proposition 15.3 below, so as to not interrupt the main argument of the proof of the theorem.

In the present case the joint continuity implies that there exist constants $c > 0$ and $N \in \mathbf{Z}_{\ge 0}$ such that

$$|B(\phi, \psi)| \leq c \, \|\phi\|_{C^N} \, \|\psi\|_{C^N}. \tag{15.12}$$

Note that this is, in effect, the estimate (15.4), which has thus been recovered.

Let U and V be relatively compact open subsets of X and Y, and select $\rho \in C_0^\infty(X)$ and $\sigma \in C_0^\infty(Y)$ such that $\rho = 1$ on U and $\sigma = 1$ on V, respectively. Guided by (15.10), we introduce $k_{\rho,\sigma} \in \mathcal{D}'(X \times Y)$ by

$$k_{\rho,\sigma}(\chi) = \frac{1}{(2\pi)^{n+m}} \int_{\mathbf{R}^n \times \mathbf{R}^m} \mathcal{F}\chi(\xi, \eta) \, B(\rho e_{i\xi}, \sigma e_{i\eta}) \, d(\xi, \eta), \tag{15.13}$$

for all $\chi \in C_0^\infty(X \times Y)$. Then $k_{\rho,\sigma}$ is well-defined. Indeed, (15.12) implies the existence of constants $c > 0$ and $N \in \mathbf{Z}_{\geq 0}$ such that

$$|B(\rho e_{i\xi}, \sigma e_{i\eta})| \leq c \, (1 + \|\xi\|)^N (1 + \|\eta\|)^N \qquad ((\xi, \eta) \in \mathbf{R}^n \times \mathbf{R}^m).$$

Combination of this estimate with the fact that $\mathcal{F}\chi \in \mathcal{S}(\mathbf{R}^n \times \mathbf{R}^m)$ yields the convergence of the integral in (15.13).

Let $\phi \in C_0^\infty(U)$ and $\psi \in C_0^\infty(V)$. Then $\mathcal{F}(\phi \otimes \psi)(\xi, \eta) = \mathcal{F}\phi(\xi) \, \mathcal{F}\psi(\eta)$. Hence application of (15.13) with $\chi = \phi \otimes \psi$ implies

$$\begin{aligned}
k_{\rho,\sigma}(\phi \otimes \psi) &= \frac{1}{(2\pi)^{n+m}} \int_{\mathbf{R}^n \times \mathbf{R}^m} \mathcal{F}\phi(\xi) \, \mathcal{F}\psi(\eta) \, \mathcal{K}(\sigma e_{i\eta})(\rho e_{i\xi}) \, d(\xi, \eta) \\
&= \mathcal{K}\Big(\sigma \frac{1}{(2\pi)^m} \int_{\mathbf{R}^m} \mathcal{F}\psi(\eta) \, e_{i\eta} \, d\eta\Big)\Big(\rho \frac{1}{(2\pi)^n} \int_{\mathbf{R}^n} \mathcal{F}\phi(\xi) \, e_{i\xi} \, d\xi\Big) \\
&= \mathcal{K}(\sigma\psi)(\rho\phi) = \mathcal{K}\psi(\phi).
\end{aligned}$$

Phrased differently, ${}^t\iota_{X \times Y \, U \times V} \circ k_{\rho,\sigma} \in \mathcal{D}'(U \times V)$ is the distribution kernel of the sequentially continuous linear mapping ${}^t\iota_{XU} \circ \mathcal{K} \circ \iota_{YV} : C_0^\infty(V) \to \mathcal{D}'(U)$, which is independent of ρ and σ. Here $\iota_{XU} : C_0^\infty(U) \to C_0^\infty(X)$ is the linear injection defined in (7.1) (note that its transpose equals $\rho_{UX} : \mathcal{D}'(X) \to \mathcal{D}'(U)$, as in (7.2)). Since we have already established the uniqueness of distribution kernels, it follows that $k_{\rho,\sigma}$ is independent of the choices of ρ and σ, but it still depends on the choices of U and V. Thus, we may write $k^{U,V}$ instead of $k_{\rho,\sigma}$. Invoking the uniqueness once more, we see that $k^{U,V} = k^{U',V'}$ on $(U \cap U') \times (V \cap V')$. On account of Theorem 7.6, the $k^{U,V}$ have a common extension to $k \in \mathcal{D}'(X \times Y)$ that satisfies (15.9). This verifies the existence part of the theorem. □

Next, we come to the proposition announced in the proof above.

Proposition 15.3. *Let the notation be as in the preceding proof. The restriction of B to $C_0^\infty(K) \times C_0^\infty(K')$ is a jointly continuous bilinear form.*

Proof. Consider arbitrary sequences $(\phi_j)_{j \in \mathbf{N}}$ in $C_0^\infty(X)$ and $(\psi_j)_{j \in \mathbf{N}}$ in $C_0^\infty(Y)$ such that $\phi_j \to 0$ in $C_0^\infty(X)$ and $\psi_j \to 0$ in $C_0^\infty(Y)$, respectively, as $j \to \infty$. Define the linear functional u_j on $C_0^\infty(X)$ by $u_j(\phi) = B(\phi, \psi_j)$. Then u_j is continuous, and so $u_j \in \mathcal{D}'(X)$. Furthermore, $\lim_{j \to \infty} u_j(\phi) = \lim_{j \to \infty} B(\phi, \psi_j) = 0$, for all $\phi \in C_0^\infty(X)$; hence u_j converges in $\mathcal{D}'(X)$ to 0. Accordingly, Theorem 5.5.(ii) implies $B(\phi_j, \psi_j) = u_j(\psi_j) \to 0$ as $j \to \infty$. In other words, B is jointly sequentially continuous, and a fortiori this is also the case for the restriction of B to $C_0^\infty(K) \times C_0^\infty(K')$. Now both factors in this product can be provided with a countable separating collection of seminorms. But then Lemma 8.7 implies that the restriction is jointly continuous. □

The preceding theorem immediately leads to the following result; Remark 11.12 and Theorem 16.17 contain different proofs.

Theorem 15.4. *Let the notation be as in Example 11.11. Then $C_0^\infty(X) \otimes C_0^\infty(Y)$ is a dense linear subspace of $C_0^\infty(X \times Y)$.*

Proof. Suppose that $k \in \mathcal{D}'(X \times Y)$ vanishes on $C_0^\infty(X) \otimes C_0^\infty(Y)$. If $\mathcal{K} : C_0^\infty(Y) \to \mathcal{D}'(X)$ denotes the mapping generated by k, then (15.9) implies that $\mathcal{K} = 0$ and this in turn gives $k = 0$, by the uniqueness of distribution kernels. Using the Hahn–Banach Theorem, Theorem 8.12, we now obtain the desired conclusion. □

A continuous linear mapping $\mathcal{K} : C_0^\infty(Y) \to C^\infty(X)$ is evidently also a continuous linear mapping $C_0^\infty(Y) \to \mathcal{D}'(X)$. By Theorem 15.2 it is therefore generated by a distribution kernel; a rule for its computation is provided by the next theorem. The formulation of the result requires some notation.

Define

$$I \otimes \mathcal{K} : C_0^\infty(X \times Y) \to C^\infty(X \times X) \quad \text{by} \quad (I \otimes \mathcal{K})\chi(x, x') = \mathcal{K}(\chi(x, \cdot))(x'),$$

for $\chi \in C_0^\infty(X \times Y)$ and x and $x' \in X$. Then $\chi(x, \cdot) \in C_0^\infty(Y)$, so that $\mathcal{K}(\chi(x, \cdot)) \in C^\infty(X)$. The notation for $I \otimes \mathcal{K}$ derives its justification from the fact that

$$\begin{aligned} (I \otimes \mathcal{K})(\phi \otimes \psi)(x, x') &= (\phi \otimes \mathcal{K}\psi)(x, x') = \phi(x)\, \mathcal{K}\psi(x') \\ &= \mathcal{K}(\phi(x)\, \psi)(x') = \mathcal{K}((\phi \otimes \psi)(x, \cdot))(x'), \end{aligned} \tag{15.14}$$

taken in conjunction with Theorem 15.4.

Theorem 15.5. *The distribution kernel $k \in \mathcal{D}'(X \times Y)$ of a continuous linear mapping $\mathcal{K} : C_0^\infty(Y) \to C^\infty(X)$ satisfies, in the notation discussed in Example 15.1,*

$$k = {}^t(I \otimes \mathcal{K})\, \Delta_* 1_X = (I \otimes {}^t\mathcal{K})\, \Delta_* 1_X, \tag{15.15}$$

or more explicitly

$$k(\chi) = \int_X \mathcal{K}(\chi(x, \cdot))(x)\, dx \qquad (\chi \in C_0^\infty(X \times Y)).$$

Proof. On account of Theorem 15.4 it is sufficient to verify formula (15.15) for all χ of the form $\phi \otimes \psi$, where $\phi \in C_0^\infty(X)$ and $\psi \in C_0^\infty(Y)$. It is not self-evident that $\Delta_* 1_X \in \mathcal{D}'(X \times X)$ may be applied to $\phi \otimes \mathcal{K}\psi \in C_0^\infty(X) \otimes C^\infty(X)$. For showing that this is the case, select $\theta \in C_0^\infty(X)$ such that $\theta = 1$ on a neighborhood of supp ϕ. Then $\phi\otimes\mathcal{K}\psi - \phi\otimes\theta\,\mathcal{K}\psi = \phi\otimes(1-\theta)\mathcal{K}\psi$ is a function in $C^\infty(X\times X)$ with support disjoint from the diagonal $\Delta(X)$, which is supp $\Delta_* 1_X$. It follows that we may define $\Delta_* 1_X(\phi \otimes \mathcal{K}\psi)$ as $\Delta_* 1_X(\phi \otimes \theta\,\mathcal{K}\psi)$ and that this definition is independent of the choice of θ.

Thus, we have

$$\begin{aligned} k(\phi \otimes \psi) = \mathcal{K}\psi(\phi) &= \int_X \phi(x)\,\mathcal{K}\psi(x)\,dx = \int_X (I \otimes \mathcal{K})(\phi \otimes \psi)(x,x)\,dx \\ &= \Delta_* 1_X((I \otimes \mathcal{K})(\phi \otimes \psi)) = {}^t(I \otimes \mathcal{K})\,\Delta_* 1_X(\phi \otimes \psi). \end{aligned}$$

Next, applying Theorem 15.4 once again and using Corollary 11.7, we see that the linear subspace $C_0^\infty(X) \otimes C_0^\infty(Y)$ is dense in $\mathcal{D}'(X \times Y)$. Hence the identity ${}^t(I \otimes \mathcal{K}) = I \otimes {}^t\mathcal{K}$ is valid on $\mathcal{D}'(X \times Y)$ as soon as it holds on $C_0^\infty(X) \otimes C_0^\infty(Y)$. Now observe, for ζ and $\phi \in C_0^\infty(X)$ and η and $\psi \in C_0^\infty(Y)$, that

$$\begin{aligned} {}^t(I \otimes \mathcal{K})(\zeta \otimes \eta)(\phi \otimes \psi) &= (\zeta \otimes \eta)(\phi \otimes \mathcal{K}\psi) = \zeta(\phi)\,\eta(\mathcal{K}\psi) = \zeta(\phi)\,{}^t\mathcal{K}\eta(\psi) \\ &= (\zeta \otimes {}^t\mathcal{K}\eta)(\phi \otimes \psi) = (I \otimes {}^t\mathcal{K})(\zeta \otimes \eta)(\phi \otimes \psi). \end{aligned}$$

This leads to the desired formula. □

Example 15.6. Take X open in $\mathbf{R}^n$, and consider a linear partial differential operator in X of order m with C^∞ coefficients

$$P(\partial) = \sum_{|\alpha| \le m} a_\alpha \partial^\alpha; \tag{15.16}$$

here $a_\alpha \in C^\infty(X)$, for $|\alpha| \le m$. Then we actually have a continuous linear mapping $P(\partial) : C_0^\infty(X) \to C_0^\infty(X)$. Recall that ${}^tP(\partial)$, the transpose operator of $P(\partial)$, is given by

$${}^tP(\partial) = \sum_{|\alpha| \le m} (-\partial)^\alpha \circ a_\alpha.$$

Applying (15.15), we obtain for the distribution kernel associated to $P(\partial)$

$$(I \otimes {}^tP(\partial))\,\Delta_* 1_X \in \mathcal{D}'(X \times X). \tag{15.17}$$

Note that this assertion agrees with the following one, derived from (15.9):

$$(I \otimes {}^tP(\partial))\,\Delta_* 1_X(\phi \otimes \psi) = \Delta_* 1_X(\phi \otimes P(\partial)\psi) = P(\partial)\psi(\phi),$$

for ϕ and $\psi \in C_0^\infty(X)$.

For instance, if $m = 0$ and $P(\partial) = \psi \in C^\infty(X)$, then $P(\partial)$ reduces to the multiplication map $\phi \mapsto \psi\,\phi$, with kernel $(I \otimes \psi)\,\Delta_* 1_X$. In particular, the distribution $\Delta_* 1_X$, which satisfies supp $\Delta_* 1_X$ = sing supp $\Delta_* 1_X = \Delta(X)$, is the distribution kernel of the identity mapping; see Example 15.1. Furthermore, if $P(\partial) = \partial_j$, for $1 \le j \le n$, then its kernel is $(I \otimes -\partial_j)\, \Delta_* 1_X$. ⊘

In applications, one often encounters (semi)regular kernels; these have images in spaces consisting of smooth functions instead of distributions, as we saw in the preceding example.

Definition 15.7. Let k and $\mathcal{K}$ be as in the Kernel Theorem, Theorem 15.2. Then the distribution kernel k is said to be *left semiregular* if $\mathcal{K}$ sends $C_0^\infty(Y)$ to $C^\infty(X)$ and if this mapping is continuous. Similarly, k is said to be *right semiregular* if the preceding condition holds with X and Y as well as $\mathcal{K}$ and ${}^t\mathcal{K}$ interchanged. ⊘

Theorem 15.8. *Suppose that the distribution kernel $k \in \mathcal{D}'(X \times Y)$ is right semiregular; in other words, the mapping ${}^t\mathcal{K}$ generated by $\tau_* k$ is a continuous mapping $C_0^\infty(X) \to C^\infty(Y)$. Then the mapping $\mathcal{K}$ extends to a linear mapping $\mathcal{E}'(Y) \to \mathcal{D}'(X)$ that is sequentially continuous in the following sense: if a sequence $(u_j)_{j\in\mathbf{N}}$ converges in $\mathcal{D}'(Y)$ to $u \in \mathcal{E}'(Y)$ and the supports of the u_j are contained in a fixed compact set, then* $\lim_{j\to\infty} \mathcal{K}u_j = \mathcal{K}u$ *in $\mathcal{D}'(X)$.*

Proof. Define

$$\widetilde{\mathcal{K}} : \mathcal{E}'(Y) \to \mathcal{D}'(X) \qquad \text{by} \qquad \widetilde{\mathcal{K}}u = u \circ {}^t\mathcal{K} \qquad (u \in \mathcal{E}'(Y)). \tag{15.18}$$

Being the composition of the continuous linear mappings ${}^t\mathcal{K} : C_0^\infty(X) \to C^\infty(Y)$ and $u : C^\infty(Y) \to \mathbf{C}$, the mapping $\widetilde{\mathcal{K}}u$ is a distribution. The sequential continuity of $\widetilde{\mathcal{K}} : \mathcal{E}'(Y) \to \mathcal{D}'(X)$ is also immediate from (15.18).

To show that $\widetilde{\mathcal{K}}$ extends $\mathcal{K} : C_0^\infty(Y) \to \mathcal{D}'(X)$, consider $u = \psi \in C_0^\infty(Y)$. We then have, from (15.18),

$$\widetilde{\mathcal{K}}\psi(\phi) = \psi({}^t\mathcal{K}\phi) = \mathcal{K}\psi(\phi) \qquad (\phi \in C_0^\infty(X)),$$

which proves the claim. We can now obviously drop the tilde, and write $\mathcal{K}$ for the map defined by (15.18). □

Definition 15.9. The case most frequently encountered in applications is that of a kernel that is both left and right semiregular. We then speak of a *regular* distribution kernel. ⊘

For instance, kernels belonging to $C_0^\infty(X \times Y)$ are regular. Theorem 15.8 has the following consequence.

Corollary 15.10. *If $k \in \mathcal{D}'(X \times Y)$ is a regular distribution kernel, then the maps $\mathcal{K}$ and ${}^t\mathcal{K}$ extend to sequentially continuous maps $\mathcal{E}'(Y) \to \mathcal{D}'(X)$ and $\mathcal{E}'(X) \to \mathcal{D}'(Y)$, respectively.*

Example 15.11. Consider $u \in \mathcal{D}'(\mathbf{R}^n)$. On account of Theorem 11.2 we then have the linear mapping

$$\mathcal{U} := u\,* : C_0^\infty(\mathbf{R}^n) \to C^\infty(\mathbf{R}^n) \qquad \text{with} \qquad \mathcal{U}\,\psi(x) = u(T_x \circ S\psi),$$

while the continuity of $\mathcal{U}$ is established in Theorem 11.3. The distribution kernel of $\mathcal{U}$ equals $d^*u \in \mathcal{D}'(\mathbf{R}^n \times \mathbf{R}^n)$, where d is the difference mapping from (10.24).

Indeed, on the strength of Example 11.9 we conclude, for ϕ and $\psi \in C_0^\infty(\mathbf{R}^n)$,

$$d^*u(\phi \otimes \psi) = u * \psi(\phi).$$

In particular, with $u = \delta$ one immediately obtains (15.8).

In order to determine ${}^t\mathcal{U}$, it is sufficient to compute $\tau_*(d^*u)$, in view of (15.6). It is straightforward to verify that $\tau^{-1} = \tau$ and $d\,\tau = S\,d : \mathbf{R}^n \times \mathbf{R}^n \to \mathbf{R}^n \times \mathbf{R}^n$. Theorem 10.8 and Corollary 11.7 then imply $\tau_* = \tau^*$ on $\mathcal{D}'(\mathbf{R}^n \times \mathbf{R}^n)$ and so

$$\tau_* d^*u = \tau^* d^*u = (d\,\tau)^*u = (S\,d)^*u = d^*(Su).$$

Accordingly, ${}^t\mathcal{U} = Su\,*$ (compare with Problem 11.11), which implies that ${}^t\mathcal{U} : C_0^\infty(\mathbf{R}^n) \to C^\infty(\mathbf{R}^n)$ is continuous too. It follows that k is a regular distribution kernel, and Corollary 15.10 then tells us that $\mathcal{U}$ can be extended to a sequentially continuous mapping $\mathcal{E}'(\mathbf{R}^n) \to \mathcal{D}'(\mathbf{R}^n)$. Thus we have recovered a major case of the existence part of Theorem 11.17. ⊘

Example 15.12. As already pointed out in Chap. 12, fundamental solutions of differential operators with constant coefficients play an important part in the theory of such operators. In general, a similar part is played by fundamental kernels.

Let $P(\partial)$ be as in (15.16). A distribution $E \in \mathcal{D}'(X \times X)$ is said to be a *right fundamental kernel* of $P(\partial)$ if the mapping $E : C_0^\infty(X) \to \mathcal{D}'(X)$ generated by it is a right inverse of $P(\partial)$, that is,

$$P(\partial)E = I. \tag{15.19}$$

Likewise, $E' \in \mathcal{D}'(X \times X)$ is said to be a *left fundamental kernel* of $P(\partial)$ if the corresponding mapping is a left inverse of $P(\partial)$, that is,

$$E'P(\partial) = I.$$

Right fundamental kernels lead to existence theorems; left fundamental kernels immediately lead to uniqueness theorems.

Let us consider (15.19) in more detail; in view of (15.9) and (10.6) it implies

$$\begin{aligned}(P(\partial) \otimes I)E(\phi \otimes \psi) &= E({}^tP(\partial)\phi \otimes \psi) = E\psi({}^tP(\partial)\phi) = P(\partial)E\,\psi(\phi)\\ &= \psi(\phi) = \Delta_* 1_X(\phi \otimes \psi) \qquad (\phi, \psi \in C_0^\infty(X)).\end{aligned}$$

Therefore, on account of Theorem 15.4,

$$(P(\partial) \otimes I)E = \Delta_* 1_X. \tag{15.20}$$

Conversely, this identity leads to (15.19). Therefore (15.20) is an alternative characterization of right fundamental kernels of $P(\partial)$.

A case frequently encountered is that of a right fundamental kernel that is regular. Suppose that E satisfies (15.19) and that both E and ${}^tE$ are continuous mappings $C_0^\infty(X) \to C^\infty(X)$. Then, given arbitrary $v \in \mathcal{E}'(X)$, one can obtain a solution $u \in \mathcal{D}'(X)$ of the inhomogeneous equation $P(\partial)u = v$. Indeed, one has only to set $u = v \circ {}^tE$. This defines a distribution, and on account of Corollary 15.10 one may conclude

$$P(\partial)u(\phi) = u({}^tP(\partial)\phi) = v({}^tE\,{}^tP(\partial)\phi) = v(\phi) \qquad (\phi \in C_0^\infty(X)). \qquad \oslash$$

Remark 15.13. In linear algebra, when considering linear spaces of finite dimension, one studies the null space and the image space of a linear operator, and one determines its inverse if the operator is bijective. In that theory, the operator and its inverse are quite similar in nature. In analysis, however, objects of major interest are differential operators, which act in linear spaces of infinite dimension. Under suitable conditions these operators are inverted by integral operators, which are given by classical kernels. It is only in the framework of distribution theory that the differential operators are also given by (distribution) kernels. As a consequence, in distribution theory both types of operators can be studied from a unified point of view.

It should be noted, however, that the class of operators defined by distribution kernels is very large. For many aspects of the theory of differential operators, it is advantageous to retain important properties of the latter class, for instance the fact that they preserve the class of infinitely differentiable functions. In addition, a differential operator is a *local* continuous linear mapping, that is, it can only decrease the support of a function or distribution to which it is applied. The integral operators of interest do not have this property, but they are *pseudo-local*; this means that they decrease singular supports.

Definition 15.14. A sequentially continuous linear mapping $\mathcal{K} : C_0^\infty(X) \to \mathcal{D}'(X)$ is said to be *pseudo-local* if it has a sequentially continuous linear extension $\mathcal{K} : \mathcal{E}'(X) \to \mathcal{D}'(X)$ (recall that according to Theorem 15.8 this is the case if the kernel $k \in \mathcal{D}'(X \times X)$ of $\mathcal{K}$ is right semiregular) and $\operatorname{sing\,supp} \mathcal{K}u \subset \operatorname{sing\,supp} u$, for all $u \in \mathcal{E}'(X)$. $\oslash$

A sequentially continuous linear mapping $\mathcal{K} : C_0^\infty(X) \to \mathcal{D}'(X)$ having a sequentially continuous linear extension $\mathcal{K} : \mathcal{E}'(X) \to \mathcal{D}'(X)$ is pseudo-local if and only if $\operatorname{sing\,supp} k \subset \Delta(X)$ (for the analogous characterization of locality, see Problem 15.17). In fact, suppose that $\mathcal{K}$ is pseudo-local and select $(x, y) \in X \times X \setminus \Delta(X)$. Then there exist disjoint open neighborhoods U of x and V of y in X such that $(U \times V) \cap \Delta(X) = \emptyset$. Let $\iota_{XU} : \mathcal{E}'(U) \to \mathcal{E}'(X)$ be the natural injection and $\rho_{VX} : \mathcal{D}'(X) \to \mathcal{D}'(V)$ the restriction mapping. If $u \in \mathcal{E}'(U)$, then $\operatorname{sing\,supp} (\rho_{VX} \circ \mathcal{K} \circ \iota_{XU})u \subset U \cap V = \emptyset$; in other words, we have the continuous linear mapping $\rho_{VX} \circ \mathcal{K} \circ \iota_{XU} : \mathcal{E}'(U) \to C^\infty(V)$. This is equivalent to the assertion that the restriction of k to $U \times V$ belongs to $C^\infty(U \times V)$. The reverse implication is left to the reader.

In this vein one obtains *pseudo-differential operators* (see Remarks 17.8 and 18.10) and their generalizations, the *Fourier integral operators*; see [6] or [14, Sects. VII.6 and VIII.6].

A starting point for these constructions is the formula, valid for $P(D)$ as in (15.16) with ∂ replaced by D and for $\phi \in \mathcal{S}(\mathbf{R}^n)$,

$$P(D)\phi(x) = \frac{1}{(2\pi)^n} \int_{\mathbf{R}^n \times \mathbf{R}^n} e^{i\langle x-y,\xi\rangle} P(x,\xi)\phi(y)\,d(y,\xi).$$

In order to obtain the generalizations of the differential operators $P(D)$, one considers larger classes of functions P and more general phase functions in the second integral.

Under some conditions, pseudo-differential operators can be composed; these operators then form an algebra. One of the fundamental aspects of their theory is that it is possible, in a canonical fashion, to associate with such an operator a certain function, called the *symbol* (for example, see (17.2)), with the property that algebraic operations on the operator get translated into differential-algebraic operations on its symbol. This is a sort of converse of the operation of *quantization*. There the problem is to associate to suitable functions f on the phase space $\mathbf{R}^{2n}$ linear mappings $\mathcal{K}_f$ acting in $L^2(\mathbf{R}^n)$ such that the coordinate functions correspond to the momentum and position operators D_j and x_j, for $1 \le j \le n$, and such that the properties of the functions f (algebra structure, positivity, boundedness, etc.) are reflected in some sense in the properties of the mappings $\mathcal{K}_f$. For more details, see for instance [9]. ⊘

Problems

15.1. *(Generalization of Example 15.1.)* Consider $u \in \mathcal{D}'(X)$ and $\mathcal{K} : C_0^\infty(X) \to \mathcal{D}'(X)$ given by $\mathcal{K}\phi = \phi\, u$. Prove that $\Delta_* u \in \mathcal{D}'(X \times X)$ is the distribution kernel of $\mathcal{K}$.

15.2. *(Kernel of a composition.)* Let $X \subset \mathbf{R}^n$, $Y \subset \mathbf{R}^m$, and $Z \subset \mathbf{R}^p$ be open subsets, and let $k \in C_0^\infty(X \times Y)$ and $\widetilde{k} \in C_0^\infty(Y \times Z)$. Prove that k and $\widetilde{k}$ define sequentially continuous linear mappings $\mathcal{K} : C_0^\infty(Y) \to C_0^\infty(X)$ and $\widetilde{\mathcal{K}} : C_0^\infty(Z) \to C_0^\infty(Y)$. Show that $\mathcal{K} \circ \widetilde{\mathcal{K}} : C_0^\infty(Z) \to C_0^\infty(X)$ is a sequentially continuous linear mapping with kernel

$$k_{\mathcal{K}\circ\widetilde{\mathcal{K}}} = \int_Y k_{\mathcal{K}}(\cdot, y)\, k_{\widetilde{\mathcal{K}}}(y, \cdot)\, dy.$$

The right-hand side is said to be the *Volterra composition product* of $k_{\widetilde{\mathcal{K}}}$ and $k_{\mathcal{K}}$.

In this situation, define the diagonal mapping Δ and the projection π by

$$\Delta : X \times Y \times Z \to X \times Y \times Y \times Z \qquad \text{and} \qquad \pi : X \times Y \times Z \to X \times Z,$$

$$\text{where}\quad \Delta(x,y,z) \mapsto (x,y,y,z) \quad\text{and}\quad \pi(x,y,z) \mapsto (x,z),$$

respectively. Verify that $k_{\mathcal{K}\circ\widetilde{\mathcal{K}}} = \pi_*\Delta^*(k_{\mathcal{K}} \otimes k_{\widetilde{\mathcal{K}}})$.

15.3.* *(Kernel and mappings of test functions.)* Consider $\mathcal{K}$ as in the Kernel Theorem, Theorem 15.2.

(i) Let $\Phi\colon C_0^\infty(X) \to C_0^\infty(X)$ and $\Psi\colon C_0^\infty(Y) \to C_0^\infty(Y)$ be continuous linear mappings. Prove that ${}^t\Phi\,\mathcal{K}\,\Psi\colon C_0^\infty(Y) \to \mathcal{D}'(X)$ is a continuous linear mapping satisfying

$$k_{{}^t\Phi\,\mathcal{K}\,\Psi} = ({}^t\Phi \otimes {}^t\Psi)k_{\mathcal{K}}.$$

(ii) Now suppose that $\Phi\colon X \to X$ and $\Psi\colon Y \to Y$ are C^∞ diffeomorphisms. Show that $\Phi_*\mathcal{K}\,\Psi^*\colon C_0^\infty(Y) \to \mathcal{D}'(X)$ is a continuous linear mapping satisfying

$$k_{\Phi_*\mathcal{K}\,\Psi^*} = (\Phi_* \otimes \Psi_*)k_{\mathcal{K}}.$$

(iii) In particular, suppose that $X = Y$ and $\Phi = \Psi$. Then demonstrate, in the notation (10.7), that

$$k_{\Phi^*\mathcal{K}\,(\Phi^{-1})^*} = (j_\Phi \otimes 1)(\Phi^* \otimes \Phi^*)k_{\mathcal{K}}.$$

(iv) Verify that $\mathcal{K}$ commutes with Φ^* if and only if

$$k_{\mathcal{K}} = (j_\Phi \otimes 1)(\Phi^* \otimes \Phi^*)k_{\mathcal{K}}.$$

(v) Finally, let $(\Phi_t)_{t\in\mathbf{R}}$ be a one-parameter family of volume-preserving C^∞ diffeomorphisms with velocity vector field v and assume that $\mathcal{K}$ commutes with all ${\Phi_t}^*$, for $t \in \mathbf{R}$. Then prove, in the notation of (10.17), that

$$\sum_{j=1}^{n}(v_j\,\partial_j \otimes I + I \otimes v_j\,\partial_j)k_{\mathcal{K}} = 0.$$

15.4. Apply Problem 15.3.(i) in order to derive (15.17) from Example 15.1.

15.5.* *(Kernel of Fourier transform.)* Verify that the Fourier transform $\mathcal{F} : C_0^\infty(\mathbf{R}^n) \to C^\infty(\mathbf{R}^n)$ is a continuous linear mapping. Prove that the distribution kernel $k_{\mathcal{F}}$ of $\mathcal{F}$ is given by $e^{-i\langle\cdot,\cdot\rangle} \in C^\infty(\mathbf{R}^n \times \mathbf{R}^n)$. Show that the symmetry $\tau_*k_{\mathcal{F}} = k_{\mathcal{F}}$ implies that $\mathcal{F} = {}^t\mathcal{F}$ and derive (14.19) and (14.20). Furthermore, obtain the identities from (14.10), (14.11), (14.29), and Problem 14.30.(i) by studying $k_{\mathcal{F}}$. (We note that (14.21) may also be obtained in this manner, but that Remark 14.17 contains an argument that is both shorter and more appropriate in this context.) Solve Problem 14.30.(iv). Finally, show that $\mathcal{F}$ extends to a sequentially continuous mapping $\mathcal{F} : \mathcal{E}'(\mathbf{R}^n) \to \mathcal{D}'(\mathbf{R}^n)$ satisfying $\mathcal{F}\delta = 1$.

15.6. *(Mellin transform.)* The linear integral operator $\mathcal{M} : C_0^\infty(\mathbf{R}) \to C^\infty(\mathbf{R}_{>0})$ defined by the kernel $k(a,x) = H(x)\,x^{a-1}$ is said to be the *Mellin transform*. (The identity $k(a,\cdot) = \Gamma(a)\,\chi_a^+$ and (13.7) can be used to extend this definition.) Prove that $\mathcal{M} \circ (-x\,\partial_x) = a \circ \mathcal{M}$, which expresses the fact that the Mellin transform diagonalizes the Euler operator $x\,\partial_x$. In addition, show, for $c > 0$, that

$$\mathcal{M} \circ x^c = T_c{}^* \circ \mathcal{M}, \qquad \mathcal{M} \circ \log x = \partial \circ \mathcal{M}, \qquad \mathcal{M} \circ c^* = c^{-a} \circ \mathcal{M}.$$

Finally, derive the second identity from the first by means of (10.15). See [7, Exercise 6.100] for more information about the Mellin transform.

15.7.* Let $X \subset \mathbf{R}^n$ and $Y \subset \mathbf{R}^m$ be open subsets and consider $u \in \mathcal{D}'(X \times Y)$.

(i) Suppose that Y is connected and that u satisfies $(I \otimes \partial_j)u = 0$, for $1 \le j \le m$. Prove the existence of $v \in \mathcal{D}'(X)$ such that $u = v \otimes 1_Y$.
(ii) Suppose that $0 \in Y$ and that u satisfies $(I \otimes x_j)u = 0$, for $1 \le j \le m$. Prove the existence of $v \in \mathcal{D}'(X)$ satisfying $u = v \otimes \delta^Y$, where δ^Y denotes the Dirac measure on Y supported at 0.

For different proofs in special cases, see Problems 11.19 and 11.18.

15.8.* *(Distribution supported by linear submanifold.)* Let $X \subset \mathbf{R}^n$ and $Y \subset \mathbf{R}^m$ be open subsets and assume that $0 \in Y$. Suppose $u \in \mathcal{E}'(X \times Y)$ satisfies $\operatorname{supp} u \subset X \times \{0\}$. Let δ^Y be the Dirac measure on Y supported at 0 and $\iota : X \to X \times Y$ the natural embedding with $\iota(x) = (x, 0)$. Prove the existence of $u_\alpha \in \mathcal{E}'(X)$, with $\alpha \in A := \{0\} \times (\mathbf{Z}_{\ge 0})^m$, such that

$$u = \sum_{\alpha \in (\mathbf{Z}_{\ge 0})^m} u_\alpha \otimes \partial^\alpha \delta^Y = \sum_{\alpha \in (\mathbf{Z}_{\ge 0})^m} (I \otimes \partial^\alpha)\iota_* u_\alpha = \sum_{\alpha \in A} \partial^\alpha (\iota_* u_\alpha),$$

where the sum is actually finite. See Problem 11.20 for another proof in a special case.

Observe that for $k \in \mathbf{N}$, any C^k submanifold in $\mathbf{R}^p$ of dimension $0 \le n < p$ is locally C^k diffeomorphic to a set of the form $X \times \{0\}$, with X open in $\mathbf{R}^n$; see [7, Theorem 4.7.1.(iv)].

15.9.* *(Another proof of the Fourier inversion formula.)* Let I and k_I be as in Example 15.1 and $k = k_{\mathcal{G}}$, with $\mathcal{G}$ as in Lemma 14.12. Show that

$$(D_j \otimes I + I \otimes D_j)k = 0 \qquad \text{and} \qquad (x_j \otimes I - I \otimes x_j)k = 0 \qquad (1 \le j \le n).$$

Define the automorphism

$$\Phi : \mathbf{R}^n \times \mathbf{R}^n \to \mathbf{R}^n \times \mathbf{R}^n \qquad \text{by} \qquad \Phi(x, y) = (x, d(x, y)) = (x, x - y)$$

and set $\widetilde{k} = \Phi^* k \in \mathcal{D}'(\mathbf{R}^n \times \mathbf{R}^n)$. For $1 \le j \le n$, deduce that

$$(D_j \otimes I)\widetilde{k} = \Phi^*(D_j \otimes I + I \otimes D_j)k = 0 \qquad \text{and} \qquad (I \otimes x_j)\widetilde{k} = 0.$$

Conclude that $\widetilde{k} = 1_{\mathbf{R}^n} \otimes c\,\delta$, for some $c \in \mathbf{C}$ and δ the Dirac measure on $\mathbf{R}^n$ and prove that $k = \Phi_* \Phi^* k = \Phi_*(1_{\mathbf{R}^n} \otimes c\,\delta) = c\,d^*\delta = c\,k_I$. Finally, show that $\mathcal{G} = (2\pi)^n I$.

Observe that an adaptation of the preceding arguments leads to another solution of Problem 14.7. The present approach is quite similar to the one in Problem 13.12.

15.10.* Verify the formula for $d_*\chi$ in (10.25) in a manner different from that in Example 10.20.

15.11.* *(Another proof of Theorem 11.3.)* Let $\mathcal{U}$ be as in Theorem 11.3 and prove the characterization of $\mathcal{U}$ as in the theorem by computing $k_{\mathcal{U}}$.

15.12.* Let $u \in \mathcal{D}'(\mathbf{R}^n)$ and define

$$\mathcal{U} : C_0^\infty(\mathbf{R}^n) \to \mathcal{D}'(\mathbf{R}^n) \quad \text{by} \quad k_{\mathcal{U}} = d^*u.$$

For any $\psi \in C_0^\infty(\mathbf{R}^n)$, prove that $\mathcal{U}\psi = u * \psi$ as in (11.1). Furthermore, verify that T_a and ∂_j commute with $\mathcal{U}$, for $a \in \mathbf{R}^n$ and $1 \le j \le n$.

15.13.* *(Characterization of Fourier transform.)* Let $\mathcal{K} : C_0^\infty(\mathbf{R}^n) \to \mathcal{D}'(\mathbf{R}^n)$ be a sequentially continuous linear mapping. Consider the following list of six conditions (compare with (14.10), (14.11), (14.19), (14.12), and Problem 14.30):

$$\text{(i)} \quad \mathcal{K}\,D_j = \xi_j\,\mathcal{K}, \qquad \text{(ii)} \quad \mathcal{K}\,x_j = -D_j\,\mathcal{K}, \qquad \text{(iii)} \quad {}^t\mathcal{K} = \mathcal{K},$$
$$\text{(iv)} \quad \mathcal{K}\,T_a{}^* = e_{ia}\,\mathcal{K}, \qquad \text{(v)} \quad \mathcal{K}\,c^* = c_*\,\mathcal{K}, \qquad \text{(vi)} \quad \mathcal{K}\,A^* = A^*\,\mathcal{K},$$

for every $1 \le j \le n$, $a \in \mathbf{R}^n$, $c > 0$, and every rotation A in $\mathbf{R}^n$.

(a) Prove that $\mathcal{K}$ equals the Fourier transform $\mathcal{F}$ if $\mathcal{K}$ satisfies only condition (i) but can be extended to a continuous linear mapping $\mathcal{K} : \mathcal{E}'(\mathbf{R}^n) \to \mathcal{D}'(\mathbf{R}^n)$ that satisfies $\mathcal{K}(\delta) = 1$.

Verify that $\mathcal{K}$ equals a multiple of $\mathcal{F}$ if $\mathcal{K}$ satisfies any of the following:

(b) conditions (i) and (ii);
(c) conditions (i) and (iii);
(d) conditions (iv)–(vi);
(e) conditions (iv) and (v), if, in addition, the range space of $\mathcal{K}$ consists of locally integrable functions that are continuous at 0.

15.14.* In Problem 14.42 it is proved that the collection of functions $(\phi)_{j\in\mathbf{Z}_{\ge 0}}$ as in (14.47) is a basis in $L^2(\mathbf{R})$. Therefore one may define a continuous linear mapping $\widetilde{\mathcal{F}} : L^2(\mathbf{R}) \to L^2(\mathbf{R})$ by means of $\widetilde{\mathcal{F}}\phi_j = (-1)^j\phi_j$, for $j \in \mathbf{Z}_{\ge 0}$. Prove that this new definition of $\widetilde{\mathcal{F}}$ coincides with the familiar one of normalized Fourier transform.

15.15.* *(Characterization of Hilbert transform.)* Let $\mathcal{K} : C_0^\infty(\mathbf{R}) \to \mathcal{D}'(\mathbf{R})$ be a sequentially continuous linear mapping. Suppose that $\mathcal{K}$ commutes with all translations and positive dilations in $\mathbf{R}$ and anticommutes with reflection. Prove that $\mathcal{K}$ is a multiple of the Hilbert transform $\mathcal{H}$ from Problem 14.49. Compare with Problem 14.52.

15.16.* Suppose that X is an open subset of $\mathbf{R}^n$ and that $\mathcal{K} : C_0^\infty(X) \to \mathcal{D}'(X)$ is given by

$$\mathcal{K} = \sum_{\alpha \in (\mathbf{Z}_{\geq 0})^n} u_\alpha \circ \partial^\alpha, \qquad \text{that is,} \qquad \mathcal{K}\psi = \sum_{\alpha \in (\mathbf{Z}_{\geq 0})^n} (\partial^\alpha \psi)\, u_\alpha,$$

where $u_\alpha \in \mathcal{D}'(X)$ and the sum is locally finite. (We may think of $\mathcal{K}$ as a linear partial differential operator in X of locally constant order, with distributions as coefficients.) Show that $\mathcal{K}$ is a local operator. Prove that the distribution kernel k of $\mathcal{K}$ is given by

$$k = \sum_{\alpha \in (\mathbf{Z}_{\geq 0})^n} (I \otimes (-\partial)^\alpha) \Delta_* u_\alpha \qquad \text{and that} \qquad \operatorname{supp} k \subset \Delta(X).$$

Conversely, suppose that $\mathcal{K} : C_0^\infty(X) \to \mathcal{D}'(X)$ has a distribution kernel k that is supported by $\Delta(X)$. Then prove that k, and therefore $\mathcal{K}$, are of the form given above.

15.17.* Consider a sequentially continuous linear mapping $\mathcal{K} : C_0^\infty(X) \to \mathcal{D}'(X)$ and denote its distribution kernel by $k \in \mathcal{D}'(X \times X)$. Show that $\mathcal{K}$ is local if and only if $\operatorname{supp} k \subset \Delta(X)$.

15.18. *(Weak form of Peetre's Theorem.)* Let $\mathcal{K} : C_0^\infty(X) \to \mathcal{D}'(X)$ be a local and sequentially continuous linear mapping. Prove that $\mathcal{K}$ is of the form given in Problem 15.16. In particular, if $\mathcal{K}$ maps into $C^\infty(X)$, then $\mathcal{K}$ is locally a linear partial differential operator in X with C^∞ coefficients. Hint: in order to demonstrate that the coefficients are of class C^∞, use $\psi_{\alpha,V} \in C_0^\infty(X)$ that coincide with monomials $x \mapsto x^\alpha$ when restricted to a relatively compact open subset V of X.

Peetre [17] proved that the latter conclusion holds even for local linear mappings; the continuity follows from the restriction on the supports. Let $\mathcal{K} : C_0^\infty(X) \to \mathcal{D}'(X)$ be a local linear mapping. For this case, Peetre showed that for every $x \in X$ there exists a neighborhood U of x such that the restriction of $\mathcal{K}$ to U is a sequentially continuous linear mapping $C_0^\infty(U) \to \mathcal{D}'(U)$, but that this is not true for x belonging to a locally finite set $\Lambda \subset X$. Next, applying results from Problems 15.17 and 15.16, he obtained the existence of $u_\alpha \in \mathcal{D}'(X)$ that are uniquely determined on $X \setminus \Lambda$ such that $\mathcal{K}\psi$ is as in Problem 15.16 plus some distribution supported by Λ, for any $\psi \in C_0^\infty(X)$.

Chapter 16
Fourier Series

The theory of Fourier series is a far-reaching analog of writing $x \in \mathbf{R}^n$ as

$$x = \sum_{j=1}^{n} \langle x, e_j \rangle e_j,$$

the decomposition of a vector into a sum of projections along the vectors e_j, which form a complete system of orthonormal vectors $(e_1, \dots, e_n)$ in $\mathbf{R}^n$. Indeed, the theory of Fourier series asserts that a sufficiently smooth function u (of class C^2 is sufficient, see Problem 16.18) on $\mathbf{R}$ that is periodic with period $a > 0$ can be written as a uniformly convergent series

$$u(x) = \sum_{n \in \mathbf{Z}} c_n \, e^{in\omega x}. \tag{16.1}$$

Here $\omega = \frac{2\pi}{a} > 0$ and c_n, the nth *Fourier coefficient* of u, is given by

$$c_n = \frac{1}{a} \int_s^{s+a} e^{-in\omega x} \, u(x) \, dx \qquad (n \in \mathbf{Z},\ s \in \mathbf{R}).$$

For a locally integrable periodic function f with period a, the integral

$$M(f) := \frac{1}{a} \int_s^{s+a} f(x) \, dx \tag{16.2}$$

is said to be the *mean* of f. Note that the periodicity of f implies that $f(x + na) = f(x)$, for all $n \in \mathbf{Z}$, and consequently

$$M(f) = \frac{1}{n\,a} \int_s^{s+na} f(x) \, dx \qquad (n \in \mathbf{N}).$$

From this it follows in turn that

J.J. Duistermaat and J.A.C. Kolk, *Distributions: Theory and Applications*,
Cornerstones, DOI 10.1007/978-0-8176-4675-2_16,

$$M(f) = \lim_{t-s\to\infty} \frac{1}{t-s} \int_s^t f(x)\,dx.$$

This result can be verified by choosing an $n \in \mathbf{N}$ such that $t \leq s + n\,a < t + a$, then writing

$$M(f) - \frac{1}{t-s} \int_s^t f(x)\,dx$$

$$= \Big(\frac{1}{n\,a} - \frac{1}{t-s}\Big)\, n \int_s^{s+a} f(x)\,dx + \frac{1}{t-s} \int_t^{s+n\,a} f(x)\,dx,$$

and then using $|\frac{1}{a} - \frac{n}{t-s}| = |\frac{t-s-n\,a}{a\,(t-s)}| \leq \frac{1}{t-s}$. This further shows that (16.2) is independent of the choice of s, as one also verifies directly.

Our aim now is to reconstruct the theory of *Fourier series* from the theory of the Fourier transform of Chap. 14. In doing so, we will obtain (16.1), with convergence in the sense of distributions, even for arbitrary periodic distributions with period a. We begin by formulating an alternative definition of the mean $M(u)$ of a periodic distribution u. This definition is required because the formula

$$\int_s^{s+a} u(x)\,dx = u(1_{[s,\,s+a]})$$

cannot be generalized: the characteristic function $1_{[s,\,s+a]}$ of the interval $[s,\, s+a]$ is not a C^∞ function.

Lemma 16.1. *For every $a > 0$ there exists $\mu \in C_0^\infty(\mathbf{R})$ such that $\sum_{n\in\mathbf{Z}} T_{na}\mu = 1$. If $u \in \mathcal{D}'(\mathbf{R})$ and $T_a u = u$, then $u(\nu) = u(\mu)$ if in addition $\nu \in C_0^\infty(\mathbf{R})$ and $\sum_{n\in\mathbf{Z}} T_{na}\nu = 1$. If u is locally integrable, then $u(\mu) = a\,M(u)$.*

Proof. Begin by choosing $\chi \in C_0^\infty(\mathbf{R})$ such that $\chi \geq 0$ and $\chi > 0$ on $[s,\, s+a\,[$. Then $\theta(x) := \sum_{n\in\mathbf{Z}} \chi(x - na)$ defines a C^∞ function on $\mathbf{R}$. Indeed, note that for every bounded open interval I in $\mathbf{R}$ there exists a finite subset E of $\mathbf{Z}$ with $\chi(x - na) = 0$ whenever $x \in I$ and $n \notin E$. This gives $\sum_{n\in\mathbf{Z}} \chi(x - na) = \sum_{n\in E} \chi(x - na)$, for $x \in I$, and from this immediately follow both the convergence and the fact that $\theta \in C^\infty(\mathbf{R})$.

Furthermore, θ is periodic with period a. For every $x \in \mathbf{R}$ there exists an $n \in \mathbf{Z}$ with $x - na \in [s,\, s+a\,[$, and therefore $\theta(x) = \theta(x - na) \geq \chi(x - na) > 0$. This gives $\mu := \chi/\theta \in C_0^\infty(\mathbf{R})$ and

$$\sum_{n\in\mathbf{Z}} T_{na}\mu = \sum_{n\in\mathbf{Z}} T_{na}\Big(\frac{\chi}{\theta}\Big) = \sum_{n\in\mathbf{Z}} \frac{T_{na}\chi}{T_{na}\theta} = \sum_{n\in\mathbf{Z}} \frac{T_{na}\chi}{\theta} = \frac{\theta}{\theta} = 1.$$

Using the periodicity of u we obtain, by a summation that is in fact finite,

$$u(\nu) = u\Big(\nu \sum_{n\in\mathbf{Z}} T_{na}\mu\Big) = \sum_{n\in\mathbf{Z}} T_{-na}u(\mu\, T_{-na}\nu) = \sum_{n\in\mathbf{Z}} u(\mu\, T_{-na}\nu) = u(\mu). \tag{16.3}$$

To obtain the latter assertion we apply the foregoing to $\mu = 1_{[s,s+a]}$, which is permitted if u is locally integrable. Naturally, in this case $u(\mu) = a\, M(u)$. □

Definition 16.2. Let $a > 0$. For an arbitrary $u \in \mathcal{D}'(\mathbf{R})$ satisfying $T_a u = u$, the number $M(u) := \frac{1}{a}\, u(\mu)$, with μ as in Lemma 16.1, is said to be the *mean* of u. ⊘

Theorem 16.3. *Every periodic distribution on* $\mathbf{R}$ *is tempered.*

Proof. For $u \in \mathcal{D}'(\mathbf{R})$ we have an estimate of the form (3.4), with $K = \operatorname{supp} \mu$ and μ as in Lemma 16.1. Formula (16.3), with ϕ instead of v, implies, for every $\phi \in C_0^\infty(\mathbf{R})$,

$$|u(\phi)| = \Big|\sum_{n\in\mathbf{Z}} u(\mu\, T_{-na}\phi)\Big| \le c \sum_{n\in\mathbf{Z}} \sup_{|\alpha|\le k,\, x\in K} |\partial^\alpha \phi(x+na)|.$$

There exists a constant $b > 0$ with $1 + |n| \le b\,(1 + |x+na|)$, for all $n \in \mathbf{Z}$ and $x \in K$. Now choose $d > 1$ such that $s := \sum_{n\in\mathbf{Z}} (1+|n|)^{-d} < \infty$. This yields

$$|u(\phi)| \le c\, b^d\, s \sup_{|\alpha|\le k,\, x\in K,\, n\in\mathbf{Z}} (1+|x+na|)^d\, |\partial^\alpha \phi(x+na)| \le c'\, \|\phi\|_{\mathcal{S}(k,d)},$$

for a constant $c' > 0$; see (14.8). □

A sequence $(c_n)_{n\in\mathbf{Z}}$ in $\mathbf{C}$ is said to be *of moderate growth* if positive constants c and N exist such that

$$|c_n| \le c\, |n|^N \qquad (n \in \mathbf{Z} \setminus \{0\}). \tag{16.4}$$

Lemma 16.4. *Let* $\omega > 0$. *The distribution*

$$v = \sum_{n\in\mathbf{Z}} c_n\, \delta_{n\omega}$$

on $\mathbf{R}$ *is tempered if and only if the sequence* $(c_n)_{n\in\mathbf{Z}}$ *of coefficients in* $\mathbf{C}$ *is of moderate growth. In that case*

$$\lim_{j,k\to\infty} \sum_{n=-j}^{k} c_n\, \delta_{n\omega} = v$$

and

$$\lim_{j,k\to\infty} \sum_{n=-j}^{k} c_n\, e_{in\omega} = u := S \circ \mathcal{F} v, \tag{16.5}$$

both with convergence in $\mathcal{S}'$.

Proof. We have $v \in \mathcal{S}'$ if and only if there exist $c > 0$, $k \in \mathbf{Z}_{\ge 0}$, and $N \in \mathbf{Z}_{\ge 0}$ such that $|v(\phi)| \le c\, \|\phi\|_{\mathcal{S}(k,N)}$, for all $\phi \in C_0^\infty(\mathbf{R})$. See (14.8), Definition 14.19, and (14.25).

First, suppose that $v \in \mathcal{S}'$. Let $\chi \in C_0^\infty(\mathbf{R})$, supp $\chi \subset \,]-\omega, \omega\,[$, and $\chi(0) = 1$. For $n \in \mathbf{Z}$ and $\phi = T_{n\omega}\chi$ we obtain $v(\phi) = c_n$, while on the other hand, $\|\phi\|_{\mathcal{S}(k,N)} \leq d\,(1 + |n|)^N$, where d is a constant that is independent of n but contains the C^k norm of χ. This implies that $(c_n)_{n\in\mathbf{Z}}$ is of moderate growth if $v \in \mathcal{S}'$.

Conversely, if the sequence $(c_n)_{n\in\mathbf{Z}}$ satisfies (16.4), we obtain, for every $b > 1$,

$$\begin{aligned}|v(\phi)| &= \Big|\sum_{n\in\mathbf{Z}} c_n\,\phi(n\omega)\Big| \leq c \sum_{n\in\mathbf{Z}}(1 + |n|)^N\,|\phi(n\omega)| \\ &\leq c\left(\sum_{n\in\mathbf{Z}}(1 + |n|)\right)^{-b} \sup_{n\in\mathbf{Z}}(1 + |n|)^{N+b}\,|\phi(n\omega)|,\end{aligned}$$

for all $\phi \in C_0^\infty(\mathbf{R})$, which implies that $v \in \mathcal{S}'$. The convergence in $\mathcal{S}'$ follows from analogous estimates, with v replaced by the difference of v and the finite sum. The latter assertion follows from (14.35) and the continuity of $S \circ \mathcal{F}$ from $\mathcal{S}'$ to $\mathcal{S}'$; see Theorem 14.24. □

Definition 16.5. When (16.5) holds, u is said to be the *distributional Fourier series* with coefficients $(c_n)_{n\in\mathbf{Z}}$. One uses the notation (see also Problem 5.10)

$$u = \sum_{n\in\mathbf{Z}} c_n\, e_{in\omega} \quad \text{in} \quad \mathcal{S}'. \qquad \oslash$$

Lemma 16.6. *The derivative of a distributional Fourier series can be obtained by termwise differentiation; more exactly, if the sequence $(c_n)_{n\in\mathbf{Z}}$ in $\mathbf{C}$ is of moderate growth, then*

$$\partial \sum_{n\in\mathbf{Z}} c_n\, e_{in\omega} = \sum_{n\in\mathbf{Z}} in\omega\, c_n\, e_{in\omega} \quad \textit{in} \quad \mathcal{S}'.$$

Proof. Application of Lemma 5.9 yields

$$\partial \sum_{n\in\mathbf{Z}} c_n\, e_{in\omega} = \lim_{j,k\to\infty} \partial \sum_{n=-j}^{k} c_n\, e_{in\omega} \quad \text{in} \quad \mathcal{S}'.$$

But from Lemma 16.4 we deduce, considering that the sequence $(in\omega\, c_n)_{n\in\mathbf{Z}}$ is of moderate growth,

$$\lim_{j,k\to\infty} \partial \sum_{n=-j}^{k} c_n\, e_{in\omega} = \sum_{n\in\mathbf{Z}} in\omega\, c_n\, e_{in\omega} \quad \text{in} \quad \mathcal{S}'. \qquad \square$$

Theorem 16.7. *Let $a, \omega > 0$ and $a\,\omega = 2\pi$. The mapping $S \circ \mathcal{F}$ then is a linear bijection from the linear space of distributions $\sum_{n\in\mathbf{Z}} c_n \delta_{n\omega}$, where the sequence*

$(c_n)_{n\in\mathbf{Z}}$ in $\mathbf{C}$ is of moderate growth, onto the linear space of periodic distributions with period a. If

$$u = \sum_{n\in\mathbf{Z}} c_n\, e_{in\omega} \quad \textit{in} \quad \mathcal{S}'$$

then

$$c_n = M(e_{-in\omega}\, u) \qquad (n \in \mathbf{Z}). \tag{16.6}$$

Proof. If $u \in \mathcal{D}'(\mathbf{R})$ and $T_a u = u$, then $u \in \mathcal{S}'$ on the strength of Theorem 16.3. We obtain $v := \frac{1}{2\pi}\mathcal{F}u \in \mathcal{S}'$ and according to (14.29) the equation $T_a u = u$ is equivalent to $e_{-ia}\, v = v$, that is, $(e_{ia} - 1)\, v = 0$. On account of Theorem 9.5 we conclude that $v = \sum_{n\in\mathbf{Z}} c_n\, \delta_{n\omega}$, for a sequence (c_n) in $\mathbf{C}$ that can be of moderate growth at most growth, because $v \in \mathcal{S}'$.

In reaching this conclusion we have also seen that conversely, $u := S \circ \mathcal{F} v$ is periodic with period a if $v = \sum_{n\in\mathbf{Z}} c_n\, \delta_{n\omega} \in \mathcal{S}'$, where we have used Theorem 14.24. The assertion in (16.5) concerning the convergence follows from Lemma 16.4 and the fact that $\mathcal{F}$ is continuous from $\mathcal{S}'$ to $\mathcal{S}'$; see Theorem 14.24.

For (16.6) we write

$$M(e_{-in\omega}\, u) = M\Big(\sum_{k\in\mathbf{Z}} c_k\, e_{ik\omega}\, e_{-in\omega}\Big) = \sum_{k\in\mathbf{Z}} c_k\, M(e_{i(k-n)\omega}) = c_n.$$

The convergence does not pose a problem, because M is a continuous linear form on $\mathcal{S}'$, as one straightforwardly reads from the definition of the mean. □

Example 16.8. Let a and ω be positive numbers satisfying $a\,\omega = 2\pi$ and consider

$$u_\omega = \sum_{n\in\mathbf{Z}} \delta_{n\omega} \in \mathcal{S}'(\mathbf{R}).$$

Then $\mathcal{F} u_\omega = S \circ \mathcal{F} u_\omega = \sum_{n\in\mathbf{Z}} e_{in\omega}$ on the basis of (14.35). A substitution of the index of summation implies $T_\omega u_\omega = u_\omega$; hence (14.29) leads to $(e_{i\omega} - 1)\, \mathcal{F} u_\omega = 0$. As in the proof of Theorem 16.7, we deduce the existence of a sequence (c_n) in $\mathbf{C}$ of moderate growth such that $\mathcal{F} u_\omega = \sum_{n\in\mathbf{Z}} c_n\, \delta_{na}$. But $e_{ia}\, \delta_{n\omega} = \delta_{n\omega}$, for all $n \in \mathbf{Z}$, implies $e_{ia}\, u_\omega = u_\omega$. Applying (14.29) once again, we obtain that $\mathcal{F} u_\omega$ is a Radon measure invariant under T_a; therefore $\mathcal{F} u_\omega$ has the same mass, say c, at every point. In other words, $\sum_{n\in\mathbf{Z}} e_{in\omega} = c \sum_{n\in\mathbf{Z}} \delta_{na}$. See Problem 10.13 for another, strongly related, proof. We now conclude from (16.6) that

$$1 = M(c\, u_a) = \frac{c}{a}\, u_a(\mu) = \frac{c}{a}\, \delta\Big(\sum_{k\in\mathbf{Z}} T_{ka}\mu\Big) = \frac{c}{a},$$

and therefore $c = a = 2\pi/\omega$. In explicit form, the formula becomes (see Fig. 16.1), for $\phi \in \mathcal{S}(\mathbf{R})$,

$$\sum_{n\in\mathbf{Z}} e_{in\omega} = a \sum_{k\in\mathbf{Z}} \delta_{ka}, \qquad \text{that is,} \qquad \sum_{n\in\mathbf{Z}} \mathcal{F}\phi(n\,\omega) = a \sum_{k\in\mathbf{Z}} \phi(k\,a). \tag{16.7}$$

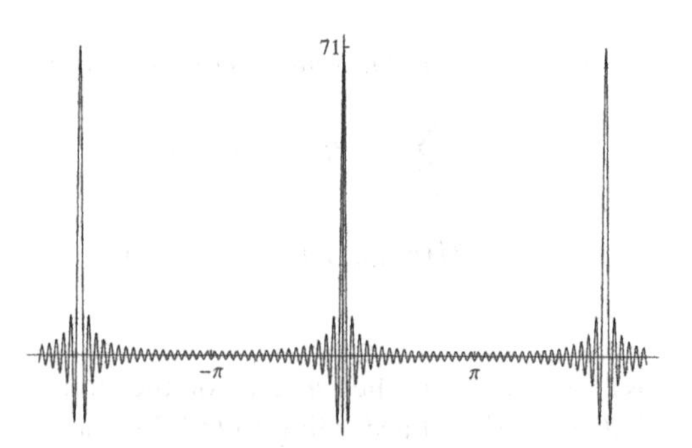

Fig. 16.1 Illustration of (16.7). Graph of $\sum_{n=-35}^{35} e_{in}$

This result is known as *Poisson's summation formula* (see Example 1.4 for a heuristic discussion, Exercise 16.17 for another proof, and (14.32) for the analogous formula in the aperiodic case).

Next, apply the summation formula with $T_x{}^*\phi$ instead of ϕ and note that $\mathcal{F}(T_x{}^*\phi) = e_{ix}\,\mathcal{F}\phi$ on account of (14.12); this leads to

$$a\sum_{k\in\mathbf{Z}} T_{k\,a}\phi = \sum_{n\in\mathbf{Z}} \mathcal{F}\phi(n\,\omega)\,e_{in\omega} \qquad (\phi \in \mathcal{S}(\mathbf{R})). \tag{16.8}$$

Observe that the formula gives the Fourier series of the *periodization* with period a of a Schwartz function on $\mathbf{R}$. The identity of distributions in (16.7) is the quintessential Fourier series expansion, because from it one can obtain other such expansions, by testing the identity; this is what is expressed by (16.8) and (16.9) below. It is quite remarkable that many expansions of commonly encountered distributions can be directly derived from (16.7) by algebraic operations, translation, differentiation, or integration; see Example16.24 and Problems 16.2, 16.8, 16.17, and 16.22. One might say that the Fourier expansion of a distribution is determined by its singularities, similarly to Riemann's key theme of the determination of a (complex-differentiable) function from its singularities. Furthermore, the identity (16.7) relates geometry of the circle $\mathbf{R}/a\,\mathbf{Z}$, specifically, the lengths (belonging to $a\,\mathbf{Z}$) of its closed geodesics, to analysis on $\mathbf{R}/a\,\mathbf{Z}$, more precisely, the (multiplicities of) the eigenvalues $-n^2\omega^2$ and eigenfunctions $e_{in\omega}$ of the Laplace operator ∂^2 associated to the circle.

The identity (16.8) is also valid for a much wider class of functions. This may be proved by convolving ϕ by a suitable function f and then sending ϕ to δ in the usual way. Alternatively, the next proof applies to continuous functions f satisfying $f(x) = O\left(\frac{1}{x^2}\right)$, $x \to \infty$:

$$M\left(e_{-in\omega}\, a \sum_{k\in\mathbf{Z}} T_{k\,a} f\right) = a \sum_{k\in\mathbf{Z}} M(e_{-in\omega}\, T_{-k\,a} f)$$
$$= \sum_{k\in\mathbf{Z}} \int_0^a e^{-in\omega x}\, f(x+k\,a)\, dx = \sum_{k\in\mathbf{Z}} \int_{k\,a}^{(k+1)a} e^{-in\omega x}\, f(x)\, dx \qquad (16.9)$$
$$= \int_{\mathbf{R}} e^{-in\omega x}\, f(x)\, dx = \mathcal{F} f(n\,\omega).$$

The distribution u_ω can also be used to determine the constant in the Fourier inversion formula (14.17) in an alternative way. In the proof of (16.6), and therefore also of (16.7), formula (14.17) is not used. We now obtain, with $a\,\omega = 2\pi$,

$$\mathcal{F} \circ \mathcal{F} u_\omega = a\, \mathcal{F} u_a = a\,\omega\, u_\omega;$$

thus, $S \circ \mathcal{F} \circ \mathcal{F} = c\, I$ is seen to act on u_ω as the scalar multiplication by $a\,\omega = 2\pi$. Because $u_\omega \neq 0$, it follows that $c = 2\pi$. ⊘

A distribution of the form $\sum c_n\, \delta_{n\omega}$ reminds one of the functions on a lattice

$$\Omega := \{n\omega \in \mathbf{R} \mid n \in \mathbf{Z}\}$$

used in numerical mathematics. By taking the limit as $\omega \downarrow 0$ one can use these to approximate arbitrary distributions on $\mathbf{R}$; compare Problem 5.12. Conversely, by application of the convergence in $\mathcal{S}$ one derives the Fourier transform in $\mathbf{R}$ from the theory of Fourier series.

There also exists a Fourier series variant of Parseval's formula. To formulate this on the natural maximal domain space, we introduce the following Hermitian inner product in the space of periodic functions with period a:

$$(f, g)_{\mathbf{R}/a\mathbf{Z}} = M(f\,\overline{g}).$$

The corresponding L^2 norm is

$$\|f\| = \|f\|_{L^2(\mathbf{R}/a\mathbf{Z})} = \left((f, f)_{\mathbf{R}/a\mathbf{Z}}\right)^{1/2}.$$

The linear space of locally integrable periodic functions f on $\mathbf{R}$ with period a such that $\|f\| < \infty$ is denoted by $L^2(\mathbf{R}/a\mathbf{Z})$. This space is *complete* (a Hilbert space) and contains the space $C^\infty(\mathbf{R}/a\mathbf{Z})$ of C^∞ periodic functions with period a as a dense linear subspace.

In the space of sequences $c = (c_n)_{n\in\mathbf{Z}}$ in $\mathbf{C}$ one defines the Hermitian inner product

$$(c, d) := \sum_{n\in\mathbf{Z}} c_n\, \overline{d_n}.$$

The corresponding norm is

$$\|c\| = \|c\|_{l^2} = \left(\sum_{n\in\mathbf{Z}} |c_n|^2\right)^{1/2},$$

and the space of sequences c with $\|c\| < \infty$ is denoted by $l^2 = l^2(\mathbf{Z})$. This, too, is a Hilbert space; the sequences (c_n) of which only finitely many c_n differ from zero (the sequences with compact support) lie dense in this space.

Theorem 16.9. *Consider a and $\omega > 0$ with $a\,\omega = 2\pi$. The mapping F_ω that assigns to $(c_n)_{n\in\mathbf{Z}} \in l^2(\mathbf{Z})$ the distributional Fourier series $u = \sum_{n\in\mathbf{Z}} c_n\, e_{in\omega}$ is a linear bijection from $l^2(\mathbf{Z})$ to $L^2(\mathbf{R}/a\mathbf{Z})$. We have Parseval's formula*

$$\|(c_n)_{n\in\mathbf{Z}}\|^2_{l^2(\mathbf{Z})} = \sum_{n\in\mathbf{Z}} |c_n|^2 = M(|u|^2) = \|u\|^2_{L^2(\mathbf{R}/a\mathbf{Z})}.$$

Proof. For a finite Fourier series, Parseval's formula holds on account of

$$M(|u|^2) = M(u\,\overline{u}) = \sum_{n,m} c_m\,\overline{c_m}\,M(e_{i(n-m)\omega}) = \sum_n c_n\,\overline{c_n}.$$

By continuous extension we immediately see that F_ω is a linear isometry from $l^2(\mathbf{Z})$ to $L^2(\mathbf{R}/a\mathbf{Z})$; in particular, it is injective.

Furthermore, this implies that the image is complete with respect to the norm on $L^2(\mathbf{R}/a\mathbf{Z})$, and therefore constitutes a closed subset of $L^2(\mathbf{R}/a\mathbf{Z})$. For every $\phi \in C^\infty(\mathbf{R}/a\mathbf{Z})$ and every $N \in \mathbf{Z}_{\geq 0}$, repeated integration by parts yields

$$c_n = M(e_{-in\omega}\phi) = O((1+|n|)^{-N}) \quad \text{as} \quad |n| \to \infty.$$

From this we deduce $c \in l^2(\mathbf{Z})$, and so $\phi \in F_\omega(l^2(\mathbf{Z}))$. Approximating an arbitrary $u \in L^2(\mathbf{R}/a\mathbf{Z})$ with respect to the L^2 norm by means of such ϕ, and using the fact that $F_\omega(l^2(\mathbf{Z}))$ is closed, we see that $u \in F_\omega(l^2(\mathbf{Z}))$. The conclusion is that the mapping F_ω is also surjective from $l^2(\mathbf{Z})$ to $L^2(\mathbf{R}/a\mathbf{Z})$. □

Example 16.10. For $n \in \mathbf{Z}_{\geq 0}$, we have the (finite) Fourier series

$$(1+e_i)^n = \sum_{k=0}^{n} \binom{n}{k} e_{ik} \qquad \text{and} \qquad 4^n \cos^{2n} = \sum_{k=0}^{2n} \binom{2n}{k} e_{2i(n-k)}$$

of functions in $L^2(\mathbf{R}/2\pi\mathbf{Z})$. Since $|1+e_i(x)|^2 = 4\cos^2\frac{x}{2}$, Parseval's formula now leads to the *Wallis integrals* of even order:

$$\begin{aligned}\int_0^\pi \cos^{2n} x\,dx = \int_0^\pi \sin^{2n} x\,dx &= \frac{\pi}{4^n}\sum_{k=0}^{n}\binom{n}{k}^2 = \frac{\pi}{4^n}\binom{2n}{n}\\ &= \pi\,\frac{1\cdot 3\cdots(2n-1)}{2\cdot 4\cdots 2n} =: \pi\,\frac{(2n-1)!!}{(2n)!!}.\end{aligned}$$

For an evaluation of the integrals by means of Euler's Beta function, use (13.38). Furthermore, note that we have obtained a binomial identity. ⊘

Remark 16.11. The extension of the theory of Fourier series to the n-dimensional variant below does not involve any real difficulties; our only reason for starting with the one-dimensional theory was to avoid being confronted with too many different aspects at the same time.

Let $a(j)$, for $1 \le j \le n$, be a basis for $\mathbf{R}^n$. A distribution $u \in \mathcal{D}'(\mathbf{R}^n)$ is said to be *n-fold periodic* with periods $a(j)$ if $T_{a(j)}u = u$, for all $1 \le j \le n$. From this one immediately derives by mathematical induction that $T_a u = u$, for all $a \in A$, where

$$A := \Big\{ \sum_{j=1}^{n} k_j\, a(j) \mid k \in \mathbf{Z}^n \Big\}$$

is the *lattice* in $\mathbf{R}^n$ *generated* by the vectors $a(j)$. The distribution u is also said to be *A-invariant* and the space of A-invariant distributions is denoted by $\mathcal{D}'(\mathbf{R}^n/A)$.

As in Lemma 16.1, one finds a $\mu \in C_0^\infty(\mathbf{R}^n)$ such that $\sum_{a\in A} T_a\mu = 1$. If, furthermore, $\nu \in C_0^\infty(\mathbf{R}^n)$ with $\sum_{a\in A} T_a\nu = 1$, then $u(\mu) = u(\nu)$, for every $u \in \mathcal{D}'(\mathbf{R}^n/A)$. The *mean* of $u \in \mathcal{D}'(\mathbf{R}^n/A)$ is defined as

$$M(u) = \frac{1}{j_A}\, u(\mu),$$

where j_A equals the n-dimensional volume of the parallelepiped spanned by the vectors $a(j)$, a *fundamental domain* for the lattice A. This corresponds to the usual mean if u is locally integrable. As in Theorem 16.3, we can use this function μ to show that every A-invariant distribution is tempered.

To study the Fourier transforms we need the *dual lattice* Ω in $\mathbf{R}^n$, generated by vectors $\omega(k)$, for $1 \le k \le n$, that are determined by the equations

$$\langle a(j),\, \omega(k)\rangle = 2\pi\, \delta_{jk}.$$

The $\omega(k)$ also form a basis for $\mathbf{R}^n$; the $\frac{1}{2\pi}\omega(k)$ form the so-called *dual basis* of the $a(j)$; see Problem 9.11.

A multisequence $(c_k)_{k\in\mathbf{Z}^n}$ in $\mathbf{C}$ is said to be *of moderate growth* if positive constants c and N exist such that $|c_k| \le c\,(1 + \|k\|)^N$, for all $k \in \mathbf{Z}^n$. As in Lemma 16.4, we obtain

$$v := \sum_{k\in\mathbf{Z}^n} c_k\, \delta_{\omega(k)} \in \mathcal{S}'(\mathbf{R}^n)$$

if and only if the multisequence $(c_k)_{k\in\mathbf{Z}^n}$ of coefficients in $\mathbf{C}$ is of moderate growth; the

$$\omega(k) := \sum_{j=1}^{n} k_j\, \omega(j)$$

run through the points of the lattice Ω.

The n-dimensional version of Theorem 16.7 asserts that the mapping $S \circ \mathcal{F}$ is a bijection from the space of these $v \in \mathcal{S}'(\mathbf{R}^n)$ to the space of A-invariant distributions in $\mathbf{R}^n$. The relevant coefficients c_k in the distributional Fourier series

$$u = \sum_{k\in\mathbf{Z}^n} c_k\, e_{i\omega(k)}$$

are given by the formula

$$c_k = M(e_{-i\omega(k)}\, u) \qquad (k \in \mathbf{Z}^n). \tag{16.10}$$

The *n-dimensional Poisson summation formula* takes the following form:

$$\mathcal{F}(\delta_\Omega) = j_A\, \delta_A \quad \text{if} \quad \delta_\Omega = \sum_{\omega\in\Omega} \delta_\omega \quad \text{and} \quad \delta_A := \sum_{a\in A} \delta_a.$$

This also gives $\mathcal{F}^2(\delta_\Omega) = j_A\, j_\Omega\, \delta_\Omega = (2\pi)^n\, \delta_\Omega$, an alternative calculation of the factor in the Fourier inversion formula.

Finally, we have Parseval's formula. It asserts that the mapping that assigns the corresponding distributional Fourier series to a multisequence in $l^2(\mathbf{Z}^n)$, is a bijective isometry from $l^2(\mathbf{Z}^n)$ to $L^2(\mathbf{R}^n/A)$. ⊘

Remark 16.12. The notation $\mathcal{D}'(\mathbf{R}^n/A)$ and $L^2(\mathbf{R}^n/A)$ derives from the following. The quotient space $\mathbf{R}^n/A$ is defined as the space of cosets $x + A$ with $x \in \mathbf{R}^n$. A function f is A-invariant if and only if f is constant on every coset relative to A, that is, if $f(x) = g(x + A)$, for all $x \in \mathbf{R}^n$, where g denotes a uniquely determined function on $\mathbf{R}^n/A$. To use yet another formulation, if $\pi : x \mapsto x + A$ is the canonical projection from $\mathbf{R}^n$ to $\mathbf{R}^n/A$, the mapping π^* is bijective from the space of functions on $\mathbf{R}^n/A$ to the space of A-invariant functions on $\mathbf{R}^n$. It is customary to identify f and g with each other.

$\mathbf{R}^n/A$ can be equipped with the structure of an n-dimensional C^∞ manifold in such a way that π^* creates a correspondence between the C^∞ functions on this manifold and the A-invariant C^∞ functions on $\mathbf{R}^n$, which explains the notation $C^\infty(\mathbf{R}^n/A)$ for the space of these functions.

This manifold can in fact be obtained as a submanifold of $\mathbf{R}^{2n}$. Let

$$C := \{\, z \in \mathbf{C} \mid |z| = 1 \,\} \tag{16.11}$$

be the unit circle in the complex plane $\mathbf{C} \simeq \mathbf{R}^2$; this is a compact, C^∞, one-dimensional submanifold of $\mathbf{R}^2$. The n-fold Cartesian product

$$T := C \times C \times \cdots \times C$$

is a compact, C^∞, n-dimensional submanifold of $\mathbf{R}^{2n}$, known as an n-dimensional *torus*. With respect to the complex multiplication we can treat C, and therefore also T, as a group. This implies that

$$\Phi : x \mapsto \left(e^{i\langle x,\omega(1)\rangle}, \dots, e^{i\langle x,\omega(n)\rangle}\right)$$

is a homomorphism from the additive group $\mathbf{R}^n$ to the multiplicative group T, with kernel equal to the period lattice A. This means that Φ induces a bijective mapping from $\mathbf{R}^n/A$ to T, by which we can transfer the manifold structure of T to $\mathbf{R}^n/A$. On the basis of this identification, $\mathbf{R}^n/A$ may also be referred to as an n-dimensional torus.

It is possible on manifolds to integrate with respect to densities; considering, on $T = \mathbf{R}^n/A$, the density that locally corresponds to the standard density on $\mathbf{R}^n$, we have

$$u(\mu) = \int_T u(x)\,dx \qquad \text{and} \qquad M(u) = \frac{1}{j_A}\int_T u(x)\,dx.$$

In Remark 10.13 we have noted that it is possible to define distributions on arbitrary manifolds. Thus we now see that distributions on T correspond to the A-invariant distributions on $\mathbf{R}^n$. The point of Remark 16.11 is that there also exists a Fourier transform defined for distributions on the torus that leads to the space of functions on $\mathbf{Z}^n$, the multisequences.

In the case of compact manifolds X, like the torus, we have the simplification that every distribution on X automatically has compact support; in this case, therefore, $C^\infty(X) = C_0^\infty(X)$ and $\mathcal{D}'(X) = \mathcal{E}'(X)$. By way of example, for the torus this yields a convolution product $u * v$ of arbitrary u and $v \in \mathcal{D}'(T)$ such that $\mathcal{F}(u*v) = \mathcal{F}u\,\mathcal{F}v$.

For *Lie groups* G, groups that also possess a manifold structure, there exists a convolution product, which, however, is not commutative when G is not commutative. For compact Lie groups there is a theory of Fourier series that generalizes the theory outlined above for tori. For noncompact, noncommutative Lie groups, Fourier analysis is much more difficult. Here, Harish-Chandra, see [10], has done groundbreaking work. $\oslash$

Until now we have considered convergence of Fourier series only in $\mathcal{S}'$; for the sake of completeness, we discuss some results on the pointwise convergence of Fourier series on $\mathbf{R}^n$. In order to see that the former convergence is really weaker than the latter, derive from (16.8), for $x \in \mathbf{R}$,

$$u_x := 2\pi \sum_{n\in\mathbf{Z}} \delta_{x+2\pi n} = \sum_{n\in\mathbf{Z}} e^{ixn}\, e_{-in}.$$

The distribution u_x has $\{x\} + 2\pi\mathbf{Z}$ as its support (and therefore equals a continuous function on the complement of a countable discrete set), while its distributional Fourier series is nowhere on $\mathbf{R}$ pointwise convergent. We will study some subclasses in $\mathcal{S}'(\mathbf{R}^n)$ possessing better behavior, to wit, pointwise convergence of the (weighted) Fourier series of their elements.

We need some preparation. For $0 < r < 1$, define the function $P_r : \mathbf{R} \to \mathbf{C}$ by $P_r = \sum_{n\in\mathbf{Z}} r^{|n|} e_{in}$. Summing two infinite geometric series, we obtain, for $x \in \mathbf{R}$,

$$P_r(x) = \sum_{n\in\mathbf{N}} r^n e^{-inx} + \sum_{n\in\mathbf{Z}_{\geq 0}} r^n e^{inx} = \frac{1-r^2}{1+r^2-2r\cos x}. \tag{16.12}$$

The family $(P_r)_{0<r<1}$ is said to be the *Poisson kernel* (see Problem 16.12 for the relation between this Poisson kernel and the one occurring in Problem 14.51). It follows that

$$\text{(i)}\quad \frac{1}{2\pi}\int_{-\pi}^{\pi} P_r(x)\,dx = 1; \qquad \text{(ii)}\quad P_r > 0; \qquad \text{(iii)}\quad \lim_{r\uparrow 1} P_r(x) = 0 \tag{16.13}$$

uniformly on $\{ x \in \mathbf{R} \mid \delta \le |x| \le \pi \}$, for every $0 < \delta < \pi$. Indeed, as $r \uparrow 1$, the numerator of $P_r(x)$ tends to 0 and the denominator to a nonzero limit, except at $x \in \mathbf{R} \setminus 2\pi\mathbf{Z}$. On account of (16.7) we have the following equality in $\mathcal{S}'(\mathbf{R})$:

$$\lim_{r\uparrow 1} P_r = \sum_{n\in\mathbf{Z}} e_{in} = 2\pi \sum_{k\in\mathbf{Z}} \delta_{2\pi k}.$$

Set

$$P_{n,r}(x) = \prod_{j=1}^{n} P_r(x_j) \qquad (x \in \mathbf{R}^n).$$

Suppose f is a locally integrable periodic function on $\mathbf{R}^n/2\pi\mathbf{Z}^n$ with distributional Fourier series $\sum_{k\in\mathbf{Z}^n} c_k\, e_{ik}$. Then

$$c_k = M(f\, e_{-ik}) = \frac{1}{(2\pi)^n}\int_{\mathbf{R}^n/2\pi\mathbf{Z}^n} f(y)\, e^{-i\langle y,k\rangle}\,dy.$$

Introduce

$$A_r f = \sum_{k\in\mathbf{Z}^n} r^{|k|} c_k\, e_{ik} \qquad \text{with} \qquad |k| = \sum_{j=1}^{n} |k_j|$$

and consider $x \in \mathbf{R}^n/2\pi\mathbf{Z}^n$. Then we obtain, because $0 < P_{n,r} \le \frac{1+r}{1-r}$,

$$\begin{aligned} A_r f(x) &= \sum_{k\in\mathbf{Z}^n} r^{|k|} \frac{1}{(2\pi)^n}\int_{\mathbf{R}^n/2\pi\mathbf{Z}^n} f(y)\, e^{i\langle x-y,k\rangle}\,dy \\ &= \frac{1}{(2\pi)^n}\int_{\mathbf{R}^n/2\pi\mathbf{Z}^n} f(y)\, P_{n,r}(x-y)\,dy \qquad (16.14) \\ &= \frac{1}{(2\pi)^n}\int_{\mathbf{R}^n/2\pi\mathbf{Z}^n} (T_y f)(x)\, P_{n,r}(y)\,dy. \qquad (16.15) \end{aligned}$$

Theorem 16.13. *Let $f \in C(\mathbf{R}^n/2\pi\mathbf{Z}^n)$. Then we have, uniformly on $\mathbf{R}^n/2\pi\mathbf{Z}^n$,*

$$\lim_{r\uparrow 1} A_r f = f. \tag{16.16}$$

We say that the Fourier series of f is summable in the sense of Abel *to f. Furthermore, for any $1 \le p < \infty$ and $f \in L^p(\mathbf{R}^n/2\pi\mathbf{Z}^n)$, the same limit is valid in $L^p(\mathbf{R}^n/2\pi\mathbf{Z}^n)$.*

Proof. For (16.16), note that (16.14) and (16.13).(i) and (ii) imply, for $x \in \mathbf{R}^n/2\pi\mathbf{Z}^n$,

$$|A_r f(x) - f(x)| \le \frac{1}{(2\pi)^n}\int_{\mathbf{R}^n/2\pi\mathbf{Z}^n} |f(x-y) - f(x)|\, P_{n,r}(y)\,dy. \tag{16.17}$$

With slight modifications the argument is now similar to that in the solution to Problem 5.1 on account of (16.13).(iii).

For the second assertion observe that (16.15) jointly with the translation invariance of the L^p norm implies, for $f \in L^p(\mathbf{R}^n/2\pi\mathbf{Z}^n)$,

$$\|A_r f\|_{L^p} \le \frac{1}{(2\pi)^n} \int_{\mathbf{R}^n/2\pi\mathbf{Z}^n} \|T_y f\|_{L^p} P_{n,r}(y)\, dy = \|f\|_{L^p}.$$

Next use that $C(\mathbf{R}^n/2\pi\mathbf{Z}^n)$ is dense in $L^p(\mathbf{R}^n/2\pi\mathbf{Z}^n)$; see Remark 20.43. Given $\epsilon > 0$ arbitrarily, one therefore obtains the existence of $g \in C(\mathbf{R}^n/2\pi\mathbf{Z}^n)$ such that $\|g - f\|_{L^p} < \epsilon$. In turn, this leads to

$$\|A_r f - f\|_{L^p} \le \|A_r(f-g)\|_{L^p} + \|A_r g - g\|_{L^p} + \|g - f\|_{L^p} \le 2\epsilon + c\,\|A_r g - g\|_{L^\infty},$$

for a suitable constant $c > 0$. By (16.16) the middle term is less than $\frac{\epsilon}{c}$ if r is close enough to 1. □

Corollary 16.14. *A continuous periodic function on* $\mathbf{R}^n$ *is uniquely determined by its Fourier series.*

Example 16.15. Consider the case of $n = 1$ and identify $\mathbf{R}/2\pi\mathbf{Z}$ with C as in (16.11) via $x \leftrightarrow e^{ix}$. Similarly, write $C(\mathbf{R}/2\pi\mathbf{Z}) \ni \widetilde{f} \leftrightarrow f \in C(C)$ if $\widetilde{f}(x) = f(e^{ix})$. Suppose $\widetilde{f} \in C(\mathbf{R}/2\pi\mathbf{Z})$ has Fourier series $\sum_{k\in\mathbf{Z}} c_k\, e_{ik}$ and define the two following series:

$$g_+(z) = \sum_{k\in\mathbf{Z}_{\ge 0}} c_k\, z^k \qquad \text{and} \qquad g_-(z) = \sum_{k\in\mathbf{N}} c_{-k}\, z^{-k} \qquad (z \in \mathbf{C}).$$

Write $D_\pm = \{z \in \mathbf{C} \mid |z| \lessgtr 1\}$. Then $C = \partial D_\pm$. Observe that the bound $|c_k| \le M(|\widetilde{f}|)$, for all $k \in \mathbf{Z}$, implies that

$$g_\pm \text{ define complex-analytic functions on } D_\pm; \text{ and } \lim_{|z|\to\infty} g_-(z) = 0. \qquad (16.18)$$

We also obtain

$$A_r \widetilde{f}(x) = \sum_{\pm} g_\pm(r^{\pm 1}\, e^{ix}) \qquad (0 < r < 1,\ x \in \mathbf{R}).$$

Therefore it is a direct consequence of Theorem 16.13 that

$$f(e^{ix}) = \widetilde{f}(x) = \lim_{r\uparrow 1} \sum_{\pm} g_\pm(r^{\pm 1}\, e^{ix}) \qquad \text{uniformly for } x \in \mathbf{R}. \qquad (16.19)$$

In other words, the functions $z \mapsto \sum_\pm g_\pm(r^{\pm 1}\, z)$ on C converge uniformly on C to f as $r \uparrow 1$. Furthermore, the function f on C is the *uniform boundary value* of the complex-analytic function g_+ on D_+ if and only if $c_k = 0$, for every $k < 0$.

In addition, note that $z \in C$ if and only if $\frac{1}{z} = \bar{z}$. Define the functions h_- and $u : D_+ \to \mathbf{C}$ by means of

$$h_-(z) = \sum_{k \in \mathbf{N}} c_{-k}\, \bar{z}^k \qquad \text{and} \qquad u = g_+ + h_-.$$

Then it is obvious that $\partial_{\bar{z}} g_+(z) = 0$ and $\partial_z h_-(z) = 0$ on D_+, in the notation of (12.12). On account of Problem 12.11 we know that the Laplacian Δ equals $4\partial_z \partial_{\bar{z}} = 4\partial_{\bar{z}} \partial_z$, which implies that $\Delta u = 0$. In other words, u is harmonic on D_+. Moreover, we verify directly, for $0 \le r < 1$ and $x \in \mathbf{R}$,

$$u(re^{ix}) = A_r \tilde{f}(x) = \frac{1-r^2}{2\pi} \int_{-\pi}^{\pi} \frac{f(e^{iy})}{1 + r^2 - 2r\cos(x-y)}\, dy.$$

Now note that

$$|re^{ix} - e^{iy}|^2 = |r - e^{i(y-x)}|^2 = (r - e^{i(y-x)})(r - e^{-i(y-x)}) = 1 + r^2 - 2r\cos(x-y).$$

Hence, writing $x \in D_+$ instead of re^{ix} and $y \in \partial D_+$ instead of e^{iy}, we obtain

$$u(x) = \frac{1-|x|^2}{2\pi} \int_{\partial D_+} \frac{f(y)}{|x-y|^2}\, dy \qquad (x \in D_+), \tag{16.20}$$

where dy denotes integration with respect to the Euclidean density on the circle ∂D_+. Phrased differently, for every $f \in C(C)$, *Poisson's integral formula* (16.20) defines a harmonic function u on D_+ having f as its uniform boundary value, or u solves the Dirichlet problem for Δ on the disk D_+ given the boundary value f (see Problem 12.4). For the natural generalization of Poisson's integral formula to the open unit ball in $\mathbf{R}^n$, see Problem 16.14. ⊘

Theorem 16.16. *If $f \in C^l(\mathbf{R}^n/2\pi\mathbf{Z}^n)$ with $l > \frac{n}{2}$, then the distributional Fourier series $\sum_{k\in\mathbf{Z}^n} c_k\, e_{ik}$ of f is absolutely and uniformly convergent on $\mathbf{R}^n/2\pi\mathbf{Z}^n$. In addition, the Fourier inversion formula from Theorem 16.7 holds pointwise for f.*

Proof. With $c_k = M(f\, e_{-ik})$, we have on account of the Cauchy–Schwarz inequality and with suitable constants c and $c' > 0$,

$$\begin{aligned}
\Big(\sum_{k\in\mathbf{Z}^n} |c_k|\Big)^2 &= \Big(\sum_{k\in\mathbf{Z}^n} (1+|k|)^{-l}\,(1+|k|)^l |c_k|\Big)^2 \\
&\le \Big(\sum_{k\in\mathbf{Z}^n} (1+|k|)^{-2l}\Big)\Big(\sum_{k\in\mathbf{Z}^n} (1+|k|)^{2l} |c_k|^2\Big) \\
&\le c \sum_{k\in\mathbf{Z}^n} (1+|k|)^{2l} |c_k|^2 \le c' \sum_{|\alpha|\le l} \sum_{k\in\mathbf{Z}^n} |k^\alpha c_k|^2.
\end{aligned}$$

Parseval's formula from Theorem 16.9 asserts that the right-hand side can be dominated by

$$c' \sum_{|\alpha|\le l} \|D^\alpha f\|_{L^2} \le c'' \|f\|_{C^l},$$

for suitable $c'' > 0$.

Because of the uniform convergence of the Fourier series, its sum defines a function $\widetilde{f} \in C(\mathbf{R}^n/2\pi\mathbf{Z}^n)$. In such a case, one a fortiori has

$$\lim_{r\uparrow 1} A_r f(x) = \lim_{r\uparrow 1} \sum_{k\in\mathbf{Z}^n} r^{|k|} c_k\, e^{i\langle x,k\rangle} = \widetilde{f}(x) \qquad (x \in \mathbf{R}^n/2\pi\mathbf{Z}^n).$$

On the other hand, it follows from Theorem 16.13 that $\lim_{r\uparrow 1} A_r f(x) = f(x)$. □

As an application, we give an elementary proof of the result discussed in Remark 11.12; for another argument see Theorem 15.4 below. We use the tensor product notation from (11.8).

Theorem 16.17. *Let $X \subset \mathbf{R}^n$ and $Y \subset \mathbf{R}^m$ be open sets and write $C_0^\infty(X) \otimes C_0^\infty(Y)$ for the linear subspace of $C_0^\infty(X \times Y)$ consisting of finite linear combinations of functions $\phi \otimes \psi$, where $\phi \in C_0^\infty(X)$ and $\psi \in C_0^\infty(Y)$. Then $C_0^\infty(X) \otimes C_0^\infty(Y)$ is dense in $C_0^\infty(X \times Y)$.*

Proof. It has to be shown that if $\phi \in C_0^\infty(X \times Y)$, then there exists a sequence $(\phi_m)_{m\in\mathbf{N}}$ of functions in $C_0^\infty(X) \otimes C_0^\infty(Y)$ that converges in $C_0^\infty(X \times Y)$ to ϕ as $m \to \infty$. By a partition of unity, the proof of this can be reduced to the case that $\operatorname{supp}\phi$ is contained in a cube. This, in turn, follows from the case that $\operatorname{supp}\phi$ is a subset of the unit cube in $\mathbf{R}^n \times \mathbf{R}^m$. But then it is clear that the theorem is implied by the following lemma. □

Lemma 16.18. *Define $I = \,]\,0,\,1\,[\, \subset \mathbf{R}$, let $n \in \mathbf{N}$, and denote by I^n the open unit cube in $\mathbf{R}^n$. Suppose that $\phi \in C_0^\infty(I^n)$. Then one can find functions*

$$\phi_m \in \bigotimes_{1\le j\le n} C_0^\infty(I) \qquad (m \in \mathbf{N}) \tag{16.21}$$

such that the sequence $(\phi_m)_{m\in\mathbf{N}}$ converges in $C_0^\infty(I^n)$ to ϕ.

Proof. Extend ϕ to all of $\mathbf{R}^n$ as a C^∞ function $\widetilde{\phi}$ that is periodic with respect to $\mathbf{Z}^n$. One can expand $\widetilde{\phi}$ as a distributional Fourier series, see (16.10),

$$\widetilde{\phi} = \sum_{k\in\mathbf{Z}^n} \widehat{\phi}(k)\, e_{2\pi i k} \qquad \text{with} \qquad \widehat{\phi}(k) = M(e_{-2\pi i k}\widetilde{\phi}).$$

According to Theorem 16.16 this series converges in $C^\infty(\mathbf{R}^n)$ to $\widetilde{\phi}$, and thus in $C^\infty(\overline{I^n})$ to ϕ.

Since ϕ is supported in the open cube I^n, there exists $\delta > 0$ such that $\operatorname{supp}\phi$ is contained in the closed cube $[\,2\delta, 1-2\delta\,]^n$. Choose $\rho \in C_0^\infty(I)$ such that $\rho = 1$ on $]\,\delta,\, 1-\delta\,[$. Furthermore, note that

$$e_{2\pi i k}(x) = \prod_{1\le j\le n} e_{2\pi i k_j}(x_j) = \Big(\bigotimes_{1\le j\le n} e_{2\pi i k_j}\Big)(x) \qquad (k \in \mathbf{Z}^n, x \in \mathbf{R}^n).$$

Now define

$$\phi_m = \sum_{\{k \in \mathbf{Z}^n \mid |k_j| \le m\ (1 \le j \le n)\}} \widehat{\phi}(k) \bigotimes_{1 \le j \le n} e_{2\pi i k_j}\,\rho.$$

Then it is clear from the convergence in $C^\infty(\overline{I^n})$ of the Fourier series to ϕ that these functions ϕ_m, which are of the desired form (16.21), converge in $C_0^\infty(I^n)$ to ϕ; and so the assertion is proved. □

In applications, the pointwise convergence of Fourier series on $\mathbf{R}$ under conditions weaker than the one in Theorem 16.16 is an important issue. Define the following functions on $\mathbf{R}$:

$$D_n = \sum_{k=-n}^{n} e_{ik} \qquad \text{and} \qquad F_n = \frac{1}{n} \sum_{k=0}^{n-1} D_k. \tag{16.22}$$

The sequence $(D_n)_{n \in \mathbf{Z}_{\ge 0}}$ is called the *Dirichlet kernel* and $(F_n)_{n \in \mathbf{N}}$ the *Fejér kernel*. We evaluate these sums explicitly. To this end, recall the summation formula $\sum_{n=0}^{n} z^n = \frac{1-z^{n+1}}{1-z}$ for the geometric series, valid for all $z \in \mathbf{C}$. Accordingly,

$$\sum_{k=-n}^{n} e^{ikx} = e^{-inx}\,\frac{1 - e^{i(2n+1)x}}{1 - e^{ix}} = \frac{e^{i(n+\frac{1}{2})x} - e^{-i(n+\frac{1}{2})x}}{e^{i\frac{x}{2}} - e^{-i\frac{x}{2}}} = \frac{\sin(n+\frac{1}{2})x}{\sin\frac{x}{2}}.$$

Furthermore, on account of the identity $2 \sin p \,\sin q = \cos(p-q) - \cos(p+q)$ we obtain from the preceding equality

$$\begin{aligned} n\,F_n(x)\,\sin^2\frac{x}{2} &= \sum_{k=0}^{n-1} \sin\left(k + \frac{1}{2}\right)x\,\sin\frac{x}{2} = \frac{1}{2}\sum_{k=0}^{n-1}(\cos kx - \cos(k+1)x) \\ &= \frac{1}{2}(1 - \cos nx) = \sin^2\frac{nx}{2}. \end{aligned}$$

Thus (see Figs. 16.2 and 16.3)

$$D_n(x) = \frac{\sin(n+\frac{1}{2})x}{\sin\frac{x}{2}} \qquad \text{and} \qquad F_n(x) = \frac{\sin^2\frac{nx}{2}}{n\,\sin^2\frac{x}{2}}. \tag{16.23}$$

Note the following properties:

$$\text{(i)}\quad \frac{1}{2\pi}\int_{-\pi}^{\pi} \begin{Bmatrix} D_n \\ F_n \end{Bmatrix}(x)\,dx = 1; \qquad \text{(ii)}\quad F_n(x) \ge 0 \qquad (x \in \mathbf{R});$$

$$\text{(iii)}\quad 0 \le F_n(x) \le \frac{\pi^2}{n x^2} \qquad (0 < |x| \le \pi); \qquad \text{(iv)}\quad \lim_{n\to\infty} F_n(x) = 0 \tag{16.24}$$

uniformly on $\{\, x \in \mathbf{R} \mid \delta \le |x| \le \pi \,\}$, for every $0 < \delta < \pi$. In property (i) we used (16.22), in (iii) that $\sin\frac{x}{2} \ge \frac{x}{\pi}$ on $[0, \pi\,]$, and in (iv) the existence of a constant $c > 0$ such that

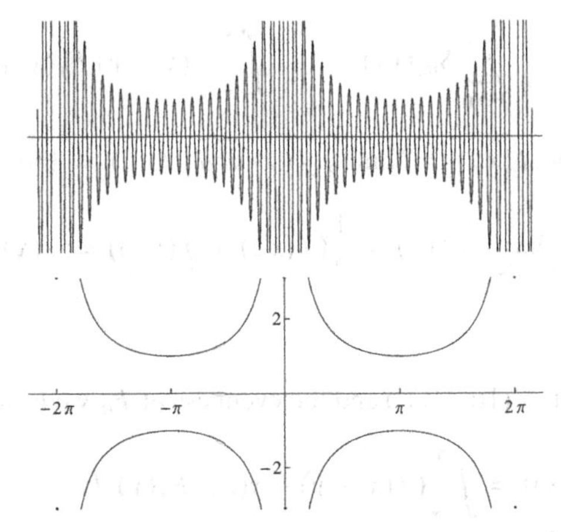

Fig. 16.2 Illustration of (16.23). Truncated graphs of D_{20} and of $x \mapsto \frac{\pm 1}{\sin\frac{x}{2}}$

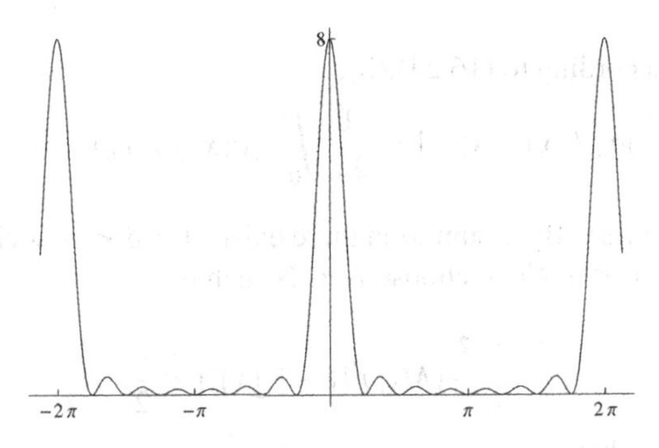

Fig. 16.3 Illustration of (16.23). Graph of F_8.

$$|F_n(x)| \leq \frac{1}{n\,\sin^2\frac{x}{2}} \leq \frac{c}{n\,\delta^2} \qquad (\delta \leq |x| \leq \pi).$$

Furthermore, observe that the partial sums D_n nowhere have a pointwise limit as $n \to \infty$, because their graphs all oscillate with increasing frequency between the graphs of two fixed functions.

We begin by proving that the distributional Fourier series of a periodic distribution of the form test f is *summable in the sense of Cesàro* at points where the function f satisfies a mild extra condition.

Theorem 16.19. *Suppose f is a locally integrable periodic function on* **R** *of period 2π and consider $x \in$* **R**. *Introduce, in the notation of Definition 16.2,*

$$S_n f(x) = \sum_{k=-n}^{n} M(f\,e_{-ik})e^{ikx} = \frac{1}{2\pi}\int_{-\pi}^{\pi} f(x-y)\,D_n(y)\,dy,$$

$$\sigma_n f(x) = \frac{1}{n}\sum_{k=0}^{n-1} S_n f(x) = \frac{1}{2\pi}\int_{-\pi}^{\pi} f(x-y)\, F_n(y)\, dy.$$

Suppose $x \in \mathbf{R}$ *and* $f(x_-) := \lim_{y\uparrow x} f(y)$ *and* $f(x_+) := \lim_{y\downarrow x} f(y)$ *exist. Then*

$$\lim_{n\to\infty} \sigma_n f(x) = \frac{1}{2}(f(x_-) + f(x_+)) =: s(x). \tag{16.25}$$

Proof. On account of (16.24).(i) and the evenness of F_n we have

$$\begin{aligned}2\pi(\sigma_n f(x) - s(x)) &= \int_{-\pi}^{\pi} (f(x-y) - s(x))\, F_n(y)\, dy,\\ &= \int_0^{\pi} (f(x-y) + f(x+y) - 2s(x))\, F_n(y)\, dy =: \int_0^{\pi} g(x,y)\, F_n(y)\, dy,\end{aligned}$$

which implies, according to (16.24).(ii),

$$|\sigma_n f(x) - s(x)| \le \frac{1}{2\pi}\int_0^{\pi} |g(x,y)|\, F_n(y)\, dy. \tag{16.26}$$

Let $\epsilon > 0$ be arbitrary. By assumption there exists $0 < \delta < \pi$ such that $|g(x,y)| < \frac{\epsilon}{2}$ whenever $0 < y < \delta$. Next, choose $N \in \mathbf{N}$ such that

$$\frac{2\pi^2}{N\,\delta^2}(M(|f|) + |s(x)|) < \frac{\epsilon}{2}.$$

It follows directly that

$$\frac{1}{2\pi}\int_0^{\delta} |g(x,y)|\, F_n(y)\, dy \le \frac{1}{2\pi}\int_0^{\delta} \frac{\epsilon}{2}\, F_n(y)\, dy \le \frac{\epsilon}{2}.$$

Furthermore, in view of (16.24).(iii),

$$\begin{aligned}\int_{\delta}^{\pi} |g(x,y)|\, F_n(y)\, dy &\le \int_{\delta}^{\pi} |g(x,y)|\, \frac{\pi^2}{n\,y^2}\, dy\\ &\le \frac{\pi^2}{n\,\delta^2}\int_{\delta}^{\pi} (|f(x-y)| + |f(x+y)| + 2|s(x)|)\, dy\\ &\le \frac{\pi^2}{n\,\delta^2}(2\pi M(|f|) + 2\pi M(|f|) + 2\pi|s(x)|) \le \frac{4\pi^3}{n\,\delta^2}(M(|f|) + |s(x)|)\\ &< \pi\,\epsilon.\end{aligned}$$

We conclude that we have $|\sigma_n f(x) - s(x)| < \epsilon$, for all $n \ge N$. □

The following result is known as *Fejér's Theorem*. Its proof closely parallels that of Theorem 16.13; see, however, Problem 16.15.

Theorem 16.20. *Suppose that f is a continuous periodic function on* **R** *of period 2π. In the notation of Theorem 16.19 we then have that $\lim_{n\to\infty} \sigma_n f = f$ uniformly on* **R**.

We mention an immediate corollary. A continuous periodic function can be uniformly approximated by *trigonometric polynomials*, that is, by functions of the form $\sum_{k=-n}^{n} c_k\, e_{ink}$. In turn, this result leads via uniform convergence of the exponential series to the approximation of a continuous function by polynomials uniformly on compacta; compare with Weierstrass's Approximation Theorem from Example 14.31.

Now we come to results that imply pointwise convergence of the distributional Fourier series of a periodic distribution test f.

Theorem 16.21. *Let f be as in Theorem 16.19. At points $x \in \mathbf{R}$ such that $y \mapsto \frac{f(y)-f(x)}{y-x}$ is integrable in a neighborhood of x, in particular, at points x of differentiability of f, we have pointwise convergence, that is,*

$$\lim_{m,n\to\infty} \sum_{k=-m}^{n} M(f\, e_{-ik})\, e^{ikx} = f(x).$$

Proof. Indeed, as usual we may suppose that $x = 0$ and $f(x) = 0$; just shift the origin and subtract a constant from f. By assumption, the function $g : x \mapsto \frac{f(x)}{e^{ix}-1}$ is integrable near 0. Furthermore, we have

$$f(x) = (e^{ix} - 1)g(x), \qquad \text{so} \qquad M(f\, e_{-ik}) = M(g\, e_{-i(k-1)}) - M(g\, e_{-ik}).$$

Therefore the Fourier series of f at 0 is a telescoping series. Indeed,

$$\sum_{k=-m}^{n} M(f\, e_{-ik}) = M(g\, e_{i(m+1)}) - M(g\, e_{-in}),$$

and this tends to $0 = f(0)$ by the Riemann–Lebesgue Theorem, Theorem 14.2, applied to $g\, 1_{[-\pi,\pi]}$, which is integrable on **R**. □

We can also deal with a jump discontinuity.

Theorem 16.22. *Let f be as in Theorem 16.19. Suppose that f has left-hand and right-hand limits at x and, in addition, that it has one-sided slopes at x, that is, $h \mapsto \frac{f(x\pm h)-f(x_\pm)}{h}$ converge to limits as $h \downarrow 0$. Then the symmetric partial sums converge; more precisely,*

$$\lim_{n\to\infty} S_n f(x) = \frac{1}{2}(f(x_-) + f(x_+)).$$

Proof. Using a translation we may suppose that $x = 0$. Subtract a constant such that $f(0_-) = -f(0_+)$. We now must show that the $S_n f(0)$ converge to 0, where

$$S_n f(0) = \frac{1}{2\pi} \int_{-\pi}^{\pi} f(y)\, D_n(y)\, dy.$$

But we need only observe that D_n is an even function. Hence

$$S_n f(0) = \frac{1}{2\pi} \int_{-\pi}^{\pi} \frac{1}{2}(f(-y) + f(y))\, D_n(y)\, dy.$$

Now we simply apply Theorem 16.21 to the function $x \mapsto \frac{1}{2}(f(-x) + f(x))$. □

Note that near a jump discontinuity the convergence of the symmetric partial sums cannot be uniform. Otherwise, the function would be continuous at that point.

Example 16.23. Consider arbitrary $z \in \mathbf{C}$ and let $f_z : \mathbf{R} \to \mathbf{C}$ be the periodic function of period 1 that on $\left]-\frac{1}{2}, \frac{1}{2}\right[$ is given by $f_z(x) = e^{2\pi i z x}$. A straightforward computation (based, for instance, on the generalization of (16.8)) gives the distributional Fourier series

$$f_z = \frac{\sin \pi z}{\pi} \sum_{n \in \mathbf{Z}} \frac{(-1)^n}{z - n} e_{2\pi i n}, \tag{16.27}$$

where a term with $n = z$ should be read as $e_{2\pi i n}$. In particular, application of the preceding theorem with $x = 0$ yields the *partial-fraction decomposition* of the cosecant (see Problem 16.21.(iii) for another proof)

$$\frac{\pi}{\sin \pi z} = \frac{1}{z} + 2z \sum_{n \in \mathbf{N}} \frac{(-1)^n}{z^2 - n^2} \qquad (z \in \mathbf{C} \setminus \mathbf{Z}). \tag{16.28}$$

Furthermore, using the preceding theorem with $x = \frac{1}{2}$ leads to the partial-fraction decomposition of the cotangent (see [7, Exercises 0.13.(i) and 0.21.(ii)] for other proofs)

$$\pi \cot \pi z = \frac{1}{z} + 2z \sum_{n \in \mathbf{N}} \frac{1}{z^2 - n^2} \qquad (z \in \mathbf{C} \setminus \mathbf{Z}). \tag{16.29}$$

In particular, we obtain the sum of *Euler's series*

$$\frac{\pi^2}{6} = \lim_{z \to 0} \frac{1 - \pi z \cot \pi z}{2z^2} = \sum_{n \in \mathbf{N}} \frac{1}{n^2}.$$

Parseval's formula from Theorem 16.9 applied to f_z or differentiation of (16.29) with respect to z implies (compare with Problem 16.10 and [7, Exercise 0.12.(ii)])

$$\frac{\pi^2}{\sin^2 \pi z} = \sum_{n \in \mathbf{Z}} \frac{1}{(z - n)^2} \qquad (z \in \mathbf{C} \setminus \mathbf{Z}). \tag{16.30}$$

See [7, Exercise 0.12] for further consequences of the identity, for instance, values of Riemann's zeta function at positive even points (see also (16.35) below) and

MacLaurin's series of the tangent. Finally, one may rewrite (16.30) in the form (see Problems 16.10 or 18.9 for another proof)

$$1 = \sum_{n \in \mathbf{Z}} T_{\pi n} \operatorname{sinc}^2 \quad \text{on} \quad \mathbf{C}, \tag{16.31}$$

which according to Problem 16.19 leads to $\int_{\mathbf{R}} \operatorname{sinc} x \, dx = \pi$ (compare with Problems 14.44 and 16.21).

We give an application of the theory above. Let $0 < a < 1$. Using the change of variables $x = \frac{1}{y}$, we get

$$\int_{\mathbf{R}_{>0}} \frac{x^{a-1}}{1+x} \, dx = \int_0^1 \frac{x^{a-1}}{1+x} \, dx + \int_0^1 \frac{y^{(1-a)-1}}{1+y} \, dy.$$

Furthermore, the following series is uniformly convergent for $x \in [\epsilon, 1-\epsilon']$, where ϵ and $\epsilon' > 0$ are arbitrary:

$$\frac{x^{a-1}}{1+x} = \sum_{k \in \mathbf{Z}_{\geq 0}} (-1)^k x^{a+k-1}.$$

Finally, apply termwise integration and (16.28) to obtain (see [7, Exercise 6.58.(v)] for another proof)

$$\int_{\mathbf{R}_{>0}} \frac{x^{a-1}}{x+1} \, dx = \frac{\pi}{\sin \pi a} \qquad (0 < a < 1). \tag{16.32}$$

$\oslash$

Example 16.24. (Fourier series of Bernoulli functions.) The sequence of polynomial functions $(b_n)_{n \in \mathbf{Z}_{\geq 0}}$ on $\mathbf{R}$ is uniquely determined by the following conditions:

$$b_0 = 1, \qquad b_n' = n b_{n-1}, \qquad \int_0^1 b_n(x) \, dx = 0 \qquad (n \in \mathbf{N}).$$

The b_n are known as the *Bernoulli polynomials* and the $B_n := b_n(0) \in \mathbf{R}$ as the *Bernoulli numbers*. It follows that $b_n(0) = b_n(1)$, for $n \geq 2$. The $b_n(x)$, for $1 \leq n \leq 4$, are respectively given by

$$x - \frac{1}{2}, \qquad x^2 - x + \frac{1}{6}, \qquad x^3 - \frac{3}{2}x^2 + \frac{1}{2}x, \qquad x^4 - 2x^3 + x^2 - \frac{1}{30}.$$

Introduce the *Bernoulli functions* $\overline{b}_n : \mathbf{R} \to \mathbf{R}$ by means of $\overline{b}_n(x) = b_n(x - [x])$; see Fig. 16.4. Then $\overline{b}_n \in \mathcal{S}'(\mathbf{R})$, for all $n \in \mathbf{N}$; in fact, $\overline{b}_n \in C^{n-2}(\mathbf{R})$, for $n \geq 2$.

Observe that $\overline{b}_1$ equals $s - \frac{1}{2}$, where s is the sawtooth function from Example 4.2. Using the formula for s' from that example and Poisson's summation formula, we see that the distributional Fourier series of s' is given by $-\sum_{k \in \mathbf{Z} \setminus \{0\}} e_{2\pi i k}$. This leads to the following identity in $\mathcal{S}'(\mathbf{R})$:

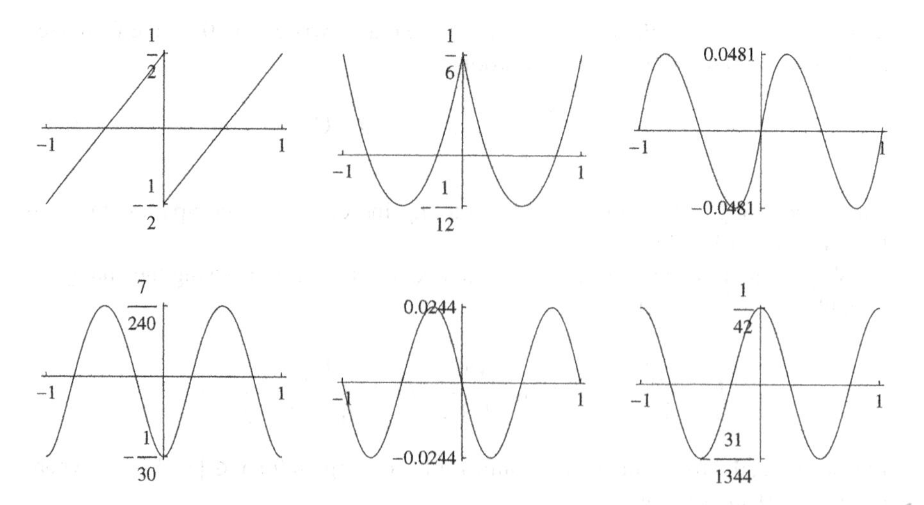

Fig. 16.4 Example 16.24. Graphs of Bernoulli functions $\overline{b}_n$, for $1 \leq n \leq 6$, with different vertical scales and ordered in rows

$$\overline{b}_1 = s - \frac{1}{2} = -\sum_{k \in \mathbf{Z} \setminus \{0\}} \frac{e_{2\pi i k}}{2\pi i k}. \tag{16.33}$$

Here the mean has been used to fix the constant of integration; in fact, $M(s) = \int_0^1 x\, dx = \frac{1}{2}$. It is also easy to verify (16.33) by direct computation of the distributional Fourier series of s. On account of Theorem 16.21 we even obtain the pointwise convergence of the series on $\mathbf{R} \setminus \mathbf{Z}$; hence (compare with [7, Exercise 0.18.(i)] and Fig. 16.5)

$$\overline{b}_1(x) = x - [x] - \frac{1}{2} = -\sum_{k \in \mathbf{N}} \frac{\sin 2\pi k x}{\pi k} \qquad (x \in \mathbf{R} \setminus \mathbf{Z}). \tag{16.34}$$

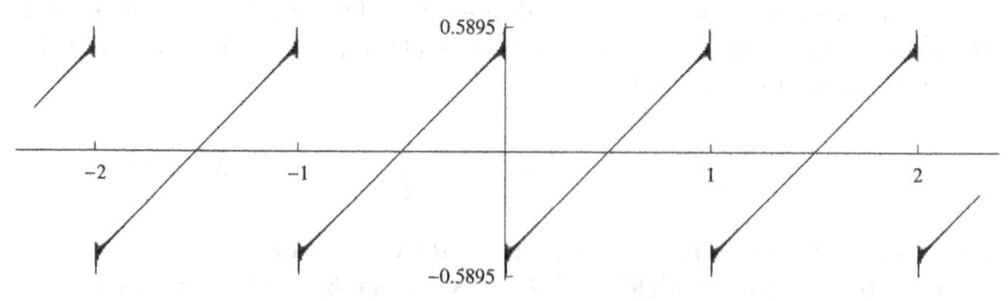

Fig. 16.5 Illustration of (16.34). Graph of $x \mapsto -\sum_{k=1}^{100} \frac{\sin(2\pi k x)}{\pi k}$

We would like to integrate (16.34) termwise in order to obtain the Fourier series of $\overline{b}_2$, but the series in (16.34) is uniformly convergent only on intervals of the form

$[\delta, 1-\delta]$, for every $0 < \delta < \frac{1}{2}$ and not on $[0, 1]$. On the other hand, $\sum_{k=1}^{n} \frac{\sin kx}{k}$ is uniformly bounded for $n \in \mathbf{N}$ and $x \in [-\pi, \pi]$. Indeed, in the notation of (16.22),

$$\sum_{k=1}^{n} \frac{\sin kx}{k} = \frac{1}{2i} \sum_{0<|k|\leq n} \frac{e^{ikx}}{k} = \frac{1}{2} \int_0^x (D_n(t) - 1)\, dt.$$

Now

$$D_n(t) = \frac{\sin(n+\frac{1}{2})t}{\sin\frac{t}{2}} = \sin nt \, \cot\frac{t}{2} + \cos nt.$$

Note that $\cot\frac{t}{2} = \frac{2}{t} + O(t)$ as $t \to 0$, and therefore $g : [-\pi, \pi] \to \mathbf{R}$ defined by $\cot\frac{t}{2} = \frac{2}{t} + g(t)$ is a continuous function. This implies

$$D_n(t) = 2n \,\operatorname{sinc} nt + g(t)\sin nt + \cos nt =: 2n \,\operatorname{sinc} nt + h_n(t),$$

where $\int_0^x h_n(t)\, dt$ is bounded, uniformly for $n \in \mathbf{N}$ and $x \in [-\pi, \pi]$. Furthermore, the convergence of $\int_{\mathbf{R}} \operatorname{sinc} t\, dt$ (see Problem 16.19) gives that $p \mapsto \int_0^p \operatorname{sinc} t\, dt$ is a bounded function on $\mathbf{R}$. Therefore

$$\int_0^x n \,\operatorname{sinc} nt\, dt = \int_0^{nx} \operatorname{sinc} t\, dt$$

is bounded, uniformly for $n \in \mathbf{N}$ and $x \in [-\pi, \pi]$.

It follows from the above that the conditions of the Dominated Convergence Theorem of Arzelà (see [7, Theorem 6.12.3]) or that of Lebesgue (see Theorem 20.26.(iv)) are satisfied; hence we obtain by means of termwise integration the following identity of functions on $\mathbf{R}$:

$$\frac{\overline{b}_2}{2!} = -\sum_{k\in\mathbf{Z}\setminus\{0\}} \frac{e_{2\pi ik}}{(2\pi ik)^2}.$$

Since this series is uniformly convergent on $\mathbf{R}$, repeated termwise integration leads to the identities

$$\frac{\overline{b}_n}{n!} = -\sum_{k\in\mathbf{Z}\setminus\{0\}} \frac{e_{2\pi ik}}{(2\pi ik)^n} \qquad (n \in \mathbf{Z}_{>1})$$

of functions on $\mathbf{R}$. In particular,

$$\frac{\overline{b}_{2n}(x)}{(2n)!} = 2(-1)^{n-1} \sum_{k\in\mathbf{N}} \frac{\cos 2\pi kx}{(2\pi k)^{2n}} \qquad (n \in \mathbf{N},\ x \in \mathbf{R}),$$

$$\frac{\overline{b}_{2n+1}(x)}{(2n+1)!} = 2(-1)^{n-1} \sum_{k\in\mathbf{N}} \frac{\sin 2\pi kx}{(2\pi k)^{2n+1}} \qquad (n \in \mathbf{N},\ x \in \mathbf{R}),$$

$$\zeta(2n) := \sum_{k\in\mathbf{N}} \frac{1}{k^{2n}} = (-1)^{n-1}\frac{1}{2}(2\pi)^{2n}\frac{B_{2n}}{(2n)!} \qquad (n \in \mathbf{N}). \tag{16.35}$$

Here $\zeta : \{s \in \mathbf{C} \mid \operatorname{Re} s > 1\} \to \mathbf{C}$ denotes Riemann's *zeta function* defined by $\zeta(s) = \sum_{k\in\mathbf{N}} \frac{1}{k^s}$. Finally, note that $B_{2n+1} = 0$, for $n \in \mathbf{N}$. ⊘

Problems

16.1.* Without computing Fourier coefficients by means of integration, determine the Fourier series of the 2π-periodic even continuous function $\arccos \circ \cos$ on $\mathbf{R}$ from Problem 1.7; that is, show that

$$\arccos \circ \cos = \frac{\pi}{2} - \frac{2}{\pi} \sum_{n\in\mathbf{Z}} \frac{1}{(2n-1)^2} e_{i(2n-1)} = \frac{\pi}{2} - \frac{4}{\pi} \sum_{n\in\mathbf{N}} \frac{\cos(2n-1)\cdot}{(2n-1)^2}.$$

Deduce that $\sum_{n\in\mathbf{N}} \frac{1}{(2n-1)^2} = \frac{\pi^2}{8}$ and *Euler's series* $\zeta(2) = \sum_{n\in\mathbf{N}} \frac{1}{n^2} = \frac{\pi^2}{6}$, and furthermore $\sum_{n\in\mathbf{N}} \frac{1}{(2n-1)^4} = \frac{\pi^4}{96}$ and $\zeta(4) = \sum_{n\in\mathbf{N}} \frac{1}{n^4} = \frac{\pi^4}{90}$. (Compare with (16.35) and [7, Exercise 0.12.(iv)].)

In addition, derive for $f = \arcsin \circ \sin$ (see Fig. 16.6),

$$f = \frac{4}{\pi} \sum_{n\in\mathbf{N}} \frac{(-1)^{n-1} \sin(2n-1)\cdot}{(2n-1)^2}, \qquad f' = \frac{4}{\pi} \sum_{n\in\mathbf{N}} \frac{(-1)^{n-1} \cos(2n-1)\cdot}{2n-1}.$$

Deduce *Leibniz's series* $\sum_{n\in\mathbf{N}} \frac{(-1)^{n-1}}{2n-1} = \frac{\pi}{4}$.

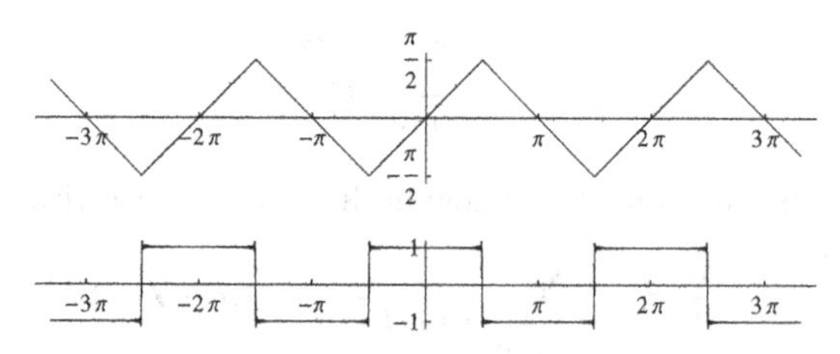

Fig. 16.6 Illustration of Problem 16.1. Graphs of $\frac{4}{\pi}\sum_{n=1}^{30} \frac{(-1)^{n-1}\sin(2n-1)\cdot}{(2n-1)^2}$ and $\frac{4}{\pi}\sum_{n=1}^{100} \frac{(-1)^{n-1}\cos(2n-1)\cdot}{2n-1}$

16.2.* In the preceding problem new Fourier series were obtained from known ones by means of differentiation. The converse method of integration is effective as well.

Generalize Poisson's summation formula (16.7) as follows:

$$a \sum_{k\in\mathbf{Z}} \delta_{h+ka} = \sum_{n\in\mathbf{Z}} e^{-in\omega h} e_{in\omega} \qquad (a, \omega, h \in \mathbf{R},\ a\,\omega = 2\pi).$$

Next deduce $(\arcsin\circ\sin)'' = \frac{4}{\pi}\sum_{n\in\mathbf{N}}(-1)^n \sin(2n-1)\cdot$ in $\mathcal{S}'(\mathbf{R})$ (see Fig. 16.6) and integrate termwise.

16.3. Show that $\sum_{k\in\mathbf{Z}}\delta_{\sqrt{2\pi}\,k}$ is an eigendistribution for the Fourier transform with eigenvalue $\sqrt{2\pi}$; compare with Problem 14.9.

16.4. Demonstrate that Poisson's summation formula may also be applied to $\phi : x\mapsto \frac{1}{1+x^2}$. Calculate

$$R_h = h\sum_{n\in\mathbf{Z}}\frac{1}{1+h^2n^2},$$

a Riemann sum approximation of the integral of ϕ. Determine the limit, as $h\downarrow 0$, of $e^{2\pi/h}(R_h-\pi)$.

16.5. Show that $\sum_{n\in\mathbf{N}}(-1)^{n-1}$ is summable in the sense of Abel and of Cesàro to $\frac{1}{2}$ in both cases.

16.6. For $x\in\mathbf{R}\setminus 2\pi\mathbf{Z}$ prove the following identities on summation of series in the sense of Abel:

$$\sum_{n\in\mathbf{Z}_{\geq 0}}(n+1)\cos nx = \frac{1}{2}-\frac{1}{4\sin^2(\frac{x}{2})},\qquad \sum_{n\in\mathbf{N}}(n+1)\sin nx = \frac{1}{2}\cot\Big(\frac{x}{2}\Big),$$

$$\sum_{n\in\mathbf{N}} n\cos nx = -\frac{1}{4\sin^2(\frac{x}{2})},\qquad \sum_{n\in\mathbf{N}} n\sin nx = 0.$$

For $x=\pi$ the third series takes the form $\sum_{n\in\mathbf{N}}(-1)^n n$. Show that its corresponding sequence of Cesàro means has two limit points: 0 and $-\frac{1}{2}$. In other words, the series is summable in the sense of Abel to $-\frac{1}{4}$ (the average of these limit points), but is not summable in the sense of Cesàro.

Conversely, suppose the series $\sum_{n\in\mathbf{N}} c_n$ of complex numbers is summable in the sense of Cesàro to s and prove that it is summable in the sense of Abel to s. Hint: apply the identity $\sum_{n\in\mathbf{N}} c_n\, r^n = (1-r)^2\sum_{n\in\mathbf{N}} n\sigma_n\, r^n$ and assume $s=0$.

16.7.* With $n\in\mathbf{N}$ and F_n as in (16.22), demonstrate that

$$F_n = \sum_{|k|\leq n-1}\Big(1-\frac{|k|}{n}\Big)e_{ik}\qquad\text{and}\qquad \mathcal{F}F_n = 2\pi\sum_{|k|\leq n-1}\Big(1-\frac{|k|}{n}\Big)\delta_k.$$

16.8.* Calculate $\sum_{n=0}^{N}\cos n\,\cdot$ and $\sum_{n=1}^{N}\sin n\,\cdot$, and discuss the convergence in $\mathcal{S}'(\mathbf{R})$ as $N\to\infty$. Prove that in $\mathcal{S}'(\mathbf{R})$ we have the following identity (compare with Example 1.4):

$$\sum_{n\in\mathbf{Z}_{\geq 0}}\cos n\,\cdot = \frac{1}{2}+\pi\sum_{n\in\mathbf{Z}}\delta_{2\pi n}.$$

In Problem 16.17.(v) we will use the Fourier transform to show

$$u := \sum_{n\in\mathbf{N}} \sin n \cdot = \frac{1}{4}\sum_{\pm} \cot\left(\frac{\cdot}{2} \pm i0\right).$$

In this problem, compute only $\mathcal{F}u$.
Remark: it is also possible to prove, without using the Fourier transform (see, for example, [7, Exercise 0.18]) the formula above for $\sum_{n\in\mathbf{Z}_{\geq 0}} \cos n \cdot$ and furthermore that the restriction of u to $\mathbf{R} \setminus 2\pi\mathbf{Z}$ equals the analytic function $x \mapsto \frac{\sin x}{2(1-\cos x)} = \frac{1}{2}\cot\frac{x}{2}$; see Fig. 16.7 and compare with Problem 16.6. For x near 0 we may write the last expression as $\frac{\cos\frac{x}{2}}{\operatorname{sinc}\frac{x}{2}}\frac{1}{x}$, where the first factor is a C^∞ function. On account of the 2π-periodicity of u, it has similar, not absolutely integrable behavior at all integer multiples of 2π. For a test function ϕ with support in the interval $]-2\pi, 2\pi[$ one has

$$\begin{aligned} u(\phi) &= \frac{\cos\frac{x}{2}}{\operatorname{sinc}\frac{x}{2}}\left(\mathrm{PV}\frac{1}{x}\right)(\phi) = \frac{1}{2}\int_{\mathbf{R}_{>0}} (\phi(x) - \phi(-x)) \cot\frac{x}{2}\, dx \\ &= \int_{\mathbf{R}} \frac{(\phi(x) - \phi(-x)) \sin x}{4(1-\cos x)}\, dx = -\frac{1}{2}\int_{\mathbf{R}} \log(1-\cos x)\, \phi'(x)\, dx, \end{aligned}$$

where all integrals are absolutely convergent. Also, u equals the distributional derivative of the locally integrable function $\frac{1}{2}\log(1-\cos)$. Using Lemma 16.6

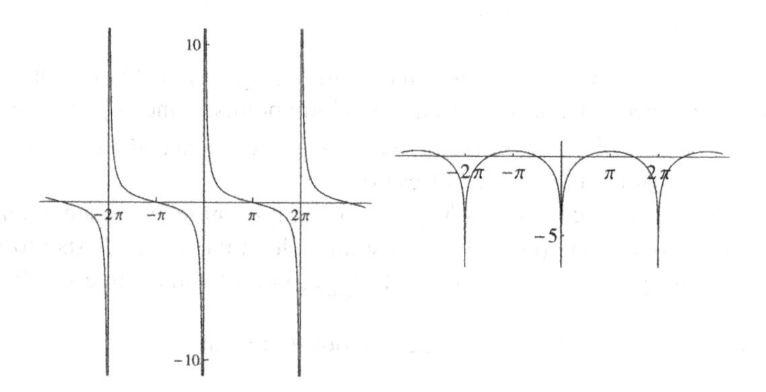

Fig. 16.7 Illustration of Problem 16.8. Graphs of $\frac{1}{2}\cot\frac{\cdot}{2}$ and $\frac{1}{2}\log(1-\cos)$

and Problem 13.6 one obtains the following identity in $\mathcal{D}'(]-2\pi, 2\pi[)$ (with some abuse of notation):

$$u'(x) = \sum_{n\in\mathbf{N}} n\cos nx = -\frac{1}{4}\frac{x-\sin x}{\sin^2\frac{x}{2}}\,\mathrm{PV}\frac{1}{x} - \frac{\cos(\frac{x}{2})}{\operatorname{sinc}(\frac{x}{2})}|x|^{-2}.$$

Note that the restriction of u' to $\mathbf{R}\setminus 2\pi\mathbf{Z}$ equals the function $x \mapsto -\frac{1}{4\sin^2(\frac{x}{2})}$; compare with Problem 16.6

16.9. Let u_0 be a periodic distribution on $\mathbf{R}$ with period $a > 0$. Prove that there exists exactly one differentiable family $t \mapsto u_t : \mathbf{R}_{>0} \to \mathcal{D}'(\mathbf{R})$ with the following properties (compare with Problem 17.1):

(i) For every $t \in \mathbf{R}_{>0}$, the distribution u_t is periodic with period a,
(ii) $\frac{d}{dt} u_t = \partial^2 u_t$,
(iii) $\lim_{t \downarrow 0} u_t = u_0$ in $\mathcal{D}'(\mathbf{R})$.

Calculate $\mathcal{F} u_t$ and u_t.

16.10.* Compute $\mathcal{F}\, \mathrm{sinc}^2$, and using Poisson's summation formula deduce that $\sum_{k \in \mathbf{Z}} T_{\pi k}\, \mathrm{sinc}^2 = 1$ on $\mathbf{R}$. Compare with (16.31) and [7, Exercise 0.14]. Derive (16.30).

16.11.* *(Relation between Fejér kernel for line and unit circle.)* For $t > 0$ define $F_t(x) = \frac{t}{2\pi} \mathrm{sinc}^2 \frac{tx}{2}$. Verify that $\mathcal{F} F_t(\xi) = \max\{0, 1 - \frac{|\xi|}{t}\}$, for every $\xi \in \mathbf{R}$. Conclude that $\lim_{t \to \infty} F_t = \delta$ in $\mathcal{S}'(\mathbf{R})$. Furthermore, for every $f \in \mathcal{S}(\mathbf{R})$ and $x \in \mathbf{R}$, deduce that

$$F_t * f(x) = \frac{1}{2\pi} \int_{-t}^{t} e^{ix\xi} \Big(1 - \frac{|\xi|}{t}\Big) \mathcal{F} f(\xi)\, d\xi.$$

In this problem, write $\overline{F}_n$ for the Fejér kernel F_n from (16.23), for $n \in \mathbf{N}$. Prove in two different ways the following identity of functions on $\mathbf{R}$:

$$\frac{1}{2\pi} \overline{F}_n = \sum_{k \in \mathbf{Z}} T_{2\pi k} F_n \qquad \text{and} \qquad \lim_{n \to \infty} \overline{F}_n = 2\pi \sum_{k \in \mathbf{Z}} \delta_{2\pi k}.$$

The passage from F_n to $\frac{1}{2\pi} \overline{F}_n$ is said to be the *periodization* of F_n.

16.12. *(Relation between Poisson kernel for half-space and unit disk.)* Define the Poisson kernel for the half-space $P_t : \mathbf{R} \to \mathbf{R}$, for $t > 0$, as in Problem 14.51, and the Poisson kernel for the unit disk $\overline{P}_r : \mathbf{R}/2\pi\mathbf{Z} \to \mathbf{R}$ by (16.12) (here the bar is used temporarily in order to avoid confusion). Using that $\mathcal{F} P_t = e^{-t|\cdot|}$, demonstrate that

$$\overline{P}_r = 2\pi \sum_{k \in \mathbf{Z}} T_{2\pi k} P_t, \qquad \text{where} \qquad r = e^{-t}.$$

Deduce that $\lim_{r \uparrow 1} \overline{P}_r = 2\pi \sum_{k \in \mathbf{Z}} \delta_{2\pi k}$ in two different ways.

16.13. Define C and $D_\pm$ as in Example 16.15. Let g_+ be complex-analytic on D_+ and write $u_r(x) := g_+(r\, e^{ix})$, for $0 \le r < 1$ and $x \in \mathbf{R}$. Verify that as $r \uparrow 1$, u_r converges in $\mathcal{D}'(\mathbf{R})$ if and only if the sequence of terms $c_n := g_+^{(n)}(0)/n!$, for $n \in \mathbf{Z}_{\ge 0}$, is of moderate growth. If this is the case, the limit f is said to be the distributional boundary value of g_+, via the parametrization $x \mapsto e^{ix} : \mathbf{R} \to C$. Prove that f is periodic with period 2π. Does every complex-analytic function on D_+ have a distributional boundary value? Does every 2π-periodic distribution on $\mathbf{R}$ equal the distributional boundary value of an analytic function on D_+?

Let g_- be complex-analytic on D_-. It is known that $g_-(z) \to 0$ as $|z| \to \infty$ if and only if $g_-(z) = h_+(\frac{1}{z})$ for an analytic function h_+ on D_+ with $h_+(0) = 0$. Write $v_r(x) := g_-(r\, e^{ix})$, for $r > 1$ and $x \in \mathbf{R}$. Prove that as $r \downarrow 1$, v_r converges in $\mathcal{D}'(\mathbf{R})$ if and only if h_+ possesses a distributional boundary value h. If this is the case, the limit g is said to be the distributional boundary value of g_-. Prove that $g = Sh$.

Show that every 2π-periodic distribution on $\mathbf{R}$ can be written uniquely as $f + g$, with f the distributional boundary value of a complex-analytic function g_+ on D_+ and g the distributional boundary value of a complex-analytic function g_- on D_- with $g_-(\infty) = 0$. Determine f and g for $u = \sum_{n \in \mathbf{Z}} \delta_{2\pi n}$.

16.14.* *(Poisson integral formula for open unit ball in* $\mathbf{R}^n$*.)* Write $B^n = \{ x \in \mathbf{R}^n \mid \|x\| < 1 \}$ and denote, as usual, by c_n the $(n-1)$-dimensional volume of the unit sphere $S^{n-1} = \partial B^n$ from (13.37). Consider $f \in C(S^{n-1})$ and prove that

$$u(x) = \frac{1 - \|x\|^2}{c_n} \int_{S^{n-1}} \frac{f(y)}{\|x - y\|^n}\, dy \qquad (x \in B^n)$$

defines a harmonic function u on B^n that satisfies $\lim_{r \uparrow 0} u(r\, x) = f(x)$, uniformly for all $x \in S^{n-1}$.

16.15.* Give a direct proof of Theorem 16.20.

16.16.* Consider $\phi \in \mathcal{S}(\mathbf{R})$. Prove that $\sum_{k \in \mathbf{Z}} |T_{2\pi k} \mathcal{F} \phi|^2 = 1$ on $\mathbf{R}$, if and only if

$$(T_l \phi,\, T_m \phi) = \delta_{lm} \qquad (l,\, m \in \mathbf{Z}),$$

where δ_{lm} denotes the Kronecker delta. Equation (16.31) and Example 14.1 show that the equivalence also may hold in the case of functions belonging to $L^2(\mathbf{R})$.

16.17.* *(Poisson summation formula.)* Define $u \in \mathcal{D}'(\mathbf{R})$ by

$$u = \frac{1}{2}\delta + \sum_{k \in \mathbf{N}} \delta_{2\pi k}.$$

(i) Verify that $u \in \mathcal{S}'(\mathbf{R})$. Deduce the following equality of distributions in $\mathcal{S}'(\mathbf{R})$:

$$\mathcal{F} u = \frac{1}{2} + \sum_{n \in \mathbf{N}} e_{-2\pi i n}.$$

(ii) Prove that $u = \lim_{\epsilon \downarrow 0} e^{-\epsilon x} u$ in $\mathcal{S}'(\mathbf{R})$ and furthermore that

$$\mathcal{F} u = \lim_{\epsilon \downarrow 0} \Big(\frac{1}{2} + \sum_{n \in \mathbf{N}} e^{-\epsilon 2\pi n} e_{-2\pi i n} \Big).$$

Show by means of summation of a geometric series that

$$\mathcal{F}u = \lim_{\epsilon\downarrow 0} \frac{1}{2i} \cot \pi(\cdot - i\epsilon) =: \frac{1}{2i} \cot \pi(\cdot - i0).$$

Now use addition of the resulting identities to obtain

$$2i \sum_{n\in\mathbf{Z}} e_{2\pi in} = \cot \pi(\cdot - i0) - \cot \pi(\cdot + i0).$$

(iii) In addition, show that the complex-analytic function $z \mapsto \pi \cot \pi z$ on $\mathbf{C}$ has a simple pole of residue 1 at every point $k \in \mathbf{Z}$. Conclude by means of the Plemelj–Sokhotsky jump relation $\frac{1}{\xi - i0} - \frac{1}{\xi + i0} = 2\pi i \delta$ (see Problem 1.3) that one has

$$\cot \pi(\cdot - i0) - \cot \pi(\cdot + i0) = 2i \sum_{k\in\mathbf{Z}} \delta_k; \qquad \text{deduce} \qquad \sum_{n\in\mathbf{Z}} e_{2\pi in} = \sum_{k\in\mathbf{Z}} \delta_k,$$

where the latter identity of tempered distributions is the Poisson summation formula known from (16.7) and Problem 10.13.

(iv) Prove that (compare with Problem 16.6)

$$4\pi \sum_{n\in\mathbf{N}} n \sin 2\pi n \cdot = -\sum_{k\in\mathbf{Z}} \delta_k'.$$

(v) Also derive from part (ii) the following identity in $\mathcal{S}'(\mathbf{R})$ (compare with Problem 16.8):

$$\sum_{n\in\mathbf{N}} \sin n \cdot = \frac{1}{4} \sum_{\pm} \cot\Big(\frac{\cdot}{2} \pm i0\Big).$$

In particular, show for x near 0 that

$$\sum_{n\in\mathbf{N}} \sin n\, x = \frac{\cos \frac{x}{2}}{\operatorname{sinc} \frac{x}{2}} \Big(\mathrm{PV}\, \frac{1}{x}\Big).$$

(vi) Compute $\mathcal{F} \cot(\cdot - i0)$ and $\mathcal{F} \cot^2(\cdot - i0)$.

Now define

$$s(x) = \sum_{n\in\mathbf{N}} \frac{\cos nx}{1+n^2} \qquad (x \in \mathbf{R}).$$

(vii) Show that

$$s(x) = \frac{1}{2}\Big(\pi \frac{\cosh(x-\pi)}{\sinh \pi} - 1\Big) \qquad (0 \le x \le 2\pi).$$

To obtain this result, note that Poisson's summation formula implies the following identities in $\mathcal{S}'(\mathbf{R})$:

$$s - s'' = \sum_{n\in\mathbf{N}} \cos n \cdot = -\frac{1}{2} + \pi \sum_{k\in\mathbf{Z}} \delta_{2\pi k}.$$

Now solve the differential equation for s using the fact that s is even and periodic. Conclude that (alternatively, apply (16.27) with $z = \pm i$ and add, or compare with [7, Exercise 6.91.(ii) and (iii)])

$$\cosh x = \frac{\sinh\pi}{\pi} + \frac{2\sinh\pi}{\pi}\sum_{n\in\mathbf{N}}\frac{(-1)^n\cos nx}{1+n^2} \qquad (-\pi \le x \le \pi),$$

in particular $$\sum_{n\in\mathbf{N}}\frac{1}{1+n^2} = x\frac{1}{2}(-1+\pi\coth\pi) = 1.076\,674\,047\,468\cdots.$$

16.18.* Suppose that f is a C^2 periodic function on $\mathbf{R}$ of period 2π. Show that the Fourier series of f converges absolutely and uniformly on $\mathbf{R}$ to f, and that in fact $S_n f - f = O(\frac{1}{n})$ uniformly as $n \to \infty$.

16.19. Integrate (16.31) over $[0,\pi]$ to obtain

$$\pi = \sum_{n\in\mathbf{Z}}\int_{n\pi}^{(n+1)\pi}\operatorname{sinc}^2 x\,dx = \int_{\mathbf{R}}\operatorname{sinc}^2 x\,dx.$$

Integration by parts yields, for any $a > 0$,

$$\int_0^a \operatorname{sinc}^2 x\,dx = -\frac{\sin^2 a}{a} + \int_0^{2a}\operatorname{sinc} x\,dx.$$

Deduce the convergence of the improper integral $\int_{\mathbf{R}}\operatorname{sinc} x\,dx$ as well as the evaluation $\int_{\mathbf{R}}\operatorname{sinc} x\,dx = \pi$ (see Problem 14.44 or 16.21 and [7, Example 2.10.14 or Exercises 6.60 and 8.19] for other proofs).

16.20.* Successively prove the following formulas:

(i) $$\frac{\pi}{\sqrt{2}} = 2 + 4\sum_{n\in\mathbf{N}}\frac{(-1)^{n-1}}{16n^2-1},$$

(ii) $$\frac{\pi}{2\cos\frac{\pi}{2}z} = 2\sum_{n\in\mathbf{N}}(-1)^n\frac{2n-1}{z^2-(2n-1)^2} \qquad \left(\frac{z-1}{2}\in\mathbf{C}\setminus\mathbf{Z}\right),$$

(iii) $$\frac{\pi}{4} = \sum_{n\in\mathbf{N}}\frac{(-1)^{n-1}}{2n-1},$$

(iv) $$\frac{\pi^2}{\cos^2\pi z} = 4\sum_{n\in\mathbf{Z}}\frac{1}{(2z-2n-1)^2} \qquad \left(\frac{z-1}{2}\in\mathbf{C}\setminus\mathbf{Z}\right),$$

(v) $$\pi\tan\pi z = 8z\sum_{n\in\mathbf{N}}\frac{1}{(2n-1)^2-4z^2} \qquad \left(\frac{z-1}{2}\in\mathbf{C}\setminus\mathbf{Z}\right),$$

(vi) $$\frac{d}{dz}\log\operatorname{sinc}\pi z = \pi\cot\pi z - \frac{1}{z} \qquad (z\in\mathbf{C}\setminus\mathbf{Z}),$$

(vii) $$\sin\pi z = \pi z\prod_{n\in\mathbf{N}}\left(1-\frac{z^2}{n^2}\right) \qquad (z\in\mathbf{C}),$$

(viii) $$\frac{1}{2} = \prod_{n=2}^{\infty}\left(1-\frac{1}{n^2}\right),$$

$$\text{(ix)} \qquad \frac{2}{\pi} = \prod_{n\in\mathbf{N}}\Bigl(1-\frac{1}{4n^2}\Bigr), \qquad \text{or} \qquad \lim_{n\to\infty}\frac{(2^n n!)^2}{(2n)!\,\sqrt{2n+1}} = \sqrt{\frac{\pi}{2}},$$

$$\text{(x)} \quad \cos\pi z = \prod_{n\in\mathbf{N}}\Bigl(1-\frac{4z^2}{(2n-1)^2}\Bigr) \qquad (z\in\mathbf{C}).$$

Note *Leibniz's series* in (iii) and *Wallis's product* in (ix).

16.21.* Consider arbitrary $z \in \mathbf{C}\setminus\mathbf{Z}$ and let $f_z : \mathbf{R} \to \mathbf{C}$ be the periodic function of period 2π that on $]\,0, 2\pi\,[$ is given by $f_z(x) = e^{-izx}$. Interpreting, if necessary, the sums over $\mathbf{Z}$ as limits of the symmetric partial sums, prove the following identity of functions on $\mathbf{R}\setminus 2\pi\mathbf{Z}$:

$$f_z = e^{-i\pi z}\,\frac{\sin\pi z}{\pi}\sum_{n\in\mathbf{Z}}\frac{e_{in}}{z+n}; \qquad \text{or} \qquad \frac{\pi\, e^{iz(\pi-x)}}{\sin\pi z} = \sum_{n\in\mathbf{Z}}\frac{e^{inx}}{z+n},$$

for $0 < x < 2\pi$. For $0 < x < 2\pi$, deduce that

$$\text{(i)} \qquad \frac{\pi\, e^{\pm i\pi z}}{\sin\pi z} = \sum_{n\in\mathbf{Z}}\frac{e^{\pm i(z+n)x}}{z+n},$$

$$\text{(ii)} \qquad \pi\cot\pi z = \sum_{n\in\mathbf{Z}}\frac{\cos(z+n)x}{z+n},$$

$$\text{(iii)} \qquad \pi = \sum_{n\in\mathbf{Z}}\frac{\sin(z+n)x}{z+n},$$

$$\text{(iv)} \qquad 1 = \sum_{n\in\mathbf{Z}}\operatorname{sinc}(z+\pi n) \qquad (z\in\mathbf{C}),$$

$$\text{(v)} \qquad \frac{\cos z(\pi-x)}{\operatorname{sinc}\pi z} - 1 = 2z^2\sum_{n\in\mathbf{N}}\frac{\cos nx}{z^2-n^2},$$

$$\text{(vi)} \qquad -\frac{\sin z(\pi-x)}{\operatorname{sinc}\pi z} = 2z\sum_{n\in\mathbf{N}}\frac{n\,\sin nx}{z^2-n^2}.$$

Observe that the identities (i) and (iii) are not valid for $x = 0$. On the other hand, (ii) with $x = 0$ leads to the partial-fraction decomposition of the cotangent in (16.29); hence, (ii) is true in this case. Setting $x = \pi$ in (iii) gives the partial-fraction decomposition of the cosecant in (16.28).

Use identity (iv) to prove $\int_{\mathbf{R}}\operatorname{sinc} x\,dx = \pi$. To this end, use the method from Problem 16.19 and estimates similar to those in Example 16.24 to justify the interchange of integration and summation. (See Problem 14.44 or 16.19 and [7, Example 2.10.14 or Exercises 6.60 and 8.19] for other proofs.)

16.22.* Consider arbitrary $z \in \mathbf{C}$ and let $g_z : \mathbf{R} \to \mathbf{C}$ be the periodic function of period 2π that on $[-\pi, \pi\,]$ is given by $g_z(x) = \cos z\,x$.

(i) Prove the following identity in $\mathcal{S}'(\mathbf{R})$:

$$g_z'' + z^2 g_z = 2z \sin \pi z \sum_{k\in\mathbf{Z}} \delta_{(2k+1)\pi}.$$

Verify that we have the following distributional Fourier series, where a term with $n = z$ should be read as $\frac{1}{2}e_{in}$:

$$g_z = \frac{z \sin \pi z}{\pi} \sum_{n\in\mathbf{Z}} \frac{(-1)^n}{z^2 - n^2} e_{in}; \qquad \text{show} \qquad 2\pi \sum_{k\in\mathbf{Z}} \delta_{(2k+1)\pi} = \sum_{n\in\mathbf{Z}} (-1)^n e_{in}.$$

Derive the latter identity also directly from Poisson's summation formula.

(ii) Define $f : \mathbf{R} \to \mathbf{R}$ by $f(x) = \arctan \circ \tan(\frac{x}{2})$ if $x \in \mathbf{R} \setminus (\pi + 2\pi\mathbf{Z})$ and $f(x) = 0$, for the remaining x. Verify that f is an odd sawtooth function, more precisely, that $f(x) = \frac{x}{2} - \pi\,[\frac{x+\pi}{2\pi}] = \frac{1}{2i} \log e^{ix}$, for all $x \in \mathbf{R}$, if log is defined on $\mathbf{C}$ by its principal value (compare with Example 4.2). (Note the slight abuse of notation: the equality is not one of functions but of distributions, because the functions differ on $\pi + 2\pi\mathbf{Z}$.) Furthermore, show that

$$f' = \frac{1}{2} - \pi \sum_{k\in\mathbf{Z}} \delta_{(2k+1)\pi} = \frac{1}{2} - \frac{1}{2} \sum_{n\in\mathbf{Z}} (-1)^n e_{in} = \sum_{n\in\mathbf{N}} (-1)^{n+1} \cos n\, \cdot.$$

Verify by integration of the preceding identity that $f = \sum_{n\in\mathbf{N}} \frac{(-1)^{n+1}}{n} \sin n\, \cdot$.

(iii) As another application of the Fourier series of g_z, deduce by taking $z = \frac{1}{2}$ that

$$|\cos| = \frac{2}{\pi} + \frac{4}{\pi} \sum_{n\in\mathbf{N}} \frac{(-1)^{n-1} \cos 2n\, \cdot}{4n^2 - 1}, \qquad |\sin| = \frac{2}{\pi} - \frac{4}{\pi} \sum_{n\in\mathbf{N}} \frac{\cos 2n\, \cdot}{4n^2 - 1}.$$

Derive

$$\sum_{n\in\mathbf{N}} \frac{(-1)^{n-1}}{4n^2 - 1} = \frac{\pi}{4} - \frac{1}{2}, \qquad \sum_{n\in\mathbf{N}} \frac{1}{4n^2 - 1} = \frac{1}{2}, \qquad \sum_{n\in\mathbf{N}} \frac{1}{(4n-2)^2 - 1} = \frac{\pi}{8},$$

$$|\cos| = \frac{4}{\pi} - 1 + \frac{8}{\pi} \sum_{n\in\mathbf{N}} \frac{(-1)^{n-1} \cos^2 n\, \cdot}{4n^2 - 1}, \qquad |\sin| = \frac{8}{\pi} \sum_{n\in\mathbf{N}} \frac{\sin^2 n\, \cdot}{4n^2 - 1}.$$

Since $\frac{1}{4n^2-1} = \frac{1}{2}(\frac{1}{2n-1} - \frac{1}{2n+1})$, one recognizes the first and third numerical series as simple modifications of Leibniz's series from Problem 16.1 and sees that the second one is telescoping. Finally, prove that

$$\sum_{k\in\mathbf{Z}} T_{k\pi}\big(1_{]0,\pi[} \cos\big) = \frac{8}{\pi} \sum_{n\in\mathbf{N}} \frac{n \sin 2n\, \cdot}{4n^2 - 1},$$

where the left-hand side is defined to attain the value 0 on $\pi\,\mathbf{Z}$. In other words, the Fourier series of the even periodic extension to $\mathbf{R}$ of $1_{[0,\pi]} \sin$ and the odd extension of $1_{]0,\pi[} \cos$ have been obtained; see Fig. 16.8.

16.23. *(Gamma distribution, Lipschitz formula, and Eisenstein series.)* Let $\alpha, \lambda > 0$. In statistics, the function $f_{\alpha,\lambda} : \mathbf{R} \to \mathbf{R}$ with

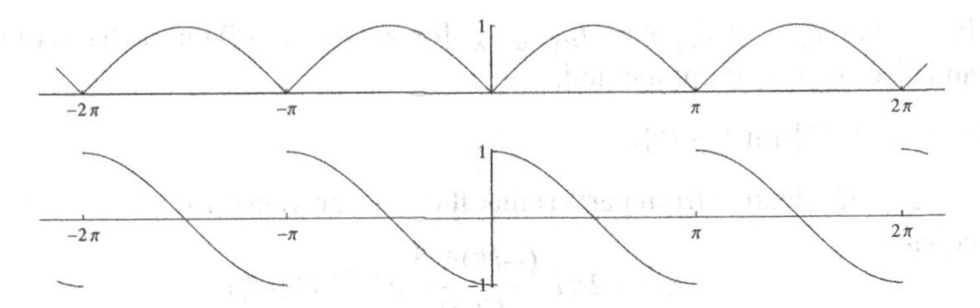

Fig. 16.8 Illustration of Problem 16.22.(iii). Graphs of the even extension of sin and the odd extension of cos

$$f_{\alpha,\lambda}(x) = \lambda^\alpha \, \chi_+^\alpha(x) \, e^{-\lambda x}$$

is said to be the *probability density of the Gamma distribution* of order α and parameter λ. In particular, for $n \in \mathbf{N}$, the function

$$f_{\frac{n}{2},\frac{1}{2}}(x) = \frac{1}{2^{\frac{n}{2}} \Gamma(\frac{n}{2})} x^{\frac{n}{2}-1} e^{-\frac{1}{2}x} \, H(x)$$

is called the *probability density of Pearson's χ^2 distribution with n degrees of freedom.*

(i) Using Problem 14.39 or by expansion of $e^{-ix\xi}$ into a power series and using [7, Exercise 0.11.(iii)] for the binomial series of $(1 + \frac{i\xi}{\lambda})^{-\alpha}$, prove that

$$\mathcal{F} f_{\alpha,\lambda}(\xi) = \left(\frac{\lambda}{\lambda + i\xi}\right)^\alpha \qquad (\xi \in \mathbf{R}).$$

For $k \in \mathbf{Z}_{\geq 0}$, introduce the *shifted factorial*, or *Pochhammer symbol*,

$$(\alpha)_k = \frac{\Gamma(\alpha + k)}{\Gamma(\alpha)} = \alpha(\alpha+1)\cdots(\alpha+k-1).$$

(ii) Show that $x^k f_{\alpha,\lambda}(x) = \frac{(\alpha)_k}{\lambda^k} f_{\alpha+k,\lambda}(x)$. Use this identity or (14.28) in order to deduce from part (i) that

$$\mathcal{F}(x \mapsto x^k f_{\alpha,\lambda}(x))(\xi) = \frac{(\alpha)_k \, \lambda^\alpha}{(\lambda + i\xi)^{\alpha+k}}, \qquad \int_{\mathbf{R}_{>0}} x^k f_{\alpha,\lambda}(x) \, dx = \frac{(\alpha)_k}{\lambda^k}.$$

The latter integral is called the kth *moment* of the function $f_{\alpha,\lambda}$. Conclude that

$$\mu := \int_{\mathbf{R}_{>0}} x \, f_{\alpha,\lambda}(x) \, dx = \frac{\alpha}{\lambda}, \qquad \sigma^2 := \int_{\mathbf{R}_{>0}} (x-\mu)^2 f_{\alpha,\lambda}(x) \, dx = \frac{\alpha}{\lambda^2}.$$

In statistical terms, μ is the *expectation* and σ^2 the *variance* of the Gamma distribution.

(iii) Prove that $f_{\alpha_1,\lambda} * f_{\alpha_2,\lambda} = f_{\alpha_1+\alpha_2,\lambda}$, for $\alpha_1, \alpha_2, \lambda > 0$ on the basis of part (i) and also by a different method.

Set $H = \{ z \in \mathbf{C} \mid \operatorname{Im} z > 0 \}$.

(iv) For $z \in H$, deduce from part (i) that the Fourier transform of $x \mapsto (x-z)^{-\alpha}$ equals

$$\xi \mapsto 2\pi i \, \frac{(-i\xi)^{\alpha-1}}{\Gamma(\alpha)} e^{-iz\xi} \, H(-\xi).$$

(v) Now use Poisson's summation formula to derive from part (iv) *Lipschitz's formula*:

$$\sum_{n\in\mathbf{Z}} \frac{1}{(z+n)^k} = \frac{(-2\pi i)^k}{(k-1)!} \sum_{n\in\mathbf{N}} n^{k-1} e^{2\pi i n z} \qquad (z \in H,\ k \in \mathbf{Z}_{\geq 2}).$$

(vi) On account of (16.29) we have the following identity, where $'$ denotes symmetrical summation:

$$\sum_{n\in\mathbf{Z}}{}' \frac{1}{z+n} = \pi \cot \pi z = -\pi i - 2\pi i \, \frac{e^{2\pi i z}}{1-e^{2\pi i z}} \qquad (z \in H).$$

Lipschitz's formula also can be obtained by expanding the right-hand side of the identity as a geometric series in $e^{2\pi i z}$ (this is where the condition $z \in H$ is necessary) and differentiating $k-1$ times with respect to z. Conversely, the partial-fraction decomposition of the cotangent can be derived from Lipschitz's formula.

Let $\zeta(2k)$, for $k \in \mathbf{N}$, be as in (16.35). The *Eisenstein series* G_k *of index* $k > 1$ is the function $G_k : H \to \mathbf{C}$ given by

$$G_k(z) = \sum_{(0,0)\neq(m,n)\in\mathbf{Z}\times\mathbf{Z}} \frac{1}{(mz+n)^{2k}} = 2\zeta(2k) + 2 \sum_{m\in\mathbf{N}} \sum_{n\in\mathbf{Z}} \frac{1}{(mz+n)^{2k}}.$$

(vii) Apply Lipschitz's formula with z replaced by mz to get the so-called Fourier expansion of G_k at infinity

$$\begin{aligned} G_k(z) &= 2\zeta(2k) + \frac{2(-2\pi i)^{2k}}{(2k-1)!} \sum_{m\in\mathbf{N}} \sum_{n\in\mathbf{N}} n^{2k-1} e^{2\pi i m n z} \\ &= 2\zeta(2k) + \frac{2(2\pi i)^{2k}}{(2k-1)!} \sum_{j\in\mathbf{N}} \sigma_{2k-1}(j) e^{2\pi i j z} \\ &= 2\zeta(2k)\Big(1 - \frac{4k}{B_{2k}} \sum_{j\in\mathbf{N}} \sigma_{2k-1}(j) e^{2\pi i j z}\Big). \end{aligned}$$

Here the coefficient $\sigma_{2k-1}(j)$ denotes the sum of the $(2k-1)$th powers of positive divisors of j.

Chapter 17
Fundamental Solutions and Fourier Transform

The Fourier transform in $\mathcal{S}'(\mathbf{R}^n)$ is a very useful tool in the analysis of linear partial differential operators

$$P(D) = \sum_{|\alpha| \leq m} c_\alpha \, D^\alpha$$

in $\mathbf{R}^n$ with constant coefficients $c_\alpha \in \mathbf{C}$. If $u \in \mathcal{S}'(\mathbf{R}^n)$, then (14.10) implies

$$\mathcal{F}(P(D)u) = P(\xi)\, \mathcal{F}u. \tag{17.1}$$

Here

$$P(\xi) = \sum_{|\alpha| \leq m} c_\alpha \, \xi^\alpha \tag{17.2}$$

is the polynomial in ξ obtained by substituting ξ_j for D_j everywhere. $P(\xi)$ is said to be the *symbol* of $P(D)$. In other words, on the Fourier transform side the differential operator $P(D)$ changes over into multiplication by the polynomial $P(\xi)$. Clearly, this is the expansion in eigenvectors that we discussed in the introduction to Chap. 14.

Remark 17.1. A distribution $u \in \mathcal{S}'(\mathbf{R}^n)$ is a solution of the equation $P(D)u = 0$ if and only if $\mathcal{F}u \in \mathcal{S}'(\mathbf{R}^n)$ satisfies $P(\xi)\, \mathcal{F}u = 0$. If $P(\xi) \neq 0$, for all $\xi \in \mathbf{R}^n$, this leads to $\mathcal{F}u = 0$ and therefore $u = 0$.

In general, one has

$$\operatorname{supp} \mathcal{F}u \subset N := \{\, \xi \in \mathbf{R}^n \mid P(\xi) = 0 \,\}.$$

If the total derivative $DP(\xi)$ is nonzero for all $\xi \in N$, then N is a C^∞ (even analytic) submanifold in $\mathbf{R}^n$ of dimension $n - 1$ according to the Submersion Theorem. In this case $\mathcal{F}u$ is given by a "distribution on N"; but for a proper formulation we have to know what distributions on submanifolds are. This therefore represents a natural occasion to introduce these. An example of an entirely different nature was given by the distributions on the torus, in connection with Fourier series; see Remark 16.12. $\oslash$

J.J. Duistermaat and J.A.C. Kolk, *Distributions: Theory and Applications*,
Cornerstones, DOI 10.1007/978-0-8176-4675-2_17,

Suppose that $E \in \mathcal{D}'(\mathbf{R}^n)$ is a fundamental solution of $P(D)$ and, in addition, that $E \in \mathcal{S}'(\mathbf{R}^n)$. Using (17.1), we conclude from (14.30) that

$$1 = \mathcal{F}(P(D)E) = P(\xi)\,\mathcal{F}E. \tag{17.3}$$

Conversely, if $Q \in \mathcal{S}'$ is a solution of the equation $P(\xi)\,Q = 1$, then $E := \mathcal{F}^{-1}Q \in \mathcal{S}'$ is a fundamental solution of $P(D)$. Indeed, in that case one has $\mathcal{F}(P(D)E) = P(\xi)\,Q = 1$; therefore $P(D)E = \delta$. It goes without saying that this result makes use of Theorem 14.24.

Equation (17.3) implies that on the complement $C = \{\,\xi \in \mathbf{R}^n \mid P(\xi) \neq 0\,\}$ in $\mathbf{R}^n$ of the zero-set N of the polynomial function P, the distribution $Q = \mathcal{F}E$ equals the C^∞ function $1/P$. In particular, a possible tempered fundamental solution of $P(D)$ is uniquely determined if P does not have real zeros. Even in this case it is not obvious, a priori, whether $1/P$ is a tempered distribution. This might lead to problems in cases in which $P(\xi)$ were to converge to 0 too fast in too large subsets of $\mathbf{R}^n$, as $\|\xi\| \to \infty$. At the present stage we do not have sufficient background on polynomials in n variables to reach general conclusions on this point. Instead, we now concentrate on a subclass of operators.

Definition 17.2. The operator $P(D)$ of order m is said to be *elliptic* if the homogeneous part

$$P_m(\xi) := \sum_{|\alpha|=m} c_\alpha\, \xi^\alpha$$

of degree m of the symbol $P(\xi)$ of $P(D)$ does not have any real zeros ξ except $\xi = 0$. That is,

$$\xi \in \mathbf{R}^n \quad \text{and} \quad \xi \neq 0 \qquad \Longrightarrow \qquad P_m(\xi) \neq 0.$$

$P_m(\xi)$ is said to be the *principal symbol* of $P(D)$. ⊘

Example 17.3. The Laplace operator Δ is elliptic. The heat operator $\partial_t - \Delta_x$ and the wave operator $\partial_t^2 - \Delta_x$ are not elliptic. ⊘

Note that the condition of ellipticity applies only to the highest-order part $P_m(D)$ of $P(D)$. Thus, if $P(D)$ is elliptic, the operator $P(D) + Q(D)$ is elliptic for every $Q(D)$ of order lower than m. The condition of ellipticity becomes effective with the estimate in the following lemma.

Lemma 17.4. *The differential operator $P(D)$ in $\mathbf{R}^n$ is elliptic of order m if and only if there exist constants $c > 0$ and $R \geq 0$ such that*

$$|P(\xi)| \geq c\,\|\xi\|^m \qquad (\xi \in \mathbf{R}^n,\ \|\xi\| \geq R). \tag{17.4}$$

Proof. Let $P(D)$ be elliptic. The continuous function $|P_m|$ attains its minimum on the compact subset $S = \{\,\xi \in \mathbf{R}^n \mid \|\xi\| = 1\,\}$ of $\mathbf{R}^n$. That is, there exists an

$\eta \in S$ such that $|P_m(\xi)| \geq |P_m(\eta)|$, for all $\xi \in S$. Now $\eta \neq 0$, and therefore $\mu := |P_m(\eta)| > 0$.

For arbitrary $\xi \in \mathbf{R}^n \setminus \{0\}$ one has $\|\xi\|^{-1} \xi \in S$, and therefore

$$\mu \leq \left|P_m\Big(\frac{1}{\|\xi\|}\,\xi\Big)\right| = \|\xi\|^{-m}\,|P_m(\xi)|.$$

Furthermore, $Q(\xi) = P(\xi) - P_m(\xi)$ is a polynomial of degree $\leq m-1$, which implies the existence of a constant $d > 0$ such that

$$|Q(\xi)| \leq d\,\|\xi\|^{m-1} \qquad (\|\xi\| \geq 1).$$

Combining the estimates, we obtain, for $\|\xi\| \geq 1$,

$$|P(\xi)| \geq |P_m(\xi)| - |Q(\xi)| \geq \mu\,\|\xi\|^m - d\,\|\xi\|^{m-1} = \Big(\mu - \frac{d}{\|\xi\|}\Big)\,\|\xi\|^m.$$

This yields the desired estimate, with $c = \mu - d/R$, whenever $R > d/\mu$ and $R \geq 1$.

Conversely, if $P(D)$ is not elliptic, there exists $\xi \in \mathbf{R}^n$ such that $\xi \neq 0$ and $P_m(\xi) = 0$. This implies $P_m(t\,\xi) = t^m\,P_m(\xi) = 0$, and consequently $P(t\,\xi) = P_m(t\,\xi) + Q(t\,\xi) = Q(t\,\xi)$ is a polynomial in t of degree $< m$. This is in contradiction to (17.4). □

The estimate (17.4) implies that for any $c \in \mathbf{R}_{\geq 0}$, the set of the $\xi \in \mathbf{R}^n$ with $|P(\xi)| \leq c$ is bounded.

Lemma 17.5. *If $P(D)$ is elliptic, it has a parametrix E (see (12.4)) satisfying* sing supp $E \subset \{0\}$.

Proof. Let R and $c > 0$, as in (17.4). Choose $\chi \in C_0^\infty(\mathbf{R}^n)$ with $\chi(\xi) = 1$ for $\|\xi\| \leq R$. The function $1/P$ is C^∞ on a suitable open neighborhood U of supp $(1-\chi)$. Therefore $v \in C^\infty(\mathbf{R}^n)$ if

$$v(\xi) := \begin{cases} \dfrac{1-\chi(\xi)}{P(\xi)} & \text{if } \ \xi \in U, \\ 0 & \text{on the interior of the set where } \chi = 1. \end{cases}$$

Furthermore, $v(\xi) = 1/P(\xi)$ if $\xi \notin$ supp χ, that is, if $\|\xi\|$ is sufficiently large. On account of (17.4), one there has $|v(\xi)| \leq \frac{1}{c}\,\|\xi\|^{-m}$, if $P(D)$ is assumed to be of order m. It follows that v is certainly bounded and defines a tempered distribution on $\mathbf{R}^n$.

If we now choose $E := \mathcal{F}^{-1}v$, then $\mathcal{F}(P(D)E) = P(\xi)\mathcal{F}E = P(\xi)v = 1-\chi$. Applying $\mathcal{F}^{-1}$, we obtain $P(D)E = \delta - \mathcal{F}^{-1}\chi$. Now $\mathcal{F}^{-1}\chi = (2\pi)^{-n}\,S \circ \mathcal{F}\chi$ is a C^∞ function on $\mathbf{R}^n$, even with a complex-analytic extension to $\mathbf{C}^n$; it follows that E is a parametrix of P.

To prove that sing supp $E \subset \{0\}$, we begin by deriving an auxiliary estimate, guided by the knowledge that the way in which v decreases at infinity improves under differentiation. We observe that by mathematical induction on l,

$$D_j{}^l\Big(\frac{1}{P}\Big)(\xi) = \frac{Q_l(\xi)}{P(\xi)^{l+1}},$$

if $P(\xi) \neq 0$. Here $Q_0 = 1$, and $Q_l = P\, D_j Q_{l-1} - l\, Q_{l-1}\, D_j P$ is a polynomial function of degree $\leq l(m-1)$. Because $l(m-1) - (1+l)m = -l - m$, this yields, in combination with (17.4),

$$\xi^\alpha\, D_j{}^l v = O\big(\|\xi\|^{|\alpha|-m-l}\big), \qquad \|\xi\| \to \infty.$$

For given $k \in \mathbf{N}$ we choose $l > n+k-m$, or $k-m-l < -n$; then $\xi^\alpha D_j{}^l v \in L^1$, for all multi-indices α with $|\alpha| \leq k$. But this means that we may differentiate

$$(-x_j)^l\, E(x) = \mathcal{F}^{-1}(D_j{}^l v)(x) = \frac{1}{(2\pi)^n} \int_{\mathbf{R}^n} e^{i\langle x,\xi\rangle}\, (D_j{}^l v)(\xi)\, d\xi$$

to order k under the integral sign. In other words, $x_j{}^l\, E$ is of class C^k, that is, $E \in C^k$ on the complement C_j of the hyperplane $\{\, x \in \mathbf{R}^n \mid x_j = 0\,\}$. Because this holds for all k, we have $E \in C^\infty$ on C_j, and because this in turn is true for all $1 \leq j \leq n$, it follows that $E \in C^\infty$ on $\bigcup_{j=1}^n C_j = \mathbf{R}^n \setminus \{0\}$. □

From the proof it will be clear that we could have chosen $E = \mathcal{F}^{-1}(1/P)$ if P had no zeros in $\mathbf{R}^n$. This applies, for example, if $P(D) = \Delta - \lambda\, I$, with $\lambda \in \mathbf{C}$ and not $\lambda \leq 0$. In that case, E also is a fundamental solution of $P(D)$, with the additional properties that $E \in \mathcal{S}'$ and sing supp $E \subset \{0\}$.

We obtain the following theorem by combining Lemma 17.5 and Theorem 12.4.

Theorem 17.6. *Every elliptic operator $P(D)$ is hypoelliptic.*

This theorem is a classical regularity result; as a consequence, operators having the regularity property were called hypoelliptic.

Remark 17.7. For the parametrix E in the proof of Lemma 17.5, $P(D)E - \delta$ equals the Fourier transform of a function with compact support, and is therefore analytic. By replacing the integration over $\mathbf{R}^n$ in the formula for E by an integral over a suitable submanifold in $\mathbf{C}^n$, which, on the strength of Cauchy's Integral Theorem (see (12.9)), leaves the result unchanged, one also proves that E is analytic on $\mathbf{R}^n \setminus \{0\}$. Thus, it follows from Theorem 12.15 that for every elliptic operator $P(D)$, the distribution u is analytic wherever $P(D)u = 0$. ⊘

Remark 17.8. An elliptic operator $P(D)$ with variable (C^∞) coefficients has a parametrix K as discussed in Remark 12.6. Its construction is a generalization of the proof of Lemma 17.5, and the result is a so-called *pseudo-differential operator* K. This has a kernel k (see Theorem 15.2 below) that is smooth off the diagonal, and the conclusion is that elliptic operators with C^∞ coefficients, too, are hypoelliptic. This theory can be found in Hörmander [11, Chap. 18], for example. ⊘

Example 17.9. We will prove the principle that every derivative can be majorized by a suitable power of the Laplacian. More specifically, suppose that ϕ belongs to

$C^\infty(\mathbf{R}^n)$ and $\Delta^j\phi$ is of polynomial growth at infinity, for $0 \le j \le k$, with $2k > n$. Then $\partial^\alpha\phi$ also is at most of polynomial growth at infinity, for all multi-indices α with $|\alpha| \le l := 2k - n - 1 \in \mathbf{Z}_{\ge 0}$.

Indeed, define $f_k = (1 + \|\cdot\|^2)^{-k} \in C^\infty(\mathbf{R}^n)$. Then $f_k \in L^1(\mathbf{R}^n)$. According to the Riemann–Lebesgue Theorem, Theorem 14.2, the fundamental solution

$$E_k := \mathcal{F}^{-1} f_k = \frac{1}{(2\pi)^n}\mathcal{F} f_k \tag{17.5}$$

of the differential operator $(I - \Delta)^k$ is a continuous function vanishing at infinity. Furthermore, Theorem 17.6 below and Theorem 12.4 imply that $\operatorname{sing\,supp} E_k = \{0\}$. Now $|\alpha| \le l$ leads to $|\alpha| - 2k \le -n - 1$, which gives the convergence of $\int_{\mathbf{R}^n} |\xi^\alpha f_k(\xi)|\,d\xi$. In turn, this implies that E_k is of class C^l and that

$$D^\alpha E_k(x) = \frac{1}{(2\pi)^n}\int_{\mathbf{R}^n} e^{-i\langle x,\xi\rangle}(-\xi)^\alpha f_k(\xi)\,d\xi.$$

Furthermore, $D^\alpha E_k$ is rapidly decreasing, as one sees using integration by parts, because all derivatives of $\xi \mapsto \xi^\alpha f_k(\xi)$ are linear combinations of functions $\xi \mapsto \xi^\beta D^\gamma f_k(\xi)$ with $|\beta| + |\gamma| \le |\alpha|$. As in the proof of Lemma 17.5 one obtains that the latter functions are $O(\|\xi\|^{|\alpha|-2k}) = O(\|\xi\|^{-n-1})$ as $\|\xi\| \to \infty$. Finally,

$$\partial^\alpha\phi = (I - \Delta)^k E_k * \partial^\alpha\phi = \partial^\alpha E_k * (I - \Delta)^k\phi,$$

where the convolution on the right-hand side is given by the usual integral. Furthermore, this convolution is at most of polynomial growth.

In the (minimal) case of $2k = n + 1$ (assuming for the moment n to be odd), Problem 18.8 below shows that $E_k = c\,e^{-\|\cdot\|}$, for some constant $c > 0$. This means that this E_k is actually of exponential decay at infinity. By applying the Fourier transform to the following identity, one sees that, for $2k > n$,

$$(I - \Delta)^{\frac{2k-n-1}{2}} E_k = c\,e^{-\|\cdot\|}.$$

This makes it plausible that all such E_k decay exponentially fast; for a proof see Sect. 17.1 below. ⊘

As another application of the preceding technique, in Example 17.9 we give a global characterization of tempered distributions; compare with Theorem 13.1 and Example 18.2.

Theorem 17.10. *A distribution in $\mathcal{D}'(\mathbf{R}^n)$ belongs to $\mathcal{S}'(\mathbf{R}^n)$ if and only if it is a finite sum of derivatives of a function in $C(\mathbf{R}^n)$ that is at most of polynomial growth at infinity.*

Proof. Suppose that $u \in \mathcal{S}'(\mathbf{R}^n)$. Then there exist $c > 0$ and $k, N \in \mathbf{Z}_{\ge 0}$ such that for all $\phi \in \mathcal{S}(\mathbf{R}^n)$,

$$|u(\phi)| \leq c\, \|\phi\|_{\mathcal{S}(k,N)} = c \sup_{|\alpha|\leq k,\, |\beta|\leq N,\, x\in \mathbf{R}^n} |x^\beta\, \partial^\alpha \phi(x)|. \tag{17.6}$$

Select $l > \frac{n}{2} + k$ and consider the fundamental solution $E_l \in \mathcal{S}'(\mathbf{R}^n)$ of $(I - \Delta)^l$ as in (17.5). Then E_l belongs to $C^k(\mathbf{R}^n)$, with all of its derivatives rapidly decreasing on the strength of Example 17.9.

We may define the function f on $\mathbf{R}^n$ by (compare with (11.1))

$$f = u * E_l \qquad \text{with} \qquad u * E_l(x) = u(T_x E_l), \tag{17.7}$$

because u is of order $\leq k$ and E_l possesses the properties established above. Since $x \mapsto T_x E_l$ is a continuous linear mapping with respect to the $\mathcal{S}(k, N)$ norm, the function f belongs to $C(\mathbf{R}^n)$. Furthermore, to prove that f is at most of polynomial growth, apply (17.7) and (17.6). In addition, use that all derivatives of E_l of order at most k are rapidly decreasing. As a consequence, $f \in \mathcal{S}'(\mathbf{R}^n)$.

On the other hand, let $\phi \in \mathcal{S}(\mathbf{R}^n)$ be arbitrary. Since E_l is absolutely integrable on $\mathbf{R}^n$, the proof of Lemma 2.18 can be modified to give $E_l * \phi \in C^\infty(\mathbf{R}^n)$. Furthermore, as in the proof of Theorem 14.16, we see that $E_l * \phi$ is rapidly decreasing. It follows that $E_l * \phi \in \mathcal{S}(\mathbf{R}^n)$. (This conclusion is also suggested by the fact that $\mathcal{F} E_l\, \mathcal{F}\phi = (1 + \|\cdot\|^2)^{-l} \mathcal{F}\phi$ belongs to $\mathcal{S}(\mathbf{R}^n)$. Actually, for every $l \in \mathbf{Z}$, the mapping $\phi \mapsto (1 + \|\cdot\|^2)^{-l}\phi$ defines a linear isomorphism from $\mathcal{S}(\mathbf{R}^n)$ onto itself, because $(1 + \|\cdot\|^2)^{-l} \in C^\infty(\mathbf{R}^n)$, while all of its derivatives are at most of polynomial growth.) Hence $u(E_l * \phi)$ is well-defined.

A variation of the proof of Theorem 14.33 shows that the convolution in (17.7) is a continuous extension of convolution of distributions belonging to $\mathcal{E}'(\mathbf{R}^n)$ and $\mathcal{S}'(\mathbf{R}^n)$. With $\epsilon^*\chi$ as in Lemma 14.7, we have $\epsilon^*\chi\, u \in \mathcal{E}'(\mathbf{R}^n)$ and thus we obtain from Lemma 14.7 and (11.23), for all $\phi \in C_0^\infty(\mathbf{R}^n)$,

$$u(E_l * \phi) = \lim_{\epsilon\downarrow 0} \epsilon^*\chi\, u(E_l * \phi) = \lim_{\epsilon\downarrow 0}(\epsilon^*\chi\, u * E_l)(\phi) = u * E_l(\phi) = f(\phi).$$

In view of (11.4) this implies

$$\begin{aligned} u(\phi) &= u(\delta * \phi) = u((I - \Delta)^l E_l * \phi) = u(E_l * (I - \Delta)^l \phi) \\ &= (u * E_l)((I - \Delta)^l \phi) = (I - \Delta)^l f(\phi). \end{aligned}$$

On account of Lemma 14.7 we now obtain the identity $u = (I - \Delta)^l f$ in $\mathcal{S}'(\mathbf{R}^n)$. □

We return to the properties of elliptic operators.

Theorem 17.11. *Every elliptic operator $P(D)$ in $\mathbf{R}^n$ has a fundamental solution E in $\mathcal{D}'(\mathbf{R}^n)$.*

Proof. Suppose $P(D)$ is of order m and let $R > 0$ be as in Lemma 17.4. Choose $\eta \in \mathbf{C}^n$ such that $P_m(\eta) \neq 0$; for example, $\eta \in \mathbf{R}^n \setminus \{0\}$. For $z \in \mathbf{C}$ we write $P(\xi + z\eta)$ as a polynomial in z:

$$P(\xi + z\eta) = P_m(\eta)\, z^m + \sum_{l=0}^{m-1} R_l(\xi)\, z^l,$$

where the coefficients $R_l(\xi) = R_l(\xi, \eta)$ are polynomials in ξ. Introducing $c_l = \sup_{\|\xi\| \le R} |R_l(\xi)|$, we obtain the estimate

$$|P(\xi + z\eta)| \ge r^m \left(|P_m(\eta)| - \sum_{l=0}^{m-1} \frac{c_l}{r^{m-l}} \right) \qquad (|z| = r,\ \|\xi\| \le R).$$

Because the quantity within parentheses converges to $|P_m(\eta)| > 0$ as $r \to \infty$, we conclude that there exist $\epsilon > 0$ and $r > 0$ such that

$$|P(\xi + z\eta)| \ge \epsilon \qquad (z \in \mathbf{C},\ |z| = r,\ \xi \in \mathbf{R}^n,\ \|\xi\| \le R).$$

Write

$$B = \{\, \xi \in \mathbf{R}^n \mid \|\xi\| \le R \,\} \qquad \text{and} \qquad C = \mathbf{R}^n \setminus B.$$

We now define $E = F + G : \mathbf{R}^n \to \mathbf{C}$, with $F = \mathcal{F}^{-1}(1_C/P)$ and G the analytic function given by

$$G(x) = \frac{1}{(2\pi)^n} \int_B \frac{1}{2\pi i} \int_{|z|=r} \frac{e^{i\langle x, \xi + z\eta\rangle}}{P(\xi + z\eta)}\, \frac{1}{z}\, dz\, d\xi.$$

Here the integral over z is the complex line integral, whose value can be regarded as the sum of all residues in the disk $\{\, z \in \mathbf{C} \mid |z| < r \,\}$ of the meromorphic function $z \mapsto e^{i\langle x,\xi+z\eta\rangle}/(z\, P(\xi + z\eta))$.

This yields $P(D)F = \mathcal{F}^{-1} 1_C$, while

$$\begin{aligned} P(D)G(x) &= \frac{1}{(2\pi)^n} \int_B \frac{1}{2\pi i} \int_{|z|=r} e^{i\langle x,\xi+z\eta\rangle}\, \frac{1}{z}\, dz\, d\xi \\ &= \frac{1}{(2\pi)^n} \int_B e^{i\langle x,\xi\rangle} d\xi = \mathcal{F}^{-1} 1_B(x), \end{aligned}$$

where in deriving the latter identity we have used Cauchy's integral formula (12.11).

The conclusion is $P(D)E = \mathcal{F}^{-1} 1_C + \mathcal{F}^{-1} 1_B = \mathcal{F}^{-1} 1 = \delta$. □

Note that sing supp $E \subset \{0\}$ for every fundamental solution of an elliptic operator; this follows from Theorem 17.6.

Remark 17.12. The *Theorem of Ehrenpreis–Malgrange* asserts the existence of a fundamental solution in $\mathcal{D}'(\mathbf{R}^n)$ for an arbitrary linear partial differential operator $P(D) \ne 0$ with constant coefficients. The construction given above is a simple variant of Hörmander's proof [12, Thm. 7.3.10] of that theorem. The simplification is made possible by the ellipticity of $P(D)$, which ensures that no problems arise due to $P(\xi)$ becoming small for large $\|\xi\|$.

By combining the Fourier transform with deep results from algebraic geometry, Bernstein and Gel'fand [2] and Atiyah [1] were able to prove that every $P(D)$

as above has a fundamental solution belonging to $\mathcal{S}'(\mathbf{R}^n)$. However, in the nonelliptic case these solutions often have local properties less favorable than those of Hörmander.

Remarkably enough, we now are prepared enough to establish the Theorem of Ehrenpreis–Malgrange in full generality. They proved the result in [8] and [16], respectively, by an approach different from the one below; to be more specific, by application of the Hahn–Banach Theorem, Theorem 8.11. The proof presented here is extremely efficient as well as explicit, but at this stage of the theory it might be difficult to see how to find it yourself. For a more "natural" verification, see the proof of Theorem 18.4 below. ⊘

Theorem 17.13. *Every linear partial differential operator $P(D)$ in $\mathbf{R}^n$ with constant coefficients (not all of them equal to 0) possesses a fundamental solution in $\mathcal{D}'(\mathbf{R}^n)$.*

More specifically, if $P(D)$ is of order m and the polynomial function P_m is the principal symbol as in Definition 17.2, there exists $\eta \in \mathbf{R}^n$ such that $P_m(\eta) \neq 0$. Then $E \in \mathcal{D}'(\mathbf{R}^n)$ defined by

$$E(x) = \frac{1}{2\pi i\, \overline{P_m(\eta)}} \int_C z^{m-1}\, e^{z\langle x,\eta\rangle}\, \mathcal{F}^{-1}\Big(\frac{\overline{P(\cdot + z\eta)}}{P(\cdot + z\eta)}\Big)(x)\, dz$$

is a distribution having the desired property. Here C denotes the unit circle in $\mathbf{C}$, and the integration with respect to z is the complex line integral over C.

Proof. The set of zeros of the nontrivial polynomial function P on $\mathbf{R}^n$ is an algebraic hypersurface in $\mathbf{R}^n$ and therefore a set of measure 0; this can be deduced from the structure theory in Whitney [25] or apply [13, Problem VI.10]. Given $z \in C$, this implies that $\xi \mapsto \frac{\overline{P(\xi+z\eta)}}{P(\xi+z\eta)}$ is an essentially bounded function on $\mathbf{R}^n$ and therefore defines an element of $\mathcal{S}'(\mathbf{R}^n)$. Furthermore, the distribution-valued mapping

$$C \to \mathcal{S}'(\mathbf{R}^n) \qquad \text{given by} \qquad z \mapsto \frac{\overline{P(\cdot + z\eta)}}{P(\cdot + z\eta)}$$

is continuous on account of Lebesgue's Dominated Convergence Theorem, Theorem 20.26.(iv), applied to $\mathbf{R}^n$. A continuous distribution-valued function may be integrated over a compact set; hence E is well-defined.

Successively using Problem 14.2, the identity $P(D) \circ \mathcal{F}^{-1} = \mathcal{F}^{-1} \circ P(\xi)$, and (14.33), one obtains with $D = D_x$,

$$\begin{aligned}
P(D)E(x) &= \frac{1}{2\pi i\, \overline{P_m(\eta)}} \int_C z^{m-1}\, P(D)\Big(e^{z\langle x,\eta\rangle}\, \mathcal{F}^{-1}\Big(\frac{\overline{P(\cdot + z\eta)}}{P(\cdot + z\eta)}\Big)(x)\Big)\, dz \\
&= \frac{1}{2\pi i\, \overline{P_m(\eta)}} \int_C z^{m-1} e^{z\langle x,\eta\rangle}\, P(D + z\eta)\, \mathcal{F}^{-1}\Big(\frac{\overline{P(\cdot + z\eta)}}{P(\cdot + z\eta)}\Big)(x)\, dz \\
&= \frac{1}{2\pi i\, \overline{P_m(\eta)}} \int_C z^{m-1} e^{z\langle x,\eta\rangle}\, \mathcal{F}^{-1}\overline{P(\cdot + z\eta)}(x)\, dz
\end{aligned}$$

$$= \frac{1}{2\pi i\,\overline{P_m(\eta)}} \int_C z^{m-1} e^{z\langle x,\eta\rangle}\, \overline{P(D+z\eta)}\,\delta\,dz,$$

with $\delta \in \mathcal{D}'(\mathbf{R}^n)$ the Dirac measure. Next, expansion of the polynomial $\overline{P(D+z\eta)}$ in the variable z and the equality $\overline{z} = \frac{1}{z}$ for $z \in C$ lead to

$$\overline{P(D+z\eta)} = \frac{1}{z^m}\overline{P_m(\eta)} + \sum_{j=0}^{m-1} \frac{1}{z^j} Q_j(D,\eta),$$

for certain polynomial functions Q_j. The integral of the leading term on the right-hand side yields δ on account of Cauchy's integral formula (12.11), while the integral of the sum vanishes by Cauchy's Integral Theorem in (12.9). □

17.1 Appendix: Fundamental Solution of $(I - \Delta)^k$

The fundamental solution E_k from (17.5), for $k > \frac{n}{2}$, deserves closer study on account of its useful properties; for instance, see Remark 18.8 below. The method used here is a vast generalization of the one employed in Problem 14.15 and uses a differential equation satisfied by E_k.

By means of a change of variables one sees that E_k is invariant under all rotations about the origin in $\mathbf{R}^n$. Special functions with this property often satisfy a differential equation of Bessel type; see (17.9) below.

In order to derive this equation, first rewrite E_k as a Fourier transform of a function of one variable. Note that $E_k(x) = E_k(\|x\|\,e_1)$, where e_1 is the first standard basis vector in $\mathbf{R}^n$. Now one has, writing $\xi = (\xi_1, \xi') \in \mathbf{R} \times \mathbf{R}^{n-1}$ and using the change of variables $\xi' = \sqrt{1+\xi_1^2}\,\eta$ in $\mathbf{R}^{n-1}$ as well as [7, Exercise 7.23],

$$\begin{aligned} E_k(x) &= \frac{1}{(2\pi)^n}\int_{\mathbf{R}^n} \frac{e^{-i\|x\|\xi_1}}{(1+\xi_1^2+\|\xi'\|^2)^k}\,d\xi \\ &= \frac{1}{(2\pi)^n}\int_{\mathbf{R}} \frac{e^{-i\|x\|\xi_1}}{(1+\xi_1^2)^{k-\frac{n-1}{2}}}\,d\xi_1 \int_{\mathbf{R}^{n-1}} \frac{1}{(1+\|\eta\|^2)^k}\,d\eta \\ &= \frac{\Gamma(k-\frac{n-1}{2})}{2^n\pi^{\frac{n+1}{2}}\,\Gamma(k)} \int_{\mathbf{R}} \frac{e^{-i\|x\|\xi}}{(1+\xi^2)^{k-\frac{n-1}{2}}}\,d\xi. \end{aligned}$$

It is a direct verification that for x and $\xi \in \mathbf{R}$,

$$\begin{aligned} \partial^2 \circ \mathcal{F} &= -\mathcal{F} \circ \xi^2, & x^2 \circ \partial^2 \circ \mathcal{F} &= \mathcal{F} \circ \partial^2 \circ \xi^2, \\ x \circ \partial \circ \mathcal{F} &= -\mathcal{F} \circ \partial \circ \xi, & x^2 \circ \mathcal{F} &= -\mathcal{F} \circ \partial^2. \end{aligned}$$

Next introduce

$$\kappa = k - \frac{n-1}{2} \qquad \text{and} \qquad F_\kappa = \mathcal{F} a \qquad \text{with} \qquad a(\xi) = \frac{1}{(1+\xi^2)^\kappa}.$$

Now try to determine constants p, q, and $r \in \mathbf{C}$ such that

$$\begin{aligned} (x^2\partial^2 + p\,x\,\partial + q + r\,x^2)F_\kappa &= 0, && \text{i.e.,} \\ \mathcal{F}\,(\partial^2(\xi^2 a) - p\,\partial(\xi a) + q\,a - r\,\partial^2 a) &= 0; && \text{hence} \\ \partial^2((\xi^2 - r)a) - p\,\partial(\xi a) + q\,a &= 0 && \text{by Fourier inversion.} \end{aligned}$$

The left-hand side of the last equation is a rational function in the variable ξ with $(1+\xi^2)^{\kappa+2}$ in the denominator and $c_4\xi^4 + c_2\xi^2 + c_0$ in the numerator, where

$$c_4 = (2\kappa - 1)p + q + 4\kappa^2 - 6\kappa + 2,$$
$$c_2 = 2((\kappa - 1)p + q - (2\kappa^2 + \kappa)r + 2 - 5\kappa), \qquad c_0 = -p + q + 2\kappa\,r + 2.$$

Equate the c_i to 0 and note that this inhomogeneous linear system for p, q, and r is nonsingular, because $\kappa > 0$. Its solution is $p = 2 - 2\kappa$, $q = 0$, and $r = -1$; hence one obtains the following ordinary differential equation for the function F_κ:

$$(x\,\partial^2 + 2(1-\kappa)\partial - x)F_\kappa = 0. \tag{17.8}$$

In fact, on account of the theory of the *indicial equation*, see [13, p. 223], one may predict the vanishing of q, because E_k is nonsingular at 0. In that case, the equations $c_4 = c_0 = 0$ take the simple form $p + 2\kappa - 2 = 0$ and $-p + 2\kappa\,r + 2 = 0$.

Now it is a straightforward to compute that for any $\lambda \in \mathbf{C}$ and C^2 function u defined on $\mathbf{R}$,

$$x^{-\lambda}(x^2\partial^2(x^\lambda u) + x\,\partial(x^\lambda u) - \lambda^2(x^\lambda u)) = x^2\partial^2 u + (2\lambda + 1)x\,\partial u.$$

Therefore select λ such that

$$2\lambda + 1 = 2 - 2\kappa = 2 - 2k + n - 1, \qquad \text{that is,} \qquad \lambda = \frac{n}{2} - k.$$

It follows that

$$v(x) := x^{\frac{n}{2}-k} E_k(x) \qquad \text{satisfies} \qquad x^2\,v'' + x\,v - (x^2 + \lambda^2)v = 0. \tag{17.9}$$

This is a differential equation for Bessel functions of purely imaginary argument, which differs from *Bessel's equation* $x^2\,u'' + x\,u' + (x^2 - \lambda^2)\,u = 0$ only in the coefficient of u; see Watson [23, § 3.7] for more information. As we will see in a moment, any two linearly independent solutions to the former equation have the property that one is exponentially increasing at infinity and the other exponentially decreasing. Hence, this differential equation satisfies the following principle: a solution that is a tempered distribution is of exponential decay at infinity. A tempered solution is given by the *Macdonald function* K_λ of order $\lambda > 0$ that is normalized by the following condition, which is a natural one in the theory of Bessel functions:

$$\lim_{x\downarrow 0}\left(\frac{x}{2}\right)^{\lambda} K_\lambda(x)=\frac{\Gamma(\lambda)}{2}.$$

K_λ tends exponentially to 0 at infinity through positive values. On account of (13.35) one obtains

$$E_k(x)=\frac{\|x\|^{k-\frac{n}{2}}}{2^{\frac{n}{2}+k-1}\pi^{\frac{n}{2}}\Gamma(k)}\,K_{k-\frac{n}{2}}(\|x\|)\qquad (x\in\mathbf{R}^n). \tag{17.10}$$

For this reason the (E_k) are said to be the *Bessel kernel*.

As a byproduct of this identity and [7, Exercise 7.30.(iii)], which reduces the Fourier transform of a function on $\mathbf{R}^n$ that is invariant under rotations to an integral over $\mathbf{R}_{>0}$ involving ordinary Bessel functions, we mention (compare with [23, § 13.6 (2)], which, however, is not correct)

$$\int_{\mathbf{R}_{>0}}\frac{r^{\frac{n}{2}}}{(1+r^2)^k}\,J_{\frac{n}{2}-1}(x\,r)\,dr=\frac{x^{k-1}}{2^{k-1}\Gamma(k)}\,K_{k-\frac{n}{2}}(x)\qquad (x\in\mathbf{R}).$$

Conversely, this equality can be applied to derive (17.10).

Finally, we derive formulas for E_k that are more explicit than the one in (17.10). It is straightforward to verify that

$$F_\kappa^1(x)=\int_x^\infty e^{-\xi}(\xi^2-x^2)^{\kappa-1}\,d\xi\qquad\text{and}\qquad F_\kappa^2(x)=\int_{-x}^x e^{-\xi}(x^2-\xi^2)^{\kappa-1}\,d\xi \tag{17.11}$$

define functions on $\mathbf{R}$ annihilated by the differential operator in (17.8). These integrals are related to Schläfli's integral; see [23, §6.15 (4)]. Furthermore, F_κ^1 is bounded on $\mathbf{R}$, whereas F_κ^2 is unbounded. Hence, there exists a constant $c\in\mathbf{C}$ such that $F_\kappa=c\,F_\kappa^1$. In this case, one again computes the constant by taking the limit as $x\downarrow 0$. Using Legendre's duplication formula (13.39) one derives $c=\frac{\pi}{4^{\kappa-1}\Gamma(\kappa)^2}$.

Combination of the preceding results leads to, for $x\in\mathbf{R}^n$,

$$E_k(x)=\frac{1}{2^{2k-1}\pi^{\frac{n-1}{2}}\Gamma(k-\frac{n-1}{2})\Gamma(k)}\int_{\|x\|}^\infty e^{-\xi}(\xi^2-\|x\|^2)^{k-\frac{n+1}{2}}\,d\xi.$$

In particular (compare with (18.18) and Problem 18.8.(i), and furthermore [23, § 3.71 (13)]),

$$E_{\frac{n+1}{2}}(x)=\frac{e^{-\|x\|}}{2^n\pi^{\frac{n-1}{2}}\Gamma(\frac{n+1}{2})}\qquad (x\in\mathbf{R}^n). \tag{17.12}$$

Using the change of variables $\xi=x+t$ in (17.11) one obtains

$$\begin{aligned}F_\kappa^1(x)&=e^{-x}\int_{\mathbf{R}_{>0}}e^{-t}(2xt+t^2)^{\kappa-1}\,dt\\&=(2x)^{\kappa-1}e^{-x}\int_{\mathbf{R}_{>0}}e^{-t}\,t^{\kappa-1}(1+\frac{t}{2x})^{\kappa-1}\,dt\\&\sim\Gamma(\kappa)(2x)^{\kappa-1}e^{-x}\Big(1+O\Big(\frac{1}{x}\Big)\Big),\quad x\to+\infty.\end{aligned} \tag{17.13}$$

Hence (note the agreement with [23, § 7.23 (1)])

$$E_k(x) \sim \frac{\|x\|^{k-\frac{n+1}{2}}\, e^{-\|x\|}}{2^{\frac{n-1}{2}+k}\pi^{\frac{n-1}{2}}\,\Gamma(k)}\Big(1+O\Big(\frac{1}{\|x\|}\Big)\Big), \qquad \|x\| \to \infty.$$

In (17.13) we restrict ourselves to deriving a crude asymptotic estimate. The binomial series expansion of $(1+\frac{t}{2x})^{\kappa-1}$ is convergent only as long as $0 < t < 2|x|$. Therefore interchange of integration and summation is not admissible. In fact, a formal interchange leads to the divergent series

$$(2x)^{\kappa-1}e^{-x}\sum_{j=0}^{\infty}\frac{\Gamma(\kappa)\Gamma(\kappa+j)}{j!\,\Gamma(\kappa-j)}\,\frac{1}{(2x)^j}.$$

Instead, one uses a finite number of terms plus a remainder in a correct analysis.

In the special case of $\kappa \in \mathbf{N}$, observe that the same change of variables and application of the Binomial Theorem to $(2x+t)^{\kappa-1}$ imply (the comparable final formula in [23, § 6.33] is erroneous)

$$F_\kappa^1(x) = (\kappa-1)!\,e^{-x}\sum_{j=0}^{\kappa-1}\frac{(2\kappa-2-j)!}{j!\,(\kappa-1-j)!}(2x)^j \qquad (x \in \mathbf{R}).$$

Therefore, if $k - \frac{n+1}{2} \in \mathbf{Z}_{\geq 0}$ and $x \in \mathbf{R}^n$ (see [23, § 3.71 (12)]), then

$$E_k(x) = \frac{e^{-\|x\|}}{2^{2k-1}\pi^{\frac{n-1}{2}}\,\Gamma(k)}\sum_{j=0}^{k-\frac{n+1}{2}}\frac{(2k-(n+1)-j)!}{j!\,(k-\frac{n+1}{2}-j)!}(2\|x\|)^j.$$

Problems

17.1. *(Heat equation on* $\mathbf{R}$ *and* $\mathbf{R}/2\pi\mathbf{Z}$*).* For $t > 0$, define $u_t \in \mathcal{S}(\mathbf{R})$ as in Problem 5.5. Prove that $\mathcal{F}u_t(\xi) = e^{-t\xi^2}$, for $\xi \in \mathbf{R}$, both by direct computation and by solving the following initial value problem on $\mathbf{R}$:

$$\frac{d}{dt}u_t = \Delta u_t \qquad (t > 0) \qquad \text{and} \qquad \lim_{t\downarrow 0} u_t = \delta.$$

Show that the solution of the following initial value problem on $\mathbf{R}$:

$$\frac{d}{dt}v_t = \Delta v_t \qquad (t > 0) \qquad \text{and} \qquad \lim_{t\downarrow 0} v_t = \sum_{n\in\mathbf{Z}}\delta_{2\pi n},$$

is given by the periodic function

$$v_t = \frac{1}{2\pi}\sum_{n\in\mathbf{Z}} e^{-t\,n^2}\, e_{in}.$$

Verify that

$$v_t(x) = \sum_{n\in\mathbf{Z}} u_t(x + 2\pi\, n) \qquad (x \in \mathbf{R}).$$

(Compare with Problem 16.9.) In other words, the solution to the initial value problem on the circle $\mathbf{R}/2\pi\mathbf{Z}$ equals the periodization (in the sense of Problem 16.12) of the solution to the initial value problem on $\mathbf{R}$.

17.2. *(Fundamental solution of heat operator.)* Let $a \in \mathbf{R}$. Prove that a fundamental solution of the differential operator $\partial + a\,I$ in $\mathbf{R}$ is given by $e^{-a\cdot}\,H$; see Example 14.30.

Next, use the partial Fourier transform with respect to the variable $x \in \mathbf{R}^n$ to show that a fundamental solution $E \in \mathcal{S}'(\mathbf{R}^n \times \mathbf{R})$ of the heat operator $\partial_t - \Delta_x$, for $(x, t) \in \mathbf{R}^n \times \mathbf{R}$, is given by the locally integrable function (see Problem 8.5 for another proof)

$$E : (x, t) \mapsto u_t(x)\, H(t) = \frac{H(t)}{(4\pi t)^{\frac{n}{2}}}\, e^{-\frac{\|x\|^2}{4t}}.$$

Here $u_t : \mathbf{R}^n \to \mathbf{R}$, for $t > 0$, is as in Problem 5.5.

17.3.* For what linear partial differential operators $P(D)$ in $\mathbf{R}^n$, with constant coefficients and of order > 0, does the equation $P(D)u = 0$ possess a solution $u \neq 0$ in (i) $\mathcal{D}'()$, (ii) $\mathcal{S}'$, (iii) $\mathcal{E}'()$, (iv) L^1, (v) C_0^∞, and (vi) $\mathcal{S}$? (Compare with Problem 14.3.)

17.4.* Prove that every harmonic and tempered distribution u on $\mathbf{R}^n$ is a polynomial. Prove that u is constant if u is a bounded harmonic function. Prove that for the Laplace operator Δ in $\mathbf{R}^n$, with $n \geq 3$, the potential of δ as in (12.3) is the only fundamental solution E with the property that $E(x) \to 0$ as $\|x\| \to \infty$.

17.5. *(Liouville's Theorem.)* Suppose that u is a tempered distribution on $\mathbf{C} \simeq \mathbf{R}^2$ and is complex-differentiable in the sense that $\partial u/\partial \bar{z} = 0$; then prove that u is a complex polynomial. Verify Liouville's Theorem, which asserts that any bounded complex-analytic function u on $\mathbf{C}$ is constant.

Generalize this result as follows. Let $P(D)$ be a partial differential operator in $\mathbf{R}^n$ with constant coefficients satisfying $P(\xi) \neq 0$ for all $\xi \in \mathbf{R}^n \setminus \{0\}$. Show that u is a polynomial function on $\mathbf{R}^n$ if $u \in \mathcal{S}'(\mathbf{R}^n)$ and $P(D)u = 0$. Conclude that u is constant if it is a bounded function. Verify that the heat operator $\partial_t - \Delta_x$ in $\mathbf{R}^{n+1}$ is an example of an operator as discussed here.

17.6.* Let $P(D)$ be an elliptic operator of order m. Prove the following assertions.

(i) If $P(D)$ possesses a homogeneous fundamental solution E, then $P(D) = P_m(D)$, that is, P is homogeneous of degree m. Furthermore, E is homogeneous of degree $m - n$.

(ii) Suppose that P is homogeneous and that E is a homogeneous fundamental solution E of $P(D)$. On account of Theorem 14.34, E is tempered. Prove that $\widetilde{E}$ is a tempered fundamental solution of P if and only if there exists a polynomial function f such that $\widetilde{E} = E + f$ and $P f = 0$. Furthermore, if $m \geq n$, then $\widetilde{E}$ is a homogeneous fundamental solution for P if and only if there exists a homogeneous polynomial function f of degree $m - n$ such that $\widetilde{E} = E + f$ and $P f = 0$. If $m < n$ and if $\widetilde{E}$ is a homogeneous fundamental solution of P, then $\widetilde{E} = E$.

(iii) If P is homogeneous and $m < n$, then $1/P$ is locally integrable on $\mathbf{R}^n$ and P has exactly one homogeneous fundamental solution.

(iv) The Laplace operator in $\mathbf{R}^2$ has no homogeneous fundamental solution.

17.7. *(Fundamental solution of Schrödinger operator.)* Prove that $E : \mathbf{R}^n \times \mathbf{R} \to \mathbf{C}$ is a fundamental solution of the *Schrödinger operator* on $\mathbf{R}^n \times \mathbf{R}$

$$D_t - \Delta_x \qquad \text{if} \qquad E(x,t) = H(t)\frac{1}{(4\pi t)^{\frac{n}{2}}}\, e^{i\left((2-n)\frac{\pi}{4} + \frac{\|x\|^2}{4t}\right)}.$$

Hint: apply Problem 14.41.

17.8.* Let $\lambda \in \mathbf{C}$ and $\lambda \notin \mathbf{R}$. Show that $P(D) = D - \lambda$ has exactly one tempered fundamental solution E_0. Calculate the fundamental solution E from the proof of Theorem 17.11. Show that E is not a tempered distribution on $\mathbf{R}$ whenever $r > (|\lambda| + R)/|\eta|$.

17.9. Fill in the details of the following arguments.

In the notation of Sect. 17.1 we have on account of (17.10), for $k > \frac{n}{2}$,

$$E_k = \frac{1}{(2\pi)^n}\mathcal{F}\Big(\frac{1}{1 + \|\cdot\|^2)^k}\Big), \qquad \text{and so} \qquad \int_{\mathbf{R}^n} E_k(x)\,dx = 1;$$

$$\text{hence} \qquad \int_{\mathbf{R}^n} \|x\|^\lambda K_\lambda(\|x\|)\,dx = 2^{\lambda+n-1}\pi^{\frac{n}{2}}\Gamma\Big(\lambda + \frac{n}{2}\Big) \qquad (\lambda > 0);$$

see [23, § 13.21 (8)].

Using the Fourier transform prove that $E_k * E_l = E_{k+l}$ and $(I - \Delta)E_k = E_{k-1}$, for admissible values of k and l. Investigate whether these properties enable the construction of a complex-analytic family $(E_k)_{k\in\mathbf{C}}$ of distributions; see Fig. 17.1.

Derive from Definition 13.30 and Example 14.11, respectively,

$$\frac{1}{(1 + \|\xi\|^2)^k} = \frac{1}{\Gamma(k)}\int_{\mathbf{R}_{>0}} e^{-t}e^{-t\|\xi\|^2}t^{k-1}\,dt, \qquad \mathcal{F}e^{-t\|\cdot\|^2} = \pi^{\frac{n}{2}}e^{-\frac{\|\cdot\|^2}{4t}}t^{-\frac{n}{2}}.$$

Deduce that

$$E_k(x) = \frac{1}{2^n\pi^{\frac{n}{2}}\Gamma(k)}\int_{\mathbf{R}_{>0}} e^{-t-\frac{\|x\|^2}{4t}}t^{k-\frac{n}{2}-1}\,dt \qquad (x \in \mathbf{R}^n),$$

$$\int_{\mathbf{R}_{>0}} e^{-t-\frac{r^2}{4t}}t^{\lambda-1}\,dt = 2\Big(\frac{r}{2}\Big)^\lambda K_\lambda(r) \qquad (r > 0,\ \lambda \in \mathbf{R}).$$

For the last identity, compare with [23, § 6.22 (15)] and [7, Exercise 2.87.(v)] for the elementary case of $\lambda = \frac{1}{2}$. In addition, the penultimate identity allows one to define E_k explicitly, for $0 < k \le \frac{n}{2}$. Show by direct computation that in this case one also has $\int_{\mathbf{R}^n} E_k(x)\,dx = 1$.

In the last identity above involving $E_k(x)$ introduce a new variable s by means of $t = \frac{\|x\|^2}{s}$. Next, suppose $0 < k < \frac{n}{2}$ and use dominated convergence to show that $E_k(x) = \widetilde{R}^{2k}(x)(1 - \tau(x))$, where $\widetilde{R}^{2k}$ is as in (14.51) and $\lim_{x \to 0} \tau(x) = 0$. This shows that there is an intimate connection between E_k and $\widetilde{R}^{2k}$, as is to be expected.

$f(t) := e^{-t-\frac{\|x\|^2}{4t}}$ attains its maximum $e^{-\|x\|}$ on $\mathbf{R}_{>0}$ at $t = \frac{\|x\|}{2}$. Further, if $\|x\| \ge 1$, then $f(t) \le e^{-(t+\frac{1}{4t})}$. Combine the two estimates in order to obtain $0 \le f(t) \le e^{-\frac{\|x\|}{2}} e^{-\frac{1}{2}(t+\frac{1}{4t})}$ and thus prove $E_k(x) = O(e^{-\frac{\|x\|}{2}}), \quad x \to \infty$.

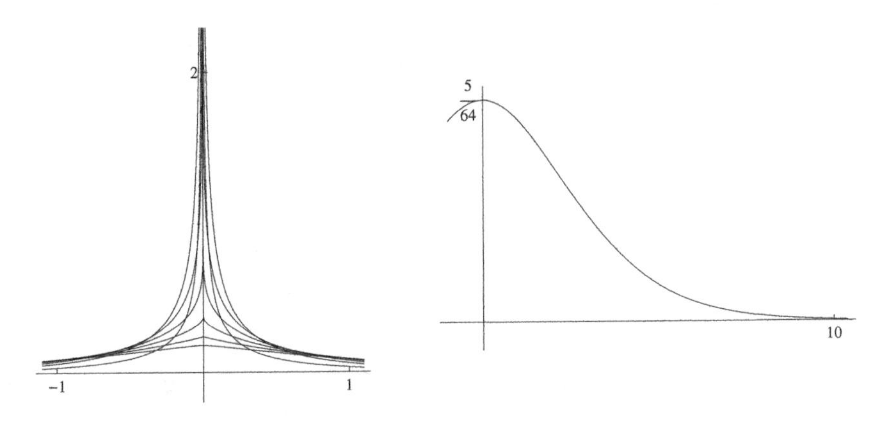

Fig. 17.1 Graphs of $E_{\frac{k}{5}}$, for $1 \le k \le 7$, and of E_4

Chapter 18
Supports and Fourier Transform

In Lemma 14.4 we have proved that the Fourier transform $\mathcal{F}u$ of a distribution with compact support can be extended to a complex-analytic function U on $\mathbf{C}^n$, called the Fourier–Laplace transform of u. In this chapter we deal with questions like, What complex-analytic functions U occur as $\mathcal{F}u$ for some $u \in \mathcal{E}'$? and How can U provide information on the support of u?

For a compact subset K of $\mathbf{R}^n$ we define

$$s_K : \mathbf{R}^n \to \mathbf{R} \qquad \text{by} \qquad s_K(\eta) = \sup_{x \in K} \langle x, \eta \rangle. \tag{18.1}$$

Theorem 18.1. *Let $K \subset \mathbf{R}^n$ be compact and $u \in \mathcal{E}'(\mathbf{R}^n)$ with* supp $u \subset K$. *If u is of order $k \in \mathbf{Z}_{\geq 0}$, there exists a constant $c > 0$ such that*

$$|\mathcal{F}u(\zeta)| \leq c\,(1 + \|\zeta\|)^k\, e^{s_K(\operatorname{Im}\zeta)} \qquad (\zeta \in \mathbf{C}^n). \tag{18.2}$$

If, in addition, u is of class C^k with $k \in \mathbf{Z}_{\geq 0}$, there exists a constant $c_k > 0$ such that

$$|\mathcal{F}u(\zeta)| \leq c_k\,(1 + \|\zeta\|)^{-k}\, e^{s_K(\operatorname{Im}\zeta)} \qquad (\zeta \in \mathbf{C}^n). \tag{18.3}$$

Proof. We are going to apply Theorem 8.8; this requires some preparation.

For given $\epsilon > 0$ we can, on the strength of Corollary 2.4, cover the compact set K by a finite number of open balls whose union U is contained in the $\epsilon/3$-neighborhood $K_{\epsilon/3}$. Applying Lemma 2.19 with this set U and with ϵ and δ replaced by $\epsilon/3$ and $2\epsilon/3$, respectively, we conclude that it is possible to find $\chi_\epsilon \in C_0^\infty(\mathbf{R}^n)$ with $\chi_\epsilon = 1$ on an open neighborhood of K, while supp $\chi_\epsilon \subset K_\epsilon$. Then there exists, for every multi-index β, a constant $c_\beta > 0$ such that

$$\sup_x |\partial^\beta \chi_\epsilon(x)| \leq c_\beta\, \epsilon^{-|\beta|}.$$

Writing $\zeta = \xi + i\eta$ with ξ and $\eta \in \mathbf{R}^n$, we have, for every multi-index γ,

J.J. Duistermaat and J.A.C. Kolk, *Distributions: Theory and Applications*,
Cornerstones, DOI 10.1007/978-0-8176-4675-2_18,

$$|\partial_x{}^\gamma e^{-i\langle x,\zeta\rangle}| \leq |\zeta^\gamma|\, e^{\langle x,\eta\rangle}.$$

On account of $\langle y, \eta\rangle \leq \|y\|\,\|\eta\|$, the Cauchy–Schwarz inequality, we get

$$e^{\langle x,\eta\rangle} \leq e^{s_K(\eta)+\epsilon\,\|\eta\|} \qquad (x \in K_\epsilon).$$

If we now apply Theorem 8.8 with $\chi = \chi_\epsilon$, and work out an explicit form of the C^k norm of $\chi_\epsilon\, e_{-i\zeta}$ by means of Leibniz's rule, we obtain an estimate of the form

$$|\mathcal{F}u(\zeta)| = |u(e_{-i\zeta})| \leq c' \sup_{|\beta|+|\gamma|\leq k} \epsilon^{-|\beta|}\, |\zeta^\gamma|\, e^{s_K(\eta)+\epsilon\,\|\eta\|}.$$

With the choice $\epsilon = 1/\|\eta\|$ we now deduce (18.2), where we have used

$$\epsilon^{-|\beta|}\, |\zeta^\gamma| = \|\eta\|^{|\beta|}\, |\zeta^\gamma| \leq (1+\|\zeta\|)^{|\beta|+|\gamma|},$$

while $e^{\epsilon\,\|\eta\|} = e$ may be included with the constant.

As regards (18.3), we observe that

$$|\mathcal{F}u(\zeta)| = \left|\int e^{-i\langle x,\zeta\rangle}\, u(x)\, dx\right| \leq \int_{\operatorname{supp} u} e^{\langle x,\eta\rangle}\, |u(x)|\, dx \leq e^{s_K(\eta)}\, \|u\|_{L^1},$$

whenever $u \in \mathcal{E}'(\mathbf{R}^n)$ is integrable. On account of (14.28) this yields, for all multi-indices α with $|\alpha| \leq k$ and for all u of class C^k,

$$|\zeta^\alpha \mathcal{F}u(\zeta)| \leq e^{s_K(\eta)}\, \|D^\alpha u\|_{L^1}.$$

This proves (18.3). □

For $u \in C_0^\infty(\mathbf{R}^n)$ we thus find that an estimate of the form (18.3) obtains for every $k \in \mathbf{Z}_{\geq 0}$. To account for the fact that the constant c will depend on k, it is written with subscript k.

Example 18.2. Let $u \in \mathcal{E}'(\mathbf{R}^n)$. Then there exist a continuous function f on $\mathbf{R}^n$ and $N \in \mathbf{Z}_{\geq 0}$ such that $u = (I - \Delta)^N f$. Indeed, every distribution with compact support is of finite order k, and Theorem 18.1 therefore yields an estimate of the form

$$|\mathcal{F}u(\xi)| \leq c\,(1+\|\xi\|)^k \qquad (\xi \in \mathbf{R}^n).$$

Now choose N sufficiently large to ensure that $2N - k > n$. One then has $g \in L^1(\mathbf{R}^n)$ if

$$g(\xi) := \left(1+\|\xi\|^2\right)^{-N} \mathcal{F}u(\xi) \qquad (\xi \in \mathbf{R}^n).$$

On account of the Riemann–Lebesgue Theorem, Theorem 14.2, $f = \mathcal{F}^{-1} g$ is a continuous function on $\mathbf{R}^n$ with $f(x) \to 0$ as $\|x\| \to \infty$. Furthermore,

$$\mathcal{F}\left((I-\Delta)^N f\right)(\xi) = \left(1+\|\xi\|^2\right)^N g(\xi) = \mathcal{F}u(\xi),$$

It follows that $(I - \Delta)^N f = u$.

Using a smooth cut-off function we can deduce, for every distribution u on an open subset X of $\mathbf{R}^n$ and for every compact subset K of X, the existence of a continuous function f on X and an $N \in \mathbf{Z}_{\geq 0}$ such that $u = (I - \Delta)^N f$ on a neighborhood of K. This implies that every distribution can locally be written as a linear combination of derivatives of a continuous function. In Theorem 13.1 we gave a different proof. ⊘

Now we come to another proof of the Ehrenpreis–Malgrange Theorem, Theorem 17.13; it is based on Theorem 18.1. We begin by considering a special case.

Proposition 18.3. *Let* $c \neq 0$ *and*

$$P(D) = c\, D_n^m + \sum_{j=0}^{m-1} P_j(D')\, D_n^j \qquad \textit{with} \qquad D' = (D_1, \dots, D_{n-1})$$

be a linear partial differential operator in $\mathbf{R}^n$ *of order* $m \in \mathbf{N}$ *with constant coefficients. Then* $P(D)$ *has a fundamental solution in* $\mathcal{D}'(\mathbf{R}^n)$.

Proof. The symbol of $P(D)$ is

$$P(\xi) = c\, \xi_n^m + \sum_{j=0}^{m-1} P_j(\xi')\, \xi_n^j, \qquad \text{where} \qquad \xi' = (\xi_1, \dots, \xi_{n-1}) \in \mathbf{R}^{n-1}$$

and the P_j are polynomial functions on $\mathbf{R}^{n-1}$. For $\eta' \in \mathbf{R}^{n-1}$, the polynomial function $\zeta_n \mapsto P(\eta', \zeta_n)$ on $\mathbf{C}$ has m zeros counted with multiplicities. Denote by $D(\eta')$ the set of those zeros with repetitions if they have positive multiplicities. Then there exists $\tau(\eta') \in \mathbf{R}$ such that

$$|\tau(\eta')| < m + 1 \qquad \text{and} \qquad |\tau(\eta') - \operatorname{Im} \lambda(\eta')| > 1 \qquad (\lambda(\eta') \in D(\eta')). \tag{18.4}$$

The roots of a polynomial with constant highest coefficient depend continuously on the remaining coefficients (see *Rouché's Theorem* from complex function theory or [7, Example 8.11.11], both of which are concerned with winding numbers, that is, degrees of mappings). Therefore one can find an open neighborhood $U(\eta')$ of η' in $\mathbf{R}^{n-1}$ such that (18.4) also is valid with η' replaced by arbitrary $\xi' \in U(\eta')$. One may take $U(\eta')$ to be an open cube in $\mathbf{R}^{n-1}$ parallel to the coordinate axes and centered at η'. These cubes form an open cover of $\mathbf{R}^{n-1}$. The Heine–Borel Theorem, Theorem 2.2.(c), allows one to extract a countable and locally finite subcover of $\mathbf{R}^{n-1}$, whose elements we denote by $U_k = U(\eta'_k)$, for $k \in \mathbf{N}$. Next, one can define sets $V_k \subset U_k$ by

$$V_1 = U_1, \qquad V_k = U_k \setminus \bigcup_{j=1}^{k-1} V_j \qquad (k \in \mathbf{Z}_{>1}).$$

Then the V_k are mutually disjoint, their union equals $\mathbf{R}^{n-1}$, and any compact subset of $\mathbf{R}^{n-1}$ meets only a finite number of them. Furthermore, define the (real) n-

dimensional subsets Z_k and Z of $\mathbf{R}^{n-1} \times \mathbf{C}$ by

$$Z_k = V_k \times (\mathbf{R} + i\,\tau(\eta'_k)) \qquad (k \in \mathbf{N}) \qquad \text{and} \qquad Z = \bigcup_{k \in \mathbf{N}} Z_k. \tag{18.5}$$

Obviously, for $\zeta = (\xi', \zeta_n) \in \mathbf{R}^{n-1} \times \mathbf{C}$,

$$|P(\zeta)| = |c| \prod_{\lambda(\xi') \in D(\xi')} |\zeta_n - \lambda(\xi')| \geq |c| \prod_{\lambda(\xi') \in D(\xi')} |\operatorname{Im} \zeta_n - \operatorname{Im} \lambda(\xi')|.$$

By construction, (18.4) holds for each ξ' belonging to some U_k. Therefore it is clear from (18.5) that

$$|P(\zeta)| \geq |c| \qquad (\zeta \in Z). \tag{18.6}$$

Now define $E \in \mathcal{D}'(\mathbf{R}^n)$ by setting, for $\phi \in C_0^\infty(\mathbf{R}^n)$,

$$E(\phi) = \frac{1}{(2\pi)^n} \int_Z \frac{\mathcal{F}\phi(\zeta)}{P(-\zeta)}\, d\zeta = \frac{1}{(2\pi)^n} \sum_{k \in \mathbf{N}} \int_{Z_k} \frac{\mathcal{F}\phi(\zeta)}{P(-\zeta)}\, d\zeta. \tag{18.7}$$

This sum converges on account of (18.6) (with P replaced by SP) as well as (18.3), which leads to the existence of constants $c > 0$ and $d > 0$ such that, for $\zeta = (\xi', \xi_n + i\,\tau(\eta'_k)) \in Z_k$ with arbitrary k,

$$|\mathcal{F}\phi(\zeta)| \leq c\,(1 + \|(\xi', \xi_n)\|)^{-n-1}\, e^{d\,|\tau(\eta'_k)|} \leq c\,(1 + \|(\xi', \xi_n)\|)^{-n-1}\, e^{d(m+1)}.$$

Note that the estimate on the right-hand side is independent of k and thus applies to all $\zeta \in Z$. Furthermore, c and d can be taken to be the same for all $\phi \in C_0^\infty(\mathbf{R}^n)$ supported in a fixed compact set. Therefore $E \in \mathcal{D}'(\mathbf{R}^n)$.

Finally, first applying Cauchy's Integral Theorem from (12.9) to the integration with respect to the variable ζ_n and then Fourier inversion, one obtains

$$\begin{aligned} P(D)E(\phi) = E(P(-D)\phi) &= \frac{1}{(2\pi)^n} \int_Z \frac{P(-\zeta)\,\mathcal{F}\phi(\zeta)}{P(-\zeta)}\, d\zeta \\ &= \frac{1}{(2\pi)^n} \int_Z \mathcal{F}\phi(\zeta)\, d\zeta = \frac{1}{(2\pi)^n} \int_{\mathbf{R}^n} \mathcal{F}\phi(\xi)\, d\xi = \phi(0) = \delta(\phi). \end{aligned}$$

□

Theorem 18.4. *Every linear partial differential operator in $\mathbf{R}^n$ with constant coefficients (not all of them equal to 0) possesses a fundamental solution in $\mathcal{D}'(\mathbf{R}^n)$.*

Proof. Denote the differential operator of order $m \in \mathbf{N}$ by $P(D)$ and its principal symbol by $P_m(\xi)$; see Definition 17.2. Let $A : \mathbf{R}^n \to \mathbf{R}^n$ be a linear isomorphism and write A_n for the last column of the matrix of A. A straightforward computation shows that the coefficient of ξ_n^m in the symbol $P(A\,\xi)$ of the differential operator $P(A\,D)$ equals $P_m(A_n)$. One can clearly choose A as above such that $P_m(A_n) \neq 0$. The preceding proposition then yields the existence of $F \in \mathcal{D}'(\mathbf{R}^n)$ such that $P(A\,D)F = \delta$. Hence, application of (10.2) with A replaced by $B := {}^tA$ and of

Example 10.12 leads to

$$P(D)(B^* F) = B^* \circ P(A\,D)F = B^*\delta = |\det B|^{-1}\delta.$$

Phrased differently, $|\det B|\, B^* F \in \mathcal{D}'(\mathbf{R}^n)$ is a fundamental solution of $P(D)$. $\square$

Next, we present a converse of Theorem 18.1. For a real-valued function s on a subset Y of $\mathbf{R}^n \setminus \{0\}$ we define the subset K_s of $\mathbf{R}^n$ by

$$K_s := \bigcap_{\eta\in Y} \{\, x \in \mathbf{R}^n \mid \langle x, \eta\rangle \le s(\eta) \,\}. \tag{18.8}$$

Before proceeding further, we give a characterization of the sets of the form K_s.

Theorem 18.5. *For every $Y \subset \mathbf{R}^n \setminus \{0\}$ and $s : Y \to \mathbf{R} \cup \{\infty\}$, the subset K_s of $\mathbf{R}^n$ as defined by (18.8) is closed and convex. Conversely, if K is a closed and convex subset of $\mathbf{R}^n$, then $K = K_s$ if $s = s_K$ is defined by (18.1).*

Proof. A subset K of a linear space is said to be *convex* if x and $y \in K$ implies that the line segment $\{\, x + t\,(y - x) \in \mathbf{R}^n \mid 0 \le t \le 1 \,\}$ from x to y is contained in K. For $\eta \in \mathbf{R}^n \setminus \{0\}$ and $c \in \mathbf{R}$, it is evident that the half-space

$$H(\eta, c) := \{\, x \in \mathbf{R}^n \mid \langle x, \eta\rangle \le c \,\}$$

is convex. Because the intersection of a collection of convex sets is also convex, we deduce the convexity of K_s, the intersection of all $H(\eta, s(\eta))$, where $\eta \in Y$. Furthermore, K_s is closed, being the intersection of the closed subsets $H(\eta, s(\eta))$ of $\mathbf{R}^n$.

With respect to the converse assertion we begin by observing that $K \subset K_s$, for every $K \subset \mathbf{R}^n$, if $s = s_K$; this immediately follows from the definitions. What remains to be shown, therefore, is that if K is convex and closed, and if $y \notin K$, it follows that $y \notin K_s$. That is, there exists $\eta \in \mathbf{R}^n$ such that $s_K(\eta) < \langle y, \eta\rangle$.

Because K is closed, there is $a \in K$ satisfying

$$\|y - a\| = d(y, K) := \inf_{x\in K} \|y - x\|.$$

Indeed, there exists a sequence $(x_j)_{j\in\mathbf{N}}$ in K such that $\|y - x_j\| \to d(x, K)$. This sequence is bounded, and hence it contains a subsequence that converges to some a; then $\|y - a\| = d(y, K)$. Because K is closed, $a \in K$.

Now let $x \in K$. Because K is convex, we see, for every $0 \le t \le 1$, that $a + t\,(x - a) \in K$; as a result,

$$\|y - a\|^2 \le \|y - (a + t\,(x - a))\|^2 = \|y - a\|^2 - 2t\,\langle y - a,\, x - a\rangle + t^2\,\|x - a\|^2.$$

This implies that the derivative of the right-hand side with respect to t cannot be negative at $t = 0$, that is,

$$\langle x,\, y-a\rangle \le \langle a,\, y-a\rangle \qquad (x\in K).$$

In other words, $s_K(y-a) = \langle a,\, y-a\rangle < \langle y,\, y-a\rangle$, where the inequality is a consequence of the assumption that $y \notin K$; therefore, $y \neq a$. This is the desired inequality, with $\eta = y - a$. □

Theorem 18.6. *Let $Y \subset \mathbf{R}^n \setminus \{0\}$ and $s : Y \to \mathbf{R}$ such that K_s is a bounded (and therefore compact) subset of $\mathbf{R}^n$. Let $N \in \mathbf{R}$. Suppose that U is a complex-analytic function on $\mathbf{C}^n$ with the property that for every $\eta \in Y$ there exists a constant $c(\eta) > 0$ such that (compare with (18.2))*

$$|U(\xi + i\tau\eta)| \le c(\eta)\,(1 + \|\xi\|)^N\, e^{\tau\, s(\eta)} \qquad (\xi\in\mathbf{R}^n,\ \tau\ge 0). \tag{18.9}$$

Then $U = \mathcal{F}u$ for some $u \in \mathcal{E}'(\mathbf{R}^n)$ satisfying supp $u \subset K_s$. *If, in addition, $N < -n-k$, then u is of class C^k.*

Proof. Denote the restriction of U to $\mathbf{R}^n$ by v. First assume that $N < -n-k$. In that case, the functions $\xi^\alpha v$ are integrable on $\mathbf{R}^n$ if $|\alpha| \le k$, and consequently, $u := \mathcal{F}^{-1}v$ is a bounded C^k function on $\mathbf{R}^n$.

Now let $\eta \in Y$. Because $\eta \neq 0$, there exists a linear mapping $A : \mathbf{R}^{n-1} \to \mathbf{R}^n$ such that $B : (\sigma, \mu) \mapsto \sigma\,\eta + A\mu$ is a bijective linear mapping from $\mathbf{R}^n$ to $\mathbf{R}^n$; the latter has a complex-linear extension $B : \mathbf{C}^n \to \mathbf{C}^n$. The change of variables $\xi = B(\sigma, \mu)$ yields

$$u(x) = |\det B| \int_{\mathbf{R}^{n-1}} \int_{\mathbf{R}} F\circ B(\sigma,\mu)\, d\sigma\, d\mu,$$

$$\text{where} \qquad F(\xi) = \frac{1}{(2\pi)^n} e^{i\langle x,\xi\rangle}\, U(\xi).$$

On account of Cauchy's Integral Theorem (see (12.9)), applied to the complex-analytic function $F \circ B(\rho, \mu)$ of $\rho = \sigma + i\tau \in \mathbf{C}$, the integration over the real σ-axis may be replaced by integration over $\sigma \mapsto \sigma + i\tau$. Here the estimate (18.9) and the condition on N guarantee that the contributions over $t \mapsto \sigma + it\tau$, for $t \in [0, 1]$, converge to 0 as $|\sigma| \to \infty$. Returning to the old variables, we get

$$u(x) = \frac{1}{(2\pi)^n} \int_{\mathbf{R}^n} e^{i\langle x,\xi + i\tau\eta\rangle}\, U(\xi + i\tau\,\eta)\, d\xi.$$

Since $|e^{i\langle x,\xi+i\tau\eta\rangle}| = e^{-\tau\langle x,\eta\rangle}$, substitution of the estimate (18.9) gives

$$|u(x)| \le c\, e^{-\tau\langle x,\eta\rangle}\, e^{\tau\, s(\eta)},$$

with a constant $c = c(\eta) > 0$ that is independent of $\tau \ge 0$. If $\langle x, \eta\rangle > s(\eta)$, the right-hand side converges to 0 as $\tau \to \infty$; in that case, therefore, $|u(x)| \le 0$, implying $u(x) = 0$. The conclusion is that $\langle x, \eta\rangle \le s(\eta)$, for all $x \in$ supp u. Because this is true for all $\eta \in Y$, we have proved supp $u \subset K_s$.

If we drop the assumption $N < -n - k$, we still have $v \in \mathcal{S}'$; so $v = \mathcal{F}u$, for some $u \in \mathcal{S}'$. Choose $\phi \in C_0^\infty(\mathbf{R}^n)$ with $\phi(x) = 0$ if $\|x\| > 1$ and $1(\phi) = 1$. For $\phi_\epsilon(x) := \epsilon^{-n}\,\phi(\epsilon^{-1}x)$ we find that $u_\epsilon := u * \phi_\epsilon \to u$ in $\mathcal{S}'$ as $\epsilon \to 0$. Furthermore, one has, on the strength of Theorem 14.33,

$$v_\epsilon(\xi) := \mathcal{F}(u_\epsilon)(\xi) = \mathcal{F}u(\xi)\,\mathcal{F}\phi_\epsilon(\xi) = U(\xi)\,\mathcal{F}\phi(\epsilon\,\xi).$$

The right-hand side has an extension to a complex-analytic function on $\mathbf{C}^n$, which we denote by U_ϵ. Using the Cauchy–Schwarz inequality we obtain $s_K(\eta) = \|\eta\|$ if

$$K = \{x \in \mathbf{R}^n \mid \|x\| \le 1\}.$$

Combining the estimate (18.3) for ϕ with (18.9) we get, for $\zeta = \xi + i\,\tau\eta$,

$$|U_\epsilon(\zeta)| \le c(\eta, k)\,(1 + \epsilon\|\zeta\|)^{-k}\,e^{\epsilon\tau\|\eta\|}\,(1 + \|\xi\|)^N\,e^{\tau\,s(\eta)}.$$

Choosing $k > N + n$ we may apply the theorem, with k as in the theorem set to 0, to conclude that $\operatorname{supp} u_\epsilon$ is contained in the half-space

$$H_\epsilon(\eta) := \{\,x \in \mathbf{R}^n \mid \langle x,\,\eta\rangle \le s(\eta) + \epsilon\|\eta\|\,\}.$$

On account of $\operatorname{supp} u_\delta \subset H_\delta(\eta) \subset H_\epsilon(\eta)$, whenever $0 < \delta \le \epsilon$ and $u = \lim_{\delta\downarrow 0} u_\delta$, this implies $\operatorname{supp} u \subset H_\epsilon(\eta)$. Because this is true for every $\epsilon > 0$, we conclude that $\operatorname{supp} u \subset H_0(\eta)$. Finally, since this applies for all $\eta \in Y$, it follows that $\operatorname{supp} u \subset K_s$. □

Theorems of this kind are called *Paley–Wiener Theorems*, and the estimates for the function U that they contain are the *Paley–Wiener estimates*. Theorems 18.1 and 18.6 in essentially this form are due to Schwartz; Paley and Wiener studied the Fourier transformation of integrable functions on **R** with support in a half-line.

Example 18.7. It is a classical idea, due to Cauchy, to solve the wave equation

$$\frac{d^2}{dt^2}u_t = \Delta u_t$$

in (x, t)-space by treating t as a parameter in the Fourier transform with respect to the variable x. We will use the notation from Remark 13.4. To initiate this treatment; we assume that $t \mapsto u_t$ is a C^2 family in $\mathcal{S}'(\mathbf{R}^n)$. Because $\mathcal{F} : \mathcal{S}' \to \mathcal{S}'$ is a continuous linear mapping, $t \mapsto \mathcal{F}u_t$ in that case also is a C^2 family in $\mathcal{S}'$ and

$$\frac{d^2}{dt^2}\mathcal{F}u_t = \mathcal{F}\Big(\frac{d^2}{dt^2}u_t\Big) = \mathcal{F}(\Delta u_t) = -\|\xi\|^2\,\mathcal{F}u_t,$$

in view of Theorem 14.24. But this implies

$$\mathcal{F}u_t(\xi) = p(\xi)\,\cos t\,\|\xi\| + q(\xi)\,\frac{\sin t\,\|\xi\|}{\|\xi\|}, \tag{18.10}$$

with

$$p = \mathcal{F} u_0 \qquad \text{and} \qquad q = \frac{d}{dt} \mathcal{F} u_t \Big|_{t=0} = \mathcal{F} \Big(\frac{d}{dt} u_t \Big|_{t=0} \Big), \tag{18.11}$$

or $p = \mathcal{F} a$ and $q = \mathcal{F} b$ in the notation of Remark 13.4.

This causes us to turn to the functions

$$A_t(\xi) := \cos t \, \|\xi\| \qquad \text{and} \qquad B_t(\xi) := \frac{\sin t \, \|\xi\|}{\|\xi\|}, \tag{18.12}$$

the solutions for $p = 1$ and $q = 0$, and for $p = 0$ and $q = 1$, respectively. Both functions possess an extension to a complex-analytic function on $\mathbf{C}^n$, owing to the fact that

$$\cos z = C(z^2) \qquad \text{and} \qquad \operatorname{sinc} z = S(z^2),$$

where C and S are complex-analytic functions on $\mathbf{C}$; this can be seen from the power series, for example. Accordingly, we may write

$$A_t(\xi) = C(t^2 \, \|\xi\|^2) \qquad \text{and} \qquad B_t(\xi) = t \, S(t^2 \, \|\xi\|^2);$$

here $\xi \mapsto \|\xi\|^2$ has the complex-analytic extension

$$\xi + i\eta = \zeta \mapsto \sum_{j=1}^{n} {\zeta_j}^2 = \|\xi\|^2 - \|\eta\|^2 + 2i \langle \xi, \eta \rangle.$$

In order to estimate $A_t(\zeta)$, we observe that $|\cos z| \le e^{|y|}$ if $z = x + iy$ with x, $y \in \mathbf{R}$. Solving y^2 from

$$x^2 - y^2 + 2ixy = z^2 = t^2 \, (\|\xi\|^2 - \|\eta\|^2 + 2i \langle \xi, \eta \rangle),$$

we obtain

$$y^2 = \frac{t^2}{2} \, (\|\eta\|^2 - \|\xi\|^2 \pm \sqrt{(\|\xi\|^2 - \|\eta\|^2)^2 + 4 \langle \xi, \eta \rangle^2}) \le t^2 \, \|\eta\|^2.$$

Here we have used the Cauchy–Schwarz inequality. This leads to

$$|A_t(\xi + i\eta)| \le e^{|t| \, \|\eta\|} \qquad (\xi, \eta \in \mathbf{R}^n).$$

Now $\langle x, \eta \rangle \le |t| \, \|\eta\|$, for all $\eta \in \mathbf{R}^n$, if and only if $\|x\| \le |t|$. Similar Paley–Wiener estimates hold for B_t if it is observed that $\frac{d}{dt} \sin tz = z \cos tz$, which implies $|\sin z| = |z \int_0^1 \cos tz \, dt| \le |z| \, e^{|y|}$. On the basis of Theorem 18.6 we thus find that $A_t = \mathcal{F} a_t$ and $B_t = \mathcal{F} b_t$, for distributions a_t and b_t in $\mathbf{R}^n$ with supp a_t and supp b_t both contained in the ball about 0 of radius $|t|$. Moreover, these are C^2 families in $\mathcal{E}'(\mathbf{R}^n)$ with

$$\frac{d^2}{dt^2} a_t = \Delta a_t \qquad \text{and} \qquad \frac{d^2}{dt^2} b_t = \Delta b_t$$

and initial conditions

$$a_0 = \delta, \qquad \frac{d}{dt} a_t \bigg|_{t=0} = 0 \qquad \text{and} \qquad b_0 = 0, \qquad \frac{d}{dt} b_t \bigg|_{t=0} = \delta. \tag{18.13}$$

The fact that a_t and b_t are distributions with compact support enables us to convolve them with arbitrary distributions. The conclusion is that for every a and $b \in \mathcal{D}'(\mathbf{R}^n)$, the distributions

$$u_t := a_t * a + b_t * b \qquad (t \in \mathbf{R})$$

give a C^2 family $t \mapsto u_t$ in $\mathcal{D}'(\mathbf{R}^n)$ that forms a solution of the *Cauchy problem*

$$\frac{d^2}{dt^2} u_t = \Delta u_t \qquad \text{with} \qquad u_0 = a \qquad \text{and} \qquad \frac{d}{dt} u_t \bigg|_{t=0} = b. \tag{18.14}$$

Combining this with the uniqueness from Remark 13.4, we conclude that the Cauchy problem for the wave equation has a unique solution, even for arbitrary distributional initial values.

Next, we consider the wave equation with an inhomogeneous term $f \in \mathcal{D}'(\mathbf{R} \times \mathbf{R}^n)$ as in (13.28). Once again, we assume f to be associated with a family f_t of distributions in $\mathcal{D}'(\mathbf{R}^n)$ via the formula $f(\phi) = \int_{\mathbf{R}} f_t(\phi_t)\, dt$, for all $\phi \in C_0^\infty(\mathbf{R}^{n+1})$, with the notation $\phi_t : x \mapsto \phi(t, x)$. Furthermore, we suppose that the $f_t \in \mathcal{D}'(\mathbf{R}^n)$ depend continuously on the parameter $t \in \mathbf{R}$. Recall that this means that the function $t \mapsto f_t(\phi)$ is continuous, for every $\phi \in C_0^\infty(\mathbf{R}^n)$. (The principle of uniform boundedness is required in showing that one really defines a distribution in this manner.)

In order to get an idea about the solution, we initially consider the situation in which $t \mapsto f_t$ is continuous with values in $\mathcal{S}'(\mathbf{R}^n)$. In that case, $t \mapsto \mathcal{F} f_t$ is also continuous with values in $\mathcal{S}'(\mathbf{R}^n)$. Upon Fourier transformation the inhomogeneous wave equation takes the form

$$\frac{d^2}{dt^2} \mathcal{F} v_t = -\|\xi\|^2\, \mathcal{F} v_t + \mathcal{F} f_t.$$

Moreover, let us suppose that $\mathcal{F} f_t(\xi)$ depends continuously on t and ξ. Under that assumption we may, for fixed ξ, consider the equation as an inhomogeneous second-order equation with constant coefficients and continuous inhomogeneous term. As in the proof of Theorem 9.4, we write it as a first-order inhomogeneous system

$$\frac{d}{dt} V_t(\xi) = L(\xi)\, V_t(\xi) + F_t(\xi), \qquad \text{where}$$

$$V_t = \begin{pmatrix} \mathcal{F} v_t \\ \frac{d}{dt} \mathcal{F} v_t \end{pmatrix}, \qquad L(\xi) = \begin{pmatrix} 0 & 1 \\ -\|\xi\|^2 & 0 \end{pmatrix}, \qquad \text{and} \qquad F_t = \begin{pmatrix} 0 \\ \mathcal{F} f_t \end{pmatrix}.$$

Let $\Phi_t(\xi)$ be the fundamental matrix determined by $A_t(\xi)$ and $B_t(\xi)$ for the homogeneous system, more precisely,

$$\Phi_t(\xi) = \begin{pmatrix} A_t & B_t \\ \frac{d}{dt} A_t & \frac{d}{dt} B_t \end{pmatrix} = \exp t\, L(\xi) \qquad \text{and} \qquad \Phi_0(\xi) = I.$$

Next, we use Lagrange's method of variation of constants to obtain a solution of the inhomogeneous system having the form $\Phi_t(\xi)\, W_t(\xi)$. This leads to

$$V_t(\xi) = \Phi_t(\xi) \int_{t_0}^{t} \Phi_{-s}(\xi)\, F_s(\xi)\, ds = \int_{t_0}^{t} \Phi_{t-s}(\xi)\, F_s(\xi)\, ds,$$

for some arbitrary $t_0 \in \mathbf{R}$. Thus,

$$\begin{aligned} \mathcal{F} v_t &= (1,0)\, V_t(\xi) = (1,0) \int_{t_0}^{t} \Phi_{t-s}\, F_s\, ds = \int_{t_0}^{t} B_{t-s}\, \mathcal{F} f_s\, ds \\ &= \int_{t_0}^{t} \mathcal{F} b_{t-s}\, \mathcal{F} f_s\, ds = \mathcal{F}\left(\int_{t_0}^{t} b_{t-s} * f_s\, ds \right), \end{aligned}$$

and hence

$$v_t = \int_{t_0}^{t} b_{t-s} * f_s\, ds. \tag{18.15}$$

The formula for v_t is a version of *Duhamel's principle*. Recall that $\operatorname{supp} b_{t-s}$ is contained in the closed ball around 0 of radius $|t-s|$. Therefore (18.15) also makes sense for a continuous family $t \mapsto f_t$ of distributions in $\mathcal{D}'(\mathbf{R}^n)$. The family $t \mapsto v_t$ thus defined is a C^2 family in $\mathcal{D}'(\mathbf{R}^n)$.

A direct verification that the family defined in (18.15) provides a solution to the inhomogeneous equation is also feasible. Indeed,

$$\frac{d}{dt} v_t = b_{t-s} * f_s|_{s=t} + \int_{t_0}^{t} \frac{d}{dt} b_{s-t} * f_s\, ds = \int_{t_0}^{t} \frac{d}{dt} b_{s-t} * f_s\, ds$$

in view of (18.13). Using these conditions once more, we obtain

$$\begin{aligned} \frac{d^2}{dt^2} v_t &= \frac{d}{dt} b_{t-s} * f_s \Big|_{s=t} + \int_{t_0}^{t} \frac{d^2}{dt^2} b_{s-t} * f_s\, ds = \delta * f_t + \int_{t_0}^{t} \Delta b_{s-t} * f_s\, ds \\ &= f_t + \Delta v_t. \end{aligned}$$

Furthermore, the initial values can be adjusted by adding to v_t a solution of the Cauchy problem (18.14) with suitably chosen constants a and b.

If f_t is a family of distributions with $f_t = 0$ for $t < t_0$, then the lower limit t_0 may be replaced by $-\infty$. For the corresponding distribution v we thus obtain the following formula, where $\phi \in C_0^\infty(\mathbf{R}^{n+1})$:

$$v(\phi) = \int_{\mathbf{R}} v_t(\phi_t)\, dt = \int_{\mathbf{R}} \int_{s \le t} b_{t-s} * f_s(\phi_t)\, ds\, dt = \int_{\mathbf{R}} \int_{t \ge s} b_{t-s}(S f_s * \phi_t)\, dt\, ds.$$

If we take $f = \delta^{\mathbf{R}^{n+1}}$, the Dirac measure on $\mathbf{R}^{n+1}$, then certainly $t \mapsto f_t$ is not a continuous family. Yet we have obtained a formula that makes sense, namely

$$v(\phi) = \int_{\mathbf{R}_{\ge 0}} b_{t-0}(S\delta^{\mathbf{R}^n} * \phi_t)\, dt = \int_{\mathbf{R}_{>0}} b_t(\phi_t)\, dt.$$

This formula really defines a fundamental solution. In fact,

$$\begin{aligned}\Box\, v(\phi) = v(\Box\,\phi) &= \int_{\mathbf{R}_{>0}} b_t\Big(\frac{d^2}{dt^2}\phi_t\Big)\,dt - \int_{\mathbf{R}_{>0}} b_t(\Delta\phi_t)\,dt\\ &= \int_{\mathbf{R}_{>0}} b_t\Big(\frac{d^2}{dt^2}\phi_t\Big)\,dt - \int_{\mathbf{R}_{>0}} \Delta b_t(\phi_t)\,dt\\ &= \int_{\mathbf{R}_{>0}} \Big(b_t\Big(\frac{d^2}{dt^2}\phi_t\Big) - \frac{d^2}{dt^2}b_t(\phi_t)\Big)\,dt.\end{aligned}$$

Using integration by parts we now deduce

$$\Box\, v(\phi) = \Big[b_t\Big(\frac{d}{dt}\phi_t\Big) - \frac{d}{dt}b_t(\phi_t)\Big]_0^\infty = \delta^{\mathbf{R}^n}(\phi_0) = \phi(0) = \delta^{\mathbf{R}^{n+1}}(\phi).$$

Denoting the fundamental solution thus obtained by E_+, we may write

$$E_+(\phi) = \int_{\mathbf{R}_{>0}} b_t(\phi_t)\,dt \qquad (\phi \in C_0^\infty(\mathbf{R}^{n+1})).$$

We note that supp b_t is contained in the closed ball around 0 of radius $|t|$. Therefore $E_+ = 0$ on the union of the set in $\mathbf{R}\times\mathbf{R}^n$ where $t < 0$ and the set where $\|x\| > t$. In other words, supp E_+ is contained in the solid cone

$$\overline{C_+} = \{\,(t,x) \in \mathbf{R}\times\mathbf{R}^n \mid \|x\| \le t\,\}.$$

On account of the uniqueness result from Remark 13.4, this fundamental solution equals the one from Theorem 13.3. This implies that b_t is analytic on $\{\,x \in \mathbf{R}^n \mid \|x\| < t\,\}$ and even vanishes on that set for odd $n > 1$.

More precisely, we obtain the formula

$$b_t = \frac{1}{2\pi^{\frac{n-1}{2}}}\, {q_t}^*\Big(\chi_+^{\frac{3-n}{2}}\Big) \qquad (t > 0).$$

Here

$$q_t = q \circ i_t : x \mapsto t^2 - \|x\|^2 \quad : \quad \mathbf{R}^n \to \mathbf{R},$$

in other words, the composition of q by the mapping $i_t : x \mapsto (x,t)$ from $\mathbf{R}^n$ to $\mathbf{R}^{n+1}$. Furthermore, q_t is a submersion from $\mathbf{R}^n \setminus \{0\}$ to $\mathbf{R}$; there is not a real problem at $x = 0$, however, since $b_t \in C^\infty$ at $x = 0$ and $\chi_+^{\frac{3-n}{2}} \in C^\infty$ at $q = t$. One hesitates to write $b_t = {i_t}^*(E_+)$, because i_t is not a submersion. For b_t with $t < 0$ one uses

$$b_{-t} = -b_t,$$

while a_t can be expressed in terms of b_t by means of the formula

$$a_t = \frac{d}{dt}b_t. \tag{18.16}$$

The analyticity of a_t and b_t in $\|x\| < t$ can also be demonstrated directly if the paths of integration in the integral formulas for $a_t = \mathcal{F}^{-1} A_t$ and $b_t = \mathcal{F}^{-1} B_t$, respectively, are slightly rotated in suitably chosen complex directions. $\oslash$

Remark 18.8. It is natural to try to derive explicit formulas for the distributions a_t and b_t, whose respective Fourier transforms A_t and B_t are given by (18.12). In view of (18.16), only b_t needs to be determined. A uniform formula for all dimensions n is given by

$$b_t(x) = p_n \lim_{\epsilon \downarrow 0} \operatorname{Im} \left(\|x\|^2 - (t + i\epsilon)^2 \right)^{-(n-1)/2} \quad \text{with} \quad p_n = \frac{\Gamma(\frac{n+1}{2})}{\pi^{\frac{n+1}{2}} \, (n-1)}. \tag{18.17}$$

If $n = 2$, then b_t equals the locally integrable function

$$b_t = \begin{cases} \operatorname{sgn} t \, (2\pi)^{-1} \, (t^2 - \|x\|^2)^{-1/2} & \text{if} \quad \|x\| < |t|, \\ 0 & \text{if} \quad \|x\| > |t|, \end{cases}$$

as can be seen from (18.17) for $n = 2$, for which some extra work is required, however. If $n = 3$, then $b_t(\phi)$ equals t times the mean of ϕ over the sphere of radius $|t|$ about the origin. For $n = 3$, the derivation of this classical description of b_t from (18.17) involves further complications. For $n \geq 4$ it is useful to observe that

$$\left(\|x\|^2 - (t + i\epsilon)^2 \right)^{-(n-1)/2} = (n-3)^{-1} (t + i\epsilon)^{-1} \frac{d}{dt} \left(\|x\|^2 - (t + i\epsilon)^2 \right)^{-(n-3)/2},$$

which shows that the right-hand side of (18.17) can be identified with a constant times the differential operator $(t^{-1} \frac{d}{dt})^{p-1}$ of order $p - 1$ applied to the distribution

$$\lim_{\epsilon \downarrow 0} \operatorname{Im} \left(\|x\|^2 - (t + i\epsilon)^2 \right)^{-1/2} \qquad \text{or} \qquad \lim_{\epsilon \downarrow 0} \operatorname{Im} \left(\|x\|^2 - (t + i\epsilon)^2 \right)^{-1},$$

depending on whether $n = 2p$ or $n = 2p + 1$, respectively. To within a constant factor, the latter distributions equal the integral of $\left(t^2 - \|x\|^2 \right)^{-1/2} \phi(x)$ over the open ball in $\mathbf{R}^n$ of radius $|t|$ about 0, and t^{n-2} times the mean of the test function ϕ over the sphere of radius $|t|$ about 0, respectively.

Equation (18.17) may be verified as follows. The function

$$v(x, t) = \frac{1}{\sqrt{\pi t}} e^{-\frac{x^2}{4t}} \qquad (x \in \mathbf{R}, \, t > 0)$$

satisfies the heat equation $\partial_t v = \partial_x^2 v$ (see Problem 5.5), and furthermore,

$$\int_{\mathbf{R}_{>0}} v(x, t) \, dx = 1.$$

Under the assumption $x > 0$, all derivatives $\partial_t^k v(x, t)$ converge to 0 as $t \downarrow 0$ and also as $t \to \infty$. This enables us to conclude that the integral

$$I(x) := \int_{\mathbf{R}_{>0}} e^{-t}\, v(x,t)\, dt$$

converges and that

$$\begin{aligned} I''(x) &= \int_{\mathbf{R}_{>0}} e^{-t}\, \partial_x^2 v(x,t)\, dt = \int_{\mathbf{R}_{>0}} e^{-t}\, \partial_t\, v(x,t)\, dt \\ &= -\int_{\mathbf{R}_{>0}} v(x,t)\, \partial_t\, e^{-t}\, dt = I(x). \end{aligned}$$

Indeed, the first identity follows by differentiating twice under the integral sign, the second from the heat equation, and the third through integration by parts. By solving the differential equation we obtain the existence of a constant $c \in \mathbf{C}$ such that $I(x) = c\, e^{-x}$, because $I(x) \to 0$ as $x \to \infty$. As a result we have

$$\begin{aligned} c &= \int_{\mathbf{R}_{>0}} I(x)\, dx = \int_{\mathbf{R}_{>0}} \int_{\mathbf{R}_{>0}} e^{-t}\, v(x,t)\, dt\, dx \\ &= \int_{\mathbf{R}_{>0}} e^{-t} \int_{\mathbf{R}_{>0}} v(x,t)\, dx\, dt = \int_{\mathbf{R}_{>0}} e^{-t}\, dt = 1. \end{aligned}$$

This proves the *subordination identity* (see [7, Exercise 2.87.(v)] for another proof)

$$e^{-x} = \frac{1}{\sqrt{\pi}} \int_{\mathbf{R}_{>0}} \frac{e^{-t}}{\sqrt{t}}\, e^{-\frac{x^2}{4t}}\, dt \qquad (x > 0).$$

Replacing x by $\|\xi\|$ with $\xi \in \mathbf{R}^n$, then multiplying by $(2\pi)^{-n} e^{i\langle x, \xi\rangle}$, integrating over $\xi \in \mathbf{R}^n$, changing the order of integration, and finally using Example 14.11, we obtain the following formula for the inverse Fourier transform of the function $f(\xi) = e^{-\|\xi\|}$ (see (17.12) for another proof):

$$\begin{aligned} \mathcal{F}^{-1} f(x) &= \frac{1}{(2\pi)^n} \int_{\mathbf{R}_{>0}} \frac{(4\pi\, t)^{\frac{n}{2}}}{(\pi\, t)^{\frac{1}{2}}}\, e^{-t\,(1+\|x\|^2)}\, dt \\ &= \frac{\Gamma(\frac{n+1}{2})}{\pi^{\frac{n+1}{2}}}\, \frac{1}{(1+\|x\|^2)^{\frac{n+1}{2}}}. \end{aligned} \tag{18.18}$$

If we now define $f_\tau(\xi) = e^{-\tau\,\|\xi\|}$, then

$$\mathcal{F}^{-1} f_\tau(x) = \frac{\Gamma(\frac{n+1}{2})}{\pi^{\frac{n+1}{2}}}\, \frac{\tau}{(\|x\|^2 + \tau^2)^{\frac{n+1}{2}}}.$$

If $n > 1$, which we assume from now on, then $F_\tau(\xi) := e^{-\tau\,\|\xi\|}/\|\xi\|$ defines a locally integrable function of ξ whose derivative with respect to τ equals $-f_\tau(\xi)$. This leads to

$$G_\tau(x) := \mathcal{F}^{-1} F_\tau(x) = \frac{p_n}{(\|x\|^2 + \tau^2)^{\frac{n-1}{2}}}, \tag{18.19}$$

where we have used the fact that both sides converge to zero as $\tau \to \infty$, and where the constant p_n is as in (18.17).

Now F_τ is a complex-differentiable function of τ with values in $\mathcal{S}'(\mathbf{R}^n)$, defined on the complex half-plane $\{ \tau \in \mathbf{C} \mid \operatorname{Re} \tau > 0$. Furthermore, this $\mathcal{S}'(\mathbf{R}^n)$-valued function is continuous on $\{ \tau \in \mathbf{C} \mid \operatorname{Re} \tau \geq 0 \}$. Because $\mathcal{F}^{-1}$ is a continuous operator, we arrive at the same conclusions, with $G_\tau = \mathcal{F}^{-1} F_\tau$ instead of F_τ, where formula (18.19) remains valid for $\operatorname{Re} \tau > 0$. If we write $\tau = \epsilon - it = -i(t + i\epsilon)$, with $t \in \mathbf{R}$ and $\epsilon > 0$, and let $\epsilon \downarrow 0$, this leads to

$$\mathcal{F}^{-1}\Big(\frac{e^{it\|\cdot\|}}{\|\cdot\|}\Big)(x) = p_n \lim_{\epsilon \downarrow 0} \big(\|x\|^2 - (t + i\epsilon)^2\big)^{-(n-1)/2}. \tag{18.20}$$

Because $F_\tau(-\xi) = F_\tau(\xi)$, the complex conjugate of $\mathcal{F}^{-1} F_\tau$ equals the inverse Fourier transform of the complex conjugate of $F_{-\tau}$; therefore, we also obtain that $\operatorname{Im} \mathcal{F}^{-1} F_\tau = \mathcal{F}^{-1} (\operatorname{Im} F_\tau)$. On account of $B_t = \operatorname{Im} F_{it}$, (18.17) now follows from (18.20). ⊘

Remark 18.9. The theory of the wave operator $\square$ has a generalization to a wide class of linear partial differential operators $P(D)$ with constant coefficients. Let $P(\xi)$ be the symbol of P and $P_m(\xi)$ the homogeneous part of degree m of $P(\xi)$, where m is the order of P. Let $\eta \in \mathbf{R}^n \setminus \{0\}$. The operator P is said to be *hyperbolic with respect to* η if for every $\xi \in \mathbf{R}^n$, the polynomial $\sigma \mapsto P_m(\xi + \sigma\eta)$ on $\mathbf{C}$ of degree m has only real zeros. For $n > 1$, this excludes the possibility that P is elliptic, because in that case $\sigma \mapsto P_m(\xi + \sigma\eta)$ does not have any real zeros if ξ and $\eta \in \mathbf{R}^n$ and ξ is not a multiple of η.

Hyperbolicity is equivalent to the assertion that $P_m(\xi + i\tau\eta) \neq 0$ whenever $\xi \in \mathbf{R}^n$, $\tau \in \mathbf{R}$, and $\tau \neq 0$. Using the homogeneity of P_m, we obtain

$$|P_m(\xi + i\tau\eta)| \geq c\,(\|\xi\| + |\tau|)^m \qquad (\xi \in \mathbf{R}^n,\ \tau \in \mathbf{R}),$$

where c is the infimum of the left-hand side for $\|\xi\| + |\tau| = 1$; that is, $c > 0$. Because the other terms in $P(\xi)$ are of degree $< m$, there exist $\tau_0 \geq 0$ and a constant $c' > 0$ such that

$$\frac{1}{|P(\xi + i\tau\eta)|} \leq \frac{c'}{(\|\xi\| + |\tau|)^m} \qquad (\xi \in \mathbf{R}^n,\ \tau > \tau_0).$$

This allows the definition of $E \in \mathcal{D}'(\mathbf{R}^n)$ by means of the formula

$$E(\phi) = \int_{\mathbf{R}^n} \frac{\mathcal{F}^{-1}\phi(\xi + i\tau\eta)}{P(\xi + i\tau\eta)}\, d\xi \qquad (\phi \in C_0^\infty(\mathbf{R}^n)),$$

where we have chosen $\tau > \tau_0$. For $\mathcal{F}^{-1}\phi(\zeta) = (2\pi)^{-n} \mathcal{F}\phi(-\zeta)$ we use Paley–Wiener estimates of the form (18.3). In view of the identity $\mathcal{F}^{-1}(P(-D)\phi)(\zeta) = P(\zeta)\, \mathcal{F}^{-1}\phi(\zeta)$, translation of the path of integration leads to

$$P(D)E(\phi) = E(P(-D)\phi) = \int_{\mathbf{R}^n} \mathcal{F}^{-1}\phi(\xi + i\tau\eta)\, d\xi = \int_{\mathbf{R}^n} \mathcal{F}^{-1}\phi(\xi)\, d\xi = \phi(0).$$

That is, $P(D)E = \delta$; in other words, E is a fundamental solution of P.

In addition, by Cauchy's Integral Theorem (see (12.9)), the definition of E is independent of the choice of $\tau \geq \tau_0$. Using the Paley–Wiener estimates for $\mathcal{F}\phi(-\zeta)$ we see that the limit as $\tau \to \infty$ equals 0 if $\langle x, \eta \rangle > 0$ for all $x \in \operatorname{supp} \phi$. That is,

$$\operatorname{supp} E \subset \{ x \in \mathbf{R}^n \mid \langle x, \eta \rangle \leq 0 \}.$$

We obtain a fundamental solution with support in the opposite half-space by substituting $-\eta$ for η.

Moreover, the Cauchy problem with initial values on the hyperplane $\{ x \in \mathbf{R}^n \mid \langle x, \eta \rangle = 0 \}$ can be shown to have a unique solution. In this text, we will not elaborate the theory any further; the main purpose of the remarks above was to demonstrate that the ideas underlying the Paley–Wiener Theorems can find application in quite a few situations. ⊘

Remark 18.10. The constructions of parametrices and solutions of the Cauchy problem of hyperbolic differential equations with **variable** coefficients find their natural context in the class of the so-called *Fourier integral operators.* These form an extension of the class of pseudo-differential operators mentioned in Remark 17.8. See, for example, Hörmander [11, Chap. 25] or Duistermaat [6]. ⊘

Suppose we want to reconstruct a function $f \in \mathcal{S}(\mathbf{R})$ from its values *sampled* at the points of the lattice $\omega\mathbf{Z}$ in $\mathbf{R}$, for $\omega > 0$. The process of sampling may be represented by

$$f_{\text{sam}} := \sum_{k \in \mathbf{Z}} f(k\omega)\, \delta_{k\omega} = \sum_{k \in \mathbf{Z}} f(k\omega)\, T_{k\omega}\delta \in \mathcal{S}'(\mathbf{R}).$$

Then, using (14.35) and applying (16.8) with x replaced by $-\xi$ and ϕ by $\mathcal{F}^{-1} f = \frac{1}{2\pi} S \mathcal{F} f$, we obtain

$$\mathcal{F} f_{\text{sam}}(\xi) = \sum_{k \in \mathbf{Z}} f(k\omega)\, e^{-ik\omega\xi} = \frac{a}{2\pi} \sum_{k \in \mathbf{Z}} \mathcal{F} f(\xi + ka) \qquad (\xi \in \mathbf{R}). \tag{18.21}$$

Here, as usual, $a\,\omega = 2\pi$. For a suitable function u that is still to be determined, we introduce the reconstructed function by

$$f_{\text{rec}} := f_{\text{sam}} * u, \qquad \text{and therefore} \qquad \mathcal{F} f_{\text{rec}} = \mathcal{F} f_{\text{sam}}\, \mathcal{F} u. \tag{18.22}$$

Furthermore, assume f to be *band-limited,* more precisely that $\operatorname{supp} \mathcal{F}^{-1} f \subset [-\rho, \rho]$. Then

$$a = 2\rho \qquad \text{and} \qquad \omega = \frac{\pi}{\rho}. \tag{18.23}$$

Select also u such that, see Example 14.1,

$$\frac{a}{2\pi} \mathcal{F} u = 1_{[-\rho, \rho]} \in \mathcal{E}'(\mathbf{R}), \qquad \text{that is,} \qquad u(x) = \operatorname{sinc} \rho x. \tag{18.24}$$

Under these assumptions, we obtain, by combination of (18.22), (18.21), and (18.24),

$$\mathcal{F} f_{\rm rec}(\xi) = 1_{[-\rho,\rho]}(\xi) \sum_{k\in\mathbf{Z}} \mathcal{F} f(\xi + ka) = \mathcal{F} f(\xi).$$

Hence, by Fourier inversion we obtain the following *cardinal series expansion* of f, valid for all $x \in \mathbf{R}$:

$$\begin{aligned} f(x) = f_{\rm rec}(x) &= \sum_{k\in\mathbf{Z}} f(k\omega)\, u(x - k\omega) = \sum_{k\in\mathbf{Z}} f(k\omega)\, \mathrm{sinc}\, \rho(x - k\omega) \\ &= \sum_{k\in\mathbf{Z}} f(k\omega)\, \mathrm{sinc}(\rho x - k\pi) = \frac{\sin \rho x}{\rho} \sum_{k\in\mathbf{Z}} \frac{(-1)^k f(k\omega)}{x - k\omega}. \end{aligned} \tag{18.25}$$

The identity (18.25) is the *Sampling Theorem*, which is due to Whittaker, Kotel'nikov, and Shannon, in the case that

$$g := \mathcal{F}^{-1} f \in C_0^\infty(\mathbf{R}).$$

It asserts that perfect reconstruction of the band-limited function $f = \mathcal{F} g$ is possible. Of course, this fact is related to the special nature of f: it can be extended to $\mathbf{C}$ as a complex-analytic function satisfying estimates as in (18.3), for every $k \in \mathbf{Z}_{\geq 0}$.

Remark 18.11. Note that (18.25) is not true for arbitrary distributions g satisfying $\mathrm{supp}\, g \subset [-\rho, \rho]$. For instance, if $f(x) = \sin \rho x$, then $g = \frac{i}{2}(\delta_\rho - \delta_{-\rho})$, but $f(k\omega) = \sin k\rho\omega = \sin k\pi = 0$, for all $k \in \mathbf{Z}$. The inverse Fourier transform of a nonconstant polynomial even has support in $\{0\}$, while the cardinal series of such a polynomial is pointwise divergent.

On the other hand, (16.27) implies that the Fourier series of the function e_{it} on $]-\pi, \pi[$, for $t \in \mathbf{R}$, is given by

$$e^{itx} = \frac{\sin \pi t}{\pi} \sum_{k\in\mathbf{Z}} \frac{(-1)^k}{t-k} e^{ikx} \qquad (-\pi < x < \pi). \tag{18.26}$$

However, for fixed $-\pi < x < \pi$, the function $f = e_{ix}$ on $\mathbf{R}$ satisfies $g = \delta_{-x}$ and thus $\mathrm{supp}\, g \subset [-\pi, \pi]$. Therefore (18.26) may be recognized as the cardinal series of f. Here we have to interpret the series as a limit of symmetric partial sums. ⊘

By expanding the *periodization* of g into a Fourier series, we now extend (18.25) to the more general case of $g \in \mathcal{E}'(\mathbf{R})$ satisfying the extra condition (18.28) below. In Remark 18.13 we will discuss properties of g or f that imply this condition.

Theorem 18.12. *Let $\rho > 0$ and $g \in \mathcal{E}'(\mathbf{R})$ with $\mathrm{supp}\, g \subset [-\rho, \rho]$ and define $f = \mathcal{F} g \in C^\infty(\mathbf{R})$. Select $\phi \in C_0^\infty(\mathbf{R})$ with $1(\phi) = 1$ and define ϕ_ϵ as in Lemma 2.19. Put*

$$\chi_\epsilon = \phi_\epsilon * 1_{[-\rho,\rho]} \in C_0^\infty(\mathbf{R}) \tag{18.27}$$

and suppose

$$g = \lim_{\epsilon \downarrow 0} \chi_\epsilon (g - T_a\, g) \quad \text{in} \quad \mathcal{E}'(\mathbf{R}). \tag{18.28}$$

Then we have

$$f = \sum_{k \in \mathbf{Z}} f(k\omega)\,(T_{k\omega} \circ \rho^*)\,\mathrm{sinc} \quad \mathit{in} \quad \mathcal{S}'(\mathbf{R}). \tag{18.29}$$

Proof. We employ the usual notation as in (18.23). Consider $h = \sum_{k \in \mathbf{Z}} T_{ka}\, g$ in $\mathcal{D}'(\mathbf{R})$. Then h is periodic of period a and so belongs to $\mathcal{S}'(\mathbf{R})$ in view of Theorem 16.3. On account of its periodicity we may expand h into its Fourier series, while Lemma 16.4 implies

$$h = \sum_{k \in \mathbf{Z}} c_k\, e_{ik\omega} = S\mathcal{F}\, v, \qquad \text{where} \qquad v = \sum_{k \in \mathbf{Z}} c_k\, \delta_{k\omega}.$$

By means of Theorem 16.6, Definition 16.2, and Lemma 16.1 we derive

$$\begin{aligned} c_k &= M(e_{-ik\omega}\, h) = \frac{1}{a} e_{-ik\omega}\, h(\mu) = \frac{1}{a} h(e_{-ik\omega}\mu) = \frac{1}{a} \sum_{n \in \mathbf{Z}} T_{na}\, g(e_{-ik\omega}\mu) \\ &= \frac{1}{a} g\Big(\sum_{n \in \mathbf{Z}} T_{-na}(e_{-ik\omega}\mu)\Big) = \frac{1}{a}\, g\Big(e_{-ik\omega} \sum_{n \in \mathbf{Z}} T_{-na}\mu\Big) = \frac{1}{a}\, g(e_{-ik\omega}) \\ &= \frac{1}{a} \mathcal{F} g(k\omega) = \frac{1}{a} f(k\omega). \end{aligned}$$

Next, Fourier inversion leads to

$$\mathcal{F} h = 2\pi\Big(\frac{1}{2\pi} \mathcal{F}\, S\mathcal{F}\Big) v = 2\pi\, v = \omega \sum_{k \in \mathbf{Z}} f(k\omega)\, \delta_{k\omega}.$$

Furthermore, condition (18.28) implies

$$g = \lim_{\epsilon \downarrow 0} \chi_\epsilon\, h \quad \text{in} \quad \mathcal{S}'(\mathbf{R}), \qquad \text{hence} \qquad f = \mathcal{F} g = \lim_{\epsilon \downarrow 0} \mathcal{F}(\chi_\epsilon\, h) \quad \text{in} \quad \mathcal{S}'(\mathbf{R}).$$

On the other hand, we have

$$\mathcal{F}(\chi_\epsilon\, h) = \frac{1}{2\pi} \mathcal{F} \chi_\epsilon * \mathcal{F} h = \frac{1}{2\pi} \mathcal{F}(\phi_\epsilon * 1_{[-\rho,\rho]}) * \mathcal{F} h = \frac{1}{2\pi} (\mathcal{F} \phi_\epsilon\, \mathcal{F} 1_{[-\rho,\rho]}) * \mathcal{F} h.$$

Write $\psi = \mathcal{F}\phi$, then $\mathcal{F}\phi_\epsilon = \epsilon^*\psi$ in view of Problem 14.30.(ii), while $\mathcal{F} 1_{[-\rho,\rho]} = a\, \rho^*\, \mathrm{sinc}$ as follows from (18.24). Accordingly

$$\begin{aligned} \mathcal{F}(\chi_\epsilon\, h) &= \frac{a\omega}{2\pi} ((\epsilon^*\psi)\,(\rho^*\,\mathrm{sinc})) * \sum_{k \in \mathbf{Z}} f(k\omega)\, \delta_{k\omega} \\ &= \sum_{k \in \mathbf{Z}} f(k\omega)\, T_{k\omega}((\epsilon^*\psi)\,(\rho^*\,\mathrm{sinc})). \end{aligned}$$

Now $\psi(0) = 1(\phi) = 1$. On account of Theorem 18.1 we know that f is at most of polynomial growth and that ψ is rapidly decreasing on $\mathbf{R}$; therefore the series on the right-hand side converges absolutely and uniformly on $\mathbf{R}$, which leads to

$$f = \lim_{\epsilon \downarrow 0} \sum_{k \in \mathbf{Z}} f(k\omega)\, T_{k\omega}((\epsilon^* \psi)\, (\rho^* \operatorname{sinc})) = \sum_{k \in \mathbf{Z}} f(k\omega)\, (T_{k\omega} \circ \rho^*) \operatorname{sinc}$$

in $\mathcal{S}'(\mathbf{R})$. This finishes the proof. □

Remark 18.13. We discuss some conditions on the function g that imply that (18.28) is satisfied. If supp g is contained in the interior of $[-\rho, \rho]$, then there exists $\epsilon_0 > 0$ such that $g = \chi_\epsilon(g - T_a\, g)$ as soon as $0 < \epsilon \le \epsilon_0$. Furthermore, (18.28) holds if there exist neighborhoods of the points $\pm\rho$ on which g is a locally integrable function.

Of course, we are also interested in conditions that guarantee the pointwise convergence of (18.29) and that are weaker than those for (18.25). Therefore, let us suppose in addition to (18.28) that

$$\sum_{k \in \mathbf{Z}} \frac{|f(k\omega)|}{1 + |k|} < \infty. \tag{18.30}$$

Then the series on the right-hand side in (18.29) converges absolutely and uniformly on $\mathbf{R}$, which implies that its sum is a continuous function. On the other hand, (18.29) is an equality in the weak sense, but then pointwise too, because both sides in (18.29) are continuous.

Because the cardinal series of f is uniformly convergent on $\mathbf{R}$ in this case, termwise integration of the series is admissible. Problem 14.44 asserts

$$\int_{\mathbf{R}} \operatorname{sinc}(\rho x - k\pi)\, dx = \int_{\mathbf{R}} \operatorname{sinc} \rho x\, dx = \frac{\pi}{\rho} = \omega;$$

in consequence, we obtain the following integration formula:

$$\int_{\mathbf{R}} f(x)\, dx = \omega \sum_{k \in \mathbf{Z}} f(k\omega). \tag{18.31}$$

Observe that (18.30) is certainly the case if g is sufficiently smooth; in fact, Hölder continuity is enough according to Problem 14.11. In addition, (18.30) is also satisfied if $g \in L^2(\mathbf{R})$, or equivalently on account of Theorem 14.32 if $f \in L^2(\mathbf{R})$. Indeed, g then is locally integrable, and considering it as a function that is periodic of period a, we have its Fourier series

$$g = \frac{1}{a} \sum_{k \in \mathbf{Z}} f(k\omega)\, e_{ik\omega}; \qquad \text{and so} \qquad \sum_{k \in \mathbf{Z}} |f(k\omega)|^2 = \int_{-\rho}^{\rho} |g(x)|^2\, dx < \infty$$

according to Parseval's formula from Theorem 16.9. The Cauchy–Schwarz inequality in $l^2(\mathbf{Z})$ then leads to

$$\Big(\sum_{k \in \mathbf{Z} \setminus \{0\}} \frac{|f(k\omega)|}{|k|} \Big)^2 \le 2 \sum_{k \in \mathbf{Z}} |f(k\omega)|^2 \sum_{k \in \mathbf{N}} \frac{1}{k^2} < \infty. \tag{18.32}$$

Finally, consider $f \in L^1(\mathbf{R})$. On account of the Riemann–Lebesgue Theorem, Theorem 14.2, we have $g \in C(\mathbf{R})$, which implies $g \in L^2(\mathbf{R})$, and therefore $f \in L^2(\mathbf{R})$.

Example 18.14. Now we study the cardinal series (18.25) from the point of view of Hilbert space theory. Consider $g \in L^2(\mathbf{R})$ satisfying $\operatorname{supp} g \subset [-\rho, \rho]$ and write $f = \mathcal{F} g$. Then f is a complex-analytic function on $\mathbf{C}$ according to Lemma 14.4, while there exist constants $c > 0$ and $k \in \mathbf{Z}_{\geq 0}$ such that

$$f(\zeta) \leq c(1 + |\zeta|)^k\, e^{\rho|\operatorname{Im}\zeta|} \qquad (\zeta \in \mathbf{C}); \qquad \text{in addition,} \qquad f \in L^2(\mathbf{R}).$$

Denote the linear space of all such functions f by $\mathrm{PW}^2_\rho(\mathbf{C})$ and introduce an inner product on $\mathrm{PW}^2_\rho(\mathbf{C})$ by $(f, h)_{\mathrm{PW}} = \int_{\mathbf{R}} f(\xi)\, \overline{h(\xi)}\, d\xi$. In fact, the complex analyticity of f guarantees that this is an inner product. Denote by $\|\cdot\|_{\mathrm{PW}}$ the corresponding norm on $\mathrm{PW}^2_\rho(\mathbf{C})$. Now Theorems 14.32 and 18.6 imply that

$$\widetilde{\mathcal{F}}^{-1} = S\widetilde{\mathcal{F}} = \frac{1}{\sqrt{2\pi}} S\mathcal{F} : \mathrm{PW}^2_\rho(\mathbf{C}) \to L^2([-\rho, \rho]) \subset L^2(\mathbf{R})$$

is a unitary isomorphism. It follows that $\mathrm{PW}^2_\rho(\mathbf{C})$ is a Hilbert space, which is called the *Paley–Wiener space* of type ρ.

Define $e^+_{ik\omega} = \frac{1}{\sqrt{a}}\, e_{ik\omega}\, 1_{[-\rho,\rho]} \in \mathcal{E}'(\mathbf{R})$, for $k \in \mathbf{Z}$. Then $(e^+_{ik\omega})_{k\in\mathbf{Z}}$ forms an orthonormal basis in $L^2([-\rho, \rho])$. Using Example 14.1 and (14.29) one easily computes, for z and $\zeta \in \mathbf{C}$,

$$\widetilde{\mathcal{F}}^{-1} e^+_{iz\omega} = -s_{-z}, \qquad \text{where} \qquad s_z(\zeta) = \frac{1}{\sqrt{\omega}} \operatorname{sinc} \rho(\zeta - z\,\omega). \tag{18.33}$$

It follows that $(s_k)_{k\in\mathbf{Z}}$ is an orthonormal basis in $\mathrm{PW}^2_\rho(\mathbf{C})$. In particular, in the case of $\rho = \pi$, the identity $\|s_0\|_{\mathrm{PW}} = 1$ implies $\int_{\mathbf{R}} \operatorname{sinc}^2 x\, dx = \pi$; compare with Problem 14.44. Furthermore, one finds that each function $f \in \mathrm{PW}^2_\rho(\mathbf{C})$ has a unique expansion of the form

$$f = \sum_{k\in\mathbf{Z}} c_k\, s_k \qquad \text{with} \qquad \sum_{k\in\mathbf{Z}} |c_k|^2 < \infty. \tag{18.34}$$

The convergence of the series on the left-hand side is understood with respect to the norm $\|\cdot\|_{\mathrm{PW}}$. But convergence in $\mathrm{PW}^2_\rho(\mathbf{C})$ implies uniform convergence on each horizontal strip in $\mathbf{C}$. Indeed, on account of the Cauchy–Schwarz inequality we obtain, for ξ and $\eta \in \mathbf{R}$,

$$\begin{aligned}|f(\xi + i\eta)| &= \left|\int_{-\rho}^{\rho} e^{-i(\xi+i\eta)x} g(x)\, dx\right| \leq e^{\rho|\eta|} \int_{-\rho}^{\rho} |g(x)|\, dx \leq \sqrt{a}\, e^{\rho|\eta|}\, \|g\| \\ &= \frac{e^{\rho|\eta|}}{\sqrt{\omega}} \|f\|_{\mathrm{PW}}.\end{aligned}$$

If we evaluate the identity in (18.34) at $k\omega$, we see that $c_k = \sqrt{\omega}\, f(k\omega)$; and as a consequence we obtain another proof of the cardinal series for f:

$$f(x) = \sqrt{\omega} \sum_{k \in \mathbf{Z}} f(k\omega)\, s_k(x) = \sum_{k \in \mathbf{Z}} f(k\omega)\ \operatorname{sinc} \rho(x - k\omega) \qquad (x \in \mathbf{R}).$$

In particular, Parseval's formula in $\mathrm{PW}^2_\rho(\mathbf{C})$ leads to the following integration formula:

$$\int_{\mathbf{R}} |f(x)|^2\, dx = \omega \sum_{k \in \mathbf{Z}} |f(k\omega)|^2, \tag{18.35}$$

which as in (18.32) corroborates the pointwise convergence of the cardinal series.

Furthermore, since $\widetilde{\mathcal{F}}^{-1}$ is unitary, we have, denoting the complex conjugate of $\zeta \in \mathbf{C}$ by $\overline{\zeta}$ and using (18.33),

$$\begin{aligned} f(\zeta) &= \int_{-\rho}^{\rho} e^{-ix\zeta}\, g(x)\, dx = \sqrt{a} \int_{\mathbf{R}} g(x)\, \frac{1}{\sqrt{a}}\, e^{-i\frac{\zeta}{\omega}\omega x}\, 1_{[-\rho,\,\rho]}(x)\, dx \\ &= \sqrt{a}\left(g,\, e^{+}_{i\frac{\overline{\zeta}}{\omega}\omega}\right) = \sqrt{a}\left(\widetilde{\mathcal{F}}^{-1} g,\, \widetilde{\mathcal{F}}^{-1} e^{+}_{i\frac{\overline{\zeta}}{\omega}\omega}\right)_{\mathrm{PW}} = -\frac{1}{\omega}\Big(Sf,\, \operatorname{sinc} \rho(\,\cdot + \overline{\zeta})\Big)_{\mathrm{PW}} \\ &= \frac{1}{\omega} \int_{\mathbf{R}} f(x)\ \operatorname{sinc} \rho(\zeta - x)\, dx = f * k(\zeta) \qquad \text{with} \qquad k(z) = \frac{1}{\pi}\, \frac{\sin \rho z}{z}. \end{aligned}$$

In other words, f is a solution to *Bateman's integral equation* for the unknown function g:

$$f(\zeta) = \frac{1}{\pi} \int_{\mathbf{R}} g(x)\, \frac{\sin \rho(\zeta - x)}{\zeta - x}\, dx.$$

In addition, we obtain the integration formula

$$\int_{\mathbf{R}} f(x)\ \operatorname{sinc} \rho(\zeta - x)\, dx = \omega \sum_{k \in \mathbf{Z}} f(k\omega)\ \operatorname{sinc} \rho(\zeta - k\omega).$$

Finally, by writing

$$K(z, \zeta) = k(z - \overline{\zeta}) = \frac{1}{\pi}\, \frac{\sin \rho(z - \overline{\zeta})}{z - \overline{\zeta}} \qquad ((z, \zeta) \in \mathbf{C} \times \mathbf{C}),$$

we obtain the *reproducing kernel* K for the Hilbert space $\mathrm{PW}^2_\rho(\mathbf{C})$ satisfying

$$f(\zeta) = (f,\, K(\,\cdot\,, \zeta))_{\mathrm{PW}} \qquad (\zeta \in \mathbf{C}). \qquad \oslash$$

Problems

18.1.* *(Fast convergence of Riemann sums for test functions.)* Let $\phi \in C_0^\infty(\mathbf{R})$ and let $k \in \mathbf{Z}_{>1}$ be arbitrary. Prove that, for all $N \in \mathbf{N}$,

$$\int_{\mathbf{R}} \phi(x)\, dx = \frac{1}{N} \sum_{n\in\mathbf{Z}} \phi\Big(\frac{n}{N}\Big) + O\Big(\frac{1}{N^k}\Big), \qquad N \to \infty.$$

Here the implied constant may depend on ϕ and k.

18.2. (Sequel to Problem 14.47.) Prove that there exist constants $c > 0$ and $k \in \mathbf{Z}_{\geq 0}$ such that

$$|u(z)| \leq c\,(1 + \|z\|)^k\, e^{\sqrt{-\lambda}\,\|\operatorname{Im} z\|} \qquad (z \in \mathbf{C}^n).$$

18.3. Suppose that $Y \subset \mathbf{R}^n$ consists of a basis $\eta(j)$ of $\mathbf{R}^n$, where $1 \leq j \leq n$, supplemented by a vector $\eta(n+1) = \sum_{j=1}^n c_j\, \eta(j)$, with $c_j < 0$ for all j. Further suppose that U is a complex-analytic function on $\mathbf{C}^n$ and that for every $\epsilon > 0$ there exist constants $c > 0$ and $N \in \mathbf{R}$ with

$$|U(\xi + i\tau\,\eta)| \leq c\,(1 + \|\xi\|)^N\, e^{\epsilon\,\tau} \qquad (\xi \in \mathbf{R}^n,\ \eta \in Y,\ \tau \geq 0).$$

Prove that U is a polynomial.

18.4. *(Klein–Gordon operator.)* Discuss fundamental solutions and the Cauchy problem for $P = \Box + \lambda$, where $\lambda \in \mathbf{C}$ is a given constant. (For $\lambda = m^2 > 0$ this is the *Klein–Gordon operator*.)

18.5. Let u be an integrable function on $\mathbf{R}$. Let $H = \{\zeta \in \mathbf{C} \mid \operatorname{Im}\zeta < 0\}$ and denote by $\overline{H}$ the closure of H in $\mathbf{C}$. Prove that $\operatorname{supp} u \subset \mathbf{R}_{\geq 0}$ if and only if $\mathcal{F}u$ has an extension to a bounded continuous function U on $\overline{H}$ that is complex-analytic on H.

18.6.* *(Distributions as boundary values of holomorphic functions.)* We will prove a result converse to that in Problem 12.14. In doing so, we use the notation and results from this problem. Consider $u \in \mathcal{E}'(\mathbf{R})$.

(i) Prove that $f_\pm : H_\pm \to \mathbf{C}$ are well-defined and belong to $f_\pm \in \mathcal{O}(H_\pm)$ if

$$f_\pm(z) = \frac{1}{2\pi} \int_{\mathbf{R}_{\gtrless 0}} e^{iz\,\xi}\, \mathcal{F}u(\xi)\, d\xi \qquad (z \in H_\pm).$$

(ii) Show the existence of $N \in \mathbf{Z}_{\geq 0}$ and $c > 0$ such that $|\operatorname{Im} z|^N\, |f_\pm(z)| \leq c$, for $|\operatorname{Im} z|$ sufficiently small.

(iii) Verify that $u = \sum_\pm \beta_\pm(f_\pm)$.

18.7. *(Conservation of energy for wave equation.)* The *energy* $E(t)$ of a solution u_t of the wave equation as in Example 18.7 is defined as

$$E(t) = \frac{1}{2} \int_{\mathbf{R}^n} \Big(\Big|\frac{d}{dt} u_t(x)\Big|^2 + \|\operatorname{grad} u_t(x)\|^2 \Big)\, dx,$$

under the assumption of convergence of the integral for all $t \in \mathbf{R}$ (see Problem 19.3.(ii)). The first term is *kinetic energy* and the second *potential energy*. Show that

$$E(t) = \frac{1}{2(2\pi)^n} \int_{\mathbf{R}^n} \left(\left| \frac{d}{dt} \mathcal{F} u_t(\xi) \right|^2 + \|\xi\|^2 |\mathcal{F} u_t(\xi)|^2 \right) d\xi$$

and use (18.10) to prove that

$$E(t) = \frac{1}{2(2\pi)^n} \int_{\mathbf{R}^n} \left(\|\xi\|^2 |\mathcal{F} u_0(\xi)|^2 + \left| \mathcal{F} \frac{d}{dt} u_t \Big|_{t=0} (\xi) \right|^2 \right) d\xi,$$

which is independent of t.

18.8. *(Fourier transform of $x \mapsto e^{-\|x\|}$ and Poisson's integral formula for half-space – higher-dimensional generalization of Problem 14.51.)* For $t > 0$, let $f_t : \mathbf{R}^n \to \mathbf{R}$ be given by $f_t(x) = e^{-t\,\|x\|}$.

(i) Prove that $f_1 \in \mathcal{S}'(\mathbf{R})$ and also (compare with Problems 14.14 and 14.15)

$$\mathcal{F} f_1(\xi) = 2^n \pi^{\frac{n-1}{2}} \Gamma\left(\frac{n+1}{2}\right) \frac{1}{(1 + \|\xi\|^2)^{\frac{n+1}{2}}} \qquad (\xi \in \mathbf{R}^n).$$

Deduce, for all $\xi \in \mathbf{R}^n$ and $t > 0$, that

$$\mathcal{F} f_t(\xi) = 2^n \pi^{\frac{n-1}{2}} \Gamma\left(\frac{n+1}{2}\right) \frac{t}{(\|\xi\|^2 + t^2)^{\frac{n+1}{2}}}.$$

Let $\mathbf{R}^{n+1}_* = \mathbf{R}^{n+1} \setminus \{0\}$. Define *Poisson's kernel* $P : \mathbf{R}^{n+1}_* \to \mathbf{R}$ by

$$P(x,t) = \frac{\Gamma(\frac{n+1}{2})}{\pi^{\frac{n+1}{2}}} \frac{|t|}{\|(x,t)\|^{n+1}}.$$

Note that $P(x,-t) = P(x,t)$. In statistics the function $P(\cdot, 1) : \mathbf{R}^n \to \mathbf{R}$ is said to be the *probability density of the Cauchy distribution*.

(ii) Conclude that $\int_{\mathbf{R}^n} P(x,t)\, dx = 1$, for all $t \neq 0$. More generally, verify that

$$\int_{\mathbf{R}^n} \frac{\cos\langle\, \xi, x\,\rangle}{(\|x\|^2 + t^2)^{\frac{n+1}{2}}}\, dx = \frac{\pi^{\frac{n+1}{2}}}{\Gamma(\frac{n+1}{2})} \frac{e^{-t\|\xi\|}}{t} \qquad (\xi \in \mathbf{R}^n,\ t > 0),$$

and show that one therefore has the following, known as *Laplace's integrals*:

$$\int_{\mathbf{R}_{>0}} \frac{\cos \xi x}{x^2 + t^2}\, dx = \frac{\pi}{2} \frac{e^{-t\xi}}{t} \qquad (\xi \geq 0,\ t > 0), \tag{18.36}$$

$$\int_{\mathbf{R}_{>0}} \frac{x \sin \xi x}{x^2 + t^2}\, dx = \frac{\pi}{2} e^{-t\xi} \qquad (\xi > 0,\ t \geq 0).$$

Use the identity $\int_{\mathbf{R}_{>0}} \operatorname{sinc} x\, dx = \frac{\pi}{2}$ from, for instance, Problem 14.44 to obtain the validity of the latter formula for $t = 0$.

(iii) Deduce from part (i) that we have the following *semigroup property*, for all t_1 and $t_2 > 0$:

$$P(\cdot\,, t_1 + t_2) = P(\cdot\,, t_1) * P(\cdot\,, t_2).$$

(iv) Let $\delta > 0$ be arbitrary. Check that $\int_{\{x\in\mathbf{R}^n \mid \|x\|>\delta\}} \|x\|^{-n-1}\,dx < \infty$, and use this to prove

$$\lim_{t\to 0}\int_{\{x\in\mathbf{R}^n \mid \|x\|>\delta\}} P(x,t)\,dx = 0.$$

Define $\mathbf{R}^{n+1}_{\pm} = \{\,(x,t)\in\mathbf{R}^{n+1} \mid \pm t > 0\,\}$.

(v) From Example 12.3 we know that the functions

$$(x,t)\mapsto \log\|(x,t)\| \qquad (n=1); \qquad (x,t)\mapsto \frac{1}{\|(x,t)\|^{n-1}} \qquad (n>1),$$

are harmonic on $\mathbf{R}^{n+1}_{*}$. Verify that, leaving scalars aside, P on $\mathbf{R}^{n+1}_{\pm}$ is the partial derivative with respect to t of the preceding function, and conclude that P is a harmonic function on $\bigcup_{\pm}\mathbf{R}^{n+1}_{\pm}$.

Let $h\in C(\mathbf{R}^n)$ be bounded and define *Poisson's integral* $\mathcal{P}h : \bigcup_{\pm}\mathbf{R}^{n+1}_{\pm}\to\mathbf{R}$ of h by

$$(\mathcal{P}h)(x,t) = (h * P(\cdot,t))(x) = \int_{\mathbf{R}^n} h(x-y)P(y,t)\,dy.$$

(vi) Prove that $\mathcal{P}h$ is a well-defined harmonic function on $\bigcup_{\pm}\mathbf{R}^{n+1}_{\pm}$. Verify by means of parts (ii) and (iv) that, uniformly for x in compact sets in $\mathbf{R}^n$,

$$\lim_{\pm t\downarrow 0}(\mathcal{P}h)(x,t) = h(x).$$

Evidently, in the terminology of Problem 12.4, the function $f = \mathcal{P}h$ is a solution of the following Dirichlet problem on $U = \mathbf{R}^{n+1}_{+}$ or $\mathbf{R}^{n+1}_{-}$:

$$\Delta f = 0 \qquad \text{with} \qquad f|_{\partial U} = h.$$

(vii) Show that, with c_n as in (13.37),

$$(\mathcal{P}h)(x,t) = \frac{2}{c_n}\int_{\mathbf{R}^n}\frac{h(x+ty)}{\|(y,1)\|^{n+1}}\,dy \qquad ((x,t)\in\mathbf{R}^{n+1}_{\pm}).$$

(viii) Write $h = \frac{1}{2}h - (-\frac{1}{2}h)$ and prove the existence of a harmonic function u on $\bigcup_{\pm}\mathbf{R}^{n+1}_{\pm}$ with

$$\lim_{t\downarrow 0}u(x,t) - \lim_{t\uparrow 0}u(x,t) = h(x) \qquad (x\in\mathbf{R}^n).$$

Evidently, a bounded continuous function on $\mathbf{R}^n$ can be represented by means of the jump along $\mathbf{R}^n\times\{0\}$ made by a suitably chosen harmonic function on $\bigcup_{\pm}\mathbf{R}^{n+1}_{\pm}$.

(ix) What can be said if $h\in C(\mathbf{R}^n)$ above is replaced by $u\in\mathcal{E}'(\mathbf{R}^n)$?

18.9.* Let $t\in\mathbf{R}$. For $f:\mathbf{R}\to\mathbf{R}$ given by

$$f(x) = \operatorname{sinc}(\pi x + t) \qquad \text{show that} \qquad \mathcal{F}f = e_{-i\frac{t}{\pi}}\,1_{[-\pi,\pi]}.$$

Deduce that $\operatorname{supp} \mathcal{F} f = [-\pi, \pi]$ and derive $\sum_{k\in\mathbf{Z}} \operatorname{sinc}^2(\pi k + t) = 1$; compare with (16.31). Similarly, verify that

$$\mathcal{F} f^2(\xi) = \frac{e^{-it\frac{\xi}{\pi}}}{2} \max\left\{0, 2 - \frac{|\xi|}{\pi}\right\}; \qquad \text{deduce} \qquad \sum_{k\in\mathbf{Z}} \operatorname{sinc}^2\left(\frac{\pi}{2}k + t\right) = 2.$$

18.10. Verify that $\mathrm{PW}^2_\rho(\mathbf{C})$ is closed under differentiation, for every $\rho > 0$.

18.11. For $k \in \mathbf{Z}$, define $t_k : \mathbf{C} \to \mathbf{C}$ by

$$t_k(\zeta) = \frac{1}{\sqrt{\omega}} \frac{\sin^2 \frac{\rho}{2}(\zeta + k\omega)}{\frac{\rho}{2}(\zeta + k\omega)}, \qquad \text{and show that} \qquad \widetilde{\mathcal{F}}^{-1}(-i \operatorname{sgn} \cdot e^+_{ik\omega}) = t_k.$$

In particular, deduce that $(t_k)_{k\in\mathbf{Z}}$ forms an orthonormal basis in $\mathrm{PW}^2_\rho(\mathbf{C})$. Using Parseval's formula and the notation and results from Problem 14.49, prove that, for $f \in \mathrm{PW}^2_\rho(\mathbf{C})$,

$$\int_{\mathbf{R}} f(x)\, t_k(x)\, dx = -\sqrt{\omega}\, \mathcal{H} f(-k\omega), \qquad \mathcal{H} f = \sum_{k\in\mathbf{Z}} f(k\omega) \frac{\sin^2 \frac{1}{2}(\rho \cdot -k\pi)}{\frac{1}{2}(\rho \cdot -k\pi)}.$$

Verify the truth of the latter formula in the case of $f = \frac{\sin \rho \cdot}{\cdot}$ by means of Problem 14.49.(vii).

Chapter 19
Sobolev Spaces

The degree of differentiability of a function u with compact support cannot easily be seen from the Fourier transform $\mathcal{F}u$. For instance, there is a difference of more than n between the exponents in the estimates for $\mathcal{F}u$ in Theorems 18.1 and 18.6, which are necessary and sufficient, respectively, for u to be of class C^k. The question how to decide by inspection whether a function is the Fourier transform of a bounded continuous function is very difficult to answer in general.

In contrast, the situation for the space $L^2(\mathbf{R}^n)$ and the corresponding L^2-norm is much simpler, because the mapping $\widetilde{\mathcal{F}} = (2\pi)^{-\frac{n}{2}} \mathcal{F}$ is a unitary isomorphism from $L^2(\mathbf{R}^n)$ to $L^2(\mathbf{R}^n)$ according to Theorem 14.32. Now $\mathcal{F} D^\alpha u = \xi^\alpha \mathcal{F} u$, see (14.28), implying that u is an L^2 function whose derivatives of order $\leq k$ all belong to $L^2(\mathbf{R}^n)$ if and only if $(1 + \|\xi\|)^k \mathcal{F} u \in L^2(\mathbf{R}^n)$. Hence, in this case the degree of differentiability can be derived from the Fourier transform. The space of these functions u is denoted by $H_{(k)} = H_{(k)}(\mathbf{R}^n)$. The following extension of this definition proves very flexible.

Definition 19.1. For every real number s we define the *Sobolev space* $H_{(s)} = H_{(s)}(\mathbf{R}^n)$ *of order* s as the space of all $u \in \mathcal{S}'(\mathbf{R}^n)$ such that $(1 + \|\cdot\|^2)^{s/2} \mathcal{F} u \in L^2(\mathbf{R}^n)$. This space is provided with the norm

$$\|u\|_{(s)} := \left(\int_{\mathbf{R}^n} |\widetilde{\mathcal{F}} u(\xi)|^2 \, (1 + \|\xi\|^2)^s \, d\xi \right)^{1/2}. \tag{19.1}$$

$\oslash$

The norm has been chosen such that $H_{(0)} = L^2$. The convention to use the weight factor $(1 + \|\xi\|^2)^{1/2}$, instead of $1 + \|\xi\|$, derives from the fact that $1 + \|\xi\|^2$ is the symbol of $I - \Delta$, where Δ is the Laplace operator. The operators

$$(I - \Delta)^z := \mathcal{F}^{-1} \circ (1 + \|\xi\|^2)^z \circ \mathcal{F}$$

can be defined for all $z \in \mathbf{C}$; they form a complex-analytic family of operators, analogous to the families of Riesz from Chap. 13 (see also Sect. 17.1). Thus, under this convention $H_{(s)}$ can also be characterized as the space of $u \in \mathcal{S}'$ such that

J.J. Duistermaat and J.A.C. Kolk, *Distributions: Theory and Applications*, Cornerstones, DOI 10.1007/978-0-8176-4675-2_19,

$v := (I - \Delta)^{s/2} u \in L^2$, and the $H_{(s)}$ norm of u as the L^2 norm of v. For estimates the norm convention is hardly of importance, considering that

$$1 + \|\xi\|^2 \leq (1 + \|\xi\|)^2 \leq 2\,(1 + \|\xi\|^2).$$

Part (a) of the following result is a version of *Sobolev's Embedding Theorem.*

Theorem 19.2. *Let $k \in \mathbf{Z}_{\geq 0}$.*

(a) If $u \in H_{(s)}$ and $s > \frac{n}{2} + k$, then u is of class C^k and $\partial^\alpha u(x) \to 0$ as $\|x\| \to \infty$, for every multi-index α with $|\alpha| \leq k$.
(b) For every $u \in \mathcal{E}'$, we have that u is of finite order, say k, and $u \in H_{(s)}$ if $s < -\frac{n}{2} - k$.

Proof. **(a).** $s > \frac{n}{2} + k$ means $2(k - s) < -n$, that is, $(1 + \|\xi\|)^{k-s} \in L^2$. The Cauchy–Schwarz inequality for the integral inner product yields the integrability of the product of two L^2 functions. Thus we see, for every multi-index α with $|\alpha| \leq k$, that the function

$$\xi \mapsto (\mathcal{F}(D^\alpha u))(\xi) = \xi^\alpha\,(1 + \|\xi\|)^{-s}\,(1 + \|\xi\|)^s\,\mathcal{F}u(\xi)$$

is integrable. In view of the Riemann–Lebesgue Theorem, Theorem 14.2, this proves (a).
(b). Let $u \in \mathcal{E}'$. From Theorem 8.8 we know that u is of finite order. From (18.2) we read that there exists $c > 0$ such that

$$|\mathcal{F}u(\xi)| \leq c\,(1 + \|\xi\|)^k \qquad (\xi \in \mathbf{R}^n).$$

This implies that the square of $(1 + \|\xi\|)^s\,|\mathcal{F}u(\xi)|$ can be estimated by the function $(1 + \|\xi\|)^{2(s+k)}$, which is integrable if $2(s + k) < -n$, that is, $s < -\frac{n}{2} - k$. $\square$

Corollary:

$$\bigcap_{s \in \mathbf{R}} H_{(s)} \subset C^\infty \qquad \text{and} \qquad \mathcal{E}' \subset \bigcup_{s \in \mathbf{R}} H_{(s)},$$

but the assertions in Theorem 19.2 are more detailed, of course.

Example 19.3. In Example 1.5 we discussed functions v defined on an open interval $]\,a, b\,[$ such that v and its derivative v' are square-integrable. If ϕ belongs to $C_0^\infty(\,]\,a, b\,[\,)$, we find by means of Leibniz's rule that $\phi\, v \in H_{(1)}$; therefore, on account of Theorem 19.2.(a), the function $\phi\, v$ is continuous on $\mathbf{R}$. Because this holds for every $\phi \in C_0^\infty(\,]\,a, b\,[\,)$, we conclude that v is continuous on $]\,a, b\,[$.

Furthermore, one has the estimate (1.9), for every $a < x < y < b$. Indeed, let $\phi \in C_0^\infty(\mathbf{R})$ with $\phi \geq 0$ and $\int \phi(z)\,dz = 1$. As usual, we introduce $\phi_\epsilon(z) = \frac{1}{\epsilon}\,\phi(\frac{z}{\epsilon})$ and define the function ψ_ϵ in terms of $\phi_\epsilon(z)$ by (see Fig. 19.1)

$$\psi_\epsilon(z) = \int_{-\infty}^{z} \big(\phi_\epsilon(\zeta - x) - \phi_\epsilon(\zeta - y)\big)\,d\zeta. \tag{19.2}$$

If ϵ is sufficiently small, one has $\psi_\epsilon \in C_0^\infty(\,]\,a,\,b\,[\,)$. Furthermore, $v(x) - v(y)$

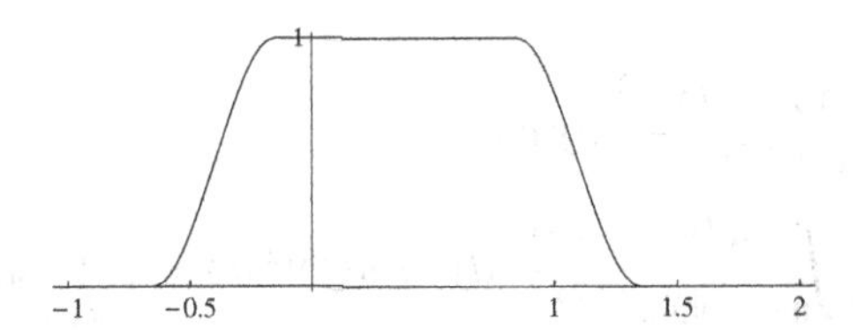

Fig. 19.1 Illustration of (19.2). Graph of ψ_ϵ with $\epsilon = 1/5$, $a = -1$, $x = -1/2$, $y = 1$, and $b = 2$

equals the limit of the $v(\psi_\epsilon')$ as $\epsilon \downarrow 0$, that is, of the $-v'(\psi_\epsilon)$. On the strength of the Cauchy–Schwarz inequality, the absolute value of $-v'(\psi_\epsilon)$ is smaller than or equal to the product of the L^2 norm of v' and the L^2 norm of ψ_ϵ. Now, ψ_ϵ converges, in the L^2-sense, to the characteristic function of $[\,x,\,y\,]$, and so the L^2 norm of ψ_ϵ converges to $(y-x)^{1/2}$ as $\epsilon \downarrow 0$. This proves (1.9).

Applying the Cauchy criterion for convergence we now also see that $v(x)$ converges as $x \downarrow a$ or $x \uparrow b$; write the corresponding limits as $v(a)$ and $v(b)$. Thus we have extended v to a continuous function on $[\,a,\,b\,]$. It now makes sense to speak of the space $H_{(1)}^{\alpha,\beta}$ of all $v \in L^2(\,]\,a,\,b\,[\,)$ such that $v' \in L^2(\,]a,\,b[\,)$ and for which $v(a) = \alpha$ and $v(b) = \beta$. What is more, v is Hölder continuous of order $1/2$ (see the following definition), because (1.9) now holds for all $x,\,y \in [\,a,\,b\,]$. $\oslash$

Definition 19.4. Let $X \subset \mathbf{R}^n$ be open. The function f is said to be *Hölder continuous* on X of order $0 < \alpha \le 1$, with notation $f \in C^\alpha(X)$, if for every compact set $K \subset X$,

$$\sup_{x,\,y\in K,\,x\neq y} \frac{|f(x)-f(y)|}{\|x-y\|^\alpha}$$

is finite. $\oslash$

In a certain sense, Hölder continuity may be viewed as fractional differentiability.

The Sobolev spaces $H_{(s)}(\mathbf{R}^n)$, for $0 < s < 1$, can be characterized in terms of integrated Hölder conditions.

Theorem 19.5. *Consider* $0 < s < 1$. *A function* $u \in L^2(\mathbf{R}^n)$ *belongs to* $H_{(s)}(\mathbf{R}^n)$ *if and only if*

$$\int_{\mathbf{R}^n}\int_{\mathbf{R}^n} \frac{|u(x+y)-u(x)|^2}{\|y\|^{n+2s}}\,dy\,dx < \infty.$$

Proof. Let $y \in \mathbf{R}^n$. In view of (14.29) we obtain $\mathcal{F}(T_y{}^*u - u) = (e_{iy} - 1)\mathcal{F}u$, and thus Theorem 14.32 implies

$$\int_{\mathbf{R}^n} |u(x+y)-u(x)|^2\,dx = \frac{1}{(2\pi)^n}\int_{\mathbf{R}^n} |e^{i\langle y,\xi\rangle}-1|^2|\mathcal{F}u(\xi)|^2\,d\xi.$$

Using the nonnegativity of the integrands as well as Lebesgue's Dominated Convergence Theorem, Theorem 20.26.(iv), we may interchange the order of integration twice to obtain

$$\int_{\mathbf{R}^n}\int_{\mathbf{R}^n}\frac{|u(x+y)-u(x)|^2}{\|y\|^{n+2s}}\,dy\,dx$$
$$=\frac{1}{(2\pi)^n}\int_{\mathbf{R}^n}\int_{\mathbf{R}^n}\frac{|e^{i\langle y,\xi\rangle}-1|^2}{\|y\|^{n+2s}}\,dy\,|\mathcal{F}u(\xi)|^2\,d\xi=c\int_{\mathbf{R}^n}\|\xi\|^{2s}\,|\mathcal{F}u(\xi)|^2\,d\xi,$$

on account of Problem 10.14, for some constant $c>0$. Finally, it is easy to see that $\|u\|_{(s)}<\infty$ if and only if $\mathcal{F}u$ and $\|\cdot\|^s\,\mathcal{F}u$ belong to $L^2(\mathbf{R}^n)$. □

Theorem 19.6. *Let $P(D)$ be a linear partial differential operator in $\mathbf{R}^n$ of order m with constant coefficients. One then has, for every $s\in\mathbf{R}$,*

(a) $P(D)$ is a continuous linear mapping from $H_{(s)}$ to $H_{(s-m)}$.
(b) If $P(D)$ is elliptic, $u\in\mathcal{E}'$, and $P(D)u\in H_{(s-m)}$, then $u\in H_{(s)}$.

Proof. (**a**). One only needs to observe that $\xi\mapsto(1+\|\xi\|)^{-m}\,P(\xi)$ is bounded, so that the L^2 norm of $(1+\|\xi\|)^{s-m}\,P(\xi)\,\mathcal{F}u(\xi)$ can be estimated as a constant times the L^2 norm of $(1+\|\xi\|)^s\,\mathcal{F}u(\xi)$.
(**b**). To prove this we use (17.4). If $\|\xi\|\geq R$ we have

$$U(\xi):=(1+\|\xi\|)^s\,\mathcal{F}u(\xi)=\frac{(1+\|\xi\|)^m}{P(\xi)}\,(1+\|\xi\|)^{s-m}\,\mathcal{F}(P(D)u)(\xi).$$

Write $B=\{\,\xi\in\mathbf{R}^n\mid\|\xi\|<R\,\}$. The function $|U(\xi)|^2$ is integrable on $\mathbf{R}^n\setminus B$, because the factor $(1+\|\xi\|)^m/P(\xi)$ is bounded on $\mathbf{R}^n\setminus B$ and the other factor in $U(\xi)$ is square-integrable on $\mathbf{R}^n$. The conclusion $\int_B|U(\xi)|^2\,d\xi<\infty$ follows from the fact that U is continuous (and even analytic). □

The definition of $H_{(s)}(\mathbf{R}^n)$ makes use of the global operation of the Fourier transform, which precludes a similar definition of a Sobolev space $H_{(s)}(X)$, for arbitrary open $X\subset\mathbf{R}^n$. Still, there exists a localized version of the Sobolev spaces.

Definition 19.7. Let X be an open subset of $\mathbf{R}^n$ and $u\in\mathcal{D}'(X)$ a distribution on X. The distribution u is said *locally to belong to* $H_{(s)}$ if $\phi\,u\in H_{(s)}(\mathbf{R}^n)$ for every $\phi\in C_0^\infty(X)$. The space of all such u is denoted by $H_{(s)}^{\mathrm{loc}}(X)$. ⊘

Theorem 19.8. *Let $k\in\mathbf{Z}_{\geq0}$ and $s\in\mathbf{R}$.*

(a) $H_{(s)}^{\mathrm{loc}}(X)\subset C^k(X)\subset H_{(k)}^{\mathrm{loc}}(X)$ for $s>\frac{n}{2}+k$.
(b) If u is a distribution of order $\leq k$ in X and $s<-\frac{n}{2}-k$, then $u\in H_{(s)}^{\mathrm{loc}}(X)$. If $u\in H_{(-k)}^{\mathrm{loc}}(X)$, it follows that u is a distribution of order k.

Proof. **(a).** Let $u \in H^{\rm loc}_{(s)}(X)$ and $s > \frac{n}{2} + k$. For every $\phi \in C_0^\infty(X)$ one then has $\phi\, u \in H_{(s)}(\mathbf{R}^n)$, and therefore $\phi\, u \in C^k(\mathbf{R}^n)$ on account of Theorem 19.2.(a). This implies $u = \frac{1}{\phi}\,\phi\, u \in C^k$, where $\phi(x) \neq 0$. For every $a \in X$, there exists a $\phi \in C_0^\infty(X)$ with $\phi(x) \neq 0$ for all x in an open neighborhood of a; this implies that $u \in C^k(X)$. The second inclusion follows from the fact that a continuous function with compact support is square-integrable.
(b). As regards the first assertion, we note that $\phi\, u \in \mathcal{E}'(X)$ is of order $\leq k$; hence Theorem 19.2.(b) implies $\phi\, u \in H_{(s)}(\mathbf{R}^n)$. Because this holds for every $\phi \in C_0^\infty(X)$, we conclude that $u \in H^{\rm loc}_{(s)} X$. The proof of the second assertion is left to the reader; it may be seen in relation to the continuous inclusion of $C^k(X)$ in $H^{\rm loc}_{(k)}(X)$. □

Theorem 19.9.

(a) Let X be an open subset of $\mathbf{R}^n$ and let $P(x, D)$ be a linear partial differential operator of order m with C^∞ coefficients on X. If $u \in H^{\rm loc}_{(s)}(X)$, one has $P(x, D)u \in H^{\rm loc}_{(s-m)}(X)$.
(b) In addition, suppose that $P(D)$ is an elliptic operator with constant coefficients and let $P(D)u \in H^{\rm loc}_{(s-m)}(X)$. Then $u \in H^{\rm loc}_{(s)}(X)$.

Proof. **(a)** follows if the following two assertions are combined:

(i) If $u \in H^{\rm loc}_{(s)}(X)$, one has $D_j u \in H^{\rm loc}_{(s-1)}(X)$, for every $1 \leq j \leq n$.
(ii) If $u \in H^{\rm loc}_{(s)}(X)$ and $\psi \in C^\infty(X)$, one has $\psi\, u \in H^{\rm loc}_{(s)}(X)$.

In the arguments below, ϕ is an arbitrary element of $C_0^\infty(X)$.

As regards (i) we note that $\phi\, D_j u = D_j(\phi\, u) - (D_j\phi)\, u$. Here $\phi\, u \in H_{(s)}(\mathbf{R}^n)$, and hence $D_j(\phi\, u) \in H_{(s-1)}(\mathbf{R}^n)$ on the strength of Theorem 19.6.(a). Because the second term lies in $H_{(s)} \subset H_{(s-1)}$, the conclusion is that $\phi\, D_j u \in H_{(s-1)}$, for all $\phi \in C_0^\infty(X)$; that is, $D_j u \in H^{\rm loc}_{(s-1)}(X)$. With respect to (ii) we note that $\phi\,\psi \in C_0^\infty(X)$, and therefore $\phi\,(\psi\, u) = (\phi\,\psi)\, u \in H_{(s)}$.
(b). Let K be a compact subset of X. On account of Theorem 8.8, there exists a $k \in \mathbf{Z}_{\geq 0}$ such that $\psi\, u$ is of order $\leq k$ whenever $\psi \in C_0^\infty(X)$ and $\operatorname{supp}\psi \subset K$. Choose $l \in \mathbf{Z}$ such that $s - l < -\frac{n}{2} - k$. In view of Theorem 19.2.(b) we then have $\psi\, u \in H_{(s-l)}$.

Now write

$$P(D)(\phi\, u) = \phi\, P(D)u + \sum_{\alpha \neq 0,\ |\alpha| \leq m} P_{(\alpha)}(D)(D^\alpha \phi\, u), \tag{19.3}$$

where $P_{(\alpha)}(D)$ are operators of order $< m$. This is true for every operator $P(D)$ of order $\leq m$. The proof proceeds by first considering the case of $P(D) = D^\alpha$, by mathematical induction on $|\alpha|$.

Now suppose that $\phi \in C_0^\infty(X)$ and $\operatorname{supp}\phi \subset K$. In that case, the support of $\psi := D^\alpha \phi$ is contained in K, and so $\psi\, u \in H_{(s-l)}$. Using Theorem 19.6.(a), we find that the terms following the summation sign in (19.3) belong to $H_{(s-l-m+1)}$. Because it is given that $\phi\, P(D)u \in H_{(s-m)}$, we can conclude that $P(D)(\phi\, u) \in H_{(s-l-m+1)}$, that is, so long as $l \geq 1$. If we now apply Theorem 19.6.(b), we find that $\phi\, u \in H_{(s-l+1)}$. By descending mathematical induction on l we arrive at the result that $\phi\, u \in H_{(s)}$, whenever $\phi \in C_0^\infty(X)$ and $\operatorname{supp} \phi \subset K$. The fact that this holds for every compact subset K of X implies $u \in H_{(s)}^{\text{loc}}(X)$. □

Remark 19.10. An important fact that justifies the name $H_{(s)}^{\text{loc}}$ is that

$$H_{(s)}(\mathbf{R}^n) \subset H_{(s)}^{\text{loc}}(\mathbf{R}^n).$$

This means that for every $\phi \in C_0^\infty(\mathbf{R}^n)$, the operator

$$A := (I - \Delta)^{s/2} \circ \phi \circ (I - \Delta)^{-s/2}$$

maps the space $L^2(\mathbf{R}^n)$ to itself.

The main step in the argument is to show that $\phi\, u \in H_{(-s)}(\mathbf{R}^n)$ if $\phi \in C_0^\infty(\mathbf{R}^n)$ and $u \in H_{(-s)}(\mathbf{R}^n)$. We treat the case of $s \geq 0$; the other one is handled similarly. Indeed, for all ξ and $\eta \in \mathbf{R}^n$, we have $\|\xi\| \leq 1 + \|\xi\|^2$. Hence

$$\begin{aligned} 1 + \|\xi - \eta\|^2 &\leq 1 + (\|\xi\| + \|\eta\|)^2 = 1 + \|\xi\|^2 + 2\|\xi\|\,\|\eta\| + \|\eta\|^2 \\ &\leq 1 + \|\xi\|^2 + 2\|\eta\| + 2\|\eta\|\,\|\xi\|^2 + \|\eta\|^2 + \|\xi\|^2\,\|\eta\|^2 \\ &= (1 + \|\xi\|^2)(1 + \|\eta\|)^2. \end{aligned}$$

Hence

$$(1 + \|\xi\|^2)^{-\frac{s}{2}} \leq (1 + \|\eta\|)^s (1 + \|\xi - \eta\|^2)^{-\frac{s}{2}}.$$

On account of $\mathcal{F}\phi \in \mathcal{S}(\mathbf{R}^n)$ we may extend the identity in (14.22) to the case of the product of ϕ and u. Next, observe that

$$\begin{aligned} f(\xi) &:= (1 + \|\xi\|^2)^{-\frac{s}{2}} |\mathcal{F}\phi * \mathcal{F}u(\xi)| \\ &\leq \int_{\mathbf{R}^n} (1 + \|\eta\|)^s |\mathcal{F}\phi(\eta)|\,(1 + \|\xi - \eta\|^2)^{-\frac{s}{2}} |\mathcal{F}u(\xi - \eta)|\, d\eta = g * h(\xi) \end{aligned}$$

with

$$g = (1 + \|\cdot\|)^s |\mathcal{F}\phi| \qquad \text{and} \qquad h = (1 + \|\cdot\|^2)^{-\frac{s}{2}} |\mathcal{F}u|.$$

Using the rapid decrease of $\mathcal{F}\phi$ once more we get $g \in L^1(\mathbf{R}^n)$, while $h \in L^2(\mathbf{R}^n)$ by the assumption on u. Therefore $f \in L^2(\mathbf{R}^n)$, because Problem 11.22 implies $\|f\|_{L^2} \leq \|g * h\|_{L^2} \leq \|g\|_{L^1}\, \|h\|_{L^2}$.

By means of pseudo-differential operators one proves that the assertion in Theorem 19.9.(b) remains valid when an elliptic operator $P(x, D)$ with C^∞ coefficients on X is substituted for $P(D)$. ⊘

Remark 19.11. The fact that Sobolev norms are better suited to partial differential operators than C^k norms is also illustrated by the fact that there exist continuous functions f with compact support in $\mathbf{R}^n$ whose potential is not C^2, at least not if $n > 1$. For example, let

$$f(x) = \begin{cases} -\dfrac{\chi(x)\, x_1^2}{\|x\|^2 \log \|x\|} & \text{if} \quad 0 < \|x\| < 1, \\ 0 & \text{for } x \text{ in all other cases.} \end{cases}$$

Here $\chi \in C_0^\infty(\mathbf{R}^n)$, $\chi(0) > 0$, and $\chi(x) = 0$ if $\|x\| \geq r$, where $0 < r < 1$. Then f is continuous if we take $f(0) = 0$. Furthermore, f has compact support and $\operatorname{sing\,supp} f \subset \{0\}$. For the potential $u = E * f$ of f we obtain $\Delta u = f$, and therefore $\operatorname{sing\,supp} u \subset \{0\}$. However, further computation yields

$$\lim_{x \to 0,\, x \neq 0} \partial_1^2 u(x) = \infty,$$

in other words, $\partial_1^2 u$ is certainly not continuous at 0.

If, however, f is Hölder continuous of order α, the potential u of f is actually C^2, and all second-order derivatives of u are also Hölder continuous of order α. Furthermore, Sobolev norms are meaningful for general distributions. However, the notion of "C^k plus Hölder" is preferred if uniform estimates for function values are required. ⊘

Problems

19.1. Let $P(D)$ be an elliptic operator of order m and E the parametrix of $P(D)$ constructed in Lemma 17.5. Prove that $E* : u \mapsto E * u$ is a continuous linear mapping from $H_{(s-m)}$ to $H_{(s)}$.

19.2. Let E be the potential of δ in $\mathbf{R}^n$, for $n \geq 3$. Is $E * u \in H_{(s)}$ for every $u \in H_{(s-2)}$?

19.3.* Let a_t and b_t be the solutions of the Cauchy problem for the wave equation from Example 18.7. Prove, for arbitrary $s \in \mathbf{R}$,

(i) $b_t *$, $a_t *$ and $\frac{d}{dt} a_t *$ are continuous linear mappings from $H_{(s)}$ to $H_{(s+1)}$, to $H_{(s)}$, and to $H_{(s-1)}$, respectively.

(ii) If u_t is a solution of the homogeneous wave equation with $u_0 \in H_{(s)}$ and $\frac{d}{dt} u_t \big|_{t=0} \in H_{(s-1)}$, one has, for every $t \in \mathbf{R}$, that $u_t \in H_{(s)}$ and $\frac{d}{dt} u_t \in H_{(s-1)}$.

19.4. Examine the following functions on $\mathbf{R}$ and decide in which Sobolev spaces they are contained, and in which they are not (parts (iii) and (iv) are difficult):

(i) the characteristic function of the interval $]\,2,\, 5\,]$,

(ii) $x \mapsto e^{-x} H(x)$,

(iii) $x \mapsto H(x)/(x+1)$,
(iv) $x \mapsto x\, H(x)$,
(v) $x \mapsto 1 - x^2$ for $|x| \leq 1$, and zero elsewhere.

19.5.* Let $s > \frac{n}{2}$ and $0 < \alpha < 1$. If $s = \frac{n}{2} + \alpha$, then $H_{(s)}(\mathbf{R}^n) \subset C^\alpha(\mathbf{R}^n)$. More generally, if $s = \frac{n}{2} + k + \alpha$, where $k \in \mathbf{N}$, then $H_{(s)}(\mathbf{R}^n)$ is contained in

$$\Big\{ f \in C^k(\mathbf{R}^n) \mid \lim_{\|x\| \to \infty} \partial^\beta f(x) = 0,\ \|\beta| \leq k;\ \partial^\beta f \in C^\alpha(\mathbf{R}^n),\ |\beta| = k \Big\}.$$

19.6.* Suppose $P(D)$, s, and m are as in Theorem 19.6.(b). If $u \in L^2(\mathbf{R}^n)$ and $P(D)u \in H_{(s-m)}(\mathbf{R}^n)$, prove that $u \in H_{(s)}(\mathbf{R}^n)$. Furthermore, prove the existence a constant $c > 0$ such that for all such u,

$$\|u\|_{(s)} \leq c\,(\|u\|_{(0)} + \|P(D)u\|_{(s-m)}).$$

Observe that some condition on the growth of u is necessary in view of examples such as $u : \mathbf{R}^2 \to \mathbf{R}$ with $u(x) = e^{x_1^2 - x_2^2} \cos 2x_1 x_2$. This u satisfies $\Delta u = 0$, but it does not belong to $H_{(s)}$ for any $s \in \mathbf{R}$. Furthermore, the assumption $u \in L^2(\mathbf{R}^n)$ may be weakened to $u \in H_{(-t)}(\mathbf{R}^n)$, for some $t > 0$.

19.7. For all ξ and $\eta \in \mathbf{R}$, prove that

$$0 < \frac{1 + \|\xi - \eta\|^2}{(1 + \|\xi\|^2)(1 + \|\eta\|^2)} \leq \frac{4}{3}.$$

Show that the maximum is attained if $\eta = -\xi$ and $\|\xi\| = \frac{1}{\sqrt{2}}$, while the infimum is approached if and only if both $\|\xi\|$ and $\|\eta\|$ tend to ∞.

19.8.* *(Fourth-order ordinary differential operator.)* Define the differential operator $P(\partial)$ acting in $\mathcal{D}'(\mathbf{R})$ by $P(\partial)u = \partial^4 u + u$.

(i) Prove that $P(\partial)u \in C^\infty(\mathbf{R})$ implies $u \in C^\infty(\mathbf{R})$. Conclude that $N \subset C^\infty(\mathbf{R})$ if N is the solution space of the homogeneous equation $P(\partial)u = 0$.
(ii) Verify that N is spanned by the following four functions:

$$x \mapsto e^{w(\pm 1 \pm i)x} \qquad \text{with} \qquad w = \frac{1}{2}\sqrt{2},$$

in which all combinations of the $+$ and $-$ signs occur. Conclude that we have $N \cap \mathcal{S}'(\mathbf{R}) = \{0\}$.
(iii) Consider a fundamental solution E of $P(\partial)$. Show that E does not belong to $\mathcal{E}'(\mathbf{R})$, but that there exists a unique E such that $E \in \mathcal{S}'(\mathbf{R})$.

Suppose $u \in \mathcal{S}'(\mathbf{R})$. We now study the properties of u in the case that this distribution satisfies the inhomogeneous equation $P(\partial)u = f \in \mathcal{D}'(\mathbf{R})$.

(iv) Show that $f \in \mathcal{S}'(\mathbf{R})$ is a necessary condition under this assumption.

Next, consider the special case of $f \in L^2(\mathbf{R})$.

(v) Prove that $\partial^j u \in L^2(\mathbf{R})$, for $0 \le j \le 4$. Show that in addition, $u \in C^3(\mathbf{R})$, but that $u \notin C^4(\mathbf{R})$ if $f \notin C(\mathbf{R})$. See Fig. 19.2.

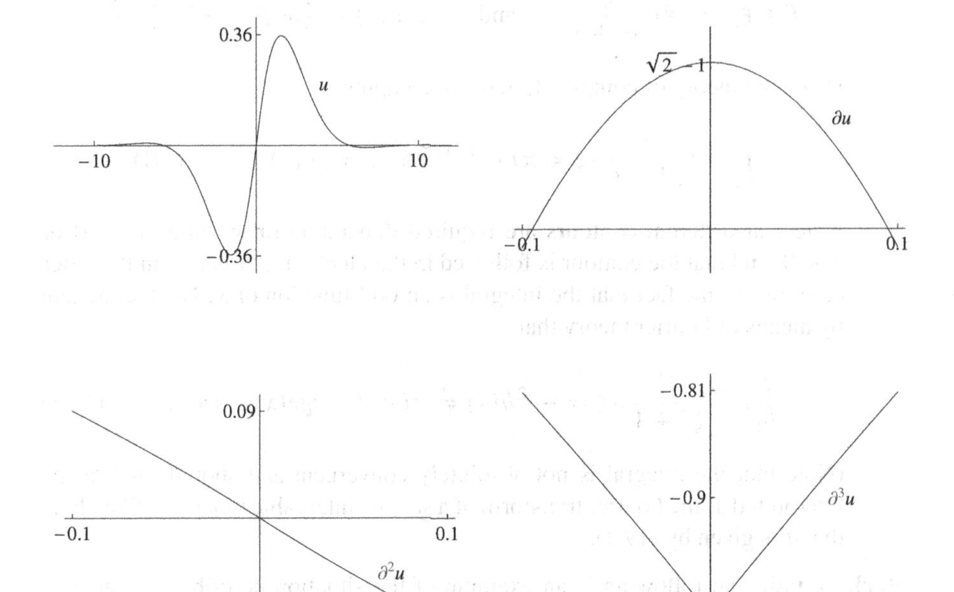

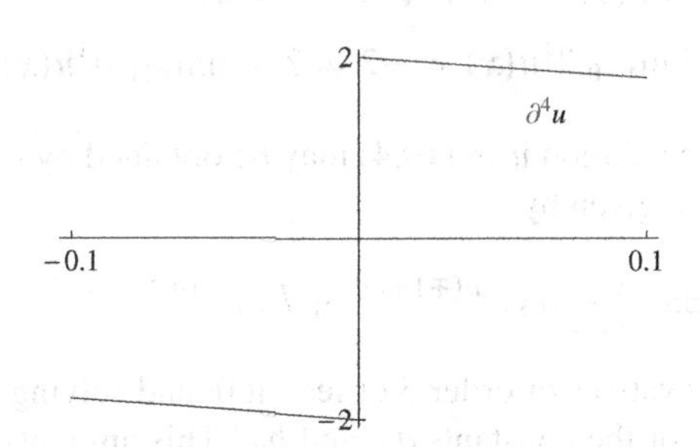

Fig. 19.2 Problem 19.8.(v). Graphs of $\partial^j u$, for $0 \le j \le 4$, ordered in rows

(vi) Verify that $g \in L^2(\mathbf{R})$ if $g(\xi) = \frac{\mathcal{F} f(\xi)}{1+\xi^4}$, and prove that a solution $u \in L^2(\mathbf{R})$ of $P(\partial)u = f$ is uniquely determined and is given by $u = \mathcal{F}^{-1} g$.

Finally, define $f \in L^2(\mathbf{R})$ by $f(x) = 2e^{-|x|} \operatorname{sgn} x$; observe that f is discontinuous at 0. We now prove that the solution $u \in L^2(\mathbf{R})$ of $P(\partial)u = f$ is given by

$$u(x) = e^{-|x|} \operatorname{sgn}(x) + e^{-w|x|} (\sin wx - \operatorname{sgn}(x) \cos wx). \tag{19.4}$$

(vii) By an a priori argument show that sing supp $u \subset \{0\}$.

(viii) Determine the solution u from (19.4) by the method of part (vi). To do so, prove (by means of Problem 14.17 or [7, Exercise 0.8]) that

$$\mathcal{F} f(\xi) = -4i \,\frac{\xi}{\xi^2+1} \qquad \text{and} \qquad \mathcal{F} u(\xi) = \frac{1}{2}\mathcal{F} f(\xi) - 2i\,\frac{\xi - \xi^3}{\xi^4+1}.$$

Using the theory of complex functions, compute

$$\int_{\mathbf{R}} e^{ix\xi}\,\frac{\xi}{\xi^4+1}\,d\xi = \pi i\, e^{-w|x|} \sin wx =: h(x) \qquad (x \in \mathbf{R}).$$

Note that different contours are required depending on whether $x > 0$ or $x < 0$, and that the contour is followed in the clockwise direction in the latter case; or use the fact that the integral is an odd function of x. Next, conclude by means of Fourier theory that

$$\int_{\mathbf{R}} e^{ix\xi}\,\frac{\xi^3}{\xi^4+1}\,d\xi = -\partial^2 h(x) = \pi i\, e^{-w|x|}\,\mathrm{sgn}(x) \cos wx. \tag{19.5}$$

(Note that the integral is not absolutely convergent and should, in fact, be interpreted as the Fourier transform of a square-integrable function.) Conclude that u is given by (19.4).

Background. The following is an example of the situation described in part (v). The function u from (19.4) belongs to $C^3(\mathbf{R})$ but not to $C^4(\mathbf{R})$ (that is, u is not a classical solution). Indeed (see Fig. 19.2 below),

$$u(0) = 0, \qquad \partial u(0) = -1 + \sqrt{2}, \qquad \partial^2 u(0) = 0, \qquad \partial^3 u(0) = -1,$$
$$\lim_{x\uparrow 0} \partial^4 u(x) = -2 \neq 2 = \lim_{x\downarrow 0} \partial^4 u(x).$$

Alternatively, the solution u in (19.4) may be obtained by requiring that the L^2-solutions $u_\pm$ on $\mathbf{R}_{\gtrless 0}$ given by

$$u_\pm(x) = \pm e^{-x} + a_\pm e^{w(\mp 1+i)x} + b_\pm e^{w(\mp 1-i)x} \qquad (x \in \mathbf{R}_{\gtrless 0}),$$

have coinciding derivatives of order 3 or less at 0, and solving the resulting system of linear equations for the constants $a_\pm$ and $b_\pm$. This amounts to a computation of the following integrals without complex analysis:

$$\int_{\mathbf{R}_{>0}} \frac{\xi \cos x\xi}{\xi^4+1}\,d\xi = \frac{\pi}{2}\, e^{-w|x|} \sin wx,$$
$$\int_{\mathbf{R}_{>0}} \frac{\xi^3 \cos x\xi}{\xi^4+1}\,d\xi = \frac{\pi}{2}\, e^{-w|x|}\,\mathrm{sgn}(x) \cos wx.$$

Chapter 20
Appendix: Integration

As we have seen, in Theorems 3.15 and 3.18 above, continuous linear forms on $C_0(X)$, for X an open subset of $\mathbf{R}^n$, arise naturally in the theory of distributions. The alternative name of Radon measure for such a form finds its origin in the existence of a bijective correspondence that associates a complex-valued measure on X to the linear form. More specifically, we have the following *(Frigyes)*[1] *Riesz Representation Theorem.*

Theorem 20.1. *Suppose $X \subset \mathbf{R}^n$ is an open subset and $u \in \mathcal{D}'(X)$. Then u is of order zero if and only if there exist positive measures μ_j on X such that every $\phi \in C_0^\infty(X)$ is integrable with respect to all μ_j, for $1 \leq j \leq 4$, and*

$$u(\phi) = \int \phi(x)\,\mu_1(dx) - \int \phi(x)\,\mu_2(dx) + i\int \phi(x)\,\mu_3(dx) - i\int \phi(x)\,\mu_4(dx).$$

We do not assume the reader to be familiar with the notions of a positive measure and integrability with respect to it; the precise formulation of the theorem is included here for the benefit of those who are. In this appendix we will show that certain aspects of the theory of distributions are related to this result in a natural way. We also give a condensed but complete proof of the theorem, which will, however, be brought to a close only after Theorem 20.35 has been verified.

It should be noted that the notation and nomenclature of measure theory, which are needed in this proof, are not completely standardized in the literature. Moreover, the usual treatments of measure theory start from functions acting on sets, rather than linear forms acting on continuous functions; see for example, Knapp [13] or Stroock [21]. This is another reason why we survey the theory of Lebesgue integration with respect to a measure from a perspective that corresponds as closely as possible to that of the theory of distributions.

The next lemma is stated here in order to motivate the subsequent development.

[1] Marcel Riesz, see Chap. 13, was his younger brother.

J.J. Duistermaat and J.A.C. Kolk, *Distributions: Theory and Applications*, Cornerstones, DOI 10.1007/978-0-8176-4675-2_20,

Lemma 20.2. *A continuous linear form* $u : C_0(X) \to \mathbf{C}$ *possesses the* monotone convergence property, *in the sense that for a sequence* $(\phi_k)_{k\in\mathbf{N}}$ *in* $C_0(X)$,

$$\phi_k \downarrow 0 \qquad \Longrightarrow \qquad \lim_{k\to\infty} u(\phi_k) = 0. \tag{20.1}$$

Here $\phi_k \downarrow 0$ *means that* $(\phi_k)_{k\in\mathbf{N}}$ *is a nonincreasing sequence of real-valued functions with pointwise limit* 0 *as* $k \to \infty$.

Proof. We have $\operatorname{supp} \phi_k \subset K := \operatorname{supp} \phi_1$, for every $k \in \mathbf{N}$, where K is a compact subset of X. On account of Dini's Theorem (see [7, Theorem 1.8.19]) the convergence to 0 of the sequence (ϕ_k) is uniform on K. Hence, the continuity of the linear form u on $C_0(X)$ leads to $\lim_{k\to\infty} u(\phi_k) = 0$; that is, u satisfies (20.1). □

In *measure theory*, one usually begins by introducing measures of sets and subsequently defines an operation of integration with respect to these measures that acts on suitable functions, which are then said to be *integrable*. The introduction of the integral of step functions is often a first step in this construction.

Integration is the counterpart in analysis of the arithmetic operation of computing a weighted average of a finite collection of numbers. Thus, integrating a function, which in general assumes countably or even uncountably many values, means averaging its values. In this process different weights can be assigned to different values: the weight assigned to a value can depend on properties such as location and size of the set of points where the value is attained.

By and large, customary requirements in a theory of integration of functions are the following:

(i) Integrable functions are not necessarily compactly supported and bounded.
(ii) On integrable functions integration acts as a linear form.
(iii) The integral of a nonnegative integrable function is nonnegative.
(iv) The class of integrable functions is closed with respect to taking specific kinds of limits. More precisely, pointwise limits of monotone sequences of integrable functions are integrable, too, and the same holds for arbitrary sequences of integrable functions that are all dominated by an integrable function.
(v) The integral of such a limit is the limit of the integrals.

Conditions (ii) and (iii) relate to algebraic properties, while (i), (iv), and (v) refer to analytic ones. More specifically, (v) demands that integration be continuous on certain sequences of functions. Observe that (i) is a necessary consequence of (iv).

The monotone convergence property as in Lemma 20.2 implies that $u \in \mathcal{D}'(X)$ of order 0 acting on the space $C_0(X)$ of functions satisfies conditions (ii) and (iv), mutatis mutandis. If one wants to consider u as an integral, one needs to show that u can be extended to a class of integrable functions, which properly contains $C_0(X)$, in such a way that (i)–(v) are valid. Preferably, the class of integrable functions should include the characteristic functions of a large class of subsets of X. Note that no characteristic function of a nonempty subset of X belongs to $C_0(X)$ itself.

The fact that we impose conditions (i) and (iv) disqualifies *Riemann integration* as the context for the extension of u. Indeed, Riemann integration primarily applies to bounded functions with compact support. Moreover, a pointwise limit of a sequence of Riemann integrable functions need not be Riemann integrable. For instance, the characteristic function 1_Q of the set $Q := \mathbf{Q} \cap [0, 1]$ of rational numbers in $[0, 1]$ is not Riemann integrable. On the other hand, if $(q_j)_{j \in \mathbf{N}}$ is an enumeration of the elements in Q, then 1_Q is the pointwise limit of the nondecreasing sequence $\big(\sum_{1 \leq j \leq n} 1_{\{q_j\}}\big)_{n \in \mathbf{N}}$ of functions, all of which have Riemann integral equal to 0. Accordingly, we will develop the more appropriate tool of *Lebesgue integration* with respect to a measure.

Just as with the Riemann integral, the Lebesgue integral is defined by means of an approximation process. In the case of the Riemann integral, the process is to use upper sums and lower sums, which capture an approximate value of an integral by adding contributions influenced by proximity in the **domain** of the integrand. The process is qualitatively different for the Lebesgue integral, which captures an approximate value of an integral by adding contributions based on what happens in the **image** of the integrand. As a consequence, the Lebesgue integral is more suitable for handling functions of rapid variation: it leads to a theory of integration of wider applicability and to a simpler treatment of singular behavior of functions.

The theories of Lebesgue and Riemann have similar permanence properties concerning algebraic operations, but in Lebesgue's theory the properties concerning limits are superior. On the other hand, a complete treatment is technically more demanding for Lebesgue integration than for Riemann integration. In addition, there exists a theory of conditionally convergent Riemann integrals, such as $\int_{\mathbf{R}} \frac{\sin x}{x}\, dx = \pi$ (see Problem 14.44), which has no counterpart in Lebesgue integration. More explicitly, this integral is not absolutely convergent, which would be the case for a Lebesgue integrable function.

Nevertheless, the distinction between the two theories rarely plays a role for the functions encountered in this book. More specifically, in practically all situations in which the interchange of a limit and integration is usually justified by an appeal to *Lebesgue's Dominated Convergence Theorem*, Theorem 20.26.(iv) below, one may as well apply *Arzelà's Dominated Convergence Theorem* (see [7, Theorem 6.12.3] for a simple proof) for Riemann integration. This is because the Riemann integrability of the limit function is seldom an issue. However, the limit properties of Lebesgue integration established in the former theorem are indispensable for demonstrating the remarkable fact that completions of $C_0(X)$ with respect to natural integral (semi)norms are again spaces of **functions** and do not consist of objects not encountered before. Furthermore, interchanging the order of integration in multiple integrals under minimal conditions on the integrands can better be handled in Lebesgue's theory.

In the remainder of this appendix we will develop Lebesgue's theory by means of a method due to Daniell, in which functions, instead of measures, are the initial objects of investigation and (semi)norms play an important role. For these reasons, the method naturally suggests itself in the context of distribution theory. If one starts with a ring of subsets and a positive measure on it, and then applies Daniell's extension method to the corresponding step functions, the method **equals** the one used in the standard theory of Lebesgue integration, but it applies in more general settings. We consider the case of a linear form $u : E \to \mathbf{R}$, where E will be specified in (20.8) below. The properties of E are abstracted from those of $C_0(X)$.

Example 20.3. A most important example of a linear form u on $C_0(X)$ is provided by the Riemann integral $1_X = \text{test}\, 1_X$, with $1_X(\phi) = \int_X \phi(x)\, dx$. In this case, the Daniell extension produces the *Lebesgue integral* acting on the (large) space of functions *integrable with respect to Lebesgue measure.* ⊘

For a characterization of Riemann integrable functions on $\mathbf{R}^n$ in the context of Lebesgue integration, see [13, Theorem 6.31].

We begin our treatment of integration theory by discussing some properties and definitions that are intimately connected with the ordering that $\mathbf{R}$ possesses but that is not present in $\mathbf{C}$.

Denote by E the real-linear subspace $C_0(X, \mathbf{R})$ of real-valued functions in $C_0(X)$. Since the real and imaginary parts of a function in $C_0(X)$ belong to E, any complex-valued u as above is uniquely determined by its restriction to E. Accordingly, in the following it will be sufficient to study $u : E \to \mathbf{C}$. Furthermore, we have

$$u = v_1 + i\, v_2 \qquad \text{with unique real-linear forms} \qquad v_1,\, v_2 : E \to \mathbf{R}. \tag{20.2}$$

Note that E and $C_0^\infty(X, \mathbf{R})$ both are linear spaces over $\mathbf{R}$. In contrast to $C_0^\infty(X, \mathbf{R})$, the space E possesses the *Riesz property*, in the sense that

$$\begin{matrix}\max\\ \min\end{matrix}(f, g) = \frac{1}{2}(f + g \pm |f - g|) \in E \qquad (f, g \in E). \tag{20.3}$$

The function $\max(f, g)$ is the smallest element in E that is $\geq f$ and g; likewise, $\min(f, g)$ is the largest element in E that is $\leq f$ and g. In particular, then,

$$0 \leq f_\pm := \max(\pm f, 0) \in E, \qquad f = f_+ - f_-, \qquad |f| = f_+ + f_- \in E. \tag{20.4}$$

Recall that a linear form $u : E \to \mathbf{R}$ is said to be *positive* if

$$u(f) \geq 0 \qquad (f \in E,\ f \geq 0). \tag{20.5}$$

According to Theorem 20.8 below, every real-valued distribution of order 0 can be written as the difference of two positive distributions. This result requires some preparation.

Definition 20.4. The linear form $u : E \to \mathbf{R}$ is of *bounded variation* if

$$|u|(\phi) := \sup\{\, |u(\psi)| \mid \psi \in E,\ |\psi| \leq \phi \,\} < \infty \qquad (0 \leq \phi \in E). \tag{20.6}$$

If this is the case, then the mapping

$$|u| : E_{\geq 0} := \{\phi \in E \mid \phi \geq 0\} \to \mathbf{R}_{\geq 0}$$

is called the *variation* of u. ⊘

Note that a positive linear form $u : E \to \mathbf{R}$ is of bounded variation, and so is the difference of two positive linear forms. Therefore, the linear form $u : E \to \mathbf{R}$ being of bounded variation is a necessary condition for the validity of Theorem 20.8.

Theorem 20.5. *A linear form $u : E \to \mathbf{R}$ defines a distribution of order* 0 *if and only if u is of bounded variation.*

Proof. $\Rightarrow$. Consider $0 \leq \phi \in E$ and set $K = \text{supp}\, \phi$. Applying (3.4) with $k = 0$, we find a constant $c > 0$ such that $|u(\psi)| \leq c\, \|\psi\|_{C^0}$, for all $\psi \in C_0(K)$. In particular, if $\psi \in E$ with $|\psi| \leq \phi$, then supp $\psi \subset K$; hence, $|u(\psi)| \leq c\, \|\psi\|_{C^0} \leq c\, \|\phi\|_{C^0}$. This implies $|u|(\phi) \leq c\, \|\phi\|_{C^0}$, which proves that u is of bounded variation.

$\Leftarrow$. Suppose that u is not continuous. As in the proof of Theorem 3.8, we find a compact set $K \subset X$ and a sequence (ϕ_j) in E satisfying

$$\text{supp}\, \phi_j \subset K \quad \text{and} \quad \|\phi_j\|_{C^0} < \frac{1}{2^j}, \qquad \text{while} \qquad |u(\phi_j)| = 1 \qquad (j \in \mathbf{N}).$$

Consider $\psi_j = \pm\phi_j$, where the sign is chosen such that $u(\psi_j) = 1$. Select $\chi \in C_0(X)$ satisfying $\chi = 1$ on K. The functions $\chi_k = \sum_{j=1}^{k} \psi_j \in E$ have the property that $|\chi_k| \leq \chi$, whereas $|u(\chi_k)| = k$, for all $k \in \mathbf{N}$. This contradicts Definition 20.4 of bounded variation of u. □

The notation in Definition 20.4 anticipates part (ii) of the following result.

Proposition 20.6. *Suppose $u : E \to \mathbf{R}$ is a linear form of bounded variation.*

(i) The mapping $|u|$ is monotone, nonnegative-homogeneous, and additive on the subset of nonnegative elements in E.
(ii) There exists a unique extension of $|u|$ to a positive linear form on E, which will again be denoted by $|u| : E \to \mathbf{R}$.

Proof. **(i).** The monotonicity and nonnegative-homogeneity of $|u|$ are straightforward. The additivity of $|u|$ on elements $0 \leq \phi_j \in E$, for $j = 1, 2$, is best shown in two steps.

Firstly, $|u|$ is subadditive. Indeed, consider arbitrary $\psi \in E$ with $|\psi| \leq \phi_1 + \phi_2$. Introduce $\psi_1 = \min(\phi_1, \psi + \phi_2)$ and $\psi_2 = \max(-\phi_2, \psi - \phi_1)$, which both belong to E. Combination of the identities $\min(\phi_1, \psi + \phi_2) = \min(\phi_1 - \phi_2, \psi) + \phi_2$ and $\max(-\phi_2, \psi - \phi_1) = \max(\phi_1 - \phi_2, \psi) - \phi_1$ with $\min(\phi_1 - \phi_2, \psi) + \max(\phi_1 - \phi_2, \psi) = \phi_1 - \phi_2 + \psi$ leads to $\psi_1 + \psi_2 = \psi$ and $|\psi_j| \leq \phi_j$, for $j = 1, 2$. Hence

$$|u(\psi)| = |u(\psi_1 + \psi_2)| \leq |u(\psi_1)| + |u(\psi_2)| \leq |u|(\phi_1) + |u|(\phi_2).$$

By taking the supremum over all such ψ, we deduce $|u|(\phi_1 + \phi_2) \leq |u|(\phi_1) + |u|(\phi_2)$.

Secondly, $|u|$ is superadditive. Since we are concerned with $\sup\{\, |u(\psi_j)| \mid \psi_j \in E,\ |\psi_j| \leq \phi_j \,\}$, we may without loss of generality consider $\psi_j \in E$ with $|\psi_j| \leq \phi_j$ and $u(\psi_j) \geq 0$. This comes down to changing the sign of ψ_j if necessary. Then $\psi_1 + \psi_2 \in E$ with $|\psi_1 + \psi_2| \leq |\psi_1| + |\psi_2| \leq \phi_1 + \phi_2$, and

$$|u(\psi_1)| + |u(\psi_2)| = u(\psi_1) + u(\psi_2) = u(\psi_1 + \psi_2) \leq |u|(\phi_1 + \phi_2).$$

Taking the two suprema on the left-hand side, we obtain that $|u|(\phi_1 + \phi_2) \geq |u|(\phi_1) + |u|(\phi_2)$ as well, and with it the additivity on the nonnegative elements in E.

(ii). If $|u|$ is the restriction of a positive linear form on all of E, then necessarily

$$|u|(\phi) = |u|(\phi_1) - |u|(\phi_2), \tag{20.7}$$

whenever $E \ni \phi = \phi_1 - \phi_2$, for $0 \leq \phi_j \in E$. Therefore the extension of $|u|$ to all of E is unique. Conversely, since every $\phi \in E$ is the difference of two nonnegative functions, we may use (20.7) to define the extension of $|u|$ to E. This definition is sound; for if $\phi = \phi'_1 - \phi'_2$, with $0 \leq \phi'_j \in E$, then $\phi_1 + \phi'_2 = \phi_2 + \phi'_1$ and a direct computation now gives $|u|(\phi_1) - |u|(\phi_2) = |u|(\phi'_1) - |u|(\phi'_2)$. □

Note that the preceding proof also applies if E is a linear space over **R** of functions on X that has the Riesz property (20.3).

Definition 20.7. Suppose $u : E \to \mathbf{R}$ is a linear form of bounded variation. Define

$$u_\pm = \frac{1}{2}(|u| \pm u). \qquad \text{Then} \qquad u = u_+ - u_- \quad \text{and} \quad |u| = u_+ + u_-.$$

The expression of u as a difference of these two positive linear forms is known as the *Jordan decomposition*. ⊘

The preceding results immediately imply the following theorem. Observe that it is an analog of the classical fact that a real-valued function of bounded variation is the difference of two nondecreasing functions.

Theorem 20.8. *Let E be a linear space of real-valued functions on X having the Riesz property (20.3) and let $u : E \to \mathbf{R}$ be a linear form on E that has the monotone convergence property (20.1) and is of bounded variation. Then there exist linear forms $u_\pm$ on E that satisfy (20.5) and (20.1) with u replaced by $u_\pm$ and such that $u = u_+ - u_-$.*

In Definitions 20.4 and 20.7 and Proposition 20.6 only very general properties of E have been used. Therefore the results above imply that the study of complex-valued distributions u on X of order 0 can be reduced to the study of positive linear forms $u : E \to \mathbf{R}$ satisfying

$$E \text{ is linear space of functions } X \to \mathbf{R} \text{ having the Riesz property (20.3),} \\ u : E \to \mathbf{R} \text{ is a linear form on } E \text{ satisfying (20.5) and (20.1).} \tag{20.8}$$

Daniell's extension method produces a space $\overline{E}$ of functions on X having values in

$$\overline{\mathbf{R}} := \mathbf{R} \cup \{-\infty\} \cup \{\infty\}$$

that contains E and also an extension of u to a real-valued function on $\overline{E}$ denoted by the same letter u. Then $\overline{E}$ and $u : \overline{E} \to \mathbf{R}$ have all the properties of the space of integrable functions with respect to a measure determined by u and Lebesgue integration, respectively.

We now discuss Daniell's method in more detail. In the remainder of the chapter, E and u are as in (20.8). The construction of the Daniell upper extension of u proceeds in two steps.

In the first step it is defined only for functions $X \to \mathbf{R} \cup \{\infty\}$ that are pointwise suprema of countable families in E. Let $E^{\uparrow}$ denote their collection: a function g belongs to $E^{\uparrow}$ if there exists a sequence $(\phi_n)_{n \in \mathbf{N}}$ of functions in E whose pointwise supremum equals g. Replacing ϕ_n by $\max_{1 \le m \le n} \phi_m$, we see that such a sequence can be chosen to be nondecreasing. It is understood that a function in $E^{\uparrow}$ may take the value ∞. Summarizing:

$$E^{\uparrow} = \{ g : X \to \mathbf{R} \cup \{\infty\} \mid \text{there exist } E \ni \phi_n \uparrow g \text{ as } n \to \infty \}.$$

The extension $u^{\uparrow}$ of u to $E^{\uparrow}$ is given by

$$u^{\uparrow}(g) = \sup\{ u(\phi) \mid \phi \in E \text{ and } \phi \le g \}.$$

If the set above is unbounded $u^{\uparrow}(g)$ is indicated by the symbol ∞; hence, we obtain $u^{\uparrow} : E^{\uparrow} \to \mathbf{R} \cup \{\infty\}$.

In the second step, we introduce

$$\overline{F} = \{ f : X \to \overline{\mathbf{R}} \} \qquad \text{and} \qquad \overline{u}(f) = \inf\{ u^{\uparrow}(g) \mid g \in E^{\uparrow} \text{ and } f \le g \}, \tag{20.9}$$

for all $f \in \overline{F}$. If the set on the right-hand side is empty, i.e., if f is not majorized by any function in $E^{\uparrow}$, $\overline{u}(f)$ is indicated by the symbol ∞. We refer to this way of defining $\overline{u} : \overline{F} \to \mathbf{R} \cup \{\infty\}$ as *Daniell's up-and-down method*.

Furthermore, in Definition 20.19 below, the space $\overline{E} = \overline{E}^u$ is defined by

$$\overline{E} = \left\{ f \in \overline{F} \mid \text{there are } \phi_n \in E \text{ with } \lim_{n \to \infty} \overline{u}(|f - \phi_n|) = 0 \right\}. \tag{20.10}$$

In other words, and with some inaccuracy, $\overline{E}$ is the closure of E in the space $\overline{F}$ if $\overline{u}(|f - \phi|)$ is viewed as the distance in $\overline{F}$ between f and $\phi \in E$. Actually, this distance is a seminorm, as discussed in Chap. 8. $\overline{E}$ is called the *Daniell closure* of E.

Following Definition 20.25 below, we prove that

$$|u(\phi) - u(\psi)| \leq \overline{u}(|f - \phi|) + \overline{u}(|f - \psi|) \qquad (\phi,\ \psi \in E,\ f \in \overline{E}).$$

If $(\phi_n)_{n \in \mathbf{N}}$ is a sequence in E satisfying $\lim_{n\to\infty} \overline{u}(|f - \phi_n|) = 0$, the preceding estimate implies that $(u(\phi_n))$ is a Cauchy sequence in $\mathbf{R}$; hence, the latter sequence possesses a limit, say $v(f) \in \mathbf{R}$. This leads to the existence of a unique $v : \overline{E} \to \mathbf{R}$ such that $|v(f) - u(\phi)| \leq \overline{u}(|f - \phi|)$ whenever $f \in \overline{E}$ and $\phi \in E$, which means that $v|_E = u$; additionally, this fact allows us to write $v = u$. Later on, in Theorem 20.26, we verify that the extended form u is real-linear and positive on $\overline{E}$. More importantly, $\overline{E}$ and $u : \overline{E} \to \mathbf{R}$ enjoy the limit properties described in (iv) in the list of requirements following the proof of Lemma 20.2. By contrast, the initial space E and the linear form $u : E \to \mathbf{R}$ are not closed in most cases under taking the limits of arbitrary monotone sequences (ϕ_n) for which the corresponding sequence $(u(\phi_n))$ is bounded.

We consider the behavior of $u^{\uparrow}$ on $E^{\uparrow}$.

Lemma 20.9. *Suppose g and $g' \in E^{\uparrow}$ satisfy $g \leq g'$. Let $(\phi_n)_{n\in\mathbf{N}}$ and $(\phi'_n)_{n\in\mathbf{N}}$ be nondecreasing sequences in E such that $\lim_{n\to\infty} \phi_n = g$ and $\lim_{n\to\infty} \phi'_n = g'$, respectively.*

(i) Then $(u(\phi_n))_{n\in\mathbf{N}}$ is a nondecreasing sequence in $\mathbf{R}$; hence it converges to an element in $\mathbf{R} \cup \{\infty\}$ as $n \to \infty$. If $\psi \in E$ and $\psi \leq g'$, then $u(\psi) \leq \lim_{n\to\infty} u(\phi'_n)$.
(ii) We have $\lim_{n\to\infty} u(\phi_n) \leq \lim_{n\to\infty} u(\phi'_n)$. In the case $g = g'$, one obtains that $\lim_{n\to\infty} u(\phi_n) = \lim_{n\to\infty} u(\phi'_n)$.
(iii) $u^{\uparrow}$ is monotone on $E^{\uparrow}$, that is, $u^{\uparrow}(g) \leq u^{\uparrow}(g')$.
(iv) The restriction of $u^{\uparrow}$ to E coincides with u.

Proof. **(i)**. The monotonicity of the sequence $(u(\phi_n))$ follows from (20.1) and the linearity of u. For the second statement, set $\chi_n = \min(\psi, \phi'_n)$. Equation (20.3) then implies $\chi_n \in E$, while we have $\chi_n \uparrow \psi$; therefore the first statement leads to $u(\chi_n) \uparrow u(\psi)$. Since $\chi_n \leq \phi'_n$ entails $u(\chi_n) \leq u(\phi'_n)$, for $n \in \mathbf{N}$, the conclusion follows.
(ii). Part (i) with ψ replaced by ϕ_m implies $u(\phi_m) \leq \lim_{n\to\infty} u(\phi'_n)$, for every m. Taking the limit for $m \to \infty$ yields the first inequality. If $g = g'$ we may interchange the sequences (ϕ_n) and (ϕ'_n), which gives the inequality in the other direction.
(iii). Direct consequence of part (ii).
(iv). The constant sequence (ϕ_n) given by $\phi_n = \phi \in E$ converges nondecreasingly to ϕ. □

Lemma 20.10. *$E^{\uparrow}$ and $u^{\uparrow} : E^{\uparrow} \to \mathbf{R} \cup \{\infty\}$ satisfy the following properties:*

(i) $E^{\uparrow}$ is closed under addition, multiplication by positive scalars, and taking countable suprema and finite infima.

(ii) For any nondecreasing sequence $(g_n)_{n\in\mathbf{N}}$ in $E^\uparrow$ with pointwise supremum g, which belongs to $E^\uparrow$ by virtue of part (i), we have

$$u^\uparrow(g) = \sup_{n\in\mathbf{N}} u^\uparrow(g_n) = \lim_{n\to\infty} u^\uparrow(g_n). \tag{20.11}$$

(iii) $u^\uparrow$ is additive and nonnegative-homogeneous on $E^\uparrow$.

Proof. **(i)**. Let $\{ g_n \mid n \in \mathbf{N} \}$ be a collection of functions in $E^\uparrow$. Every g_n is the supremum of a countable collection in E, say

$$g_n = \sup\{ \phi_{n,k} \in E \mid k \in \mathbf{N} \} \qquad (n \in \mathbf{N}). \tag{20.12}$$

Then $g_1 + g_2$ is the supremum of $\{ \phi_{1,k} + \phi_{2,l} \in E \mid k, l \in \mathbf{N} \}$, while $r\,g_1$ is the supremum of $\{ r\,\phi_{1,k} \in E \mid k \in \mathbf{N} \}$, provided $r > 0$. Next, $\sup_{n\in\mathbf{N}} g_n$ is the pointwise supremum of the collection $\{ \phi_{n,k} \in E \mid n, k \in \mathbf{N} \}$. Lastly, observe that g_n is the pointwise limit of the nondecreasing sequence $(\widetilde{\phi}_{n,k})_{k\in\mathbf{N}}$ with

$$\widetilde{\phi}_{n,k} = \max_{1\le j\le k} \phi_{n,j} \in E.$$

On account of the preceding argument, $\min(g_1, g_2) = \sup_{k\in\mathbf{N}} \min(\widetilde{\phi}_{1,k}, \widetilde{\phi}_{2,k})$ belongs to $E^\uparrow$.
(ii). According to Lemma 20.9.(iii), the limit in (20.11) equals the supremum. The inequality

$$u^\uparrow(g) \ge \sup_{n\in\mathbf{N}} u^\uparrow(g_n)$$

is straightforward. For the reverse inequality, consider any $r \in \mathbf{R}$ with $r < u^\uparrow(g)$. There exist $E \ni \phi \le g$ with $u(\phi) > r$ and collections $\{ \phi_{n,k} \in E \}$ as in (20.12). The sequence $(\widetilde{\phi}_n)$ of functions

$$E \ni \widetilde{\phi}_n = \min\Big(\phi, \max_{1\le m,k\le n} \phi_{m,k}\Big) \le g_n$$

is pointwise nondecreasing to ϕ. From (20.1) it follows that $\lim_{n\to\infty} u(\widetilde{\phi}_n) > r$. Since $u^\uparrow(g_n) \ge u(\widetilde{\phi}_n)$, we have $\sup_{n\in\mathbf{N}} u^\uparrow(g_n) > r$ and thus the desired inequality

$$\sup_{n\in\mathbf{N}} u^\uparrow(g_n) \ge u^\uparrow(g).$$

(iii). The nonnegative-homogeneity is obvious. Now let g and $g' \in E^\uparrow$. There exist sequences (ϕ_n) and (ϕ'_n) of functions in E that are pointwise nondecreasing to g and g', respectively. Clearly $(\phi_n + \phi'_n)$ is pointwise nondecreasing to $g + g'$. According to part (ii),

$$u^\uparrow(g + g') = \lim_{n\to\infty} u(\phi_n + \phi'_n) = \lim_{n\to\infty} u(\phi_n) + u(\phi'_n) = u^\uparrow(g) + u^\uparrow(g'). \qquad \square$$

Proposition 20.11. *The function $\overline{u} : \overline{F} \to \mathbf{R} \cup \{\infty\}$ as in (20.9) has the following properties:*

(i) $\overline{u}$ is monotone.
(ii) The restriction of $\overline{u}$ to $E^{\uparrow}$ coincides with $u^{\uparrow}$; in particular, $\overline{u}$ agrees with u on E.
(iii) $\overline{u}$ is nonnegative-homogeneous.
(iv) $\overline{u}$ is countably subadditive; that is to say, for any sequence $(f_n)_{n\in\mathbf{N}}$ of nonnegative functions in $\overline{F}$

$$\overline{u}\Big(\sum_{n\in\mathbf{N}} f_n\Big) \le \sum_{n\in\mathbf{N}} \overline{u}(f_n).$$

Proof. **(i)**. Consider f and $f' \in \overline{F}$ with $f \le f'$. If $g \in E^{\uparrow}$ satisfies $f' \le g$, then $f \le g$; hence $\overline{u}(f) \le u^{\uparrow}(g)$. Taking the infimum over all such g, we obtain $\overline{u}(f) \le \overline{u}(f')$.
(ii). If $f \in E^{\uparrow}$, then $u^{\uparrow}(f) \le u^{\uparrow}(g)$ for all $f \le g \in E^{\uparrow}$ according to Lemma 20.9.(iii). Hence $u^{\uparrow}(f) \le \overline{u}(f)$, whereas taking $g = f$ we obtain that $\overline{u}(f) \le u^{\uparrow}(f)$.
(iii). If $f \in \overline{F}$, $g \in E^{\uparrow}$, and $c > 0$, then $f \le g$ if and only if $c\,f \le c\,g$, while $u^{\uparrow}(c\,g) = c\,u^{\uparrow}(g)$ in view of Lemma 20.10.(iii).
(iv). If $\sum_{n\in\mathbf{N}} \overline{u}(f_n) < r$ for some $r \in \mathbf{R}$, there exist functions $f_n \le g_n \in E^{\uparrow}$ with $\sum_{n\in\mathbf{N}} u^{\uparrow}(g_n) < r$. We now successively apply parts (i) and (ii) and Lemma 20.10.(ii) and (iii), to derive

$$\begin{aligned}\overline{u}\Big(\sum_{n\in\mathbf{N}} f_n\Big) &\le u^{\uparrow}\Big(\sum_{n\in\mathbf{N}} g_n\Big) = u^{\uparrow}\Big(\lim_{n\to\infty}\sum_{k=1}^{n} g_k\Big) = \lim_{n\to\infty} u^{\uparrow}\Big(\sum_{k=1}^{n} g_k\Big)\\ &= \lim_{n\to\infty}\sum_{k=1}^{n} u^{\uparrow}(g_k) < r.\end{aligned}$$

Since this holds for all admissible r, the desired inequality follows. □

The countable subadditivity is the only one of these features of $\overline{u}$ that the Riemann upper integral does not share. All limit theorems of the integral constructed by means of the Daniell extension are consequences of this.

Definition 20.12. On the set $\overline{F}$ we define the *Daniell seminorm* $\|\cdot\| = \|\cdot\|_u$ associated with the extension $\overline{u}$ of u to $\overline{F}$ by

$$\|f\| = \overline{u}(|f|) \in \mathbf{R} \cup \{\infty\} \qquad (f \in \overline{F}).$$

Denote the space of all functions $f : X \to \mathbf{R}$ by F. We then say that the sequence $(f_n)_{n\in\mathbf{N}}$ in F *converges with respect to the Daniell seminorm* or *in seminorm* to $f \in \overline{F}$ if $\lim_{n\to\infty} \|f - f_n\| = 0$; compare with Chap. 8. The terminology is justified by the fact that $\|\cdot\|$ satisfies conditions (a)–(c) in Definition 8.5, as will be shown in the following theorem. ⊘

Theorem 20.13. *The Daniell seminorm* $\|\cdot\| : \overline{F} \to \mathbf{R} \cup \{\infty\}$ *has the following properties:*

(i) $\|\cdot\|$ *is* solid, *that is,* $|f| \leq |g|$ *implies* $\|f\| \leq \|g\|$, *for any* f *and* $g \in \overline{F}$. *Furthermore,* $\|\cdot\|$ *is absolute-homogeneous as well as countably subadditive; that is to say, for any sequence* $(f_n)_{n\in\mathbf{N}}$ *of nonnegative functions in* $\overline{F}$,

$$\Big\|\sum_{n\in\mathbf{N}} f_n\Big\| \leq \sum_{n\in\mathbf{N}} \|f_n\|.$$

(ii) $\|\cdot\|$ *is finite on all of* E*; in addition, it dominates* $|\,u\,|$ *on* E*; that is,* $|u(\phi)| \leq \|\phi\|$, *for all* $\phi \in E$.

(iii) For every sequence $(\phi_n)_{n\in\mathbf{N}}$ *of nonnegative functions in* E,

$$\sup_{m\in\mathbf{N}} \Big\|\sum_{n=1}^{m} \phi_n\Big\| < \infty \qquad \Longrightarrow \qquad \lim_{n\to\infty} \|\phi_n\| = 0.$$

Proof. **(i).** This is immediate from Proposition 20.11.(i), (iii), and (iv).
(ii). Consequence of Proposition 20.11.(ii) and (i).
(iii). If $\|\sum_{n=1}^{m} \phi_n\| = \sum_{n=1}^{m} u(\phi_n)$ admits a bound independent of m, then obviously $\lim_{n\to\infty} \|\phi_n\| = \lim_{n\to\infty} u(\phi_n) = 0$. □

Definition 20.14. A function $f \in \overline{F}$ is said to be *negligible* if $\|f\| = 0$. A subset A of X is *negligible* if its characteristic function 1_A as defined in Definition 2.17 is negligible. A property of the points of X is said to hold *almost everywhere*, or *a.e.* for short, if the subset of points in X where it fails to hold is negligible. ⊘

For f and $g \in \overline{F}$, we define the following subsets of X:

$$\begin{aligned} (f = g) &= \{\, x \in X \mid f(x) = g(x)\,\}, \\ (f \neq g) &= \{\, x \in X \mid f(x) \neq g(x)\,\}. \end{aligned} \tag{20.13}$$

Proposition 20.15. *Negligibility has the following permanence properties.*

(i) The sum of countably many nonnegative negligible functions is negligible. The union of countably many negligible sets is negligible. Any subset of a negligible set is negligible. The empty set $\emptyset$ *is negligible.*

(ii) A function $f \in \overline{F}$ *is negligible if and only if it vanishes almost everywhere, that is to say, if and only if the set* $(f \neq 0)$ *is negligible.*

(iii) If f *and* $f' \in \overline{F}$ *agree almost everywhere, then* $\|f\| = \|f'\|$.

(iv) A function with finite seminorm assumes finite values almost everywhere.

Proof. **(i).** If $f_n \in \overline{F}$ are nonnegative negligible functions, Theorem 20.13.(i) leads to

$$\Big\|\sum_{n\in\mathbf{N}} f_n\Big\| \leq \sum_{n\in\mathbf{N}} \|f_n\| = \sum_{n\in\mathbf{N}} 0 = 0.$$

The second claim is a particular instance of this, since characteristic functions of sets are nonnegative. The third claim is immediate from the solidity of $\|\cdot\|$. Finally, $1_\emptyset = 0$ and $\overline{u}(0) = 0$ by Theorem 20.13.(i).
(ii). For the implication $\Rightarrow$ note that $1_{(f\neq 0)} \leq \sum_{n\in\mathbf{N}} |f|$. Hence, if $\|f\| = 0$, then

$$\|1_{(f\neq 0)}\| \leq \sum_{n\in\mathbf{N}} \|f\| = \sum_{n\in\mathbf{N}} 0 = 0.$$

Conversely, $|f| \leq \sum_{n\in\mathbf{N}} 1_{(f\neq 0)}$; accordingly $\|1_{(f\neq 0)}\| = 0$ implies

$$\|f\| \leq \sum_{n\in\mathbf{N}} \|1_{(f\neq 0)}\| = \sum_{n\in\mathbf{N}} 0 = 0.$$

(iii). Write A and B for the sets $(f = f')$ and $(f \neq f')$, respectively. Theorem 20.13.(i) and part (ii) then imply

$$\|f\| = \|1_A\,|f| + 1_B\,|f|\,\| \leq \|1_A\,f\| + \|1_B\,f\| = \|1_A\,f\| + 0 \leq \|f\|,$$
$$\text{so} \qquad \|f\| = \|1_A\,f\| = \|1_A\,f'\| = \|f'\|.$$

(iv). We have $n\,1_{(|f|=\infty)} \leq |f|$ and so $n\,\|1_{(|f|=\infty)}\| \leq \|f\|$, for all $n \in \mathbf{N}$. If $(|f| = \infty)$ is not negligible, $\|f\|$ must be infinite. □

The only functions of interest for the purpose at hand are, of course, those with finite seminorm. We should like to argue that the sum of any two of them also has finite seminorm, in view of the subadditivity of $\|\cdot\|$. A technical difficulty appears: even if f and g have finite seminorm, there may be points $x \in X$ where $f(x) = \infty$ and $g(x) = -\infty$ or vice versa; $f(x) + g(x)$ is not defined for such x. A similar problem arises with $0\,f(x)$. The solution is to note that according to Proposition 20.15.(iv) such ambiguities can occur in a negligible set of points x at the most. We simply extend $\|\cdot\|$ to functions that are defined only almost everywhere, as follows.

Definition 20.16. A function $f \in \overline{F}$ is said to be *defined almost everywhere* on X if the complement in X of the domain of f is negligible. For such f, set $\|f\| = \|f'\|$, where f' is any function defined everywhere on X and coinciding with f almost everywhere at the points where f is defined. ⊘

Proposition 20.15.(iii) entails that this definition is sound: it does not matter which function f' we choose to agree almost everywhere with f; any two will differ negligibly and thus have the same seminorm. Given two functions f and g with finite seminorm that are defined almost everywhere, we declare their sum $f+g$ to equal $f(x)+g(x)$ at points x where both $f(x)$ and $g(x)$ are finite. This function is defined almost everywhere, because the set of points where f and/or g are either infinite or not defined is negligible on account of Proposition 20.15.(iv). It is clear how to define the maximum, minimum, scalar multiples, etc. of functions that are defined almost everywhere, as well as how to read $f \leq g$ almost everywhere etc. when f and g are defined almost everywhere.

Definition 20.17. A function $f \in \overline{F}$ defined almost everywhere is said to have *finite seminorm*, or to be *finite in seminorm*, if $\|f\| < \infty$. Let $\overline{F}_{\|\ \|}$ denote the collection of functions defined almost everywhere with finite seminorm. ⊘

We now derive a crucial technical result.

Theorem 20.18. *(i) $\overline{F}_{\|\ \|}$ is closed under taking finite linear combinations, and finite maxima and minima. In addition, $\|\cdot\|$ is a solid and countably subadditive seminorm on $\overline{F}_{\|\ \|}$.*

(ii) Every Cauchy sequence in $\overline{F}_{\|\ \|}$ with respect to the seminorm has a subsequence that converges pointwise almost everywhere to a limit in $\overline{F}_{\|\ \|}$ with respect to the seminorm.

(iii) The space $(\overline{F}_{\|\ \|}, \|\cdot\|)$ is complete, *that is, every Cauchy sequence in it is convergent to an element contained in the space.*

Proof. **(i)**. For finite sums, apply $|f + g| \leq |f| + |g|$ and Theorem 20.13.(i) to obtain $\|f + g\| \leq \|f\| + \|g\|$. Next, for the maxima and minima use (20.3) plus the fact that $|f - g| \leq |f| + |g|$. Solidity is obvious and countable subadditivity is proved by applying similar arguments as for finite sums.
(ii). Let (f_n) be a Cauchy sequence in seminorm in $\overline{F}_{\|\ \|}$; that is to say, $\|f_n - f_m\| \to 0$ as n and $m \to \infty$. For every $n \in \mathbf{N}$, there exists a function f'_n that is defined and finite everywhere and agrees with f_n almost everywhere. Let N_n denote the negligible set of points where f_n is not defined or does not agree with f'_n. There is a nondecreasing sequence $(n_k)_{k\in\mathbf{N}}$ of indices such that

$$\|f'_n - f'_{n_k}\| < \frac{1}{2^{k+1}} \qquad (n \geq n_k).$$

Clearly $\|f'_{n_{k+1}} - f'_{n_k}\| < \frac{1}{2^{k+1}}$, so that $g := \sum_{k\in\mathbf{N}} |f'_{n_{k+1}} - f'_{n_k}|$

has finite seminorm, by Theorem 20.13.(i). Therefore the set $B = \bigcup_{n\in\mathbf{N}} N_n \cup (g = \infty)$ is negligible according to Proposition 20.15.(i) and (iv). If x belongs to $A = X \setminus B$, then $g(x) < \infty$ and

$$f(x) := f'_{n_1}(x) + \sum_{k\in\mathbf{N}} (f'_{n_{k+1}} - f'_{n_k})(x) = \lim_{k\to\infty} f'_{n_k}(x) \in \mathbf{R},$$

because the series above converges absolutely.

Next, applying Theorem 20.13.(i) once again we obtain, for any $m \in \mathbf{N}$,

$$\|f - f_{n_m}\| = \|f - f'_{n_m}\| = \|1_A(f - f'_{n_m})\| = \Big\|1_A \sum_{k=m}^{\infty} (f'_{n_{k+1}} - f'_{n_k})\Big\|$$
$$\leq \sum_{k=m}^{\infty} \|f'_{n_{k+1}} - f'_{n_k}\| < \frac{1}{2^m}.$$

Hence the subsequence $(f'_{n_k})_{k\in\mathbf{N}}$ converges to f pointwise on A, as well as in seminorm. So does the subsequence $(f_{n_k})_k$ of the original sequence (f_n).

(iii). Given $\epsilon > 0$, let N be so large that for n_k, n, and m larger than N we have

$$\|f - f_{n_k}\| = \|f - f'_{n_k}\| < \frac{\epsilon}{2} \quad \text{and} \quad \|f_n - f_m\| < \frac{\epsilon}{2};$$
$$\text{hence} \quad \|f - f_n\| \leq \|f - f_{n_k}\| + \|f_{n_k} - f_n\| < \epsilon.$$

This shows that $\lim_{n\to\infty} f_n = f$ in seminorm, while $\lim_{k\to\infty} f_{n_k} = f$ almost everywhere as well as in seminorm. □

Definition 20.19. A function in $\overline{F}$ that is defined almost everywhere is said to be *integrable* if there exists a sequence of functions in E converging to it in seminorm. In other words, a function is integrable if it belongs to the closure of E in $\overline{F}_{\| \|}$ with respect to the seminorm. The collection of integrable functions is denoted by $\overline{E}$, the *Daniell closure* of E as in (20.10). Phrased differently,

$$\overline{E} = \left\{ f \in \overline{F} \mid \text{there are } \phi_n \in E \text{ with } \lim_{n\to\infty} \|f - \phi_n\| = 0 \right\} \subset \overline{F}_{\| \|}. \qquad \oslash$$

One does not want to apply the definition to decide whether a given function f is integrable. Rather, one establishes the permanence properties of integrability and checks that f is made up of functions known to be integrable via constructions that preserve integrability. It is expedient to do this first, and then to define and examine the extension of u to $\overline{E}$ with these tools at one's disposal.

Proposition 20.20. *Let f and f' be functions in $\overline{E}$ and $r \in \mathbf{R}$. Then $f + f'$, $r\,f$, $|f|$, $\max(f, f')$, and $\min(f, f')$ belong to $\overline{E}$. In particular, $\overline{E}$ is a linear space.*

Proof. According to Theorem 20.18.(i) we obtain, for any two functions ϕ and $\phi' \in E$,

$$|f + f' - (\phi + \phi')| \leq |f - \phi| + |f' - \phi'|,$$
$$\text{so} \quad \|f + f' - (\phi + \phi')\| \leq \|f - \phi\| + \|f' - \phi'\|.$$

Since the right-hand side can be made as small as desired by the choice of ϕ and ϕ', so can the left-hand side. This implies that $f + f' \in \overline{E}$. The same argument applies to the inequalities

$$|r\,f - r\,\phi| \leq |r|\,|f - \phi| \quad \text{and} \quad \big|\,|f| - |\phi|\,\big| \leq |f - \phi|.$$

For $\max(f, f')$ and $\min(f, f')$ use the preceding results in combination with (20.3). □

We have the following result on completeness of the space $\overline{E}$ of integrable functions with respect to its seminorm.

Theorem 20.21. *(i) If $(f_n)_{n\in\mathbf{N}}$ is a sequence of functions in $\overline{E}$ converging in seminorm to $f \in F_{\|\ \|}$, then $f \in \overline{E}$.*
(ii) $\overline{E}$ is complete with respect to the seminorm. Moreover, every Cauchy sequence in seminorm $(f_n)_{n\in\mathbf{N}}$ in $\overline{E}$ has a subsequence $(f_{n_k})_{k\in\mathbf{N}}$ that converges almost everywhere to a seminorm limit of $(f_n)_{n\in\mathbf{N}}$.
(iii) If $\overline{E} \ni f_n \to f$ in seminorm and $f = f'$ almost everywhere, then $f_n \to f'$ in seminorm as $n \to \infty$.

Proof. **(i).** For every $n \in \mathbf{N}$, select $\phi_n \in E$ with $\|f_n - \phi_n\| < \frac{1}{n}$. Now $f \in \overline{E}$, because the subadditivity of $\|\cdot\|$ leads to

$$\|f - \phi_n\| \le \|f - f_n\| + \|f_n - \phi_n\| < \|f - f_n\| + \frac{1}{n} \to 0 \quad \text{as} \quad n \to \infty.$$

(ii). Immediate from Theorem 20.18.(iii) and (ii) and part (i) above.
(iii). Use $\|f_n - f'\| \le \|f_n - f\| + \|f - f'\|$. □

Lemma 20.22. *(i) A nonnegative function in $\overline{E}$ is the seminorm limit of nonnegative functions in E.*
(ii) The property in Theorem 20.13.(iii) extends to any sequence $(f_n)_{n\in\mathbf{N}}$ of nonnegative functions in $\overline{E}$.

Proof. **(i).** Suppose arbitrary $0 \le f \in \overline{E}$ and $\epsilon > 0$ are given. There exists $\phi \in E$ with $\|f - \phi\| < \epsilon$. Since $|f - |\phi|| = ||f| - |\phi|| \le |f - \phi|$, Theorem 20.18.(i) leads to $\|f - |\phi|\,\| < \epsilon$, where $|\phi| \in E$.
(ii). For all $n \in \mathbf{N}$, we have the existence of $0 \le \phi_n \in E$ with $\|f_n - \phi_n\| < 2^{-n}$ on the basis of part (i). Since

$$\begin{aligned}\sup_{m\in\mathbf{N}} \Big\| \sum_{n=1}^{m} \phi_n \Big\| &\le \sup_{m\in\mathbf{N}} \Big\| \sum_{n=1}^{m} (\phi_n - f_n) \Big\| + \sup_{m\in\mathbf{N}} \Big\| \sum_{n=1}^{m} f_n \Big\| \\ &\le \sum_{m\in\mathbf{N}} \frac{1}{2^m} + \sup_{m\in\mathbf{N}} \Big\| \sum_{n=1}^{m} f_n \Big\| < \infty,\end{aligned}$$

we obtain $\lim_{n\to\infty} \|f_n\| = \lim_{n\to\infty} \|\phi_n\| = 0$ on account of Theorem 20.13.(iii). □

The following result is known as the *Monotone Convergence Theorem.*

Theorem 20.23. *Let $(f_n)_{n\in\mathbf{N}}$ be a nondecreasing or nonincreasing sequence of functions in $\overline{E}$ whose seminorms form a bounded set in $\mathbf{R}$. Then $(f_n)_{n\in\mathbf{N}}$ converges to its pointwise limit f in seminorm; in particular, $f \in \overline{E}$.*

Proof. Since the sequence $(f_n(x))$ in $\overline{\mathbf{R}}$ is monotone, it has a limit $f(x)$, possibly equal to $\pm\infty$. First we consider the case that (f_n) is nondecreasing. Then (f_n) is a Cauchy sequence in seminorm. Indeed, suppose it is not. Then there would exist an $\epsilon > 0$ and a subsequence $(f_{n_k})_k$ with $\|f_{n_{k+1}} - f_{n_k}\| > \epsilon$. But the sequence $(f_{n_{k+1}} - f_{n_k})_k$ consists of nonnegative functions in $\overline{E}$ satisfying

$$\sup_{m\in\mathbf{N}} \left\| \sum_{k=1}^{m} (f_{n_{k+1}} - f_{n_k}) \right\| = \sup_{m\in\mathbf{N}} \|f_{n_m} - f_{n_1}\| \le \|f_{n_1}\| + \sup_n \|f_n\| < \infty.$$

By Lemma 20.22.(ii), though, $(f_{n_{k+1}} - f_{n_k})_k$ must converge to 0 in seminorm.

Now that we know that (f_n) is a Cauchy sequence, we may employ Theorem 20.21.(ii). Thus we find a limit in seminorm $f' \in \overline{E}$ and a subsequence $(f_{n_k})_k$ such that $f_{n_k}(x) \to f'(x)$ for all x outside some negligible set N. For every $x \in X$, however, $f_n(x) \to f(x)$. Therefore

$$f(x) = \lim_{n\to\infty} f_n(x) = \lim_{k\to\infty} f_{n_k}(x) = f'(x) \qquad (x \notin N).$$

Hence f is equal almost everywhere to the limit in seminorm f'. Consequently, it is a limit in seminorm itself, on account of Theorem 20.21.(iii).

If (f_n) is nonincreasing rather than nondecreasing, then $(-f_n)$ is nondecreasing pointwise, and therefore, by the above, nondecreasing in seminorm, to $-f$; again $\lim_{n\to\infty} f_n = f$ in seminorm. □

Next comes the *Lebesgue Dominated Convergence Theorem*. It is a central result in integration theory: many other results can be derived from it.

Theorem 20.24. *Let $(f_n)_{n\in\mathbf{N}}$ be a sequence of functions in $\overline{E}$ and assume:*

(i) $\lim_{n\to\infty} f_n =: f$ pointwise almost everywhere;
(ii) there exists a function g with finite seminorm such that $|f_n| \le g$, for all $n \in \mathbf{N}$.

Then $(f_n)_{n\in\mathbf{N}}$ converges to f in seminorm, and consequently $f \in \overline{E}$.

Proof. As in the proof of the preceding theorem, we begin by showing that (f_n) has the Cauchy property. To this end we consider the nonnegative function

$$g_k = \sup_{k\le n,m} |f_n - f_m| = \lim_{l\to\infty} \max_{k\le n,m\le l} |f_n - f_m| \le 2g \qquad (k \in \mathbf{N}).$$

By Proposition 20.20 and Theorem 20.23, we get $g_k \in \overline{E}$. Moreover, the sequence $(g_k(x))_{k\in\mathbf{N}}$ converges nonincreasingly to 0 at all points x where $(f_n(x))$ converges, that is, almost everywhere. A simple modification of the preceding theorem then leads to $\lim_{k\to\infty} \|g_k\| = 0$. Now $\|f_n - f_m\| \le \|g_k\|$, for m and $n \ge k$, so that (f_n) is indeed a Cauchy sequence in seminorm.

By Theorem 20.21.(ii), the sequence possesses a limit in seminorm f' and a subsequence $(f_{n_k})_k$ that converges pointwise almost everywhere to f'. Since $(f_{n_k})_k$ also converges to f almost everywhere, the functions f and f' agree almost everywhere, namely, at all points x where both $(f_n(x))$ and $(f_{n_k}(x))_k$ converge. Thus $\lim_{n\to\infty} \|f_n - f\| = \lim_{n\to\infty} \|f_n - f'\| = 0$. □

Definition 20.25. We are now in a position to define the extension $u : \overline{E} \to \mathbf{R}$ of the linear form $u : E \to \mathbf{R}$. Namely, for any $f \in \overline{E}$, select an arbitrary sequence $(\phi_n)_{n\in\mathbf{N}}$ of functions in E converging in seminorm to f and set $u(f) = \lim_{n\to\infty} u(\phi_n)$. ⊘

The extension is well-defined. First, on the strength of Theorem 20.13.(ii) and (i),

$$|u(\phi_n) - u(\phi_m)| = |u(\phi_n - \phi_m)| \le \|\phi_n - \phi_m\| \le \|\phi_n - f\| + \|f - \phi_m\| \to 0,$$

as n and $m \to \infty$. Hence, $(u(\phi_n))$ is a Cauchy sequence in $\mathbf{R}$ and does have a limit.

Next, if (ϕ'_n) is another sequence of functions in E converging in seminorm to f, then

$$|u(\phi_n) - u(\phi'_n)| = |u(\phi_n - \phi'_n)| \le \|\phi_n - \phi'_n\| \le \|\phi_n - f\| + \|f - \phi'_n\| \to 0,$$

as $n \to \infty$. Phrased differently, $(u(\phi_n))$ and $(u(\phi'_n))$ have the same limit.

Next, we present the fundamental properties of the extension. Part (iv) below is also referred to as the *Lebesgue Dominated Convergence Theorem.*

Theorem 20.26. *(i) The extension $u : \overline{E} \to \mathbf{R}$ is linear and monotone; that is to say, for any f and $g \in \overline{E}$ and r and $s \in \mathbf{R}$,*

$$u(r\,f + s\,g) = r\,u(f) + s\,u(g), \qquad \text{and} \qquad f \le g \implies u(f) \le u(g).$$

(ii) The extension is still majorized by the seminorm. More specifically,

$$|u(f)| \le u(|f|) = \|f\| \qquad (f \in \overline{E}).$$

In particular, u assigns equal values to functions in $\overline{E}$ that are equal almost everywhere.

(iii) If the sequence $(f_n)_{n\in\mathbf{N}}$ in $\overline{E}$ converges in seminorm to $f \in F$, then $f \in \overline{E}$ and $\lim_{n\to\infty} u(f_n) = u(f)$.

(iv) The conclusion in part (iii) holds if the sequence $(f_n)_{n\in\mathbf{N}}$ in $\overline{E}$ converges almost everywhere to f and $|f_n| \le g$, for all $n \in \mathbf{N}$ and for some function $g \in \overline{E}$. In particular, the choice $g = \sup_{n\in\mathbf{N}} |f_n|$ is admissible if

$$\|\sup_{n\in\mathbf{N}} |f_n|\,\| < \infty.$$

Proof. **(i)**. Let (ϕ_n) and (ψ_n) be sequences in E converging in seminorm to f and $g \in \overline{E}$, respectively. Then $(r\,\phi_n + s\,\psi_n)$ converges to $r\,f + s\,g$ in seminorm, because

$$\|r\,f + s\,g - (r\,\phi_n + s\,\psi_n)\| \le |r|\,\|f - \phi_n\| + |s|\,\|g - \psi_n\| \to 0 \quad \text{as} \quad n \to \infty.$$

Thus

$$u(r\,f+s\,g)=\lim_{n\to\infty}u(r\,\phi_n+s\,\psi_n)=\lim_{n\to\infty}(r\,u(\phi_n)+s\,u(\psi_n))=r\,u(f)+s\,u(g).$$

This shows that the extension is linear.

On account of the linearity of the extension its monotonicity follows if $f\geq 0$ implies $u(f)\geq 0$, for all $f\in\overline{E}$. By Lemma 20.22.(i), there exist nonnegative functions $\phi_n\in E$ converging in seminorm to f. Then $u(\phi_n)\geq 0$ implies $u(f)=\lim_{n\to\infty}u(\phi_n)\geq 0$.

(ii). Suppose (ϕ_n) in E satisfies $\lim_{n\to\infty}\|f-\phi_n\|=0$. Then the fact that $u(f)=\lim_{n\to\infty}u(\phi_n)$ leads to $|u(f)|=\lim_{n\to\infty}|u(\phi_n)|$. In addition, since $|\,|f|-|\phi_n|\,|\leq|f-\phi_n|$, it follows from Theorem 20.13.(i) that $\lim_{n\to\infty}\|\,|f|-|\phi_n|\,\|=0$; hence

$$u(|f|)=\lim_{n\to\infty}u(|\phi_n|)=\lim_{n\to\infty}\|\phi_n\|=\|f\|.$$

The monotonicity of u on E leads to $|u(\phi_n)|\leq u(|\phi_n|)$. Combination of the latter two results yields the desired

$$|u(f)|=\lim_{n\to\infty}|u(\phi_n)|\leq\lim_{n\to\infty}u(|\phi_n|)=u(|f|)=\|f\|.$$

(iii). Consequence of part (ii).

(iv). Evident from part (iii) and Lebesgue's Dominated Convergence Theorem, Theorem 20.24. □

Theorem 20.26 completes the construction of the Daniell closure as well as the description of its properties.

Example 20.27. Consider the family of functions $(A_r)_{0\leq r\leq 1}$ from Example 1.4. We recall the argument at the end of the example that $(A_r)_{0\leq r\leq 1}$ does not admit an integrable majorant on $[-\pi,\pi]$; see Fig. 1.4. ⊘

We now introduce some concepts from measure theory.

Definition 20.28. Let X be a set. A *ring* of subsets of X is a collection $\mathcal{A}$ of subsets of X such that $\emptyset\in\mathcal{A}$ and

$$A\cup B\in\mathcal{A},\qquad A\cap B\in\mathcal{A},\qquad A\setminus B\in\mathcal{A}\qquad (A,\,B\in\mathcal{A}).$$

Note the redundancy in this definition; for instance, $\emptyset=A\setminus A$ and $A\cap B=A\setminus(A\setminus B)$.

A *positive measure* on $\mathcal{A}$, or on X, is a set function $\mu:\mathcal{A}\to\mathbf{R}$ that is positive in the sense that

$$\mu(A)\geq 0\qquad (A\in\mathcal{A}),$$

(finitely) additive in the sense that

$$A,\,B\in\mathcal{A}\ \text{ and }\ A\cap B=\emptyset\quad\Longrightarrow\quad\mu(A\cup B)=\mu(A)+\mu(B),$$

and has the *monotone convergence property* in the sense that

$$(A_n)_{n\in\mathbf{N}} \text{ nonincreasing in } \mathcal{A} \text{ with } \bigcap_{n=1}^{\infty} A_n = \emptyset \quad \Longrightarrow \quad \lim_{n\to\infty} \mu(A_n) = 0. \tag{20.14}$$

Given the ring $\mathcal{A}$ of subsets of X, a function $f : X \to \mathbf{R}$ is called *$\mathcal{A}$-elementary* if $f(X) \subset \mathbf{R}$ is a finite subset and $f^{-1}(\{c\}) \in \mathcal{A}$, for every $c \in \mathbf{R} \setminus \{0\}$. Denote the set of all $\mathcal{A}$-elementary functions by $E_{\mathcal{A}}$.

For any $f \in E_{\mathcal{A}}$, the *integral* of f with respect to the positive measure μ on $\mathcal{A}$ is defined as

$$\mathrm{I}_\mu(f) := \int f(x)\,\mu(dx) := \sum_{c\in f(X)\setminus\{0\}} c\,\mu(f^{-1}(\{c\})). \tag{20.15}$$

⊘

For arbitrary $A \subset X$, recall the characteristic function 1_A of A as defined in Definition 2.17.

Definition 20.29. Given E and $u : E \to \mathbf{R}$ as in (20.8) with Daniell extension $u : \overline{E} \to \mathbf{R}$, introduce the collection $\mathcal{A}$ of subsets of X and the function μ on $\mathcal{A}$ by

$$\mathcal{A} = \mathcal{A}_u = \{\, A \subset X \mid 1_A \in \overline{E}\,\},$$
$$\mu = \mu_u : \mathcal{A} \to \mathbf{R} \quad \text{by} \quad \mu(A) = u(1_A) \quad (A \in \mathcal{A}).$$

⊘

Lemma 20.30. *In the notation as above, we have the following properties:*

(i) $\mathcal{A}$ is a ring of subsets of X.
(ii) μ is a positive measure on $\mathcal{A}$.
(iii) $E_{\mathcal{A}}$ and I_μ possess the properties of E and u listed in (20.8). In particular, Daniell's extension method is applicable to $\mathrm{I}_\mu : E_{\mathcal{A}} \to \mathbf{R}$. Furthermore, $E_{\mathcal{A}} \subset \overline{E}$ and I_μ equals the restriction of u to $E_{\mathcal{A}}$.

Proof. **(i).** $1_\emptyset = 0$. For A and $B \in \mathcal{A}$, the following functions all belong to $\overline{E}$ in view of Proposition 20.20:

$$1_{A\cup B} = \max(1_A, 1_B), \qquad 1_{A\cap B} = \min(1_A, 1_B), \qquad 1_{A\setminus B} = 1_A - 1_{A\cap B}.$$

(ii). We have $\mu(A) = u(1_A) \geq u(0) = 0$ because $1_A \geq 0$. In addition,

$$\mu(A \cup B) = u(1_{A\cup B}) = u(1_A + 1_B) = u(1_A) + u(1_B) = \mu(A) + \mu(B)$$

if A and $B \in \mathcal{A}$ and $A \cap B = \emptyset$.

Finally, let (A_n) be a nondecreasing sequence in $\mathcal{A}$ such that $(\mu(A_n))$ is bounded from above, and let A be the union of all A_n. Then $\overline{E} \ni 1_{A_n} \uparrow 1_A$ and $(u(1_{A_n}))$ is bounded from above. Therefore the Monotone Convergence Theorem, Theorem 20.23, implies that $1_A \in \overline{E}$ and $u(1_A) = \lim_{n\to\infty} u(1_{A_n})$; hence, $A \in \mathcal{A}$ and $\mu(A) = \lim_{n\to\infty} \mu(A_n)$. An analogous statement can be made for nonincreasing sequences in $\mathcal{A}$, which implies (20.14).

(iii). Provided with the pointwise addition and scalar multiplication of functions, the space F of all functions $f : X \to \mathbf{R}$ is a linear space over $\mathbf{R}$, and $E_{\mathcal{A}}$ is a linear subspace of F. Also, in F we have the partial ordering defined by $f \leq g$. The ring property of $\mathcal{A}$ implies that $E_{\mathcal{A}}$ possesses the Riesz property (20.3).

For $f \in E_{\mathcal{A}}$, we have $f = \sum_{c \in f(X) \setminus \{0\}} c\, 1_{f^{-1}(\{c\})} \in \overline{E}$, because the summation is finite. On account of (20.15) and Definition 20.29 this leads to

$$\mathrm{I}_\mu(f) = \sum_{c \in f(X) \setminus \{0\}} c\, \mu(f^{-1}(\{c\})) = u\Big(\sum_{c \in f(X) \setminus \{0\}} c\, 1_{f^{-1}(\{c\})} \Big) = u(f).$$

Consequently I_μ is linear and positive.

Any sequence of functions $\overline{E} \supset E_{\mathcal{A}} \ni f_n \downarrow 0$ is majorized by $f_1 \in \overline{E}$; therefore application of Theorem 20.26.(iv) entails

$$\lim_{n \to \infty} \mathrm{I}_\mu(f_n) = \lim_{n \to \infty} u(f_n) = u(0) = 0.$$

This means that I_μ has the monotone convergence property as in (20.1). □

Example 20.31. Application of the preceding lemma to the positive linear form from Definition 20.3 gives rise to the *Lebesgue measure* on the ring of *Lebesgue measurable sets* in $X \subset \mathbf{R}^n$. ⊘

Definition 20.32. Let the notation be as above. Functions belonging to $\overline{E_{\mathcal{A}}}^{\mathrm{I}_\mu}$, the Daniell closure of $E_{\mathcal{A}}$ with respect to the measure μ, are said to be *integrable with respect to* μ or μ*-integrable*. ⊘

Observe that $E_{\mathcal{A}}$ is not contained in E. On the other hand, $E_{\mathcal{A}} \subset \overline{E}$ and I_μ equals u on $E_{\mathcal{A}}$. When applying the extension method with $u : E \to \mathbf{R}$ replaced by $\mathrm{I}_\mu : E_{\mathcal{A}} \to \mathbf{R}$, it is natural to ask whether $E \subset \overline{E_{\mathcal{A}}}^{\mathrm{I}_\mu}$ and whether u equals the restriction to E of the extension of I_μ to $\overline{E_{\mathcal{A}}}^{\mathrm{I}_\mu}$. This is not always the case. Indeed, for arbitrary $f \in E$, the value $u(f)$ cannot always be recovered as the integral of f with respect to the measure μ. The following example, admittedly somewhat artificial, shows this convincingly. In other words, the theory of the Daniell closure is a true generalization of the theory of Lebesgue integration with respect to a measure.

Example 20.33. Let X be a subset of $\mathbf{R}_{>0}$ having at least two elements. Let E be the set of all restrictions to X of linear functions on $\mathbf{R}$. That is, $f \in E$ if and only if there exists $c \in \mathbf{R}$ such that $f(x) = c\,x$ for every $x \in X$. Define $u(f) = c$ if $f(x) = c\,x$ for every $x \in X$. Phrased differently, if we choose $x_0 \in X$, then $u(f) = f(x_0)/x_0$ for every $f \in E$. It follows that E is a vector space of functions on X having the Riesz property (20.3) and that u is a linear form on E satisfying (20.5) and (20.1). And yet one has $E^\uparrow = E$, $\overline{E} = E$, $\mathcal{A}_u = \{\emptyset\}$, $E_{\mathcal{A}_u} = \{0\}$, $\mu_u = 0$, and $\mathrm{I}_{\mu_u} = 0$. ⊘

Under an additional assumption, however, one obtains a positive result.

Theorem 20.34. *Suppose that E is a linear space of real-valued functions on a set X having the Riesz property (20.3) and that $u : E \to \mathbf{R}$ is a linear form on E satisfying (20.5) and (20.1). Furthermore, assume that E has the* Stone property, *that is,* $\min(\phi, 1_X) \in E$ *for every* $\phi \in E$. *Then*

$$\overline{E} = \overline{E_{\mathcal{A}}}^{\,\mathrm{I}_\mu} \qquad \textit{and} \qquad u(f) = \int f(x)\,\mu(dx) \qquad (f \in \overline{E}).$$

Proof. First we prove that $\phi \in \overline{E_{\mathcal{A}}}^{\,\mathrm{I}_\mu}$ and $u(\phi) = \int \phi(x)\,\mu(dx)$, for any $\phi \in E$. We will do this by approximating ϕ in a monotone way by means of functions in $E_{\mathcal{A}}$.

In view of (20.4), with f replaced by ϕ, we may assume that $\phi \geq 0$. For any $c > 0$, we have $\min(\phi, c\,1_X) = c\,\min(\frac{1}{c}\,\phi, 1_X) \in E$. Accordingly we also obtain

$$\psi_{c,c'} := \frac{1}{c - c'}\,(\min(\phi, c\,1_X) - \min(\phi, c'\,1_X)) \in E \qquad (0 < c' < c).$$

Now $\psi_{c,c'}(x)$ takes the following values: 0 if $\phi(x) \leq c'$; $\psi(\phi(x) - c')/(c - c') \in [0, 1]$ if $c' \leq \phi(x) \leq c$; and 1 if $c \leq \phi(x)$. Let (c_n) be a sequence in $\mathbf{R}_{>0}$ such that $0 < c_n < c$ for every n and $c_n \uparrow c$ as $n \to \infty$. Then $\chi_n := \min_{1 \leq m \leq n} \psi_{c,c_m} \in E$ and also $\chi_n \downarrow 1_{(\phi \geq c)}$, in the notation (20.13). The Monotone Convergence Theorem, Theorem 20.23, for $\overline{E}$ implies $1_{(\phi \geq c)} \in \overline{E}$; this in turn leads to

$$(\phi \geq c) \in \mathcal{A} \qquad \text{and} \qquad 1_{(\phi \geq c)} \in E_{\mathcal{A}}.$$

Let $F \subset \mathbf{R}_{>0}$ be nonempty and finite. Set $c'_F = \min\{\, c' \in F \mid c' > c \,\}$ for every $c \in F$, where $c'_F = \infty$ if $c = \max F$. In view of the fact that $(c'_F > \phi \geq c) = (\phi \geq c) \setminus (\phi \geq c'_F) \in \mathcal{A}$, we obtain

$$f_F := \sum_{c \in F} c\,1_{(c'_F > \phi \geq c)} \in E_{\mathcal{A}} \subset \overline{E}.$$

In view of $\phi \geq f_F$ and $(c'_F > \phi \geq c) = f_F^{-1}(\{c\})$ and applying Lemma 20.30.(iii) we deduce

$$u(\phi) \geq u(f_F) = \int f_F(x)\,\mu(dx).$$

If (F_n) is a nondecreasing sequence of finite subsets of $\mathbf{R}_{>0}$ such that the union of all F_n is dense in $\mathbf{R}_{>0}$, then $f_{F_n} \uparrow \phi$ as $n \to \infty$. Since the $u(f_{F_n})$ are bounded from above by $u(\phi)$, the Monotone Convergence Theorem for $u : \overline{E} \to \mathbf{R}$ leads to

$$u(\phi) = \lim_{n \to \infty} u(f_{F_n}) = \lim_{n \to \infty} \int f_{F_n}(x)\,\mu(dx).$$

On the other hand, the Monotone Convergence Theorem for I_μ implies that $\phi \in \overline{E_{\mathcal{A}}}^{\,\mathrm{I}_\mu}$ and

$$\lim_{n\to\infty} \int f_{F_n}(x)\,\mu(dx) = \int \phi(x)\,\mu(dx).$$

Combination of the two identities gives the desired equality $u(\phi) = \int \phi(x)\,\mu(dx)$.

Thus we have established that $E \subset \overline{E_{\mathcal{A}}}^{\mathrm{I}_\mu}$ and $u|_E = \mathrm{I}_\mu|_E$. Since the Daniell closure of I_μ equals I_μ, the Daniell closure of $u|_E = \mathrm{I}_\mu|_E$ is contained in the Daniell closure of I_μ, that is, $\overline{E} \subset \overline{E_{\mathcal{A}}}^{\mathrm{I}_\mu}$ and $u|_{\overline{E}} = \mathrm{I}_\mu|_{\overline{E}}$. On the other hand, since $E_{\mathcal{A}} \subset \overline{E}$ and $u|_{\overline{E}}$ is Daniell closed, we also have $\overline{E_{\mathcal{A}}}^{\mathrm{I}_\mu} \subset \overline{E_{\mathcal{A}}}^{u} \subset \overline{E}$, which implies $\overline{E} = \overline{E_{\mathcal{A}}}^{\mathrm{I}_\mu}$ and $u = \mathrm{I}_\mu$ on $\overline{E}$. □

As a direct consequence we derive the following important result. Recall that on account of Theorem 3.18, a positive distribution is a distribution of order 0.

Theorem 20.35. *Let $X \subset \mathbf{R}^n$ be an open subset. Then u is a positive distribution on X if and only if there exists a positive measure μ on X such that ϕ is integrable with respect to μ and $u(\phi) = \int \phi(x)\,\mu(dx)$, for every $\phi \in C_0^\infty(X)$.*

Proof. $\Leftarrow$. Lemma 20.30.(iii) implies that $u = \mathrm{I}_\mu$ is a linear form on $C_0^\infty(X)$. If K is an arbitrary compact subset of X we have, for all $\phi \in C_0^\infty(K)$,

$$|u(\phi)| = \Big|\int \phi(x)\,\mu(dx)\Big| \le \int |\phi(x)|\,\mu(dx) \le \sup_{x\in X} |\phi(x)|\,\mu(K) = \mu(K)\|\phi\|_{C^0}.$$

In view of Theorem 3.8 this implies that $u \in \mathcal{D}'(X)$. The positivity of u is a consequence of the positivity of μ.

$\Rightarrow$. From Theorem 3.18 and Lemma 20.2 it follows that u possesses the property in (20.1). We conclude that the space $E = C_0(X, \mathbf{R})$ and the linear form $u : E \to \mathbf{R}$ satisfy (20.3), the Stone property, (20.5), and (20.1). Since E contains $C_0^\infty(X, \mathbf{R})$, the desired conclusion now follows from Theorem 20.34. □

We have now completed the preparations required for the proof of Riesz's Representation Theorem, Theorem 20.1.

Proof. $\Leftarrow$. This implication follows as in the proof of Theorem 20.35.
$\Rightarrow$. Apply (20.2) and Theorems 20.8 and 20.35. □

The condition imposed on the measure μ, namely that every $\phi \in C_0^\infty(X)$ be integrable with respect to μ, is not very direct in terms of the measure μ as a set function itself. This is remedied by the following proposition.

Proposition 20.36. *Let μ be as in Theorem 20.35. Then the following assertions are equivalent:*

(i) Every $\phi \in C_0^\infty(X)$ is μ-integrable.
(ii) 1_U is μ-integrable for every relatively compact *(i.e., having compact closure) open subset U of X.*
(iii) 1_K is μ-integrable for every compact subset K of X.

(iv) Every $\phi \in C_0(X)$ is μ-integrable.

Proof. Write $E = E_{\mathcal{A}}$, $\overline{E} = \overline{E_{\mathcal{A}}}^{\mathrm{I}_\mu}$ and $u = \mathrm{I}_\mu$. Then $f \in \overline{E}$ if and only if f is μ-integrable.
(i) ⇒ (ii). Use Lemma 8.2 and Corollary 2.16 to choose $\chi \in C_0^\infty(X)$ and an increasing sequence (χ_n) in $C_0^\infty(X)$ such that $\chi_n \uparrow 1_U \leq \chi$. Because $u(\chi_n) \leq u(\chi)$ for every n, it follows from the Monotone Convergence Theorem that $1_U \in \overline{E}$.
(ii) ⇒ (iii). For $K \subset X$ compact, Corollary 2.4 implies the existence of a relatively compact open $U \subset X$ with $K \subset U$. Then $K = U \setminus (U \setminus K)$ exhibits K as a difference of two relatively compact open subsets; hence K belongs to the ring $\mathcal{A}$ and so $1_K \in \overline{E}$.
(ii) ⇐ (iii). Application of the preceding argument with the roles of K and U interchanged leads to the conclusion $1_U \in \overline{E}$.
(ii) ⇒ (iv). In view of (20.4) it is sufficient to consider $\phi \in C_0(X)$ satisfying $\phi \geq 0$. Then $(\phi > c)$ is a relatively compact open subset of X for every $c > 0$; hence $1_{(\phi > c)} \in \overline{E}$. Since $(\phi \geq c)$ is compact, (iii) implies that $1_{(\phi \geq c)} \in \overline{E}$. The ring property of $\mathcal{A}$ leads to $1_{(a \leq \phi < b)} \in \overline{E}$, for any a and $b \in \mathbf{R}$ with $0 < a < b$. Approximating ϕ in a monotone way by a sequence of linear combinations of functions $1_{(a \leq \phi < b)}$ as in the proof of Theorem 20.34, we conclude that $\phi \in \overline{E}$.
(iv) ⇒ (i) is trivial. □

The proposition above justifies the following definition.

Definition 20.37. Let μ be as in Theorem 20.35 and $f \in \overline{F}$. Then f is said to be a *locally μ-integrable function* if the product $f\, 1_U$ is μ-integrable for every relatively compact open subset U of X. In particular, if μ is the Lebesgue measure on X, we speak of a *locally integrable function* f. ⊘

The next two results clarify the relation between a locally integrable function and the distribution defined by it.

Theorem 20.38. *Let $X \subset \mathbf{R}^n$ be open, $E = C_0(X)$, and u a positive distribution acting on $\overline{E}$ with corresponding Daniell seminorm $\|\cdot\|$. For any $f \in \overline{E}$, we have*

$$\|f\| = \sup\{\, |u(\phi\, f)| \in \mathbf{R}_{\geq 0} \mid \phi \in E,\ \|\phi\|_{C^0} \leq 1\,\}.$$

Proof. Consider any $\phi \in E$. We obviously have

$$|u(\phi\, f)| \leq \|\phi\|_{C^0}\, \|f\|. \tag{20.16}$$

Furthermore, let $K = \operatorname{supp} \phi \subset X$ and define $K(\delta) = \{\, x \in K \mid |\phi(x)| \geq \delta\,\}$, for $\delta > 0$. Now select $\chi_\delta \in E$ such that $\chi_\delta = 1$ on $K(2\delta)$, $\chi_\delta = 0$ on $X \setminus K(\delta)$, and $0 \leq \chi_\delta(x) \leq 1$, for all $x \in X$. Set

$$\phi_\delta = \chi_\delta \frac{|\phi|}{\phi}. \qquad \text{Then} \qquad \phi_\delta \in E, \qquad \|\phi_\delta\|_{C^0} \leq 1,$$
$$0 \leq |\phi| - \phi_\delta\,\phi = (1-\chi_\delta)|\phi| \leq 2\delta, \qquad \text{so} \qquad |\,\|\phi\| - |u(\phi_\delta\,\phi)|\,| \leq 2\delta\,\mu_u(K). \tag{20.17}$$

Consider arbitrary $\epsilon > 0$. There exists $\phi \in E$ such that $\|f - \phi\| < \epsilon$. For $\delta > 0$ sufficiently small, we then obtain from (20.17) and (20.16), plus the subadditivity of $|\cdot|$,

$$\begin{aligned}|\,\|f\| - |u(\phi_\delta\,f)|\,| &\leq |\,\|f\| - \|\phi\|\,| + |\,\|\phi\| - |u(\phi_\delta\,\phi)|\,| \\ &\quad + |\,|u(\phi_\delta\,\phi)| - |u(\phi_\delta\,f)|\,| < 2\epsilon + \|\phi_\delta\|_{C^0}\,\|f - \phi\| < 3\epsilon.\end{aligned}$$

□

Corollary 20.39. *Suppose the function f is locally integrable with respect to the Lebesgue measure on an open set X in $\mathbf{R}^n$ and $f = 0$ in $\mathcal{D}'(X)$, that is,* test $f = 0$. *Then $f = 0$ almost everywhere on X.*

It is by now evident that two functions in $\overline{E}$ defined almost everywhere that differ only negligibly are the same for all practical purposes. From Definition 20.14 and Theorem 20.18.(i) it is clear that "equality" of two such functions is an equivalence relation. Hence, we may identify them by the formation of equivalence classes. For any function f defined almost everywhere with $f \in \overline{E}$ almost everywhere, we denote by $[f]$ the *equivalence class* of all functions in $\overline{E}$ that agree with f almost everywhere, that is,

$$[f] = \{\, f' \in \overline{E} \mid f' = f \text{ almost everywhere} \,\}. \tag{20.18}$$

Proposition 20.15.(iv) implies that every class contains a function that is defined everywhere and assumes only finite values.

The sum and scalar product of classes are defined in the obvious way: $[f]+[g]$ is the class of $f+g$, and $r\,[f]$ is the class of $r\,f$. These classes do not depend on the choice of the representatives $f \in [f]$ and $g \in [g]$. For instance, if $f = f'$ almost everywhere and $g = g'$ almost everywhere, then the class of $f+g$ is clearly the same as the class of $f' + g'$. In addition, the Daniell **seminorm** of functions in $\overline{E}$ now leads to the **norm** of a class by means of

$$\|\,[f]\,\| = \|f\|,$$

where $f \in [f]$; again, this number will not depend on the choice of $f \in [f]$ according to Definition 20.16.

Let us denote the collection of equivalence classes of functions in $\overline{E}$ defined almost everywhere by

$$L^1(X).$$

Observe that $L^1(X)$ is the quotient of $\overline{E}$ by its linear subspace of negligible functions, while $\|\cdot\|$ is the quotient norm. By application of Theorem 20.21.(ii) one immediately establishes the following result.

Theorem 20.40. *$(L^1(X), \|\cdot\|)$, the space of Lebesgue integrable functions on X, is a* Banach space, *i.e., a normed linear space that is complete. In addition, the linear form u is continuous with respect to the topology of $L^1(X)$ defined by the norm.*

There is an additional structure on $\overline{E}$, the order. The question arises whether it is inherited by the space $L^1(X)$. It is. Let us say that the class $[f]$ is smaller than the class $[g]$ if a representative of $[f]$ is smaller than a representative of $[g]$ almost everywhere. In other words, $[f] \leq [g]$ if and only if $f \leq g$ almost everywhere, for any, and then clearly all, representatives $f \in [f]$ and $g \in [g]$.

Finally, we define the spaces $L^p(X)$, for $1 < p < \infty$ and any open subset $X \subset \mathbf{R}^n$. We denote by $\|\cdot\| = \|\cdot\|_1$ the Daniell seminorm corresponding to the Lebesgue measure on X and introduce $\|\cdot\|_p$ by

$$\|\phi\|_p := \|\phi\|_{L^p(X)} := \|\,|\phi|^p\,\|^{1/p} = \left(\int_X |\phi(x)|^p\,dx\right)^{1/p} \tag{20.19}$$

for $\phi \in C_0(X)$. *Minkowski's inequality*, see Remark 20.42 below or [7, Exercise 6.73.(iii)],

$$\|\phi + \psi\|_p \leq \|\phi\|_p + \|\psi\|_p \qquad (\phi, \psi \in C_0(X)),$$

then expresses the subadditivity of $\|\cdot\|_p$ on $C_0(X)$.

Theorem 20.41. *Let the notation be as in (20.19). Then all assertions of Theorem 20.13 remain valid if the seminorm $\|\cdot\|$ acting on $E = C_0(X, \mathbf{R})$ is replaced by $\|\cdot\|_p$, but for the fact that $\|\cdot\|_p$ dominates $|u|$ as in part (ii) of the theorem.*

Proof. **(i)**. The solidity of $\|\cdot\|_p$ follows from the fact that $r \mapsto r^p$ is strictly increasing on $\mathbf{R}_{\geq 0}$ and the solidity of $\|\cdot\|$. The absolute homogeneity of $\|\cdot\|_p$ is a consequence of the same property of $\|\cdot\|$. If (ϕ_k) is a nonincreasing sequence of functions with pointwise limit 0, then $(|\phi_k|^p)$ has the same property. Since $\|\cdot\|$ is monotone decreasing, we now obtain that this is also true of $\|\cdot\|_p$. The countable subadditivity follows easily from this and the subadditivity. Indeed, for any sequence (ϕ_k) of nonnegative functions in E we obtain

$$\Big\|\sum_{k\in\mathbf{N}} \phi_k\Big\|_p = \lim_{l\to\infty}\Big\|\sum_{k=1}^{l} \phi_k\Big\|_p \leq \lim_{l\to\infty}\sum_{k=1}^{l}\|\phi_k\|_p = \sum_{k\in\mathbf{N}}\|\phi_k\|_p.$$

(ii). The finiteness of $\|\cdot\|_p$ on E is implied by the corresponding property of $\|\cdot\|$.
(iii). Let (ϕ_k) be a sequence of nonnegative functions in E and set

$$\psi_l = \sum_{k=1}^{l}\phi_k \in E; \qquad \text{then, by assumption} \qquad \sup_{l\in\mathbf{N}}\|\psi_l\|_p < \infty.$$

Now $(\psi_l^p)_l$ is a nondecreasing sequence in E, while $\sup_l \|\,\psi_l^p\,\| < \infty$. The Monotone Convergence Theorem, Theorem 20.23, then implies that (ψ_l^p) is a Cauchy sequence with respect to $\|\cdot\|$. In particular, $\lim_{l\to\infty}\|\psi_l^p - \psi_{l-1}^p\| = 0$. Furthermore $\phi_l^p \le \psi_l^p - \psi_{l-1}^p$, by virtue of the fact that $1 + t^p \le (1+t)^p$, for $t \ge 0$. (The latter assertion follows from the observation that we have equality for $t = 0$, while the derivative with respect to t of the left-hand side is dominated by that of the right-hand side.) Thus we obtain $\lim_{l\to\infty}\|\phi_l^p\| = \lim_{l\to\infty}\|\phi_l\|_p^p = 0$ and this gives the desired conclusion. □

The arguments and definitions following Theorem 20.13 that underlie Theorem 20.21 can now be copied, and they lead to the linear space $\overline{E}$ consisting of the *p-integrable functions*, which is *complete* with respect to the seminorm $\|\cdot\|_p$. Analogous to Theorem 20.40 for $L^1(X)$, one obtains the Banach space $L^p(X)$.

Remark 20.42. In order to give a direct proof of *Minkowski's inequality*, for f and g belonging to $L^p(X)$, write $a = \|f\|_p$ and $b = \|g\|_p$. To avoid trivialities, we may suppose that $a > 0$ and $b > 0$. By scaling, it suffices to consider the case that $a+b = 1$. Let $\tilde{f} = \frac{|f|}{a}$ and $\tilde{g} = \frac{|g|}{b}$. Then $\int_X \tilde{f}(x)^p\,dx = \int_X \tilde{g}(x)^p\,dx = 1$. Now $|f+g| \le |f| + |g| = a\tilde{f} + b\tilde{g}$. The second derivative of the function $r \mapsto r^p$ is nonnegative on $\mathbf{R}_{\ge 0}$, which implies that this function is convex on $\mathbf{R}_{\ge 0}$. Therefore we have $|f+g|^p \le a\tilde{f}^p + b\tilde{g}^p$, and so $\int_X |f+g|^p(x)\,dx \le a+b = 1$. This leads to $\|f+g\|_p \le 1 = \|f\|_p + \|g\|_p$. ⊘

Remark 20.43. We list some more properties of the spaces $L^p(X)$.

$C_0(X)$ is dense in $L^p(X)$ with respect to $\|\cdot\|_p$. Hölder's inequality from Problem 11.21 immediately implies that any $f \in L^p(X)$ is locally integrable. The inclusion mapping $L^p(X) \to \mathcal{D}'(X)$ given by $u \mapsto \text{test}\, u$ is injective, as a consequence of Remark 3.7. On account of Hölder's inequality we also obtain, for arbitrary $u \in L^p(X)$ and $\phi \in C_0^\infty(X)$,

$$|u(\phi)| = \left|\int_X u(x)\phi(x)\,dx\right| \le \|u\|_{L^p}\,\|\phi\|_{L^q}.$$

Here q is the real number such that $\frac{1}{p} + \frac{1}{q} = 1$. This entails that we have a

$$\text{continuous injection} \qquad L^p(X) \to \mathcal{D}'(X). \tag{20.20}$$

As an application, consider a sequence $(u_j)_{j\in\mathbf{N}}$ in $C_0^\infty(X)$ that has the Cauchy property with respect to the L^p norm and satisfies $\lim_{j\to\infty} u_j = 0$ in $\mathcal{D}'(X)$. Then we claim that $\lim_{j\to\infty}\|u_j\|_{L^p} = 0$. Indeed, in view of the completeness of $L^p(X)$, there exists $u \in L^p(X) \subset \mathcal{D}'(X)$ such that $\lim_{j\to\infty} u_j = u$ in $L^p(X)$. Then the continuity in (20.20) gives $0 = \lim_{j\to\infty} u_j = u$ in $\mathcal{D}'(X)$ and therefore $u = 0$ in $L^p(X)$ by the injectivity in (20.20). This in turn leads to the desired conclusion. ⊘

Finally we give a useful result on the continuity of pullback under translation acting in $L^p(\mathbf{R}^n)$.

Lemma 20.44. *Let* $1 \le p < \infty$ *and* $f \in L^p(\mathbf{R}^n)$. *Then*

$$\lim_{h \to 0} \|(T_h{}^* - I)f\|_p = 0.$$

Proof. Let $\epsilon > 0$ be arbitrary. Applying Theorem 20.41, choose $\phi \in C_0(\mathbf{R}^n)$ such that $\|f - \phi\|_p < \frac{\epsilon}{3}$. Then it is clear from the invariance under translation of the Lebesgue measure that $\|T_h{}^* f - T_h{}^* \phi\|_p = \|f - \phi\|_p < \frac{\epsilon}{3}$, for all $h \in \mathbf{R}^n$.

Furthermore, there exists $R > 0$ such that $\phi(x) = 0$, for $\|x\| \geq R$; write $B = \{ x \in \mathbf{R}^n \mid \|x\| \leq R + 1 \}$. Since ϕ is uniformly continuous on $\mathbf{R}^n$, we may select $0 < \delta < 1$ such that

$$|\phi(x+h) - \phi(x)| < \frac{\epsilon}{3}\Big(\frac{1}{\mathrm{vol}_n(B)}\Big)^{\frac{1}{p}} \qquad (\|h\| < \delta).$$

Hence

$$\int_{\mathbf{R}^n} |\phi(x+h) - \phi(x)|^p \, dx = \int_B |\phi(x+h) - \phi(x)|^p \, dx < \Big(\frac{\epsilon}{3}\Big)^p \qquad (\|h\| < \delta),$$

which is to say that $\|T_h{}^*\phi - \phi\|_p < \frac{\epsilon}{3}$. Accordingly, we obtain, for $\|h\| < \delta$,

$$\|T_h{}^* f - f\|_p \leq \|T_h{}^* f - T_h{}^* \phi\|_p + \|T_h{}^*\phi - \phi\|_p + \|\phi - f\|_p < \epsilon. \qquad \square$$

Chapter 21
Solutions to Selected Problems

1.3 Let ϕ be continuously differentiable and suppose $\phi(x) = 0$ if $|x| \geq m > 0$. We introduce some notation. For $\epsilon \geq 0$ and $x \in \mathbf{R}$, set

$$\Phi_\epsilon(x) = \frac{\phi(x) - \phi(0)}{x + i\epsilon} \qquad \text{and} \qquad I_a = \,]-a, a\,[\qquad (a \in \mathbf{R}).$$

Note that Φ_ϵ is continuous on $\mathbf{R}$ (see [7, Proposition 2.2.1.(ii)]), for all $\epsilon \geq 0$. Now we get, writing $\phi(x) = \phi(0) + (\phi(x) - \phi(0))$,

$$\int_{\mathbf{R}} \frac{\phi(x)}{x + i\epsilon}\, dx = \int_{I_m} \frac{\phi(x)}{x + i\epsilon}\, dx = \phi(0) \int_{I_m} \frac{x - i\epsilon}{x^2 + \epsilon^2}\, dx + \int_{I_m} \Phi_\epsilon(x)\, dx.$$

Since $x \mapsto x/(x^2 + \epsilon^2)$ is an odd function on $\mathbf{R}$ and I_m is symmetric about the origin, we obtain, by means of the change of variables $x = \epsilon\, y$,

$$\int_{I_m} \frac{x - i\epsilon}{x^2 + \epsilon^2}\, dx = -i \int_{I_m} \frac{\epsilon}{x^2 + \epsilon^2}\, dx = -i \int_{I_{\frac{m}{\epsilon}}} \frac{1}{y^2 + 1}\, dy$$
$$= -2i \arctan \frac{m}{\epsilon} \to -\pi i \quad \text{as} \quad \epsilon \downarrow 0.$$

Furthermore, we have

$$\lim_{\epsilon \downarrow 0} \int_{I_m} \Phi_\epsilon(x)\, dx = \int_{I_m} \Phi_0(x)\, dx. \tag{21.1}$$

Postponing for the moment the proof of this identity, we apply it to obtain

$$\lim_{\epsilon \downarrow 0} \int_{I_m} \Phi_\epsilon(x)\, dx = \lim_{\delta \downarrow 0} \Big(\int_{-m}^{-\delta} \Phi_0(x)\, dx + \int_{\delta}^{m} \Phi_0(x)\, dx \Big) = \Big(\mathrm{PV}\, \frac{1}{x} \Big)(\phi).$$

With the notation $\delta(\phi) = \phi(0)$, combination of these identities leads to the following *Plemelj–Sokhotsky jump relations* (see Example 14.30 or Problem 12.14.(ix) for other proofs):

J.J. Duistermaat and J.A.C. Kolk, *Distributions: Theory and Applications*,
Cornerstones, DOI 10.1007/978-0-8176-4675-2_21,

$$\frac{1}{x \pm i\,0} \pm \pi i\,\delta = \mathrm{PV}\,\frac{1}{x}, \qquad \delta = \frac{1}{2\pi i}\Big(\frac{1}{x - i\,0} - \frac{1}{x + i\,0}\Big).$$

Formula (21.1) follows from the *Dominated Convergence Theorem of Arzelà*, see [7, Theorem 6.12.3], or that of *Lebesgue*, see Theorem 20.26.(iv). Indeed, we have

$$|\Phi_\epsilon(x)| = \Big|\frac{x}{x + i\epsilon}\,\Phi_0(x)\Big| = \frac{|x|}{\sqrt{x^2+\epsilon^2}}\,|\Phi_0(x)| \le |\Phi_0(x)|.$$

A more elementary proof of (21.1) is as follows. The integrand on the left-hand side is a continuous function that admits a majorant that is independent of ϵ. Accordingly, for arbitrary $\eta > 0$ there exists $m > \delta > 0$ such that for every $\epsilon \ge 0$,

$$\Big|\int_{I_\delta} \Phi_\epsilon(x)\,dx\Big| < \frac{\eta}{3}.$$

Next, observe that for all $\epsilon > 0$ and $x \in I'_\delta := [-m, m\,] \setminus I_\delta$,

$$|\Phi_\epsilon(x) - \Phi_0(x)| = \Big|\frac{1}{x + i\epsilon} - \frac{1}{x}\Big|\,|\phi(x) - \phi(0)| = \epsilon\,\frac{|\Phi_0(x)|}{|x + i\epsilon|} \le \epsilon\,\frac{|\Phi_0(x)|}{\delta}.$$

In other words, the convergence of Φ_ϵ to Φ_0 is uniform on I'_δ as $\epsilon \downarrow 0$. This implies the existence of $\epsilon_0 > 0$ such that for all $0 < \epsilon \le \epsilon_0$,

$$\Big|\int_{I'_\delta} (\Phi_\epsilon(x) - \Phi_0(x))\,dx\Big| < \frac{\eta}{3}.$$

As a consequence we obtain, for all $0 < \epsilon \le \epsilon_0$,

$$\Big|\int_{I_m} \Phi_\epsilon(x)\,dx - \int_{I_m} \Phi_0(x)\,dx\Big|$$

$$\le \Big|\int_{I_\delta} \Phi_\epsilon(x)\,dx\Big| + \Big|\int_{I_\delta} \Phi_0(x)\,dx\Big| + \Big|\int_{I'_\delta} (\Phi_\epsilon(x) - \Phi_0(x))\,dx\Big| < 3\,\frac{\eta}{3} = \eta.$$

Finally note that for $x = \epsilon = \frac{1}{n}$ with $n \in \mathbf{N}$, we have $\frac{1}{|x+i\epsilon|} = \frac{n}{\sqrt{2}}$; this implies that the convergence of Φ_ϵ to Φ_0 is **not** uniform on all of I_m.

A second proof can be based on integration by parts. For $x \in \mathbf{R}$ and $\epsilon > 0$, we have $\frac{d}{dx}\log|x + i\epsilon| = \frac{x}{x^2+\epsilon^2}$ and $\frac{d}{dx}\arctan\frac{\epsilon}{x} = -\frac{\epsilon}{x^2+\epsilon^2}$, and therefore

$$\frac{d}{dx}\log(x + i\epsilon) = \frac{x - i\epsilon}{x^2+\epsilon^2} = \frac{1}{x + i\epsilon}.$$

Hence integration by parts leads to

$$\begin{aligned}\int_{\mathbf{R}} \frac{\phi(x)}{x+i\epsilon}\,dx &= \int_{\mathbf{R}} \phi(x)\frac{d}{dx}\log(x+i\epsilon)\,dx \\ &= -\int_{\mathbf{R}} \phi'(x)\log(x+i\epsilon)\,dx + [\phi(x)\log(x+i\epsilon)]_{-\infty}^{\infty} \\ &= -\int_{\mathbf{R}} \phi'(x)\log|x+i\epsilon|\,dx - i\int_{\mathbf{R}} \phi'(x)\arg(x+i\epsilon)\,dx.\end{aligned}$$

Now $\lim_{\epsilon\downarrow 0}\arg(x+i\epsilon)$ equals 0 for $x>0$, and π for $x<0$. Furthermore, we have $|\arg(x+i\epsilon)|\le 2\pi$. By dominated convergence we obtain

$$\lim_{\epsilon\downarrow 0}\int_{\mathbf{R}} \phi'(x)\arg(x+i\epsilon)\,dx = \pi\int_{-\infty}^{0}\phi'(x)\,dx = \pi\,\phi(0) = \pi\,\delta(\phi).$$

Assuming $\epsilon<1$ we see that

$$|\log|x+i\epsilon|| = \begin{cases} \log|x+i\epsilon| \le \dfrac{1}{2}\log(x^2+1), & |x+i\epsilon|\ge 1; \\ \log\dfrac{1}{|x+i\epsilon|} \le \log\dfrac{1}{|x|} = |\log|x||, & |x+i\epsilon|<1. \end{cases}$$

So another application of dominated convergence leads to

$$\lim_{\epsilon\downarrow 0}\int_{\mathbf{R}} \frac{\phi(x)}{x+i\epsilon}\,dx = -\int_{\mathbf{R}} \phi'(x)\log|x|\,dx - \pi i\,\delta(\phi).$$

1.4 Consider $\tilde{p}(x) = -(x-1)^3(x+1)^3 = 1-3x^2+3x^4-x^6$. Then $\int_{-1}^{1}\tilde{p}(x)\,dx = 32/35$ and therefore $p(x) = \frac{35}{32}(1-x^2)^3$. That ϕ is twice continuously differentiable on some neighborhood of ± 1, respectively, is an easy verification.

1.5 Integration by parts leads to $(|\cdot| * \psi)'' = 2\psi$. Indeed,

$$\begin{aligned}(|\cdot| * \psi)''(x) = (|\cdot| * \psi'')(x) &= \int_{\mathbf{R}} |y|\,\psi''(x-y)\,dy \\ &= -\int_{-\infty}^{0} y\,\psi''(x-y)\,dy + \int_{0}^{\infty} y\,\psi''(x-y)\,dy \\ &= -\int_{-\infty}^{0} \psi'(x-y)\,dy + \int_{0}^{\infty} \psi'(x-y)\,dy = 2\psi(x).\end{aligned}$$

Now select $\psi = \phi_\epsilon$. Then integration implies

$$g'(x) = -1 + 2\int_{-1}^{x/\epsilon} p(t)\,dt \qquad \text{and} \qquad g(x) = -x + 2\int_{-1}^{x/\epsilon}\int_{-1}^{\xi/\epsilon} p(t)\,dt\,d\xi.$$

1.6 (i). $f * g$ equals 0 outside $[-2,2]$. Indeed, $\int_{\mathbf{R}} f(x-y)\,g(y)\,dy \ne 0$ implies that there is $y\in[-1,1]$ with $-1\le x-y\le 1$, which leads to $-2\le y-1\le x\le y+1\le 2$.

(ii). Take both f and g equal to the characteristic function $1_{[-1,1]}$. Then $f * g(x) = \int_{-1}^{1} 1_{[-1,1]}(x-y)\, dy$, where the interval of integration may actually be restricted to $[x-1, x+1]$. Hence, for $-2 \le x \le 0$ and $0 \le x \le 2$ we find that $f * g(x)$ equals

$$\int_{-1}^{x+1} dy = 2 + x \qquad \text{and} \qquad \int_{x-1}^{1} dy = 2 - x, \qquad \text{respectively.}$$

It follows that $f * g(x) = \max\{0, 2 - |x|\}$, for all $x \in \mathbf{R}$.
(iii). The functions from part (ii) satisfy all demands in (iii).
(iv). Following the hint, we find by means of the substitution $y = x\,t$ that

$$f * g(x) = \int_0^x (x-y)^\alpha\, y^\alpha\, dy = x^{2\alpha+1} \int_0^1 (1-t)^\alpha\, t^\alpha\, dt,$$

where the latter integral converges. $f * g$ is discontinuous at 0 if $2\alpha + 1 < 0$, that is, if $-1 < \alpha < -1/2$.

2.1 There exists $z \in U$ such that $d(y, U) = \|y - z\|$. Therefore we obtain

$$d(x, U) \le \|x - z\| \le \|x - y\| + \|y - z\|,$$

which implies $d(x, U) - d(y, U) \le \|x - y\|$. Similarly $d(y, U) - d(x, U) \le \|x - y\|$.

2.2 Note that $\operatorname{supp} \phi^{(k)} \subset \operatorname{supp} \phi$, for all $k \in \mathbf{Z}_{\ge 0}$.
(i). Suppose there exists a compact set $K \subset \mathbf{R}$ such that $\{j\} + \operatorname{supp} \phi = \operatorname{supp} \phi_j \subset K$, for all $j \in \mathbf{N}$. Then $j \in K + (-\operatorname{supp} \phi)$, which is a compact set; hence no convergence in $C_0^\infty(\mathbf{R})$. On the other hand, for fixed k and arbitrary j and x,

$$|\phi_j^{(k)}(x)| = \frac{1}{j} |\phi^{(k)}(x - j)| \le \frac{1}{j} \sup_{x \in \mathbf{R}} |\phi^{(k)}(x)|,$$

which shows that $\lim_{j\to\infty} \phi_j^{(k)} = 0$ converges uniformly on $\mathbf{R}$.
(ii). ϕ cannot be a polynomial, owing to the compactness of $\operatorname{supp} \phi$; consequently, there exists $x_0 \in \mathbf{R}$ with $\phi^{(p+1)}(x_0) \ne 0$. Now set $x_j = x_0/j$. Then we have, as $j \to \infty$,

$$|\phi_j^{(p+1)}(x_j)| = j^{p+1-p} \left|\phi^{(p+1)}\Big(j\, \frac{x_0}{j}\Big)\right| = j\, |\phi^{(p+1)}(x_0)| \to \infty.$$

This implies that $(\phi_j^{(p+1)})_{j\in\mathbf{N}}$ is not uniformly convergent on $\mathbf{R}$, and therefore $(\phi_j)_{j\in\mathbf{N}}$ is not convergent in $C_0^\infty(\mathbf{R})$. Now consider $x \in \mathbf{R}$ fixed. Since 0 does not belong to the compact set $\operatorname{supp} \phi$, there is $N = N(x)$ such that $jx \notin \operatorname{supp} \phi$, for all $j \ge N$. Accordingly, for such j and fixed $k \in \mathbf{Z}_{\ge 0}$, we have $\phi_j^{(k)}(x) = j^{k-p}\, \phi^{(k)}(jx) = 0$ and thus $\lim_{j\to\infty} \phi_j^{(k)}(x) = 0$.
(iii). Select $a < 0 < b$ such that $\operatorname{supp} \phi \subset [a, b]$. Then $\operatorname{supp} \phi_j \subset \frac{1}{j} \operatorname{supp} \phi \subset [a, b]$, for all j. Fix k arbitrarily. Then we have, for all j and x,

$$|\phi_j^{(k)}(x)| = j^k e^{-j}\, |\phi^{(k)}(jx)| \le j^k e^{-j} \sup_{x\in\mathbf{R}} |\phi^{(k)}(x)|,$$

whereas $\lim_{j\to\infty} j^k e^{-j} = 0$. It follows that $\lim_{j\to\infty} \phi_j^{(k)} = 0$ uniformly on $\mathbf{R}$, and this holds for every k. As a consequence, we have convergence to 0 in $C_0^\infty(\mathbf{R})$.

2.3 supp ϕ_ϵ is contained in the closed ball $\overline{B(0;\epsilon)}$, while supp ψ is a compact subset of the open set X. According to Corollary 2.4, there exist $\epsilon_0 > 0$ and a compact set $K \subset X$ such that the ϵ-neighborhood of supp ψ is contained in K, for $0 < \epsilon \le \epsilon_0$. Now Lemma 2.18 implies, for $0 < \epsilon < \epsilon_0$,

$$\operatorname{supp}(\psi * \phi_\epsilon) \subset \operatorname{supp}\psi + \overline{B(0;\epsilon)} \subset K \subset X,$$

which shows that condition (a) in Definition 2.13 is satisfied. According to the proof of Lemma 2.18 one has $\partial^\alpha(\psi * \phi_\epsilon) = \partial^\alpha\psi * \phi_\epsilon$, for arbitrary multi-index α. On account of Lemma 1.6, it follows that $\partial^\alpha\psi * \phi_\epsilon$ converges uniformly to $\partial^\alpha\psi$ on K as $\epsilon \downarrow 0$; this leads to the uniform convergence of $\partial^\alpha(\psi * \phi_\epsilon)$ to $\partial^\alpha\psi$ on X. Therefore condition (b) in Definition 2.13 is satisfied too.

2.5 We have $-y\,\gamma_\epsilon(y) = \frac{\epsilon^2}{2}\gamma_\epsilon'(y)$ and furthermore $\int_{-\infty}^{-x} \gamma_\epsilon(y)\,dy = \int_x^\infty \gamma_\epsilon(y)\,dy$, because γ_ϵ is an even function. Hence, for any $x \in \mathbf{R}^n$,

$$\begin{aligned}
(|\cdot| * \gamma_\epsilon)(x) &= \int_{\mathbf{R}} |x-y|\,\gamma_\epsilon(y)\,dy = \Big(\int_{-\infty}^{x} - \int_x^\infty\Big)(x-y)\,\gamma_\epsilon(y)\,dy \\
&= x\Big(\int_{-\infty}^{x} - \int_x^\infty\Big)\gamma_\epsilon(y)\,dy + \frac{\epsilon^2}{2}\big([\gamma_\epsilon]_{-\infty}^{x} - [\gamma_\epsilon]_x^\infty\big) \\
&= x\int_{-x}^{x} \gamma_\epsilon(y)\,dy + \epsilon^2\,\gamma_\epsilon(x) = \frac{x}{\sqrt{\pi}}\int_{-x/\epsilon}^{x/\epsilon} e^{-y^2}\,dy + \epsilon^2\,\gamma_\epsilon(x),
\end{aligned}$$

where we have obtained the last integral by a change of variables. In view of (2.14),

$$\lim_{\epsilon\downarrow 0} \epsilon^2\,\gamma_\epsilon(x) = 0 \qquad\text{and}\qquad \lim_{\epsilon\downarrow 0}\frac{1}{\sqrt{\pi}}\int_{-x/\epsilon}^{x/\epsilon} e^{-y^2}\,dy = \begin{cases} \pm 1, & x \gtrless 0; \\ 0, & x = 0. \end{cases}$$

Accordingly

$$\lim_{\epsilon\downarrow 0}(|\cdot| * \gamma_\epsilon)(x) = |x|.$$

A standard computation now gives

$$(|\cdot| * \gamma_\epsilon)'(x) = \frac{1}{\sqrt{\pi}}\int_{-x/\epsilon}^{x/\epsilon} e^{-y^2}\,dy \qquad\text{and}\qquad (|\cdot| * \gamma_\epsilon)''(x) = 2\gamma_\epsilon(x).$$

3.2 Suppose there exists a neighborhood U of 0 in $\mathbf{R}$ such that u is of order $< k$ on U. Then for every compact subset $K \subset U$, there exists $c > 0$ such that

$$|\phi^{(k)}(0)| = |u(\phi)| \le c\,\|\phi\|_{C^{k-1}} \qquad (\phi \in C_0^\infty(K)).$$

For arbitrary $\psi \in C_0^\infty(\mathbf{R})$ satisfying $\psi(0) \neq 0$ and $\operatorname{supp} \psi \subset U$ and $t > 0$, consider $\phi_t \in C_0^\infty(\mathbf{R})$ given by $\phi_t(x) = e^{itx}\psi(x)$. Then $\phi_t{}^{(k)}(0) = (it)^k\,\psi(0) + O(t^{k-1})$ on account of Leibniz's formula (2.8), while $\|\phi_t\|_{C^{k-1}} = O(t^{k-1})$ as $t \to \infty$. This leads to a contradiction and therefore u is of order k on U.

3.5 (i). Both functions are locally integrable on $\mathbf{R}^2$ and therefore define distributions, of order 0.
(ii).(a). A straightforward estimate yields that $|u(\phi)| \leq 2\pi\,\|\phi\|_{C^1}$, for all $\phi \in C_0^\infty(\mathbf{R}^2)$; this implies that u is a distribution of order ≤ 1.
The integrand in the definition of $u(\phi)$ is equal to $\partial_r\phi(r\cos t, r\sin t)|_{r=1}$. Therefore, if we take $\phi(x) = \psi(\|x\|)$ for some $\psi \in C_0^\infty(\mathbf{R}_{>0})$, then $u(\phi) = \pi\,\psi'(1)$, whereas on the other hand $\sup_{x\in\mathbf{R}^2}|\phi(x)| = \sup_{r>0}|\psi(r)|$. Hence, if u were of order 0, there would exist $c > 0$ such that $|\psi'(1)| \leq c\,\sup_{r>0}|\psi(r)|$, for all $\psi \in C_0^\infty(\mathbf{R}_{>0})$; this is in contradiction to Problem 3.2 for $k = 1$. We conclude that u has order 1.
(ii).(b). Note that $(-\sin t\,\partial_1 + \cos t\,\partial_2)\phi(\cos t, \sin t) = \frac{d}{dt}\phi(\cos t, \sin t)$, which implies that v is of order 0.

4.1 For any $\phi \in C_0^\infty(\mathbf{R})$, one finds by integration by parts that

$$
\begin{aligned}
|\cdot|'(\phi) &= -|\cdot|(\phi') = \int_{-\infty}^0 x\,\phi'(x)\,dx - \int_0^\infty x\,\phi'(x)\,dx \\
&= -\int_{-\infty}^0 \phi(x)\,dx + \int_0^\infty \phi(x)\,dx = \int_{\mathbf{R}} \operatorname{sgn}(x)\,\phi(x)\,dx = \operatorname{sgn}(\phi).
\end{aligned}
$$

Similarly,

$$
\begin{aligned}
|\cdot|''(\phi) &= \operatorname{sgn}'(\phi) = -\operatorname{sgn}(\phi') = \int_{-\infty}^0 \phi'(x)\,dx - \int_0^\infty \phi'(x)\,dx = 2\phi(0) \\
&= 2\delta(\phi).
\end{aligned}
$$

4.2 This is a direct consequence of (1.3).

4.6 Without any restriction in generality we may assume that $p = 0$, i.e., $v(x) = \frac{1}{\|x\|^n}\,x$. Introduce spherical coordinates $(r, y) \in \mathbf{R}_{>0} \times S^{n-1}$ in $\mathbf{R}^n$ via $x = \Psi(r, y) = r\,y$; one then has $|\det D\Psi(r, y)| = r^{n-1}\omega(y)$, where ω denotes the Euclidean $(n-1)$-dimensional density on S^{n-1}. The local integrability of the v_j over $\mathbf{R}^n$, for $1 \leq j \leq n$, then follows from

$$
\begin{aligned}
\int_{B(0;\epsilon)} |v_j(x)|\,dx &= \int_0^\epsilon r^{-n+1+n-1}\,dr \int_{S^{n-1}} |y_j|\,\omega(y)\,dy \\
&= \epsilon \int_{S^{n-1}} |y_j|\,\omega(y)\,dy < \infty \qquad (\epsilon > 0).
\end{aligned}
$$

For any $\phi \in C_0^\infty(\mathbf{R}^n)$, we have

$$
\begin{aligned}
(\operatorname{div} v)(\phi) &= -\sum_{j=1}^{n} v_j(\partial_j \phi) = -\int_{\mathbf{R}^n} \sum_{j=1}^{n} v_j(x)\, \partial_j \phi(x)\, dx \\
&= -\int_{\mathbf{R}^n} \langle\, v(x),\, \operatorname{grad} \phi(x)\,\rangle\, dx = -\int_{\mathbf{R}^n} D\phi(x) v(x)\, dx.
\end{aligned}
$$

Upon the introduction of spherical coordinates in $\mathbf{R}^n$ this leads to

$$
\begin{aligned}
&(\operatorname{div} v)(\phi) = -\int_{S^{n-1}} \int_{\mathbf{R}_{>0}} r^{n-1} D\phi(r\,y) v(r\,y)\, dr\, \omega(y)\, dy \\
&= -\int_{S^{n-1}} \int_{\mathbf{R}_{>0}} D\phi(r\,y) y\, dr\, \omega(y)\, dy = -\int_{S^{n-1}} \int_{\mathbf{R}_{>0}} \frac{d}{dr} \phi(r\,y)\, dr\, \omega(y)\, dy \\
&= c_n\, \phi(0) = c_n\, \delta(\phi).
\end{aligned}
$$

4.7 The local integrability of E over $\mathbf{R}^n$ follows as in the preceding problem. For any $\phi \in C_0^\infty(\mathbf{R}^n)$ and $n \neq 2$ we obtain, through integration by parts,

$$
\begin{aligned}
(\partial_j E)(\phi) &= -E(\partial_j \phi) = -\int_{\mathbf{R}^n} E(x)\, \partial_j \phi(x)\, dx \\
&= \frac{1}{(n-2)c_n} \int_{\mathbf{R}^{n-1}} \int_{\mathbf{R}} \|x\|^{2-n}\, \partial_j \phi(x)\, dx_j\, dx_{\hat{j}} \\
&= \frac{1}{c_n} \int_{\mathbf{R}^{n-1}} \int_{\mathbf{R}} \|x\|^{1-n} \frac{x_j}{\|x\|}\, \phi(x)\, dx_j\, dx_{\hat{j}} = \frac{1}{c_n} \int_{\mathbf{R}^n} \frac{x_j}{\|x\|^n}\, \phi(x)\, dx \\
&= \frac{1}{c_n} v_j(\phi).
\end{aligned}
$$

Finally, $\Delta E = \operatorname{div}(\operatorname{grad} E) = \frac{1}{c_n} \operatorname{div} v = \frac{1}{c_n}\, c_n\, \delta = \delta$. The case of $n = 2$ is treated in a similar way.

5.1 For any $0 \neq \phi \in C_0^\infty(\mathbf{R}^n)$ and $r > 0$, we have, in view of condition (a),

$$
\begin{aligned}
&(\operatorname{test} f_j)(\phi) - \delta(\phi) = \int_{\mathbf{R}^n} f_j(x)\, (\phi(x) - \phi(0))\, dx \\
&= \int_{\|x\| \le r} f_j(x)(\phi(x) - \phi(0))\, dx + \int_{\|x\| \ge r} f_j(x)\, (\phi(x) - \phi(0))\, dx =: \sum_{i=1}^{2} I_i.
\end{aligned}
$$

Now select $\epsilon > 0$ arbitrarily. In view of the continuity of ϕ there exists $r > 0$ such that $|\phi(x) - \phi(0)| < \epsilon/2$, for all $\|x\| \le r$. Since $f_j \ge 0$, condition (a) implies

$$
|I_1| \le \int_{\|x\| \le r} f_j(x) |\phi(x) - \phi(0)|\, dx < \frac{\epsilon}{2} \int_{\|x\| \le r} f_j(x)\, dx \le \frac{\epsilon}{2} \qquad (j \in \mathbf{N}).
$$

Next, write $m = \sup_{x \in \mathbf{R}^n} |\phi(x)| > 0$. On account of condition (b) there exists $N \in \mathbf{N}$ such that for all $j \ge N$,

$$0 \le \int_{\|x\|\ge r} f_j(x)\,dx < \frac{\epsilon}{4m}, \qquad \text{so} \qquad |I_2| \le 2m \int_{\|x\|\ge r} f_j(x)\,dx < \frac{\epsilon}{2}.$$

Combination of the estimates leads to the desired conclusion.

5.2 Consider $\operatorname{Re} f$ and $\operatorname{Im} f$ separately. Since $f = \sum_{\pm}(f \pm |f|)/2$, we see that f may be assumed to be nonnegative and integrable on $\mathbf{R}^n$. Write $f_\epsilon^\circ = \frac{1}{c} f_\epsilon$ if $c \neq 0$. Then $(f_\epsilon^\circ)_{\epsilon>0}$ satisfies conditions (a) and (b) from Problem 5.1. Indeed,

$$\int_{\|x\|\ge r} f_\epsilon^\circ(x)\,dx = \int_{\|x\|\ge \frac{r}{\epsilon}} f^\circ(x)\,dx \qquad \text{with} \qquad \lim_{\epsilon\downarrow 0} \frac{r}{\epsilon} = \infty.$$

Thus the conclusion follows from the same problem.

Alternatively, one can proceed as follows. For any $\phi \in C_0^\infty(\mathbf{R}^n)$, one has

$$(\text{test } f_\epsilon)(\phi) = \int_{\mathbf{R}^n} f_\epsilon(x)\phi(x)\,dx = \int_{\mathbf{R}^n} f(x)\phi(\epsilon\, x)\,dx.$$

Now $|f(x)\phi(\epsilon\, x)| \le (\sup_{x\in\mathbf{R}^n} |\phi(x)|)\,|f(x)|$, for all $x \in \mathbf{R}^n$ and all $\epsilon > 0$, with a dominating function that is integrable on $\mathbf{R}^n$. Furthermore, $\lim_{\epsilon\downarrow 0} f(x)\phi(\epsilon\, x) = \phi(0)\, f(x)$. As one can see by applying either the Dominated Convergence Theorem of Arzelà (see [7, Theorem 6.12.3]) or that of Lebesgue (see Theorem 20.26.(iv)), one has $\lim_{\epsilon\downarrow 0}(\text{test } f_\epsilon)(\phi) = \int_{\mathbf{R}^n} f(x)\,dx\,\delta(\phi)$.

5.5 Note that $u_t(x) = f_{\sqrt{t}}(x)$ in the notation of Problem 5.2, where

$$f(x) = (4\pi)^{-\frac{n}{2}} e^{-\frac{\|x\|^2}{4}} \qquad \text{and} \qquad \int_{\mathbf{R}^n} f(x)\,dx = 1.$$

Application of that same problem therefore implies $\lim_{t\downarrow 0} u_t(x) = \delta$ in $\mathcal{D}'(\mathbf{R}^n)$. The following formulas show that u_t satisfies the heat equation:

$$\frac{d}{dt}u_t(x) = \Big(\frac{\|x\|^2}{4t^2} - \frac{n}{2t}\Big)u_t(x), \qquad \operatorname{grad}_x u_t(x) = -\frac{u_t(x)}{2t}\,x,$$

$$\Delta u_t(x) = -\frac{u_t(x)}{2t}\operatorname{div}_x x - \frac{1}{2t}\langle \operatorname{grad}_x u_t(x),\, x\,\rangle = \Big(\frac{\|x\|^2}{4t^2} - \frac{n}{2t}\Big)u_t(x),$$

where we have used that $\Delta = \operatorname{div}\operatorname{grad}$. Consequently, $\lim_{t\downarrow 0}\frac{d}{dt}u_t = \Delta\delta$ in $\mathcal{D}'(\mathbf{R}^n)$. Indeed, for any $\phi \in C_0^\infty(\mathbf{R}^n)$,

$$\lim_{t\downarrow 0}\frac{d}{dt}u_t(\phi) = \lim_{t\downarrow 0}\Delta u_t(\phi) = \lim_{t\downarrow 0} u_t(\Delta\phi) = \delta(\Delta\phi) = \Delta\delta(\phi).$$

5.6 Here we evaluate only the integral. Use the change of variables $t = \frac{\|x\|^2}{4u}$ to obtain

$$E(x) = -\frac{1}{(4\pi)^{\frac{n}{2}}}\int_{\mathbf{R}_{>0}} t^{-\frac{n}{2}}\, e^{-\frac{\|x\|^2}{4t}}\,dt = -\frac{\|x\|^{2-n}}{4\pi^{\frac{n}{2}}}\int_{\mathbf{R}_{>0}} u^{\frac{n}{2}-2}\,e^{-u}\,du.$$

Note that according to (13.30) and (13.31) the integral on the right-hand side equals

$$\Gamma\Big(\frac{n}{2}-1\Big)=\frac{2\Gamma(\frac{n}{2})}{n-2}; \qquad \text{so} \qquad E(x)=\frac{1}{(2-n)\,c_n\,\|x\|^{n-2}}$$

in view of (13.37).

6.2 We discuss two different methods to prove the result.

Consider arbitrary $\phi\in C_0^\infty(\mathbf{R}^n)$ and define $f:\mathbf{R}_{>0}\to\mathbf{C}$ by $f(t)=u_t(\phi)$. From Problem 5.5 we obtain $\lim_{t\downarrow 0}f(t)=\phi(0)$, and more generally,

$$f^{(j)}(t)=\frac{d^j}{dt^j}u_t(\phi)=\Delta^j u_t(\phi)=u_t(\Delta^j\phi)\to\Delta^j\phi(0) \quad \text{as} \quad t\downarrow 0.$$

As a consequence, $f\in C^\infty(\mathbf{R}_{>0})$ and the $f^{(j)}(t)$ have limits as $t\downarrow 0$, for all $j\in\mathbf{Z}_{\geq 0}$. Therefore we may use Taylor expansion to derive

$$f(t)=\sum_{j=0}^{k-1}\frac{t^j}{j!}f^{(j)}(+0)+\int_0^t\frac{(t-s)^{k-1}}{(k-1)!}f^{(k)}(s)\,ds \qquad (t>0).$$

Hence, by means of the change of variables $s=tu$ in the integral, we obtain, for $k\in\mathbf{Z}_{\geq 0}$,

$$\begin{aligned}\lim_{t\downarrow 0}\frac{1}{t^k}\Big(u_t-\sum_{j=0}^{k-1}\frac{t^j}{j!}\Delta^j\delta\Big)(\phi)&=\lim_{t\downarrow 0}\int_0^1\frac{(1-u)^{k-1}}{(k-1)!}f^{(k)}(s)\Big|_{s=tu}du\\&=\frac{1}{k!}f^{(k)}(+0)=\frac{1}{k!}\Delta^k\delta(\phi).\end{aligned}$$

Next, the second method. According to Problem 6.1 we may apply Proposition 6.3 to the function u_t. Indeed, if $2t=\epsilon^2$, then $u_t(x)=\frac{1}{\epsilon^n}f(\frac{1}{\epsilon}x)$ where $f(x)=(2\pi)^{-n/2}e^{-\|x\|^2/2}$. In the notation of the proposition and on account of [7, Exercise 6.51], we have, for all $\alpha\in(\mathbf{Z}_{\geq 0})^n$,

$$\begin{aligned}c_\alpha&=\int_{\mathbf{R}^n}x^\alpha f(x)\,dx=\frac{1}{(2\pi)^{\frac{n}{2}}}\prod_{j=1}^n\int_{\mathbf{R}}x_j^{\alpha_j}e^{-\frac{1}{2}x_j^2}\,dx_j\\&=\begin{cases}0, & \text{if there exists } j \text{ with } \alpha_j \text{ odd};\\ \dfrac{\pi^{\frac{n}{2}}}{(2\pi)^{\frac{n}{2}}}\dfrac{2^n}{2^{\frac{n}{2}}}\displaystyle\prod_{j=1}^n\frac{(\alpha_j)!}{(\frac{\alpha_j}{2})!\,2^{\frac{\alpha_j}{2}}}=\frac{\alpha!}{2^{\frac{|\alpha|}{2}}\,(\frac{\alpha}{2})!}, & \text{for } \alpha\in(2\mathbf{Z}_{\geq 0})^n.\end{cases}\end{aligned}$$

Therefore we have, with $j\in\mathbf{Z}_{\geq 0}$,

$$u_j=\sum_{|\alpha|=j}\frac{c_\alpha}{\alpha!}\partial^\alpha\delta=\begin{cases}0, & j \text{ odd};\\ \dfrac{1}{2^{\frac{j}{2}}}\displaystyle\sum_{|\alpha|=j,\,\alpha\in(2\mathbf{Z}_{\geq 0})^n}\frac{1}{(\frac{\alpha}{2})!}\partial^\alpha\delta, & j \text{ even}.\end{cases}$$

Next apply Proposition 6.3 with k replaced by $2k$. This leads to

$$0=\lim_{\epsilon\downarrow 0}\frac{1}{\epsilon^{2k}}\Big(f_\epsilon-\sum_{j=0}^{k}\epsilon^{2j}\,u_{2j}\Big)=\frac{1}{2^k}\lim_{t\downarrow 0}\frac{1}{t^k}\Big(u_t-\sum_{j=0}^{k}t^j\sum_{|\alpha|=j}\frac{1}{\alpha!}\,\partial^{2\alpha}\delta\Big).$$

The desired formula now follows by application of the following Multinomial Theorem (see [7, Exercise 2.52.(ii)]):

$$\sum_{|\alpha|=j}\frac{x^{2\alpha}}{\alpha!}=\frac{(\sum_{i=1}^{n}x_i^2)^j}{j!},\qquad\text{that is,}\qquad\sum_{|\alpha|=j}\frac{1}{\alpha!}\,\partial^{2\alpha}=\frac{1}{j!}\,\Delta^j.$$

6.3 (i). Introducing spherical coordinates in $\mathbf{R}^n$, we see that

$$\begin{aligned}u_t(\phi)&=(4\pi t)^{-\frac{n}{2}}\int_{\mathbf{R}_{>0}}e^{-\frac{r^2}{4t}}r^{n-1}\int_S\phi(r\,y)\,dy\,dr\\&=c_n(4\pi t)^{-\frac{n}{2}}\int_{\mathbf{R}_{>0}}e^{-\frac{r^2}{4t}}r^{n-1}\,S_\phi(r)\,dr.\end{aligned}$$

(ii). Expand S_ϕ in its Taylor series at 0 and next apply the substitution of variables $r=\sqrt{4ts}$ as well as (13.30). Thus we get, for any $k\in\mathbf{Z}_{\geq 0}$,

$$\begin{aligned}u_t(\phi)&=\frac{c_n}{(4\pi t)^{\frac{n}{2}}}\sum_{j=0}^{k}\frac{S_\phi^{(j)}(0)}{j!}\int_{\mathbf{R}_{>0}}e^{-\frac{r^2}{4t}}r^{n-1+j}\,dr+O(t^{k+1}),\quad t\downarrow 0\\&=\frac{c_n}{\pi^{\frac{n}{2}}}\sum_{j=0}^{k}2^{j-1}\Gamma\Big(\frac{n+j}{2}\Big)S_\phi^{(j)}(0)\,\frac{t^{\frac{j}{2}}}{j!}+O(t^{k+1}),\quad t\downarrow 0.\end{aligned}$$

On the other hand, according to Problem 6.2 we have, for any $k\in\mathbf{Z}_{\geq 0}$,

$$u_t(\phi)=\sum_{j=0}^{k}\Delta^j\phi(0)\,\frac{t^j}{j!}+O(t^{k+1}),\quad t\downarrow 0.$$

Equate the coefficients of like powers of t to find that $S_\phi^{(j)}(0)=0$ if j is odd. Therefore, suppose that $j=2k$; then

$$\frac{c_n}{\pi^{\frac{n}{2}}}\,\frac{2^{2k-1}\,k!}{(2k)!}\,\Gamma\Big(\frac{n}{2}+k\Big)\,S_\phi^{(2k)}(0)=\Delta^k\phi(0).$$

On account of (13.31) this leads to

$$S_\phi^{2k}(0)=\frac{(2k)!}{2^k\,k!\,\prod_{j=0}^{k-1}(n+2j)}\Delta^k\phi(0).$$

(iii). Multiply $y\mapsto y^\alpha$ by a suitable cut-off function and apply part (ii).

7.4 See [7, Formula (7.44)].

7.5 See [7, Exercise 6.95].

7.6 **(i)** and **(ii)**. For these results, see the beginning of Chap. 13.
(ii). Consider $k \in \mathbf{Z}$ such that $-\mathrm{Re}\, a - 1 < k < -\mathrm{Re}\, a$. It is easy to show that x_+^a has order $\leq k$. We will show that x_+^a cannot have order $< k$. If this were the case, there would exist $c > 0$ such that

$$|x_+^a(\phi)| \leq c\, \|\phi\|_{C^{k-1}} \qquad (\phi \in C_0^\infty([\,0,1\,])).$$

For $\phi \in C_0^\infty([\,0,1\,])$ and $0 < \epsilon \leq 1$, define $\phi_\epsilon(x) = \phi(\frac{x}{\epsilon})$. Routine computation shows that

$$x_+^a(\phi_\epsilon) = \epsilon^{a+1} x_+^a(\phi).$$

Now select $\phi \in C_0^\infty([\,0,1\,])$ such that $x_+^a(\phi) \neq 0$. Such ϕ exist, because otherwise $x_+^a = 0$ on $]\,0,1\,[$, which by Theorem 4.3 would imply that on $]\,0,1\,[$ the distribution x_+^a equals some polynomial of degree $\leq k-1$, which cannot be the case. Therefore we have, for all $0 < \epsilon \leq 1$,

$$\epsilon^{\mathrm{Re}\, a+1}|x_+^a(\phi)| = |x_+^a(\phi_\epsilon)| \leq c\, \|\phi_\epsilon\|_{C^{k-1}} = c \sup_{l<k,\, 0\leq x\leq 1} \epsilon^{-l}\, |\phi^{(l)}(x)|.$$

Hence we derive, for all $0 < \epsilon \leq 1$,

$$0 < |x_+^a(\phi)| \leq c \sup_{l<k,\, 0\leq x\leq 1} \epsilon^{-l-\mathrm{Re}\, a-1} |\phi^{(l)}(x)| \leq c\, \epsilon^{-k-\mathrm{Re}\, a} \sup_{l<k,\, 0\leq x\leq 1} |\phi^{(l)}(x)|.$$

Since $-k - \mathrm{Re}\, a > 0$, the right-hand side in the inequalities above will tend to 0 as $\epsilon \downarrow 0$; thus, we obtain a contradiction.
(iii). Obviously l_+ is of order 0. Suppose $k \geq 1$ and select $\phi \in C_0^\infty(\mathbf{R})$ with $\mathrm{supp}\, \phi \subset [\,-1,1\,]$ and $\partial^{k-1}\phi(0) = 1$. Now define

$$\phi_j(x) = \frac{1}{j^{k-1} \log j}\, \phi(j\, x) \qquad (j \in \mathbf{Z}_{>1}).$$

Then

$$\begin{aligned}(-1)^k\, \partial^k l_+(\phi_j) &= \int_{\mathbf{R}_{>0}} \partial^k \phi_j(x)\, \log x\, dx = \frac{j}{\log j} \int_{\mathbf{R}_{>0}} \partial^k \phi(j\, x)\, \log x\, dx \\ &= \frac{1}{\log j} \int_{\mathbf{R}_{>0}} \partial^k \phi(x)\, \log \frac{x}{j}\, dx = \frac{1}{\log j} \int_{\mathbf{R}_{>0}} \partial^k \phi(x)\, \log x\, dx + \partial^{k-1}\phi(0).\end{aligned}$$

As a consequence, $\lim_{j\to\infty}(-1)^k\, \partial^k l_+(\phi_j) = 1$. On the other hand, for $0 \leq l < k$ we obtain

$$\|\partial^l \phi_j\|_{C^0} = O\Big(\frac{1}{j^{k-l-1} \log j}\Big) = o(1), \quad j \to \infty.$$

It follows that $\partial^k l_+$ is of order k on $\mathbf{R}$, by arguments similar to those in the solution to Problem 3.2.

8.2 If $C = X \setminus K$, then C open in $\mathbf{R}^n$. For any $x \in C$, there exists an open neighborhood U of x with $K \cap U = \emptyset$. Now consider $\phi \in C_0^\infty(U)$. Since $\operatorname{supp} u_j \cap \operatorname{supp} \phi \subset K \cap U$, formula (7.3) says that $u_j(\phi) = 0$ and thus $u(\phi) = \lim_{j\to\infty} u_j(\phi) = 0$. In other words, $u = 0$ on U and then Theorem 7.1 asserts that $u = 0$ on C; that is, $\operatorname{supp} u \subset K$. This implies $u \in \mathcal{E}'(X)$, in view of Theorem 8.8. Next, select $\chi \in C_0^\infty(X)$ with $\chi = 1$ on a neighborhood of K and let $\psi \in C^\infty(X)$ be arbitrarily chosen. Then $\operatorname{supp}(\psi - \chi\,\psi) \cap K = \emptyset$, and so

$$u_j(\psi - \chi\,\psi) = 0 \qquad (j \in \mathbf{N}) \qquad \text{and} \qquad u(\psi - \chi\,\psi) = 0.$$

Because $\lim_{j\to\infty} u_j = u$ in $\mathcal{D}'(X)$ and $\chi\,\psi \in C_0^\infty(X)$, we obtain

$$\lim_{j\to\infty} u_j(\psi) = \lim_{j\to\infty} u_j(\chi\,\psi) = u(\chi\,\psi) = u(\psi),$$

which implies $\lim_{j\to\infty} u_j = u$ in $\mathcal{E}'(X)$.

The final assertion is a direct consequence of the preceding results.

8.5 The function $(x,t) \mapsto E(x,t)$ is locally integrable on $\mathbf{R}^{n+1}$ because it equals 0 if $t \le 0$, while $\int_{\mathbf{R}^n} E(x,t)\,dx = 1$, if $t > 0$, which defines a locally integrable function on $\mathbf{R}$. Further, $\operatorname{sing\,supp} E = \{0\}$. In order to see this, we have only to consider a neighborhood of a point $(x, 0)$ with $x \neq 0$. In fact, we have to show that all partial derivatives of $E(x,t)$ tend to 0 as $t \downarrow 0$.

We have, for $\phi \in C_0^\infty(\mathbf{R}^{n+1})$,

$$\begin{aligned} v(\phi) &= (\partial_t E - \Delta_x E)(\phi) = -E(\partial_t\phi + \Delta_x\phi) \\ &= -\int_{\mathbf{R}^n}\int_{\mathbf{R}_{>0}} E(x,t)(\partial_t\phi + \Delta_x\phi)(x,t)\,dt\,dx =: \lim_{\epsilon\downarrow 0} I_\epsilon + J_\epsilon, \end{aligned}$$

with

$$I_\epsilon = -\int_{\mathbf{R}^n}\int_\epsilon^\infty (E\,\partial_t\phi)(x,t)\,dt\,dx, \quad J_\epsilon = -\int_\epsilon^\infty\int_{\mathbf{R}^n} (E\,\Delta_x\phi)(x,t)\,dx\,dt.$$

Integration by parts with respect to the variable t gives

$$I_\epsilon = \int_{\mathbf{R}^n} (E\,\phi)(x,\epsilon)\,dx + \int_{\mathbf{R}^n}\int_\epsilon^\infty (\phi\,\partial_t E)(x,t)\,dt\,dx.$$

Apply Green's second identity (see [7, Example 7.9.6]) to J_ϵ. There is no boundary term, because ϕ has compact support; hence we see that

$$J_\epsilon = -\int_\epsilon^\infty\int_{\mathbf{R}^n} (\phi\,\Delta_x E)(x,t)\,dx\,dt.$$

Since E satisfies the heat equation on $\mathbf{R}^n \times \,]\,\epsilon, \infty\,[$, we therefore obtain, by means of the change of variables $x = 2\sqrt{\epsilon}\,y$,

$$I_\epsilon + J_\epsilon = \int_{\mathbf{R}^n} (E\,\phi)(x,\epsilon)\,dx = \pi^{-\frac{n}{2}} \int_{\mathbf{R}^n} e^{-\|y\|^2}\phi(2\sqrt{\epsilon}\,y, \epsilon)\,dy.$$

On account of the Mean Value Theorem and the compactness of supp ϕ we can find $c > 0$ such that for all $0 < \epsilon \le 1$ and $y \in \mathbf{R}^n$,

$$|\phi(2\sqrt{\epsilon}\, y, \epsilon) - \phi(0)| \le c\sqrt{\epsilon}\, \|y\|.$$

Hence the equality $\phi(2\sqrt{\epsilon}\, y, \epsilon) = \phi(0) + \phi(2\sqrt{\epsilon}\, y, \epsilon) - \phi(0)$ leads to

$$I_\epsilon + J_\epsilon = \phi(0)\, \pi^{-\frac{n}{2}} \int_{\mathbf{R}^n} e^{-\|y\|^2}\, dy + R_\epsilon,$$

$$\text{where} \quad |R_\epsilon| \le \sqrt{\epsilon}\, c\pi^{-\frac{n}{2}} \int_{\mathbf{R}^n} \|y\|\, e^{-\|y\|^2}\, dy = O(\sqrt{\epsilon}), \quad \epsilon \downarrow 0.$$

It follows that $v = \delta \in \mathcal{D}'(\mathbf{R}^{n+1})$, in view of

$$v(\phi) = \lim_{\epsilon \downarrow 0} I_\epsilon + J_\epsilon = \phi(0) + \lim_{\epsilon \downarrow 0} R_\epsilon = \phi(0) = \delta(\phi).$$

9.2 For every $\phi \in C_0^\infty(\mathbf{R}^n)$, we have on account of Leibniz's rule (2.8),

$$\begin{aligned}
&(\psi\, \partial^\alpha \delta)(\phi) = (\partial^\alpha \delta)(\psi\, \phi) = (-1)^{|\alpha|} \partial^\alpha (\psi\, \phi)(0) \\
&= (-1)^{|\alpha|} \sum_{\beta \le \alpha} \binom{\alpha}{\beta}\, \partial^{\alpha-\beta} \psi(0)\, \partial^\beta \phi(0) = \sum_{\beta \le \alpha} \binom{\alpha}{\beta}\, \partial^{\alpha-\beta} \psi(0)\, (-1)^{|\alpha|-|\beta|}\, \partial^\beta \delta\, \phi \\
&= \sum_{\beta \le \alpha} (-1)^{|\alpha-\beta|} \binom{\alpha}{\beta}\, \partial^{\alpha-\beta} \psi(0)\, \partial^\beta \delta\, \phi.
\end{aligned}$$

9.3 Application of Problem 9.2 implies

$$\begin{aligned}
x^k\, \partial^m \delta &= \sum_{l=0}^{m} \binom{m}{l} (-1)^{m-l} (\partial^{m-l} x^k)(0)\, \partial^l \delta \\
&= \begin{cases} \displaystyle\sum_{l=0}^{m} \binom{m}{l} (-1)^{m-l} k(k-1)\cdots(k-m+l+1) x^{k-m+l}\big|_{x=0} \partial^l \delta, & m-l \le k; \\ 0, & m-l > k. \end{cases}
\end{aligned}$$

Therefore $x^k\, \partial^m \delta \ne 0$ only if there exists an l such that $l \ge 0$ and $k - m + l = 0$, that is, $l = m - k \ge 0$. If this is the case, we obtain

$$x^k\, \partial^m \delta = (-1)^k \frac{m!}{(m-k)!} \partial^{m-k} \delta.$$

Now consider $u \in \mathcal{D}'(\mathbf{R})$ satisfying $x^k u = 0$. On account of Theorem 9.5 we have supp $u \subset \{ x \in \mathbf{R} \mid x^k = 0 \} = \{0\}$, while Theorem 8.10 then implies $u = \sum_m c_m\, \partial^m \delta$, with $c_m \in \mathbf{C}$. Now

$$0 = x^k\, u = \sum_m c_m\, x^k \partial^m \delta = \sum_{m \geq k} c_m (-1)^k \frac{m!}{(m-k)!} \partial^{m-k} \delta.$$

Accordingly $c_m = 0$, for $m \geq k$, and so $u = \sum_{m=0}^{k-1} c_m \partial^m \delta$ is the general solution for $x^k\, u = 0$.

9.4 According to Problem 9.3 we have $u' = c\,\delta$ with $c \in \mathbf{C}$ arbitrary. In view of Example 4.2 the latter equation has the particular solution $u = c\,H$, while Theorem 4.3 says that $u = c\,H + d$ is the general solution, where $d \in \mathbf{C}$.

9.5 Since $x\,\delta = 0$, the desired formula may be easily deduced from the formula $\frac{1}{x+i\,0} = -\pi i\,\delta + \mathrm{PV}\,\frac{1}{x}$, which occurs in the solution of Problem 1.3. The problem of solving $x^k\, u = 1$ can be reduced to finding all solutions u_0 to the homogeneous problem $x^k\, u_0 = 0$, which is done in Problem 9.3, plus a particular solution to the inhomogeneous problem $x^k\, u = 1$. In the case $k = 1$, we already know that $\mathrm{PV}\,\frac{1}{x}$ solves the latter problem. Next, suppose that u_k satisfies $x^k\, u_k = 1$. This implies

$$0 = x\,(x^k\, u_k)' = k x^k\, u_k + x^{k+1}\, u_k' = k + x^{k+1}\, u_k',$$

$$\text{so} \qquad x^{k+1}\, u_{k+1} := x^{k+1}\Big(-\frac{1}{k} u_k'\Big) = 1.$$

Therefore, the general solution to $x^k\, u = 1$ is given by, see Problem 4.2,

$$u = \frac{(-1)^{k-1}}{(k-1)!}\, \partial^{k-1}\, \mathrm{PV}\,\frac{1}{x} + \sum_{m=0}^{k-1} c_m \partial^m \delta = \frac{(-1)^{k-1}}{(k-1)!} \log^{(k)} |x| + \sum_{m=0}^{k-1} c_m \partial^m \delta.$$

9.6 Let $\phi \in C_0^\infty(\mathbf{R})$ and suppose $\phi(x) = 0$ if $|x| \geq m > 0$. We have

$$\Big(e^{itx}\, \mathrm{PV}\,\frac{1}{x}\Big)(\phi) = \lim_{\epsilon \downarrow 0} J_\epsilon,$$

where, by means of a change of variables,

$$J_\epsilon := \int_{\epsilon \leq |x| \leq m} \frac{e^{itx}\phi(x)}{x}\, dx = \int_\epsilon^m \Big(\frac{e^{itx}\phi(x) - e^{-itx}\phi(-x)}{x}\Big)\, dx.$$

Writing $\phi(x) = \phi(0) + x\psi(x)$ for all $x \in \mathbf{R}$, with ψ continuously differentiable on $\mathbf{R}$, we get

$$\begin{aligned} J_\epsilon &= \phi(0) \int_\epsilon^m \frac{e^{itx} - e^{-itx}}{x}\, dx + \int_\epsilon^m (e^{itx}\psi(x) + e^{-itx}\psi(-x))\, dx \\ &= 2i\,\delta(\phi) \int_\epsilon^{tm} \frac{\sin x}{x}\, dx + \int_\epsilon^m (e^{itx}\psi(x) + e^{-itx}\psi(-x))\, dx. \end{aligned}$$

Both integrands on the right-hand side are continuous, which implies

$$\lim_{\epsilon \downarrow 0} J_\epsilon = 2i\,\delta(\phi) \int_0^{tm} \frac{\sin x}{x}\,dx + K_t,$$

where

$$K_t = \int_0^m (e^{itx}\psi(x) + e^{-itx}\psi(-x))\,dx.$$

Now we have the well-known evaluation $\int_{\mathbf{R}_{>0}} \frac{\sin x}{x}\,dx = \frac{\pi}{2}$, see Problems 14.44, 16.19, or 16.21 or [7, Example 2.10.14 or Exercises 0.14, 6.60, 8.19]. Furthermore, the identity $e^{itx} = \frac{1}{it}\partial_x e^{itx}$ and integration by parts lead to, for $t > 0$,

$$|K_t| = \left| \frac{i}{t} \int_0^m (e^{itx}\psi'(x) + e^{-itx}\psi(-x))\,dx \right| \leq \frac{1}{t} \int_{-m}^m |\psi'(x)|\,dx. \qquad (21.2)$$

We now immediately obtain (9.5). The second identity follows by adding together (9.5) and the analogous identity obtained by replacing x by $-x$.

By combination of Problem 1.3 and (9.5) it follows that

$$\lim_{t\to\infty} e^{itx} \frac{1}{x \pm i\,0} = \lim_{t\to\infty} \Big(\mp\pi i\, e^{itx}\,\delta + e^{itx}\,\mathrm{PV}\frac{1}{x} \Big) = \mp\pi i\,\delta + \pi i\,\delta,$$

and this proves (i) and (ii). Assertion (iii) follows from the fact that the inner limit is already equal to 0, as can be demonstrated in a way similar to the proof of (21.2).

9.7 Theorem 9.5 implies that supp u is contained in the hyperplane $H = \{x \in \mathbf{R}^n \mid x_n = 0\}$. Select $\chi \in C_0^\infty(\mathbf{R})$ with $\chi = 1$ on an open neighborhood of 0 and define $\tilde\chi \in C^\infty(\mathbf{R}^n)$ by $\tilde\chi(x_1, \dots, x_n) = \chi(x_n)$. Then $1 - \tilde\chi = 0$ on an open neighborhood of H; hence $u = \tilde\chi\, u$. Now consider $\phi \in C_0^\infty(\mathbf{R}^n)$. Then

$$\begin{aligned}\phi(x_1, \dots, x_n) &= \phi(x_1, \dots, x_{n-1}, 0) + x_n \int_0^1 \partial_n\phi(x_1, \dots, x_{n-1}, t x_n)\,dt \\ &= \phi(\pi(x)) + x_n\,\eta(x),\end{aligned}$$

where $\pi : x \mapsto (x_1, \dots, x_{n-1}, 0)$ and $\eta \in C_0^\infty(\mathbf{R}^n)$. This leads to

$$u(\phi) = u(\tilde\chi\,\phi) = u(\tilde\chi\,(\phi\circ\pi)) + u(x_n\,\tilde\chi\,\eta) = u(\tilde\chi\,(\phi\circ\pi)).$$

Now $\tilde\chi(x)\,(\phi\circ\pi)(x) = \phi(x_1, \dots, x_{n-1}, 0)\,\chi(x_n) = \iota^*\phi \otimes \chi(x)$, so $u(\phi) = u(\iota^*\phi \otimes \chi)$. Finally, we note that $\psi \mapsto \psi \otimes \chi$ defines a continuous linear mapping $C_0^\infty(\mathbf{R}^{n-1}) \to C_0^\infty(\mathbf{R}^n)$. Hence, $v : \psi \mapsto u(\psi \otimes \chi)$ belongs to $\mathcal{D}'(\mathbf{R}^{n-1})$ and satisfies $u(\phi) = v(\iota^*\phi)$.

9.14 On account of the data, the solution I should be supported by $\mathbf{R}_{\geq 0}$, that is, it has to contain the Heaviside function H as a factor. Note that $H' = \delta$. Furthermore, the restriction of I to $\mathbf{R}_{>0}$ should satisfy the corresponding homogeneous differential equation $L\,I'' + R\,I' + \frac{1}{C}\,I = 0$. Now the solutions of the latter equation are given by, for $c_\pm \in \mathbf{C}$,

$$I_0(t) = \sum_{\pm} c_{\pm} e^{t(-A \pm B)} \qquad \text{with} \qquad a = \frac{R}{2L} \qquad \text{and} \qquad B = \frac{1}{2L}\sqrt{R^2 - \frac{4L}{C}}$$

under the assumption of $B \neq 0$. Hence, consider $I = I_0\, H$; then by Leibniz's rule and Example 9.1,

$$I' = I_0(0)\,\delta + I_0'\,H \qquad \text{and} \qquad I'' = I_0(0)\,\delta' + I_0'(0)\,\delta + I_0''\,H.$$

As a consequence, the condition

$$\begin{aligned} L\,I'' + R\,I' + \frac{1}{C}\,I &= L(I_0(0)\,\delta' + I_0'(0)\,\delta) + R\,I_0(0)\,\delta \\ &= L\,I_0(0)\,\delta' + (L\,I_0'(0) + R\,I_0(0))\delta = \delta \end{aligned}$$

leads to $I_0(0) = 0$ and $L\,I_0'(0) = 1$, whence

$$-c_- = c_+ = \frac{1}{2LB} = \frac{1}{\sqrt{R^2 - \frac{4L}{C}}}.$$

In the case of $B = 0$, one uses $I_0(t) = c_- e^{-tA} + c_+ t\, e^{-tA}$ and derives similar equations as before. These imply $c_- = 0$ and $L\,c_+ = 1$.

10.1 Taking the transpose of (10.1) and using Lemma 10.1 as well as the fact that the transpose of ∂_k is $-\partial_k$ leads to the desired identity.

10.7 Successively applying the chain rule and the transposition of a matrix, we obtain, for $\phi \in C_0^\infty(X)$,

$$\begin{aligned} &D(\phi \circ \Psi) = (D\phi) \circ \Psi\, D\Psi, \implies D(\Psi^*\phi) = \Psi^*(D\phi)\, D\Psi, \implies \\ &\operatorname{grad}(\Psi^*\phi) = {}^t D\Psi\, \Psi^*(\operatorname{grad}\phi), \implies \Psi^* \circ \operatorname{grad}(\phi) = ({}^t D\Psi)^{-1} \circ \operatorname{grad} \circ \Psi^*(\phi). \end{aligned}$$

10.8 The mapping $t \mapsto \det D\Phi_t$ is continuous and nonvanishing on the connected set $\mathbf{R}$, while $\det D\Phi_0 = \det I = 1$. Therefore, $\det D\Phi_t > 0$, for all $t \in \mathbf{R}$. On account of (10.10) and the fact that $(\Phi_{-t})^{-1} = \Phi_t$, one obtains, on $\mathcal{D}'(X)$,

$$\frac{d}{dt}(\Phi_{-t})_*\Big|_{t=0} = \frac{d}{dt}\big(((\Phi_{-t})^{-1})^*\,(\det D(\Phi_{-t}))^{-1}\big)\Big|_{t=0} = \frac{d}{dt}(\Phi_t{}^*\,\det D\Phi_t)\Big|_{t=0}.$$

According to (10.18) and (10.17), this implies

$$\begin{aligned} \sum_{j=1}^n \partial_j \circ v_j &= \frac{d}{dt}\Phi_t{}^*\Big|_{t=0} \det DI + I^*\,\frac{d}{dt}\det D\Phi_t\Big|_{t=0} \\ &= \sum_{j=1}^n v_j \circ \partial_j + \frac{d}{dt}\det D\Phi_t\Big|_{t=0}. \end{aligned}$$

Hence,

$$\frac{d}{dt}\det D\Phi_t\Big|_{t=0} = \sum_{j=1}^{n}(\partial_j \circ v_j - v_j \circ \partial_j) = \sum_{j=1}^{n}\partial_j v_j = \operatorname{div} v.$$

In fact, one can prove that

$$\frac{d}{dt}\det D\Phi_t = (\operatorname{div} v)\circ \Phi_t \,\det D\Phi_t \qquad (t \in \mathbf{R}).$$

In view of $\det D\Phi_0 = 1$, solving the differential equation (compare with [7, Formula (5.32)]) yields

$$\det D\Phi_t(x) = e^{\int_0^t \operatorname{div} v(\Phi_\tau(x))\,d\tau} \qquad ((t,x) \in \mathbf{R}\times X).$$

10.9 δ_x is invariant under $(\Phi_t)_{t\in\mathbf{R}}$ as a distributional density iff $(\Phi_t)_*\delta_x = \delta_x$ for all $t \in \mathbf{R}$, and in view of Example 10.4 this is the case iff $\delta_{\Phi_t(x)} = \delta_x$, that is, iff $\Phi_t(x) = x$, which is equivalent to $v(x) = 0$. Furthermore, assume that δ_x is invariant under $(\Phi_t)_{t\in\mathbf{R}}$ as a generalized function. On account of Example 10.12, we have

$$\Phi_t{}^*\delta_x = \frac{1}{|\det D\Phi_t(\Phi_{-t}(x))|}\,\delta_{\Phi_{-t}(x)} \qquad (t \in \mathbf{R}). \tag{21.3}$$

By considering the supports of the distributions occurring in (21.3), it follows that the assumption implies $x = \Phi_{-t}(x)$, or equivalently $\Phi_t(x) = x$, for all $t \in \mathbf{R}$. In turn, (21.3) then leads to $\det D\Phi_t(x) = 1$, for all $t \in \mathbf{R}$. Upon differentiation with respect to t at $t = 0$ and application of Problem 10.8 we obtain the desired conclusion.

10.10 Writing Φ_t for the rotation by the angle t, we obtain $v(x) = \frac{d}{dt}\Phi_t(x)\big|_{t=0} = (-x_2, x_1)$. On account of Theorem 10.16, the distributional density $u \in \mathcal{D}'(\mathbf{R}^2)$ is invariant under all of the Φ_t if and only if

$$\sum_{j=1}^{2}\partial_j(v_j\,u) = -\partial_1(x_2\,u) + \partial_2(x_1\,u) = -x_2\,\partial_1 u + x_1\,\partial_2 u = 0.$$

Furthermore, $u \in \mathcal{D}'(\mathbf{R}^2)$ is invariant as a generalized function if and only if $\sum_{j=1}^{2} v_j\,\partial_j u = -x_2\,\partial_1 u + x_1\,\partial_2 u = 0$.

10.13 Using the identity for the exponential function, integration by parts twice, and the convergence of $\sum_{n\in\mathbf{N}}\frac{1}{n^2}$, one obtains the desired estimate. Theorem 3.8 then implies that $u \in \mathcal{D}'(\mathbf{R})$.

Part (i) follows by a shift in the index of summation. For (ii), consider, with $\phi \in C_0^\infty(\mathbf{R})$,

$$(T_{\frac{2\pi}{\omega}})_*u(\phi) = u(T_{\frac{2\pi}{\omega}}{}^*\phi) = {}_{n\in\mathbf{Z}}e_{in\omega}(T_{\frac{2\pi}{\omega}}{}^*\phi) = \sum_{n\in\mathbf{Z}}\int_{\mathbf{R}} e^{in\omega x}\phi\Big(x+\frac{2\pi}{\omega}\Big)\,dx$$

$$= \sum_{n\in\mathbf{Z}}\int_{\mathbf{R}} e^{in\omega x}\phi(x)\,dx = u(\phi).$$

Next, $e^{i\omega x} - 1 = 0$ only if $x \in \frac{2\pi}{\omega}\mathbf{Z}$, while $\partial(e_{i\omega} - 1) = i\omega \neq 0$ at such x. Hence, Theorem 9.5 implies the existence of constants $c_k \in \mathbf{C}$ such that $u = \sum_{k\in\mathbf{Z}} c_k\, \delta_{k\,\frac{2\pi}{\omega}}$. The periodicity of u and Example 10.4 now lead to $c_k = c_{k+1}$, and so $c_k = c$, for all $k \in \mathbf{Z}$.

10.14 The integral is convergent at 0 because near 0 the numerator can be estimated by $\|x\|^2$ while $s < 1$, and at ∞ because the numerator is bounded by 4 and $s > 0$. The integral defines a function on $\mathbf{R}^n$ homogeneous of degree $2s$, as follows by the change of variables $x = \frac{1}{t}y$. Finally, the integral is invariant under every rotation A about the origin on account of the substitution $x = Ay$.

10.15 **(i)**. According to Theorem 10.17, we have $\sum_{k=1}^n x_k\, \partial_k u_j = a\, u_j$, for all j. Since $\partial_k : \mathcal{D}'(\mathbf{R}^n) \to \mathcal{D}'(\mathbf{R}^n)$ is a continuous linear mapping, on account of (8.2), it follows that $\partial_k u_j \to \partial_k u$ in $\mathcal{D}'(\mathbf{R}^n)$; by taking the limit as $j \to \infty$, we therefore obtain $\sum_{k=1}^n x_k\, \partial_k u = a\, u$, which implies $u \in \mathcal{H}_a$.
(ii). Applying ∂_j to the last identity in (i), we see that

$$\sum_{k\neq j} x_k\, \partial_j \partial_k u + \partial_j u + x_j\, \partial_j^2 u = \sum_k x_k\, \partial_k \partial_j u + \partial_j u = a\, \partial_j u;$$

in other words, $\partial_j u \in \mathcal{H}_{a-1}$. The assertion for $x_k\, u$ follows by a similar argument.
(iii). If $a \in \mathbf{R}$, then $t \mapsto t^a$ belongs to $C^\infty(\mathbf{R})$ if and only if $a \in \mathbf{Z}_{\geq 0}$. For any $a \in \mathbf{C} \setminus \mathbf{R}$,

$$t^a = t^{\operatorname{Re} a}\, e^{i(\operatorname{Im} a)\log t} = t^{\operatorname{Re} a}\big(\cos((\operatorname{Im} a)\log t) + i\,\sin((\operatorname{Im} a)\log t)\big),$$

where $\operatorname{Im} a \neq 0$. Hence, $t \mapsto t^a$ does not belong to $C^\infty(\mathbf{R})$ in this case. Now suppose $x \in \mathbf{R}^n$ and $a \in \mathbf{Z}_{\geq 0}$. Then $\psi(t\,x) = t^a\, \psi(x)$ is an identity of functions in $C^\infty(\mathbf{R})$. Now differentiate this identity a times with respect to t at $t = 0$; the right-hand side then gives $a!\, \psi(x)$, while on account of the chain rule the left-hand side leads to

$$D^a\psi(tx)(x, \dots, x)|_{t=0} = D^a\psi(0)(x, \dots, x) =: p(x),$$

with a copies of x, and where p is a polynomial function on $\mathbf{R}^n$. Accordingly, $\psi = \frac{1}{a!}\, p$.

According to [7, Exercise 0.11.(iv)], properties of the binomial series imply

$$\sum_{a\in\mathbf{Z}_{\geq 0}} |\{\alpha \in (\mathbf{Z}_{\geq 0})^n \mid |\alpha| = a\}|\, t^a = \sum_{\alpha\in(\mathbf{Z}_{\geq 0})^n} \prod_{j=1}^n t^{\alpha_j} = \Big(\sum_{a\in\mathbf{Z}_{\geq 0}} t^a\Big)^n$$
$$= (1-t)^{-n} = \sum_{a\in\mathbf{Z}_{\geq 0}} \binom{a+n-1}{a} t^a.$$

(iv). Set $g(x) = \|x\|^a\, f(\frac{1}{\|x\|}\, x)$. The local integrability of g can be proved using spherical coordinates. And $g \in \mathcal{H}_a$, because according to Theorem 10.8 we have, for any $\phi \in C_0^\infty \mathbf{R}^n$,

$$c^* g(\phi) = g(c_* \phi) = g(c^{-n}(c^{-1})^* \phi) = \int_{\mathbf{R}^n} g(x)\, \phi(c^{-1}x)\, c^{-n}\, dx$$
$$= \int_{\mathbf{R}^n} g(cx)\, \phi(x)\, dx = c^a\, g(\phi).$$

10.16 On the basis of Theorem 10.8 we see, for all $c > 0$ and $\phi \in C_0^\infty(\mathbf{R}^n)$, that

$$c^* \delta(\phi) = \delta(c_* \phi) = \delta(c^{-n}(c^{-1})^* \phi) = c^{-n}\, \delta(\phi).$$

This proves that δ is homogeneous of degree $-n$. Using Problem 10.15.(ii) and mathematical induction over $|\alpha|$, we now find that $\partial^\alpha \delta$ is homogeneous of degree $-n - |\alpha|$.

10.18 Problem 10.1 implies that

$$\Phi_* \circ \partial_j = \sum_{k=1}^{p} \partial_k \circ \Phi_* \circ \partial_j \Phi_k \quad : \quad C_0^\infty(X) \to C_0^\infty(Y)$$

is an identity of continuous linear mappings. On account of Theorem 10.18, the pullback $\Phi^* : \mathcal{D}'(Y) \to \mathcal{D}'(X)$ is well-defined as the transpose of $\Phi_* : C_0^\infty(X) \to C_0^\infty(Y)$, the mapping Φ being a C^∞ submersion. Therefore the desired identity can be obtained by transposition.

10.20 **(i).** From $\Phi(x) = \langle Ax, x \rangle$ one obtains $\operatorname{grad} \Phi(x) = 2Ax$. Because A is invertible, its kernel consists of 0 only. This proves the assertion.
(ii). Apply Problem 10.18 with $p = 1$ to obtain, since $\Phi(x) = y$ and y denotes the variable in $\mathbf{R}$,

$$(\partial_j \circ \Phi^*) v = (\partial_j \Phi)\, \Phi^* v' \quad \text{in} \quad \mathcal{D}'(\mathbf{R}^n \setminus \{0\}).$$

Using this identity leads to

$$((\partial_i \partial_j) \circ \Phi^*) v = \partial_i (\partial_j \circ \Phi^*) v = (\partial_i \partial_j \Phi)\, \Phi^* v' + (\partial_j \Phi)(\partial_i \circ \Phi^*) v'$$
$$= (\partial_i \partial_j \Phi)\, \Phi^* v' + (\partial_j \Phi)(\partial_i \Phi)\, \Phi^* v''.$$

Part (i) now implies $\partial_j \Phi(x) = 2 \sum_{k=1}^n A_{jk}\, x_k$ and $\partial_i \partial_j \Phi = 2A_{ji}$. Hence, in $\mathcal{D}'(\mathbf{R}^n \setminus \{0\})$,

$$(P \circ \Phi^*) v = \sum_{i,j=1}^{n} B_{ij} ((\partial_i \partial_j) \circ \Phi^*) v$$
$$= \sum_{i,j=1}^{n} B_{ij} 2A_{ji}\, \Phi^* v' + \sum_{i,j,k,l=1}^{n} B_{ij} 4 A_{jk} A_{il}\, x_k x_l\, \Phi^* v''$$
$$= 2 \sum_{i=1}^{n} \delta_{ii}\, \Phi^* v' + 4y\, \Phi^* v'' = \Phi^* (2n\, v' + 4y\, v''),$$

because

$$\sum_{i,j,k,l=1}^{n} B_{ij} A_{jk} A_{il}\, x_k x_l = \sum_{i,k,l=1}^{n} \delta_{ik} A_{il}\, x_k x_l = \sum_{k,l=1}^{n} A_{kl}\, x_k x_l = y.$$

(iii). Application of the first identity in part (ii) and of part (i) leads to

$$\begin{aligned}\sum_{j=1}^{n} x_j\, (\partial_j \circ \Phi^*) v &= \sum_{j=1}^{n} x_j\, (\partial_j \Phi)\, \Phi^* v' = 2 \sum_{j=1}^{n} x_j\, (Ax)_j\, \Phi^* v' \\ &= 2y\, \Phi^* v' = 2\Phi^*(y\, v') = 2a\, \Phi^* v.\end{aligned}$$

Thus, the claim is a consequence of Theorem 10.17.
(iv). The first assertion also follows from Theorem 10.17. If v is homogeneous of degree $1 - \frac{n}{2}$, then u is homogeneous of degree $2 - n$.
(v). The former identity follows upon restriction of the identity in part (ii) from $\mathcal{D}'(\mathbf{R})$ to $C^\infty(\mathbf{R})$, while the latter is obtained by means of transposition.
(vi). In this case, the matrix B is the identity matrix on $\mathbf{R}^2$ and so is A. Hence Φ as in this problem coincides with Φ as in Example 10.5. Furthermore,

$$(\Phi_*\, \delta_V)(\phi) = \int_{-\pi}^{\pi} \phi(1)\, d\alpha = 2\pi\, \delta_1(\phi).$$

Since $y\, \delta_1 = \delta_1$ according to Example 9.1, one derives

$$\Phi_*\, \Delta\, \delta_V = 4(\partial_y^2 \circ y - \partial_y) 2\pi\, \delta_1 = 8\pi(\delta_1{}'' - \delta_1{}').$$

10.21 We use the notation of the proof of Theorem 10.18 and Remark 10.19. Given $y \in Y$, write

$$V(y) = \Pi^{-1}(\{y\}) \cap V = \{\, z \in \mathbf{R}^q \mid (y, z) \in V \,\}.$$

In view of formulas (10.21) and (10.22), we obtain

$$(\Phi_* f)(y) = \int_{V(y)} j_\Lambda(y, z)\, f(\Lambda(y, z))\, dz \qquad (f \in C_0(U),\ y \in Y). \tag{21.4}$$

We will rewrite this integral in a more intrinsic manner, as an integral with respect to Euclidean integration over the fiber of Φ over the value $y \in Y$, which possesses the following parametrization:

$$\Lambda_y : V(y) \to U \cap \Phi^{-1}(\{y\}) \qquad \text{given by} \qquad \Lambda_y(z) = \Lambda(y, z).$$

To this end, we derive the identity (21.9) below for $j_\Lambda(y, z)$.

Identify $x' \in \mathbf{R}^n$ with $(y, z) \in \mathbf{R}^p \times \mathbf{R}^q$ and write $x = \Lambda(x') = \Lambda(y, z)$, for $(y, z) \in V$. Suppose $x = \Lambda(x') \in U$. Then $\Phi(x) = \Pi \circ K \circ \Lambda(x') = \Pi(x') = y$; in other words, $x \in \Phi^{-1}(\{y\})$. On account of the Submersion Theorem, near x the set $\Phi^{-1}(\{y\})$ is a C^∞ submanifold of $\mathbf{R}^n$ of dimension $n - p = q$. According to [7,

Theorem 5.1.2], the tangent space of $\Phi^{-1}(\{y\})$ at x is the following linear subspace of $\mathbf{R}^n$ of dimension q:

$$T := T_x\Phi^{-1}(\{y\}) = \ker\phi \qquad \text{where} \qquad \phi = D\Phi(x) \in \operatorname{Lin}(\mathbf{R}^n, \mathbf{R}^p). \quad (21.5)$$

Application of the chain rule to the identity $\Phi \circ \Lambda = \Pi$ on V gives the following identity in $\operatorname{Lin}(\mathbf{R}^n, \mathbf{R}^p)$:

$$D\Phi(x) \circ D\Lambda(x') = D\Pi(x') = \Pi \qquad (x' \in V).$$

For the moment we keep x and x' fixed and rewrite this equality more concisely as

$$\phi\,\lambda = \phi(\lambda^{\vdash}\,\lambda^{\parallel}) = (\phi\,\lambda^{\vdash}\,\phi\,\lambda^{\parallel}) = \Pi. \quad (21.6)$$

Here $\lambda = D\Lambda(x') \in \operatorname{Aut}(\mathbf{R}^n)$, while for $1 \le j \le n$,

$$\lambda_j = D_j\Lambda(x') \in \mathbf{R}^n, \qquad \lambda^{\vdash} = (\lambda_1 \cdots \lambda_p), \qquad \lambda^{\parallel} = (\lambda_{p+1} \cdots \lambda_n) = D\Lambda_y(z).$$

As is customary, we identify a linear mapping and its matrix with respect to the standard basis vectors. It is a consequence of (21.6) and (21.5) that T is spanned by the vectors that occur in $\lambda^{\parallel}$; these are the linearly independent vectors λ_k, for $p < k \le n$. Furthermore, (21.6) implies that $\phi\,\lambda_j = e_j$, for $1 \le j \le p$. In particular, the vectors that occur in $\lambda^{\vdash}$ are transversal to T in view of (21.5).

Now consider the parallelepiped in $\mathbf{R}^n$ spanned by all λ_j, for $1 \le j \le n$. Its volume equals

$$J := |\det D\Lambda(x')| = j_\Lambda(y,z) = |\det(\lambda_1 \cdots \lambda_n)| = |\det\lambda\,|.$$

Denote the orthocomplement in $\mathbf{R}^n$ of T by $T^\perp$. Then $\dim T^\perp = p$. The value of J does not change if we replace $\lambda_j \in \mathbf{R}^n$ by its projection $\pi_j \in T^\perp$ along T, for all $1 \le j \le p$. Using the notation $\pi = (\pi_1 \cdots \pi_p)$ this leads to

$$\begin{aligned} J^2 &= (\det\lambda)^2 = (\det(\lambda_1 \cdots \lambda_n))^2 = (\det(\pi_1 \cdots \pi_p\,\lambda_{p+1} \cdots \lambda_n))^2 \\ &= (\det(\pi\,\lambda^{\parallel}))^2 = \det({}^t(\pi\,\lambda^{\parallel})\,(\pi\,\lambda^{\parallel})). \end{aligned}$$

The vectors in π belong to $T^\perp$, while those in $\lambda^{\parallel}$ belong to T; hence

$${}^t(\pi\,\lambda^{\parallel})\,(\pi\,\lambda^{\parallel}) = {}^t\pi\,\pi \otimes {}^t\lambda^{\parallel}\,\lambda^{\parallel},$$

where the former factor on the right-hand side belongs to $\operatorname{Aut}(\mathbf{R}^p)$ and the latter one to $\operatorname{Aut}(\mathbf{R}^q)$. This implies

$$J^2 = \det({}^t\pi\,\pi)\,\det({}^t\lambda^{\parallel}\,\lambda^{\parallel}). \quad (21.7)$$

Next, we evaluate $\det({}^t\pi\,\pi)$. From linear algebra it is well-known that $T^\perp$ may be described by

$$T^{\perp} = (\ker \phi)^{\perp} = \operatorname{im} {}^t\phi, \qquad \text{in particular,} \qquad \mathbf{R}^n = \ker \phi \oplus \operatorname{im} {}^t\phi. \tag{21.8}$$

(Observe that (21.8) proves the identity $\operatorname{rank} {}^t\phi = \operatorname{rank} \phi$; see [7, Rank Lemma 4.2.7].) Indeed, for $p \in \mathbf{R}^n$,

$$\begin{aligned} p \in \ker \phi \quad &\Longleftrightarrow \quad \phi\, p = 0 \quad \Longleftrightarrow \quad \langle \phi\, p, q \rangle = 0 \text{ for all } q \in \mathbf{R}^p \\ &\Longleftrightarrow \quad \langle p, {}^t\phi\, q \rangle = 0 \text{ for all } q \in \mathbf{R}^p \quad \Longleftrightarrow \quad p \in (\operatorname{im} {}^t\phi)^{\perp}, \end{aligned}$$

which shows that $(\ker \phi)^{\perp} = (\operatorname{im}({}^t\phi))^{\perp\perp} = \operatorname{im}({}^t\phi)$. Now (21.8) yields the existence of vectors $\eta_1, \dots, \eta_p \in \mathbf{R}^p$ such that $\pi = {}^t\phi\, \eta$. Since $\lambda_j - \pi_j \in T$, for $1 \leq j \leq p$, it follows from (21.5) that $\phi(\lambda^{\vdash} - \pi) = 0$. Hence, we obtain from (21.6),

$$I_p = \phi\, \lambda^{\vdash} = \phi\, \pi = \phi\, {}^t\phi\, \eta = \gamma\, \eta, \qquad \text{where} \qquad \gamma = \phi\, {}^t\phi.$$

Note that $\gamma \in \operatorname{Aut}(\mathbf{R}^p)$ because it is Gram's matrix associated with the linearly independent **row** vectors of ϕ. As a consequence, $\eta = \gamma^{-1}$, and thus $\pi = {}^t\phi\, \gamma^{-1}$. In turn, this leads to

$${}^t\pi\, \pi = ({}^t\gamma)^{-1} (\phi\, {}^t\phi) \gamma^{-1} = \gamma^{-1}\, \gamma\, \gamma^{-1} = \gamma^{-1}.$$

On account of (21.7) we therefore obtain

$$\det({}^t\pi\, \pi) = \frac{1}{\det \gamma} \qquad \text{and} \qquad J^2 = \frac{\det({}^t\lambda^{\parallel}\, \lambda^{\parallel})}{\det(\phi\, {}^t\phi)}.$$

More explicitly, this means that

$$j_{\Lambda}(y, z) = \frac{\sqrt{\det({}^tD\Lambda_y(z) \circ D\Lambda_y(z))}}{\sqrt{\det\big(D\Phi(\Lambda(y, z)) \circ {}^tD\Phi(\Lambda(y, z))\big)}} = \frac{\sqrt{\det({}^tD\Lambda_y(z) \circ D\Lambda_y(z))}}{(\operatorname{gr} \Phi)(\Lambda(y, z))}. \tag{21.9}$$

On account of the definition of Euclidean q-dimensional integration (see [7, Sect. 7.3]), one now deduces from (21.4), for $f \in C_0(U)$ and $y \in Y$, that

$$\begin{aligned} (\Phi_* f)(y) &= \int_{V(y)} \frac{f(\Lambda_y(z))}{(\operatorname{gr} \Phi)(\Lambda_y(z))} \sqrt{\det({}^tD\Lambda_y(z) \circ D\Lambda_y(z))}\, dz \\ &= \int_{U \cap \Phi^{-1}(\{y\})} \frac{f(x)}{(\operatorname{gr} \Phi)(x)}\, dx. \end{aligned} \tag{21.10}$$

10.22 (i). The first identity follows from $(\Phi_*\phi)(y) = \delta_y(\Phi_*\phi) = (\Phi^*\delta_y)(\phi)$ and $\delta_y = \partial\, T_y H$, which is a consequence of

$$\phi(y) = -\int_0^{\infty} \partial\phi(x + y)\, dx = -H(T_{-y}\, \partial\phi) = \partial\, T_y H(\phi).$$

The second may be obtained from (10.1).
(ii). Select $\chi \in C^\infty(\mathbf{R})$ satisfying $0 \le \chi \le 1$, $\chi = 0$ on $]-\infty, -1[$, and $\chi = 1$ on $]1, \infty[$. For $t > 0$, write $(t^*\chi)(y) = \chi(ty)$. Then, for $x \in X$,

$$\Phi^*(T_y\, t^*\chi)(x) = (T_y\, t^*\chi)(\Phi(x)) = \chi(t(\Phi(x) - y)),$$

and so

$$\lim_{t\to\infty} \Phi^*(T_y\, t^*\chi)(x) = H(\Phi(x) - y) = 1_{\Phi^{-1}(]y,\infty[)}(x).$$

the Dominated Convergence Theorem of Arzelà, see [7, Theorem 6.12.3], or that of Lebesgue, see Theorem 20.26.(iv), now leads to

$$\begin{aligned}\Phi^*(T_y\, H)(\phi) &= \lim_{t\to\infty} \Phi^*(T_y\, t^*\chi)(\phi) = \lim_{t\to\infty} \int_X \Phi^*(T_y\, t^*\chi)(x)\, \phi(x)\, dx \\ &= \int_X 1_{\Phi^{-1}(]y,\infty[)}(x)\, \phi(x)\, dx = \int_{\Phi^{-1}(]y,\infty[)} \phi(x)\, dx.\end{aligned}$$

(iii). Part (ii) and the Theorem on Integration of a Total Derivative imply

$$\begin{aligned}\partial_j(\Phi^*(T_y H))(\phi) &= -(\Phi^*(T_y H))(\partial_j\phi) = -\int_{\Phi^{-1}(]y,\infty[)} \partial_j\phi(x)\, dx \\ &= \int_{\Phi^{-1}(\{y\})} \phi(x)\, \frac{\partial_j\Phi(x)}{\|\operatorname{grad}\Phi(x)\|}\, dx.\end{aligned}$$

Here we have used that $\operatorname{grad}\Phi(x)$ points inward in $\Phi^{-1}(]y, \infty[)$.
(iv). Parts (i) and (iii) lead to

$$\begin{aligned}\frac{(\partial_j\Phi)^2}{\|\operatorname{grad}\Phi\|^2}\, \Phi^*\delta_y(\phi) &= \partial_j(\Phi^*(T_y H))\Big(\phi\, \frac{\partial_j\Phi}{\|\operatorname{grad}\Phi\|^2}\Big) \\ &= \int_{\Phi^{-1}(\{y\})} \phi(x)\, \frac{(\partial_j\Phi(x))^2}{\|\operatorname{grad}\Phi(x)\|^3}\, dx,\end{aligned}$$

and so

$$(\Phi_*\phi)(y) = \frac{\sum_{j=1}^n (\partial_j\Phi)^2}{\|\operatorname{grad}\Phi\|^2}\, \Phi^*\delta_y(\phi) = \int_{\Phi^{-1}(\{y\})} \frac{\phi(x)}{\|\operatorname{grad}\Phi(x)\|}\, dx.$$

10.23 **(i)**. The first identity is a direct consequence of (10.23). The second follows from the first by noting that $(\Phi^*\delta_y)(\phi) = \delta_y(\Phi_*\phi) = (\Phi_*\phi)(y)$.
(ii). From part (i) we obtain, for any $y \in \mathbf{R}$,

$$\begin{aligned}\int_{-\infty}^{y} (\Phi_*\phi)(t)\, dt &= \int_{-\infty}^{y} \int_{\Phi^{-1}(\{t\})} \frac{\phi(x)}{\|\operatorname{grad}\Phi(x)\|}\, dx\, dt \\ &= \int_{\Phi^{-1}(]-\infty,\, y[)} \phi(x)\, dx.\end{aligned}$$

In view of the Fundamental Theorem of Integral Calculus on **R** this implies

$$(\Phi^*\delta_y)(\phi) = (\Phi_*\phi)(y) = \partial_y \int_{-\infty}^{y} (\Phi_*\phi)(t)\,dt = \partial_y \int_{\Phi^{-1}(]-\infty,\,y[)} \phi(x)\,dx.$$

(iii). We have

$$\begin{aligned}
\psi(\Phi_*\phi) = (\Phi^*\psi)(\phi) &= \int_{\mathbf{R}^n} \phi(x)\,\psi(\Phi(x))\,dx \\
&= -\int_{\mathbf{R}^n} \phi(x) \int_{\Phi(x)}^{\infty} \partial\psi(y)\,dy\,dx = -\int_{\mathbf{R}} \partial\psi(y) \int_{\Phi^{-1}(]-\infty,\,y[)} \phi(x)\,dx\,dy \\
&= \int_{\mathbf{R}} \psi(y)\,\partial_y \int_{\Phi^{-1}(]-\infty,\,y[)} \phi(x)\,dx\,dy.
\end{aligned}$$

10.24 Combining of Problem 10.18 and $\Phi^*\delta = \frac{1}{\|\operatorname{grad}\Phi\|}\,\delta_{\partial\Omega}$ leads to

$$\partial_j(\Phi^*H) = \partial_j\Phi\;\Phi^*(\partial H) = \partial_j\Phi\;\Phi^*\delta = \frac{\partial_j\Phi}{\|\operatorname{grad}\Phi\|}\,\delta_{\partial\Omega}.$$

For arbitrary $\phi \in C_0^\infty(X)$ we now obtain

$$\partial_j(\Phi^*H)(\phi) = -(H\circ\Phi)(\partial_j\phi) = -\int_{\Phi^{-1}(\mathbf{R}_{>0})} \partial_j\phi(x)\,dx = -\int_{\Omega} \partial_j\phi(x)\,dx,$$

while

$$\frac{\partial_j\Phi}{\|\operatorname{grad}\Phi\|}\,\delta_{\partial\Omega}(\phi) = \int_{\partial\Omega} \phi(y)\,\frac{\partial_j\Phi(y)}{\|\operatorname{grad}\Phi(y)\|}\,dy = -\int_{\partial\Omega} \phi(y)\,\nu_j(y)\,dy.$$

Here we have used that $\partial_j\Phi(y)/\|\operatorname{grad}\Phi(y)\|$ equals the jth component of the normalized gradient vector of Φ at y, which points inward in Ω, since the latter is the inverse image $\Phi^{-1}(\mathbf{R}_{>0})$; phrased differently, it equals $-\nu_j(y)$, where $\nu_j(y)$ is the outer normal to $\partial\Omega$ at y. The desired equality is now obvious.

11.1 **(i).** $u = \phi = 0$.
(ii). Select $u = 1 \in C^\infty(\mathbf{R})$ and $\phi \in C_0^\infty(\mathbf{R})$ satisfying $\int_{\mathbf{R}} \phi(y)\,dy = 1$. Then

$$(u*\phi)(x) = \int_{\mathbf{R}} u(x-y)\phi(y)\,dy = \int_{\mathbf{R}} \phi(y)\,dy = 1.$$

(iii). Select $u \in C^\infty(\mathbf{R})$ with $u(x) = x$ and ϕ as in part (ii) and, in addition, assume that ϕ is an even function. Then

$$(u*\phi)(x) = \int_{\mathbf{R}} (x-y)\phi(y)\,dy = x\int_{\mathbf{R}} \phi(y)\,dy - \int_{\mathbf{R}} y\,\phi(y)\,dy = x.$$

(iv). Select $u(x) = \sin x$ and an even $\phi \in C_0^\infty(\mathbf{R})$ with $\int_{\mathbf{R}} \phi(y)\cos y\,dy = 1$. Then

$$(u * \phi)(x) = \int_{\mathbf{R}} \sin(x - y)\,\phi(y)\,dy$$
$$= \sin x \int_{\mathbf{R}} \phi(y) \cos y\, dy - \cos x \int_{\mathbf{R}} \phi(y)\, \sin y\, dy = \sin x.$$

11.3 The implication (ii) $\Rightarrow$ (i) follows from (11.4). For the reverse implication, note that for every $\phi \in C_0^\infty(\mathbf{R}^n)$ and $x \in \mathbf{R}^n$,

$$\partial_{a_j} T_a\phi(x) = \partial_{a_j}(a \mapsto \phi(x-a)) = -\partial_j\phi(x-a) = -T_a\partial_j\phi(x) = -\partial_j T_a\phi(x),$$

because ∂_j commutes with T_a. Application of the chain rule to

$$a \mapsto (a, a) \mapsto T_a \circ \mathcal{A} \circ T_{-a}\phi(x) \quad : \quad \mathbf{R}^n \to \mathbf{R}^n \times \mathbf{R}^n \to \mathbf{R},$$

taken in conjunction with assumption (i), implies

$$\partial_{a_j}(T_a \circ \mathcal{A} \circ T_{-a})\phi = (\partial_{a_j} T_a) \circ \mathcal{A} \circ T_{-a}\phi + T_a \circ \mathcal{A} \circ (\partial_{a_j} T_{-a})\phi$$
$$= -T_a \circ \partial_j \circ \mathcal{A} \circ T_{-a}\phi + T_a \circ \mathcal{A} \circ \partial_j \circ T_{-a}\phi = 0.$$

Apparently $a \mapsto T_a \circ \mathcal{A} \circ T_{-a}$ is a constant mapping, with value $\mathcal{A}$, for $a = 0$; or $T_a \circ \mathcal{A} = \mathcal{A} \circ T_a$. The desired conclusion now follows from Theorem 11.3.

11.5 Consider $\phi \in C_0^\infty(\mathbf{R}^n)$ with $1(\phi) = 1$. On the basis of Problem 5.2 we then have $\lim_{\epsilon\downarrow 0} \phi_\epsilon = \delta$. Accordingly, $u * \phi_\epsilon \in C^\infty(\mathbf{R}^n)$ satisfies $\partial_j(u * \phi_\epsilon) = \partial_j u * \phi_\epsilon = 0$, for $1 \le j \le n$. Thus there exists $c_\epsilon \in \mathbf{C}$ with $u * \phi_\epsilon = c_\epsilon$, and this leads to $u = \lim_{\epsilon\downarrow 0} u * \phi_\epsilon = \lim_{\epsilon\downarrow 0} c_\epsilon$ in $\mathcal{D}'(\mathbf{R}^n)$. This in turn implies $c := u(\phi) = \lim_{\epsilon\downarrow 0} c_\epsilon\, 1(\phi) = \lim_{\epsilon\downarrow 0} c_\epsilon$ in $\mathbf{C}$ and therefore $u = c \in \mathbf{C}$.

11.6 Set $m = \dim L$ and let $u = u_1, \dots, u_m$ be some enumeration of all the $\partial^\alpha u \in \mathcal{D}'(\mathbf{R}^n)$, for suitable multi-indices α, and denote by U the vector in $(\mathcal{D}'(\mathbf{R}^n))^m$ having these distributions as components. Then there exist $m \times m$ matrices A_j with coefficients in $\mathbf{C}$ such that $\partial_j U = A_j\, U$, where the differentiation is componentwise; hence $\partial_j(e^{-\sum_{i=1}^n x_i A_i}\, U) = 0$, for all $1 \le j \le n$ (see [7, Example 2.4.10] for the exponential of a matrix). But then Problem 11.5 implies the existence of some $C \in \mathbf{C}^m$ such that $U = e^{\sum_{i=1}^n x_i A_i}\, C$. The assertion follows by considering the first component of U.

11.7 f is characterized by the condition $f^{(m+1)} = 0$ on account of Example 4.4.

11.8 In view of (11.11) and Example 11.13 we obtain

$$(\delta_a * \delta_b)(\phi) = (\delta_a \otimes \delta_b)(\Sigma^*\phi) = \delta_{(a,b)}(\phi \circ \Sigma) = \phi(a+b) = \delta_{a+b}(\phi).$$

11.10 From (11.19) and (11.25) we deduce $Pv = \delta * Pv = (P\delta) * v$. The second assertion is a direct consequence of (11.25).

11.11 According to Theorem 11.2 the image under the mapping is $C_0^\infty(\mathbf{R}^n)$, while Theorem 11.17 implies that it is continuous. The desired equality follows from (11.23), in view of ${}^t(Sv*)u(\phi) = u(Sv * \phi) = (u * v)(\phi)$, for all $\phi \in C_0^\infty(\mathbf{R}^n)$.

11.14 A short proof of the formula for $(\phi H)^{(k)}$ may be given by mathematical induction on k. On the other hand, one might apply Leibniz's rule (9.1) and obtain

$$\begin{aligned}(\phi H)^{(k)} &= \sum_{i=0}^{k-1}\binom{k}{i+1}\phi^{(k-i-1)}\,\delta^{(i)} + \phi^{(k)} H\\ &= \sum_{i=0}^{k-1}\binom{k}{i+1}\sum_{j=0}^{i}(-1)^{i-j}\binom{i}{j}\phi^{(k-j-1)}(0)\,\delta^{(j)} + \phi^{(k)} H\\ &= \sum_{j=0}^{k-1}\Big(\sum_{i=j}^{k-1}(-1)^{i-j}\binom{i}{j}\binom{k}{i+1}\Big)\phi^{(k-j-1)}(0)\,\delta^{(j)} + \phi^{(k)} H.\end{aligned}$$

In the second equality we used that for all $\psi \in C_0^\infty(\mathbf{R})$,

$$(\phi^{(k-i-1)}\delta^{(i)})(\psi) = \delta^{(i)}(\phi^{(k-i-1)}\psi) = (-1)^i(\phi^{(k-i-1)}\psi)^{(i)}(0),$$

as well as Leibniz's rule once again. Comparison of the two right-hand sides implies that the inner sum equals 1.

11.15 On account of (11.25), $1 * \delta' = 1' * \delta = 0$ and so $(1 * \delta') * H = 0$. On the other hand, $\delta' * H = \delta * H' = \delta * \delta = \delta$ and so $1 * (\delta' * H) = 1$. This result does not violate (11.22), because neither of the distributions 1 and H has compact support.

11.19 **(i)**. This assertion follows by means of (10.15).
(ii). Select $\chi \in C_0^\infty(\mathbf{R})$ such that $1_{\mathbf{R}}(\chi) = 1$. Since $u(\phi) = u(T_{-t}{}^*\phi) = u(T_t\phi)$, for all $t \in \mathbf{R}$ and $\phi \in C_0^\infty(\mathbf{R}^n)$, we see that

$$u(\phi) = \int_{\mathbf{R}}\chi(t)\,dt\,u(\phi) = \int_{\mathbf{R}}\chi(t)\,u(T_t\phi)\,dt = u\Big(\int_{\mathbf{R}}\chi(t)\,T_t\phi\,dt\Big). \tag{21.11}$$

For any $s \in \mathbf{R}$, consider the embedding $\iota_s : \mathbf{R}^{n-1} \to \mathbf{R}^n$ satisfying $x' \mapsto (x', s)$. Thus, we have

$$\begin{aligned}\Big(\int_{\mathbf{R}}\chi(t)\,T_t\phi\,dt\Big)(x) &= \int_{\mathbf{R}}\chi(t)\,\phi(x', x_n - t)\,dt = \int_{\mathbf{R}}\phi(x', s)\,\chi(x_n - s)\,ds\\ &= \int_{\mathbf{R}}\iota_s{}^*\phi(x')\,T_s\chi(x_n)\,ds = \Big(\int_{\mathbf{R}}\iota_s{}^*\phi\otimes T_s\chi\,ds\Big)(x)\\ &= \Big(\int_{\mathbf{R}}T_s(\iota_s{}^*\phi\otimes\chi)\,ds\Big)(x).\end{aligned} \tag{21.12}$$

Note that in the fourth and fifth expressions, T_s denotes a translation acting in $\mathbf{R}$. Hence, combining (21.11) and (21.12) as well as using the invariance of u under $\mathbf{T}$ once more, we may write

$$u(\phi) = \int_{\mathbf{R}}u(T_s(\iota_s{}^*\phi\otimes\chi))\,ds = \int_{\mathbf{R}}u(\iota_s{}^*\phi\otimes\chi)\,ds = u\Big(\int_{\mathbf{R}}\iota_s{}^*\phi\,ds\otimes\chi\Big)$$

$$= u(\pi_*\phi \otimes \chi) = v(\pi_*\phi) = (\pi^* v)(\phi),$$

if we use (10.21) in the fourth equality and introduce $v \in \mathcal{D}'(\mathbf{R}^{n-1})$ by $v(\psi) = u(\psi \otimes \chi)$, for $\psi \in \mathcal{D}'(\mathbf{R}^{n-1})$. This demonstrates that $u = \pi^* v$. Finally, according to (11.14),

$$(v \otimes 1_{\mathbf{R}})(\phi) = v(x' \mapsto 1_{\mathbf{R}}(x_n \mapsto \phi(x', x_n))) = v(\pi_*\phi) = \pi^* v(\phi).$$

11.20 The proof is a generalization of that of Theorem 8.10. We describe its new aspects. u is of finite order, say $\leq k$, according to Theorem 8.8. Next, consider the Taylor expansion of the function on $\mathbf{R}^q$ given by $z \mapsto \phi(y, z)$ to order k at 0, with $y \in \mathbf{R}^p$ as a parameter,

$$\phi(y, z) = \sum_{\alpha \in A,\, |\alpha| \leq k} \partial^\alpha \phi(y, 0) \frac{z^\alpha}{\alpha!} + R(y, z).$$

Here $R(y, z)$ is a finite sum of terms of the form $\widetilde{\phi}(y, z)\, z^\alpha$ with $\widetilde{\phi} \in C^\infty(\mathbf{R}^n)$ and $\alpha \in A$ satisfying $|\alpha| = k + 1$. This implies $u(R) = 0$. Note that $\phi(y, 0) = \phi(\iota(y)) = \iota^*\phi(y)$, where $\iota^*\phi \in C_0^\infty(\mathbf{R}^p)$. Furthermore, p_α is a polynomial function on $\mathbf{R}^q$ if $p_\alpha(z) = \frac{z^\alpha}{\alpha!}$. Replacing A by its finite subset $\{\, \alpha \in A \mid |\alpha| \leq k \,\}$, we may write

$$\phi = \sum_{\alpha \in A} \iota^* \partial^\alpha \phi \otimes p_\alpha + R.$$

Next define

$$u_\alpha \in \mathcal{E}'(\mathbf{R}^p) \qquad \text{by} \qquad u_\alpha(\psi) = (-1)^{|\alpha|} u(\psi \otimes p_\alpha) \qquad (\psi \in C_0^\infty(\mathbf{R}^p)).$$

Since $(-1)^{|\alpha|} u_\alpha(\iota^* \partial^\alpha \phi) = (\partial^\alpha \circ \iota_*) u_\alpha(\phi)$, we obtain

$$u(\phi) = \sum_{\alpha \in A} (\partial^\alpha \circ \iota_*) u_\alpha(\phi) + u(R) = \sum_{\alpha \in A} (\partial^\alpha \circ \iota_*) u_\alpha(\phi).$$

12.2 Successively using (11.4), (11.18), and (11.3), we obtain, for the potential $u \in \mathcal{D}'(\mathbf{R}^3)$ of the dipole $f \in \mathcal{E}'(\mathbf{R}^3)$,

$$u = \Big(\sum_j v_j \partial_j \delta_a\Big) * E = \sum_j v_j\, T_a \delta * \partial_j E = T_a\Big(\sum_j v_j \delta * \partial_j E\Big)$$

$$= T_a\Big(\sum_j v_j\, \partial_j E\Big), \qquad \text{where} \qquad E(x) = -\frac{1}{4\pi} \frac{1}{\|x\|}.$$

Now

$$\partial_j E(x) = \frac{x_j}{4\pi \|x\|^3}, \qquad \text{so} \qquad u(x) = T_a\Big(x \mapsto \frac{\langle v, x\rangle}{4\pi \|x\|^3}\Big) = \frac{1}{4\pi} \frac{\langle v, x - a\rangle}{\|x - a\|^3},$$

for $x \neq a$. In particular, for $a = 0$, $v = e_1$, and x in the (x_1, x_2)-plane provided with polar coordinates (r, α), we see that

$$u(x) = \frac{1}{4\pi}\frac{x_1}{(x_1^2+x_2^2)^{3/2}} = \frac{1}{4\pi}\frac{\cos\alpha}{r^2}.$$

Note that the equipotential lines satisfy a polynomial equation of degree 6.

In the case of $n = 2$,

$$E(x) = \frac{1}{2\pi}\log\|x\|, \qquad \partial_j E(x) = \frac{1}{2\pi}\frac{x_j}{\|x\|^2}, \qquad \text{so} \qquad u(x) = \frac{1}{2\pi}\frac{x_1}{\|x\|^2}.$$

The equipotential lines now become circles centered on the x_1-axis and tangent to the x_2-axis; they satisfy $(x_1 - c)^2 + x_2^2 = c^2$, where $c \in \mathbf{R}$.

12.3 In order to obtain concise formulas, we introduce some notation. Denote by $S \subset \mathbf{R}^2$ the strip parallel to the main diagonal, given by

$$\begin{aligned} S &= \{\,(x_1,\xi_1) \in \mathbf{R}^2 \mid x_1 \in \mathbf{R},\ x_1 - a \le \xi_1 \le x_1 + a\,\} \\ &= \{\,(x_1,\xi_1) \in \mathbf{R}^2 \mid \xi_1 \in \mathbf{R},\ \xi_1 - a \le x_1 \le \xi_1 + a\,\}; \end{aligned}$$

furthermore,

$$\rho(x) = \|(0, x_2, x_3)\| \quad \text{and} \quad \beta(a, x) = \frac{x_1 + a}{\rho(x)} \qquad (x \in \mathbf{R}^3).$$

Moreover, recall the antiderivative $\int \frac{dt}{\sqrt{1+t^2}} = \log(t + \sqrt{1+t^2})$. Then, for any $\phi \in C_0^\infty(\mathbf{R}^3)$, we obtain, by changing the order of integration over S,

$$\begin{aligned} -4\pi\, u_a(\phi) &= -4\pi\,(E * f)(\phi) = -4\pi \int_{\mathbf{R}^3\times\mathbf{R}^3} E(x) f(y)\,\phi(x+y)\,d(x,y) \\ &= \int_{\mathbf{R}^3} \frac{1}{\|x\|}\int_{-a}^{a} \phi(x_1 + y_1, x_2, x_3)\,dy_1\,dx \\ &= \int_{\mathbf{R}^3} \frac{1}{\|x\|}\int_{x_1-a}^{x_1+a} \phi(\xi_1, x_2, x_3)\,d\xi_1\,dx \\ &= \int_{\mathbf{R}^2}\int_S \frac{1}{\|x\|}\,\phi(\xi_1, x_2, x_3)\,d(x_1,\xi_1)\,d(x_2,x_3) \\ &= \int_{\mathbf{R}^2}\int_{\mathbf{R}}\int_{\xi_1-a}^{\xi_1+a} \frac{1}{\sqrt{x_1^2 + \rho(x)^2}}\,dx_1\,\phi(\xi_1, x_2, x_3)\,d\xi_1\,d(x_2,x_3) \\ &= \int_{\mathbf{R}^3} [\log(t + \sqrt{1+t^2})]_{\beta(-a,x)}^{\beta(a,x)}\,\phi(x)\,dx. \end{aligned}$$

Hence

$$\begin{aligned} \exp(-4\pi\, u_a(x)) &= \frac{\beta(a,x) + \sqrt{1 + \beta(a,x)^2}}{\beta(-a,x) + \sqrt{1+\beta(-a,x)^2}} \\ &= \frac{1 + x_1 a^{-1} + \sqrt{\rho(x)^2 a^{-2} + (1 + x_1 a^{-1})^2}}{-1 + x_1 a^{-1} + \sqrt{\rho(x)^2 a^{-2} + (-1 + x_1 a^{-1})^2}}. \end{aligned}$$

For the purpose of studying the behavior of u_a as $a \to \infty$, we note that $\sqrt{1+h} = 1 + \frac{h}{2} - \frac{h^2}{8} + O(h^3)$ as $h \to 0$. This implies, as $a \to \infty$,

$$\begin{aligned}\sqrt{\rho(x)^2 a^{-2} + (\pm 1 + x_1 a^{-1})^2} &= \sqrt{1 \pm 2x_1 a^{-1} + \|x\|^2 a^{-2}} \\ &= 1 \pm x_1 a^{-1} + \frac{1}{2}\rho(x)^2 a^{-2} + O(a^{-3}).\end{aligned}$$

Therefore, as $a \to \infty$,

$$\exp(-4\pi\, u_a(x)) = \frac{2 + 2x_1 a^{-1} + O(a^{-2})}{\frac{1}{2}\rho(x)^2 a^{-2} + O(a^{-3})} = \frac{4a^2}{\rho(x)^2}(1 + x_1 a^{-1} + O(a^{-2})).$$

The unnormalized potential u_a becomes unbounded as $a \to \infty$. To remedy this, consider $v_a = u_a - u_a(e_3)$. Then we have, for $x \in \mathbf{R}^3$ with $\rho(x) > 0$ and as $a \to \infty$,

$$\begin{aligned}\exp(-4\pi\, v_a(x)) &= \frac{\exp(-4\pi\, u_a(x))}{\exp(-4\pi\, u_a(e_3))} = \frac{4a^2(1 + x_1 a^{-1} + O(a^{-2}))}{4a^2 \rho(x)^2 (1 + O(a^{-1}))} \\ &= \frac{1}{\rho(x)^2}(1 + O(a^{-1})).\end{aligned}$$

This leads to

$$v_a(x) = \frac{1}{2\pi} \log \rho(x) - \frac{1}{4\pi} \log(1 + O(a^{-1})) = \frac{1}{2\pi} \log \rho(x) + O(a^{-1}),$$

and accordingly

$$\lim_{a\to\infty} v_a(x) = \frac{1}{2\pi} \log \rho(x) = \frac{1}{2\pi} \log \|(0, x_2, x_3)\|.$$

Observe that in this manner we have obtained the fundamental solution as in (12.3) of the Laplacian Δ acting on $\mathbf{R}^2$.

12.4 The first assertions follow in a straightforward manner.

Green's second identity (see [7, Example 7.9.6]) asserts that for $u \in C^\infty(X)$ and $\phi \in C_0^\infty(\mathbf{R}^n)$ and in the notation of Example 7.3,

$$\int_{\mathbf{R}^n} (1_U\, u\, \Delta\phi - 1_U\, \phi\, \Delta u)(x)\, dx = \int_{\mathbf{R}^n} (\delta_{\partial U}\, u\, \partial_\nu \phi - \delta_{\partial U}\, \phi\, \partial_\nu u)(x)\, dx.$$

This implies the following identity in $\mathcal{E}'(\mathbf{R}^n)$:

$$\Delta(u\, 1_U) - (\Delta u)\, 1_U = -\partial_\nu(u\, \delta_{\partial U}) - (\partial_\nu u)\, \delta_{\partial U}.$$

The function u being harmonic on X, it satisfies $\Delta u = 0$; therefore one obtains the identity $\Delta(u\, 1_U) = -\partial_\nu(u\, \delta_{\partial U}) - (\partial_\nu u)\, \delta_{\partial U}$ in $\mathcal{E}'(\mathbf{R}^n)$. Note that the fundamental solution $E \in \mathcal{D}'(\mathbf{R}^n)$ of Δ as in (12.3) is a locally integrable function on $\mathbf{R}^n$ with $\operatorname{sing\,supp} E = \{0\}$. Convolution of the preceding identity with E is well-defined on account of (11.31); alternatively, one might apply Theorems 14.34 and 14.33.

Furthermore, U is disjoint with sing supp $E * (u\, \delta_{\partial U})$. Therefore, one deduces

$$\begin{aligned} u\, 1_U &= (\Delta E) * (u\, 1_U) = E * \Delta(u\, 1_U) = -E * \partial_\nu(u\, \delta_{\partial U}) - E * (\partial_\nu u)\, \delta_{\partial U} \\ &= -\partial_\nu E * (u\, \delta_{\partial U}) - E * (\partial_\nu u)\, \delta_{\partial U}. \end{aligned}$$

Hence

$$u(x) = \int_{\partial U} \Big(\partial_\nu (y \mapsto E(x-y))\, u(y) - E(x-y)\, \partial_\nu u(y) \Big)\, dy \qquad (x \in U).$$

The vanishing of the integral on $X \setminus \overline{U}$ can also be derived from the formula above.

12.6 According to Problem 12.5 the mean value of $y \mapsto P(x+y)$ over a sphere centered at the origin equals $P(x)$; the function $e^{-\frac{1}{2}\|\cdot\|^2}$ is constant on such spheres, while its total integral over $\mathbf{R}^n$ is $(2\pi)^{\frac{n}{2}}$ on the strength of Problem 2.7.

12.7 The direct approach to prove that $a = \frac{1}{2}$ is as follows. Write, for arbitrary $\phi \in C_0^\infty(\mathbf{R}^2)$,

$$\begin{aligned} \Box 1_V(\phi) = 1_V(\Box \phi) &= \int_V (\partial_t^2 - \partial_x^2)\, \phi(x,t)\, d(x,t) \\ &= \int_{\mathbf{R}} \int_{|x|}^{\infty} \partial_t^2 \phi(x,t)\, dt\, dx - \int_{\mathbf{R}_{>0}} \int_{-t}^{t} \partial_x^2 \phi(x,t)\, dx\, dt. \end{aligned}$$

Continuing this evaluation is straightforward but leads to tedious calculations.

Instead, we discuss two alternative, more conceptual, approaches. First, note that the rotation Ψ as described in the problem satisfies $V = \Psi(\mathbf{R}^2_{>0})$ and is given by the matrix $\Psi = \frac{1}{\sqrt{2}}\begin{pmatrix} 1 & -1 \\ 1 & 1 \end{pmatrix} \in \mathbf{SO}(2, \mathbf{R})$. In particular, $({}^t D\Psi(y))^{-1} = D\Psi(y) = \Psi$, and so $\det D\Psi(y) = 1$, for all $y \in \mathbf{R}^2$. By application of Problem 10.7 we obtain, for $1 \le j \le 2$,

$$\begin{aligned} &\Psi^*(\partial_j^2 \phi) = \frac{1}{\sqrt{2}}(\partial_1 + (-1)^j \partial_2)\Psi^*(\partial_j \phi) = \frac{1}{2}(\partial_1 + (-1)^j \partial_2)^2\, \Psi^*\phi, \\ \Longrightarrow \quad &\Psi^*(\Box \phi) = \frac{1}{2}\big((\partial_1 + \partial_2)^2 - (\partial_1 - \partial_2)^2\big)\, \Psi^*\phi = 2\, \partial_1 \partial_2(\Psi^*\phi). \end{aligned}$$

Hence, the Change of Variables Theorem (see [7, Theorem 6.6.1]) implies

$$\begin{aligned} \Box 1_V(\phi) &= 1_V(\Box\phi) = \int_V \Box\phi(x)\, dx = \int_{\Psi(\mathbf{R}^2_{>0})} \Box\phi(x)\, dx \\ &= \int_{\mathbf{R}^2_{>0}} \Psi^*(\Box\phi)(y)\, j_\Psi(y)\, dy = 2 \int_{\mathbf{R}_{>0}} \int_{\mathbf{R}_{>0}} \partial_1(\partial_2(\Psi^*\phi))(y_1, y_2)\, dy_1\, dy_2 \\ &= -2 \int_{\mathbf{R}_{>0}} \partial_2(\Psi^*\phi)(0, y_2)\, dy_2 = 2\, \Psi^*\phi(0) = 2\,\phi(\Psi(0)) = 2\,\phi(0) = 2\,\delta(\phi). \end{aligned}$$

Next, we describe the second method. Consider the vector field

$$v = -S\,\mathrm{grad}\,\phi : \mathbf{R}^2 \to \mathbf{R}^2 \qquad \text{with} \qquad S = -\begin{pmatrix} 0 & 1 \\ 1 & 0 \end{pmatrix} \in \mathrm{Mat}(2, \mathbf{R}).$$

Introducing the notation $J = \Psi^2$, the rotation in $\mathbf{R}^2$ by $\pi/2$, and $\overline{(g_1, g_2)} = (g_1, -g_2)$ for a vector field g, we have

$$S = \begin{pmatrix} 0 & -1 \\ 1 & 0 \end{pmatrix}\begin{pmatrix} 1 & 0 \\ 0 & -1 \end{pmatrix} = J \circ \overline{}.$$

Furthermore, $J \in \mathbf{SO}(2, \mathbf{R})$ implies

$$\mathrm{curl}\, v = \mathrm{div}({}^tJ\, v) = -\,\mathrm{div}({}^tJJ\, \overline{\mathrm{grad}\,\phi}) = -\,\mathrm{div}(\,\overline{\mathrm{grad}\,\phi}\,) = (\partial_2^2 - \partial_1^2)\phi = \Box\,\phi.$$

It is a straightforward verification that a positive parametrization $y : \mathbf{R} \to \partial V$ is given by $y(s) = (s,\ \mathrm{sgn}(s)\, s)$, for $s \in \mathbf{R}$. Differentiation then leads to the following formula for $Dy(s)$, where we note that sgn is a locally constant function:

$$Dy(s) = \begin{pmatrix} 1 \\ \mathrm{sgn}(s) \end{pmatrix} \qquad \text{and} \qquad Dy(s) = \mathrm{sgn}(s) S\, Dy(s) \qquad (s \in \mathbf{R} \setminus \{0\}).$$

Next we obtain, on account of the chain rule and ${}^tS = S$, for $s \in \mathbf{R} \setminus \{0\}$,

$$\begin{aligned}(\phi \circ y)'(s) &= D\phi(y(s))\, Dy(s) = \mathrm{sgn}(s)\langle \mathrm{grad}\ \phi(y(s)),\ S Dy(s)\,\rangle \\ &= \mathrm{sgn}(s)\langle\, (S\,\mathrm{grad}\,\phi) \circ y(s),\ Dy(s)\,\rangle = \mathrm{sgn}(-s)\langle\, v \circ y,\ Dy\,\rangle(s).\end{aligned}$$

On the basis of Green's Integral Theorem and the compact support of ϕ we obtain

$$\begin{aligned}\int_V \Box\,\phi(x)\,dx &= \int_V \mathrm{curl}\, v(x)\,dx = \int_{\partial V} \langle\, v(y), d_1 y\,\rangle = \int_{\mathbf{R}} \langle\, v \circ y, Dy\,\rangle(s)\,ds \\ &= \int_{\mathbf{R}} \mathrm{sgn}(-s)\,(\phi \circ y)'(s)\,ds = \int_{-\infty}^{0} (\phi \circ y)'(s)\,ds \\ &\quad - \int_0^{\infty} (\phi \circ y)'(s)\,ds = \phi(y(0)) - (-\phi(y(0))) = 2\phi(0).\end{aligned}$$

It is straightforward that $\mathrm{supp}\,E = \overline{V}$ and $\mathrm{sing\,supp}\,E = \partial V$. For the last two assertions, we verify that the condition of Theorem 11.17 is satisfied with $A = \overline{V}$ and $B = H$ or, differently phrased, that the sum mapping $\Sigma : \overline{V} \times H \to \mathbf{R}^2$ is proper. Indeed, consider arbitrary $(x, t) \in \overline{V}$ and $(x', t') \in H$ and suppose that $(x + x', t + t')$ belongs to a compact subset in $\mathbf{R}^2$. Then $x + x'$ and $t + t'$ belong to compact subsets in $\mathbf{R}$. From $t \geq 0$ and $t' \geq t_0$ we get that both t and t' belong to a compact subset of $\mathbf{R}$. But now $|x| \leq t$ implies that x belongs to a compact subset of $\mathbf{R}$, which finally implies that x' belongs to a compact subset of $\mathbf{R}$. This proves the claim. The theorem says that u is a well-defined element of $\mathcal{D}'(\mathbf{R}^2)$, while (11.25) leads to $\Box\,u = (\Box\,E) * f = f$.

12.9 Since $\Delta u = 0$, it follows from Example 12.8 that $u \in C^\infty(X)$. But for smooth functions u the result is classical.

12.10 Since u is homogeneous of degree 0 as well as invariant under rotations, it follows from Theorem 10.17 and Problem 10.11.(ii) that u satisfies the system of partial differential equations

$$\sum_{j=1}^{n} x_j\,\partial_j u = 0 \qquad \text{and} \qquad (x_j\,\partial_k - x_k\,\partial_j)u = 0 \qquad (1 \le j < k \le n).$$

Hence, for all $1 \le j \le k \le n$, we have $x_k x_j\,\partial_j u = x_j^2\,\partial_k u$, which implies

$$\|x\|^2\,\partial_k u = 0. \qquad \text{Thus} \qquad \partial_k u = 0 \quad \text{on} \quad \mathbf{R}^n \setminus \{0\} \qquad (1 \le k \le n).$$

On account of Problem 12.9 and Theorem 8.10 we obtain $c \in \mathbf{C}$ and a polynomial function p on $\mathbf{R}^n$ such that $u = c + p(\partial)\,\delta$. Problem 10.16 then leads to $p = 0$.

12.11 For the first assertion, iterate (12.5).

Note that on $\mathcal{D}'(\mathbf{R}^2)$,

$$\partial_z\,\partial_{\bar z} = \frac{1}{4}(\partial_x - i\,\partial_y)(\partial_x + i\,\partial_y) = \frac{1}{4}(\partial_x^2 + \partial_y^2) = \frac{1}{4}\Delta.$$

Hence

$$\begin{aligned}\delta &= \frac{1}{2\pi}\,\Delta(\log\|(x,y)\|) = \frac{4}{2\pi}\,\partial_{\bar z}\partial_z\frac{1}{2}\log(x^2+y^2)\\ &= \frac{1}{\pi}\,\partial_{\bar z}\frac{1}{2}\Big(\partial_x - i\,\partial_y\Big)\log(x^2+y^2) = \frac{1}{\pi}\partial_{\bar z}\Big(\frac{x-iy}{x^2+y^2}\Big) = \partial_{\bar z}\Big(\frac{1}{\pi z}\Big) = \partial_z\Big(\frac{1}{\pi \bar z}\Big).\end{aligned}$$

12.12 The idea underlying the solution is that, formally speaking, the equation $\operatorname{grad} f = g$ in $(\mathcal{D}'(\mathbf{R}^n))^n$ leads to $\Delta f = \operatorname{div} g$, which in turn implies that $f = E * \Delta f = E * \operatorname{div} g \in \mathcal{D}'(\mathbf{R}^n)$.

Select $\phi \in C_0^\infty(\mathbf{R}^n)$ such that $\phi = 1$ on an open neighborhood of the closed ball around 0 of radius r. Now introduce

$$\tilde g_j = \phi\,g_j \qquad \text{and} \qquad \tilde f = \sum_{j=1}^{n}\partial_j E * \tilde g_j.$$

Note that $\tilde f$ is well-defined, because the $\tilde g_j$ have compact support. On account of the identity $\Delta E = \delta$ as well as the integrability conditions satisfied by the g_j, this leads to

$$\begin{aligned}\partial_k \tilde f &= \sum_{j=1}^{n}\partial_k\partial_j E * \tilde g_j = \sum_{j=1}^{n}\partial_j^2 E * \tilde g_k + \sum_{j=1}^{n}(\partial_k\partial_j E * \tilde g_j - \partial_j^2 E * \tilde g_k)\\ &= \tilde g_k + \sum_{j=1}^{n}\partial_j E * (\partial_k\tilde g_j - \partial_j\tilde g_k) = \tilde g_k + \sum_{j=1}^{n}\partial_j E * ((\partial_k\phi)\,g_j - (\partial_j\phi)\,g_k)\end{aligned}$$

$$=: \tilde{g}_k + h_k.$$

Take E as in (12.3), note that $\operatorname{supp}((\partial_k\phi)\, g_j - (\partial_j\phi)\, g_k) \subset \mathbf{R}^n \setminus \overline{B(0;r)}$, and apply Theorem 11.32. One obtains $\operatorname{sing\,supp} h_k \subset \mathbf{R}^n \setminus \overline{B(0;r)}$. Furthermore, in view of $h_k = \partial_k \tilde{f} - \tilde{g}_k$, one has on $B(0;r)$, for all j and k,

$$\partial_j h_k - \partial_k h_j = \partial_k \tilde{g}_j - \partial_j \tilde{g}_k = \partial_k g_j - \partial_j g_k = 0.$$

Applying Poincaré's Lemma in the classical setting (see [7, Lemma 8.2.6]), one finds $H \in C^\infty(B(0;r))$ such that in $C^\infty(B(0;r))$,

$$\partial_k H = h_k \qquad (1 \leq k \leq n).$$

Any two such solutions H differ by an additive constant. Hence, one may assume that solutions H and H' associated with balls $B(0;r)$ and $B(0;r')$, for $r < r'$, coincide on $B(0,r)$. It follows that $f := \tilde{f} - H$ defines a distribution on $B(0;r)$ satisfying $\partial f_k = g_k$ on $B(0;r)$, for $1 \leq k \leq n$, while any two such solutions f agree on the intersection of open balls centered at 0. Therefore Theorem 7.6 implies the existence of $f \in \mathcal{D}'(\mathbf{R}^n)$ with the desired properties; indeed, it is of the form given in the problem.

12.13 (i). Because the power series expansions of g and h involve only powers of $\|x\|^2 = \sum_{j=1}^3 x_j^2$, both g and h belong to $C^\infty(\mathbf{R}^3)$. According to Problem 4.7, the function $x \mapsto \frac{1}{\|x\|}$ on $\mathbf{R}^3 \setminus \{0\}$ defines an element of $\mathcal{D}'(\mathbf{R}^3)$, and therefore $E_\pm$ does too.
(ii). By direct computation or by using a computer algebra system such as *Mathematica* one directly verifies that h is a solution of the homogeneous equation and furthermore that the identities for $\operatorname{grad} g$, $\operatorname{grad}\big(\frac{1}{\|\cdot\|}\big)$ on $\mathbf{R}^3 \setminus \{0\}$, and Δg are valid. On the basis of Problem 4.7 we have $\Delta\big(\frac{1}{\|x\|}\big) = -4\pi\,\delta$. Furthermore, Leibniz's rule (9.1) may now be used to obtain the second identity (compare with [7, Exercise 2.40.(i)]), which then leads to

$$\Delta\Big(g\frac{1}{\|\cdot\|}\Big)(x) = -4\pi\, g(x)\,\delta + 2k\,\frac{h(x)}{\|x\|} - k^2\,\frac{g(x)}{\|x\|} - 2k\,\frac{h(x)}{\|x\|} = -k^2\,\frac{g(x)}{\|x\|} - 4\pi\,\delta.$$

(iii). According to the well-known formula for the Laplacian in spherical coordinates in $\mathbf{R}^3$ (see [7, Exercise 3.8.(vi)]), one has

$$(\Delta f)_0(r) = f_0''(r) + \frac{2}{r}\, f_0'(r).$$

(iv). The function $\widetilde{f_0} : r \mapsto r\, f_0(r)$ satisfies

$$\widetilde{f_0}'(r) = r\, f_0'(r) + f_0(r), \qquad \widetilde{f_0}''(r) = r\Big(f_0''(r) + \frac{2}{r}\, f_0'(r)\Big),$$

and this implies

$$\widetilde{f_0}''(r) = r(\Delta f)_0(r) = -k^2 r\, f_0(r) = -k^2 \widetilde{f_0}(r).$$

Accordingly, there exist constants a and $b \in \mathbf{C}$ such that

$$\widetilde{f_0}(r) = r\, f_0(r) = a\, \cos kr + b\, \sin kr, \qquad \text{so} \qquad f(x) = a\, \frac{g(x)}{\|x\|} + b\, h(x).$$

(v). $h \in C^\infty(\mathbf{R}^3)$ is a classical solution of Helmholtz' equation, as has been observed in part (i). Therefore, one only needs to compute $(\Delta + k^2)\frac{g}{\|\cdot\|}$ in $\mathcal{D}'(\mathbf{R}^3)$. For arbitrary $\phi \in C_0^\infty(\mathbf{R}^3)$ one obtains

$$\begin{aligned}\Big((\Delta + k^2)\frac{g}{\|\cdot\|}\Big)(\phi) &= \frac{g}{\|\cdot\|}(\Delta + k^2)(\phi) = \int_{\mathbf{R}^3} \frac{g(x)}{\|x\|}(\Delta + k^2)\phi(x)\, dx \\ &= \lim_{\epsilon \downarrow 0} \int_{\|x\|>\epsilon} \frac{g(x)}{\|x\|}(\Delta + k^2)\phi(x)\, dx =: \lim_{\epsilon \downarrow 0} I_\epsilon.\end{aligned}$$

Since $(\Delta + k^2)(\frac{g}{\|\cdot\|}) = 0$ on the open set $\{\, x \in \mathbf{R}^3 \mid \|x\| > \epsilon\,\}$, application of Green's second identity leads to

$$I_\epsilon = \int_{\|y\|=\epsilon} \Big(\frac{g(y)}{\|y\|}\, \partial_\nu \phi(y) - \phi(y)\, \partial_\nu \frac{g(y)}{\|y\|}\Big)\, d_2 y.$$

For the evaluation of the integral, introduce spherical coordinates $(r, \omega) \in \mathbf{R}_{>0} \times S^2$ for $x \in \mathbf{R}^3$ and use $dx = r^2\, dr\, d\omega$, where $d\omega$ indicates two-dimensional integration over the unit sphere S^2. On S^2, considered as the boundary of the complement of the unit ball, the direction of the outer normal is the opposite of the radial direction; hence,

$$\partial_\nu \frac{g(y)}{\|y\|}\Big|_{y=r\omega} = -\partial_r \frac{\cos kr}{r} = \frac{kr\, \sin kr + \cos kr}{r^2}.$$

Conclude that

$$\begin{aligned} I_\epsilon &= \epsilon \cos k\epsilon \int_{S^2} \partial_\nu \phi(\epsilon\omega)\, d\omega - k\epsilon \sin k\epsilon \int_{S^2} \phi(\epsilon\omega)\, d\omega - \cos k\epsilon \int_{S^2} \phi(\epsilon\omega)\, d\omega \\ &=: \sum_{i=1}^{3} I_\epsilon^{(i)}.\end{aligned}$$

To estimate $I_\epsilon^{(1)}$, observe that for any $y \in \mathbf{R}^3$,

$$|\partial_\nu \phi(y)| = |\langle\, \operatorname{grad} \phi(y),\ \nu(y)\,\rangle| \le \sup_{x \in \mathbf{R}^3} \|\operatorname{grad} \phi(x)\| =: m.$$

Hence

$$|I_\epsilon^{(1)}| \le \epsilon |\cos k\epsilon|\, m \int_{S^2} d\omega \to 0, \quad \text{for} \quad \epsilon \downarrow 0,$$

$$|I_\epsilon^{(2)}| \le \epsilon |\sin k\epsilon| \sup_{x\in\mathbf{R}^3} |\phi(x)| \int_{S^2} d\omega \to 0, \quad \text{for} \quad \epsilon \downarrow 0,$$

$$I_\epsilon^{(3)} = -\cos k\epsilon\, \phi(0) \int_{S^2} d\omega - \cos k\epsilon \int_{S^2} (\phi(\epsilon\omega) - \phi(0))\, d\omega.$$

On account of the Mean Value Theorem one has

$$|\phi(\epsilon\omega) - \phi(0)| \le \epsilon \sup_{x\in\mathbf{R}^3} \|D\phi(x)\| =: \epsilon\, M,$$

which implies

$$\Big| \int_{S^2} (\phi(\epsilon\omega) - \phi(0))\, d\omega \Big| \le \int_{S^2} |\phi(\epsilon\omega) - \phi(0)|\, d\omega \le \epsilon\, 4\pi M.$$

Accordingly,

$$\Big((\Delta + k^2) \frac{g}{\|\cdot\|} \Big)(\phi) = \lim_{\epsilon\downarrow 0} I_\epsilon = \lim_{\epsilon\downarrow 0} I_\epsilon^{(3)} = -4\pi\phi(0) = -4\pi\, \delta(\phi).$$

(vi). The results above imply that rotation-invariant fundamental solutions are obtained only by taking $a = 1$ and $b \in \mathbf{C}$ arbitrarily in part (iv). This amounts to $c_\pm = \frac{1\mp bi}{2}$, so that $\sum_\pm c_\pm = 1$.

12.14 (i). Recall that $\lim_{y\downarrow 0} \log y = -\infty$ and $\lim_{y\downarrow 0} y\,|\log y| = 0$. Therefore the nontrivial case of the estimate occurs if $x + iy \in H_+$ is near 0. If so, we have $|\log(x^2 + y^2)| \le 2|\log y|$ and thus

$$\begin{aligned} y\,|\log_+(x + iy)| &= y\,\big|\log|x + iy| + i\arg(x + iy)\big| \\ &\le \frac{y}{2}\,|\log(x^2 + y^2)| + \pi y \le y\,|\log y| + \pi y. \end{aligned}$$

Next, $2\,|xy| \le x^2 + y^2$ leads to

$$\frac{y}{|x + iy|} = \frac{|yx - iy^2|}{x^2 + y^2} \le \frac{|xy|}{x^2 + y^2} + \frac{y^2}{x^2 + y^2} \le \frac{1}{2} + 1 = \frac{3}{2}.$$

(ii). We have

$$\begin{aligned} 2\,\partial_{\bar z}\widetilde{\phi}_N(z) &= (\partial_x + i\,\partial_y) \sum_{k=0}^{N} \frac{\phi^{(k)}(x)}{k!}(iy)^k \\ &= \sum_{k=1}^{N+1} \frac{\phi^{(k)}(x)}{(k-1)!}(iy)^{k-1} - \sum_{k=1}^{N} \frac{\phi^{(k)}(x)}{(k-1)!}(iy)^{k-1} = \frac{\phi^{(N+1)}(x)}{N!}(iy)^N. \end{aligned}$$

(iii). This follows directly from Leibniz's rule (9.1) and the fact that $\partial_{\bar z} f = 0$, for all $f \in \mathcal{O}(H_+)$, on account of (12.6).

(iv). Identify $g = g_1 + i g_2$ with the following vector field on $\mathbf{R}^2$:

$$g = \begin{pmatrix} g_1 \\ g_2 \end{pmatrix}, \quad \text{then} \quad \overline{g} = \begin{pmatrix} g_1 \\ -g_2 \end{pmatrix} \quad \text{and} \quad J\overline{g} := \begin{pmatrix} 0 & -1 \\ 1 & 0 \end{pmatrix} \begin{pmatrix} g_1 \\ -g_2 \end{pmatrix} = \begin{pmatrix} g_2 \\ g_1 \end{pmatrix}.$$

In addition,

$$\operatorname{curl}(\overline{g} + iJ\overline{g}) = \partial_x i(g_1 + ig_2) - \partial_y(g_1 + ig_2) = i(\partial_x + i\partial_y)(g_1 + ig_2) = 2i\,\partial_{\bar{z}} g.$$

Application of Green's Integral Theorem (see [7, formula (8.27) and Theorem 8.3.5]) now implies

$$\int_{\partial R} g(z)\,dz = \int_{\partial R} \langle (\overline{g} + iJ\overline{g})(z),\, d_1 z\rangle = 2i \int_R \partial_{\bar{z}} g(x+iy)\,dx\,dy.$$

Here the second integral denotes the oriented line integral of a vector field on $\mathbf{R}^2$.
(v). The function $z \mapsto \widetilde{\phi}_N(z)\, f^\epsilon(z)$ is well-defined on $R = [\,a, b\,] \times [\,0, h\,]$ and continuous on $\overline{R}$, while $\widetilde{\phi}_N(x) = \phi(x)$ and $\widetilde{\phi}_N(z) = 0$ if $z = a + iy$ or $b + iy$. Hence, successive application of parts (iv), (iii), and (ii) leads to

$$\begin{aligned} \int_a^b \phi(x)\, f^\epsilon(x)\,dx - \int_a^b \widetilde{\phi}_N(x+ih)\, f^\epsilon(x+ih)\,dx &= \int_{\partial R} (\widetilde{\phi}_N\, f^\epsilon)(z)\,dz \\ = 2i \int_R \partial_{\bar{z}}(\widetilde{\phi}_N\, f^\epsilon)(z)\,dz &= 2i \int_a^b \int_0^h R_{\phi,N}(x)\,(iy)^N\, f^\epsilon(x+iy)\,dx\,dy. \end{aligned}$$

(vi). There exist $a < b$ such that $\operatorname{supp} \phi \subset\,]\,a, b\,[$. Because the functions $x \mapsto \widetilde{\phi}_N(x+ih)\, f^\epsilon(x+ih)$ are bounded on $[\,a, b\,]$ uniformly in $0 < \epsilon \leq 1$, Arzelà's or Lebesgue's Dominated Convergence Theorem implies

$$\lim_{\epsilon \downarrow 0} \int_a^b \widetilde{\phi}_N(x+ih)\, f^\epsilon(x+ih)\,dx = \sum_{k=0}^{N} \frac{(ih)^k}{k!} \int_{\mathbf{R}} \phi^{(k)}(x)\, f(x+ih)\,dx.$$

By definition of $f \in \mathcal{O}_N(H_+)$, we have that $y + \epsilon \mapsto (y+\epsilon)^N |f(x+iy+i\epsilon)|$ is bounded, and so $y \mapsto y^N |f^\epsilon(x+iy)|$ is uniformly bounded for $0 < \epsilon \leq 1$. Accordingly, also in this case we obtain

$$\begin{aligned} &\lim_{\epsilon \downarrow 0} \int_0^h \int_a^b R_{\phi,N}(x)\,(iy)^N\, f^\epsilon(x+iy)\,dx\,dy \\ &\quad = \int_0^h \int_a^b R_{\phi,N}(x)\,(iy)^N\, f(x+iy)\,dx\,dy \\ &\quad = \frac{(ih)^{N+1}}{N!} \int_{\mathbf{R}} \phi^{(N+1)}(x) \int_0^1 t^N\, f(x+ith)\,dt\,dx. \end{aligned}$$

The assertion about the order of $\beta_+(f)$ is a straightforward estimate.
(vii). Successively use part (vi) with $N = 0$, integration by parts, $\log_+(x+i0) = \log|x| + \pi i\, H(-x)$ for $x \in \mathbf{R}$, and (1.3) in order to obtain, for $\phi \in C_0^\infty(\mathbf{R})$,

$$\frac{1}{x+i0}(\phi) = \int_{\mathbf{R}} \frac{\phi(x)}{x+ih}\,dx + \int_{\mathbf{R}} \phi'(x)(\log_+(x+ih) - \log_+(x+i0))\,dx$$
$$= -\int_{\mathbf{R}} \phi'(x)\log_+(x+i0)\,dx = -\int_{\mathbf{R}} \phi'(x)\log|x|\,dx$$
$$-\pi i \int_{-\infty}^{0} \phi'(x)\,dx = \Big(\mathrm{PV}\frac{1}{x} - \pi i\,\delta\Big)(\phi).$$

(viii). Because $f \in \mathcal{O}(H_+)$, we have $\partial_z f = (\partial_z + \partial_{\bar{z}})f = \partial_x f$ on H_+. Therefore

$$\beta_+(\partial_z f)(\phi) = \lim_{\epsilon\downarrow 0}\int_{\mathbf{R}} \phi(x)\,\partial_z f(x+i\epsilon) = \lim_{\epsilon\downarrow 0}\int_{\mathbf{R}} \phi(x)\,\partial_x f(x+i\epsilon)\,dx$$
$$= -\lim_{\epsilon\downarrow 0}\int_{\mathbf{R}} \phi'(x)\,f(x+i\epsilon)\,dx = -\beta_+(f)(\phi') = \partial_x(\beta_+(f))(\phi).$$

(ix). $\beta_+(\log_+)$ and $\beta_-(\log_-)$ agree along $\mathbf{R}_{>0}$ and differ by $2\pi i$ along $\mathbf{R}_{<0}$. Next, apply part (viii).
(x). Straightforward computation.
(xi). We have $p(r, x+iy) = p(r,x) + y^2 - 2ixy$, for $z = x+iy \in H_-$. If $\operatorname{Im} p(r,z) = 0$, then $x = 0$ and so $p(r,z) = r^2 + y^2 > 0$. In other words, $p(r,z) \in G$. Accordingly, for every $r \in \mathbf{R}$ and $a \in \mathbf{C}$, the function $f_{r,a}$ is well-defined and complex-differentiable on H_-. For $\operatorname{Re} a \le 0$ and $x \in \mathbf{R}$, we have that $\lim_{\epsilon\downarrow 0} f_{r,a}(x - i\epsilon)$ exists and is a locally integrable function of the variable x; hence it defines a distribution on $\mathbf{R}$. Next, suppose $\operatorname{Re} a > 0$. Then

$$|p(r, x+iy)|^2 = p(r,x)^2 + y^4 + 2(r^2+x^2)y^2 \qquad (x+iy \in H_-), \quad (21.13)$$

which implies $|p(r,x+iy)| \ge y^2$. Conclude from part (vi) that $f_{r,a} \in \mathcal{O}_N(H_-)$ if $N \in \mathbf{Z}_{\ge 0}$ satisfies $N \ge \operatorname{Re} a$. In turn, this means that we have the boundary value $f_{r,a}(x - i0) = \beta_-(f_{r,a}) \in \mathcal{D}'(\mathbf{R})$, which is given by

$$\beta_-(f_{r,a})(\phi) = \lim_{\epsilon\downarrow 0}\int_{\mathbf{R}} \phi(x)\,f_{r,a}(x-i\epsilon)\,dx \qquad (\phi \in C_0^\infty(\mathbf{R})).$$

In addition, it is a distribution of order at most $N+1$.

13.1 Note that $\phi = \Gamma(1-a)\,u * \chi_+^{1-a}$. According to (13.4) and (13.3) we have $\chi_+^{a-1} = \chi_+^{-1} * \chi_+^a = \delta' * \chi_+^a$; so another application of (13.4) and of (11.4) gives

$$u = (u * \chi_+^{1-a}) * \chi_+^{a-1} = \frac{1}{\Gamma(1-a)}(\phi * \delta') * \chi_+^a = \frac{1}{\Gamma(1-a)}\phi' * \chi_+^a.$$

The reflection formula from Lemma 13.5 now leads to the desired formula. That $u \in C^\infty(\mathbf{R})$ follows from Theorem 11.2.

13.2 For $\operatorname{Re} a > 0$ we have

$$\chi_-^a(\phi) = -\int_\infty^0 \frac{x^{a-1}}{\Gamma(a)}\,\phi(-x)\,dx = \int_0^\infty \frac{x^{a-1}}{\Gamma(a)}\,S\phi(x)\,dx = \chi_+^a(S\phi).$$

The identity now follows for all $a \in \mathbf{C}$ by analytic continuation. In particular, it follows from (13.3) that $\chi_-^{-k}(\phi) = \chi_+^{-k}(S\phi) = \delta^{(k)}(S\phi) = (-1)^k\delta^{(k)}(\phi)$.

13.4 Taylor expansion of ϕ around 0 leads to the existence of $\psi \in C^\infty(\mathbf{R})$ such that $\phi(x) - \phi(0) = x\,\psi(x)$. Hence

$$x^{a-1}(\phi(x) - \phi(0)) = x^a\,\psi(x).$$

This function is integrable on $\mathbf{R}_{>0}$; indeed, the right-hand side is locally integrable at 0 since $-1 < \operatorname{Re} a$, and the left-hand side is locally integrable at ∞ because $\operatorname{Re} a - 1 < -1$. On account of (13.7) one has $\chi_+^a = \frac{d}{dx}\,\chi_+^{a+1}$, where $0 < \operatorname{Re} a + 1$. Therefore, integration by parts gives

$$\begin{aligned}
\chi_+^a(\phi) &= -\chi_+^{a+1}(\phi') = -\int_{\mathbf{R}_{>0}} \frac{x^a}{\Gamma(a+1)}\,\phi'(x)\,dx \\
&= -\Big[\frac{x^a}{\Gamma(a+1)}(\phi(x)-\phi(0))\Big]_0^\infty + \int_{\mathbf{R}_{>0}} \frac{x^{a-1}}{\Gamma(a)}(\phi(x)-\phi(0))\,dx \\
&= \int_{\mathbf{R}_{>0}} \frac{x^{a-1}}{\Gamma(a)}(\phi(x)-\phi(0))\,dx.
\end{aligned}$$

More generally, for $k \in \mathbf{Z}_{\geq 0}$ and $0 < \operatorname{Re}(a+k+1) < 1$, define $\phi_{k+1} = \phi^{(k+1)}$ and also

$$\phi_j(x) = \int_0^x \phi_{j+1}(t)\,dt$$

by means of downward mathematical induction on $0 \leq j \leq k$. Using mathematical induction and $\int_0^x \phi^{(j)}(t)\,dt = \phi^{(j-1)}(x) - \phi^{(j-1)}(0)$ one proves

$$\phi_j(x) = \phi^{(j)}(x) - \sum_{l=0}^{k-j} \frac{x^l}{l!}\,\phi^{(j+l)}(0).$$

Then

$$x^{a+j}\,\phi_j(x) = \begin{cases} x^{a+j}\,O(x^{k-j}) = O(x^{a+k}) = o(1), & x \to \infty, \\ x^{a+j}\,O(x^{k-j+1}) = O(x^{a+k+1}) = o(1), & x \downarrow 0. \end{cases}$$

In the particular case of $j = 0$, the O-estimates also imply that $x \mapsto x^{a-1}\,\phi_0(x)$ is integrable on $\mathbf{R}_{\geq 0}$. On the basis of these estimates, $(k+1)$-fold integration by parts leads to

$$\begin{aligned}
\chi_+^a(\phi) &= \frac{d^{k+1}}{dx^{k+1}}\,\chi_+^{a+k+1}(\phi) = (-1)^{k+1}\chi_+^{a+k+1}(\phi^{(k+1)}) \\
&= (-1)^{k+1}\int_{\mathbf{R}_{>0}} \frac{x^{a+k}}{\Gamma(a+k+1)}\,\phi_{k+1}(x)\,dx
\end{aligned}$$

$$= \sum_{j=0}^{k}(-1)^{j+1}\Big[\frac{x^{a+j}}{\Gamma(a+j+1)}\,\phi_j(x)\Big]_0^\infty + \int_{\mathbf{R}_{>0}} \frac{x^{a-1}}{\Gamma(a)}\,\phi_0(x)\,dx$$

$$= \int_{\mathbf{R}_{>0}} \frac{x^{a-1}}{\Gamma(a)}\Big(\phi(x) - \sum_{j=0}^{k}\frac{x^j}{j!}\phi^{(j)}(0)\Big)\,dx$$

$$= \frac{1}{\Gamma(a)}\lim_{\epsilon\downarrow 0}\Big(\int_\epsilon^\infty x^{a-1}\,\phi(x)\,dx + \sum_{j=0}^{k}\frac{\phi^{(j)}(0)}{j!\,(a+j)}\,\epsilon^{a+j}\Big).$$

The last equality follows by means of a simple integration.

Finally, (13.3) implies that $\chi_+^j\big|_L = 0$ if $j \in \mathbf{Z}_{\leq 0}$, which is also the case for $(\text{test}\,\frac{x^{j-1}}{\Gamma(j)}H)\big|_L$ because of the zero of $\frac{1}{\Gamma}$ at j; see Corollary 13.6.

13.6 Indeed, for $\phi \in C_0^\infty(\mathbf{R})$,

$$\partial_x\,\mathrm{PV}\,\frac{1}{x}(\phi) = -\,\mathrm{PV}\,\frac{1}{x}(\phi') = -\lim_{\epsilon\downarrow 0}\int_{\mathbf{R}\setminus[-\epsilon,\,\epsilon]}\frac{\phi'(x)}{x}\,dx =: -\lim_{\epsilon\downarrow 0} I_\epsilon.$$

Integration by parts gives

$$I_\epsilon = \Big[\frac{\phi(x)}{x}\Big]_{-\infty}^{-\epsilon} + \Big[\frac{\phi(x)}{x}\Big]_\epsilon^\infty + \int_{\mathbf{R}\setminus[-\epsilon,\,\epsilon]}\frac{\phi(x)}{x^2}\,dx$$

$$= -\frac{1}{\epsilon}\sum_\pm \phi(\pm\epsilon) + \int_\epsilon^\infty \frac{1}{x^2}\sum_\pm \phi(\pm x)\,dx.$$

Next, Taylor expansion of ϕ about 0 implies $\phi(x) = \phi(0) + x\,\phi'(0) + x^2\,\psi(x)$, where $\psi \in C^\infty(\mathbf{R})$. Hence $\sum_\pm \phi(\pm\epsilon) = 2\phi(0) + \epsilon^2\sum_\pm \psi(\pm\epsilon)$. In view of $\int_\epsilon^\infty \frac{1}{x^2}\,dx = \frac{1}{\epsilon}$, one obtains

$$I_\epsilon = \int_\epsilon^\infty \frac{1}{x^2}\Big(\sum_\pm \phi(\pm x) - 2\phi(0)\Big)\,dx - \epsilon\sum_\pm \psi(\pm\epsilon),$$

and therefore

$$\lim_{\epsilon\downarrow 0} I_\epsilon = \int_{\mathbf{R}_{>0}} x^{-2}\Big(\sum_\pm \phi(\pm x) - 2\phi(0)\Big)\,dx = |x|^{-2}(\phi).$$

13.9 Using (13.20) and (13.21), we obtain outside $\overline{C_-}$,

$$\rho = \frac{d}{da}R_+^a\Big|_{a=0} = \frac{d}{da}\frac{\Gamma(\frac{a+1}{2})}{\pi^{\frac{n}{2}}\,\Gamma(a)}\,q^*\Big(\chi_+^{\frac{a-n+1}{2}}\Big)\Big|_{a=0}.$$

Next, we use Leibniz's rule and Corollary 13.6 to conclude that outside $\overline{C_-}$,

$$\rho = \frac{\Gamma(\frac{1}{2})}{\pi^{\frac{n}{2}}} \frac{d}{da} \frac{1}{\Gamma(a)}\Big|_{a=0} q^*\Big(\chi_+^{\frac{-n+1}{2}}\Big) = \pi^{\frac{1-n}{2}}\, q^*\Big(\chi_+^{\frac{-n+1}{2}}\Big).$$

In the particular case of $n-1$ even, (13.3) implies $\rho = \pi^{\frac{1-n}{2}}\, q^*\big(\delta^{(\frac{n-1}{2})}\big)$ outside $\overline{C_-}$, which is a distribution with support contained in ∂C_+. In view of (13.26) we have

$$q\,\rho = q\,\frac{d}{da} R_+^a\Big|_{a=0} = (-n+1)\, R_+^2 = (1-n)E_+.$$

Finally, the last equality is immediate from (13.25). This equality in turn implies that ρ is not homogeneous.

13.10 (i). Note that $q(x,t) < y$ implies $|t| < \sqrt{\|x\|^2 + y}$. Using Problem 10.23.(ii) and changing the order of differentiation and integration over $\mathbf{R}^3$, we therefore get

$$\begin{aligned} q^*(\delta)(\phi) &= \frac{d}{dy}\Big|_{y=0} \int_{\{\,(x,t)\in\mathbf{R}^4 \mid q(x,t)<y\,\}} \phi(x,t)\, d(x,t) \\ &= \frac{d}{dy}\Big|_{y=0} \int_{\mathbf{R}^3} \int_{-\sqrt{\|x\|^2+y}}^{\sqrt{\|x\|^2+y}} \phi(x,t)\, dt\, dx \\ &= \int_{\mathbf{R}^3} \frac{1}{2\sqrt{\|x\|^2+y}} \Big(\phi\big(x, \sqrt{\|x\|^2+y}\big) + \phi\big(x, -\sqrt{\|x\|^2+y}\big)\Big)\Big|_{y=0} dx \\ &= \sum_{\pm} \frac{1}{2} \int_{\mathbf{R}^3} \frac{\phi(x, \pm\|x\|)}{\|x\|}\, dx. \end{aligned}$$

(ii). Denoting the reflection by S, we have

$$(S\delta_\pm)(\phi) = \delta_\pm(S\phi) = \frac{1}{2} \int_{\mathbf{R}^3} \frac{\phi(x, \mp\|x\|)}{\|x\|}\, dx = \delta_\mp(\phi).$$

It follows from Theorem 3.18 that δ_+ is a measure on X. Setting $V = \mathbf{R}^4 \setminus \partial C_+$, we have, for arbitrary $\phi \in C_0^\infty(\mathbf{R}^4 \setminus \{0\})$ with supp $\phi \subset V$, that $\phi(x, \|x\|) = 0$ for every $x \in \mathbf{R}^3$; this implies $\delta_+(\phi) = 0$. Accordingly, supp $\delta_+ \subset \partial C_+$, and in fact supp $\delta_+ = \partial C_+$. Using the local integrability over $\mathbf{R}^3$ of the function $\|\cdot\|^{-1}$ (see Problem 4.7), we find that δ_+ can be extended to an element of $\mathcal{D}'(\mathbf{R}^4)$ and actually defines a measure.

(iii). Problem 10.20.(ii) immediately implies the identity in $\mathcal{D}'(X)$. Furthermore, $\delta' \in \mathcal{D}'(\mathbf{R})$ is homogeneous of degree -2 on account of Problem 10.16. But in view of Theorem 10.17, that amounts to $(q\,\partial_q + 2)\delta' = 0$, and so $\Box\,(q^*\delta) = 0$ on X.

(iv). From the preceding part it follows that supp $\Box\,\delta_+ \subset \{0\}$. Hence, the first statement is a consequence of Theorem 8.10. In fact, δ_+ is homogeneous of degree -2 because for $c > 0$,

$$c^{4-2}\delta_+(c^* f) = \frac{c^2}{2}\int_{\mathbf{R}^3} \frac{f(cx, c\|x\|)}{\|x\|}\,dx = \frac{1}{2}\int_{\mathbf{R}^3} \frac{f(x,\|x\|)}{\|x\|}\,dx = \delta_+(f).$$

Recalling Problem 10.15.(ii), we deduce the homogeneity of $\Box\,\delta_+$ of degree -4, and thus the argument can be finished by copying the proof of Theorem 13.3.
(v). Consider ϕ as given. Using a suitable cut-off function we can construct a function $\psi \in C_0^\infty(\mathbf{R}^4)$ that coincides with ϕ on an open neighborhood of the compact set $\operatorname{supp}\phi \cap \operatorname{supp}\delta_+$. The definition $\delta_+(\phi) := \delta_+(\psi)$ is now independent of the particular choice of ψ. For ϕ as specifically given, we have

$$c\,\phi(0) = \delta_+(\Box\,\phi) = \frac{1}{2}\int_{\mathbf{R}^3} \frac{\psi''(\|x\|)}{\|x\|}\,dx.$$

Employing spherical coordinates in $\mathbf{R}^3$ we thus obtain

$$c\,\phi(0) = \frac{1}{2}4\pi\int_{\mathbf{R}_{>0}} r\,\psi''(r)\,dr = -2\pi\int_{\mathbf{R}_{>0}} \psi'(r)\,dr = 2\pi\,\psi(0) = 2\pi\,\phi(0).$$

(vi). $u = E_+ * f$ is a solution of the inhomogeneous equation according to Theorem 12.2. In view of (11.1) and Theorem 11.2,

$$\begin{aligned} E_+ * f(x,t) &= E_+(T_{(x,t)}Sf) = \frac{1}{4\pi}\int_{\mathbf{R}^3} \frac{(T_{(x,t)}Sf)(y',\|y'\|)}{\|y'\|}\,dy' \\ &= \frac{1}{4\pi}\int_{\mathbf{R}^3} \frac{f(x-y', t-\|y'\|)}{\|y'\|}\,dy' = \frac{1}{4\pi}\int_{\mathbf{R}^3} \frac{f(y, t-\|x-y\|)}{\|x-y\|}\,dy. \end{aligned}$$

14.3 We may assume that P is a function actually depending on its first variable, that is, that $\lambda_1 \mapsto P(\lambda_1, \dots, \lambda_n)$ is a polynomial of degree at least 1. For every $(\lambda_2, \dots, \lambda_n) \in \mathbf{C}^{n-1}$, this polynomial function has at least one zero on account of the Fundamental Theorem of Algebra (see [7, Exercise 8.13], for instance). This proves that the zero-set is infinite. It follows that there are infinitely many $\lambda \in \mathbf{C}^n$ such that $P(\partial)\,e_\lambda = P(\lambda)\,e_\lambda = 0$, while the e_λ are linearly independent complex-analytic functions on $\mathbf{C}^n$.

14.5 Straightforward verification, by writing an integral over $\mathbf{R}^n$ as iterated one-dimensional integrals.

14.7 We first show that $L\psi(a) = 0$ if $\psi \in \mathcal{S}$, $a \in \mathbf{R}^n$, and $\psi(a) = 0$. Note that there exist $\psi_1, \dots, \psi_n \in \mathcal{S}$ such that

$$\psi(x) = \sum_{j=1}^{n}(x_j - a_j)\,\psi_j(x) \qquad (x \in \mathbf{R}^n). \tag{21.14}$$

Indeed, by writing $\psi(x) - \psi(a)$ as the integral from 0 to 1 of $\frac{d}{dt}\psi(a + t\,(x-a))$, we obtain (21.14), with $\widetilde{\psi}_j \in C^\infty$ instead of the $\psi_j \in \mathcal{S}$. Now choose $\chi \in C_0^\infty$ with $\chi = 1$ on a neighborhood of a. Then

$$\psi_j(x) := \chi(x)\,\widetilde{\psi}_j(x) + (1-\chi(x))\,\psi(x)\,\frac{x_j - a_j}{\|x-a\|^2} \qquad (1 \le j \le n)$$

satisfy (21.14). This implies

$$L\psi = L\Big(\sum_{j=1}^{n}(x_j - a_j)\,\psi_j\Big) = \sum_{j=1}^{n} L(x_j\,\psi_j) - L(a_j\,\psi_j) = \sum_{j=1}^{n}(x_j - a_j)\,L\psi_j,$$

from which we conclude that $L\psi(a) = 0$ if $\psi \in \mathcal{S}$ and $\psi(a) = 0$.

Now let $\gamma \in \mathcal{S}$ be chosen such that γ is nowhere zero on $\mathbf{R}^n$. For arbitrary $\phi \in \mathcal{S}$, define $\psi = \phi - \frac{\phi(a)}{\gamma(a)}\,\gamma \in \mathcal{S}$; then ψ has a zero at a, and so

$$0 = L\psi(a) = L\phi(a) - \frac{\phi(a)}{\gamma(a)}\,L\gamma(a).$$

Because this holds for all $a \in \mathbf{R}^n$, the conclusion is, for all $\phi \in \mathcal{S}$,

$$L\phi = c\,\phi \qquad \text{with} \qquad c := \frac{L\gamma}{\gamma} \in C^\infty(\mathbf{R}^n).$$

Finally, we see that c is constant, because application of this identity with ϕ successively replaced by γ and $D_j\gamma$ implies

$$\gamma\,D_j c + c\,D_j\gamma = D_j(c\,\gamma) = D_j(L\gamma) = L(D_j\gamma) = c\,D_j\gamma,$$

and so $\gamma\,D_j c = 0$; in other words, $D_j c = 0$, for every $1 \le j \le n$.

14.8 For assertion (i) see the initial part of the solution to Problem 14.7 with $a = 0$. The two assertions are just Fourier transforms of one another.

14.9 Suppose $\phi \in \mathcal{S}(\mathbf{R})$ satisfies $\mathcal{F}\phi = \lambda\phi$, for some $\lambda \in \mathbf{C}$. According to Theorem 14.13 we have $(2\pi)^2\phi = \mathcal{F}(\mathcal{F}S)(S\mathcal{F})\mathcal{F}\phi = \mathcal{F}^4\phi = \lambda^4\phi$. Thus $\lambda^4 = (2\pi)^2$, that is, $\lambda = \sqrt{2\pi}\,i^j$, for $0 \le j \le 3$. Next, set $\phi(x) = e^{-x^2/2}$ and $\lambda = \sqrt{2\pi}$. Then $\mathcal{F}\phi = \lambda\phi$ according to Example 14.11. Note that $\partial\phi = -x\,\phi$. Using this, we see that $\mathcal{F}(x\,\phi) = -D\mathcal{F}\phi = i\partial(\lambda\phi) = -i\lambda(x\,\phi)$. Finally,

$$\begin{aligned}\mathcal{F}\Big(\Big(x^2 - \frac{1}{2}\Big)\phi\Big) &= -D\mathcal{F}(x\,\phi) - \frac{1}{2}\lambda\phi = i\partial(-i\lambda x\,\phi) - \frac{1}{2}\lambda\phi \\ &= \lambda\Big(\phi - x^2\phi - \frac{1}{2}\phi\Big) = -\lambda\Big(x^2 - \frac{1}{2}\Big)\phi.\end{aligned}$$

More generally, for $j \in \mathbf{Z}_{\ge 0}$, define the *Hermite polynomials* H_j and the *Hermite functions* ψ_j on $\mathbf{R}$ (the ψ_j are also called the *quantum harmonic oscillator wave functions*) by

$$H_j(x) = (-1)^j e^{x^2}\partial_x^j e^{-x^2} \qquad \text{and} \qquad \psi_j = \frac{1}{\sqrt{2^j\,j!\sqrt{\pi}}}\,\phi\,H_j.$$

By means of integration by parts and the identity $\partial H_j = 2j\,H_{j-1}$ one sees that $(\psi_j)_{j\in\mathbf{Z}_{\geq 0}}$ forms an orthonormal system in $L^2(\mathbf{R})$. In fact, it is a complete system; for a proof, see Problem 14.42.

In addition, every ψ_j is an eigenfunction of the Fourier transform; more specifically, $\mathcal{F}\psi_j = \sqrt{2\pi}(-i)^j\,\psi_j$, for $j \in \mathbf{Z}_{\geq 0}$. Indeed, using integration by parts and Example 14.11 one obtains

$$\begin{aligned}\mathcal{F}(\phi\, H_j)(\xi) &= \int_{\mathbf{R}} e^{-x^2}\partial_x^j e^{-ix\xi+\frac{1}{2}x^2}\,dx = i^j e^{\frac{1}{2}\xi^2}\int_{\mathbf{R}} e^{-x^2}\partial_\xi^j e^{\frac{1}{2}(x-i\xi)^2}\,dx\\ &= i^j e^{\frac{1}{2}\xi^2}\partial_\xi^j e^{-\frac{1}{2}\xi^2}\int_{\mathbf{R}} e^{-ix\xi-\frac{1}{2}x^2}\,dx = \sqrt{2\pi}(-i)^j\,(\phi\, H_j)(\xi).\end{aligned}$$

14.10 By analytic continuation it follows that the identity from Problem 12.6 is valid for all $x \in \mathbf{C}^n$. In particular, for $x = -i\,\xi$ with $\xi \in \mathbf{R}^n$, we obtain

$$\int_{\mathbf{R}^n} e^{-\frac{1}{2}\|y\|^2} P(y - i\,\xi)\,dy = (2\pi)^{\frac{n}{2}}(-i)^* P(\xi).$$

Next an n-fold application of Cauchy's Integral Theorem as in (12.9) leads to the third equality in the following:

$$\begin{aligned}e^{\frac{1}{2}\|\xi\|^2}\mathcal{F}(e^{-\frac{1}{2}\|\cdot\|^2}P)(\xi) &= \int_{\mathbf{R}^n} e^{-\frac{1}{2}(\|y\|^2+2i\langle y,\xi\rangle-\|\xi\|^2)}P(y)\,dy\\ &= \int_{\mathbf{R}^n} e^{-\frac{1}{2}\|y+i\,\xi\|^2}P(y)\,dy = \int_{\mathbf{R}^n} e^{-\frac{1}{2}\|y\|^2}P(y-i\,\xi)\,dy = (2\pi)^{\frac{n}{2}}(-i)^* P(\xi).\end{aligned}$$

Observe that the argument above gives a computation of $\mathcal{F}e^{-\frac{1}{2}\|\cdot\|^2}$ different from the one in Example 14.11. In addition, in the case of $n > 1$ Hecke's formula produces infinitely many eigenfunctions similarly as in Problem 14.9 where $n = 1$.

14.11 Applying the estimate in Definition 19.4 with $K = \operatorname{supp} f$, we obtain a constant $c' > 0$ such that for arbitrary $x \in \operatorname{supp} f$ and $\xi \in \mathbf{R}^n\setminus\{0\}$ with $x - \frac{\pi}{\|\xi\|^2}\xi \in \operatorname{supp} f$,

$$|(I - T_{\frac{\pi}{\|\xi\|^2}\xi})f(x)| \leq c'\,\|\xi\|^{-\alpha}.$$

Application of (14.6) and (14.5) then leads to the desired estimate.

14.12 Prove

$$(1+\|\xi\|^2)^{\frac{n+1}{2}} \leq \Big(1 + \sum_{1\leq j\leq n}|\xi_j|\Big)^{n+1} \leq c_n' \sum_{|\alpha|\leq n+1}|\xi^\alpha| \qquad (\xi \in \mathbf{R}^n),$$

and deduce

$$\begin{aligned}(1+\|\xi\|^2)^{\frac{n+1}{2}}|\mathcal{F}f(\xi)| &\leq c_n' \sum_{|\alpha|\leq n+1}|\xi^\alpha\mathcal{F}f(\xi)| = c_n' \sum_{|\alpha|\leq n+1}|\mathcal{F}(D^\alpha f)(\xi)|\\ &\leq c_n' \sum_{|\alpha|\leq n+1}\|D^\alpha f\|_{L^1}.\end{aligned}$$

Verify $\|f\|_{L^\infty} \le (2\pi)^{-n}\|\mathcal{F} f\|_{L^1}$ by means of Theorem 14.13, and conclude that

$$\|f\|_{L^\infty} \le c_n'' \int_{\mathbf{R}^n} (1+\|\xi\|^2)^{-\frac{n+1}{2}}\, d\xi \sum_{|\alpha|\le n+1} \|D^\alpha f\|_{L^1}.$$

14.13 Interchanging the order of integration implies, for all $t > 0$,

$$\begin{aligned}\int_{-t}^{t} \mathcal{F}\phi(\xi)\, d\xi &= \int_{-t}^{t}\int_{\mathbf{R}} e^{-ix\xi}\phi(x)\, dx\, d\xi = \int_{\mathbf{R}} \phi(x) \int_{-t}^{t} e^{-ix\xi}\, d\xi\, dx \\ &= 2\int_{\mathbf{R}} \phi(x)\, \frac{\sin tx}{x}\, dx.\end{aligned}$$

The desired equality for the limit now follows on account of Definition 14.19 and Theorem 14.13.

14.14 (ii). We have $a \in \mathcal{S}$. On the other hand, the functions b and c do not belong to $\mathcal{S}$ because they are not differentiable at 0, while d is not in $\mathcal{S}$ since $x \mapsto x^3\, d(x)$ is unbounded on $\mathbf{R}$. All functions belong to L^2.
(iii). Since all functions belong to $L^1(\mathbf{R})$, their Fourier transforms are given by (14.3). Note that $a(x) = e\, e^{-(x-1)^2}$. Hence Example 14.11 leads to

$$\begin{aligned}\mathcal{F}a(\xi) &= e\int_{\mathbf{R}} e^{-ix\xi} e^{-(x-1)^2}\, dx = e\int_{\mathbf{R}} e^{-i(x+1)\xi} e^{-x^2}\, dx \\ &= e^{1-i\xi}\int_{\mathbf{R}} e^{-ix\xi} e^{-2x^2/2}\, dx = e^{1-i\xi}\sqrt{2\pi}\,\frac{1}{\sqrt{2}}\, e^{-\frac{\xi^2}{4}} = \sqrt{\pi}\, e^{-(\frac{\xi}{2}+i)^2}.\end{aligned}$$

$\mathcal{F}b(\xi) = \frac{1}{i\xi+1}$ follows from Example 14.30 or by direct computation. Further, observe that $c = b + Sb$. Since $\mathcal{F}S = S\mathcal{F}$, this means that

$$\mathcal{F}c(\xi) = \frac{1}{1+i\xi} + \frac{1}{1-i\xi} = \frac{2}{1+\xi^2}.$$

$\mathcal{F}d(\xi) = \pi\, e^{-|\xi|}$. Indeed, note that $d = \frac{1}{2}\mathcal{F}c$; therefore, Fourier inversion gives

$$\mathcal{F}d = \frac{1}{2}\mathcal{F}^2 c = \pi\Big(\frac{1}{2\pi} S\mathcal{F}\mathcal{F}\Big)(Sc) = \pi\, Sc = \pi\, c.$$

On the other hand, $\mathcal{F}d$ is given by (14.3). Combination of these two results immediately produces Laplace's integral.
(v). Part (ii) taken in conjunction with Theorem 14.32 gives that all Fourier transforms belong to L^2. Together with Theorem 14.13 it also implies that $\mathcal{F}a \in \mathcal{S}$, whereas the remaining Fourier transforms do not. Finally, $\mathcal{F}b \notin L^1$ but the remaining do belong to L^1.

14.15 On account of Problem 4.1 we have $\partial f(x) = -f(x)\,\partial|x| = -(f\,\mathrm{sgn})(x)$. Hence,

$$\partial^2 f = -\partial f\ \mathrm{sgn} - 2f\,\delta = f\,\mathrm{sgn}^2 - 2f(0)\delta = f - 2\delta.$$

So $D^2 f + f = 2\delta$. Since $\mathcal{F}\delta = 1$, this means that $(\xi^2 + 1)\mathcal{F} f(\xi) = 2$, that is, $\mathcal{F} f(\xi) = \frac{2}{1+\xi^2}$ (compare with Problem 14.14.(iii)). Next, note that $i\, D \arctan x = \frac{1}{1+x^2}$. Hence, $i\,\xi\,\mathcal{F} \arctan \xi = \pi e^{-|\xi|}$; therefore Theorem 9.5 implies the existence of $c \in \mathbf{C}$ such that

$$\mathcal{F} \arctan = \frac{\pi}{i} \,\mathrm{PV}\, \frac{e^{-|\xi|}}{\xi} + c\,\delta.$$

Since arctan is odd, so is $\mathcal{F}$ arctan, which leads to $c = 0$.

14.17 We have $f_0(x) = x\, d(x)$ and so $\mathcal{F} f_0 = i\partial\mathcal{F} d$ according to (14.28). Now $\mathcal{F} d(\xi) = \pi e^{-|\xi|}$, and so the solution to Problem 14.15 leads to the desired formula for $\mathcal{F} f_0$. Next note that $f - f_0 = g$, where $g \in L^1(\mathbf{R})$. The Riemann–Lebesgue Theorem, Theorem 14.2, then implies that $\mathcal{F} f - \mathcal{F} f_0 = \mathcal{F} g$ is continuous.

14.18 $\phi \in \mathcal{S}(\mathbf{R}^n)$ is a direct consequence of $x^\beta\, \partial^\alpha \phi(x) = \prod_{j=1}^n x^{\beta_j}\, \partial^{\alpha_j} \phi_j(x_j)$, while the identity $c_n = (c_1)^n$ results from a similar computation.

Write $f(x) = e^{-|x|}$. From Problem 14.14.(iii) or 14.15 we obtain $\mathcal{F} f(\xi) = 2/(1 + \xi^2)$; in particular, $\mathcal{F} f \in L^1(\mathbf{R})$. Select $\phi \in C_0^\infty(\mathbf{R})$ satisfying $\phi \geq 0$ and $1(\phi) = 1$, and define $\phi_\epsilon(x) = \epsilon^{-1}\phi(x/\epsilon)$. Now $\mathcal{F}\phi_\epsilon(\xi) = \mathcal{F}\phi(\epsilon\,\xi)$ and $|\mathcal{F}\phi| \leq 1(\phi) = 1$ imply $|\mathcal{F}\phi_\epsilon| \leq 1$. Furthermore,

$$|\mathcal{F}\phi_\epsilon(\xi) - 1| \leq \int_{\mathbf{R}} |e^{-i\epsilon\xi x} - 1|\, \phi(x)\, dx$$

implies $\mathcal{F}\phi_\epsilon \to 1$ uniformly on compacta as $\epsilon \downarrow 0$. This in turn leads to

$$\lim_{\epsilon \downarrow 0} \|\mathcal{F}\phi_\epsilon\, \mathcal{F} f - \mathcal{F} f\|_{L^1} = 0.$$

Indeed, split the integration into two parts: one over a suitable compact set, the other part over its complement C such that $\int_C \mathcal{F} f(\xi)\, d\xi$ is sufficiently small. Since $\phi_\epsilon \in C_0^\infty(\mathbf{R}) \subset \mathcal{E}'(\mathbf{R})$ and $f \in L^1(\mathbf{R}) \subset \mathcal{S}'(\mathbf{R})$, we have $\mathcal{F}(\phi_\epsilon * f) = \mathcal{F}\phi_\epsilon\, \mathcal{F} f$ on account of Theorem 14.33. But $\mathcal{F}\phi_\epsilon$ belongs to $\mathcal{S}(\mathbf{R})$ and the properties of $\mathcal{F} f$ now imply $\mathcal{F}\phi_\epsilon\, \mathcal{F} f \in \mathcal{S}(\mathbf{R})$ and so $\phi_\epsilon * f \in \mathcal{S}(\mathbf{R})$ according to Theorem 14.13. From the proof of that same theorem we get the existence of $c_1 > 0$ such that $\mathcal{G}(\psi) = c_1\, \psi$, for all $\psi \in C_0^\infty(\mathbf{R})$. In particular, we obtain, for all $\epsilon > 0$,

$$\int_{\mathbf{R}} \mathcal{F}\phi_\epsilon(\xi)\, \mathcal{F} f(\xi)\, d\xi = c_1(\phi_\epsilon * f)(0).$$

Now we are sufficiently equipped to take the limit in this identity as $\epsilon \downarrow 0$; thus we obtain

$$2\pi = \int_{\mathbf{R}} \frac{2}{1 + \xi^2}\, d\xi = \int_{\mathbf{R}} \mathcal{F} f(\xi)\, d\xi = c_1\, f(0) = c_1.$$

14.24 By definition of $\mathcal{S}'(\mathbf{R}^n)$ there exist, for every $u \in \mathcal{S}'(\mathbf{R}^n)$, constants $c > 0$ and $k, N \in \mathbf{Z}_{\geq 0}$ such that for all $a \in \mathbf{R}^n$ and $\phi \in \mathcal{S}(\mathbf{R}^n)$,

$$|u(T_a\phi)| \leq c\, \|T_a\phi\|_{\mathcal{S}(k,N)} = c \sup_{|\alpha| \leq k,\, |\beta| \leq N,\, x \in \mathbf{R}^n} |x^\beta (\partial^\alpha \phi)(x - a)|$$

$$
\begin{aligned}
&= c \sup_{|\alpha|\le k,\,|\beta|\le N,\,x\in\mathbf{R}^n} |(x+a)^\beta (\partial^\alpha\phi)(x)| \\
&= c \sup_{|\alpha|\le k,\,|\beta|\le N,\,x\in\mathbf{R}^n} \Big|\sum_{\gamma\le\beta}\binom{\beta}{\gamma} x^\gamma a^{\beta-\gamma}(\partial^\alpha\phi)(x)\Big| \\
&\le c'\,(1+\|a\|)^N \sup_{|\alpha|\le k,\,|\gamma|\le N,\,x\in\mathbf{R}^n} |x^\gamma(\partial^\alpha\phi)(x)| \le d\,(1+\|a\|)^N,
\end{aligned}
$$

for a suitable $d > 0$. Now let $\xi \in \mathbf{C}^n$ and suppose that $e_{i\xi} \in \mathcal{S}'(\mathbf{R}^n)$. Then $e_{i\xi}(T_a\phi) = (T_a{}^* e_{i\xi})(\phi) = e^{i\langle a,\xi\rangle}\, e_{i\xi}(\phi)$; hence we have

$$
|e_{i\xi}(T_a\phi)| \le d\,(1+\|a\|)^N \Longrightarrow e^{i\langle a,\xi\rangle} = O(\|a\|^N), \quad \|a\|\to\infty \Longrightarrow \xi\in\mathbf{R}^n.
$$

14.27 **(i).** Formula (14.36) immediately leads to the identity $\mathcal{F} e_{ia} = 2\pi\,\delta_a$ in $\mathcal{S}'(\mathbf{R})$, for all $a \in \mathbf{R}$. It follows that we have in $\mathcal{S}'(\mathbf{R})$,

$$
\begin{aligned}
\mathcal{F}\cos &= \mathcal{F}\Big(\frac{e_i + e_{-i}}{2}\Big) = \pi(\delta_{-1}+\delta_1), \\
\mathcal{F}\sin &= \mathcal{F}\Big(\frac{e_i - e_{-i}}{2i}\Big) = \pi i(\delta_{-1}-\delta_1), \\
\mathcal{F}\sin &= -i\,\mathcal{F}\,D\cos = -\pi i\,\xi(\delta_{-1}+\delta_1) = \pi i(\delta_{-1}-\delta_1), \\
\mathcal{F}(x\mapsto x\sin x) &= -D\mathcal{F}(\sin) = i\,\partial(\pi i(\delta_{-1}-\delta_1)) = -\pi(\delta'_{-1}-\delta'_1), \\
\mathcal{F}\cos^2 &= \frac{1}{2\pi}(\mathcal{F}\cos)*(\mathcal{F}\cos) = \frac{\pi}{2}(\delta_{-1}+\delta_1)*(\delta_{-1}+\delta_1) \\
&= \frac{\pi}{2}(\delta_{-2}+2\delta_0+\delta_2).
\end{aligned}
$$

In the third identity we have applied (14.28) and Example 9.1, in the fourth (14.28) once again, and in the fifth (14.22) and Theorem 14.33.
(ii). We have $\mathcal{F}\,\mathrm{sinc} = \pi\,1_{[-1,1]}$. Indeed, this follows by Fourier inversion from $\mathcal{F}(1_{[-1,1]}) = 2\,\mathrm{sinc}$; see Example 14.1. Another way to prove this is as follows. On account of Example 4.2 we get

$$
i\,\partial\,\mathcal{F}\,\mathrm{sinc} = \mathcal{F}(x\ \mathrm{sinc}) = \mathcal{F}(\sin) = i\,\pi(\delta_{-1}-\delta_1) = i\,\partial\,\pi\,1_{[-1,1]}.
$$

Therefore there exists $c \in \mathbf{C}$ such that $\mathcal{F}\,\mathrm{sinc} = \pi\,1_{[-1,1]} + c$, and the well-known evaluation $\mathcal{F}\,\mathrm{sinc}(0) = \int_{\mathbf{R}} \mathrm{sinc}\,x\,dx = \pi$ (see Problem 14.44 or 16.19 or [7, Example 2.10.14 or Exercise 0.14, 6.60, or 8.19]) then leads to $c = 0$.
(iii). Applying (14.38) and Problem 13.6, we see that

$$
\mathcal{F}(x\,H) = i\,\partial\,\mathcal{F}H = i\,\partial_\xi\Big(\pi\,\delta - i\ \mathrm{PV}\,\frac{1}{\xi}\Big) = \pi i\,\delta' - |\cdot|^{-2}.
$$

Observe that $|x| = x\ \mathrm{sgn}\,x$; hence (14.37) and Problem 13.6 lead to

$$
\mathcal{F}|\cdot| = i\,\partial\,\mathcal{F}\,\mathrm{sgn} = i(-2i)\,\partial_\xi\,\mathrm{PV}\,\frac{1}{\xi} = -2|\cdot|^{-2}.
$$

We have
$$\sin|x| = \mathrm{sgn}(x)\,\sin x = \frac{e^{ix}-e^{-ix}}{2i}\,\mathrm{sgn}(x).$$
Next, apply the first identity in (14.29) to obtain
$$\mathcal{F}\sin|\cdot| = \mathrm{PV}\,\frac{1}{\xi+1} - \mathrm{PV}\,\frac{1}{\xi-1}.$$

Note that $2\pi\,|\cdot| = S\mathcal{F}\mathcal{F}|\cdot| = \mathcal{F}S(-2|\cdot|^{-2}) = -2\mathcal{F}|\cdot|^{-2}$.

14.28 Suppose $R \neq 0$. Then application of (14.28) leads to
$$I = \mathcal{F}^{-1}\Big(\xi \mapsto \frac{1}{L\,(i\xi)^2 + R\,i\xi + \frac{1}{C}}\Big).$$

The roots of the polynomial in the variable $i\xi$ in the denominator are
$$\frac{-R \pm i\,Q}{2L}, \qquad \text{where} \qquad Q = \sqrt{\frac{4L}{C} - R^2}. \tag{21.15}$$

If these two roots are distinct, we apply partial-fraction decomposition to obtain
$$I = \frac{1}{i\,Q}\,\mathcal{F}^{-1}\Big(\xi \mapsto \frac{1}{i\xi + \frac{R}{2L} - i\frac{Q}{2L}} - \frac{1}{i\xi + \frac{R}{2L} + i\frac{Q}{2L}}\Big).$$

Thus, analytic continuation of (14.39) to $\{\,\epsilon \in \mathbf{C} \mid \mathrm{Re}\,\epsilon > 0\,\}$ immediately implies
$$I(t) = \frac{2}{Q}\,H(t)\,e^{-\frac{R}{2L}t}\sin\frac{Q\,t}{2L} = \frac{2}{\sqrt{\frac{4L}{C} - R^2}}\,H(t)\,e^{-\frac{R}{2L}t}\sin\frac{t}{2L}\sqrt{\frac{4L}{C} - R^2}.$$

In the case of $R^2 < \frac{4L}{C}$, this is the desired formula for I. If, however, $R^2 > \frac{4L}{C}$, we rewrite this expression as
$$I(t) = \frac{2}{\sqrt{R^2 - \frac{4L}{C}}}\,H(t)\,e^{-\frac{R}{2L}t}\sinh\frac{t}{2L}\sqrt{R^2 - \frac{4L}{C}}.$$

The case of $R = 0$ is obtained by taking limits for $R \to 0$ in the preceding formulas.
Finally, if the roots in (21.15) coincide, we have $R^2 = \frac{4L}{C} > 0$ and
$$I = \frac{1}{L}\,\mathcal{F}^{-1}\Big(\xi \mapsto \frac{1}{(i\xi + \frac{R}{2L})^2}\Big) = \frac{1}{L}\,\mathcal{F}^{-1}\Big(\xi \mapsto -D_\xi\,\frac{1}{i\xi + \frac{R}{2L}}\Big).$$

Hence (14.28) implies
$$I(t) = \frac{1}{L}\,H(t)\,t\,e^{-\frac{R}{2L}t}.$$

14.30 **(i)**. In view of Example 10.14 we have, for any $u \in \mathcal{S}'(\mathbf{R}^n)$ and $\phi \in \mathcal{S}(\mathbf{R}^n)$,

$$\mathcal{F}(A^*u)(\phi) = (A^*u)(\mathcal{F}\phi) = u(A_*(\mathcal{F}\phi)) = \frac{1}{|\det A|}\,u((A^{-1})^*\mathcal{F}\phi).$$

By means of the change of variable $x = B^{-1}y = {}^tAy$ we obtain, for every $\xi \in \mathbf{R}^n$,

$$\begin{aligned}(A^{-1})^*\mathcal{F}\phi(\xi) &= \mathcal{F}\phi(A^{-1}\xi) = \int e^{-i\langle x, A^{-1}\xi\rangle}\phi(x)\,dx = \int e^{-i\langle Bx,\xi\rangle}\phi(x)\,dx \\ &= \int e^{-i\langle y,\xi\rangle}\phi({}^tAy)|\det A|\,dy = |\det A|\,\mathcal{F}(({}^tA)^*\phi)(\xi).\end{aligned}$$

On the basis of Example 10.14 again, this leads to

$$\begin{aligned}\mathcal{F}(A^*u)(\phi) &= u(\mathcal{F}(({}^tA)^*\phi)) = ({}^tA)_*(\mathcal{F}u)(\phi) = \frac{1}{|\det {}^tA|}\,({}^tA^{-1})^*(\mathcal{F}u)(\phi) \\ &= |\det B|\,B^*(\mathcal{F}u)(\phi).\end{aligned}$$

(ii). The identity is a direct consequence of part (i). For the homogeneity of $\mathcal{F}u$, note that $c^*\mathcal{F}u = c^{-n}\mathcal{F}\circ(c^{-1})^*u = c^{-n}\mathcal{F}c^{-a}u = c^{-n-a}\mathcal{F}u$.
(iii). If A is orthogonal, then ${}^tAA = I$; so $B = {}^tA^{-1} = A$ and $|\det A| = 1$, which implies $\mathcal{F}\circ A^* = A^*\circ\mathcal{F}$. Using the injectivity of $\mathcal{F}$ we see that $A^*u = u$, for all orthogonal A, if and only if $A^*(\mathcal{F}u) = \mathcal{F}u$, for all such A.
(iv). Part (i) implies $A = |\det({}^tA)^{-1}|\,({}^tA)^{-1}$, which gives $|\det A|\,{}^tAA = I$. In turn this leads to $|\det A| = 1$ and thus ${}^tAA = I$.

14.32 The equivalence of (i) and (ii) follows from Problems 14.30.(ii) and (14.30), while Arzelà's or Lebesgue's Dominated Convergence Theorem may be used to prove (i).

14.33 $\mathcal{F}u = P$ is a polynomial function on $\mathbf{R}^n$ that is invariant under all orthogonal transformations in $\mathbf{R}^n$ according to (14.33) and Problem 14.30.(iii). Therefore it is the pullback under $\|\cdot\|^2$ of a polynomial function P_0 on $\mathbf{R}$, that is, $P(\xi) = P_0(\|\xi\|^2)$, for all $\xi \in \mathbf{R}^n$. Indeed, a polynomial function P on $\mathbf{R}^n$ invariant under orthogonal transformations is constant on spheres centered at the origin, and these spheres intersect $L := \mathbf{R}\times\{0\} \subset \mathbf{R}^n$ in a vector and its opposite. The restriction of P to L may be identified with a polynomial P_2 on $\mathbf{R}$ invariant under reflection. The latter condition is satisfied if and only if P_2 is a linear combination of monomials of even degree, that is, $P_2(x) = P_1(x^2)$, for some polynomial P_1 on $\mathbf{R}_{\geq 0}$. One directly verifies that the mapping $P \mapsto P_0 := SP_1$ is a linear bijection.

14.35 For $1 \leq i, j \leq n$ with $i \neq j$, define the vector field v_{ij} on $\mathbf{R}^n$ by $v_{ij}(x) = x_j\,e_i$, where $e_i \in \mathbf{R}^n$ is the ith standard basis vector; that is, the linear mapping $v_{ij} : \mathbf{R}^n \to \mathbf{R}^n$ has a matrix E_{ij} with all entries 0 with the exception of 1 at the position (i, j). Then the flow $(\Phi_t)_{t\in\mathbf{R}}$ in $\mathbf{R}^n$ generated by v_{ij} satisfies

$$\frac{d}{dt}\Phi_t = E_{ij}\,\Phi_t, \qquad \text{and therefore} \qquad \Phi_t = e^{t\,E_{ij}};$$

see [7, Example 2.4.10]. It follows that $\det\Phi_t = 1$, because of [7, Exercise 2.44.(v)]; in other words, $\Phi_t \in \mathbf{SL}(n,\mathbf{R})$. Now apply Theorem 10.16 to obtain $x_j\,\partial_i u = 0$.

For every connected component U of the complement of the union of the hyperplanes $\{ x \in \mathbf{R}^n \mid x_j = 0 \}$ for all $1 \le j \le n$, Problem 12.9 then implies the existence of $c_U \in \mathbf{C}$ such that $u = c_U 1_U$ on U. The invariance of u under rotations in $\mathbf{R}^n$ now gives $u = c_1 1_{\mathbf{R}^n}$ on $\mathbf{R}^n \setminus \{0\}$, for some $c_1 \in \mathbf{C}$. Next, according to Theorem 8.10 there exists a polynomial function p on $\mathbf{R}^n$ such that

$$u = c_1 1_{\mathbf{R}^n} + p(D)\,\delta \in \mathcal{S}'(\mathbf{R}^n). \qquad \text{Hence} \qquad \mathcal{F}u = c_1 (2\pi)^n \delta + p\, 1_{\mathbf{R}^n},$$

while $\mathcal{F}u$ is invariant under $\mathbf{SL}(n, \mathbf{R})$, too, on account of Problem 14.30.(i). On the strength of the preceding arguments we therefore obtain $\mathcal{F}u = p'(D)\,\delta + c_2 1_{\mathbf{R}^n}$. Since δ and $1_{\mathbf{R}^n}$ are homogeneous of degree $-n$ and 0 according to Problem 10.16, respectively, this gives that p equals the constant function c_2. In other words, u is as desired.

Under the final assumption, u is also invariant under dilations. Now $1_{\mathbf{R}^n}$ possesses this invariance and δ not.

Observe that $\mathbf{SL}(1, \mathbf{R}) = \{(1)\}$ and that all factors occurring in the Iwasawa decomposition of $\mathbf{GL}(n, \mathbf{R})$ (see [7, Exercise 5.49]) play a role in this solution.

14.36 (i) $\Leftrightarrow$ (ii). Approximate $\sum_{l=1}^{k} c_l\, \delta_{x_l}$ by functions $\phi \in C_0^\infty(\mathbf{R}^n)$ to show that (i) $\Rightarrow$ (ii), and approximate integrals by Riemann sums to prove the converse.
(i) and (ii) $\Rightarrow$ (iii). Note that the choices $c_1 = 1$, $c_2 = c$, $x_1 = 0$, and $x_2 = x$ in (ii) imply

$$(1 + |c|^2)\, p(0) + c\, p(x) + \overline{c}\, p(-x) \ge 0. \tag{21.16}$$

The choice $c = 0$ gives $p(0) \ge 0$. Selecting $c = 1$ and $c = i$ leads to $p(x) + p(-x) \in \mathbf{R}$ and $i(p(x) - p(-x)) \in \mathbf{R}$, respectively, which give $p(-x) = \overline{p(x)}$. Furthermore, suppose $p(x) \ne 0$ and take $c = -\frac{|p(x)|}{p(x)}$. Then we obtain that $p(0) - |p(x)| \ge 0$ from (21.16). Thus $|p(x)| \le p(0)$, for all $x \in \mathbf{R}^n$.

As a consequence, $p \in \mathcal{S}'(\mathbf{R}^n)$, and thus we may write $p = \mathcal{F}\mu$, for some $\mu \in \mathcal{S}'(\mathbf{R}^n)$. Furthermore, on account of Theorems 11.2 and 11.5 and (14.21) we have, for all $\phi \in C_0^\infty(\mathbf{R}^n)$,

$$\begin{aligned} 0 \le (p * S\phi, S\phi) &= (p * S\phi)(\overline{S\phi}) = p(\phi * \overline{S\phi}) = \mathcal{F}\mu(\phi * \overline{S\phi}) \\ &= \mu(\mathcal{F}\phi\, \mathcal{F}(\overline{S\phi})) = \mu(\mathcal{F}\phi\, \overline{\mathcal{F}\phi}) = \mu(|\mathcal{F}\phi|^2). \end{aligned} \tag{21.17}$$

By continuity this inequality holds for all $\phi \in \mathcal{S}(\mathbf{R}^n)$. Therefore we obtain $(\mu, \chi) \ge 0$, for all $\chi \in \mathcal{S}(\mathbf{R}^n)$ satisfying $\chi \ge 0$; to verify this, approximate χ by the square of $(\chi + \epsilon\, e^{-\|\cdot\|^2})^{\frac{1}{2}} \in \mathcal{S}(\mathbf{R}^n)$, for $\epsilon > 0$. The term containing ϵ is required because $\chi^{\frac{1}{2}}$ itself is nondifferentiable at zeros of χ. On account of Theorem 3.18 we find that μ is a positive Radon measure. Furthermore, if ϕ_ϵ is an even function as in Lemma 2.19, then (21.17) and Problem 14.30.(ii) imply

$$p(0) = \lim_{\epsilon \downarrow 0} (p * \phi_\epsilon, \phi_\epsilon) = \lim_{\epsilon \downarrow 0} \mu(|\epsilon^* \mathcal{F}\phi|^2) \ge \mu(1).$$

Thus, μ is of finite mass and equality holds.
(iii) ⇒ (i). In this case, the identities in (21.17) lead to $(p*\phi, \phi) = \mu(|\mathcal{F} S\phi|^2) \geq 0$, for any $\phi \in C_0^\infty(\mathbf{R}^n)$.

14.38 On account of (14.43) we have, for $\phi \in C_0^\infty(\mathbf{R})$,

$$\begin{aligned} D^2u(\phi) = u(D^2\phi) &= \int_{\mathbf{R}} \int_{\mathbf{R}} g(x,\xi)\, f(\xi)\, d\xi\, D^2\phi(x)\, dx \\ &= \int_{\mathbf{R}\times\mathbf{R}} g(x,\xi)\, (D^2 \otimes I)(\phi \otimes f)(x,\xi)\, d(x,\xi) = (D^2 \otimes I)g(\phi \otimes f) \\ &= \Delta_* 1_J(\phi \otimes f) = \int_J f(\xi)\, \phi(\xi)\, d\xi = f(\phi). \end{aligned}$$

The properties of the distribution g allow us to extend its action to locally integrable functions.

In the particular case that $f \in C_0^\infty(\mathbf{R})$ we may write

$$u(\phi) = \int_{\mathbf{R}} \int_{\mathbf{R}} g(x,\xi)\, f(\xi)\, d\xi\, \phi(x)\, dx = g(\phi \otimes f) = g((\otimes f)\phi) = {}^t(\otimes f) g(\phi).$$

Problems 4.1, 14.27.(iii) or 14.39 imply that E is the desired fundamental solution. The solution $E * \delta_\xi = T_\xi E$ of the inhomogeneous equation follows from (11.24). The uniqueness of g follows from Theorem 9.4. Solving the boundary conditions from (14.43) for a and b leads to

$$g(x,\xi) = \frac{1}{2}(x - 2\xi x + \xi - |x - \xi|).$$

Note that $g = 0$ along the boundary ∂J^2, where $x = 0$ or 1 or $\xi = 0$ or 1.

Now $\mathcal{F} E(\xi) = \frac{1}{\xi^2 + p^2}$. In the case of $p = 1$, the formula for E follows from Problem 14.15; the case of general $p > 0$ then follows from Problem 14.30.(ii). The general solution of the homogeneous equation is given by $x \mapsto ae^{px} + be^{-px}$. For the remainder, use *Mathematica*.

14.43 According to (13.25) the R_+^a for $a \in \mathbf{C}$ are homogeneous distributions. Theorem 14.34 then implies that all of the R_+^a belong to $\mathcal{S}'(\mathbf{R}^{n+1})$. Since $(R_+^a)_{a\in\mathbf{C}}$ is a complex-analytic family of distributions and $\mathcal{F}$ is continuous linear, $(\mathcal{F} R_+^a)_{a\in\mathbf{C}}$ is a similar family.

For the computation of the Fourier transforms we use a variation of the method of Example 14.30. For $\operatorname{Re} a > n + 1$, define R_ϵ^a on $\mathbf{R}^{n+1} \simeq \mathbf{R}^n \times \mathbf{R}$ by

$$R_\epsilon^a(x,t) = e^{-\epsilon t} R_+^a(x,t) \qquad (\epsilon > 0).$$

Then $R_\epsilon^a \in L^1(\mathbf{R}^n)$. Hence, for $(\xi, \tau) \in \mathbf{R}^{n+1}$ and $c(a)$ as in Lemma 13.2, the Fourier transform of this function is given by

$$\mathcal{F} R_\epsilon^a(\xi,\tau) = c(a) \int_{\mathbf{R}^{n+1}} e^{-i\langle x,\xi\rangle - (i\tau+\epsilon)t} H(t - \|x\|)(t^2 - \|x\|^2)^{\frac{a-n-1}{2}}\, d(x,t).$$

On account of the invariance of the integrand under rotations in $\mathbf{R}^n \times \{0\}$ we may assume that $\xi = \|\xi\| e_1$ with e_1 the first standard basis vector in $\mathbf{R}^n$. Upon the change to spherical coordinates in $\mathbf{R}^{n-1}$ and with $B_t^n = \{\, x \in \mathbf{R}^n \mid \|x\| \leq t \,\}$ and c_{n-1} as in (13.37), the integral becomes

$$\begin{aligned}
&\int_{\mathbf{R}_{>0}} \int_{B_t^n} e^{-i\langle x, \xi\rangle - (i\tau+\epsilon)t} (t^2 - \|x\|^2)^{\frac{a-n-1}{2}} \, dx \, dt \\
&= \int_{\mathbf{R}_{>0}} \int_{-t}^{t} \int_{B^{n-1}_{\sqrt{t^2-y^2}}} e^{-iy\|\xi\| - (i\tau+\epsilon)t} (t^2 - y^2 - \|z\|^2)^{\frac{a-n-1}{2}} \, dz \, dy \, dt \\
&= c_{n-1} \int_{\mathbf{R}_{>0}} \int_{-t}^{t} \int_0^{\sqrt{t^2-y^2}} r^{n-2} e^{-iy\|\xi\| - (i\tau+\epsilon)t} (t^2 - y^2 - r^2)^{\frac{a-n-1}{2}} \, dr \, dy \, dt.
\end{aligned}$$

Next set $y = ts$ and $r = t\sqrt{1-s^2}\,\rho$. Then the triple integral takes the form

$$\begin{aligned}
&\int_{\mathbf{R}_{>0}} \int_{-1}^{1} \int_0^1 t^{a-1} (1-s^2)^{\frac{a-2}{2}} \rho^{n-2} (1-\rho^2)^{\frac{a-n-1}{2}} e^{-(is\|\xi\| + i\tau + \epsilon)t} \, d\rho \, ds \, dt \\
&= \int_0^1 \rho^{n-2} (1-\rho^2)^{\frac{a-n-1}{2}} \, d\rho \int_{-1}^{1} (1-s^2)^{\frac{a-2}{2}} \int_{\mathbf{R}_{>0}} t^{a-1} e^{-(is\|\xi\| + i\tau + \epsilon)t} \, dt \, ds.
\end{aligned}$$

In view of (13.34) and (13.30), respectively,

$$\begin{aligned}
\int_0^1 \rho^{n-2} (1-\rho^2)^{\frac{a-n-1}{2}} \, d\rho &= \frac{1}{2} B\Big(\frac{n-1}{2}, \frac{a-n+1}{2}\Big), \\
\int_{\mathbf{R}_{>0}} t^{a-1} e^{-(is\|\xi\| + i\tau + \epsilon)t} \, dt &= (is\|\xi\| + i\tau + \epsilon)^{-a} \Gamma(a).
\end{aligned}$$

Thus we obtain, for $\mathcal{F} R_\epsilon^a(\xi, \tau)$,

$$c(a)\, c_{n-1} \, \frac{1}{2} \, B\Big(\frac{n-1}{2}, \frac{a-n+1}{2}\Big) \Gamma(a) \int_{-1}^{1} (1-s^2)^{\frac{a-2}{2}} (is\|\xi\| + i\tau + \epsilon)^{-a} \, ds.$$

The remaining integral equals

$$B\Big(\frac{a}{2}, \frac{1}{2}\Big) (\|\xi\|^2 + (i\tau + \epsilon)^2)^{-\frac{a}{2}}.$$

Here we used that

$$\int_{-1}^{1} (1-s^2)^{\frac{a-2}{2}} (i\alpha s + \beta)^{-a} \, ds = B\Big(\frac{a}{2}, \frac{1}{2}\Big) (\alpha^2 + \beta^2)^{-\frac{a}{2}},$$

which can be seen by expanding $(1 - (-i\frac{\alpha}{\beta} s))^{-a}$ in its binomial series in s, see [7, Exercise 0.11], interchanging integration and summation and summing the resulting binomial series. If necessary, we have to use analytic continuation in α and/or β. Combining the preceding results and Legendre's duplication formula (13.39), we obtain

$$\mathcal{F} R^a_\epsilon(\xi, \tau) = (\|\xi\|^2 + (i\tau + \epsilon)^2)^{-\frac{a}{2}}.$$

Application of Problem 12.14.(xi) now leads to the desired formula.

Now we give a more conceptual proof. The subgroup $\mathbf{Lo}^\circ$ of the Lorentz group $\mathbf{Lo}$ consisting of all proper orthochronous Lorentz transformations, i.e., all Lorentz transformations that preserve the (solid) forward cone C_+, acts transitively on each of the connected components of the level sets of the quadratic form q with the exception of $q^{-1}(\{0\}) = \bigcup_\pm \partial C_\pm$; $\mathbf{Lo}^\circ$ does, however, act transitively on each of the three sets $\partial C_\pm \setminus \{0\}$ and $\{0\}$.

In view of Problem 14.30.(i) and (ii) it follows that $\mathcal{F} R^a_+$ is invariant under the action of $\mathbf{Lo}^\circ$ (because $\mathbf{Lo}^\circ$ is invariant under taking transposes) and homogeneous of degree $-(n+1) - (a - n - 1) = -a$, respectively. Now suppose $\operatorname{Re} a < 0$ and let $\mathcal{X}$ denote the set of connected components of $\mathbf{R}^{n+1} \setminus \bigcup_\pm \partial C_\pm$. It follows that $\mathcal{F} R^a_+ \in \mathcal{S}'(\mathbf{R}^{n+1})$ has to be of the form

$$\mathcal{F} R^a_+ = \sum_{X \in \mathcal{X}} f_X(a) |q|^{-\frac{a}{2}} 1_X + u_a,$$

where $f_X(a) \in \mathbf{C}$ and $u_a \in \mathcal{S}'(\mathbf{R}^{n+1})$ is $\mathbf{Lo}^\circ$-invariant, homogeneous of degree $-a$ and satisfies $\operatorname{supp} u_a \subset \bigcup_\pm \partial C_\pm = q^{-1}(\{0\})$.

Define $\sigma : \mathbf{R}^n \times \mathbf{R} \to \mathbf{R}^n \times \mathbf{R}$ by $\sigma(\xi, \tau) = (\xi, -\tau)$. Now $\rho_+ \in \mathcal{S}'(\mathbf{R}^{n+1} \setminus \{0\})$ that is defined as u_a on $\mathbf{R}^n \times \mathbf{R}_{>0}$ and $\sigma^* u_a$ on $\mathbf{R}^n \times \mathbf{R}_{<0}$ is invariant under the action of the full Lorentz group $\mathbf{Lo}$. Hence $\rho_+ = q^* w$ for some $w \in \mathcal{D}'(\mathbf{R})$ on account of Problem 13.12. Since $\operatorname{supp} u_a \subset q^{-1}(\{0\})$ we have $\operatorname{supp} w \subset \{0\}$. Moreover, w is homogeneous of degree $-\frac{a}{2} > 0$, because u_a is homogeneous of degree $-a$; thus w must equal 0 on account of Problem 10.16. This implies $\rho_+ = 0$. By a similar argument $\rho_- \in \mathcal{S}'(\mathbf{R}^{n+1} \setminus \{0\})$ that equals u_a on $\mathbf{R}^n \times \mathbf{R}_{<0}$ and $\sigma^* u_a$ on $\mathbf{R}^n \times \mathbf{R}_{>0}$ must be equal to 0. Therefore u_a vanishes away from the origin. The homogeneity of u_a now implies that $u_a = 0$ in $\mathcal{S}'(\mathbf{R}^{n+1})$.

Our next goal is the evaluation of the $f_X(a)$. Note that $\mathcal{F} R^a_+$ is a continuous function, since $\operatorname{Re} a < 0$. Therefore the product $\mathcal{F} R^a_+ \, \mathcal{F} R^b_+$ is well-defined if $\operatorname{Re} a$ and $\operatorname{Re} b < 0$. Although the conditions in Theorem 14.33 are not satisfied, its conclusion $\mathcal{F}(R^a_+ * R^b_+) = \mathcal{F} R^a_+ \, \mathcal{F} R^b_+$ nevertheless holds in this case, as we will now prove. Since $C_0^\infty(\mathbf{R}^{n+1})$ is dense in $\mathcal{S}'(\mathbf{R}^{n+1})$, there exists a sequence $(\psi^b_k)_{k \in \mathbf{N}}$ in $C_0^\infty(\mathbf{R}^{n+1})$ such that $\psi^b_k \to R^b_+$ in $\mathcal{S}'(\mathbf{R}^{n+1})$ as $k \to \infty$. Let $\phi \in \mathcal{S}(\mathbf{R}^{n+1})$. Then

$$\mathcal{F}(R^a_+ * R^b_+)(\phi) = (R^a_+ * R^b_+)(\mathcal{F}\phi) = R^a_+(S R^b_+ * \mathcal{F}\phi) = R^a_+(\lim_{k\to\infty} S\psi^b_k * \mathcal{F}\phi).$$

Here S denotes reflection in the origin. Let θ be a C^∞ function that is supported on a sufficiently small open neighborhood of $\overline{C_+}$ and equals 1 on C_+. Then we have that $\theta\,(S\psi^b_k * \mathcal{F}\phi)$ converges to $\theta\,(S R^b_+ * \mathcal{F}\phi)$ in $\mathcal{S}(\mathbf{R}^{n+1})$ as $k \to \infty$. So

$$\mathcal{F}(R^a_+ * R^b_+)(\phi) = R^a_+\Big(\lim_{k\to\infty} \theta\,(S\psi^b_k * \mathcal{F}\phi)\Big) = \lim_{k\to\infty} \theta\, R^a_+(S\psi^b_k * \mathcal{F}\phi)$$

$$= \lim_{k\to\infty} R_+^a(S\psi_k^b * \mathcal{F}\phi) = \lim_{k\to\infty} (R_+^a * \psi_k^b)(\mathcal{F}\phi) = \lim_{k\to\infty} \mathcal{F}(R_+^a * \psi_k^b)(\phi).$$

Now Theorem 14.33 can be applied to obtain

$$\mathcal{F}(R_+^a * R_+^b)(\phi) = \lim_{k\to\infty} (\mathcal{F}\psi_k^b)\,\mathcal{F}R_+^a(\phi) = \mathcal{F}R_+^a\,\mathcal{F}R_+^b(\phi).$$

For the last equality, Arzelà's Dominated Convergence Theorem has been used.

As a consequence of (13.17) we now obtain

$$\mathcal{F}R_+^{a+b} = \mathcal{F}(R_+^a * R_+^b) = \mathcal{F}R_+^a\,\mathcal{F}R_+^b; \qquad \text{so} \qquad f_X(a+b) = f_X(a)\,f_X(b).$$

Note that f_X has an analytic continuation to $\mathbf{C}$; therefore, $f_X(a) = f_X(1)^a$ on account of [7, Exercise 0.2 (iii)]. In addition $\mathcal{F}R_+^{-2} = \mathcal{F}(\Box\delta) = -q\,\mathcal{F}\delta = -q$; therefore $f_X(1)^{-2} = f_X(-2) = -\operatorname{sgn} q|_X$. This leads to $f_X(1) \in \{\pm i\}$ if $X = C_\pm$ and $f_X(1) \in \{\pm 1\}$ if X equals one of the connected components of $\mathbf{R}^{n+1} \setminus \overline{\bigcup_\pm C_\pm}$.

Let $X_0 = \mathbf{R}^{n+1} \setminus \overline{\bigcup_\pm C_\pm}$. Observe that X_0 is connected if and only if $n > 1$. If $n = 1$, then X_0 consists of two connected components X_1 and X_2. Due to the $\mathbf{Lo}^\circ$-invariance, the functions f_{X_1} and f_{X_2} coincide. Now suppose $n > 1$. Since R_+^a is not invariant under reflection of t, this precludes the invariance of $\mathcal{F}R_+^1$. Since X_0 is invariant under reflection of t, it follows that $f_{C_+}(1)$ and $f_{C_-}(1)$ are distinct; hence $f_{C_+}(1) = -f_{C_-}(1)$.

From now on we will assume $n > 1$. Let $\phi : \mathbf{R}^{n+1} \to \mathbf{R}$ be the function $e^{-\frac{\|\cdot\|^2}{2}}$. Then $\Box\phi = (n-1+q)\phi$. Therefore

$$R_+^{-2}(\phi) = \delta(\Box\phi) = n - 1 > 0.$$

The function $\mathbf{R} \ni a \mapsto R_+^a(\phi)$ is real-analytic and real-valued; hence $R_+^a(\phi) > 0$ for a in a sufficiently small open neighborhood of -2 in $\mathbf{R}$. In addition, since $\mathcal{F}\phi = (2\pi)^{\frac{n}{2}}\phi$ and ϕ is invariant under reflection in t, we find for a with $\operatorname{Re} a < 0$ that

$$(2\pi)^{\frac{n}{2}} R_+^a(\phi) = R_+^a(\mathcal{F}\phi) = \mathcal{F}R_+^a(\phi) = f_{X_0}(a)(-q)^{-\frac{a}{2}} 1_{X_0}(\phi).$$

Note that $(-q)^{-\frac{a}{2}} 1_{X_0}(\phi) > 0$ for $a < 0$. Therefore $f_{X_0}(a) > 0$ for a in a sufficiently small open neighborhood of -2 in $\mathbf{R}$ and thus $f_{X_0}(a) = 1$ first for these a and then for all $a \in \mathbf{C}$ by analytic continuation.

To determine $f_{C_\pm}(1)$ we consider some integrals. For $\operatorname{Re} a > n+1$,

$$f_{C_\pm}(a) = \mathcal{F}R_+^a(0, \pm 1) = \lim_{\epsilon\downarrow 0} c(a) \int_{C_+} e^{-(\pm i+\epsilon)t}(t^2 - \|x\|^2)^{\frac{a-n-1}{2}}\, d(x,t),$$

which by the first part of the calculation in the preceding proof equals

$$\lim_{\epsilon\downarrow 0} c(a)\, c_{n-1}\frac{1}{2} B\Big(\frac{n-1}{2}, \frac{a-n+1}{2}\Big)\Gamma(a) \int_{-1}^{1} (1-s^2)^{\frac{a-2}{2}} (\pm i + \epsilon)^{-a}\, ds$$
$$= \gamma(a) e^{\mp i\frac{\pi}{2}a},$$

where γ is a positive function. Because the absolute value of $f_{C_\pm}(a)$ equals 1, γ must be equal to the constant function 1. By analytic continuation $f_{C_\pm}(1) = e^{\mp i\frac{\pi}{2}} = \mp i$.

14.44 From Example 14.1 or Exercise 14.27.(ii) we obtain $\mathcal{F}\,\text{sinc} = \pi\, 1_{[-1,1]}$. Evaluating this identity at 0 gives $\int_{\mathbf{R}} \text{sinc}\, t\, dt = \pi$, while Parseval's formula from Theorem 14.32 implies $\int_{\mathbf{R}} \text{sinc}^2\, t\, dt = \pi$, since sinc $\in L^2(\mathbf{R})$. The substitution of variables $t = ax$ leads to the desired formulas. An extension of (14.22) to the case at hand in combination with Exercise 1.6.(ii) implies $\mathcal{F}\,\text{sinc}^2(\xi) = \frac{\pi}{2}\max\{0, 2-|\xi|\}$; hence, the integral now follows by setting $\xi = 0$.

14.49 **(i)**. Note that $\mathcal{H}\phi \in C^\infty(\mathbf{R})$ on account of Theorem 11.2. From Example 14.29 it is known that PV $\frac{1}{y} \in \mathcal{S}'(\mathbf{R})$. Therefore Theorem 14.33 implies $\mathcal{H}\phi \in \mathcal{S}'(\mathbf{R})$ as well as $\mathcal{F} \circ \mathcal{H}(\phi) = \mathcal{F}(\mathcal{H}\phi) = \frac{1}{\pi}\mathcal{F}\phi\,\mathcal{F}\,\text{PV}\,\frac{1}{y} = -i\ \text{sgn} \circ \mathcal{F}\phi$, in view of (14.37).
(ii). Apply Parseval's formula (14.20) in order to obtain $\|\mathcal{F}(\mathcal{H}\phi)\| = \|\mathcal{F}\phi\| = \sqrt{2\pi}\,\|\phi\|$. This leads to $\mathcal{F}(\mathcal{H}\phi) \in L^2(\mathbf{R})$, and invoking Parseval's formula once more, one gets the desired equality. For the extension of $\mathcal{H}$ to a continuous linear mapping from $L^2(\mathbf{R})$ into itself, one copies the proof of Theorem 14.32 with $\widetilde{\mathcal{F}}$ replaced by $\mathcal{H}$. Finally, $\mathcal{H}^2 = -I$ follows by means of Fourier inversion from

$$\mathcal{F} \circ \mathcal{H}^2 = (\mathcal{F} \circ \mathcal{H}) \circ \mathcal{H} = -i\ \text{sgn} \circ \mathcal{F} \circ \mathcal{H} = -(\text{sgn})^2 \circ \mathcal{F} = -\mathcal{F}.$$

(iii). From Problem 14.15 we obtain $\mathcal{F} f = \pi\, e^{-|\cdot|}$, and Problem 14.17 then leads to $\mathcal{F} g = -i\ \text{sgn} \circ \mathcal{F} f = \mathcal{F}\mathcal{H} f$.
(iv). From Problem 14.27.(i) it is known that $\mathcal{F}\cos = \pi(\delta_{-1} + \delta_1)$ and $\mathcal{F}\sin = \pi i(\delta_{-1} - \delta_1)$; this implies $\mathcal{F}\cos = -i\ \text{sgn} \circ \mathcal{F}\sin$. On the other hand, the identity $\int_{\mathbf{R}} \text{sinc}\, y\, dy = \pi$ from Problem 14.44 leads to

$$\begin{aligned}\mathcal{H}\cos &= \frac{1}{\pi}\ \text{PV}\int_{\mathbf{R}} \frac{\cos(\cdot - y)}{y}\, dy = \text{PV}\int_{\mathbf{R}} \frac{\cos y}{y}\, dy\, \frac{\cos}{\pi} + \int_{\mathbf{R}} \frac{\sin y}{y}\, dy\, \frac{\sin}{\pi} \\ &= \sin.\end{aligned}$$

Finally, use $\mathcal{H}^2 = -I$.
(v). The first formula follows from the definition of $\mathcal{H}u$ and Problem 1.1 and the second from part (i) and (14.4), the third from the identity $\|\mathcal{H}u\|^2 = \|u\|^2$ in part (ii) and the fourth from Parseval's formula and the third.
(vi). The former equality follows by the inverse Fourier transform from the second and first in part (iv). Finally, for the latter, note that for $\phi \in \mathcal{S}(\mathbf{R})$,

$$\begin{aligned} i\int_{\mathbf{R}} \phi(x)\log\left|\frac{x-a}{x+a}\right| dx &= \mathcal{F}\Big(\frac{\sin a\,\cdot}{|\cdot|}\Big)(\phi) = \int_{\mathbf{R}} \frac{\sin a\xi}{|\xi|}\int_{\mathbf{R}} e^{-ix\xi}\phi(x)\, dx\, d\xi \\ &= \int_{\mathbf{R}} \phi(x)\int_{\mathbf{R}} e^{-ix\xi}\,\frac{\sin a\xi}{|\xi|}\, d\xi\, dx = -i\int_{\mathbf{R}} \phi(x)\int_{\mathbf{R}} \sin x\xi\,\frac{\sin a\xi}{|\xi|}\, d\xi\, dx.\end{aligned}$$

Actually, one should insert a convergence factor in the variable ξ in order to justify the interchange in the order of integrations, and use a limit argument. The resulting integral is of Frullani type; see [7, Example 2.10.8].
(vii). From $\mathcal{F}(\frac{\sin a\cdot}{\cdot}) = \pi\, 1_{[-a,a]}$ we get $\mathcal{H}(\frac{\sin a\cdot}{\cdot}) = \mathcal{F}^{-1}(-\pi i \operatorname{sgn} 1_{[-a,a]})$.

14.50 The integral equation can be written as $a\,\psi + b\,\mathcal{H}\psi = \phi$, which leads to $a\,\mathcal{H}\psi - b\,\psi = \mathcal{H}\phi$ on account of Problem 14.49.(ii). Therefore $(a^2+b^2)\psi = a\,\phi - b\,\mathcal{H}\phi$, and this proves that $\psi \in C^\infty(\mathbf{R})$.

14.52 On account of (14.29) one gets, for every $c \in \mathbf{R}$,

$$\begin{aligned}\mathcal{F}\circ T_a \circ \mathcal{H} &= e_{-ia}\circ -i \operatorname{sgn}\circ\mathcal{F} = -i \operatorname{sgn}\circ e_{-ia}\circ\mathcal{F} = -i \operatorname{sgn}\circ\mathcal{F}\circ T_a\\ &= \mathcal{F}\circ\mathcal{H}\circ T_a.\end{aligned}$$

Theorem 11.3 now implies the existence and uniqueness of u. In view of Problem 14.30.(ii) one obtains, for all $c \in \mathbf{R}$,

$$\begin{aligned}\mathcal{F}\circ c^*\circ\mathcal{H} &= c^{-1}(c^{-1})^*\circ\mathcal{F}\circ\mathcal{H} = c^{-1}(c^{-1})^*\circ -i \operatorname{sgn}\circ\mathcal{F}\\ &= -i \operatorname{sgn}\circ c^{-1}(c^{-1})^*\circ\mathcal{F} = -i \operatorname{sgn}\circ\mathcal{F}\circ c^* = \mathcal{F}\circ\mathcal{H}\circ c^*.\end{aligned}$$

On the one hand, one gets by applying (11.1), for all $\phi \in C_0^\infty(\mathbf{R})$,

$$(c^*\circ\mathcal{H})\phi(0) = \mathcal{H}\phi(0) = u*\phi(0) = u(S\phi).$$

And on the other, by Theorem 10.8,

$$\begin{aligned}(\mathcal{H}\circ c^*)\phi(0) &= \mathcal{H}(c^*\phi)(0) = u(Sc^*\phi) = u(c^*S\phi) = c_*u(S\phi)\\ &= c^{-1}(c^{-1})^*u(S\phi).\end{aligned}$$

In other words, $u = c^{-1}(c^{-1})^*u$, that is, $c^*u = c^{-1}u$. A combination of Euler's Theorem, Theorem 10.17, and Problem 9.5 now implies the desired formula for u.

14.54 (i). Replace $\epsilon > 0$ in Example 14.30 by $-iz$, where $z = x + it \in H_+$; then $\operatorname{Re} -iz = t > 0$. Hence, for ξ and $x \in \mathbf{R}$,

$$H_{-iz}(\xi) = e^{iz\xi}\,H(\xi) \qquad\text{and}\qquad \mathcal{F}H_{-iz}(x) = \frac{1}{i(x-z)}.$$

As a consequence, the desired formula comes down to $\mathcal{F}H_{-iz}(\phi) = H_{-iz}(\mathcal{F}\phi)$.
(ii). If $z \in H_-$, replace ϕ by $S\phi$ and z by $-z \in H_+$ in order to obtain

$$-\frac{1}{i}\int_{\mathbf{R}}\frac{\phi(y)}{y-z}\,dy = \int_{\mathbf{R}_{<0}} e^{iz\xi}\,\mathcal{F}\phi(\xi)\,d\xi.$$

By differentiation under the integral sign it now follows that Φ satisfies the Cauchy–Riemann equation (12.6) on $\mathbf{C}\setminus\mathbf{R}$.
(iii). Note that $\frac{1}{y-z} - \frac{1}{y-\bar z} = 2i\,\frac{t}{|y-z|^2} = 2i\,\frac{t}{(x-y)^2+t^2}$. Therefore

$$u_\phi(x,t) = \frac{1}{\pi}\int_{\mathbf{R}} \frac{t}{(x-y)^2+t^2}\,\phi(y)\,dy = P_t * \phi(x).$$

On the other hand, $iz = -t + ix$ and $i\overline{z} = t + ix$ imply

$$\begin{aligned} u_\phi(x,t) &= \frac{1}{2\pi}\int_{\mathbf{R}_{>0}} e^{-t\xi}e^{ix\xi}\,\mathcal{F}\phi(\xi)\,d\xi + \frac{1}{2\pi}\int_{\mathbf{R}_{<0}} e^{t\xi}e^{ix\xi}\,\mathcal{F}\phi(\xi)\,d\xi \\ &= \frac{1}{2\pi}\int_{\mathbf{R}} e^{-t|\xi|}\,e^{ix\xi}\,\mathcal{F}\phi(\xi)\,d\xi. \end{aligned}$$

Now one may either apply Problem 14.51.(iii) or the following argument. We have that $\lim_{t\downarrow 0} e^{-t|\xi|} = 1$ uniformly for ξ in compact sets in $\mathbf{R}$, while $0 < e^{-t|\xi|} \le 1$. Hence, by Arzelà's or Lebesgue's Dominated Convergence Theorem and Fourier inversion,

$$\lim_{t\downarrow 0}\frac{1}{2\pi}\int_{\mathbf{R}} e^{-t|\xi|}\,e^{ix\xi}\,\mathcal{F}\phi(\xi)\,d\xi = \frac{1}{2\pi}\int_{\mathbf{R}} e^{ix\xi}\,\mathcal{F}\phi(\xi)\,d\xi = \phi(x).$$

Similarly, $-i\big(\frac{1}{y-z} + \frac{1}{y-\overline{z}}\big) = -2i\,\frac{y-x}{|y-z|^2} = 2i\,\frac{x-y}{(x-y)^2+t^2}$. Therefore

$$v_\phi(x,t) = \frac{1}{\pi}\int_{\mathbf{R}} \frac{x-y}{(x-y)^2+t^2}\,\phi(y)\,dy = Q_t * \phi(x).$$

Furthermore,

$$\begin{aligned} v_\phi(x,t) &= -\frac{i}{2\pi}\int_{\mathbf{R}_{>0}} e^{-t\xi}e^{ix\xi}\,\mathcal{F}\phi(\xi)\,d\xi + \frac{i}{2\pi}\int_{\mathbf{R}_{<0}} e^{t\xi}e^{ix\xi}\,\mathcal{F}\phi(\xi)\,d\xi \\ &= -\frac{i}{2\pi}\int_{\mathbf{R}} e^{-t|\xi|}\,e^{ix\xi}\,\mathrm{sgn}(\xi)\,\mathcal{F}\phi(\xi)\,d\xi = \frac{1}{2\pi}\int_{\mathbf{R}} e^{-t|\xi|}\,e^{ix\xi}\,\mathcal{F}(\mathcal{H}\phi)(\xi)\,d\xi. \end{aligned}$$

Proceeding as before, one obtains

$$\lim_{t\downarrow 0}\frac{1}{2\pi}\int_{\mathbf{R}} e^{-t|\xi|}\,e^{ix\xi}\,\mathcal{F}(\mathcal{H}\phi)(\xi)\,d\xi = \frac{1}{2\pi}\int_{\mathbf{R}} e^{ix\xi}\,\mathcal{F}(\mathcal{H}\phi)(\xi)\,d\xi = \mathcal{H}\phi(x).$$

(iv). In this case, $\Phi_-(\overline{z}) = -\overline{\Phi_+(z)}$. Hence,

$$u_\phi = \frac{1}{2\pi}(\Phi_+ + \overline{\Phi_+}) = \frac{1}{\pi}\,\mathrm{Re}\,\Phi \qquad \text{and} \qquad v_\phi = -\frac{i}{2\pi}(\Phi_+ - \overline{\Phi_+}) = \frac{1}{\pi}\,\mathrm{Im}\,\Phi.$$

The real and imaginary parts of a complex-differentiable function are harmonic.
(v). Direct consequence from

$$\Phi(z) = \frac{1}{i}\int_{-a}^{a}\frac{1}{y-z}\,dy = i\,\log\frac{z+a}{z-a} \qquad (z \in \mathbf{C}\setminus\mathbf{R}).$$

14.58 (i). Use spherical coordinates.
(ii). Clearly, v is homogeneous of degree -2 on $\mathbf{R}^n \setminus \{0\}$. Theorem 14.34 then implies that $v \in \mathcal{S}'(\mathbf{R}^n)$.

(iii). We have

$$-\|\xi\|^2\,\mathcal{F}E(\xi) = 1 \quad\Longrightarrow\quad \mathcal{F}(\Delta E) = \mathcal{F}\delta \quad\Longrightarrow\quad \Delta E = \delta.$$

The homogeneity of E of degree $2-n$ is a consequence of Problem 14.30.(ii).
(iv) – (vi). Let E be a homogeneous fundamental solution of Δ. Then $E \in \mathcal{S}'(\mathbf{R}^n)$ and so $\mathcal{F}E(\xi) = -\frac{1}{\|\xi\|^2}$, for all $\xi \in \mathbf{R}^n \setminus \{0\}$. It follows that E is homogeneous of degree $2-n$, and E is invariant under rotations on account of Problem 14.30.(iii). The hypoellipticity of Δ guarantees that the restriction of E to $\mathbf{R}^n \setminus \{0\}$ is a C^∞ function. Contributions to E supported at 0 are ruled out by the homogeneity of E: derivatives of δ are homogeneous of degree $\le -n$; see Problem 10.16. Combination of these data leads to the existence of a constant $c \in \mathbf{R}$ such that $E : x \mapsto c\,\|x\|^{2-n}$.
(vii). Applying Parseval's formula (14.20), with $\phi = u_1$ as in Example 14.11 and $\psi = E$, we get

$$c \int_{\mathbf{R}^n} \frac{e^{-\|x\|^2/2}}{\|x\|^{n-2}}\,dx = -\frac{1}{(2\pi)^{n/2}} \int_{\mathbf{R}^n} \frac{e^{-\|\xi\|^2/2}}{\|\xi\|^2}\,d\xi.$$

Substituting spherical coordinates in $\mathbf{R}^n$ and using (13.31) and (13.37), we obtain the desired value for c.

14.59 Because $x \mapsto E(x,\cdot)$ is a C^1 family in $\mathcal{S}'(\mathbf{R})$, partial Fourier transform of E with respect to the second variable is admissible on the strength of Theorem 14.24. By interchanging differentiation and integration one obtains, for $\phi \in \mathcal{S}(\mathbf{R})$,

$$\mathcal{F}\partial_x E(\phi) = -E(\partial_x \mathcal{F}\phi) = -E(\mathcal{F}\partial_x\phi) = \partial_x \mathcal{F}E(\phi),$$

while the same Theorem 14.24 implies $\mathcal{F}(\frac{1}{i}\partial_y E) = \eta\,\mathcal{F}E$. Furthermore, $\mathcal{F}\delta(y) = 1$, on the basis of Lemma 14.4 or formula (14.30). A combination of these arguments leads to the desired equation

$$\partial_x u - \eta\, u = 2\delta \qquad \text{with} \qquad u = \mathcal{F}E(\cdot,\eta) \in \mathcal{S}'(\mathbf{R}).$$

For $\phi \in \mathcal{S}(\mathbf{R})$ one has

$$(\partial_x u - \eta\, u)(\phi) = -u(\partial_x\phi + \eta\,\phi) = -u(e^{-x\eta}\,\partial_x(e^{x\eta}\,\phi)) = e^{x\eta}\,\partial_x(e^{-x\eta}\,u)(\phi);$$

hence, the differential equation takes the form

$$e^{x\eta}\,\partial_x(e^{-x\eta}\,u) = 2\delta, \qquad \text{that is,} \qquad \partial_x(e^{-x\eta}\,u) = 2e^{-x\eta}\,\delta = 2\delta.$$

In view of Example 4.2, a solution is given by $e^{-x\eta}\,u = 2H$, and Theorem 4.3 asserts that every solution is of the form

$$x \mapsto \mathcal{F}E(x,\eta) = u(x) = 2(c(\eta) + H(x))e^{x\eta}.$$

On account of Problem 14.24, the function $\eta \mapsto e^{x\eta}$ does not define a tempered distribution on $\mathbf{R}$. With the choice $c(\eta) = -H(\eta)$, however, one obtains

$$\eta \mapsto \mathcal{F}E(x,\eta) = 2(H(x) - H(\eta))\, e^{x\eta} = \begin{cases} 0 & \text{if} \quad x\eta > 0, \\ 2e^{x\eta} & \text{if} \quad \eta < 0 < x, \\ -2e^{x\eta} & \text{if} \quad x < 0 < \eta. \end{cases}$$

This particular solution of the differential equation defines a tempered distribution on $\mathbf{R}$, as is required by assumption, and from its description one obtains, for $(x,\eta) \in \mathbf{R}^2$,

$$\mathcal{F}E(x,\eta) = -2\,\mathrm{sgn}(\eta) H(-x\eta)\, e^{x\eta} = -2\,\mathrm{sgn}(\eta)\, H(-x\,\mathrm{sgn}(\eta))\, e^{x\eta}.$$

Consider $\phi \in \mathcal{S}(\mathbf{R}^2)$. Then $y \mapsto \phi(x,y)$ belongs to $\mathcal{S}(\mathbf{R})$, for every $x \in \mathbf{R}$. Since $\eta \mapsto \mathcal{F}E(x,\eta)$ defines an element in $\mathcal{S}'(\mathbf{R})$, Theorem 14.24 now implies, for all $x \in \mathbf{R}$,

$$E(x,\cdot)(\phi(x,\cdot)) = \frac{1}{2\pi}(\mathcal{F} \circ S \circ \mathcal{F}E(x,\cdot))(\phi(x,\cdot)) = \frac{1}{2\pi}(\mathcal{F}E(x,\cdot))(S\mathcal{F}\phi(x,\cdot)).$$

$\mathcal{F}E$ being a locally integrable and rapidly decreasing function on $\mathbf{R}^2$, one obtains $\mathcal{F}E \in \mathcal{S}'(\mathbf{R}^2)$ and thus $E \in \mathcal{S}'(\mathbf{R}^2)$ too. Accordingly, the admissible interchange of the order of integration yields

$$\begin{aligned}
E(\phi) &= \frac{1}{2\pi}(\mathcal{F}E)(S\mathcal{F}\phi) = \frac{1}{2\pi}\int_{\mathbf{R}^2} (\mathcal{F}E)(x,\eta)(S\mathcal{F}\phi)(x,\eta)\, d(x,\eta) \\
&= \frac{2}{2\pi}\int_{\mathbf{R}}\int_{\mathbf{R}} -\mathrm{sgn}(\eta)\, H(-x\,\mathrm{sgn}(\eta))\, e^{x\eta} \int_{\mathbf{R}} e^{iy\eta}\phi(x,y)\, dy\, dx\, d\eta \\
&= \frac{1}{\pi}\int_{\mathbf{R}}\int_{\mathbf{R}}\int_{\mathbf{R}} -\mathrm{sgn}(\eta)\, H(-x\,\mathrm{sgn}(\eta))\, e^{(x+iy)\eta}\phi(x,y)\, d\eta\, dx\, dy \\
&= \frac{1}{\pi}\int_{\mathbf{R}}\int_{\mathbf{R}}\Big(H(x)\int_{-\infty}^{0} e^{(x+iy)\eta}\, d\eta - H(-x)\int_{0}^{\infty} e^{(x+iy)\eta}\, d\eta\Big)\phi(x,y)\, dx\, dy \\
&= \frac{1}{\pi}\int_{\mathbf{R}}\Big(\int_{0}^{\infty} \frac{\phi(x,y)}{x+iy}\, dx + \int_{-\infty}^{0} \frac{\phi(x,y)}{x+iy}\, dx\Big)\, dy \\
&= \frac{1}{\pi}\int_{\mathbf{R}^2} \frac{1}{x+iy}\,\phi(x,y)\, d(x,y) = \frac{1}{\pi}\frac{1}{x+iy}(\phi) = \frac{1}{\pi z}(\phi).
\end{aligned}$$

Furthermore, Theorem 14.34 confirms once more the fact that E belongs to $\mathcal{S}'(\mathbf{C})$. Finally, suppose $\epsilon > 0$ and consider $E(\epsilon,\xi) = \frac{1}{\pi}\frac{1}{i\xi+\epsilon}$. Then, in the notation of Example 14.30,

$$\mathcal{F}E(\epsilon,x) = -2\,\mathrm{sgn}(x)\, H(-\epsilon\,\mathrm{sgn}(x))\, e^{\epsilon x} = 2H_\epsilon(-x) = 2SH_\epsilon(x).$$

Equation (14.39) now follows by Fourier inversion; indeed,

$$\pi\, E(\epsilon,\cdot) = \frac{1}{2}\mathcal{F}S\mathcal{F}E(\epsilon,\cdot) = \mathcal{F}S^2H_\epsilon = \mathcal{F}H_\epsilon.$$

14.60 **(i).** R^a is continuous on $\mathbf{R}^n \setminus \{0\}$, for all $a \in \mathbf{C}$, and therefore it is locally integrable on this set. To prove the local integrability of R^a at 0 if $\mathrm{Re}\, a > 0$, we integrate this function over the unit ball B centered at 0, using spherical coordinates (note that $\Gamma(\frac{a}{2}) \neq 0$ if $\mathrm{Re}\, a > 0$):

$$\int_B |R^a(x)|\, dx = c(a) \int_B \|x\|^{\mathrm{Re}\, a - n}\, dx = \lim_{\epsilon \downarrow 0} \frac{2}{\Gamma(\frac{a}{2})} \int_\epsilon^1 r^{\mathrm{Re}\, a - 1}\, dr$$
$$= \frac{2}{\Gamma(\frac{a}{2})\, \mathrm{Re}\, a} \lim_{\epsilon \downarrow 0} (1 - \epsilon^{\mathrm{Re}\, a}) < \infty.$$

Obviously, R^a is homogeneous on $\mathbf{R}^n \setminus \{0\}$ if $\mathrm{Re}\, a > 0$; according to Theorem 14.34 this implies that the distribution R^a is temperate. We may therefore test it with $e^{-\|\cdot\|^2} \in \mathcal{S}(\mathbf{R}^n)$, which yields

$$R^a(e^{-\|\cdot\|^2}) = c(a) \int_{\mathbf{R}^n} \|x\|^{a-n} e^{-\|x\|^2}\, dx = \frac{2}{\Gamma(\frac{a}{2})} \int_{\mathbf{R}_{>0}} r^{a-1} e^{-r^2}\, dr$$
$$= \frac{2}{\Gamma(\frac{a}{2})} \int_{\mathbf{R}_{>0}} s^{\frac{a-1}{2}} e^{-s} \frac{1}{2} s^{-\frac{1}{2}}\, ds = \frac{1}{\Gamma(\frac{a}{2})} \int_{\mathbf{R}_{>0}} s^{\frac{a}{2}-1} e^{-s}\, ds = 1,$$

in view of (13.30).

(ii). It is straightforward that q is a C^∞ function satisfying $Dq(x) = 2\,{}^t x : \mathbf{R}^n \to \mathbf{R}$, and the latter is $\neq 0$ if $x \neq 0$. Hence $Dq(x)$ is surjective for all $x \in \mathbf{R}^n \setminus \{0\}$; in other words, $q : \mathbf{R}^n \setminus \{0\} \to \mathbf{R}$ is a C^∞ submersion. The complex-analyticity of c follows from Corollary 13.6. Comparison of (13.5) with the definition of R^a directly implies the following identity of functions on $\mathbf{R}^n \setminus \{0\}$ and hence also of distributions belonging to $\mathcal{D}'(\mathbf{R}^n \setminus \{0\})$:

$$R^a = d(a)\, q^* \Big(\chi_+^{\frac{a-n+2}{2}} \Big),$$

initially for $\mathrm{Re}\, a > 0$, but then by analytic continuation of identities for all $a \in \mathbf{C}$.

(iii). We now come to the first proof of the formula for ΔR^a. For $x \in \mathbf{R}^n \setminus \{0\}$ we have

$$\begin{aligned}\Delta R^a(x) &= c(a) \Delta(\|x\|^{a-n}) = c(a)\, \mathrm{div}((a-n)\|x\|^{a-n-2}\, x) \\ &= c(a)(n(a-n)\|x\|^{a-n-2} + (a-n)(a-n-2)\|x\|^{a-n-2}) \\ &= c(a)(a-2)(a-n)\|x\|^{a-n-2} = 2(a-n)c(a-2)\|x\|^{a-2-n} \\ &= 2(a-n)R^{a-2}(x).\end{aligned}$$

For the second proof we denote the variable in the range space $\mathbf{R}$ of q also by q. With the notation $v = \chi_+^{\frac{a-n+2}{2}}$ we get $v' = \chi_+^{\frac{a-n}{2}}$ by (13.6); on account of (13.10) this implies $q\, v'' = \frac{a-n-2}{2}\, v'$. In turn, that leads to

$$2n\, v' + 4q\, v'' = (2n + 2a - 2n - 4)\, v' = 2(a-2)\, v' = 2(a-2)\, \chi_+^{\frac{a-n}{2}}.$$

Now applying Problem 10.20 with $P = \Delta$, $B = A = I$, and $\alpha = q$, we obtain, using part (ii) and (13.31),

$$\Delta R^a = d(a)\,\Delta(q^*v) = d(a)2(a-2)\,q^*\Big(\chi_+^{\frac{a-n}{2}}\Big) = 2(a-2)\frac{d(a)}{d(a-2)}\,R^{a-2}$$
$$= 2(a-2)\frac{\Gamma(\frac{a-n}{2}+1)}{\Gamma(\frac{a}{2})}\frac{\Gamma(\frac{a}{2}-1)}{\Gamma(\frac{a-n}{2})}\,R^{a-2} = 2(a-n)\,R^{a-2}.$$

Equation (14.50) follows by repeated application of (14.49), while (14.50) makes sense since $\mathrm{Re}(a+2j-n) \le \mathrm{Re}(a+2k-n) \le 2-n < 0$; furthermore, it is a routine computation to see that the extension is actually independent of the choice of k.

(iv). $R^0 = 0$ on $\mathbf{R}^n \setminus \{0\}$, because $a \mapsto \frac{1}{\Gamma(\frac{a}{2})}$ has a zero at $a = 0$, according to Corollary 13.6; thus, $\mathrm{supp}\, R^0 \subset \{0\}$. Note that $R^a(\phi) \ge 0$ if $\phi \in C_0^\infty(\mathbf{R}^n)$ is positive, that is, the distributions R^a are positive. Theorems 8.10 and 3.18 then imply $R^0 = c_0\,\delta$. Furthermore, $1 = R^0(e^{-\|\cdot\|^2}) = c_0$ by analytic continuation of the identity in part (i); in other words, $R^0 = \delta$. The second identity now follows as a direct consequence of part (iii) and the substitution $k-j \mapsto j$ of the index in the product.

(v). Both properties follow by means of analytic continuation of identities.

(vi). Direct computation.

(vii). $\mathcal{F}R^a$ is radial and homogeneous of degree $(n-a)-n$ on account of part (v) and Problem 14.30.(ii) and (iii). Hence there exists a constant $p(a) \in \mathbf{C}$ such that $\mathcal{F}R^a = p(a)\,R^{n-a}$. Now, the Fourier transform of the function $e^{-\|\cdot\|^2}$ on $\mathbf{R}^n$ equals $\pi^{n/2}e^{-\|\cdot\|^2/4}$ according to Example 14.11. Furthermore, from the homogeneity of R^{n-a} of degree $-a$ and part (i) we obtain $R^{n-a}(e^{-\|\cdot\|^2/4}) = 2^{n-a}$. Therefore part (i) and Parseval's formula (14.20) imply

$$1 = R^a\big(e^{-\|\cdot\|^2}\big) = \frac{1}{(2\pi)^n}\mathcal{F}R^a\Big(\pi^{\frac{n}{2}}e^{-\frac{\|\cdot\|^2}{4}}\Big) = p(a)\,2^{-n}\pi^{-\frac{n}{2}}\,R^{n-a}\Big(e^{-\frac{\|\cdot\|^2}{4}}\Big)$$
$$= p(a)\,\pi^{-\frac{n}{2}}2^{-a}.$$

For the remainder, replace a by $n-a$ and next select $a = n-2$. Finally, the equivalence of (14.52) with (14.46) is shown using Legendre's duplication formula (13.39) (with a replaced by $\frac{a}{2}$) and the reflection formula from Lemma 13.5.

(viii). Take $a = -2k$ in part (vii) and apply (14.34) to the formulas in part (iv).

14.61 (i). $\|\operatorname{grad}\Phi_\omega(x)\| = \|\omega\| = 1$, for all $x \in \mathbf{R}^n$, which proves that Φ_ω is submersive. From (10.23) one finds that the value at a point of the pushforward of a function under a submersion is obtained by integrating the function over the fiber over that point.

(ii). One has

$$\mathcal{R}(T_h{}^*\phi)(\omega,t) = \int_{N(\omega,t)} \phi(x+h)\,d_{n-1}x = \int_{N(\omega,t+\langle\omega,h\rangle)} \phi(y)\,d_{n-1}y$$

$$= \mathcal{R}\phi(\omega, t + \langle \omega, h \rangle).$$

In view of (10.12) the remaining identities follow by taking limits in this formula.
(iii). The pullback $\Phi_\omega{}^* : \mathcal{D}'(\mathbf{R}) \to \mathcal{D}'(\mathbf{R}^n)$ is well-defined according to Theorem 10.18, while a combination of Problem 13.2 and part (i) yields

$$\begin{aligned} \Phi_\omega{}^*(|\chi|^{-n+1})(\phi) &= \Phi_\omega{}^*(2\delta^{(n-1)})(\phi) = 2\delta^{(n-1)}((\Phi_\omega)_*\phi) \\ &= 2(-1)^{n-1}\partial_t^{n-1}(\Phi_\omega)_*\phi(t)\big|_{t=0} = 2\partial_t^{n-1}\mathcal{R}\phi(\omega, t)\big|_{t=0}. \end{aligned}$$

For the last equality, use part (ii).
(iv). $(A^a)_{a\in\mathbf{C}}$ is a complex-analytic family of distributions because $(|\chi|^a)_{a\in\mathbf{C}}$ is one, and because $\Phi_\omega{}^*$ is a linear mapping. Changing the order of integration, one obtains

$$\begin{aligned} A^a(\phi) &= \int_{S^{n-1}} \int_{\mathbf{R}^n} \Phi_\omega{}^*(|\chi|^{a+1})(x)\phi(x)\,dx\,d\omega \\ &= \int_{S^{n-1}} \int_{\mathbf{R}^n} |\chi|^{a+1}(\Phi_\omega(x))\phi(x)\,dx\,d\omega \\ &= \int_{\mathbf{R}^n} \int_{S^{n-1}} \frac{|\langle \omega, x \rangle|^a}{\Gamma(a+1)}\,d\omega\,\phi(x)\,dx. \end{aligned}$$

The second identity is a direct consequence of part (iii).
(v). The homogeneity is obvious from the definition and the invariance follows by an orthogonal substitution of variables. As a consequence, we obtain that $A^a(x) = d(a)\,c(a+n)\|x\|^a = d(a)\,R^{a+n}(x)$, initially for $\mathrm{Re}\,a > 0$, but then for all $a \in \mathbf{C}$ by means of analytic continuation. For the evaluation of $d(a)$, we note that $d(a) = A^a(e_n)/c(a+n)$, with $c(a)$ as in Problem 14.60. Next, introduce spherical coordinates; in particular, in the notation of [7, Exercise 7.21.(vii)], set $\langle \omega, e_n \rangle = \omega_n = \sin\theta_{n-2}$ and $d\omega = d_{n-1}\omega = d_{n-2}\omega\,\cos^{n-2}\theta_{n-2}\,d\theta_{n-2}$. Thus we conclude, for $\mathrm{Re}\,a > 0$, with the value (13.37) for c_n and by means of (13.33),

$$\begin{aligned} \int_{S^{n-1}} |\langle \omega, x \rangle|^a\,d\omega &= \int_{S^{n-2}} d\omega_{n-2} \int_{-\frac{\pi}{2}}^{\frac{\pi}{2}} |\sin\theta_{n-2}|^a \cos^{n-2}\theta_{n-2}\,d\theta_{n-2} \\ &= c_{n-1}\,2\int_0^{\frac{\pi}{2}} \sin^a\theta\cos^{n-2}\theta\,d\theta = \frac{2\pi^{\frac{n-1}{2}}}{\Gamma(\frac{n-1}{2})}\,\frac{\Gamma(\frac{a+1}{2})\Gamma(\frac{n-1}{2})}{\Gamma(\frac{a+n}{2})}. \end{aligned}$$

In other words, application of Legendre's duplication formula to $\Gamma(a+1)$ leads to

$$\begin{aligned} d(a) &= \frac{2\pi^{\frac{n-1}{2}}\,\Gamma(\frac{a+1}{2})}{\Gamma(a+1)\Gamma(\frac{a+n}{2})}\,\frac{\pi^{\frac{n}{2}}\Gamma(\frac{a+n}{2})}{\Gamma(\frac{n}{2})} = \frac{2\pi^{n-\frac{1}{2}}\Gamma(\frac{a+1}{2})}{\Gamma(\frac{n}{2})2^a\pi^{-\frac{1}{2}}\Gamma(\frac{a+1}{2})\Gamma(\frac{a}{2}+1)} \\ &= \frac{2^{1-a}\pi^n}{\Gamma(\frac{n}{2})\Gamma(\frac{a}{2}+1)}. \end{aligned}$$

According to Corollary 13.6, the function $a \mapsto \frac{1}{d(a)}$ is well-defined, unless we have $-(\frac{a}{2}+1) \in \mathbf{Z}_{\geq 0}$.

(vi). From part (v) and the reflection formula one deduces

$$\frac{1}{d(-n)} = \frac{\Gamma(\frac{n}{2})\Gamma(1-\frac{n}{2})}{2^{n+1}\pi^n} = \frac{1}{2^{n+1}\pi^{n-1}\sin n\frac{\pi}{2}} = \frac{(-1)^{\frac{n-1}{2}}}{4(2\pi)^{n-1}}.$$

The formula for $\phi(0)$ follows by successive application of Theorem 13.3, the identity from part (v) with $a = -n$, and part (iv). Furthermore, the first identity from part (ii) implies the formula for $\phi(x)$, for arbitrary $x \in \mathbf{R}^n$.

(vii). The duplication formula implies

$$\Gamma\left(\frac{n}{2}\right) = \frac{\pi^{\frac{1}{2}}\Gamma(n)}{2^{n-1}\Gamma(\frac{1+n}{2})} \qquad \text{and} \qquad \frac{\Gamma(1-\frac{n}{2})}{\Gamma(1-n)} = \frac{2^n\pi^{\frac{1}{2}}}{\Gamma(\frac{1-n}{2})}.$$

Hence, the reflection formula leads to

$$\begin{aligned} b(n) := \frac{1}{d(-n)\Gamma(1-n)} &= \frac{\Gamma(n)}{2^n\pi^{n-1}\Gamma(\frac{1+n}{2})\Gamma(\frac{1-n}{2})} \\ &= \frac{(n-1)!\,\sin(n+1)\frac{\pi}{2}}{(2\pi)^n} = \frac{(-1)^{\frac{n}{2}}(n-1)!}{(2\pi)^n}. \end{aligned}$$

In this case we have

$$\begin{aligned} \frac{1}{b(n)}\phi(0) &= \frac{1}{b(n)}\delta(\phi) = \frac{1}{b(n)}R^0(\phi) = \int_{S^{n-1}} \Phi_\omega{}^*(|\cdot|^{-n})(\phi)\,d\omega \\ &= \int_{S^{n-1}} |\cdot|^{-n}(\mathcal{R}\phi(\omega,\cdot))\,d\omega \\ &= \int_{S^{n-1}}\int_{\mathbf{R}_{>0}} t^{-n}\Big(\sum_{\pm}\mathcal{R}\phi(\omega,\pm t) - 2\sum_{k=0}^{\frac{n}{2}-1}\partial_s^{2k}\mathcal{R}\phi(\omega,s)\big|_{s=0}\frac{t^{2k}}{(2k)!}\Big)\,dt\,d\omega. \end{aligned}$$

Finally, apply the first identity from part (ii) in order to get $\phi(x)$.

15.3 (i). In view of Theorem 15.4 it is sufficient to observe, for $\phi \in C_0^\infty(X)$ and $\psi \in C_0^\infty(Y)$, that

$$\begin{aligned} k_{{}^t\Phi\,\mathcal{K}\,\Psi}(\phi\otimes\psi) &= ({}^t\Phi\,\mathcal{K}\,\Psi)\psi(\phi) = \mathcal{K}(\Psi\psi)(\Phi\phi) = k_{\mathcal{K}}(\Phi\phi\otimes\Psi\psi) \\ &= k_{\mathcal{K}}((\Phi\otimes\Psi)(\phi\otimes\psi)) = {}^t(\Phi\otimes\Psi)k_{\mathcal{K}}(\phi\otimes\psi) = ({}^t\Phi\otimes{}^t\Psi)k_{\mathcal{K}}(\phi\otimes\psi). \end{aligned}$$

Here the last equality is obtained as in the proof of Theorem 15.5.

(ii). Apply part (i) with the roles of Φ and Ψ played by Φ^* and Ψ^*, respectively.

(iii). In this case, (ii) implies $k_{\Psi_*\mathcal{K}\Psi^*} = (\Psi\otimes\Psi)_* k_{\mathcal{K}}$. According to Theorem 10.8 we have $\Psi_* = j_\Phi\,\Phi^*$ if Φ now denotes Ψ^{-1}. Since $j_{\Phi\otimes\Phi} = j_\Phi\otimes j_\Phi$, the desired identity follows from

$$(1 \otimes j_\Phi) k_{\Phi^* \mathcal{K} (\Phi^{-1})^*} = (j_\Phi \otimes j_\Phi)(\Phi \otimes \Phi)^* k_{\mathcal{K}}.$$

(iv). Note that $\mathcal{K}\,\Phi^* = \Phi^* \mathcal{K}$ is equivalent to $\mathcal{K} = \Phi^* \mathcal{K}(\Phi^{-1})^*$.
(v). The identity is a direct consequence of Theorem 10.16.

15.5 By means of differentiation under the integral sign one sees that $\mathcal{F}$ maps $C_0^\infty(\mathbf{R}^n)$ to $C^\infty(\mathbf{R}^n)$. Given $\alpha \in (\mathbf{Z}_{\geq 0})^n$ and a compact $K \subset \mathbf{R}^n$, the existence of $c > 0$ such that $|\int_{\mathbf{R}^n} x^\alpha (e_{-i\xi}\phi)(x)\,dx| \leq c\,\|\phi\|_{C^0}$, for all $\phi \in C_0^\infty(K)$, leads to estimates as in Example 8.6 and, consequently, to the continuity of $\mathcal{F}$.

The formula for $k_{\mathcal{F}}$ follows from the fact that $\mathcal{F}$ is an integral operator with integral kernel $k_{\mathcal{F}}$.

Let I be as in Example 15.1. Since ${}^t D_j = -D_j$ and ${}^t x_j = x_j$ and furthermore $(I \otimes D_j)\langle \xi, x\rangle = -i\,\xi_j$ and $(I \otimes e_{ia})e^{-i\langle \xi, x\rangle} = e^{-i\langle \xi - a, x\rangle}$, Problem 15.3.(i) implies

$$\begin{aligned} k_{\mathcal{F}\,D_j} &= (I \otimes -D_j)e^{-i\langle \cdot,\cdot\rangle} = (\xi_j \otimes I)e^{-i\langle \cdot,\cdot\rangle} = k_{\xi_j\,\mathcal{F}}, \\ k_{\mathcal{F}\,x_j} &= (I \otimes x_j)e^{-i\langle \cdot,\cdot\rangle} = (-D_j \otimes I)e^{-i\langle \cdot,\cdot\rangle} = k_{-D_j\,\mathcal{F}}, \\ k_{\mathcal{F}\,e_{ia}} &= (I \otimes e_{ia})e^{-i\langle \cdot,\cdot\rangle} = (T_a \otimes I)e^{-i\langle \cdot,\cdot\rangle} = k_{T_a\,\mathcal{F}}. \end{aligned}$$

In the notation of Problem 14.30 we obtain $B = {}^t(A^{-1})$ and so $\langle B\xi, Ax\rangle = \langle \xi, x\rangle$, for all ξ and $x \in \mathbf{R}^n$; this implies $(B \otimes A)^* k_{\mathcal{F}} = k_{\mathcal{F}}$. In view of $\det(B \otimes A) = (\det B)(\det A) = 1$, we have $k_{B_*\,\mathcal{F}\,A^*} = (B \otimes A)_* k_{\mathcal{F}} = k_{\mathcal{F}}$ on account of Problem 15.3.(ii). Hence $B_*\,\mathcal{F}\,A^* = \mathcal{F}$.

Next, note that $e^{-i\langle \xi, x\rangle} = |\det A|\,e^{-i\langle A\xi, Ax\rangle}$ implies $|\det A| = 1$ and thus ${}^t AA = I$.

The last assertion follows from Theorem 15.8.

15.7 **(i)**. For the moment, consider a fixed $\phi \in C_0^\infty(X)$. For every $\psi \in C_0^\infty(Y)$, we have that $w : \psi \mapsto u(\phi \otimes \psi)$ is a distribution on Y satisfying, for $1 \leq j \leq m$,

$$\partial_j w(\psi) = -u(\phi \otimes \partial_j \psi) = -u((I \otimes \partial_j)(\phi \otimes \psi)) = (I \otimes \partial_j)u(\phi \otimes \psi) = 0;$$

that is, $\partial_j w = 0$. Hence, on account of Problem 12.9, there exists $v(\phi) \in \mathbf{C}$ with

$$u(\phi \otimes \psi) = w(\psi) = v(\phi)\,1_Y(\psi).$$

Applying this identity with a ψ satisfying $1_Y(\psi) = 1$, we obtain that v belongs to $\mathcal{D}'(X)$. In turn, this leads to $u = v \otimes 1_Y$. As a consequence of Theorem 15.4 this identity now holds on $C_0^\infty(X \times Y)$, which implies that it is an equality in $\mathcal{D}'(X \times Y)$.
(ii). In this case we have $x_j\,w(\psi) = u(\phi \otimes x_j\,\psi) = (I \otimes x_j)u(\phi \otimes \psi) = 0$. Hence, on account of Theorem 9.5, there exists $v(\phi) \in \mathbf{C}$ with

$$u(\phi \otimes \psi) = w(\psi) = v(\phi)\,\delta^Y(\psi) = v \otimes \delta^Y(\phi \otimes \psi).$$

15.8 Since u is compactly supported, it is of order $\leq k$, for some $k \in \mathbf{Z}_{\geq 0}$. For the moment, consider a fixed $\phi \in C_0^\infty(X)$. For every $\psi \in C_0^\infty(Y)$ with $0 \notin \operatorname{supp} \psi$, we have $\operatorname{supp}(\phi \otimes \psi) \cap (X \times \{0\}) = \emptyset$, and so $u(\phi \otimes \psi) = 0$. This shows that $\psi \mapsto u(\phi \otimes \psi)$ is a distribution on Y of order $\leq k$ with support contained in $\{0\}$. Hence, for $\alpha \in (\mathbf{Z}_{\geq 0})^m$, there exist $u_\alpha(\phi) \in \mathbf{C}$ such that this distribution is given by

$$\psi \mapsto u(\phi \otimes \psi) = \sum_{|\alpha| \leq k} u_\alpha(\phi)\, \partial^\alpha \delta^Y(\psi).$$

Applying this identity with $\psi = \psi_\alpha \in C_0^\infty(Y)$ that equals $y \mapsto y^\alpha$ in a neighborhood of 0, we obtain a constant c_α such that

$$u_\alpha(\phi) = c_\alpha\, u(\phi \otimes \psi_\alpha) \qquad (\phi \in C_0^\infty(X)).$$

But this implies that $u_\alpha : \phi \mapsto u_\alpha(\phi)$ belongs to $\mathcal{E}'(X)$ and is of order $\leq k$. In other words,

$$u = \sum_{|\alpha| \leq k} u_\alpha \otimes \partial^\alpha \delta^Y \quad \text{on} \quad C_0^\infty(X) \otimes C_0^\infty(Y).$$

As a consequence of Theorem 15.4 this identity now holds on $C_0^\infty(X \times Y)$, which implies that it is an equality in $\mathcal{D}'(X \times Y)$.

For the remaining equalities we note that for $(x, y) \in X \times Y$,

$$\begin{aligned} &\iota^*(I \otimes (-\partial)^\alpha)(\phi \otimes \psi)(x) = (\phi \otimes (-\partial)^\alpha \psi)(x, 0) = \phi(x)\, \partial^\alpha \delta^Y(\psi), \\ \text{so} \quad &(I \otimes \partial^\alpha)\iota_* u_\alpha(\phi \otimes \psi) = u_\alpha(\phi)\, \partial^\alpha \delta^Y(\psi) = u_\alpha \otimes \partial^\alpha \delta^Y(\phi \otimes \psi). \end{aligned}$$

15.9 On account of Problem 15.3.(i) and (14.16).(ii) and (i) we have, for $1 \leq j \leq n$,

$$\begin{aligned} (D_j \otimes I + I \otimes D_j)k &= k_{D_j \mathcal{G}} + k_{-\mathcal{G} D_j} &= 0, \\ (x_j \otimes I - I \otimes x_j)k &= k_{x_j \mathcal{G}} - k_{\mathcal{G} x_j} &= 0. \end{aligned}$$

According to Problem 10.18 we then obtain

$$\begin{aligned} (D_j \otimes I)\widetilde{k} = (D_j \otimes I)\Phi^* k &= \sum_{l=1}^{2n} D_j \Phi_l \circ \Phi^* \circ D_l k \\ &= \Phi^*(D_j \otimes I + I \otimes D_j)k = 0 \end{aligned} \tag{21.18}$$

and analogously

$$(I \otimes x_j)\widetilde{k} = 0. \tag{21.19}$$

In view of Problem 15.7.(i) we obtain from (21.18) the existence of $u \in \mathcal{D}'(\mathbf{R}^n)$ such that $\widetilde{k} = 1_{\mathbf{R}^n} \otimes u$, while Problem 15.7.(ii) and (21.19) imply that $\widetilde{k} = 1_{\mathbf{R}^n} \otimes c\,\delta$, for some $c \in \mathbf{C}$ and δ the Dirac measure on $\mathbf{R}^n$. Application of Theorem 10.8 now leads to $k = \Phi_* \Phi^* k = \Phi_*(1_{\mathbf{R}^n} \otimes c\,\delta) = c\, k_I$, see (10.25). Finally, testing against $e^{-\|\cdot\|^2/2}$ as in the proof of Theorem 14.13 gives $c = (2\pi)^n$; hence $\mathcal{G} = (2\pi)^n I$.

Observe that the preceding arguments and Problem 14.7 are strongly related.

15.10 In view of Theorem 15.4 is sufficient to compute $d_*(\phi \otimes \psi)$ with ϕ and $\psi \in C_0^\infty(\mathbf{R}^n)$. To this end, consider arbitrary $u \in \mathcal{D}'(\mathbf{R}^n)$ and successively apply Example 15.11 and Theorem 11.5 to get

$$u(d_*(\phi \otimes \psi)) = d^*u(\phi \otimes \psi) = u * \psi(\phi) = u(S\psi * \phi),$$
$$\text{with} \qquad S\psi * \phi(y) = \int_{\mathbf{R}^n} \phi \otimes \psi(x, x-y)\, dx.$$

15.11 If $\mathcal{U} = u\,*$, then $k_{\mathcal{U}} = d^*u$ in view of Example 15.11. For $h \in \mathbf{R}^n$, we have $d(T_h \otimes T_h) = d$ and so $k_{\mathcal{U}} = (T_h \otimes T_h)^* k_{\mathcal{U}}$. Now Problem 15.3.(iv) implies that $\mathcal{U}$ commutes with all translations in $\mathbf{R}^n$.

For the reverse implication, use results from the proof of Problem 15.9.

15.12 For $\phi \in C_0^\infty(\mathbf{R}^n)$, we have, according to Example 11.9 and Theorem 11.5,

$$\mathcal{U}\psi(\phi) = k_{\mathcal{U}}(\phi \otimes \psi) = d^*u(\phi \otimes \psi) = u(S\psi * \phi) = u * \psi(\phi).$$

That T_a commutes with $\mathcal{U}$ follows as in Problem 15.11; hence, (10.15) implies that ∂_j commutes with $\mathcal{U}$.

15.13 If $k \in \mathcal{D}'(\mathbf{R}^n \times \mathbf{R}^n)$ denotes the kernel of $\mathcal{K}$, we will compute it to be (a multiple of) the function $(\xi, x) \mapsto e^{-i\langle \xi, x\rangle}$ occurring in Problem 15.5.
(a). Application of Problem 15.3.(i) implies $(I \otimes -D_j)k = (\xi_j \otimes I)k$, for $1 \le j \le n$. In order to solve this system of partial differential equations, write $k = e^{-i\langle\cdot,\cdot\rangle}\,\widetilde{l}$, for some $\widetilde{l} \in \mathcal{D}'(\mathbf{R}^n \times \mathbf{R}^n)$. Then $(I \otimes \partial_j)\widetilde{l} = 0$, for $1 \le j \le n$; see the solution to Problem 15.5. According to Problem 15.7.(i) this entails $\widetilde{l} = l \otimes 1_{\mathbf{R}^n}$, for some $l \in \mathcal{D}'(\mathbf{R}^n)$; therefore

$$k = e^{-i\langle\cdot,\cdot\rangle}(l \otimes 1_{\mathbf{R}^n}). \tag{21.20}$$

Thus we have, for ψ and $\phi \in C_0^\infty(\mathbf{R}^n)$,

$$\begin{aligned}
\mathcal{K}\psi(\phi) &= k(\phi \otimes \psi) = e^{-i\langle\cdot,\cdot\rangle}(l \otimes 1_{\mathbf{R}^n})(\phi \otimes \psi) = (l \otimes 1_{\mathbf{R}^n})(e^{-i\langle\cdot,\cdot\rangle}(\phi \otimes \psi))\\
&= l(\xi \mapsto 1_{\mathbf{R}^n}(x \mapsto e^{-i\langle\xi,x\rangle}\phi(\xi)\psi(x)))\\
&= l\Big(\xi \mapsto \phi(\xi)\int_{\mathbf{R}^n} e^{-i\langle\xi,x\rangle}\psi(x)\,dx\Big) = l(\xi \mapsto \phi(\xi)\mathcal{F}\psi(\xi))\\
&= l((\mathcal{F}\psi)\phi) = ((\mathcal{F}\psi)l)(\phi).
\end{aligned}$$

Here (11.14) has been used in the fourth equality. Hence, $\mathcal{K}$ satisfies

$$\mathcal{K}\psi = (\mathcal{F}\psi)l \in \mathcal{D}'(\mathbf{R}^n) \qquad (\psi \in C_0^\infty(\mathbf{R}^n)). \tag{21.21}$$

In particular, $1 = \mathcal{K}(\delta) = (\mathcal{F}\delta)l = l$ on account of (14.30), which proves the claim in (c).
(b). Using the same method as in (a) we obtain $c \in \mathbf{C}$ such that $l = c\,1_{\mathbf{R}^n}$.
(c). Conditions (i) and (iii) imply (ii). Now apply (b).
(d). As in Remark 14.10, condition (i) follows from (iv) and so (21.20) holds. From (v) we obtain $\mathcal{K} = c_*\,\mathcal{K}\,(c^{-1})^*$. Hence Problem 15.3.(ii) and $(c \otimes c^{-1})^*\langle\cdot,\cdot\rangle = \langle\cdot,\cdot\rangle$ imply

$$\begin{aligned}
e^{-i\langle\cdot,\cdot\rangle}(l \otimes 1_{\mathbf{R}^n}) &= k = k_{c_*\mathcal{K}(c^{-1})^*} = (c_* \otimes (c^{-1})_*)k\\
&= (c \otimes c^{-1})_*(e^{-i\langle\cdot,\cdot\rangle}(l \otimes 1_{\mathbf{R}^n})) = e^{-i\langle\cdot,\cdot\rangle}(c_* l \otimes 1_{\mathbf{R}^n}).
\end{aligned}$$

This leads to $l = c_* l$, for every $c > 0$; in other words, l is invariant under positive dilations. Similarly, from (vi) and (10.8) we get $\mathcal{K} = A_* \, \mathcal{K} \, A^*$. So $(A \otimes A)^* \langle \cdot, \cdot \rangle = \langle \cdot, \cdot \rangle$ gives

$$e^{-i\langle \cdot, \cdot \rangle}(l \otimes 1_{\mathbf{R}^n}) = (A \otimes A)_*(e^{-i\langle \cdot, \cdot \rangle}(l \otimes 1_{\mathbf{R}^n})) = e^{-i\langle \cdot, \cdot \rangle}(A_* l \otimes 1_{\mathbf{R}^n}).$$

Hence, $l = A_* l$, for every rotation A; in other words, l is invariant under rotations. Finally, application of Problem 12.10 implies that l is a constant function on $\mathbf{R}^n$.
(e). Now it follows from (21.21) that l is a function continuous at 0. Since l is also homogeneous of degree 0, it has to be a constant function.

15.14 We prove that the conditions in Problem 15.13.(b) are satisfied. Since we have continuous inclusions $C_0^\infty(\mathbf{R}) \subset L^2(\mathbf{R}) \subset \mathcal{D}'(\mathbf{R})$, the restriction of $\widetilde{\mathcal{F}}$ to $C_0^\infty(\mathbf{R})$ is a sequentially continuous linear mapping to $\mathcal{D}'(\mathbf{R})$. Use the identities $\partial_x = \frac{1}{2}(a^- - a^+)$, $a^- \phi_j = 2j\, \phi_{j-1}$, $a^+ \phi_j = \phi_{j+1}$, and $x = \frac{1}{2}(a^- + a^+)$ from Problem 14.42 to conclude that

$$\begin{aligned}(\widetilde{\mathcal{F}} \circ \frac{1}{i}\partial_x)\phi_j &= \frac{1}{2}(2j(-i)^j\, \phi_{j-1} - (-i)^{j+2}\phi_{j+1}) = (-i)^j \frac{1}{2}(2j\, \phi_{j-1} + \phi_{j+1}) \\ &= (x \circ \widetilde{\mathcal{F}})\phi_j.\end{aligned}$$

This leads to the identity $\widetilde{\mathcal{F}} \, D_x = x \, \widetilde{\mathcal{F}}$ of linear mappings in $L^2(\mathbf{R})$, and so in particular in $C_0^\infty(\mathbf{R})$. Similarly, one establishes the identity $\widetilde{\mathcal{F}} \, x = -D_x \, \widetilde{\mathcal{F}}$. Finally, the identity $\widetilde{\mathcal{F}} \phi_0 = \phi_0$ determines the multiplicative constant.

15.15 Denote by $k \in \mathcal{D}'(\mathbf{R} \times \mathbf{R})$ the kernel of $\mathcal{K}$. Since $\mathcal{K}$ commutes with translations, it follows from Problem 15.11 that $k = d^* u$, for some $u \in \mathcal{D}'(\mathbf{R})$. Furthermore, for every $c > 0$, we have $c^* \mathcal{K} = \mathcal{K} \, c^*$; hence Problem 15.3.(iv) implies that $k = (c \otimes 1)(c^* \otimes c^*)k$; in other words, $(c^* \otimes c^*)k = c^{-1}k$. Now $d(c \otimes c) = c\, d$ gives that $(c^* \otimes c^*)d^* = d^* c^*$. Hence $d^* c^* u = d^* c^{-1} u$, and the injectivity of d^* leads to $c^* u = c^{-1} u$; phrased differently, u is homogeneous of degree -1. According to Problem 15.3.(i) the equality $\mathcal{K} \, S = -S \, \mathcal{K}$ entails $(I \otimes S)k = -(S \otimes I)k$; that is, $(S \otimes S)k = -k$. Since $d(S \otimes S) = S\, d$, this gives $d^* S u = d^*(-u)$, and so u is antisymmetric. Now Theorem 10.17 and Problem 9.5 imply that u is a multiple of PV $\frac{1}{x}$.

15.16 $\mathcal{K}$ is local because the support of the distribution $\mathcal{K}\psi$ is contained in the support of the function $\psi \in C_0^\infty(X)$. For the claim about k, we verify that (15.9) is satisfied. Indeed, for $\phi \in C_0^\infty(X)$ and $x \in X$,

$$\Delta^*(I \otimes \partial^\alpha)(\phi \otimes \psi)(x) = (\phi \otimes \partial^\alpha \psi)(x, x) = \phi(x)\, \partial^\alpha \psi(x) = ((\partial^\alpha \psi)\phi)(x).$$

$$\text{Hence} \qquad (I \otimes (-\partial)^\alpha)\Delta_* u_\alpha(\phi \otimes \psi) = u_\alpha((\partial^\alpha \psi)\phi) = (\partial^\alpha \psi)\, u_\alpha(\phi),$$

$$\text{and so} \qquad k(\phi \otimes \psi) = \sum_{\alpha \in (\mathbf{Z}_{\geq 0})^n} (\partial^\alpha \psi)\, u_\alpha(\phi) = \mathcal{K}\psi(\phi).$$

Now the proof of the converse statement. Suppose $k \in \mathcal{E}'(\mathbf{R}^n)$ satisfies supp $k \subset \Delta(\mathbf{R}^n)$ and let Φ be the automorphism of $\mathbf{R}^n \times \mathbf{R}^n$ as in Problem 15.9. Then

$\Phi_* k \in \mathcal{E}'(\mathbf{R}^n \times \mathbf{R}^n)$ satisfies $\operatorname{supp}(\Phi_* k) \subset \Phi(\operatorname{supp} k) \subset \mathbf{R}^n \times \{0\}$ in view of Theorem 10.6; hence, from Problem 15.8 we obtain

$$\Phi_* k = \sum_{\alpha \in (\mathbf{Z}_{\geq 0})^n} (I \otimes \partial^\alpha) \iota_* u_\alpha, \qquad \text{thus} \qquad k = \sum_{\alpha \in (\mathbf{Z}_{\geq 0})^n} \Phi_* (I \otimes \partial^\alpha) \iota_* u_\alpha,$$

because of $\Phi^2 = I$. Now Problem 10.1 and $\Phi\,\iota = \Delta$ on $\mathbf{R}^n$ imply

$$k = \sum_{\alpha \in (\mathbf{Z}_{\geq 0})^n} (I \otimes (-\partial)^\alpha) \Delta_* u_\alpha,$$

which is the kernel of $\sum_{\alpha \in (\mathbf{Z}_{\geq 0})^n} u_\alpha \circ \partial^\alpha$ according to the initial part of the problem.

15.17 Suppose that $\mathcal{K}$ is local. Select $(x, y) \in (X \times X) \setminus \Delta(X)$ arbitrarily. Then there exist disjoint open neighborhoods U of x and V of y in X such that we have $(U \times V) \cap \Delta(X) = \emptyset$. Consider any $\psi \in C_0^\infty(V)$ and $\phi \in C_0^\infty(U)$. Since $\operatorname{supp} \mathcal{K}\psi \subset \operatorname{supp} \psi \subset V$ and $\operatorname{supp} \phi \subset U$, it follows that $0 = \mathcal{K}\psi(\phi) = k(\phi \otimes \psi)$. On the basis of Theorem 15.4 this implies that the restriction of k to $U \times V$ vanishes. Therefore $(x, y) \notin \operatorname{supp} k$, which proves $\operatorname{supp} k \subset \Delta(X)$. The reverse implication follows from Problem 15.16.

16.1 According to Problem 1.7 and (16.8) we have

$$\arccos \circ \cos = \frac{1}{2\pi} \sum_{n \in \mathbf{Z}} \mathcal{F}\phi(n) e_{in} \qquad \text{with} \qquad \phi = 1_{[0,\pi]} * 1_{[0,\pi]}.$$

As a consequence of (14.21) (use approximation) and (14.4),

$$\mathcal{F}\phi(\xi) = (\mathcal{F} 1_{[0,\pi]})^2 = \left(i\, \frac{e^{-i\pi\xi} - 1}{\xi} \right)^2 = \pi^2 e^{-i\pi\xi} \operatorname{sinc}^2\left(\frac{\pi}{2}\xi\right) \qquad (\xi \in \mathbf{R}).$$

In particular,

$$\mathcal{F}\phi(n) = \begin{cases} \pi^2, & n = 0; \\ 0, & n = 2k \neq 0; \\ -\dfrac{4}{(2k-1)^2}, & n = 2k - 1. \end{cases}$$

Hence the coefficients are as desired, while we have pointwise convergence on account of Theorem 16.21. So the first numerical identity follows by evaluation at 0 of the identity of functions. For the second, split the desired series into a series of even terms and one of odd terms and use the preceding identity. Application of Parseval's formula to $\arccos \circ \cos$ on $[-\pi, \pi]$ leads to

$$\frac{\pi^2}{3} = \frac{2}{2\pi} \int_0^\pi x^2\, dx = \frac{\pi^2}{4} + \frac{8}{\pi^2} \sum_{n \in \mathbf{N}} \frac{1}{(2n-1)^4}.$$

For the final three identities, use $\arccos \circ \cos(\cdot + \frac{\pi}{2}) = \frac{\pi}{2} + \arcsin \circ \sin$. Note that the pointwise convergence of the Fourier series for f' follows from Theorem 16.22.

16.2 On account of Example 10.4 and (10.14), application of $(T_h)_*$ to the Poisson summation formula leads to the desired result. In turn, it implies

$$\begin{aligned}
(\arcsin\circ\sin)'' &= \Big(\sum_{k\in\mathbf{Z}} T_{2\pi k}(2\,1_{[-\frac{\pi}{2},\frac{\pi}{2}]} - 1_{[\pi,\pi]})\Big)' = 2\sum_{k\in\mathbf{Z}}(\delta_{-\frac{\pi}{2}+2\pi k} - \delta_{\frac{\pi}{2}+2\pi k}) \\
&= \frac{1}{\pi}\sum_{n\in\mathbf{Z}}(e^{in\frac{\pi}{2}} - e^{-in\frac{\pi}{2}})\,e_{in} = \frac{2i}{\pi}\sum_{n\in\mathbf{Z}}\sin\Big(n\frac{\pi}{2}\Big)\,e_{in} \\
&= \frac{2i}{\pi}\sum_{n\in\mathbf{Z}}(e_{i(4n+1)} - e_{i(4n+3)}) = \frac{4}{\pi}\sum_{n\in\mathbf{N}}(-1)^n\sin(2n-1)\,\cdot\,.
\end{aligned}$$

Now integrate termwise and determine the constants of integration using

$$\sum_{n\in\mathbf{N}}\frac{(-1)^{n-1}}{2n-1} = \sum_{n\in\mathbf{N}}\int_0^1(-x^2)^{n-1}\,dx = \int_0^1\frac{1}{1+x^2}\,dx = \frac{\pi}{4}$$

and the fact that $\arcsin\circ\sin$ is an odd function.

16.7 By writing F_n as a double sum and interchanging the order of summation we see that

$$n\,F_n = \sum_{k=0}^{n-1}\sum_{j=-k}^{k} e_{ij} = \sum_{j=-(n-1)}^{n-1}\sum_{k=|j|}^{n-1} e_{ij} = \sum_{|j|\le n-1}(n-|j|)e_{ij}.$$

16.8 Denote by D_N the function as in (16.22); then $\lim_{N\to\infty} D_N$ exists on account of Lemma 16.4. In combination with Poisson's summation formula (16.7) this implies

$$\lim_{N\to\infty} D_N = \sum_{n\in\mathbf{Z}} e_{in} = 2\pi\sum_{k\in\mathbf{Z}}\delta_{2\pi k}.$$

Furthermore,

$$\sum_{n=0}^{N}\cos n\cdot = \frac{1}{2}\sum_{n=0}^{N}(e_{in}+e_{-in}) = \frac{1}{2}\sum_{n=-N}^{N} e_{in} + \frac{1}{2} = \frac{1}{2}(D_N+1),$$

which implies

$$\sum_{n\in\mathbf{Z}_{\ge 0}}\cos n\cdot = \frac{1}{2} + \pi\sum_{k\in\mathbf{Z}}\delta_{2\pi k}.$$

From the solution to Problem 14.27.(i) we get $\mathcal{F}(\sin n\,\cdot) = \pi i(\delta_{-n} - \delta_n)$, so

$$\mathcal{F}\Big(\sum_{n\in\mathbf{N}}\sin n\,\cdot\Big) = \pi i\sum_{n\in\mathbf{N}}(\delta_{-n}-\delta_n).$$

We have, for $x\in\mathbf{R}\setminus 2\pi\mathbf{Z}$,

$$\sum_{n\in\mathbf{N}}\frac{\cos nx}{n} = -\log\Big|2\sin\frac{x}{2}\Big| = -\frac{1}{2}\log\Big|4\sin^2\frac{x}{2}\Big| = -\frac{1}{2}\log 2(1-\cos x).$$

On account of Lemma 5.9, termwise differentiation gives the following identity in $\mathcal{S}'(\mathbf{R} \setminus 2\pi\mathbf{Z})$:

$$\sum_{n\in\mathbf{N}} \sin n\cdot = \frac{\cos\frac{\cdot}{2}}{2\sin\frac{\cdot}{2}} = \frac{1}{2}\cot\frac{\cdot}{2} = \frac{\sin}{2(1-\cos)}.$$

Taylor expansion around 0 implies $\phi(x) - \phi(-x) = 2x\phi'(0) + O(x^2)$ as $x \to 0$. Thus, the following integral is absolutely convergent, while integration by parts yields

$$\int_{\mathbf{R}} \frac{\sin x}{1-\cos x}(\phi(x) - \phi(-x))\,dx = -\int_{\mathbf{R}} \log(1-\cos x)(\phi'(x) + \phi'(-x))\,dx.$$

Since the integrand in the latter integral is an even function, the right-hand side takes the form $-2\int_{\mathbf{R}} \log(1-\cos x)\,\phi'(x)\,dx$.

16.10 From Exercise 14.44 we obtain $\mathcal{F}\,\mathrm{sinc}^2(\xi) = \frac{\pi}{2}\max\{0, 2-|\xi|\}$. The desired formula now follows from (16.8) by taking $\phi = \mathrm{sinc}^2$, $a = \pi$, and thus $\omega = 2$. For the final identity, replace x by πx.

16.11 The formula for $\mathcal{F}F_t$ follows from Problems 14.44 and 14.30.(ii). In view of the continuity of $\mathcal{F}$ we obtain $\mathcal{F}(\lim_{t\to\infty} F_t) = \lim_{t\to\infty}\mathcal{F}F_t = 1 = \mathcal{F}\delta$ in $\mathcal{S}'(\mathbf{R})$. Next use that $F_t * f = \frac{1}{2\pi}S\mathcal{F}\mathcal{F}(F_t * f) = \frac{1}{2\pi}S\mathcal{F}(\mathcal{F}F_t\,\mathcal{F}f)$. Finally, on account of (14.29), (16.7), and Problem 16.7,

$$\begin{aligned}\sum_{k\in\mathbf{Z}} \mathcal{F}(T_{2\pi k}F_n) &= \max\Big\{0, 1-\frac{|\cdot|}{n}\Big\}\sum_{k\in\mathbf{Z}} e_{-2\pi i k} = \max\Big\{0, 1-\frac{|\cdot|}{n}\Big\}\sum_{k\in\mathbf{Z}}\delta_k\\ &= \sum_{|k|\le n-1}\Big(1-\frac{|k|}{n}\Big)\delta_k = \frac{1}{2\pi}\mathcal{F}\overline{F}_n.\end{aligned}$$

Alternatively, for $x \in \mathbf{R}$, we have in view of (16.30),

$$\sum_{k\in\mathbf{Z}}\frac{1}{(x+2\pi k)^2} = \frac{1}{4\pi^2}\sum_{k\in\mathbf{Z}}\frac{1}{(\frac{x}{2\pi}-k)^2} = \frac{1}{4\sin^2\frac{x}{2}}, \qquad \text{hence}$$

$$\sum_{k\in\mathbf{Z}} F_n(x+2\pi k) = \frac{2\sin^2\frac{nx}{2}}{\pi n}\sum_{k\in\mathbf{Z}}\frac{1}{(x+2\pi k)^2} = \frac{1}{2\pi}\,\frac{\sin^2\frac{nx}{2}}{n\,\sin^2\frac{x}{2}}.$$

Note that in this manner one may obtain another verification of (16.30).

16.14 For a given $y \in S^{n-1}$ set $v(x) = \frac{1-\|x\|^2}{\|x-y\|^n}$. Then v is harmonic on $\mathbf{R}^n \setminus \{y\}$. It suffices to prove that $w : x \mapsto v(x+y)$ is harmonic on $\mathbf{R}^n \setminus \{0\}$. In view of $1 - \|x+y\|^2 = -(2\langle x, y\rangle + \|x\|^2)$, we have

$$-w(x) = \frac{2\langle x, y\rangle}{\|x\|^n} + \frac{1}{\|x\|^{n-2}}.$$

From (12.3) we know that the second term on the right-hand side is harmonic on $\mathbf{R}^n \setminus \{0\}$. Now, applying ∂_j to a harmonic function on an open set in $\mathbf{R}^n$ gives another, so the following are harmonic on $\mathbf{R}^n \setminus \{0\}$:

$$w_j(x) = \partial_j \frac{1}{\|x\|^{n-2}} = (2-n)\frac{x_j}{\|x\|^n} \qquad (1 \le j \le n).$$

In the case of $n = 2$, we take

$$\partial_j \log \|x\| = \frac{x_j}{\|x\|^2} \qquad (1 \le j \le 2).$$

This proves the claim about v. By means of differentiation under the integral sign it now follows that u is harmonic on B^n.

Define

$$p : \mathbf{R}_{>0} \times S^{n-1} \times S^{n-1} \to \mathbf{R} \qquad \text{by} \qquad p(r,x,y) = \frac{1-r^2}{c_n\,\|r\,x - y\|^n}.$$

We claim that

$$\int_{S^{n-1}} p(r,x,y)\,dy = 1 \qquad (r > 0,\ x \in S^{n-1}).$$

By rotational invariance, this integral is clearly independent of x; hence we may denote its value by $d(r)$. We obtain

$$d(r) = \frac{1}{c_n}\int_{S^{n-1}}\int_{S^{n-1}} p(r,x,y)\,dy\,dx = \frac{1}{c_n}\int_{S^{n-1}}\int_{S^{n-1}} p(r,x,y)\,dx\,dy.$$

But $x \mapsto p(r,x,y)$ is harmonic, for $\|x\| < \frac{1}{r}$, so the Mean Value Theorem for harmonic functions from Problem 12.5 implies that the inner integral is independent of r. Evaluation at $r = 0$ gives its value as 1, which leads to $d(r) = 1$. Furthermore, $\lim_{r\uparrow 1} p(r,x,y) = 0$ uniformly for y belonging to the complement in S^{n-1} of any open subset in S^{n-1} containing x. As in Problem 5.1 we therefore obtain $\lim_{r\uparrow 1} u(r\,x) = f(x)$, uniformly for all $x \in S^{n-1}$.

16.15 From (16.24).(i) and (ii) it is obvious that

$$|\sigma_n f(x) - f(x)| \le \frac{1}{2\pi}\int_{-\pi}^{\pi} |f(x-y) - f(x)|\,F_n(y)\,dy.$$

The continuous periodic function f is uniformly continuous on $\mathbf{R}$. As a result, given $\epsilon > 0$ arbitrarily, we can find $\delta > 0$ such that for all x and all $y \in \mathbf{R}$ with $|y| < \delta$ we have $|f(x-y) - f(x)| < \frac{\epsilon}{2}$. Hence, for all $n \in \mathbf{N}$,

$$\int_{-\delta}^{\delta} |f(x-y) - f(x)|\,F_n(y)\,dy \le \frac{\epsilon}{2}\int_{-\delta}^{\delta} F_n(y)\,dy \le \frac{\epsilon}{2}.$$

Next (16.24).(iv) implies that we can find $N \in \mathbf{N}$ such that for all $n \ge N$,

$$\sum_{\pm} \pm \int_{\pm\delta}^{\pm\pi} F_n(y)\,dy \le \frac{\epsilon}{4m}, \qquad \text{where} \qquad m = \sup_{x\in\mathbf{R}} |f(x)|.$$

This entails

$$\sum_{\pm} \pm \int_{\pm\delta}^{\pm\pi} |f(x-y) - f(x)| F_n(y)\,dy \le 2m \sum_{\pm} \pm \int_{\pm\delta}^{\pm\pi} F_n(y)\,dy \le \frac{\epsilon}{2}.$$

It follows that $|\sigma_n f(x) - f(x)| \le \epsilon$, for all $n \ge N$ and $x \in \mathbf{R}$.

16.16 As in (21.17), we have

$$|\mathcal{F}\phi|^2 = \mathcal{F}(\phi * S\overline{\phi}), \qquad \text{so} \qquad \mathcal{F}|\mathcal{F}\phi|^2 = \mathcal{F}^2(\phi * S\overline{\phi}) = 2\pi\, S(\phi * S\overline{\phi}).$$

Therefore (16.8) with $a = 2\pi$ and ϕ replaced by $|\mathcal{F}\phi|^2$ leads to, for all $\xi \in \mathbf{R}$,

$$\sum_{k\in\mathbf{Z}} |\mathcal{F}\phi(\xi + 2\pi\, k)|^2 = \sum_{k\in\mathbf{Z}} S(\phi * S\overline{\phi})(k)\, e^{ik\xi} = \sum_{k\in\mathbf{Z}} (\phi * S\overline{\phi})(k)\, e^{-ik\xi}.$$

Next observe, for all $l \in \mathbf{Z}$, that

$$(\phi * S\overline{\phi})(k) = \int_{\mathbf{R}} \phi(x)\overline{\phi(x-k)}\,dx = \int_{\mathbf{R}} \phi(x-l)\overline{\phi(x-(k+l))}\,dx.$$

16.17 **(i)**. We have $u = \sum_{k\in\mathbf{Z}} \frac{1}{2}(1 + \operatorname{sgn}(k))\,\delta_{2\pi k}$, which shows that u is a distribution as in Lemma 16.4, with coefficients forming a sequence of moderate growth. According to the lemma, u belongs to $\mathcal{S}'(\mathbf{R})$ and satisfies

$$S \circ \mathcal{F}u = \frac{1}{2} + \sum_{n\in\mathbf{N}} e_{2\pi i n} \qquad \text{and therefore} \qquad \mathcal{F}u = \frac{1}{2} + \sum_{n\in\mathbf{N}} e_{-2\pi i n}.$$

(ii). For any $\epsilon > 0$ and $x \in \operatorname{supp} u = 2\pi\mathbf{Z}_{\ge 0}$, we see that $e^{-\epsilon x} \le 1$, which means that we can find a function $\chi_\epsilon \in \mathcal{S}(\mathbf{R})$ coinciding with $x \mapsto e^{-\epsilon x}$ on $]-\frac{1}{2}, \infty[$. Thus we obtain, for any $\phi \in \mathcal{S}(\mathbf{R})$,

$$e^{-\epsilon x} u(\phi) = \chi_\epsilon u(\phi) = u(\chi_\epsilon\, \phi) = u(e^{-\epsilon x}\, \phi).$$

As a consequence,

$$|u(\phi) - e^{-\epsilon x}u(\phi)| = |u(\phi) - u(e^{-\epsilon x}\phi)| \le \sum_{k\in\mathbf{N}} |(1 - e^{-2\pi\epsilon k})\phi(2\pi k)|$$

$$\le \sum_{k\in\mathbf{N}} \frac{1 - e^{-2\pi\epsilon k}}{(2\pi k)^2} \sup_{x\in\mathbf{R}} |x^2\, \phi(x)| = \frac{\|\phi\|_{\mathcal{S}(0,2)}}{4\pi^2} \sum_{k\in\mathbf{N}} \frac{1 - e^{-2\pi\epsilon k}}{k^2}.$$

The series at the right converges uniformly for all $\epsilon \ge 0$, which implies

$$\lim_{\epsilon\downarrow 0} \sum_{k\in\mathbf{N}} \frac{1 - e^{-2\pi\epsilon k}}{k^2} = \sum_{k\in\mathbf{N}} \lim_{\epsilon\downarrow 0} \frac{1 - e^{-2\pi\epsilon k}}{k^2} = 0,$$

$$\text{and so} \qquad \lim_{\epsilon \downarrow 0} |u(\phi) - e^{-\epsilon x} u(\phi)| = 0.$$

This proves $\lim_{\epsilon \downarrow 0} e^{-\epsilon x} u = u$ in $\mathcal{S}'(\mathbf{R})$.

According to Theorem 14.24 the mapping $\mathcal{F} : \mathcal{S}'(\mathbf{R}) \to \mathcal{S}'(\mathbf{R})$ is continuous; consequently $\mathcal{F}u = \lim_{\epsilon \downarrow 0} \mathcal{F}(e^{-\epsilon x} u)$. Once more on the basis of Lemma 16.4 and summing a convergent geometric series, we obtain, for $\xi \in \mathbf{R}$,

$$\begin{aligned}
\mathcal{F}(e^{-\epsilon x} u)(\xi) &= \mathcal{F}\Big(\frac{1}{2}\delta + \sum_{k \in \mathbf{N}} e^{-2\pi\epsilon k} \delta_{2\pi k}\Big)(\xi) \\
&= S\Big(\frac{1}{2} + \sum_{n \in \mathbf{N}} e^{-2\pi\epsilon n} e_{2\pi i n}(\xi)\Big) = \frac{1}{2} + \sum_{n \in \mathbf{N}} e^{-2\pi(\epsilon + i\xi)n} \\
&= \frac{1}{2} + \frac{e^{-2\pi(\epsilon+i\xi)}}{1 - e^{-2\pi(\epsilon+i\xi)}} = \frac{1}{2} \frac{1 + e^{-2\pi(\epsilon+i\xi)}}{1 - e^{-2\pi(\epsilon+i\xi)}} \\
&= \frac{1}{2} \frac{e^{\pi(\epsilon+i\xi)} + e^{-\pi(\epsilon+i\xi)}}{e^{\pi(\epsilon+i\xi)} - e^{-\pi(\epsilon+i\xi)}} = \frac{1}{2} \frac{e^{i\pi(\xi-i\epsilon)} + e^{-i\pi(\xi-i\epsilon)}}{e^{i\pi(\xi-i\epsilon)} - e^{-i\pi(\xi-i\epsilon)}} = \frac{1}{2i} \cot \pi(\xi - i\epsilon).
\end{aligned}$$

Combining the preceding results now leads to

$$\mathcal{F}u = \lim_{\epsilon \downarrow 0} \frac{1}{2i} \cot \pi(\cdot - i\epsilon) = \frac{1}{2i} \cot \pi(\cdot - i0). \tag{21.22}$$

Furthermore,

$$\begin{aligned}
2i \sum_{n \in \mathbf{Z}} e^{2\pi i n \xi} &= 2i(\mathcal{F}u + S\mathcal{F}u)(\xi) = 2i(\mathcal{F}u(\xi) + \mathcal{F}u(-\xi)) \\
&= \cot \pi(\xi - i0) - \cot \pi(\xi + i0) =: c(\xi).
\end{aligned}$$

(iii). The properties of the function $f(z) = \cot \pi z$ follow directly, as well as the fact that f is complex-analytic on $\mathbf{C} \setminus \mathbf{Z}$. As a consequence we have, for ξ belonging to the open subset $\mathbf{R} \setminus \mathbf{Z}$,

$$c(\xi) = \lim_{\epsilon \downarrow 0} (f(\xi - i\epsilon) - f(\xi + i\epsilon)) = f(\xi) - f(\xi) = 0 = 2i \sum_{k \in \mathbf{Z}} \delta_k|_{\mathbf{R} \setminus \mathbf{Z}}.$$

On the other hand, for $z \in U_k = \{ z \in \mathbf{C} \mid |z - k| < 1/2 \}$, we may write

$$f(z) = \frac{1}{\pi(z-k)} + g(z),$$

where g is complex-analytic on U_k. The Plemelj–Sokhotsky jump relation then implies, for $\xi \in \mathbf{R} \cap U_k$,

$$c(\xi) = \lim_{\epsilon \downarrow 0} \Big(\frac{1}{\pi(\xi - k - i\epsilon)} - \frac{1}{\pi(\xi - k + i\epsilon)} + g(\xi - i\epsilon) - g(\xi + i\epsilon) \Big)$$

$$= \frac{1}{\pi(\xi - k - i0)} - \frac{1}{\pi(\xi - k + i0)} = 2i\,\delta_k = 2i \sum_{l \in \mathbf{Z}} \delta_l|_{U_k}.$$

In view of $\mathbf{R} = (\mathbf{R} \setminus \mathbf{Z}) \cup \bigcup_{k \in \mathbf{Z}} (\mathbf{R} \cap U_k)$ and Theorem 7.1 we now obtain

$$\cot \pi(\cdot - i0) - \cot \pi(\cdot + i0) = 2i \sum_{k \in \mathbf{Z}} \delta_k.$$

(iv). Poisson's summation formula from the preceding part and Lemma 16.6 imply

$$\sum_{k \in \mathbf{Z}} \delta_k' = \sum_{n \in \mathbf{Z}} e_{2\pi i n}' = \sum_{n \in \mathbf{Z} \setminus \{0\}} 2\pi i n\, e_{2\pi i n} = -4\pi \sum_{n \in \mathbf{N}} n \sin 2\pi n\,\cdot\,.$$

Alternatively, differentiate the first formula in Problem 16.8.
(v). The identity from part (ii) and its transform under S imply

$$\begin{Bmatrix} \mathcal{F}u \\ S\mathcal{F}u \end{Bmatrix} = \frac{1}{2} + \sum_{n \in \mathbf{N}} e_{\mp 2\pi i n} = \pm \frac{1}{2i} \cot \pi(\cdot \mp i0).$$

So $$\sum_{n \in \mathbf{N}} \sin 2\pi n\,\cdot = \frac{1}{2i}(S\mathcal{F}u - \mathcal{F}u) = \frac{1}{4} \sum_{\pm} \cot \pi(\cdot \pm i0),$$

and hence $$\sum_{n \in \mathbf{N}} \sin n\,\cdot = \frac{1}{4} \sum_{\pm} \cot(\frac{\cdot}{2} \pm i0).$$

For x near 0 we have $\cot x = \frac{\cos x}{\operatorname{sinc} x} \frac{1}{x}$, where the first factor is a C^∞ function having value 1 at 0. So the Plemelj–Sokhotsky jump relation now leads to

$$\cot(x \pm i0) = \frac{\cos x}{\operatorname{sinc} x}\Big(\mathrm{PV}\,\frac{1}{x} \mp \pi i\,\delta\Big) = \frac{\cos x}{\operatorname{sinc} x}\,\mathrm{PV}\left(\frac{1}{x}\right) \mp \pi i\,\delta.$$

In other words, for x near 0,

$$\sum_{n \in \mathbf{N}} \sin n\,x = \frac{\cos \frac{x}{2}}{\operatorname{sinc} \frac{x}{2}}\,\mathrm{PV}\,\frac{1}{x}.$$

(vi). Define $A : \mathbf{R} \to \mathbf{R}$ by $Ax = \pi^{-1}x$. Then ${}^tA = A$. According to Problem 14.30.(i) and Theorem 14.24, and Example 10.4, respectively, we have

$$\mathcal{F} \circ A^* \circ \mathcal{F} = A_* \circ \mathcal{F}^2 = 2\pi A_* \circ S \qquad \text{and} \qquad A_*\delta_{-2\pi k} = \delta_{-2k}.$$

Applying $\mathcal{F} \circ A^*$ to (21.22) in part (ii), we now obtain

$$\mathcal{F} \cot(\cdot - i0) = 2\pi i\Big(\delta + 2\sum_{k \in \mathbf{N}} \delta_{-2k}\Big).$$

Next note that $\cot^2 = -(1 + \cot')$, and so

$$\mathcal{F}\cot^2(\cdot - i0) = -\mathcal{F}1 - ix\,\mathcal{F}\cot(\cdot - i0) = -2\pi\delta + 2\pi\, x\Big(\delta + 2\sum_{k\in\mathbf{N}}\delta_{-2k}\Big)$$
$$= -2\pi\delta - 8\pi\sum_{k\in\mathbf{N}} k\,\delta_{-2k}.$$

(vii). On the basis of Lemma 16.6 and Problem 16.8 the desired formula follows from

$$s - s'' = \sum_{n\in\mathbf{N}}\frac{\cos n\,\cdot}{1+n^2} + \sum_{n\in\mathbf{N}}\frac{n^2\cos n\,\cdot}{1+n^2} = \sum_{n\in\mathbf{N}}\cos n\,\cdot = -\frac{1}{2} + \pi\sum_{k\in\mathbf{Z}}\delta_{2\pi k}.$$

Restricting s to the open interval $]\,0,\,2\pi\,[$, we obtain for it the differential equation $s - s'' = -1/2$, which on $\mathbf{R}$ has the general solution

$$\tilde{s}(x) = a\,e^x + b\,e^{-x} - \frac{1}{2} \qquad \text{for arbitrary} \qquad a,\ b \in \mathbf{C}.$$

From the definition of s we see that it is 2π-periodic, even, and continuous on $\mathbf{R}$, because the series is uniformly convergent on $\mathbf{R}$. Consequently, the function s is completely determined on all of $\mathbf{R}$ by the choice of a and b. Periodicity implies $s(0) = s(2\pi)$, that is,

$$a + b - \frac{1}{2} = a\,e^{2\pi} + b\,e^{-2\pi} - \frac{1}{2}, \qquad \text{that is,} \qquad b = a\,\frac{e^{2\pi}-1}{1-e^{-2\pi}}.$$

In view of the uniform convergence of the series on $\mathbf{R}$ we may write

$$\int_0^{2\pi} s(x)\,dx = \int_0^{2\pi}\sum_{n\in\mathbf{N}}\frac{\cos nx}{1+n^2}\,dx = \sum_{n\in\mathbf{N}}\frac{1}{1+n^2}\int_0^{2\pi}\cos nx\,dx = 0.$$

On the other hand,

$$\int_0^{2\pi} s(x)\,dx = \int_0^{2\pi}\Big(a\,e^x + b\,e^{-x} - \frac{1}{2}\Big)\,dx = a\,e^{2\pi} - b\,e^{-2\pi} - a + b - \pi;$$

hence

$$a = \frac{\pi}{2(e^{2\pi}-1)} \qquad \text{and} \qquad b = \frac{\pi}{2(1-e^{-2\pi})}.$$

Inserting the values for a and b, we get (see Fig. 21.1)

$$\tilde{s}(x) = \frac{1}{2}\Big(\pi\Big(\frac{e^x}{e^{2\pi}-1} + \frac{e^{-x}}{1-e^{-2\pi}}\Big) - 1\Big)$$
$$= \frac{1}{2}\Big(\pi\Big(\frac{e^{x-\pi}}{e^{\pi}-e^{-\pi}} + \frac{e^{-(x-\pi)}}{e^{\pi}-e^{-\pi}}\Big) - 1\Big) = \frac{1}{2}\Big(\pi\,\frac{\cosh(x-\pi)}{\sinh\pi} - 1\Big).$$

A less elegant, but slightly more illuminating, method for obtaining a second equation for a and b is as follows. Let $0 < \epsilon_1 < \epsilon_2 < 1$ and $\phi_{\epsilon_1,\epsilon_2} \in C_0^\infty(\mathbf{R})$ be such that $\phi_{\epsilon_1,\epsilon_2}(x) = 1$ for $|x| < \epsilon_1$, and $\phi_{\epsilon_1,\epsilon_2}(x) = 0$ for $|x| > \epsilon_2$. Then we have

$$\begin{aligned}\pi &= \pi \sum_{k\in\mathbf{Z}} \delta_{2\pi k}(\phi_{\epsilon_1,\epsilon_2}) = \Big(s - s'' + \frac{1}{2}\Big)(\phi_{\epsilon_1,\epsilon_2}) \\ &= \int_{-\epsilon_2}^{\epsilon_2} \Big(s(x) - s''(x) + \frac{1}{2}\Big)\, \phi_{\epsilon_1,\epsilon_2}(x)\, dx.\end{aligned}$$

This implies

$$\begin{aligned}\pi &= \lim_{\epsilon_2\downarrow 0} \int_{-\epsilon_2}^{\epsilon_2} \Big(s(x) - s''(x) + \frac{1}{2}\Big)\, \phi_{\epsilon_1,\epsilon_2}(x)\, dx = -\lim_{\epsilon_1\downarrow 0} \int_{-\epsilon_1}^{\epsilon_1} s''(x)\, dx \\ &= \lim_{\epsilon_1\downarrow 0} (s'(-\epsilon_1) - s'(\epsilon_1)) = \tilde{s}'(2\pi) - \tilde{s}'(0) = a\, e^{2\pi} - b\, e^{-2\pi} - a + b.\end{aligned}$$

Observe that this formula implies the *jump relations* (see Fig. 21.1)

$$\lim_{x\downarrow 2\pi k} s'(x) - \lim_{x\uparrow 2\pi k} s'(x) = -\pi \qquad (k \in \mathbf{Z}).$$

The choice $x = 0$ finally leads to the desired value of the numerical series.

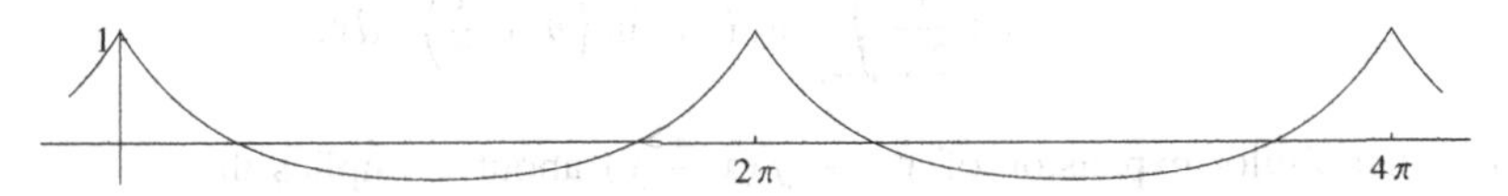

Fig. 21.1 On $[\,0,\ 2\pi\,]$ the graph of s is a catenary; therefore on $\mathbf{R}$ the graph looks like a suspension bridge. On account of the jump relations one computes the angles between the tangent lines at the points $2\pi n$ to be $\pi - 2\arctan\frac{\pi}{2} \approx 65^\circ$

Note that the Fourier transform provides another method for solving the differential equation for s; it turns out to be somewhat more laborious. We describe the main steps. Application of Theorem 14.24 and Lemma 16.4 leads to

$$\mathcal{F}s(\xi) = -\pi\delta(\xi) + \pi \sum_{n\in\mathbf{Z}} \frac{e^{-2\pi i n\xi}}{1+\xi^2},$$

and therefore

$$s(x) = -\frac{1}{2} + \frac{1}{2}\sum_{n\in\mathbf{Z}} \int_{\mathbf{R}} \frac{e^{i(x-2n\pi)\xi}}{1+\xi^2}\, d\xi.$$

On the basis of [7, Exercise 2.85 or 6.99.(iii)] or using the Residue Theorem from complex analysis, one obtains

$$\int_{\mathbf{R}} \frac{e^{i(x-2n\pi)\xi}}{1+\xi^2}\, d\xi = \pi e^{-(x-2n\pi)\,\mathrm{sgn}(x-2n\pi)}, \qquad \text{and so}$$

$$s(x) = -\frac{1}{2} + \frac{\pi}{2} \sum_{\{n\in\mathbf{Z} \mid x-2n\pi<0\}} e^{x-2n\pi} + \frac{\pi}{2} \sum_{\{n\in\mathbf{Z} \mid x-2n\pi\geq 0\}} e^{-x+2n\pi}.$$

In particular, if $x \in [\,0, 2\pi\,[$, then

$$\{\, n \in \mathbf{Z} \mid x - 2n\pi < 0 \,\} = \mathbf{Z}_{\geq 1} \qquad \text{and} \qquad \{\, n \in \mathbf{Z} \mid x - 2n\pi \geq 0 \,\} = \mathbf{Z}_{\leq 0}.$$

Thus, for $x \in [\,0, 2\pi\,]$,

$$\begin{aligned} s(x) &= -\frac{1}{2} + \frac{\pi}{2} e^{x} \sum_{n\in\mathbf{Z}_{\geq 1}} (e^{-2\pi})^n + \frac{\pi}{2} e^{-x} \sum_{n\in\mathbf{Z}_{\geq 0}} (e^{-2\pi})^n \\ &= \frac{1}{2}\Big(\pi \frac{\cosh(x-\pi)}{\sinh \pi} - 1\Big). \end{aligned}$$

Finally, it is clear that the Fourier series of the 2π-periodic function given by $x \mapsto \frac{1}{2}\big(\pi \frac{\cosh(x-\pi)}{\sinh \pi} - 1\big)$ on $[\,0, 2\pi\,]$ equals $s(x)$.

16.18 The absolute convergence of the Fourier series follows from $M(f''\, e_{-in}) = -n^2 M(f\, e_{-in})$, for all $n \in \mathbf{Z}$. Using (16.24).(i) we obtain

$$\begin{aligned} S_n f(x) - f(x) &= \frac{1}{2\pi} \int_{-\pi}^{\pi} \frac{f(x-y) - f(x)}{\sin \frac{y}{2}} \sin\Big(n + \frac{1}{2}\Big) y \, dy \\ &=: \frac{1}{2\pi} \int_{-\pi}^{\pi} g_x(y) \sin\Big(n + \frac{1}{2}\Big) y \, dy. \end{aligned}$$

Note that Taylor expansion of $y \mapsto f(x-y)$ about x implies that g_x is a C^1 function. Furthermore, there exists $m > 0$ such that $|g_x'(y)| \leq m$, for all x and $y \in \mathbf{R}$. Now apply integration by parts.

16.20 Use the results from Example 16.23 and successively apply the following arguments. (i) Set $z = \frac{1}{4}$ in (16.28). (ii) Evaluate the Fourier series of f_z at $x = \frac{1}{4}$ and take its imaginary part. (iii) Put $z = 0$ in (ii). (iv) Replace z in (16.30) by $\frac{1}{2} - z$. (v) Differentiation of this identity gives the one in (iv), while the constant of integration equals 0, as follows by taking $z = 0$. (vi) Direct verification. (vii) Use that $\frac{d}{dz} \log(1 - \frac{z^2}{n^2}) = \frac{2z}{z^2 - n^2}$. At the outset, the identity is obtained for $0 < |z| < 1$, but on account of analytic continuation, the identity holds for all $z \in \mathbf{C}$. (viii) Take the limit for $z \to 1$ in

$$-\frac{\sin \pi z - \sin \pi}{\pi(z-1)} = z(z+1) \prod_{n=2}^{\infty} \Big(1 - \frac{z^2}{n^2}\Big).$$

(ix) Set $z = \frac{1}{2}$ in (vii). (x) The identity $\sin z \, \cos z = \frac{1}{2} \sin 2z$ in combination with (vii) implies

$$\sin \pi z \, \cos \pi z = \pi z \prod_{n\in\mathbf{N}} \Big(1 - \frac{(2z)^2}{(2n)^2}\Big) \prod_{n\in\mathbf{N}} \Big(1 - \frac{(2z)^2}{(2n-1)^2}\Big).$$

16.21 The distributional Fourier series of f_z has coefficients

$$\frac{1}{2\pi}\int_0^{2\pi} e^{-i(z+n)x}\,dx = \frac{1-e^{-2\pi iz}}{2i\,\pi(z+n)} = e^{-\pi iz}\,\frac{\sin\pi z}{\pi(z+n)} \qquad (n\in\mathbf{Z}).$$

The pointwise convergence on $\mathbf{R}\setminus 2\pi\mathbf{Z}$ follows directly from Theorem 16.21. Identity (i) with the $+$ sign is a simple rewrite of the Fourier series, while the case of $-$ is obtained by replacing z by $-z$. For (ii) and (iii), apply $\sum_\pm$ and $\sum_\pm \pm$ to (i), respectively. Next, (iv) follows from (iii) by taking $x=\pi$ and replacing πz by z. For (v) and (vi) note that

$$\frac{e^{inx}}{z+n}+\frac{e^{-inx}}{z-n} = \frac{2z\cos nx}{z^2-n^2} - i\,\frac{2n\sin nx}{z^2-n^2}.$$

Next, apply this identity to the Fourier series, suppose $z\in\mathbf{R}$ for the moment, split into real and imaginary parts, and finally use analytic continuation in z. Note that (vi) is the derivative of (v) with respect to x, under the assumption that termwise differentiation is admissible.

16.22 **(i)**. We have, for arbitrary $\phi\in\mathcal{S}'(\mathbf{R})$,

$$(g_z''+z^2g_z)(\phi) = g_z(\phi''+z^2\phi) = \sum_{n\in\mathbf{Z}}\int_{(2n-1)\pi}^{(2n+1)\pi} g_z(x)(\phi''(x)+z^2\phi(x))\,dx.$$

Next, twice integrating by parts, we obtain

$$\begin{aligned}
&\int_{(2n-1)\pi}^{(2n+1)\pi} (\phi''(x)+z^2\phi(x))g_z(x)\,dx\\
&\quad= \int_{(2n-1)\pi}^{(2n+1)\pi} (\phi''(x)+z^2\phi(x))\cos z(x-2n\pi)\,dx\\
&\quad= \cos\pi z\sum_\pm \pm\phi'((2n\pm1)\pi) + z\sin\pi z\sum_\pm \phi((2n\pm1)\pi).
\end{aligned}$$

The desired formula now follows upon summation, since the sum involving the terms $\phi'((2n\pm1)\pi)$ is telescoping.

In the notation of Example 16.23 we get, for $x\in\mathbf{R}$,

$$\begin{aligned}
g_z(2\pi z\,x) &= \frac{1}{2}(f_z(x)+f_{-z}(x)) = \frac{\sin\pi z}{\pi}\sum_{n\in\mathbf{Z}}\frac{(-1)^n}{2}\Big(\frac{1}{z-n}-\frac{1}{-z-n}\Big)e^{2\pi inx}\\
&= \frac{z\sin\pi z}{\pi}\sum_{n\in\mathbf{Z}}\frac{(-1)^n}{z^2-n^2}\,e^{2\pi inx},
\end{aligned}$$

which leads to the distributional Fourier series for g_z. The final identity is deduced by means of repeated termwise differentiation of that series.

Poisson's summation formula (16.7) implies

$$\sum_{n\in\mathbf{Z}} e_{in} = 2\pi\sum_{k\in\mathbf{Z}}\delta_{2\pi k} \qquad\text{and}\qquad 2\sum_{n\in\mathbf{Z}} e_{2in} = 2\pi\sum_{k\in\mathbf{Z}}\delta_{\pi k}.$$

Subtracting the former identity from the latter, we get the desired result.
(ii). This is established by straightforward computations. The pointwise convergence of the Fourier series of f is a consequence of Theorem 16.22.
(iii). Replace x by $2x$ for $|\cos|$ and then replace x by $\frac{\pi}{2} - x$ for $|\sin|$. For the final identity, note that $|\sin| = \sum_{k\in\mathbf{Z}} T_{k\pi}(1_{[0,\pi]}\sin)$ and differentiate. The Fourier series for $|\cos|$ and $|\sin$ converge pointwise on account of Theorem 16.21, while the pointwise convergence of the last Fourier series follows from Theorem 16.22.

17.3 With respect to the cases (i), (ii), and (v), observe that $P(D)e_{i\xi} = P(\xi)e_{i\xi}$, for all $\xi \in \mathbf{C}^n$. Because the degree of P is positive, there exists $\xi \in \mathbf{C}^n$ with $P(\xi) = 0$; hence, $u = e_{i\xi} \in C^\infty(\mathbf{R}^n) \subset \mathcal{D}'(\mathbf{R}^n)$ satisfies $P(D)u = 0$. Note that according to Problem 14.24 we have $e_{i\xi} \in \mathcal{S}'(\mathbf{R}^n)$ if and only if $\xi \in \mathbf{R}^n$; therefore, in case (ii) it is necessary that the zero ξ belong to $\mathbf{R}^n$. For the cases (iii), (iv), and (vi), note that $P(D)u = 0$ implies

$$P(\xi)\,\widehat{u}(\xi) = 0 \qquad (\xi \in \mathbf{R}^n).$$

This equality holds if and only if at least one of $P(\xi)$ and $\widehat{u}(\xi)$ equals 0. Furthermore, the closed subset $N = \{\,\xi \in \mathbf{R}^n \mid P(\xi) = 0\,\}$ is nowhere dense; therefore we have $\widehat{u}(\xi) = 0$, for ξ belonging to the dense open subset $\mathbf{R}^n \setminus N$. According to Lemma 14.4, Theorem 14.2, and Lemma 14.9, depending on the case respectively, $\widehat{u}$ is continuous on $\mathbf{R}^n$. This yields $\widehat{u} = 0$, and we see that Theorem 14.24 implies $u = 0$.

17.4 Consider $u \in \mathcal{S}'(\mathbf{R}^n)$ with $\Delta u = 0$. The Fourier transform implies $\|\xi\|^2\,\mathcal{F}u = 0$, or $\operatorname{supp}\mathcal{F}u \subset \{0\}$; and so Theorem 8.10 leads to the existence of $k \in \mathbf{Z}_{\geq 0}$ and $c_\alpha \in \mathbf{C}$ such that on account of (14.33),

$$\mathcal{F}u = \sum_{|\alpha|\leq k} c_\alpha\,\partial^\alpha\delta = \frac{1}{(2\pi)^n}\sum_{|\alpha|\leq k}(-i)^{|\alpha|}\,c_\alpha\,\mathcal{F}(x^\alpha).$$

Hence, the injectivity of $\mathcal{F}$ gives that u is a polynomial. If u is a bounded harmonic function, it defines a tempered distribution. The preceding argument then gives that u is a bounded polynomial, which is possible only if it is a constant. Next, suppose $n \geq 3$. Denote by E the potential of δ as in (12.3). Now suppose $F \in \mathcal{D}'(\mathbf{R}^n)$ is a fundamental solution of Δ satisfying $F(x) \to 0$ as $\|x\| \to \infty$. In particular, $\Delta(E - F) = 0$, while Δ is hypoelliptic according to Example 12.8. Therefore $E - F$ is a bounded harmonic function, and the preceding argument gives that it is a constant. The condition at infinity implies that $E - F = 0$.

17.6 **(i).** $E \in \mathcal{S}'(\mathbf{R}^n)$ according to Theorem 14.34, while Problem 14.30.(ii) implies that $\mathcal{F}E$ is homogeneous of degree $-a - n$ if E is homogeneous of degree a. Next, write $P(\xi) = \sum_{l=0}^m P_l(\xi)$ with P_l homogeneous of degree l. Then the Fourier transform of $P(D)E = \delta$ gives $1 = \sum_{l=0}^m P_l\,\mathcal{F}E$, where $P_l\,\mathcal{F}E$ is homogeneous of degree $l - a - n$ and $P_m\,\mathcal{F}E \neq 0$. Accordingly, $P(D) = P_m(D)$ and $m - a - n = 0$, that is, $a = m - n$.
(ii). It follows immediately that $\mathcal{F}E = \mathcal{F}\widetilde{E} = 1/P$ on $\mathbf{R}^n \setminus \{0\}$ and thus $\operatorname{supp}(\mathcal{F}(E - \widetilde{E})) \subset \{0\}$. As in Problem 17.4, this implies $\widetilde{E} - E = f$, with

f a polynomial function on $\mathbf{R}^n$. If $\widetilde{E}$ is homogeneous, then f is homogeneous of degree $m-n$ according to part (i). Consequently, $f = 0$ if $m < n$, and thus $E = \widetilde{E}$ in this case. If $m \geq n$, then f is a homogeneous polynomial function of degree $m-n$.

(iii). In this case, $1/P$ is a homogeneous function on $\mathbf{R}^n \setminus \{0\}$ of degree $-m > -n$, which implies its local integrability on $\mathbf{R}^n$. From Theorem 14.34 we obtain $1/P \in \mathcal{S}'(\mathbf{R}^n)$, and therefore $E := \mathcal{F}^{-1}(1/P)$ is a homogeneous fundamental solution, of degree $-(-m) - n = m - n$. The uniqueness of E finally follows from part (ii).

(iv). According to (12.3), $\frac{1}{2\pi} \log \|x\|$ is a fundamental solution of Δ, and as a consequence of the hypoellipticity of Δ any two of its fundamental solutions differ by a function belonging to $C^\infty(\mathbf{R}^2)$. Therefore any fundamental solution possesses a logarithmic singularity at 0, which is incompatible with homogeneity.

17.8 The fundamental solution $E_0 \in \mathcal{S}'(\mathbf{R})$ is given by $\mathcal{F}^{-1}(\frac{1}{\xi-\lambda})$.

Set $\Omega = \{\, z \in \mathbf{C} \mid |z| < r \,\}$. For the solution E from the proof of Theorem 17.11 we first compute, for $|\xi| \leq R$ and $\eta \in \mathbf{R}^n \setminus \{0\}$,

$$\begin{aligned} \frac{1}{2\pi i} \int_{\partial\Omega} \frac{e^{i\langle x, \xi + z\eta\rangle}}{z\, P(\xi + z\eta)}\, dz &= \frac{e^{ix\xi}}{2\pi i} \int_{\partial\Omega} \frac{e^{ixz\eta}}{z(\xi + z\eta - \lambda)}\, dz \\ &= \frac{e^{ix\xi}}{2\pi i\eta} \int_{\partial\Omega} \frac{e^{ixz\eta}}{z(z-\alpha)}\, dz =: \frac{e^{ix\xi}}{2\pi i\eta} \int_{\partial\Omega} f(z)\, dz, \end{aligned}$$

where we have written $\alpha = \frac{\lambda - \xi}{\eta}$. For the moment, assume that r is large enough that the singular points of $f(z)$ within Ω occur at 0 and α. Then, by residue calculus,

$$\begin{aligned} \frac{1}{2\pi i} \int_{\partial\Omega} f(z)\, dz &= \operatorname*{Res}_{z=0} f(z) + \operatorname*{Res}_{z=\alpha} f(z) = \lim_{z\to 0} z\, f(z) + \lim_{z\to\alpha} (z-\alpha)\, f(z) \\ &= -\frac{1}{\alpha} + \frac{e^{ix(\lambda-\xi)}}{\alpha}. \end{aligned}$$

We obtain

$$\frac{1}{2\pi i} \int_{\partial\Omega} \frac{e^{i\langle x, \xi + z\eta\rangle}}{z\, P(\xi + z\eta)}\, dz = \frac{e^{ix\xi}}{\eta} \frac{e^{ix(\lambda-\xi)} - 1}{\alpha} = \frac{e^{ix\lambda} - e^{ix\xi}}{\lambda - \xi}.$$

Accordingly,

$$G(x) = \frac{1}{2\pi} \int_{-R}^{R} \frac{e^{ix\xi} - e^{ix\lambda}}{\xi - \lambda}\, d\xi = \frac{1}{2\pi} \int_{-R}^{R} \frac{e^{ix\xi}}{\xi - \lambda}\, d\xi - \frac{e^{ix\lambda}}{2\pi} \int_{-R}^{R} \frac{1}{\xi - \lambda}\, d\xi,$$

which is nontempered, since $\lambda \in \mathbf{C} \setminus \mathbf{R}$. For the point α to belong to Ω (while $|\xi| \leq R$), we need $|\lambda - \xi| < r|\eta|$. In particular, if $\xi = -R\lambda/|\lambda|$, we obtain $|\lambda - \xi| = |\lambda|(1 + R/|\lambda|) = |\lambda| + R$. Hence, $r|\eta| > |\lambda| + R$ is necessary. On the other hand, this condition is sufficient. Indeed, $r|\eta| > |\lambda| + R \geq |\lambda| + |\xi| \geq |\lambda - \xi|$.

18.1 Use Poisson's summation formula 16.7, the fact that $\mathcal{F}\phi(0) = 1(\phi)$, and (18.3) for estimating the remainder.

18.6 (i). We prove the results for f_+; the arguments for f_- are similar, mutatis mutandis. First note that $\mathcal{F}u \in C^\infty(\mathbf{R})$ according to Lemma 14.4. Observe that

$$2\pi\, f_+(z) = \int_{\mathbf{R}_{>0}} e^{i\,\mathrm{Re}\,z\,\xi} e^{-\mathrm{Im}\,z\,\xi}\, \mathcal{F}u(\xi)\, d\xi \qquad (z \in H_+). \tag{21.23}$$

From Theorem 18.1 we obtain the existence of constants $c' > 0$ and $N' \in \mathbf{Z}_{\geq 0}$ such that $|\mathcal{F}u(\xi)| \leq c'(1 + |\xi|)^{N'}$, for all $\xi \in \mathbf{R}$. Set $N = N' + 3$. Because $e^{-p} \leq N!\, p^{-N}$, for $p > 0$, we obtain, for all sufficiently large $\xi > 0$ and a suitable constant $d > 0$,

$$e^{-\mathrm{Im}\,z\,\xi} |\mathcal{F}u(\xi)| \leq \frac{d\,\xi^{N'}}{(\mathrm{Im}\,z)^N\, \xi^{N'+3}} = \frac{d}{(\mathrm{Im}\,z)^N} \frac{1}{\xi^3}.$$

This estimate proves that the integrand in f_+ is integrable on $\mathbf{R}_{>0}$. In addition, it shows that f_+ may be differentiated under the integral sign with respect to $x = \mathrm{Re}\,z$ and $y = \mathrm{Im}\,z$. That f_+ is complex-differentiable on H_+ follows from

$$2\partial_{\bar z} f_+(z) = (\partial_x + i\partial_y) f_+(x+iy) = \frac{1}{2\pi} \int_{\mathbf{R}_{>0}} (i\xi - i\xi) e^{ix\xi} e^{-y\xi} \mathcal{F}u(\xi)\, d\xi = 0.$$

(ii). (21.23) implies that for a sufficiently large constant $m > 0$,

$$\begin{aligned} 2\pi\, |f_+(z)| &\leq \int_0^m e^{-\mathrm{Im}\,z\,\xi} |\mathcal{F}u(\xi)|\, d\xi + \int_m^\infty e^{-\mathrm{Im}\,z\,\xi} |\mathcal{F}u(\xi)|\, d\xi \\ &\leq \int_0^m e^{-\mathrm{Im}\,z\,\xi} |\mathcal{F}u(\xi)|\, d\xi + \frac{d}{(\mathrm{Im}\,z)^N} \int_m^\infty \frac{1}{\xi^2}\, d\xi \leq d' + \frac{d''}{(\mathrm{Im}\,z)^N}. \end{aligned}$$

For $\mathrm{Im}\,z$ sufficiently small, we have $d' \leq \frac{d''}{(\mathrm{Im}\,z)^N}$, which leads to the desired estimate.

(iii). On the basis of Problem 12.14.(vi) and part (ii) we obtain the existence of $\beta_\pm(f_\pm) \in \mathcal{D}'(\mathbf{R})$. Indeed, we get, for $\phi \in C_0^\infty(\mathbf{R})$,

$$2\pi\, \beta_+(f_+)(\phi) = \lim_{y \downarrow 0} \int_{\mathbf{R}} \int_{\mathbf{R}_{>0}} e^{ix\xi} e^{-y\xi} \mathcal{F}u(\xi)\, d\xi\, \phi(x)\, dx.$$

For fixed $y > 0$, the function $(\xi, x) \mapsto e^{-y\xi} |\mathcal{F}u(\xi)|\, |\phi(x)|$ is integrable on the product $\mathbf{R} \times \mathbf{R}_{>0}$. Therefore, we may interchange the order of integration to obtain

$$\begin{aligned} 2\pi\, \beta_+(f_+)(\phi) &= \lim_{y \downarrow 0} \int_{\mathbf{R}_{>0}} e^{-y\xi} \mathcal{F}u(\xi) \int_{\mathbf{R}} e^{ix\xi} \phi(x)\, dx\, d\xi \\ &= \lim_{y \downarrow 0} \int_{\mathbf{R}_{>0}} e^{-y\xi} \mathcal{F}u(\xi)\, \mathcal{F}\phi(-\xi)\, d\xi. \end{aligned}$$

Since $\mathcal{F}u$ is at most of polynomial growth, we may take the limit under the integral sign. Thus we get $2\pi\, \beta_+(f_+)(\phi) = \int_{\mathbf{R}_{>0}} \mathcal{F}u(\xi)\, \mathcal{F}\phi(-\xi)\, d\xi$ and similarly $2\pi\, \beta_-(f_-)(\phi) = \int_{\mathbf{R}_{<0}} \mathcal{F}u(\xi)\, \mathcal{F}\phi(-\xi)\, d\xi$. Hence Theorem 14.13 leads to

$$\sum_{\pm} \beta_{\pm}(f_{\pm})(\phi) = \frac{1}{2\pi} \int_{\mathbf{R}} \mathcal{F}u(\xi)\, \mathcal{F}\phi(-\xi)\, d\xi = \frac{1}{2\pi} \mathcal{F}u(S\mathcal{F}\phi)$$
$$= u\Big(\frac{1}{2\pi}\mathcal{F}S\mathcal{F}\phi\Big) = u(\phi).$$

18.9 We have $f = \pi^* \circ T_{-t}$ sinc. Now apply Problem 14.30.(ii), formula (14.29), and Problem 16.10 for computing the Fourier transforms. For the former series apply the integration formula (18.35) with $\omega = 1$ and for the latter (18.31) with $\omega = \frac{1}{2}$.

19.3 **(ii)**. Use the inequality $|z + w|^2 \le 2(|z|^2 + |w|^2)$, valid for z and $w \in \mathbf{C}$, with $z = p(\xi) \cos t\|\xi\|$ and $w = \frac{q(\xi)}{\|\xi\|} \sin t\|\xi\|$ as in (18.10). Then we obtain from $\mathcal{F}u_t(\xi) = z + w$ and (18.11), for all $t \in \mathbf{R}$ and $\xi \in \mathbf{R}^n$ with $\|\xi\| \ge 0$,

$$|\mathcal{F}u_t(\xi)|^2 \le 2\Big(|\mathcal{F}u_0(\xi)|^2 + \frac{|\mathcal{F}(\frac{d}{dt}u_t|_{t=0})(\xi)|^2}{\|\xi\|^2}\Big).$$

This proves the claim about u_t. For $\frac{d}{dt}u_t$ apply the same argument to

$$\mathcal{F}\frac{d}{dt}u_t(\xi) = \frac{d}{dt}\mathcal{F}u_t(\xi) = \|\xi\|\Big(-p(\xi)\sin t\|\xi\| + \frac{q(\xi)}{\|\xi\|}\cos t\|\xi\|\Big).$$

19.5 Consider $u \in H_{(s)}(\mathbf{R}^n)$. Then u is continuous on account of Theorem 19.2.(a). Furthermore, Fourier inversion and the Cauchy–Schwarz inequality imply, for x and $y \in \mathbf{R}^n$,

$$|u(x+y) - u(x)| = \frac{1}{(2\pi)^n}\Big|\int_{\mathbf{R}^n} e^{i\langle x,\xi\rangle}\, \mathcal{F}u(\xi)(e^{i\langle y,\xi\rangle} - 1)\, d\xi\Big|$$
$$\le c\Big(\int_{\mathbf{R}^n} |\mathcal{F}u(\xi)|^2\, \|\xi\|^{2s}\, d\xi\Big)^{1/2}\Big(\int_{\mathbf{R}^n} |e^{i\langle y,\xi\rangle} - 1|^2\, \|\xi\|^{-2s}\, d\xi\Big)^{1/2}.$$

Write $\mathbf{R}^n$ as the union of the two subsets D_1 and D_2 determined by the condition that their elements ξ satisfy $\|\xi\| \le 2\|y\|^{-1}$ and $\|\xi\| \ge 2\|y\|^{-1}$, respectively. Note that the Mean Value Theorem implies that $|e^{ip} - 1| \le |p|$ and that we have the trivial estimate $|e^{ip} - 1| \le 2$, for $p \in \mathbf{R}$. Successively using these estimates, one obtains

$$\int_{\mathbf{R}^n} |e^{i\langle y,\xi\rangle} - 1|^2\, \|\xi\|^{-2s}\, d\xi \le \|y\|^2 \int_{D_1} \|\xi\|^{2-2s}\, d\xi + 4\int_{D_2} \|\xi\|^{-2s}\, d\xi.$$

Next, use spherical coordinates to compute these two integrals. With c_n as in (13.37), one gets

$$c_n\, \|y\|^2 \int_0^{\frac{2}{\|y\|}} r^{2-2s+n-1}\, dr + 4c_n \int_{\frac{2}{\|y\|}}^{\infty} r^{-2s+n-1}\, dr = c_n \frac{2^{1-2\alpha}}{\alpha(1-\alpha)}\, \|y\|^{2\alpha}.$$

Combining the three preceding estimates now leads to the proof.

It is natural that the coefficient of $\|y\|^{2\alpha}$ blows up as $\alpha \downarrow 0$; the fact that it also does as $\alpha \uparrow 1$ indicates that elements of $H_{(\frac{n}{2}+1)}(\mathbf{R}^n)$ need not be *Lipschitz continuous*, that is, Hölder continuous of order 1. Indeed, this can be proved.

19.6 Lemma 17.4 implies that there are $c' > 0$ and $R \geq 0$ such that $|P(\xi)|^{-2} \leq c'(1 + \|\xi\|^2)^{-m}$, for all $\|\xi\| \geq R$. Set $c^2 = (2\pi)^n \max((1+R)^{2s}, c')$ and furthermore $B = B(0; R)$ and $C = \mathbf{R}^n \setminus B$. Then

$$\int_B |\mathcal{F}u(\xi)|^2 (1 + \|\xi\|^2)^s \, d\xi \leq (1 + R^2)^s \int_B |\mathcal{F}u(\xi)|^2 \, d\xi \leq c^2 \, \|u\|_{(0)}^2,$$

$$\int_C |\mathcal{F}u(\xi)|^2 (1 + \|\xi\|^2)^s \, d\xi \leq c' \int_C |P(\xi)\mathcal{F}u(\xi)|^2 (1 + \|\xi\|^2)^{s-m} \, d\xi$$

$$\leq c' \int_{\mathbf{R}^n} |\mathcal{F}(P(D)u)(\xi)|^2 (1 + \|\xi\|^2)^{s-m} \, d\xi \leq c^2 \, \|P(D)u\|_{(s-m)}^2.$$

Addition of the estimates leads to the desired inequality.

19.8 (i). The fourth-order homogeneous part of the symbol $P(\xi)$ of $P(\partial) = P(D)$ equals ξ^4, which has no real zeros other than $\xi = 0$. Hence $P(\partial)$ is an elliptic operator, and both assertions now follow from Theorem 17.6, in other words, (12.5) applies.

(ii). From part (i) we know that all elements of N are classical solutions of $P(\partial)$. Standard facts about linear ordinary differential operators imply that N is a linear space of dimension 4. The given functions are linearly independent and satisfy the differential equation, because

$$(w(\pm 1 \pm i))^4 + 1 = w^4((\pm 1 \pm i)^2)^2 + 1 = \frac{1}{4}(\pm 2i)^2 + 1 = 0.$$

Every nontrivial function in N is of exponential growth at ∞ or at $-\infty$. This proves the second assertion.

(iii). Suppose that the fundamental solution E belongs to $\mathcal{E}'(\mathbf{R})$. According to Lemma 14.4, its Fourier transform $\mathcal{F}E$ is a complex-analytic function on $\mathbf{C}$. On the other hand, in view of $\mathcal{E}'(\mathbf{R}) \subset \mathcal{S}'(\mathbf{R})$, we obtain from (17.3) that $\mathcal{F}E(\xi) = \frac{1}{1+\xi^4} =: f(\xi)$, for all $\xi \in \mathbf{R}$. The only meromorphic extension to $\mathbf{C}$ of f is given by the same formula, for all $\xi \in \mathbf{C} \setminus \{w(\pm 1 \pm i)\}$, while this extension has simple poles at the singular points. This means that we have reached a contradiction. Next, the estimate

$$\int_{\mathbf{R}} |f(\xi)| \, d\xi \leq \int_{\mathbf{R}} \frac{1}{1+\xi^2} \, d\xi = \pi$$

and Example 14.22 together imply that $f \in L^1(\mathbf{R}) \subset \mathcal{S}'(\mathbf{R})$; furthermore, f is the unique solution of the equation $P(\xi) f(\xi) = 1$. On account of the assertion following (17.3) it now follows that $E = \mathcal{F}^{-1} f$ is the unique fundamental solution of $P(\partial)$ in $\mathcal{S}'(\mathbf{R})$.

(iv). The statement is a direct consequence of Theorem 14.21.

(v). From Theorems 14.24 and 14.32 we have $(\xi^4 + 1)\mathcal{F}u = \mathcal{F}(P(\partial)u) = \mathcal{F}f \in$

$L^2(\mathbf{R})$; hence, by a simple estimate, $\xi^j \mathcal{F}u \in L^2(\mathbf{R})$, for all $0 \le j \le 4$. Applying Theorems 14.24 and 14.32 once again, we obtain $\partial^j u \in L^2(\mathbf{R})$, for all $0 \le j \le 4$. Furthermore, $2\xi^2 \le 1 + \xi^4$ and so $(1+\xi^2)^2 = 1 + 2\xi^2 + \xi^4 \le 2(1+\xi^4)$, which implies $(1+\xi^2)^2 \mathcal{F}u \in L^2(\mathbf{R})$; in other words, u belongs to the Sobolev space $H_{(4)}(\mathbf{R})$. Since $4 > 3 + \frac{1}{2}$, Theorem 19.2.(a) then implies $u \in C^3(\mathbf{R})$. Finally, we have $u \in C(\mathbf{R})$; hence $f \notin C(\mathbf{R})$ leads to $\partial^4 u = f - u \notin C(\mathbf{R})$, that is, $u \notin C^4(\mathbf{R})$.
(vi). The estimate $|g(\xi)| \le |\mathcal{F}f(\xi)|$ implies $g \in L^2(\mathbf{R})$. The existence and uniqueness of u then follow from Theorem 14.32.
(vii). In view of the hypoellipticity of $P(\partial)$, the assertion follows from the equality $\operatorname{sing\,supp} P(\partial)u = \{0\}$.
(viii). The formula for $\mathcal{F}f$ is a direct consequence of Problem 14.17. The Fourier transform of $P(\partial)u = f$ leads to $(1+\xi^4)\mathcal{F}u(\xi) = \mathcal{F}f(\xi)$, hence

$$\begin{aligned}\mathcal{F}u(\xi) &= -4i\,\frac{\xi}{(1+\xi^2)(1+\xi^4)} = -2i\,\frac{\xi}{1+\xi^2} - 2i\,\frac{\xi-\xi^3}{1+\xi^4}\\ &= \frac{1}{2}\mathcal{F}f(\xi) - 2i\,\frac{\xi-\xi^3}{1+\xi^4}.\end{aligned}$$

Next, we introduce the function $k : \mathbf{R}\times(\mathbf{C}\setminus\{w(\pm1\pm i)\}) \to \mathbf{C}$ by $k(x,\xi) = \frac{\xi\, e^{ix\xi}}{\xi^4+1}$ and evaluate $\int_{\mathbf{R}} k(x,\xi)\,d\xi$ by means of the Residue Theorem from complex analysis. More precisely, under the assumption $x > 0$ we integrate $k(x,\cdot)$ counterclockwise along the closed path in $\mathbf{C}$ that forms the boundary of the upper half of the disk in $\mathbf{C}$ of radius R with center 0. The function $k(x,\cdot)$ decays fast enough for the contribution from the circle arc to vanish, as $R \to \infty$. Therefore, in the case of $x > 0$,

$$\begin{aligned}\int_{\mathbf{R}} k(x,\xi)\,d\xi &= 2\pi i\Big(\operatorname*{Res}_{\xi=w(1+i)} k(x,\xi) + \operatorname*{Res}_{\xi=w(-1+i)} k(x,\xi)\Big)\\ &= \frac{2\pi i\; w(1+i)\, e^{ixw(1+i)}}{2w\,2wi\,2w(1+i)} - \frac{2\pi i\; w(-1+i)\, e^{ixw(-1+i)}}{2w\,2wi\,2w(-1+i)}\\ &= \pi i\, e^{-wx}\,\frac{e^{wxi} - e^{-wxi}}{2i} = \pi i\, e^{-wx}\sin wx = h(x).\end{aligned}$$

Furthermore, if $x < 0$ one has

$$\int_{\mathbf{R}} k(x,\xi)\,d\xi = -\int_{\mathbf{R}} k(-x,\xi)\,d\xi = -h(-x) = h(x).$$

On the basis of the Fourier inversion formula we now see

$$-2i\,\frac{\xi}{\xi^4+1} = \mathcal{F}(e^{-w|\cdot|}\sin w\,\cdot)(\xi).$$

The first equality in (19.5) follows from Theorem 14.24, while computing the second-order derivative of h for $x > 0$ and $x < 0$, respectively, gives the second equality. Observe that (19.5) also holds for $x = 0$ under the convention $\operatorname{sgn}(0) = 0$, and that the first equality also may be obtained by interchanging differentiation and integration in the identity defining h.

References

1. Atiyah, M.F.: Resolution of singularities and division of distributions. *Comm. Pure Appl. Math.* **23**, 145–150 (1970).
2. Bernshtein, I.N., Gel'fand, S.I.: Meromorphic property of the function P^{λ}. *Funct. Anal. Appl.* **3**, 68–69 (1969).
3. Bochner, S.: Sur les fonctions presques périodiques de Bohr. *C. R. Acad. Sci. Paris* **180**, 1156–1158 (1925).
4. Bourbaki, N.: *Éléments de Mathématique, Livre V: Espaces Vectoriels Topologiques*. Fasc. XVII: Chap. I, II, Fasc. XVIII: Chap. III–V. Hermann, Paris (1964).
5. Duistermaat, J.J.: Selfsimilarity of "Riemann's nondifferentiable function." *Nieuw Arch. Wisk.* (4) **9**, 303–337 (1991).
6. Duistermaat, J.J.: *Fourier Integral Operators*. Birkhäuser, Boston (1996).
7. Duistermaat, J.J., Kolk, J.A.C.: *Multidimensional Real Analysis, Vols. I and II*. Cambridge University Press, Cambridge (2004).
8. Ehrenpreis, L.: Solutions of some problems of division I. *Amer. J. Math.* **76**, 883–903 (1954).
9. Folland, G.B.: *Harmonic Analysis in Phase Space*. Princeton University Press, Princeton (1989).
10. Harish-Chandra: *Collected Papers, Vols. I–IV*. Springer-Verlag, Berlin (1984).
11. Hörmander, L.: *The Analysis of Linear Partial Differential Operators, Vols. I–IV*. Springer-Verlag, Berlin (1983–85).
12. Hörmander, L.: [11, Vol. I]. Second edition. Springer-Verlag, Berlin (1990).
13. Knapp, A.W.: *Basic Real Analysis*. Birkhäuser, Boston (2005).
14. Knapp, A.W.: *Advanced Real Analysis*. Birkhäuser, Boston (2005).
15. Kolk, J.A.C., Varadarajan, V.S.: Riesz distributions. *Math. Scand.* **68**, 273–291 (1991).
16. Malgrange, B.: Existence et approximation des solutions des équations aux dérivées partielles et des équations de convolution. *Ann. Inst. Fourier (Grenoble)* **6**, 271–355 (1955–56).
17. Peetre, J.: Réctification a l'article "Une caractérisation abstraite des opérateurs différentiels". *Math. Scand.* **8**, 116–120 (1960).
18. Riesz, M.: L'intégrale de Riemann-Liouville et le problème de Cauchy. *Acta Math.* **81**, 1–223 (1949).
19. Rudin, W.: *Functional Analysis*. McGraw–Hill, New York (1973).
20. Schwartz, L.: *Théorie des Distributions, Tome I–II*. Hermann, Paris (1950–51).
21. Stroock, D.W.: *A Concise Introduction to the Theory of Integration*. Third edition. Birkhäuser, Boston (1999).
22. Varadarajan, V.S.: *Euler Through Time: A New Look at Old Themes*. Amer. Math. Soc., Providence (2006).
23. Watson, G.N.: *A Treatise on the Theory of Bessel Functions*. Second edition. Cambridge University Press, Cambridge (1944).

24. Weyl, H.: The method of orthogonal projection in potential theory. *Duke Math. J.* **7**, 411–444 (1940).
25. Whitney, H.: Elementary structure of real algebraic varieties. *Ann. of Math.* (2) **66**, 545–556 (1957).

Index of Notation

^{t} 91
$\hat{}$ 178
$\circ$ 60
$*$ 11
$\otimes$ 121
$\square$ 157
$\langle\cdot,\cdot\rangle$ 187
$(\cdot,\cdot)_{\mathrm{PW}}$ 305
$(\cdot,\cdot)_{\mathbf{R}/a\mathbf{Z}}$ 243
$\|\cdot\|_{C^k}$ 38
$\|\cdot\|_{C^k,K}$ 73
$\|\cdot\|_{L^1}$ 131
$\|\cdot\|_{L^2}$ 196
$\|\cdot\|_{L^2(\mathbf{R}/a\mathbf{Z})}$ 243
$\|\cdot\|_{L^p}$ 190, 345
$\|\cdot\|_{L^\infty}$ 190
$\|\cdot\|_{l^2}$ 243
$\|\cdot\|_{\mathrm{PW}}$ 305
$\|\cdot\|_{\mathcal{S}(k,N)}$ 181
$\|\cdot\|_{(s)}$ 311
1_U 28
$\frac{1}{x\pm i0}$ 4
∂^α 25
$\frac{\partial^\alpha}{\partial x^\alpha}$ 25
∂_z 143
$\partial_{\bar{z}}$ 143
∂_ν 147
A^a 218
A_s 100
$B(u;\delta)$ 20
$B(n, a, \epsilon)$ 73
C 36
$C(0)$ 179
C_0 24
C_0^k 21
C_0^∞ 21
$C_0(X, \mathbf{R})$ 324
$C_\pm$ 157
D 59
D_j 182
$\mathcal{D}'$ 33
$\mathcal{D}'(\cdot)_A$ 95
$\mathcal{D}'(\mathbf{R})_+$ 153
$\mathcal{D}'(\mathbf{R}^n/A)$ 246
d 20
div 2
E 34
E' 34
E_+ 162
$\mathcal{E}'$ 77
e_λ 177
$\mathcal{F}$ 178
H 46
$H_{(s)}$ 311
$H_{(s)}^{\mathrm{loc}}$ 314
$\mathcal{H}$ 211
$\mathcal{H}_a$ 109
$\mathcal{K}$ 222
k_{op} 222
I_+^a 154
L^1 131
L^2 196
$L^2(\mathbf{R}^n/A)$ 246
L^p 190
L^∞ 190
Lo 159
Lo$^\circ$ 159
l^2 244
l_+ 69
M 237, 239
$\mathcal{M}$ 156, 233
PV 4
$P(\partial)$ 68
$P(\xi)$ 271
$P(x,\partial)$ 85
$\mathbf{R}^n/A$ 246
$\overline{\mathbf{R}}$ 327
R^a 216
$\widetilde{R}^a$ 217
R_+^a 161
$R_{f,a}^k$ 61
$\mathcal{R}$ 218
S 36, 98
$\mathcal{S}$ 181
$\mathcal{S}'$ 188
sinc 178
sing supp 68
sing supp$_{\mathrm{anal}}$ 144
supp 21, 66
s_K 287
T_h 99
$T_y f$ 11
test 36
U_δ 20
$U_{-\delta}$ 20
$\overline{X}$ 2
x_+^a 69
$x_\pm^a$ 170
α 25
$|\alpha|$ 25
B 165
Γ 164
Δ 49, 96
δ 38
$\delta^{\mathbf{R}}$ 134
δ_V 39
δ_a 37
Σ 121
Φ^* 93
Φ_* 93
ϕ_ϵ 24
χ_+^a 155
χ_-^a 169
$|\chi|^a$ 169
$\binom{\alpha}{\beta}$ 25

Index

A

$\mathcal{A}$-elementary function 339
a.e. 331
Abel integral equation 169
— summable 248
— Theorem 5
absolute homogeneity 73
absorbing sequence 72
additivity, finite 338
algebra, Lie 207
almost everywhere 37
— everywhere, function defined 332
— everywhere, property 331
analytic continuation 154
— function 21
— singular support 144
annihilation operator 208
approximate identity 13
Arzelà Dominated Convergence Theorem 6, 188, 323, 350
Arzelà–Ascoli Theorem 10
associated Poisson kernel for half-space 212
averaging with weight function 10

B

backward cone 157
Banach space 131, 345
Banach–Steinhaus Theorem 52
band-limited 301
basis 74
—, dual 89, 245
Bateman integral equation 306
Bernoulli function 257
— number 257
— polynomial 257
Bessel equation 280
— kernel 281
Beta function 165
binomial coefficient 25
Bochner Theorem 204
Bolzano–Weierstrass Theorem 19
boost 108
boundary 2
— value map 151
— value, uniform 249
bounded linear mapping 75
— variation 324

C

C^k norm 38
calculus of variations 7
capacitor 99
cardinal series 302
cardinalis, sinus 178
Cauchy distribution 308
— integral equation 211
— integral formula 143
— Integral Theorem 142
— problem, existence 295
— problem, uniqueness 163
Cauchy–Riemann operator 144
Cauchy–Schwarz inequality 9
chain rule 60
Change of Variables Theorem 107
characteristic function 28
closed mapping 95
closure 2
—, Daniell's 327, 334
coefficient, Fourier 237
commutator 207
compact 17

—, relatively 342
complete 54, 131, 190, 196, 333, 346
completeness 54, 333
— of $\mathcal{D}'$ 52
— of L^1 131
— of L^2 196
— of $L^2(\mathbf{R}/a\mathbf{Z})$ 243
— of L^p 190, 346
complex derivative 141
— line integral 142
complex-analytic 143
— family of distributions 154
— function, distribution-valued 154
— in several variables 144
complex-differentiable 141
— distribution 141
component, connected 159
composition 60
— of partial differentiation and pullback 93, 110
— of pullback and partial differentiation 107
— of pushforward and partial differentiation 106
— product, Volterra's 232
cone, backward 157
—, forward 157
connected 22
— component 159
continuation, analytic 154
continuous 74
— inclusion 181
— linear form 33, 71
— linear mapping 74
—, Hölder 313
—, sequentially 74
contravariant 94
convergence in $C_0^\infty(X)$ 26
— in $\mathcal{D}'(X)$ 51
— in $C^k(X)$ 73
— in $C^\infty(X)$ 71
— in $\mathcal{E}'(X)$ 71
— in $\mathcal{S}(\mathbf{R}^n)$ 181
— in $\mathcal{S}'(\mathbf{R}^n)$ 189
— in seminorm 330
— of distributions 51
— w.r.t. Daniell seminorm 330
—, weak 75
convex 74, 291
—, locally 73
convolution 11, 115, 125
covariant 94
cover, open 17
creation operator 208
cut-off function 27

D

δ-neighborhood 20
Daniell closure 327, 334
— seminorm 330
— up-and-down method 327
decomposition, Jordan 326
—, partial-fraction 256
dense 24, 72
density of measure 42
—, Gauss 30
derivative of distribution 45
—, complex 141
—, normal 147
—, total 59
—, transversal 135
descent, Hadamard's method of 172
determinant, Jacobi's 96
diagonal 96
— mapping 96
diffeomorphism 96
difference mapping 104
differentiable family 100
diffusion equation 55
dilation 97
dipole vector 146
Dirac function 37
— measure 42
— notation 38
Dirichlet kernel 252
— problem 147
discrete 70
distribution 33
— kernel 222
— of function under measure 106
— with compact support 77
—, χ^2- 269
—, Cauchy's 308
—, complex-differentiable 141
—, derivative of 45
—, extension of 67
—, Gamma 269
—, homogeneous 102
—, local structure 156, 288
—, Lorentz-invariant 108
—, Mellin's 156
—, multiperiodic 245
—, normal 30
—, order of 40
—, periodic 108
—, positive 42
—, potential of 138

—, reflected 98
—, restriction of 65
—, Riesz's 161
—, support of 66
—, tempered 188
distribution-valued complex-analytic function 154
distributional extension 67
— sense, in 10
— solution of Cauchy problem 164
distributions, convergence of 51
divergence 2
Divergence Theorem 2
domain, fundamental 245
Dominated Convergence Theorem, Arzelà's 6, 188, 323, 350
— Convergence Theorem, Lebesgue's 6, 179, 188, 323, 336, 337, 350
— Extension Theorem 79
dual basis 89, 245
— lattice 245
—, topological 34, 75
Duhamel principle 296

E

Ehrenpreis–Malgrange Theorem 277
Eisenstein series 270
elliptic operator 272
embedding 105
energy 307
—, kinetic 307
—, potential 307
equation, Abel's integral 169
—, Bateman's integral 306
—, Bessel's 280
—, Cauchy's integral 211
—, diffusion 55
—, Euler's differential 102
—, heat 55
—, Helmholtz's 149
—, indicial 280
—, Klein–Gordon 307
—, Laplace's 138
—, Poisson's 138
—, Schrödinger's 284
—, wave 141, 157
equations of Euler–Lagrange 8
equivalence class 344
essential supremum 190
Euler differential equation 102
— series 256, 260
Euler–Lagrange, equations of 8
evaluation 38
expectation 43, 269
exponential polynomial 133, 193
extension of distribution 67

F

family of distributions, complex-analytic 154
—, differentiable 100
Fejér kernel 252
— Theorem 254
fiber, integration over 103
finite additivity 338
— in seminorm 333
— seminorm 333
flow 101
form, continuous linear 71
—, linear 33
—, monotone linear 43
formula, Cauchy's integral 143
—, Fourier's inversion 186
—, Fourier–Gel'fand 193
—, Green's 147
—, Hecke's 200
—, Legendre's duplication 167
—, Leibniz's 25
—, Lipschitz's 270
—, Parseval's 187
—, Pizzetti's 63
—, Poisson's integral 264
—, Poisson's summation 109, 242
—, Poisson's summation, in $\mathbf{R}^n$ 246
—, Pompeiu's integral 143
—, Radon's inversion 219
—, reflection, for Gamma function 167
—, Rodrigues's 208
forward cone 157
Fourier coefficient 237
— integral operator 232, 301
— inversion formula 186
— series 238
— series in $\mathcal{S}'(\mathbf{R})$ 240
— transform 178
Fourier–Gel'fand formula 193
Fourier–Laplace transform 181
function, $\mathcal{A}$-elementary 339
—, analytic 21
—, Bernoulli's 257
—, Beta 165
—, characteristic 28
—, complex-analytic 143
—, complex-analytic in several variables 144
—, complex-differentiable 141

—, cut-off 27
—, defined almost everywhere 332
—, Dirac's 37
—, distribution-valued complex-analytic 154
—, Gamma 164
—, Green's 205
—, harmonic 138
—, Heaviside's 46
—, Hermite's 208, 390
—, indicator 28
—, integrable 322, 334
—, Macdonald's 280
—, moment of 62
—, negligible 331
—, p-integrable 346
—, rapidly decreasing 181
—, real-analytic 144
—, Riemann's "nondifferentiable" 56
—, sawtooth 46
—, square-integrable 196
—, support of 21
—, test 21
—, translated 11
—, unit triangle 15
—, zeta 260
functional analysis 34
—, linear 33
fundamental domain 245
— matrix 86
— solution 137

G

Gamma distribution 269
— function 164
Gauss density 30
gliding hump, method of 52
Green formula 147
— function 205
group law 101
—, Lie 247
—, Lorentz 108, 159

H

Hadamard's method of descent 172
Hahn–Banach Theorem 79
harmonic 138
heat equation 55
— operator 140
Heaviside function 46
Hecke formula 200
Heine–Borel Theorem 19
Helmholtz equation 149
Hermite function 208, 390
— operator 209
— polynomial 208, 390
Hilbert space 196
— transform 211
Hölder continuous 313
— inequality 135
homogeneous distribution 102
Hörmander Theorem 198
Huygens principle 163
hyperbolic operator 300
hyperfunction 214
hypoelliptic 139

I

identity, approximate 13
—, subordination 299
in distributional sense 10
— weak sense 10
inclusion, continuous 181
increasing 72
independent sum of probability measures 134
indicator function 28
indicial equation 280
inequality, Cauchy–Schwarz 9
—, Hölder's 135
—, Minkowski's 345, 346
—, Sobolev's 200
—, Young's 135
integer part 46
integrability, local 343
integrable function 322, 334
— w.r.t. measure 340
— with respect to Lebesgue measure 324
—, locally 28
integral equation, Abel's 169
— equation, Bateman's 306
— equation, Cauchy's 211
— formula, Cauchy's 143
— formula, Poisson's 250, 264
— kernel 140, 221
— norm 131
— operator 221
— transform 221
— with respect to positive measure 339
—, complex line 142
—, Laplace's 201, 308
—, Lebesgue's 324
—, Poisson's 309
—, Riemann–Liouville 156
—, Wallis's 244
integration 322

Integration of a Total Derivative, Theorem on 112
integration over fiber 103
— under distribution sign 117
—, Lebesgue 28, 323
—, Riemann 28, 323
invariant, as density 102
—, as generalized function 102
Inverse Function Theorem 96

J

Jacobi determinant 96
— matrix 96
Jacobian 96
Jordan decomposition 326
jump relations 423
— relations, Plemelj–Sokhotsky's 152, 195, 349

K

Kernel Theorem 224
kernel, Bessel's 281
—, Dirichlet's 252
—, distribution 222
—, Fejér's 252
—, integral 140, 221
—, left fundamental 230
—, Mellin's 156
—, Poisson's 247
—, regular 229
—, reproducing 306
—, Riesz's 161
—, right fundamental 230
—, Schwartz's 222
—, semiregular 229
kinetic energy 307
Klein–Gordon equation 307
— operator 307

L

L^2 inner product, L^2 norm 196
L^p-(semi)norm 345
Laplace equation 138
— integral 201, 308
— operator 49, 138
— transform 181, 203
lattice 245
—, dual 245
layer 40
—, double 99
—, multiple 99
Lebesgue Dominated Convergence Theorem 6, 179, 188, 323, 336, 337, 350
— integral 324
— integration 28, 323
— measurable set 340
— measure 340
left fundamental kernel 230
Legendre duplication formula 167
Leibniz formula 25
— rule 26, 84
— series 260, 267
Lemma, Weyl's 140
Lie algebra 207
— group 247
linear form 33
— form, continuous 71
— form, monotone 43
— form, positive 324
— functional 33
— mapping, bounded 75
— mapping, continuous 74
— partial differential equation with variable coefficients 85
— partial differential operator 68
Liouville Theorem 283
Lipschitz continuous 430
— formula 270
local μ-integrability 343
— integrability 343
— mapping 231
locally convex 73
— in $H_{(s)}$ 314
— integrable 28
Lorentz group 108, 159
— group, proper orthochronous 159
Lorentz-invariant distribution 108

M

μ-integrability, local 343
Macdonald function 280
mapping, boundary value 151
—, bounded linear 75
—, closed 95
—, continuous linear 74
—, diagonal 96
—, difference 104
—, local 231
—, proper 95
—, sum 121
matrix, fundamental 86
—, Jacobi 96
maximum principle 147
mean of $u \in \mathcal{D}'(\mathbf{R}^n/A)$ 245

Mean Value Theorem 147
mean of periodic distribution 239
— of periodic function 237
measurable set 28
measure 42
— theory 322
—, density of 42
—, Dirac's 42
—, Lebesgue's 340
—, point 37
—, positive 42, 338
—, probability 43
—, Radon's 42
Mellin distribution 156
— kernel 156
— transform 156, 233
metaplectic representation 209
method of gliding hump 52
minimizing 8
Minkowski inequality 345, 346
moderate growth 239
— growth, multisequence of 245
modulo C^∞ 138
mollifier 13
moment of function 62, 269
momentum operator 182
monotone convergence property of linear form 322
— convergence property of measure 338
Monotone Convergence Theorem 335
monotone linear form 43
multi-index 25
—, order of 25
multiperiodic 245
multisequence of moderate growth 245

N

negligible function 331
— set 331
neighborhood 73
norm, integral 131
normal derivative 147
number, Bernoulli 257

O

open ball 20
— cover 17
operator commuting with translations 116
— generated by kernel 222
—, annihilation 208
—, Cauchy–Riemann 144
—, creation 208
—, elliptic 272
—, Fourier integral 232, 301
—, heat 140
—, Hermite's 209
—, hyperbolic 300
—, integral 221
—, Klein–Gordon 307
—, Laplace's 49
—, linear partial differential 68
—, momentum 182
—, position 85
—, pseudo-differential 232, 274
—, pseudo-local 231
—, Schrödinger's 284
—, singular integral 140
—, wave 141, 157
orbit 159
order of distribution 40
— of multi-index 25
oscillator representation 209

P

p-integrable function 346
Paley–Wiener space 305
— Theorem 293
parametrix (distribution) 138
—, as operator 140
Parseval formula 187
partial differential equation 85
partial-fraction decomposition 256
partition of unity 26
Pearson χ^2 distribution 269
periodic distribution 108
periodization 242, 263, 302
Pizzetti formula 63
plane wave 178
Plemelj–Sokhotsky jump relations 152, 195, 349
Pochhammer symbol 269
point measure 37
— source 2
—, stationary 69
Poisson equation 138
— integral 309
— integral formula 250, 264
— kernel 247
— kernel for half-space 212, 308
— kernel, associated, for half-space 212
— summation formula 109, 242
— summation formula in $\mathbf{R}^n$ 246
polynomial, Bernoulli's 257
—, exponential 133, 193
—, Hermite's 208, 390
—, Taylor's 61

Pompeiu integral formula 143
position operator 85
positive definite 205
— distribution 42
— linear form 324
— measure 42, 338
potential 138
— energy 307
—, retarded 171, 173
pre-sheaf 68
principal symbol 272
— value 4
principle of uniform boundedness 52
—, Duhamel's 296
—, Huygens's 163
—, maximum 147
probability density 43
— density of Cauchy distribution 308
— density of Gamma distribution 269
— density of normal distribution 30
— density of Pearson χ^2 distribution 269
— measure 43
problem, Cauchy's, existence 295
—, Cauchy's, uniqueness 163
—, Dirichlet's 147
product, tensor 121
—, Wallis's 267
proper mapping 95
— orthochronous Lorentz group 159
property of linear form, monotone convergence 322
— of measure, monotone convergence 338
—, Riesz's 324
—, sifting 38
—, Stone's 341
pseudo-differential operator 232, 274
pseudo-local operator 231
pullback, of distributions 97
—, of functions 93
pushforward, of distributions 93

Q

quantization 232
quantum harmonic oscillator wave function 390

R

Radon inversion formula 219
— measure 42
— transform 218
rapidly decreasing 181
real-analytic 144
reflection formula for Gamma function 167
—, of distribution 98
—, of function 36
regular kernel 229
Regularity Theorem 140
relatively compact 342
relativity, special 108
remainder of order k 61
representation, metaplectic 209
—, oscillator 209
—, Segal–Shale–Weil 209
reproducing kernel 306
restriction of distribution 65
retarded potential 171, 173
Riemann integration 28, 323
— "nondifferentiable" function 56
Riemann–Lebesgue Theorem 179
Riemann–Liouville integral 156
Riesz distribution 161
— kernel 161
— property 324
— Representation Theorem 321
right fundamental kernel 230
ring of subsets 338
Rodrigues formula 208
Rouché Theorem 289
rule, Leibniz's 26, 84

S

sample 301
Sampling Theorem 302
sawtooth function 46
Schrödinger operator 284
Schwartz kernel 222
Segal–Shale–Weil representation 209
semigroup property 212, 308
seminorm 73
—, Daniell's 330
—, finite 333
seminorms, separating 73
semiregular kernel 229
sense, in distributional 10
—, in weak 10
separating seminorms 73
sequence of moderate growth 239
—, absorbing 72
sequentially continuous 74
series, cardinal 302
—, Eisenstein's 270
—, Euler's 256, 260
—, Fourier's 238
—, Leibniz's 260, 267
set of sums 19

—, Lebesgue measurable 340
—, measurable 28
—, negligible 331
Shale–Weil Theorem 208
sheaf 68
shifted factorial 269
sifting property 38
signature 207
sinc 178
singular integral operator 140
— support 68
— support, analytic 144
sinus cardinalis 178
Sobolev Embedding Theorem 312
— inequality 200
— space 311
solid 331
solution of Cauchy problem, distributional 164
—, fundamental 137
source, point 2
space, Banach 131, 345
—, Hilbert 196
—, Paley–Wiener 305
—, Sobolev 311
—, topological linear 34
special relativity 108
Spectral Theorem 186
square-integrable 196
stationary point 69
Stone property 341
structure of distribution 156, 288
subadditivity 73
subcover 17
submersion 103
Submersion Theorem 31
subordination identity 299
subsequence 17
sum mapping 121
summable in the sense of Cesàro 253
—, in the sense of Abel 248
support, analytic singular 144
—, of distribution 66
—, of function 21
—, singular 68
supremum, essential 190
symbol 271
— of operator 232
—, Pochhammer's 269
—, principal 272

T

Taylor polynomial 61
tempered distribution 188
tensor product 121
test function 21
Theorem on Integration of a Total Derivative 112
— on Interchanging Differentiations 25
—, Abel's 5
—, Arzelà's Dominated Convergence 6, 188, 323, 350
—, Arzelà–Ascoli 10
—, Banach–Steinhaus 52
—, Bochner's 204
—, Bolzano–Weierstrass 19
—, Cauchy's Integral 142
—, Change of Variables 107
—, Divergence 2
—, Dominated Extension 79
—, Ehrenpreis–Malgrange 277
—, Fejér's 254
—, Hahn–Banach 79
—, Heine–Borel 19
—, Hörmander's 198
—, Inverse Function 96
—, Kernel 224
—, Lebesgue's Dominated Convergence 6, 179, 188, 323, 336, 337, 350
—, Liouville's 283
—, Mean Value 147
—, Monotone Convergence 335
—, Paley–Wiener 293
—, Regularity 140
—, Riemann–Lebesgue 179
—, Riesz's Representation 321
—, Rouché's 289
—, Sampling 302
—, Shale–Weil 208
—, Sobolev's Embedding 312
—, Spectral 186
—, Submersion 31
—, Weierstrass's Approximation 120, 196
theory, measure 322
topological dual 34, 75
— linear space 34
topology 74
torus 246
total derivative 59
transform, as density 98
—, as generalized function 98
—, Fourier 178
—, Fourier–Laplace 181
—, Hilbert 211
—, integral 221
—, Laplace 181, 203
—, Mellin 156, 233

—, Radon 218
translated function 11
translation 99
transpose 91
transversal derivative 135
trigonometric polynomials 255

U

uniform boundary value 249
— boundedness, principle of 52
unit triangle function 15
unitary 196
up-and-down method, Daniell's 327

V

value, principal 4
variance 269
variation of linear form 325
—, bounded 324
velocity vector field 101
Volterra composition product 232

W

Wallis integral 244
— product 267
wave equation 141, 157
— operator 141, 157
—, plane 178
weak convergence 51, 75
— convergence of measures 51
— sense, in 10
Weierstrass Approximation Theorem 120, 196
weight function, averaging with 11
Weyl Lemma 140

Y

Young inequality 135

Z

zeta function 260